RADIOCARBON DATES

*from samples funded by English Heritage
between 2003 and 2006*

RADIOCARBON DATES

*from samples funded by English Heritage
between 2003 and 2006*

Peter Marshall, Alex Bayliss, Christopher Bronk Ramsey,
Gordon Cook, Gerry McCormac, and Johannes van der Plicht

Published by Historic England, The Engine House, Fire Fly Avenue, Swindon SN2 2EH
www.HistoricEngland.org.uk

Historic England is a Government service championing England's heritage and giving
expert, constructive advice, and the English Heritage Trust is a charity caring for the
National Heritage Collection of more than 400 historic properties and their collections.

© Historic England 2020

Images (except as otherwise shown) © Historic England, © Crown Copyright.
HE, or Reproduced by permission of Historic England

First published 2020

ISBN 978-1-84802-384-0

British Library Cataloguing in Publication Data
A CIP catalogue record for this book is available from the British Library.

For more information about images from the Archive, contact Archives Services Team,
Historic England, The Engine House, Fire Fly Avenue, Swindon SN2 2EH; telephone
(01793) 414600.

Brought to publication by Kate Cullen, Betty Publishing, for Historic England.

Cover design and page layout by Mark Simmons
Indexed by Alan Rutter
Printed by Ashford Colour Press Ltd

Contents

*Radiocarbon Dates funded by English Heritage between
2003 and 2006 by Alex Bayliss* vii

    Introduction vii

    Sampling strategies x

    Sample selection and characterisation xi

    Laboratory methods xv

    Fractionation and radiocarbon ages xvi

    Calibration xvii

    Quality assurance xviii

    Weighted means of replicate results xxviii

    Statistical modelling xxviii

    Using the datelist xxxi

    Acknowledgements xxxi

Datelist 1

Bibliography 296

Index of laboratory codes 310

General index 317

Radiocarbon Dates funded by English Heritage between 2003 and 2006

Introduction

This volume presents a detailed catalogue of the radiocarbon dates funded by English Heritage between April 2003 and March 2006. In total, details of 1100 determinations are provided.

Only samples from sites in which English Heritage had a formal interest were eligible for dating through the Scientific Dating Team of the Centre for Archaeology. Often samples came from archaeological excavations funded, wholly or in part, by English Heritage. Most samples were from projects undertaken by external partners with funding from the Historic Environment Enabling Programme, although others came from sites in the care of English Heritage, or from sites excavated by the Centre for Archaeology, or on historic buildings where works were being undertaken with grant-aid.

Some excavations, such as the large-scale work at Yarnton, Oxfordshire (Hey 2004; Hey et al 2011; Hey et al 2016), were undertaken in advance of development or mineral extraction where permission had been granted before funding from developers became widely available, following the adoption of new planning guidance in the early 1990s (PPG16 1990).

Work also continued on the post-excavation analysis of sites that had been excavated with funding from English Heritage and its predecessors before the implementation of PPG16, such as the Iron Age promontory fort at Trevelgue

Head, Cornwall (Fig. 1; Nowakowski and Quinnell 2011) or the Neolithic long barrow at Ascott-under-Wychwood, Oxfordshire (Benson and Whittle 2007). The programme to bring such sites to publication over the preceding few years had been successful, however, and so the proportion of resources spent on the analysis of such sites had begun to decline by this period.

Resources could thus be refocused on sites that fell outside the planning process. In some cases, this was because assistance was needed for sites where archaeological remains had been encountered during development that were more significant than expected, despite the best efforts of the planning system, for example the late Bronze Age pottery production site revealed at Tinney's Lane, Sherborne, Dorset (Best and Woodward 2011), or the Anglo-Saxon settlement and cemetery at Bloodmoor Hill, Carlton Colville, Suffolk (Lucy et al 2009).

In other cases, sites were threatened by natural processes. Coastal or riverine erosion prompted beach monitoring along the Norfolk coast in the vicinity of Holme-next-the-sea (Robertson and Ames 2015) and research into human skulls recovered from the River Thames in London (Edwards et al 2009), for example. Other kinds of threats to archaeological sites also prompted research, such as the erosion damage caused by a combination of rabbits, sheep, and bracken which led to evaluation by excavation at Harehaugh hillfort, Northumberland (Carlton 2011), or a major programme to

Fig 1. *Trevelgue Head [ACS_3902] (©HER Cornwall Council)*

Fig 2. *Grey Mare's Tail Tower, Warkworth Castle, Northumberland [HAW_9387_18]*

assess the state of waterlogged sites and structures at risk through de-watering in the Somerset Levels (Brunning 2013).

At this time the first attempts to wiggle-match floating tree-ring sequences from historic buildings that could not be dated by dendrochronology were made (Bayliss 2007). These were usually undertaken to inform a programme of repair works (eg at Baguley Hall, Greater Manchester; Hamilton *et al* 2007), but could also enhance the presentation of a Property in Care (eg at Warkworth Castle, Northumberland (Arnold *et al* 2006; Fig. 2).

A number of major research projects specifically aimed at exploiting the potential for precise dating provided by the routine use of Bayesian chronological modelling were undertaken at this time. The project to date a series of Neolithic long barrows in southern England was nearing completion, and had produced a critical mass of data that allowed each site to be discussed within the context of its contemporary neighbours (Fig. 3; Bayliss and Whittle 2007).

The project to date early Anglo-Saxon burials of the period *c.* AD 580–*c* AD 720, using a combination of high-precision radiocarbon dating of the skeletons and the relative sequence provide by correspondence analysis of the associated grave-assemblages continued, but at an inevitably slow pace

because of the time needed for high-precision dating (Bayliss *et al* 2013b). Following the completion of the long barrows project and the Bayesian chronological modelling of the Neolithic monument complex on Hambledon Hill, Dorset (Mercer and Healy 2008), a major new initiative began to date causewayed enclosures across Britain and Ireland with partnership funding from the Arts and Humanities Research Board (Whittle *et al* 2011; Fig 4).

Between 2003 and 2006, English Heritage maintained collaborative research arrangements with four radiocarbon dating facilities (Fig 5). Conventional radiocarbon dating was provided by the laboratory at the Scottish Universities Environmental Research Centre, East Kilbride (GU-) using liquid scintillation spectrometry (LSS), with high-precision measurements, also undertaken using LSS, provided by the laboratory of the Queen's University, Belfast (UB-). Dating by accelerator mass spectrometry (AMS) was provided by the Oxford Radiocarbon Accelerator Unit (OxA-), the Rijkuniversiteit Groningen, the Netherlands (GrA-), and (from April 2004) the Scottish Universities Environmental Research Centre (SUERC-). In 2005, a small number of samples were kindly dated by accelerator mass spectrometry at the Christian-Albrechts-Universität zu Kiel, Germany (KIA-).

Ascott-under-Wychwood

Hazleton

West Kennet

Fussell's Lodge

Fig 3. *Varied endings for four long barrows in southern Britain in the decades around 3630 cal BC (drawing: Ian Dennis)*

Fig 4. *The Neolithic causewayed enclosure at Windmill Hill, Wiltshire [HAW_9387_18] (© Historic England Archive. Harold Wingham Collection)*

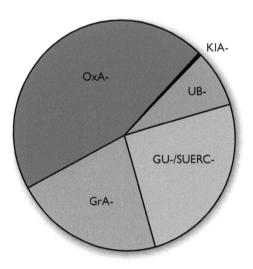

Fig 5. *Proportion of radiocarbon measurements included in this volume processed by each collaborating facility (UB-, The Queen's University, Belfast Radiocarbon Dating Laboratory; GU-/SUERC-, Scottish Universities Environmental Research Centre; OxA- Oxford Radiocarbon Accelerator Unit, GrA-, Rijksuniversiteit Groningen; KIA-, Christian-Albrechts-Universität zu Kiel).*

Multiple laboratories were now essential for undertaking the radiocarbon dating programme funded by English Heritage. Not only did this enable inter-laboratory replication to ensure that any technical problems were identified and resolved swiftly, but it also mitigated the risk of technical breakdown at any one laboratory (and the consequent delay to the reporting of results). These issues were a real and present danger, particularly because of the significant expansion of the dating programme arising from research funded by the Aggregates Levy Sustainability Fund (Bayliss *et al* 2007a; 2008).

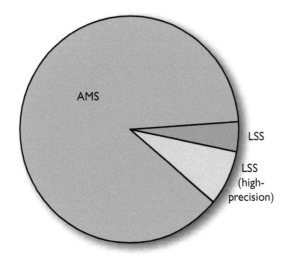

Fig 6. *Techniques of radiocarbon dating used for the measurements reported in this volume (LSS, liquid scintillation spectrometry; LSS (high-precision), high-precision liquid scintillation spectrometry; AMS, accelerator mass spectrometry).*

Between 2003 and 2006, almost 90% of all the radiocarbon dates obtained were measured by AMS (Fig 6). The precision provided by this technique was now comparable to that achieved by routine conventional dating. Measurements made by high-precision LSS at Belfast, however, were still considerably more precise than those made by AMS. Conventional high-precision dating was therefore still sometimes essential for applications where high measurement precision was required to provide chronologies that were sufficiently precise to be useful for archaeological interpretation.

A general introduction to methods of measuring the radiocarbon content of archaeological samples is provided by Bayliss *et al* (2004a).

By the time the samples covered in this volume were submitted for dating, the programme to publish the radiocarbon dates that English Heritage had funded as a series of monographs had been initiated (Jordan *et al* 1994). Consequently, almost all the information published in this volume was gathered at the time of sample submission, and during subsequent post-excavation analysis. Some additional technical information has been supplied by the dating laboratories concerned. Submitters were asked to provide interpretative comments on the overall utility of the suite of radiocarbon dates and on each individual measurement.

The majority of the radiocarbon dates included in this volume have not been published previously in datelist form, although most appear in archaeological publications on specific sites.

Sampling strategies

As the precision of AMS dating improved, the constraints on sample selection imposed by the quantity of material required for conventional dating diminished (Table 1). This remained an issue in cases where high-precision radiocarbon dating was essential, but generally the question was not whether sufficient datable material could be found, but rather which samples should be dated from the many thousands of organic items recovered on a particular site.

Table 1. Typical quantities of material required for different radiocarbon measurement techniques in 2003–2006.

Material	LSC	AMS
Charcoal	10g	1 fragment
Wood (waterlogged)	100g	1g
Peat (waterlogged)	200g	2g
Bone and antler	200g	2g

Over this period a rigorous procedure to enable the construction of Bayesian chronological models on a routine basis was forged from practice (Fig 7). This was by no means a purely mathematical process (such approaches tended to flounder in the face of the realities of sample taphonomy and radiocarbon measurement), but rather a pragmatic mix of statistical, scientific, and archaeological criteria (Bayliss and Bronk Ramsey 2004).

Statistical simulation played a role in determining the minimum number of samples that should be submitted to resolve a particular archaeological question and, crucially, in determining which archaeological problems could not be successfully addressed given the limitations of the available techniques and samples. Simulation also demonstrated how powerful vertical stratigraphy could be in obtaining precise chronologies through Bayesian statistical modelling.

In practice, however, simulation models acted as a guide in selecting an archaeologically representative set of samples that related as directly as possible to the problem or deposits in question. More samples might be needed in cases where the available material was of doubtful taphonomy, or additional samples might be submitted to address the scientific risk inherent in dating particular sample types. Replicate measurements were made largely on a judgemental basis, for example when dating both the humic acid and the humin fractions of sediment samples.

Sequential sampling strategies emerged as the most effective. The minimum number of samples needed to resolve the question at hand (as determined by simulation) was submitted as the first series of samples. When these results were returned, preliminary models with further simulated dates were constructed and a second suite of samples selected. These would address particular scientific or archaeological issues identified by the first round of radiocarbon dates and provide the additional dates needed if, for example, the site fell on a less favourable part of the calibration curve than originally anticipated. This approach maximised the cost-effectiveness of the dating programme, but could present severe challenges for project management given the turn-around time between sample submission and the reporting of results was typically several months.

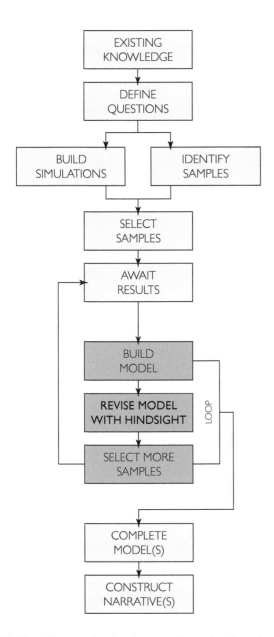

Fig 7. *Flow diagram showing the stages in routine chronology building in 2003–2006.*

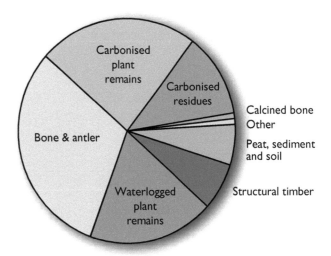

Fig 8. *Types of material dated*

bulk sample will include fragments of various ages, giving a radiocarbon measurement that is the mean of all and the age of none.

In some circumstances, however, bulk material may provide the best (or only) means of dating particular deposits (Fig 9). This is particularly true in dating peat, sediment, and soil samples. Often chemical fractions of the whole sediment are dated, these being bulk materials by definition. Such samples comprise most (66%) of the bulk samples included in this volume. The remainder of bulk samples (34%) were waterlogged, charred, or otherwise preserved plant remains that were too small individually for AMS dating. These could be bulked together to make a viable sample. Such samples were often submitted in preference to dating a bulk chemical fraction of the sediment.

Where possible all the samples reported in this volume were identified before submission for dating. In most cases identification was to age and species, but in some cases identification simply ensured that the dated material was short-lived (eg seed), and in others identification was simply to species (eg hazel) and the maturity of the sample was inferred on the basis that the relevant species does not generally grow to a great age.

Sample selection and characterisation

Once the overall sampling strategy had been designed, particular samples were identified for dating. Whilst a wide range of organic materials could be dated (Fig 8), bone and antler (31%), charcoal and other charred plant remains (24%), and waterlogged plant remains (including waterlogged wood) (19%) constituted the majority of samples. Charred residues on pottery sherds (12%), structural timbers from buildings (7%), and peat and other sediments (6%) provided most other samples, although small numbers of calcined bones, plant macrosfossils preserved by anoxic conditions in the centre of Silbury Hill, and substances that might be used in the processing of samples were also dated.

As sample size was rarely now a constraint on sample selection, the vast majority of samples (91%) consisted of material which originally derived from a single organism. This avoids the risk, highlighted by Ashmore (1999), that a

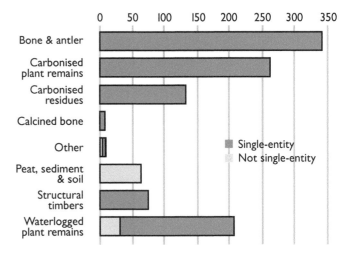

Fig 9. *Single-entity and bulk samples by material type.*

The identification of wood and charcoal samples is critical for interpreting the resultant radiocarbon date because of the old-wood effect (Bowman 1990, 51). The carbon in tree-rings is fixed from the atmosphere during the year in which the tree-ring formed. Consequently, the carbon in a twig is only a few years old when the twig is burnt and enters the archaeological record, but the rings at the centre of a long-lived tree can contain carbon that is several centuries older than the burning event. If this age-at-death offset is unknown, the radiocarbon date may be much older than archaeological activity with which the sample is associated.

Only seven samples of charcoal or waterlogged wood consisted of a species of tree which lives to some age (eg oak or ash), where the potential offset between the dated material and the outside of the tree could not be assessed. These samples could have an old-wood offset of several centuries, if wood from the centre of a mature tree was sampled, although many trees were felled before they reach such an age and, even when a mature tree was dated, the majority of the wood would have derived from the later rings rather than the centre pith. In these circumstances old-wood offsets of more than a century or two are probably rare, although the potential for an age-at-death offset in such samples means that they can only strictly be interpreted as *termini post quos* for the deposits from which they were recovered.

Bone and antler constituted the largest category of dated samples (Fig 8). Over 40% of these samples are of human bone, which overall formed the material for 13% of all the measurements reported in this volume. Although bone and antler are short-lived materials, with the turnover of carbon between ingested food and bone protein being within a decade or two at most (Hedges *et al* 2007b, 810–14), you are what you eat. This means that there is the potential for radiocarbon offsets to be transferred to the bones of terrestrial carnivores and omnivores if the dated individuals consumed a component of marine or freshwater protein. This changes the methodology needed to infer accurate chronologies from these radiocarbon dates (Bayliss *et al* 2004b).

Although, in areas such as Britain that are naturally devoid of C_4 plants that are of any dietary significance, the marine component of diet can be assessed purely on the basis of $\delta^{13}C$ values (Arneborg *et al* 1999), an input of freshwater resources may not be apparent simply on the basis of stable carbon isotopes (Lanting and van der Plicht 1998). This is a complex area that has been the subject of much research in the decades following the submission of the samples reported in this volume (eg Phillips and Gregg 2003; Hedges and Reynard 2007a; Parnell *et al* 2010; Fernandes *et al* 2014), and is still far from completely understood.

For this reason, $\delta^{15}N$ values were obtained on human bone samples submitted for dating (in addition to the $\delta^{13}C$ values obtained to allow the calculation of conventional radiocarbon ages). It was hoped that any elevated nitrogen values obtained would highlight individuals who may have consumed significant quantities of fish, and so act as a warning that the resultant chronologies should be interpreted with an appropriate degree of caution. Potentially mixed dietary sources could also be taken into account in the calibration process (Bronk Ramsey 1998; 2001).

For this to be possible, it is necessary to estimate the proportion of marine or freshwater protein in the diet of each dated individual and to estimate the reservoir age of that protein. During the period covered by this volume, diet proportions were estimated by linear interpolation from the isotopic endpoints of the potential food sources (Mays 1998, 181–90). At this time the marine component of a diet could be estimated using $\delta^{13}C$ (Arneborg *et al* 1999), and potential freshwater components by $\delta^{15}N$ (Cook *et al* 2001). It is also necessary to know the radiocarbon reservoir of the various potential dietary sources. Fortunately, the scale of the marine offset in the coastal waters around England is relatively well understood (Harkness 1983), although at the time these measurements were made there was almost no understanding of the reservoir ages of freshwater sources within England.

Figure 10 shows the carbon and nitrogen stable isotopic values for the dated human skeletons in this volume. There is no indication of enrichment in $\delta^{15}N$ values through time, although the Roman individuals from Higham Ferrers: Kings Meadow Lane are clearly more elevated in $\delta^{15}N$ than the remainder of the dataset. In comparison to the isotope values of ancient food sources, most individuals clearly fit well within the parameters for diet mainly based on terrestrial food sources.

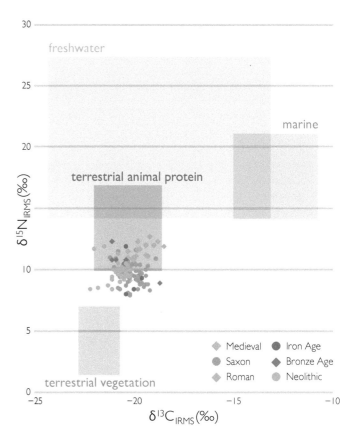

Fig 10. *Stable isotope values for the human skeletons whose radiocarbon dates are reported in this volume. The boxes are created from graphing the minimum and maximum stable isotope values of $\delta^{13}C$ and $\delta^{15}N$ from ancient food sources (vegetarian, terrestrial animal protein, freshwater fish, and marine fish), following Beavan and Mays (2013, fig 4.3). Error bars on the isotopic measurements are not shown as many of these seem not to be reproducible within the quoted error (see below).*

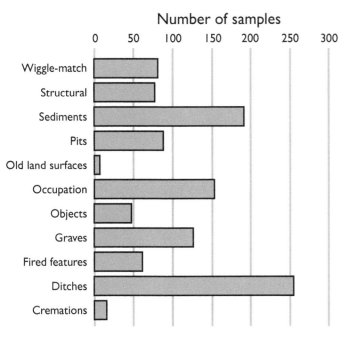

Number of samples

Context
Wiggle-match
Structural
Sediments
Pits
Old land surfaces
Occupation
Objects
Graves
Fired features
Ditches
Cremations

Fig 11. *Contexts of dated samples.*

Peat, sediment, and soil constitute the last type of sample that was frequently dated (Fig 8). These samples are rarely described more specifically. Generally, the term used to describe the deposit submitted for dating appears to reflect its perceived organic content, rather than any more technical definition.

The character of the sample material is only one criterion by which the association between a radiocarbon date and the target event that is of archaeological interest can be assessed. The importance of considering the taphonomy of dated material has been long known (Waterbolk 1971).

The types of archaeological deposits which provided the samples considered here are shown in Figure 11. The largest group is provided by samples from ditches, which make up nearly a quarter of sampled contexts, with most samples being composed of animal bone, antler, carbonised residues on pottery sherds, short-lived charred plant remains, and human bone. The association of these samples with the activity concerned is of variable security, as illustrated by material dated as part of the causewayed enclosures project (Whittle *et al* 2011). Samples such as the articulated human burial found at the base of the ditch at Offham Hill, East Sussex (GrA-27322 and OxA-14177; *see* below p00) or the possibly articulating cattle tibia and astragalus found in the inner ditch at Maiden Castle, Dorset (GrA-29107; *see* below p00; Fig 12) must be contemporary with their deposition.

Antler tools, such as one found in the lowest spit of the inner ditch at Windmill Hill, Wiltshire (GrA-29708, *see* below p00), may be functionally associated with construction. Charred plant remains may enter ditches through many different processes, although the majority of samples of this type reported in this volume appear to consist of coherent dumps of refuse such as hearth sweepings (eg GrA-29555 and OxA-14790 from close to the base of segment 1 of Kingsborough 2, Kent; *see* below p00). Pottery sherds similarly can enter ditches through a variety of processes including rubbish disposal. Only in rare cases where refitting sherds with ancient breaks are found can we be certain that the sample is close in age to the context (eg five sherds of Neolithic Bowl found in the inner ditch at Maiden Castle, Dorset; *see* below p00; Fig 13). Usually it is only the fragility of much prehistoric pottery in Britain that argues against substantive reworking of material.

Pits and other occupation deposits make up another 22% of sampled contexts, with most samples being composed of animal bone or short-lived charred plant remains. Again, the security of the association between these samples and their contexts is variable. Articulated burials and articulating groups of animal bone, such as the neonate pig recovered from a pit in the settlement at Carlton Colville: Bloodmoor Hill, Suffolk (OxA-13756; *see* below p00), do occur, but most samples appear to derive from the disposal of rubbish.

Sedimentary units produced almost 18% of samples dated. In many cases (34%) the bulk organic content of a deposit, usually peat, was itself dated. The sample is therefore

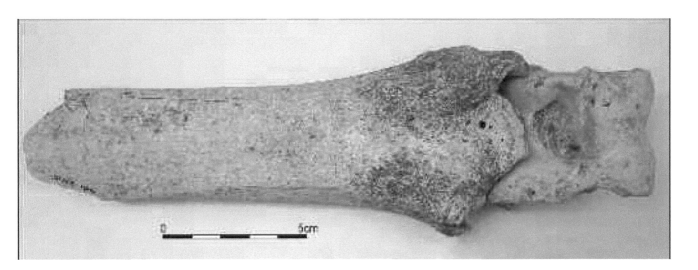

Fig 12. *Distal right cattle tibia and astragalus from the inner ditch of the causewayed enclosure at Maiden Castle, possibly articulating (although the surface condition of the two bones is different) (GrA-29107). © Alex Bayliss. Historic England.*

0   5cm

Fig 13. *Conjoining sherds of a Neolithic Bowl with internal carbonised residue, from the midden layers in the inner ditch of the causewayed enclosure at Maiden Castle. The white line denotes the area of residue required for AMS dating (GrA-29211). © Alex Bayliss. Historic England.*

composed of the unit that is of interest. In other cases, however, fragments of waterlogged plant material (including wood) were isolated from a deposit and dated. Even when dating the sediment itself, however, the relationship between the dated material and the archaeological event that is of interest has to be considered. All the material within an organic deposit does not necessarily date to the time when it formed. It could contain reworked material, for example if already waterlogged material was washed into a deposit as it was being laid down, or it could contain a component of more recent rootlets that grew down into an existing layer. Such issues can only be assessed on a case-by-case basis by consideration of the characteristics of particular deposits and by assessing the compatibility of groups of related dates (*see* below).

A further 14% of samples derive from structural contexts. In this category, there is usually a direct functional relationship between the dated material and the archaeological structure that is of interest, as almost all samples derive from the wood from which the structure was built. This material was all identified as from relatively short-lived timber (or was part of wiggle-matching, see below), but even so complications can arise. Although in the past most wood was not seasoned before use, as this makes it much harder to work, building timber was a valuable resource which could, and was, reused. Such reuse would make a radiocarbon date older than the structure from which

it was recovered. This potential issue highlights the advantages of obtaining dates from more than one timber in a structure wherever possible.

Funerary contexts form 12% of sampled contexts in this volume. A large majority of radiocarbon dates on human bone reported in this datelist come from graves containing articulated human skeletons. Here the juxtaposition of the bones provides good evidence that the individual had recently died when their corpse was interred, and so the radiocarbon date should be close in age to that of the burial. For disarticulated bones from collective burial deposits, the association is less certain, as the dated bones could come from bodies that were original deposited whole but have subsequently been dispersed or could represent bones that were already defleshed when interred. One antler artefact, buried as a grave good in an inhumation grave at Binchester, Durham (OxA-14991; *see* below p00), was also dated. This has the potential to be substantially earlier than the time of burial if it was an heirloom.

Generally, calcined bone was dated from cremation deposits, although short-lived charcoal was also dated as this probably derived from fuel used in the cremation process, and so is functionally related to the deposit from which the dated material was recovered. Since the time these measurements were made, further research has shown that calcined bone may exhibit an age-at-death offset derived from the incorporation of carbon from the pyre fuel during the cremation process (Snoek *et al* 2014). The scale of offsets of this kind is currently uncertain, as is their prevalence in the past. Most pairs of measurements on calcined bone and short-lived charcoal from the same cremation deposit undertaken so far seem to be statistically consistent (Lanting *et al* 2001), and so significant age-at-death offsets in prehistoric cremation deposits may be uncommon in practice (but see, Olsen *et al* 2012).

Functional arguments also apply to short-lived charred material recovered from fired-features, such as hearths and kilns, which provide another 6% of samples.

A small number of samples were dated from old land surfaces. Here the objective was usually to provide a *terminus post quem* for an overlying deposit (eg OxA-13476 and OxA-13609 from Groundwell Ridge, Wiltshire, *see* below p00), rather than to date the activity on the old land surface itself.

The final class of material that was submitted for dating comprises those samples which are of intrinsic interest. In these cases, the context of the find is irrelevant. Such material includes, for example, the carbonised residues on sherds of London Shellyware from Billingsgate Lorry Park, Greater London (*see* below p00; Hall *et al* 2010).

Laboratory methods

Details of the methods used for the preparation and radiocarbon dating of the samples included in this volume are provided in the references cited in this section. It is important that these technical details can be traced for every measurement as scientific methods are continuously evolving. This information is thus essential in assessing the reliability of each measurement in any future analysis.

At the Belfast Radiocarbon Dating Laboratory (UB-) samples of waterlogged wood were bleached to de-lignify and extract cellulose as described by Hoper *et al* (1998). The method of pretreatment used for bone and antler samples at Belfast was essentially that described by Longin (1971). The sample was demineralised in 2% hydrochloric acid until the bone had softened and the pH remained stable. The acid was replaced if necessary. The sample was then washed in demineralised water to remove calcium humates, and placed in slightly acid (pH 2) demineralised water, heated to 90°C for 5–18 hours, and the supernatant vacuum filtered. The sample was then evaporated dry, re-dissolved in de-ionised water, and filtered again. The supernatant was then evaporated dry before combustion.

Samples were then combusted to carbon dioxide in a positive pressure combustion stream of oxygen, converted to benzene using a chromium-based catalyst as described by Noakes *et al* (1965), and dated by LSS (Pearson 1984; McCormac 1992; McCormac *et al* 1993; 2001).

Sub-samples of bone samples dated at Belfast were sent to the Rafter Radiocarbon Laboratory, New Zealand where bone gelatin was prepared and $\delta^{15}N$, $\delta^{13}C$, %N, and %C measurements obtained at Iso trace New Zealand as described by Beavan *et al* (2011).

Samples dated at the Scottish Universities Environmental and Research Centre by LSS (GU-) were prepared as described by Stenhouse and Baxter (1983), combusted to carbon dioxide, converted to benzene using a chromium-activated catalyst, and dated as described by Noakes *et al* (1965).

The gelatin fraction of antler and bone samples was extracted and dated (Longin 1971). All other samples underwent an acid-alkali-alkali-acid pretreatment protocol (Olsson 1979). Wood samples underwent an additional stage of bleaching with a hypochlorite solution before combustion. For organic sediments, different fractions could be selected for dating: the alkali-soluble 'humic acid' fraction after either the first or second alkali pretreatment (or both together if the sample was small), or the acid- and alkali-insoluble 'humin' fraction. The chemical fraction selected for dating for each of these sediment samples is noted in the datelist.

Samples of waterlogged and charred plant remains, sediments, and charred residues on pottery sherds dated at the Scottish Universities Environmental Research Centre by AMS (SUERC-) were pretreated by the acid-base-acid protocol (Mook and Waterbolk 1985; Hall *et al* 2007). The acid and alkali insoluble fraction was dated, except for sediment samples where the acid insoluble/alkali soluble fraction (humic acid fraction) could also be dated (as specified in the datelist). Samples of bone and antler were pre-treated using a modified version of the method described by Longin (1971). Calcined bone was prepared as described by Lanting *et al* (2001), and samples of waterlogged wood and structural timber were pretreated as described by Hoper *et al* (1998).

The samples were then converted to carbon dioxide in pre-cleaned sealed quartz tubes (Vandeputte *et al* 1996), graphitised as described by Slota *et al* (1987), and measured by AMS (Freeman *et al* 2004; Xu *et al* 2004).

Fig 14. *Robert Anderson undertaking benzene synthesis at the Scottish Universities Environmental Research Centre. (© Historic England, photography by Amanda Grieve)*

A general description of the methods used by the Oxford Radiocarbon Accelerator Unit (OxA-) for producing the measurements reported in this volume follows, but full details of the methods used for each sample are available from https://c14.arch.ox.ac.uk/ (pretreatment codes can be found in Brock *et al* 2010).

Samples of bone and antler were pretreated, gelatinised, and ultrafiltered as described by Bronk Ramsey *et al* (2004a; AF) and, in many cases, also underwent solvent extraction using acetone, methanol, and chloroform (AF★). Occasionally the ultrafiltration stage was omitted (AG; AG★), and two samples (OxA-14086 and OxA-14991) were dated using re-purified excess gelatin from a first pretreatment using the original ultrafiltration protocol used at Oxford (Bronk Ramsey *et al* 2000; 2004a). Samples of calcined bone were pretreated as described by Lanting *et al* (2001).

Samples of charcoal and carbonised plant macrofossils, were generally pretreated using the acid-alkali-acid protocol described by Hedges *et al* (1989a, and *see* Brock *et al* 2010, table 1 (ZR)). Occasionally, a sample was so fragile that it would not withstand the alkali step, and so it was simply treated with acid and multiple water rinses (RR). This was the method used for the majority of carbonised residues on pottery, although occasionally a solvent extraction (RR★) or an alkali step (ZR) were used.

Waterlogged wood, wood from standing buildings, waterlogged plant remains, and organic sediments were pretreated using an acid-alkali-acid protocol, followed by a bleaching step using sodium hypochlorite (Hedges *et al* 1989a, and *see* Brock *et al* 2010, table 1 (UV or UW)). Occasionally, solvent extraction was also undertaken (UW★). For fragile samples, the bleaching step was omitted (Hedges *et al* 1989a, and *see* Brock *et al* 2010, table 1 (VV or WW)), and for particularly fragile waterlogged plant material sometimes an acid only pretreatment was undertaken (RR). The acid- and alkali-insoluble, 'humin', fraction was generally selected for dating.

Figure 15. *Fsaha Ghebru graphitising a sample at Rijkuniversiteit Groningen (© Historic England, photography by Amanda Grieve)*

No pretreatment was undertaken for the samples of evostick, calgon, and sodium bicarbonate (OxA-15714–5 and OxA-X-2128-16) measured to determine their radiocarbon age.

Samples were then converted to carbon dioxide, usually by combustion (Hedges *et al* 1992c) although the carbonate from calcined bone was devolved under vacuum using phosphoric acid (Brock *et al* 2010, 108).

Over 95% of the samples processed at Oxford and reported in this datelist were then graphitised (Dee and Bronk Ramsey 2000), although samples which yielded very little carbon were run as carbon dioxide targets (Bronk Ramsey and Hedges 1997). Dating was undertaken using the HVEE AMS (Bronk Ramsey *et al* 2004b).

Most samples dated at the Rijksuniversiteit, Groningen (GrA-) were processed using the acid/alkali/acid protocol of Mook and Waterbolk (1985), although samples of calcined bone were prepared as described by Lanting *et al* (2001) and samples of unburnt bone were prepared as described by Longin (1971). Carbonised residues on pottery sherds were pretreated by using the acid/alkali/acid method on the entire sherd, and then the alkali-soluble fraction was selected for dating (Mook and Streurman 1983).

The samples were then combusted to carbon dioxide and graphitised as described by Aerts-Bijma *et al* (1997; 2001; Fig. 15) and dated by AMS (van der Plicht *et al* 2000).

The four bone samples from Wayland's Smithy long barrow, Wiltshire (*see* below p00) dated at the Leibniz Labor für Altersbestimmung und Isotopenforschung, Christian-Albrechts Universität, Kiel were first treated with acetone, rinsed with demineralised water, and subsequently demineralised in hydrochloric acid (1%) (Grootes *et al* 2004). The demineralised bone was then treated with sodium hydroxide (1% at 20˚C for 1 hour), and again with hydrochloric acid (1% at 20˚C for 1 hour). Bone gelatin was dissolved overnight in water (at 85˚C and pH 3), filtered through a pre-combusted 0.45µm pore silver filter, and freeze dried. Combustion, graphitisation, and measurement procedures were those described by Nadeau *et al* (1997; 1998).

Fractionation and radiocarbon ages

The conventions for quoting radiocarbon ages and supporting information used here conform to the international standard known as the Trondheim Convention (Stuiver and Kra 1986).

The uncalibrated results are given as radiocarbon years before present (BP) where present has been fixed at AD 1950. These results are conventional radiocarbon ages (Stuiver and Polach 1977) and have been corrected for fractionation using measured $\delta^{13}C$ values. Three samples of peat from Exmoor Iron: Roman Lode, peat sequence, Devon (OxA-15750 and OxA-15825–6) and a fragment of charcoal from Taplow Court, Buckinghamshire (SUERC-4972) date to after AD 1950. The radiocarbon content of these samples is expressed as a fraction of modern carbon (Mook and van der Plicht 1999).

Results which are, or may be, of the same actual radiocarbon age have been tested for statistical consistency using methods described by Ward and Wilson (1978).

All $\delta^{13}C$ and $\delta^{15}N$ values in this volume were measured by Isotope Ratio Mass Spectrometry (IRMS) on sub-samples of the material combusted for dating, except for the $\delta^{13}C$ values for the four samples dated at Kiel which were measured by AMS (and so, may reflect fractionation that occurred during the dating process as well the natural isotopic composition of the sample). For conventional measurements, where open-tubed combustion was undertaken, this measurement can include a component of fractionation that occurred during laboratory processing. This measurement most closely reflects the fractionation in the dating process and is thus used for age calculation.

In order to obtain more accurate estimates of the natural isotopic composition of the dated samples, however, for conventional samples aliquots of the dated gelatin were recombusted using closed-tube combustion and repeat $\delta^{13}C$ values (and $\delta^{15}N$ values) obtained by IRMS. These are indicated as $\delta^{13}C$ (diet) and $\delta^{15}N$ (diet) in the datelist. It should be noted that, as closed-tube combustion is used in AMS dating, these measurements are equivalent to those obtained by IRMS as part of the dating process by the AMS laboratories whose measurements are included in this volume.

For some sites quality indicators of the protein extracted for dating are available. These are either C:N ratios (De Niro 1985) or amino-acid analysis (Stafford *et al* 1988).

Calibration

Radiocarbon results are not true calendar ages, but have to be converted to calendar time using a calibration curve (Pearson 1987). This is made up of radiocarbon measurements on samples whose age is known through other methods. High-precision data are currently available back to 10,600 BC, based on tree-ring samples which have been dated by dendrochronology. Beyond this a variety of archives now provide calibration back to 50,000 cal BP, although the uncertainties in this period are much greater. Reimer *et al* (2013) present the calibration curves which are presently agreed by the international radiocarbon community and provide a discussion of current understanding of the subject.

Calibrated date ranges provided in this datelist have been calculated using the maximum intercept method (Stuiver and Reimer 1986), OxCal v4.2 (Bronk Ramsey 1995; 1998; 2001; 2009a), and the dataset for terrestrial samples from the northern hemisphere published by Reimer *et al* (2013). Where appropriate, the modern results have been calibrated using the post-1950 calibration curve for the northern hemisphere atmosphere (zone 1) compiled by Hua *et al* (2013).

Calibrated date ranges are quoted in this volume in the form recommended by Mook (1986) with the end points rounded outwards to 10 years (or five years when error terms are less than ±25 BP). The date ranges for measurements which calibrate before 10,600 cal BC have been rounded outwards to 100 years to reflect the greater uncertainty on the calibration data for this early period. For the modern results, date ranges have been rounded outwards to the nearest year.

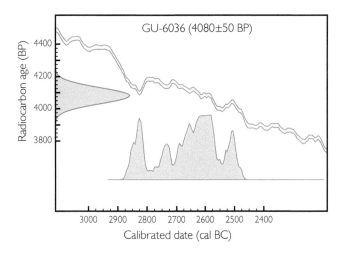

Figure 16. *Calibrated radiocarbon date for GU-6036.*

Ranges in the datelist itself are quoted at 1σ and 2σ; the calibrated date ranges referred to in the commentaries are those for 2σ unless otherwise specified.

The maximum intercept method has been used for the calibrated dates provided in this datelist and, whilst it is hoped that readers will find the calibrations provided is this volume helpful, it is necessary to recognise their limitations. First, the intercept method itself is best regarded as a 'quick and simple' way of providing an indication of the calendar date of a sample. The full complexity of the calendar age is only apparent from the probability distribution of the calibrated date. This can be illustrated by considering the calibration of GU-6036, a determination on the humic acid fraction of bulk sample of peat from MARISP: Dewar's track, Somerset (*see* below p00). This measurement (4080 ±50 BP) calibrates to 2880–2470 cal BC (at 2σ) and 2840–2490 cal BC (at 1σ) using the maximum intercept method. The calibration of this sample using the probability method (Stuiver and Reimer 1993) is shown in Figure 16. It can be seen that some parts of the calibrated range are more probable than others. It is not so much that the intercept calibration is wrong, but it does not necessarily convey the full complexity of the scientific information available.

The second limitation of the calibrated dates provided in this volume is that they are not definitive. Radiocarbon calibration is continually being refined, with updated and internationally agreed calibration curves being issued periodically (eg Stuiver and Pearson 1986; Pearson and Stuiver 1986; Stuiver *et al* 1998; Reimer *et al* 2004, Reimer *et al* 2009; and currently Reimer *et al* 2013). It is thus certain that the calibrated dates quoted here will become outmoded, and that the measurements listed here will need to be recalibrated. It is one of the major objectives of this datelist to provide easy access to the information needed for such recalibration so that these data can be used in future research. It is for this reason that it is so important that users cite both the unique laboratory identifier for each measurement and the uncalibrated radiocarbon age when using the results listed in this volume—this is a courtesy and convenience to the readers of your publications who will themselves need to recalibrate the results in due course!

Quality assurance

By the time the measurements reported in this volume were made the ongoing series of international radiocarbon inter-comparison studies had been established (Otlet *et al* 1980; International Study Group 1982; Scott *et al* 1990; Rozanski 1991; Rozanski *et al* 1992; Scott 2003).

A Fifth International Radiocarbon Inter-comparison study (VIRI) was carried out between 2004 and 2008. All of the laboratories whose measurements are reported in this datelist took part in this exercise (Scott *et al* 2007, table 1). The study provided a spot-check of operational performance of the participating laboratories at the time the inter-comparison samples were analysed. It does not measure consistent performance over a period of time, and so only the anonomised analysis of the reported results has been published Scott *et al* 2007, table 2a–d; 2010a, table 2a–e; 2010b). The study did provide valuable information to the laboratories at the time of the inter-comparison, however, which enabled them to deal with any issues identified.

Overall, 10–15% of all the results reported in this inter-comparison were identified as outliers, which is more than twice as many as would be expected on purely statistical grounds. Most of these outliers were reported by a small number of the participating laboratories. Generally, a much smaller range of results was reported by AMS laboratories (Scott *et al* 2007, fig 2).

Periodic, formal international inter-comparison exercises are only one strand in the protocols radiocarbon laboratories adopt to ensure the accuracy of the measurements they report. All the laboratories whose results are included in this datelist also maintained a continual programme of internal laboratory quality assurance procedures during the time when the reported measurements were made. The results of these internal quality control procedures are not usually published.

A summary of results on known-age tree-rings produced at the Oxford Radiocarbon Accelerator Unit including the period covered by the datelist is provided, however, by Staff *et al* (2014, fig 1), and summaries of results on a pig bone from the shipwreck of the Mary Rose (sank AD 1545) and on bison bones of background age are provided by Brock *et al* (2007, figs 2–3). Naysmith *et al* (2010) discuss results on standards and replicate materials made by AMS at the Scottish Universities Environmental Research Centre between 2003 and 2008. McCormac *et al* (2011) provide details of the quality assurance procedures undertaken as part of the Anglo-Saxon chronology project.

The variation in replicate measurements made on the same material is one of the principal methods for assessing the reproducibility of dating laboratories. This is only one of a number of reasons why repeat analyses may be undertaken, however, and many of the replicate groups reported in this volume were undertaken for other reasons. In total, there are 123 sets of replicate measurements relevant to dated samples reported in this volume, with 22 samples having more than two measurements. These results are listed in Table 2.

Table 2. Replicate radiocarbon measurements from samples included in this datelist (AMS: Accelerator Mass Spectrometry; LSS: Liquid Scintillation Spectrometry; HP: High-precision Liquid Scintillation Spectrometry), GPC: Gas Proportional Counting; entries in red are statistically significantly different at 95% confidence (Ward and Wilson 1978). [* denotes a measurement expressed as fraction modern].

Site	Material	Laboratory Number	Method	Radiocarbon Age (BP)	Ward and Wilson (1978)
Ainsbrook	human bone	SUERC-10511	AMS	1095±35	T'=0.5; T'(5%)=3.8; v=1
		SUERC-10642	AMS	1060±35	
Anglo-Saxon graves and grave goods (male graves): Edix Hill (Barrington A)	human bone	UB-4922	HP	1508±19	T'=1.2; T'(5%)=3.8; v=1
		UB-4510	HP	1479±19	
Anglo-Saxon graves and grave goods (male graves): Gally Hills	human bone	UB-4920	HP	1419±18	T'=8.0; T'(5%)=3.8; v=1
		UB-4727	HP	1487±16	
Anglo-Saxon graves and grave goods (male graves): Melbourn, Water Lane	human bone	UB-6345	HP	1516±23	T'=3.6; T'(5%)=3.8; v=1
		UB-4886	HP	1458±20	
Anglo-Saxon graves and grave goods (male graves): St Peter's Tip	human bone	UB-4924	HP	1261±16	T'=4.3; T'(5%)=3.8; v=1
		UB-6534	HP	1311±18	
Anglo-Saxon graves and grave goods (male graves): St Peter's Tip	human bone	UB-4930	HP	1414±19	T'=0.7; T'(5%)=3.8; v=1
		UB-6346	HP	1435±16	
Ascott under Wychwood: long barrow	animal bone	OxA-13316	AMS	1130±24	T'=0.5; T'(5%)=3.8; v=1
		OxA-13317	AMS	1153±24	
Aveley Marshes	humic/humin	SUERC-7354	AMS	2580±35	T'=5.9; T'(5%)=3.8; v=1
		SUERC-7355	AMS	2700±35	
Baguley Hall	wood	OxA-14586	AMS	460±29	T'=0.1; T'(5%)=3.8; v=1
		OxA-14587	AMS	472±28	
Binchester: butchery	animal bone	OxA-12370	AMS	1714±26	T'=5.7; T'(5%)=3.8; v=1
		OxA-8706	AMS	1600±40	
Binchester: butchery	animal bone	OxA-12371	AMS	1723±27	T'=5.5; T'(5%)=3.8; v=1
		OxA-8707	AMS	1610±40	

Site	Material	Laboratory Number	Method	Radiocarbon Age (BP)	Ward and Wilson (1978)
Binchester: butchery	animal bone	OxA-12372	AMS	1761±30	T'=0.3; T'(5%)=3.8; ν=1
		OxA-8711	AMS	1735±35	
Binchester: Saxon burial A1584	animal bone	OxA-14991	AMS	1637±29	T'=26.8; T'(5%)=3.8; ν=1
		OxA-9059	AMS	1380±40	
Boden Vean: Iron Age activity	charcoal	OxA-14515	AMS	2425±29	T'=0.8; T'(5%)=3.8; ν=1
		OxA-14516	AMS	2462±29	
Boden Vean: Bronze Age structure	carbonised residue	OxA-14567	AMS	2277±33	T'=229.5; T'(5%)=3.8; ν=1
		SUERC-6170	AMS	3005±35	
Bouldnor Cliff: BCII, BCIV, and BCV	waterlogged wood	OxA-15698	AMS	6956±35	T'=0.6; T'(5%)=3.8; ν=1
		OxA-15721	AMS	6915±40	
Brandon: Staunch Meadow, waterfront activity	waterlogged wood	OxA-14604	AMS	1263±29	T'=0.7; T'(5%)=3.8; ν=1
		OxA-14607	AMS	1230±28	
Callington: St Sampson's Church	human bone	OxA-14584	AMS	888±27	T'=0.6; T'(5%)=7.8; ν=3
		OxA-14808	AMS	882±31	
		SUERC-6865	AMS	880±35	
		SUERC-6932	AMS	855±35	
Carlton Colville: Bloodmoor Hill, settlement and pottery	carbonised residue	GrA-25563	AMS	1375±35	T'=2.7; T'(5%)=3.8; ν=1
		OxA-13755	AMS	1449±28	
Carlton Colville: Bloodmoor Hill, settlement and pottery	carbonised residue	GrA-25589	AMS	1385±35	T'=7.8; T'(5%)=3.8; ν=1
		OxA-13726	AMS	1509±27	
Carlton Colville: Bloodmoor Hill, settlement and pottery	carbonised residue	GrA-25590	AMS	1425±35	T'=9.9; T'(5%)=3.8; ν=1
		OxA-14019	AMS	1559±24	
Carlton Colville: Bloodmoor Hill, settlement and pottery	carbonised residue	GrA-25592	AMS	1440±35	T'=0.1; T'(5%)=3.8; ν=1
		OxA-13966	AMS	1425±27	
Carlton Colville: Bloodmoor Hill, settlement and pottery	carbonised residue	GrA-25925	AMS	1305±40	T'=0.1; T'(5%)=3.8; ν=1
		OxA-13710	AMS	1316±25	
Carlton Colville: Bloodmoor Hill, settlement and pottery	animal bone	GrA-26357	AMS	1500±35	T'=0.1; T'(5%)=3.8; ν=1
		OxA-13757	AMS	1510±26	
Carlton Colville: Bloodmoor Hill, settlement and pottery	carbonised residue	GrA-25929	AMS	1505±40	T'=2.4; T'(5%)=7.8; ν=3
		GrA-25937	AMS	1490±40	
		OxA-13754	AMS	1530±26	
		OxA-13882	AMS	1559±29	
Carlton Colville: Bloodmoor Hill, settlement and pottery	animal bone	GrA-26355	AMS	1805±35	T'=4.4; T'(5%)=6.0; ν=2
		OxA-14044	AMS	1851±28	
		UB-6185	HP	1779±20	
Causewayed Enclosures: Chalk Hill, outer ditch	carbonised residue	GrA-30888	AMS	4825±50	T'=0.6; T'(5%)=6.0; ν=2
		OxA-15509	AMS	4867±36	
		OxA-17122	AMS	4839±31	
Causewayed Enclosures: Crickley Hill, continuous ditch	charcoal	OxA-14321	AMS	4891±31	T'=0.2; T'(5%)=3.8; ν=1
		OxA-14428	AMS	4913±34	
Causewayed Enclosures: Crickley Hill, continuous ditch	animal bone	OxA-14416	AMS	4890±32	T'=2.2; T'(5%)=3.8; ν=1
		OxA-14417	AMS	4823±32	
Causewayed Enclosures: Crickley Hill, inner causewayed ditch and 'banana barrow'	animal bone	GrA-27820	AMS	4770±40	T'=1.9; T'(5%)=3.8; ν=1
		OxA-14414	AMS	4696±35	
Causewayed Enclosures: Crickley Hill, outer causewayed ditch	animal bone	GrA-27813	AMS	4830±170	T'=1.4; T'(5%)=3.8; ν=1
		GrA-30368	AMS	4625±40	
Causewayed Enclosures: Crickley Hill, outer causewayed ditch	carbonised residue	GrA-31103	AMS	4450±45	T'=1.6; T'(5%)=3.8; ν=1
		OxA-15704	AMS	4530±45	
Causewayed Enclosures: Etton, ditch	carbonised residue	GrA-29355	AMS	4225±40	T'=1.9; T'(5%)=3.8; ν=1
		OxA-14972	AMS	4300±36	
Causewayed Enclosures: Etton, ditch	waterlogged wood	GrA-29358	AMS	4765±40	T'=0.8; T'(5%)=3.8; ν=1
		BM-2890	AMS	4820±45	
Causewayed Enclosures: Etton, ditch	animal bone	OxA-14969	AMS	4809±36	T'=2.5; T'(5%)=3.8; ν=1
		BM-2765	LSS	4960±90	
Causewayed Enclosures: Knap Hill	antler	GrA-29808	AMS	4975±40	T'=4.6; T'(5%)=3.8; ν=1
		BM-205	GPC	4710±115	

Site	Material	Laboratory Number	Method	Radiocarbon Age (BP)	Ward and Wilson (1978)
Causewayed Enclosures: Knap Hill	animal bone	GrA-29809	AMS	4755±40	T′=1.1; T′(5%)=3.8; ν=1
		OxA-15200	AMS	4699±37	
Causewayed Enclosures: Maiden Castle, inner ditch	animal bone	GrA-29108	AMS	4915±40	T′=16.2; T′(5%)=3.8; ν=1
		OxA-1144	AMS	4550±80	
Causewayed Enclosures: Maiden Castle, outer ditch	carbonised residue	GrA-29213	AMS	4605±40	T′=17.2; T′(5%)=3.8; ν=1
		OxA-14793	AMS	4870±50	
Causewayed Enclosures: Maiden Castle, inner ditch	carbonised residue	GrA-29207	AMS	4935±45	T′=3.6; T′(5%)=3.8; ν=1
		OxA-14734	AMS	4830±33	
Causewayed Enclosures: Maiden Castle, inner ditch	carbonised residue	GrA-29209	AMS	4910±45	T′=1.6; T′(5%)=3.8; ν=1
		OxA-14733	AMS	4980±32	
Causewayed Enclosures: Maiden Castle, inner ditch	human bone	OxA-14832	AMS	4886±35	T′=0.8; T′(5%)=3.8; ν=1
		OxA-1148	AMS	4810±80	
Causewayed Enclosures: Offham Hill	human bone	GrA-27322	AMS	4685±45	T′=0.4; T′(5%)=3.8; ν=1
		OxA-14177	AMS	4722±32	
Causewayed Enclosures: Peak Camp	animal bone	GrA-30030	AMS	4760±40	T′=0.2; T′(5%)=3.8; ν=1
		OxA-15284	AMS	4782±31	
Causewayed Enclosures: Staines	carbonised residue	GrA-30036	AMS	3165±40	T′=205.4; T′(5%)=3.8; ν=1
		OxA-15253	AMS	3869±27	
Causewayed Enclosures: The Trundle, outer ditch	human bone	GrA-26819	AMS	2135±30	T′=0.1; T′(5%)=3.8; ν=1
		OxA-13935	AMS	2124±28	
Causewayed Enclosures: Whitehawk Hill, ditch I	antler	GrA-26962	AMS	4715±35	T′=1.6; T′(5%)=3.8; ν=1
		OxA-14126	AMS	4774±31	
Causewayed Enclosures: Whitehawk Hill, ditch II	human bone	GrA-26966	AMS	4605±40	T′=6.2; T′(5%)=3.8; ν=1
		OxA-14061	AMS	4739±36	
Causewayed Enclosures: Whitehawk Hill, ditch III	carbonised residue	GrA-26976	AMS	4710±45	T′=0.6; T′(5%)=3.8; ν=1
		OxA-14041	AMS	4820±130	
Causewayed Enclosures: Whitehawk Hill, ditch III	human bone	GrA-26977	AMS	4785±40	T′=0.0; T′(5%)=3.8; ν=1
		OxA-14063	AMS	4792±33	
Causewayed Enclosures: Whitehawk Hill, ditch III	animal bone	GrA-29363	AMS	4720±45	T′=1.3; T′(5%)=3.8; ν=1
		OxA-14062	AMS	4785±35	
Causewayed Enclosures: Whitehawk Hill, ditch IV	antler	GrA-26973	AMS	4410±35	T′=0.2; T′(5%)=3.8; ν=1
		OxA-14064	AMS	4389±32	
Causewayed Enclosures: Whitehawk Hill, ditch IV	antler	GrA-29364	AMS	4720±45	T′=1.5; T′(5%)=3.8; ν=1
		OxA-14065	AMS	4650±35	
Causewayed Enclosures: Whitesheet Hill, internal features	carbonised plant macrofossil	GrA-30072	AMS	4765±40	T′=0.2; T′(5%)=3.8; ν=1
		BM-2823	LSS	4740±35	
Causewayed Enclosures: Whitesheet Hill, internal features	carbonised plant macrofossil	OxA-15322	AMS	4797±33	T′=1.4; T′(5%)=3.8; ν=1
		BM-2823	LSS	4740±35	
Causewayed Enclosures: Whitesheet Hill	animal bone	OxA-30071	AMS	4800±45	T′=0.3; T′(5%)=3.8; ν=1
		OxA-15291	AMS	4768±33	
Causewayed Enclosures: Windmill Hill, inner ditch	carbonised residue	GrA-25391	AMS	4360±50	T′=21.4; T′(5%)=3.8; ν=1
		OxA-13732	AMS	4672±45	
Causewayed Enclosures: Windmill Hill, middle ditch	animal bone	GrA-25368	AMS	3650±50	T′=4.7; T′(5%)=3.8; ν=1
		OxA-13730	AMS	3524±30	
Causewayed Enclosures: Windmill Hill, middle ditch	animal bone	OxA-14967	AMS	4729±33	T′=2.9; T′(5%)=3.8; ν=1
		OxA-13814	AMS	4807±32	
Causewayed Enclosures: Windmill Hill, middle ditch	animal bone	GrA-29706	AMS	4700±40	T′=5.2; T′(5%)=9.5; ν=4
		UB-6186	HP	4699±20	
		OxA-15075	AMS	4717±30	
		OxA-15076	AMS	4673±30	
		OxA-15088	AMS	4770±33	
Causewayed Enclosures: Windmill Hill, outer ditch	human bone	GrA-25367	AMS	3640±50	T′=1.8; T′(5%)=3.8; ν=1
		OxA-13759	AMS	3716±28	
Causewayed Enclosures: Windmill Hill, outer ditch	human bone	GrA-29711	AMS	4615±40	T′=3.1; T′(5%)=3.8; ν=1
		OxA-14966	AMS	4521±35	

Site	Material	Laboratory Number	Method	Radiocarbon Age (BP)	Ward and Wilson (1978)
Causewayed Enclosures: Windmill Hill, outer ditch	carbonised residue	GrA-25389	AMS	4050±150	T'=391.2; T'(5%)=6.0; ν=2
		GrA-25821	AMS	3980±50	
		OxA-13561	AMS	2770±40	
Exmoor Iron: Blacklake Wood	charcoal	OxA-14511	AMS	1744±27	T'=0.9; T'(5%)=3.8; ν=1
		OxA-14512	AMS	1781±27	
Exmoor Iron: Roman Lode (peat sequence)	humic/humin	OxA-15827	AMS	2184±29	T'=2.1; T'(5%)=3.8; ν=1
		OxA-15865	AMS	2127±26	
Exmoor Iron: Roman Lode (peat sequence)	humic/humin	OxA-15750	AMS	1.065±0.003*	T'=160.2; T'(5%)=6.0; ν=2
		OxA-15825	AMS	1.066±0.003*	
		OxA-15826	AMS	1.019±0.003*	
Howick, Sea Houses Farm: Environmental	waterlogged plant macrofossils	OxA-12952	AMS	6988±37	T'=5.8; T'(5%)=3.8; ν=1
		OxA-12953	AMS	7117±39	
Long Barrows Project: Fussell's Lodge	human bone	GrA-28209	AMS	4860±50	T'=0.3; T'(5%)=3.8; ν=1
		OxA-13186	AMS	4824±39	
Longstone Edge: barrow 1, cremations	calcined human bone	GrA-26548	AMS	3555±40	T'=0.0; T'(5%)=3.8; ν=1
		OxA-14087	AMS	3560±40	
MARISP: Dewar's track	humic/humin	GU-6036	LSS	4080±50	T'=0.0; T'(5%)=3.8; ν=1
		GU-6037	LSS	4080±50	
MARISP: Glastonbury Lake Village	waterlogged plant macrofossils	SUERC-9828	AMS	2455±35	T'=98.0; T'(5%)=6.0; ν=2
	humic/humin	OxA-16233	AMS	2861±30	
		OxA-16234	AMS	2869±30	
MARISP: Glastonbury Lake Village	waterlogged plant macrofossils	SUERC-9829	AMS	2615±40	T'=28.5; T'(5%)=6.0; ν=2
	humic/humin	OxA-16235	AMS	2393±30	
		OxA-16236	AMS	2355±33	
MARISP: Harters Hill	humic/humin	SUERC-9826	AMS	2935±35	T'=1.7; T'(5%)=3.8; ν=1
		SUERC-9836	AMS	3000±35	
MARISP: Sharpham	humic/humin	SUERC-9834	AMS	2325±35	T'=1.2; T'(5%)=3.8; ν=1
		SUERC-9838	AMS	2270±35	
MARISP: Sharpham	humic/humin	SUERC-9835	AMS	2510±35	T'=0.5; T'(5%)=3.8; ν=1
		SUERC-9839	AMS	2545±35	
MARISP: Street Causeway	humic/humin	SUERC-9827	AMS	2015±35	T'=0.2; T'(5%)=3.8; ν=1
		SUERC-9837	AMS	2035±35	
Peak District Mines: Lord and Ladies Mine	charcoal	OxA-13981	AMS	161±24	T'=1.4; T'(5%)=3.8; ν=1
		OxA-13982	AMS	122±23	
Predictive modelling of archaeological sites in raised mires: Hatfield Moors	humic/humin	GU-6365	LSS	5650±50	T'=11.7; T'(5%)=3.8; ν=1
		SUERC-9646	AMS	5860±35	
Predictive modelling of archaeological sites in raised mires: Hatfield Moors	humic/humin	SUERC-9638	AMS	4225±35	T'=11.8; T'(5%)=3.8; ν=1
		SUERC-9639	AMS	4395±35	
Predictive modelling of archaeological sites in raised mires: Hatfield Moors	humic/humin	SUERC-9640	AMS	5350±35	T'=118.8; T'(5%)=3.8; ν=1
		SUERC-9641	AMS	5890±35	
Predictive modelling of archaeological sites in raised mires: Hatfield Moors	humic/humin	SUERC-9688	AMS	3880±35	T'=6.9; T'(5%)=3.8; ν=1
		SUERC-9689	AMS	4010±35	
Predictive modelling of archaeological sites in raised mires: Hatfield Moors	humic/humin	SUERC-9690	AMS	4470±35	T'=2.9; T'(5%)=3.8; ν=1
		SUERC-9691	AMS	4555±35	
Predictive modelling of archaeological sites in raised mires: Hatfield Moors	humic/humin	SUERC-9692	AMS	3205±35	T'=2.0; T'(5%)=3.8; ν=1
		SUERC-9696	AMS	3135±35	
Predictive modelling of archaeological sites in raised mires: Hatfield Moors	waterlogged plant macrofossils	SUERC-8846	AMS	5430±45	T'=919.7; T'(5%)=6.0; ν=2
	humic/humin	SUERC-9688	AMS	3880±35	
		SUERC-9689	AMS	4010±35	
Predictive modelling of archaeological sites in raised mires: Hatfield Moors	waterlogged plant macrofossils	SUERC-8848	AMS	4495±35	T'=48.7; T'(5%)=7.8; ν=3
		SUERC-8849	AMS	4745±35	
	humic/humin	SUERC-9636	AMS	4425±35	
		SUERC-9637	AMS	4500±35	
Predictive modelling of archaeological sites in raised mires: Hatfield Moors	waterlogged plant macrofossils	SUERC-8875	AMS	4895±40	T'=116.9; T'(5%)=7.8; ν=3
		SUERC-8876	AMS	4730±75	

Site	Material	Laboratory Number	Method	Radiocarbon Age (BP)	Ward and Wilson (1978)
	humic/humin	SUERC-9647	AMS	4390±35	
		SUERC-9648	AMS	4400±40	
Predictive modelling of archaeological sites in raised mires: Thorne Moors	humic/humin	OxA-15930	AMS	4205±31	T'=13.5; T'(5%)=3.8; ν=1
		OxA-15834	AMS	4374±34	
Predictive modelling of archaeological sites in raised mires: Thorne Moors	humic/humin	OxA-15946	AMS	4305±34	T'=10.7; T'(5%)=3.8; ν=1
		OxA-15835	AMS	4462±34	
Predictive modelling of archaeological sites in raised mires: Thorne Moors	waterlogged plant macrofossil	OxA-15832	AMS	3115±34	T'=12.0; T'(5%)=6.0; ν=2
	humic/humin	OxA-15833	AMS	3048±33	
		OxA-15993	AMS	2944±36	
Ripon Cathedral: RIPCSQ01	wood	SUERC-11434	AMS	140±35	T'=0.1; T'(5%)=3.8; ν=1
		SUERC-8963	AMS	100±35	
Ripon Cathedral: RIPCSQ01	wood	SUERC-11435	AMS	160±35	T'=0.1; T'(5%)=3.8; ν=1
		SUERC-8964	AMS	135±35	
Ripon Cathedral: RIPCSQ01	wood	SUERC-11439	AMS	135±35	T'=1.1; T'(5%)=3.8; ν=1
		SUERC-8965	AMS	145±35	
Ripon Cathedral: RIPCSQ01	wood	SUERC-11440	AMS	110±35	T'=0.1; T'(5%)=3.8; ν=1
		SUERC-8969	AMS	150±35	
Ripon Cathedral: RIPCSQ01	wood	SUERC-11441	AMS	85±35	T'=0.5; T'(5%)=3.8; ν=1
		SUERC-8970	AMS	155±35	
Ripon Cathedral: RIPCSQ01	wood	SUERC-11442	AMS	140±35	T'=0.5; T'(5%)=3.8; ν=1
		SUERC-8971	AMS	240±35	
Ripon Cathedral: RIPCSQ01	wood	SUERC-11443	AMS	145±35	T'=0.1; T'(5%)=3.8; ν=1
		SUERC-8972	AMS	270±35	
Ripon Cathedral: RIPCSQ01	wood	SUERC-11444	AMS	230±35	T'=0.1; T'(5%)=3.8; ν=1
		SUERC-8973	AMS	275±35	
Ripon Cathedral: RIPCSQ01	wood	SUERC-11445	AMS	140±35	T'=0.5; T'(5%)=3.8; ν=1
		SUERC-8974	AMS	245±35	
Ripon Cathedral: RIPCSQ01	wood	SUERC-11449	AMS	160±35	T'=0.5; T'(5%)=3.8; ν=1
		SUERC-8975	AMS	275±35	
Sutton Common : pollen core	humic/humin	SUERC-8018	AMS	8920±60	T'=0.7; T'(5%)=3.8; ν=1
		SUERC-7622	AMS	8860±40	
Sutton Common : pollen core	humic/humin	SUERC-8168	AMS	1730±35	T'=116.6; T'(5%)=3.8; ν=1
		SUERC-8169	AMS	2265±35	
Warkworth Castle: Grey Mare's Tail Tower	wood	SUERC-6540	AMS	700±20	T'=0.0; T'(5%)=3.8; ν=1
		SUERC-6541	AMS	695±20	
Warkworth Castle: Grey Mare's Tail Tower	wood	SUERC-6542	AMS	715±20	T'=0.0 T'(5%)=3.8; ν=1
		SUERC-6543	AMS	710±20	
Warkworth Castle: Grey Mare's Tail Tower	wood	SUERC-6547	AMS	760±20	T'=0.0; T'(5%)=3.8; ν=1
		SUERC-6548	AMS	765±20	
Warkworth Castle: Grey Mare's Tail Tower	wood	SUERC-6550	AMS	775±20	T'=0.3; T'(5%)=3.8; ν=1
		SUERC-6551	AMS	790±20	
Warkworth Castle: Grey Mare's Tail Tower	wood	SUERC-6556	AMS	800±20	T'=0.1; T'(5%)=3.8; ν=1
		SUERC-6557	AMS	810±20	
Warkworth Castle: Grey Mare's Tail Tower	wood	SUERC-6558	AMS	810±20	T'=0.0; T'(5%)=3.8; ν=1
		SUERC-6559	AMS	810±20	
Warkworth Castle: Grey Mare's Tail Tower	wood	SUERC-6563	AMS	850±20	T'=1.1; T'(5%)=3.8; ν=1
		SUERC-6564	AMS	880±20	
Warkworth Castle: Grey Mare's Tail Tower	wood	SUERC-6566	AMS	850±20	T'=0.5; T'(5%)=3.8; ν=1
		SUERC-6567	AMS	870±20	
West Heslerton: environmental	humic/humin	GU-5996	LSS	10830±120	T'=1.2; T'(5%)=3.8; ν=1
		GU-5997	LSS	11080±200	
Yarnton Neolithic and Bronze Age: activity in early Neolithic rectangular structure 3871	calcined bone	OxA-14479	AMS	4867±35	T'=3.5; T'(5%)=3.8; ν=1
		SUERC-5689	AMS	4775±35	
Yarnton Neolithic and Bronze Age: Bronze Age activity on sites 4c, 4e, 9, and 10	calcined bone	OxA-14492	AMS	3136±29	T'=2.1; T'(5%)=3.8; ν=1
		SUERC-5695	AMS	3070±35	
Yarnton Neolithic and Bronze Age: waterholes and burnt stone features, sites 17 and 21	waterlogged wood	OxA-12886	AMS	3437±35	T'=2.9; T'(5%)=3.8; ν=1
		OxA-12887	AMS	3352±36	

In total, 92 sets of replicate radiocarbon measurements are statistically consistent at 95% confidence, with a further seven groups statistically inconsistent at 95% confidence, but consistent at 99% confidence (Ward and Wilson 1978). There are, however, 24 sets of replicate radiocarbon measurements which are divergent at more than 99% confidence.

Figure 17 illustrates this reproducibility by showing offsets between replicate pairs of radiocarbon measurements on the same sample (where more than two measurements are available, each is successively compared with the first measurement in the group listed in Table 2). Although many of the data scatter around 0 (as would be expected if they exhibit only the scatter expected on statistical grounds), there

are some clear misfits. These do not occur equally across all sample types.

Thirteen of the discrepant groups include bulk fractions (acid insoluble/alkali-soluble humic acids or acid- and alkali-insoluble humin) from samples of sediment, sometimes accompanied by measurements on short-lived waterlogged terrestrial plant macrofossils from the same deposit. Overall, nearly half of all replicate groups from sediments provided divergent radiocarbon ages. This illustrates the difficulties of providing robust dating for such material, although this is clearly to some extent site specific as only two (out of seven) replicate groups from the Somerset Levels (MARISP, *see* below p00) were problematic, whereas seven (out of nine)

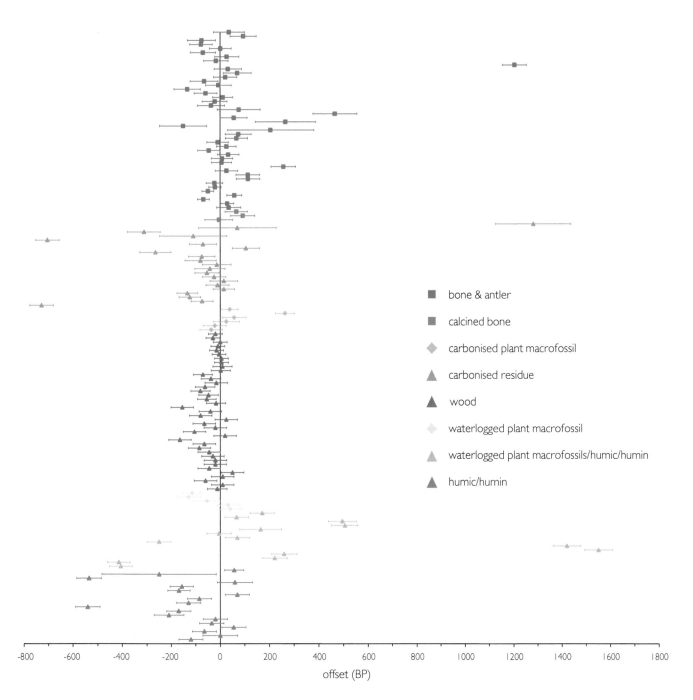

■ bone & antler

■ calcined bone

◆ carbonised plant macrofossil

▲ carbonised residue

▲ wood

◇ waterlogged plant macrofossil

▲ waterlogged plant macrofossils/humic/humin

▲ humic/humin

offset (BP)

Figure 17. *Offsets between replicate pairs of radiocarbon measurements on the same sample (where more than two measurements are available, each is successively compared with the first measurement in the group listed in Table 2).*

from Hatfield Moors (Predictive modelling of archaeological sites in raised mires, *see* below p00) were divergent.

Six of the remaining discrepant groups of radiocarbon ages are on carbonised residues on pottery sherds. In this case, contaminants from the burial environment were probably incompletely removed by the pretreatment protocols applied. Approximately one third of replicate groups on this material are divergent, which presents a substantial risk when dating such samples.

Three other divergent groups of radiocarbon ages are on bone samples. One set of replicate measurements was made to assess the effectiveness of the chemical protocols used to remove Polyvinyl Acetate PVA from a sample of human bone from the Anglo-Saxon project: Gally Hills, Bansted, Surrey (UB-4727 and UB-4920, *see* below p00) (Bayliss *et al* 2013b, 232–7). More consolidant clearly remained after pretreatment of UB-4727, and so UB-4920 is to be preferred. A measurement (GrA-29108) on a deer vertebra from the inner ditch of the causewayed enclosure at Maiden Castle, Dorset is much earlier than a measurement undertaken on the same bone at the Oxford Radiocarbon Accelerator Unit in the late 1980s (OxA-1144; *see* below p00). OxA-1144 is not compatible with the other radiocarbon dates from this ditch (Whittle *et al* 2011, 182–4) and is anomalously recent. Finally, a replicate measurement made on an antler artefact from Saxon burial A1584 at Binchester, Durham (OxA-14991, *see* below p00) is substantially later than a measurement on the same object made without ultrafiltration in 1999 (OxA-9059). The repeat measurement is much more compatible with the measurement on the skeleton from this grave (OxA-9058; 1572 ±37 BP). Furthermore, several other bone samples from this site were redated using ultrafiltration and produced earlier results which are in much closer agreement with other dates in the Binchester sequence (Marshall *et al* 2010). This includes two replicate pairs (OxA-12370 and OxA-8706; OxA-12371 and OxA-8707), which are statistically inconsistent at 95% confidence, but consistent at 99% confidence. In this case, the additional ultrafiltration

pretreatment does have a significant effect on the accuracy of the radiocarbon ages, by more successful removal of more recent contaminants (Jacobi *et al* 2006).

Removing from consideration these 24 replicate groups whose inconsistency derives from the character of the dated material rather than laboratory reproducibility, then 99 replicate groups remain. On statistical grounds alone, we expect 1 in 20 results to be more than two standard deviations away from the true value. This means that in nine or ten cases we would expect groups of results on the same sample to be statistically significantly different at 95% confidence (using the method of Ward and Wilson 1978). In fact, five groups of replicate measurements are different at this level of confidence, but statistically consistent at 99% confidence and so probably simply include one measurement that is a slight statistical outlier. Two further replicate groups, both on samples of wood from Ripon Cathedral, North Yorkshire (GrA-30761–2, SUERC-8971–2, and SUERC-11442–3; Table 2), are divergent at more than 99% confidence.

Overall, this analysis suggests that the radiocarbon measurements reported in this volume are accurate within the precision quoted, although there is likely to be a small number of measurements that are misfits well beyond statistical expectation (1–2% of the total). This is in addition, however, to inaccurate ages which may have been obtained on difficult sample types, principally organic sediments and carbonised residues on pottery sherds. Measurements on such material must be considered with caution.

From April 2002, collaborating AMS laboratories were asked to quote error estimates on the IRMS stable isotopic measurements provided. The Oxford Radiocarbon Accelerator Unit (OxA-) and the Scottish Universities Environmental Research Centre (SUERC-) both quoted errors of ±0.2‰ for $\delta^{13}C$ and ±0.3‰ for $\delta^{15}N$, and the Rijksuniversiteit Groningen (GrA-) quoted errors of ±0.1‰ for $\delta^{13}C$ and ±0.2‰ for $\delta^{15}N$. Error estimates on the stable isotopic ratios provided by the Rafter Radiocarbon Laboratory, New Zealand on bone samples dated in Belfast are given in the datelist and discussed fully by Beavan *et al* (2011).

Table 3. Replicate $\delta^{13}C$ values obtained by Isotope Ratio Mass Spectrometry from samples included in this datelist; entries in red are statistically significantly different at 95% confidence (Ward and Wilson 1978).

Site	Material	Laboratory Number	$\delta^{13}C$ (‰)	Ward and Wilson (1978)
Ainsbrook	human bone	SUERC-10511	−21.4±0.2	T′=8.0; T′(5%)=3.8; ν=1
		SUERC-10642	−20.6±0.2	
Anglo-Saxon graves and grave goods (male graves): Melbourn, Water Lane	human bone	UB-6345	−19.5±0.4	T′=1.0; T′(5%)=3.8; ν=1
		UB-4886	−20.0±0.3	
Anglo-Saxon graves and grave goods (male graves): St Peter's Tip	human bone	UB-4924	−19.7±0.3	T′=0.4; T′(5%)=3.8; ν=1
		UB-6534	−19.4±0.4	
Anglo-Saxon graves and grave goods (male graves): St Peter's Tip	human bone	UB-4930	−19.4±0.3	T′=0.6; T′(5%)=3.8; ν=1
		UB-6346	−19.0±0.4	
Ascott under Wychwood: long barrow	animal bone	OxA-13316	−20.3±0.2	T′=0.1; T′(5%)=3.8; ν=1
		OxA-13317	−20.2±0.2	
Baguley Hall	wood	OxA-14586	−24.8±0.2	T′=1.1; T′(5%)=3.8; ν=1
		OxA-14587	−25.1±0.2	

Site	Material	Laboratory Number	Radiocarbon Age (BP)	Ward and Wilson (1978)
Boden Vean: Iron Age activity	charcoal	OxA-14515	−24.3±0.2	T'=3.1; T'(5%)=3.8; ν=1
		OxA-14516	−24.8±0.2	
Boden Vean: Bronze Age structure	carbonised residue	OxA-14567	−23.4±0.2	T'=6.1; T'(5%)=3.8; ν=1
		SUERC-6170	−22.7±0.2	
Bouldnor Cliff: BCII, BCIV, and BCV	waterlogged wood	OxA-15698	−26.0±0.2	T'=0.5; T'(5%)=3.8; ν=1
		OxA-15721	−26.2±0.2	
Brandon: Staunch Meadow, waterfront activity	waterlogged wood	OxA-14604	−24.4±0.2	T'=2.0; T'(5%)=3.8; ν=1
		OxA-14607	−24.8±0.2	
Callington: St Sampson's Church	human bone	OxA-14584	−19.9±0.2	T'=18.5; T'(5%)=7.8; ν=3
		OxA-14808	−19.7±0.2	
		SUERC-6865	−20.8±0.2	
		SUERC-6932	−20.4±0.2	
Carlton Colville: Bloodmoor Hill, settlement and pottery	carbonised residue	GrA-25563	−30.0±0.1	T'=20.0; T'(5%)=3.8; ν=1
		OxA-13755	−29.0±0.2	
Carlton Colville: Bloodmoor Hill, settlement and pottery	carbonised residue	GrA-25589	−28.1±0.1	T'=1.8; T'(5%)=3.8; ν=1
		OxA-13726	−27.8±0.2	
Carlton Colville: Bloodmoor Hill, settlement and pottery	carbonised residue	GrA-25590	−29.2±0.1	T'=64.8; T'(5%)=3.8; ν=1
		OxA-14019	−27.4±0.2	
Carlton Colville: Bloodmoor Hill, settlement and pottery	carbonised residue	GrA-25592	−29.6±0.1	T'=33.8; T'(5%)=3.8; ν=1
		OxA-13966	−28.3±0.2	
Carlton Colville: Bloodmoor Hill, settlement and pottery	carbonised residue	GrA-25925	−30.3±0.1	T'=0.2; T'(5%)=3.8; ν=1
		OxA-13710	−30.4±0.2	
Carlton Colville: Bloodmoor Hill, settlement and pottery	animal bone	GrA-26357	−21.8±0.1	T'=9.8; T'(5%)=3.8; ν=1
		OxA-13757	−21.1±0.2	
Carlton Colville: Bloodmoor Hill, settlement and pottery	carbonised residue	GrA-25929	−27.5±0.1	T'=390.9; T'(5%)=7.8; ν=3
		GrA-25937	−29.7±0.1	
		OxA-13754	−27.0±0.2	
		OxA-13882	−26.4±0.2	
Carlton Colville: Bloodmoor Hill, settlement and pottery	animal bone	GrA-26355	−21.7±0.1	T'=0.0; T'(5%)=3.8; ν=1
		OxA-14044	−21.7±0.2	
Causewayed Enclosures: Chalk Hill, outer ditch	carbonised residue	GrA-30888	−30.9±0.1	T'=408.8; T'(5%)=6.0; ν=2
		OxA-15509	−27.3±0.2	
		OxA-17122	−27.5±0.2	
Causewayed Enclosures: Crickley Hill, continuous ditch	charcoal	OxA-14321	−25.9±0.2	T'=0.0; T'(5%)=3.8; ν=1
		OxA-14428	−25.9±0.2	
Causewayed Enclosures: Crickley Hill, continuous ditch	animal bone	OxA-14416	−20.3±0.2	T'=0.1; T'(5%)=3.8; ν=1
		OxA-14417	−20.4±0.2	
Causewayed Enclosures: Crickley Hill, inner causewayed ditch and 'banana barrow'	animal bone	GrA-27820	−21.9±0.1	T'=16.2; T'(5%)=3.8; ν=1
		OxA-14414	−21.0±0.2	
Causewayed Enclosures: Crickley Hill, outer causewayed ditch	animal bone	GrA-27813	−21.8±0.1	T'=0.2; T'(5%)=3.8; ν=1
		GrA-30368	−21.7±0.1	
Causewayed Enclosures: Crickley Hill, outer causewayed ditch	carbonised residue	GrA-31103	−24.4±0.1	T'=500.2; T'(5%)=3.8; ν=1
		OxA-15704	−29.4±0.2	
Causewayed Enclosures: Etton, ditch	carbonised residue	GrA-29355	−28.8±0.1	T'=9.8; T'(5%)=3.8; ν=1
		OxA-14972	−28.1±0.2	
Causewayed Enclosures: Knap Hill	animal bone	GrA-29809	−21.3±0.1	T'=5.0; T'(5%)=3.8; ν=1
		OxA-15200	−20.8±0.2	
Causewayed Enclosures: Maiden Castle, outer ditch	carbonised residue	GrA-29213	−29.5±0.1	T'=16.2; T'(5%)=3.8; ν=1
		OxA-14793	−28.6±0.2	
Causewayed Enclosures: Maiden Castle, inner ditch	carbonised residue	GrA-29207	−28.8±0.1	T'=135.2; T'(5%)=3.8; ν=1
		OxA-14734	−26.2±0.2	
Causewayed Enclosures: Maiden Castle, inner ditch	carbonised residue	GrA-29209	−29.1±0.1	T'=168.2; T'(5%)=3.8; ν=1
		OxA-14733	−26.2±0.2	
Causewayed Enclosures: Offham Hill	human bone	GrA-27322	−20.9±0.1	T'=3.2; T'(5%)=3.8; ν=1
		OxA-14177	−20.5±0.2	

Site	Material	Laboratory Number	Radiocarbon Age (BP)	Ward and Wilson (1978)
Causewayed Enclosures: Peak Camp	animal bone	GrA-30030	−22.7±0.1	T'=20.0; T'(5%)=3.8; ν=1
		OxA-15284	−21.7±0.2	
Causewayed Enclosures: Staines	carbonised residue	GrA-30036	−26.1±0.1	T'=7.2; T'(5%)=3.8; ν=1
		OxA-15253	−25.5±0.2	
Causewayed Enclosures: The Trundle, outer ditch	human bone	GrA-26819	−20.5±0.1	T'=16.2; T'(5%)=3.8; ν=1
		OxA-13935	−19.6±0.2	
Causewayed Enclosures: Whitehawk Hill, ditch I	antler	GrA-26962	−23.8±0.1	T'=16.2; T'(5%)=3.8; ν=1
		OxA-14126	−22.9±0.2	
Causewayed Enclosures: Whitehawk Hill, ditch II	human bone	GrA-26966	−21.2±0.1	T'=16.2; T'(5%)=3.8; ν=1
		OxA-14061	−20.3±0.2	
Causewayed Enclosures: Whitehawk Hill, ditch IV	antler	GrA-26973	−23.5±0.1	T'=3.2; T'(5%)=3.8; ν=1
		OxA-14064	−23.1±0.2	
Causewayed Enclosures: Whitehawk Hill, ditch III	carbonised residue	GrA-26976	−31.0±0.1	T'=24.2; T'(5%)=3.8; ν=1
		OxA-14041	−29.9±0.2	
Causewayed Enclosures: Whitehawk Hill, ditch III	human bone	GrA-26977	−21.1±0.1	T'=5.0; T'(5%)=3.8; ν=1
		OxA-14063	−20.6±0.2	
Causewayed Enclosures: Whitehawk Hill, ditch III	animal bone	GrA-29363	−20.9±0.1	T'=7.2; T'(5%)=3.8; ν=1
		OxA-14062	−21.5±0.2	
Causewayed Enclosures: Whitehawk Hill, ditch IV	antler	GrA-29364	−20.9±0.1	T'=9.8; T'(5%)=3.8; ν=1
		OxA-14065	−20.2±0.2	
Causewayed Enclosures: Whitesheet Hill	animal bone	OxA-30071	−22.2±0.2	T'=3.1; T'(5%)=3.8; ν=1
		OxA-15291	−21.7±0.2	
Causewayed Enclosures: Windmill Hill, inner ditch	carbonised residue	GrA-25391	−28.3±0.1	T'=4799.4; T'(5%)=3.8; ν=1
		OxA-13732	−12.8±0.2	
Causewayed Enclosures: Windmill Hill, middle ditch	animal bone	GrA-25368	−21.1±0.1	T'=24.2; T'(5%)=3.8; ν=1
		OxA-13730	−20.0±0.2	
Causewayed Enclosures: Windmill Hill, middle ditch	animal bone	OxA-14967	−21.4±0.2	T'=3.1; T'(5%)=3.8; ν=1
		OxA-13814	−21.9±0.2	
Causewayed Enclosures: Windmill Hill, middle ditch	animal bone	GrA-29706	−21.3±0.1	T'=15.9; T'(5%)=9.5; ν=4
		OxA-15075	−20.6±0.2	
		OxA-15076	−20.8±0.2	
		OxA-15088	−20.7±0.2	
Causewayed Enclosures: Windmill Hill, outer ditch	human bone	GrA-25367	−21.9±0.1	T'=39.2; T'(5%)=3.8; ν=1
		OxA-13759	−20.5±0.2	
Causewayed Enclosures: Windmill Hill, outer ditch	human bone	GrA-29711	−21.7±0.1	T'=7.2; T'(5%)=3.8; ν=1
		OxA-14966	−21.1±0.2	
Causewayed Enclosures: Windmill Hill, outer ditch	carbonised residue	GrA-25389	−29.1±0.1	T'=64.0; T'(5%)=6.0; ν=2
		GrA-25821	−29.9±0.1	
		OxA-13561	−28.3±0.2	
Exmoor Iron: Blacklake Wood	charcoal	OxA-14511	−25.5±0.2	T'=0.5; T'(5%)=3.8; ν=1
		OxA-14512	−25.7±0.2	
Howick, Sea Houses Farm: Environmental	waterlogged plant macrofossils	OxA-12952	−26.5±0.2	T'=2.0; T'(5%)=3.8; ν=1
		OxA-12953	−26.1±0.2	
Long Barrows Project: Fussell's Lodge	human bone	GrA-28209	−21.2±0.1	T'=12.8; T'(5%)=3.8; ν=1
		OxA-13186	−20.4±0.2	
Peak District Mines: Lord and Ladies Mine	charcoal	OxA-13981	−25.1±0.2	T'=2.0; T'(5%)=3.8; ν=1
		OxA-13982	−24.7±0.2	
Ripon Cathedral: RIPCSQ01	wood	GrA-30753	−25.7±0.1	T'=83.1; T'(5%)=6.0; ν=2
		SUERC-11434	−23.9±0.2	
		SUERC-8963	−24.0±0.2	
Ripon Cathedral: RIPCSQ01	wood	GrA-30755	−25.9±0.1	T'=150.0; T'(5%)=6.0; ν=2
		SUERC-11435	−23.5±0.2	
		SUERC-8964	−23.6±0.2	
Ripon Cathedral: RIPCSQ01	wood	GrA-30756	−26.1±0.1	T'=165.5; T'(5%)=6.0; ν=2
		SUERC-11439	−23.7±0.2	
		SUERC-8965	−23.4±0.2	

Site	Material	Laboratory Number	Radiocarbon Age (BP)	Ward and Wilson (1978)
Ripon Cathedral: RIPCSQ01	wood	GrA-30757	−25.3±0.1	T'=79.5; T'(5%)=6.0; ν=2
		SUERC-11440	−23.6±0.2	
		SUERC-8969	−23.5±0.2	
Ripon Cathedral: RIPCSQ01	wood	GrA-30635	−27.2±0.1	T'=391.1; T'(5%)=6.0; ν=2
		SUERC-11441	−23.3±0.2	
		SUERC-8970	−23.5±0.2	
Ripon Cathedral: RIPCSQ01	wood	GrA-30761	−26.3±0.1	T'=199.2; T'(5%)=6.0; ν=2
		SUERC-11442	−23.5±0.2	
		SUERC-8971	−23.7±0.2	
Ripon Cathedral: RIPCSQ01	wood	GrA-30762	−25.6±0.1	T'=136.7; T'(5%)=6.0; ν=2
		SUERC-11443	−23.3±0.2	
		SUERC-8972	−23.4±0.2	
Ripon Cathedral: RIPCSQ01	wood	GrA-30763	−25.6±0.1	T'=149.0; T'(5%)=6.0; ν=2
		SUERC-11444	−23.2±0.2	
		SUERC-8973	−23.3±0.2	
Ripon Cathedral: RIPCSQ01	wood	GrA-30765	−25.1±0.1	T'=82.6; T'(5%)=6.0; ν=2
		SUERC-11445	−23.4±0.2	
		SUERC-8974	−23.2±0.2	
Ripon Cathedral: RIPCSQ01	wood	GrA-30766	−25.4±0.1	T'=68.3; T'(5%)=6.0; ν=2
		SUERC-11449	−24.1±0.2	
		SUERC-8975	−23.3±0.2	
Ripon Cathedral: RIPCSQ02	wood	GrA-30767	−23.9±0.1	T'=0.2; T'(5%)=3.8; ν=1
		OxA-15406	−23.8±0.2	
Ripon Cathedral: RIPCSQ02	wood	GrA-30768	−24.5±0.1	T'=5.0; T'(5%)=3.8; ν=1
		OxA-15497	−24.0±0.2	
Ripon Cathedral: RIPCSQ02	wood	GrA-30770	−24.1±0.1	T'=33.8; T'(5%)=3.8; ν=1
		OxA-15407	−22.8±0.2	
Ripon Cathedral: RIPCSQ02	wood	GrA-30772	−24.5±0.1	T'=35.2; T'(5%)=3.8; ν=1
		OxA-15408	−23.1±0.2	
Ripon Cathedral: RIPCSQ02	wood	GrA-30773	−24.1±0.1	T'=16.2; T'(5%)=3.8; ν=1
		OxA-15409	−23.2±0.2	
Ripon Cathedral: RIPCSQ02	wood	GrA-30775	−24.4±0.1	T'=33.8; T'(5%)=3.8; ν=1
		OxA-15410	−23.1±0.2	
Ripon Cathedral: RIPCSQ02	wood	GrA-30776	−24.1±0.1	T'=64.8; T'(5%)=3.8; ν=1
		OxA-15411	−22.3±0.2	
Ripon Cathedral: RIPCSQ02	wood	GrA-30777	−25.4±0.1	T'=156.8; T'(5%)=3.8; ν=1
		OxA-15412	−22.6±0.2	
Ripon Cathedral: RIPCSQ02	wood	GrA-30779	−24.4±0.1	T'=28.8; T'(5%)=3.8; ν=1
		OxA-15413	−23.2±0.2	
Ripon Cathedral: RIPCSQ02	wood	GrA-30780	−23.7±0.1	T'=0.0; T'(5%)=3.8; ν=1
		OxA-15414	−23.7±0.2	
Warkworth Castle: Grey Mare's Tail Tower	wood	SUERC-6540	−25.1±0.2	T'=0.5; T'(5%)=3.8; ν=1
		SUERC-6541	−25.3±0.2	
Warkworth Castle: Grey Mare's Tail Tower	wood	SUERC-6542	−24.8±0.2	T'=2.0; T'(5%)=3.8; ν=1
		SUERC-6543	−24.4±0.2	
Warkworth Castle: Grey Mare's Tail Tower	wood	SUERC-6547	−24.0±0.2	T'=6.1; T'(5%)=3.8; ν=1
		SUERC-6548	−24.7±0.2	
Warkworth Castle: Grey Mare's Tail Tower	wood	SUERC-6550	−26.0±0.2	T'=4.5; T'(5%)=3.8; ν=1
		SUERC-6551	−25.4±0.2	
Warkworth Castle: Grey Mare's Tail Tower	wood	SUERC-6556	−25.8±0.2	T'=0.1; T'(5%)=3.8; ν=1
		SUERC-6557	−25.9±0.2	
Warkworth Castle: Grey Mare's Tail Tower	wood	SUERC-6558	−24.2±0.2	T'=0.5; T'(5%)=3.8; ν=1
		SUERC-6559	−24.4±0.2	
Warkworth Castle: Grey Mare's Tail Tower	wood	SUERC-6563	−24.3±0.2	T'=0.1; T'(5%)=3.8; ν=1
		SUERC-6564	−24.4±0.2	

Site	Material	Laboratory Number	δ¹³C (‰)	Ward and Wilson (1978)
Warkworth Castle: Grey Mare's Tail Tower	wood	SUERC-6566	−24.4±0.2	T'=0.0; T'(5%)=3.8; ν=1
		SUERC-6567	−24.4±0.2	
Yarnton Neolithic and Bronze Age: waterholes and burnt stone features, sites 17 and 21	waterlogged wood	OxA-12886	−26.1±0.2	T'=2.0; T'(5%)=3.8; ν=1
		OxA-12887	−26.5±0.2	

Table 4. Replicate δ¹⁵N values obtained by Isotope Ratio Mass Spectrometry from samples included in this datelist; entries in red are statistically significantly different at 95% confidence (Ward and Wilson 1978).

Site	Material	Laboratory Number	δ¹⁵N (‰)	Ward and Wilson (1978)
Ainsbrook	human bone	SUERC-10511	10.1±0.3	T'=12.5; T'(5%)=3.8; ν=1
		SUERC-10642	11.6±0.3	
Anglo-Saxon graves and grave goods (male graves): Melbourn, Water Lane	human bone	UB-6345	8.7±0.3	T'=0.2; T'(5%)=3.8; ν=1
		UB-4886	8.9±0.4	
Anglo-Saxon graves and grave goods (male graves): St Peter's Tip	human bone	UB-4924	8.3±0.4	T'=1.0; T'(5%)=3.8; ν=1
		UB-6534	8.8±0.3	
Anglo-Saxon graves and grave goods (male graves): St Peter's Tip	human bone	UB-4930	8.7±0.4	T'=1.4; T'(5%)=3.8; ν=1
		UB-6346	8.1±0.3	
Callington: St Sampson's Church	human bone	OxA-14584	9.9±0.3	T'=8.0; T'(5%)=7.8; ν=3
		OxA-14808	11.1±0.3	
Causewayed Enclosures: Whitehawk Hill, ditch III	human bone	GrA-26977	9.7±0.2	T'=0.3; T'(5%)=3.8; ν=1
		OxA-14063	9.9±0.3	
Causewayed Enclosures: Windmill Hill, middle ditch	animal bone	GrA-25368	3.1±0.2	T'=155.8; T'(5%)=3.8; ν=1
		OxA-13730	7.6±0.3	
Causewayed Enclosures: Windmill Hill, outer ditch	human bone	GrA-29711	11.9±0.2	T'=0.0; T'(5%)=3.8; ν=1
		OxA-14966	11.9±0.3	

The 72 replicate groups of δ¹³C values are listed in Table 3. Of these 38 are statistically consistent at 95% confidence, with three more inconsistent at 95% confidence, but consistent at 99% confidence. Over 40% of all replicate groups are divergent at more than 99% confidence. Replication of δ¹³C values within each laboratory is much better, with over 90% of same-laboratory groups statistically consistent at 95% confidence. In contrast, over 85% of inter-laboratory groups are statistically inconsistent at 95% confidence. This may suggest that the divergence arose from variations in the pretreatment protocols employed in the different laboratories or from the standard materials used (see Bayliss *et al* forthcoming). Overall reproducibility of δ¹³C values within the quoted errors is clearly poor. For this reason, the errors on these values have not been reported in the datelist.

The eight replicate groups of δ¹⁵N values are listed in Table 4. Five of these are statistically consistent at 95% confidence, although the other three pairs are divergent. Again, the error terms on the reported δ¹⁵N values have also not been reported in the datelist.

Weighted means of replicate results

Weighted means of replicate radiocarbon measurements should be taken before calibration and incorporation in statistical modelling. Most commonly these are replicate

samples from the same living organism. For example, the weighted mean of the two measurements on calcined bone from activity in early Neolithic rectangular structure 3871 at Yarnton Neolithic and Bronze Age, Oxfordshire (OxA-14479 and SUERC-5689; Table 2, see below p00) is 4821 ±25 BP, which calibrates to 3660–3530 cal BC (at 2σ), or 3650–3540 cal BC (at 1σ).

Statistical modelling

The Bayesian approach to the interpretation of archaeological chronologies has been described by Buck *et al* (1996). It is based on the principle that although the calibrated age ranges of radiocarbon measurements accurately estimate the calendar ages of the samples themselves, it is the dates of archaeological events associated with those samples that are important. Bayesian techniques can provide realistic estimates of the dates of such events by combining scientific dating evidence, such as radiocarbon dates with relative dating evidence, such as stratigraphic relationships between radiocarbon samples. These 'posterior density estimates', (which, by convention, are always expressed *in italics*) are not absolute. They are interpretative estimates, which will change as additional data become available or as the existing data are modelled from different perspectives (Fig 18).

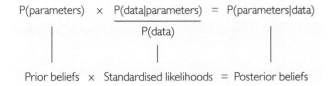

Figure 18. *Bayes' theorem as applied to chronological modelling in archaeology*

Lindley (1985) provides a user-friendly introduction to the principles of Bayesian statistics, and Bayliss *et al* (2007b) provide an introduction to the practice of chronological modelling for archaeological problems.

The technique used to implement Bayesian statistics in practice is a form of Markov Chain Monte Carlo sampling (Gilks *et al* 1996; Gelfand and Smith 1990). Almost all the models considered in this volume have been constructed using the OxCal software, usually v.1.3–v.3.10 (Bronk Ramsey 1995; 1998; 2000; 2001; Bronk Ramsey *et al* 2001), although v.4.0–v.4.2 (Bronk Ramsey 2008, 2009a–b; Bronk Ramsey and Lee 2013) has also been used more recently, particularly for sediment sequences where age-depth modelling is required or samples where mixed-source calibration is required.

An OxCal model is constructed explicitly specifying the known or assumed relative ages of the radiocarbon samples. The model structure is typically defined by the site's Harris matrix. The program calculates the probability distributions of the individual calibrated radiocarbon dates (Stuiver and Reimer 1993), and then attempts to reconcile these distributions with the relative ages of the samples, by repeatedly sampling each distribution to build up the set of solutions consistent with the model structure.

This process produces a posterior density estimate of each sample's calendar age, which occupies only part of the calibrated probability distribution (the prior distribution of the sample's calendar age). The posterior distribution is then compared to the prior distribution; an index of agreement is calculated that reflects the consistency of the two distributions. If the posterior distribution is situated in a high-probability region of the prior distribution, the index of agreement is high (sometimes 100% or more). If the index of agreement falls below 60% (a threshold value analogous to the 0.05 significance level in a χ^2 test), however, the radiocarbon date is regarded as inconsistent with the sample's calendar age, if the latter is consistent with the sample's age relative to the other dated samples. Sometimes this merely indicates that the radiocarbon result is a statistical outlier (more than 2 standard deviations from the sample's true radiocarbon age), but a very low index of agreement may mean that the sample is residual or intrusive (ie that its calendar age is different to that implied by its stratigraphic position).

An overall index of agreement is calculated from the individual agreement indices, providing a measure of the consistency between the archaeological information included in the model and the radiocarbon dates. Again, this has a threshold value of 60%. The program is also able to calculate distributions for the dates of events that have not been dated directly, such as the beginning and end of a continuous phase of activity (which is represented by several radiocarbon results), and for the durations of phases of activity or hiatuses between such phases.

By the time the samples reported in this datelist were measured, the selection and interpretation of radiocarbon dates within a Bayesian statistical framework was routinely employed for sites funded by English Heritage (Bayliss and Bronk Ramsey 2004). Overall, 82% of radiocarbon dates in this volume were interpreted using Bayesian statistics (Fig 19). All the synthetic studies were undertaken within a Bayesian framework, and over 90% of dates from site-based applications were included in Bayesian models. In contrast, only slightly over half of samples from landscape surveys were interpreted using this approach, and results from samples of intrinsic interest were simply calibrated.

At this time, a series of synthetic research projects were initiated which considered archaeological problems at a wider scale than that of the single site.

Completion of the dating programmes for the Neolithic long barrows at Ascott under Wychwood, Oxfordshire (Benson and Whittle 2006), Fussells Lodge, Wiltshire (Wysocki *et al* 2007) and Wayland's Smithy, Oxfordshire (Whittle *et al* 2007a), allowed the diversity of a group of contemporary sites in a small area of southern England to be considered (Fig 3; Whittle *et al* 2007b). A similar approach, based on providing precise dating for a group of related sites so that their origin, development, and differences could be considered together, underlay a major new initiative to date Neolithic causewayed enclosures (Whittle *et al* 2011). This was undertaken with partnership funding from the Arts and Humanities Research Board.

Research to provide robust dating for the emergence of London Sandy Shellyware (Museum of London Pottery Fabric Code SSW) arose from its unexpectedly early occurrence in Perth, Scotland (Hall *et al* 2007). Further dating from the sites of Billingsgate Lorry Park, Greater London and Bryggen, Bergen, Norway, confirmed the presence of these ceramics in all three towns substantially before the mid-twelfth century AD, in London probably from the tenth century (Hall *et al* 2010, fig 4). This project was undertaken with partnership funding from Historic Scotland.

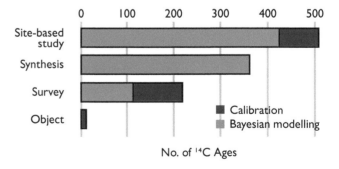

Figure 19. *The use of Bayesian chronological modelling for the interpretation of the radiocarbon dates reported in this volume.*

Another major project explored the chronology of Anglo-Saxon graves and grave goods between the later sixth century AD and the end of furnished burial in England. This combined the relative dating of grave assemblages revealed through correspondence analysis with a series of high-precision radiocarbon dates on the skeletons from the burials. This study revealed unexpected variation in the frequency of furnished burial through this period, and also suggested that the practice ended abruptly towards the end of the seventh century cal AD (Bayliss *et al* 2013b).

Models which addressed the chronology of single sites still formed the largest class of application, however. A Mesolithic structure with a contemporary environmental profile from Howick, Northumberland was dated to the earlier 8th millennium cal BC (Bayliss *et al* 2007c; Boomer *et al* 2007) and a hearth and related environmental sequence from submerged deposits at Bouldner Cliff in the Solent date to the end of the seventh millennium cal BC (Momber *et al* 2011).

Two early Neolithic structures of varying form were dated at Yarnton, Oxfordshire, along with a series of early and middle Neolithic pits, and a later Neolithic structure (Hey *et al* 2016). At Silbury Hill, Wiltshire work continued to date the construction of the mound based on the archive from the 1960s excavation, although this sampling programme was augmented by additional material retrieved during remedial works which followed the partial collapse of part of the mound in 2000 (Bayliss *et al* 2007d).

At Longstone Edge, Derbyshire (Last 2014) ephemeral Neolithic mortuary activity, preceded the construction of a barrow accompanied by both inhumation and cremation burials in the early Bronze Age. Middle Bronze Age occupation was dated at Boden Vean, Cornwall (Gossip 2013) and on two eyots in the river Thames at Yarnton, Oxfordshire (Hey *et al* 2016). An important production site for pottery in the Post Deverel-Rimbury plain ware tradition was dated to the 12th–11th centuries cal BC at Tinney's Lane, Sherborne, Dorset (Best and Woodward 2012), and a ditched enclosure at Taplow Court, Buckinghamshire fell in the early centuries of the first millennium cal BC (Allen *et al* 2009). Wooden structures dating to second half of the second millennium cal BC and the first half of the first millennium cal BC were identified in the Somerset Levels at Harters Hill, Sharpham, and Dewar's Track (Brunning 2013).

Chronological modelling was undertaken for a series of sites of the middle and later Iron Age, including a hillfort at Taplow Court, Buckinghamshire (Allen *et al* 2009), the promentary fort at Trevelue Head, Cornwall (Nowakowski and Quinnell 2011), and a fogou at Boden Vean, Cornwall (Gossip 2013). Dating of paleoenvironmental proxies in a sediment sequence adjacent to Glastonbury Lake Village, Somerset provided a landscape context for this iconic site (Brunning 2013).

Dating programmes provided more secure chronologies for cemeteries of the later Iron Age and Roman period at Duxford: Hinxton Road, Cambridgeshire (Lyons 2011) and the later Roman period at Higham Ferrers: Kings Meadow Lane, Northamptonshire (Lawrence and Smith 2009). Iron working on Exmoor, Devon at Sherracombe Ford proved

to have begun shortly after the Roman conquest and continued into the later third century cal AD, whilst the industry at Blacklake Wood was active later in the Roman period (*see* below, p00).

A range of sites of post-Roman date were subject to formal programmes of chronological modelling. The long sequence of deposits above the late Roman bath-house at Binchester, Durham proved to have been formed with unexpected rapidity (Marshall *et al* 2010) and, in addition to the national corpus of graves included in the Anglo-Saxon chronology project described above (Bayliss *et al* 2013b), key samples were also dated from the contemporary cemetery at Dover Buckland, Kent (Bayliss *et al* 2012b). Early Saxon settlement was dated at Carlton Coleville: Bloodmoor Hill, Suffolk, where a specific attempt was made to refine the dating of the ceramic sequence (Marshall *et al* 2009).

Middle Saxon occupation sites were dated at Ainsbrook, North Yorkshire (*see* below, p00) and Brandon, Suffolk (Marshall and Bayliss 2014), and a fish-trap off the suffolk coast at Holbrook Bay proved to be of seventh-century date (*see* below, p00). Using the archives of a number of sites excavated in the 1980s, we also addressed the question of when Ipswich Ware first appeared in Lundenwic (Marshall *et al* 2012). A causeway at Street in the Somerset Levels was probably of eighth- or ninth-century date (Brunning 2013). A cemetery, potentially associated with an encampment of King Harald Hardraada in AD 1066, at Riccall, North Yorkshire was probably used over a much longer period that previously thought (Hall *et al* 2008).

Between 2003 and 2006, the first attempts were made to wiggle-match tree-ring sequences from standing buildings using AMS dating. This approach had previously only been employed using conventional high-precision measurements (eg at Holme-next-the-Sea, Norfolk; Bayliss *et al* 1999), but the increasing precision of AMS measurements (Bayliss 2016, fig 1) made such applications now potentially feasible for the small samples recovered by coring from structural timbers in buildings.

Applications ranged from studies on structures in the care of English Heritage, such as Warkworth Castle, Northumberland (Fig 2; Arnold *et al* 2006) or Baguley Hall, Greater Manchester (Hamilton *et al* 2007), to buildings where information was required to guide decisions on conservation or repair (Bayliss *et al* 2006; 2014).

Multi-period age-depth models for the chronologies of sediment sequences at Sutton Common (Gearey *et al* 2009), and from Thorne and Hatfield Moors, Yorkshire (Marshall 2013) should also be noted.

Already published models in due course will be reinterpreted and remodelled. Radiocarbon dates that originally simply provided spot dates for wooden structures in the foreshore or for the organic sediments surviving at particular locations, will in time become part of chronological models for other problems (the currency of particular forms of fish-trap, or the date of past sea level, for example).

It is in the creation of new models that the detailed information contained in this datelist will prove invaluable. It will allow the necessary critical assessment of sample character and taphonomy, and measurement accuracy, to be made. This will allow informed decisions to be made about how each radiocarbon date is most realistically incorporated into a particular model.

Using the Datelist

Radiocarbon determinations are identified by a unique laboratory code. So, for example, OxA- is the code for the Oxford Radiocarbon Accelerator Unit, and OxA-15791 was the 15,791st measurement produced by the laboratory. This code is the internationally agreed identifier by which every radiocarbon determination can be traced. OxA-15791 refers to the result produced on human bone F 005 from Kinsey Cave, North Yorkshire and only to that measurement. An index of these codes is therefore provided to enable further details of dates cited elsewhere to be easily traced.

A more traditional index of key terms is also provided. This enables dates from particular sites, or of particular materials, or with particular archaeological associations to be traced (eg dates relating to the elm decline or Collared Urns).

Acknowledgments

This datelist has been compiled and edited by Kate Cullen, on the basis of information provided by the submitters of the samples dated and by the radiocarbon laboratories. We are grateful to all the submitters of the samples included in this datelist, who have generously responded to our requests for information. Bisserka Gaydarksa kindly collated details of the replicate measurements.

Design has been the responsibility of Mark Simmons, and the overall production of the volume has been overseen by John Hudson. The information has been output from the Historic England Radiocarbon Database. This has been developed over many years, successively by Paul Cheetham, Sarah Hill, Manuela Lopez, Marcos Guillen, Mike Gratton, David Head, Carlton Carver, and Gordon Mackay. Henriette Granlund Marsden kindly proofread this volume.

I am particularly indebted to Christopher Bronk Ramsey, Gordon Cook, Stephen Hoper, and Johannes van der Plicht who have all checked through the datelist and contributed materially to the accuracy of the information in this introduction. Radiocarbon dating is a complex and labour-intensive process which takes time. It would be impossible without the dedicated attention of the laboratory staff to each and every sample. We are grateful to Stephen Hoper, James McDonald, and Michelle Thompson for processing and dating the samples at the Queen's University, Belfast, to Robert Anderson, Andrew Dougans, Elaine Dunbar, Stewart Freeman, Philip Naysmith and Sheng Xu for similarly processing and dating samples dated at the Scottish Universities Environmental Research Centre, to Angela Bowles, Fiona Brock, David Chivall, Jane Davies, Peter Ditchfield, Martin Humm, Philip Leach, Celia Sykes, and Christine Tompkins for undertaking the measurements at the Oxford Radiocarbon Accelerator Unit, and to Anita Aerts-Bijma, Henk Been, Fsaha Ghebru, Bert Kers, Stef Wijma and Dicky van Zonneveld for dating the samples processed at the Rijksuniversiteit Groningen.

Alex Bayliss
Historic England, 4th Floor, Cannon Bridge House,
25 Dowgate Hill, London, EC4R 2YA

Ainsbrook, North Yorkshire

Project manager: R Finlayson and R Hall (York Archaeological Trust), 2005

Archival body: York Archaeological Trust

Description: Ainsbrook, a site in North Yorkshire, first came to attention in late 2003. Metal detectorists declared to the Finds Liaison Officer at The Yorkshire Museum for the Portable Antiquities Scheme that they had discovered a group of Viking-Age artefacts that might represent a burial. The detectorists had discovered many items of early medieval date, as well as some other material. The area extends over some 31ha.

Concerned at the rate of erosion on the site, York Archaeological Trust carried out a limited excavation of the findspot. This was designed to look for further traces of the circumstances of burial, and recover any other information and objects. They demonstrated that this was not a Viking ship burial, but revealed that there were archaeological features in the vicinity of the objects, albeit their date and purpose could not be defined.

English Heritage then undertook a geophysical survey of part of the extensive site as defined by the detectorists' pattern of findspots. This survey revealed part of a massive curving ditch-like feature.

These investigations revealed that the site contained a large, possibly D-shaped ditched enclosure of unknown date, a pattern of apparently Iron Age/Romano-British field systems and associated features, and medieval ridge and furrow. This suggested that the Ainsbrook site was potentially an archaeological resource of the first rank for the study of economy and society, power and prestige in the late first century BC and the first millennium AD in the Yorkshire region; it may well be of national or even international significance.

Objectives: to provide a date for the inhumation (SK1), and to provide a chronology for the occupation of the site.

Final comment: P Marshall (13 October 2014), the radiocarbon dating has confirmed the Anglo-Saxon date for the inhumation (SK1). This individual was interred a short time after a period of pit digging had taken place. The dating of other isolated features contributes to understanding the period over which the site was episodically used.

References: Finlayson 2006
                 Hall 2005

SUERC–10161 1270 ±35 BP

$\delta^{13}C$: -25.0‰

Sample: Sample 103; context 22036 A, submitted on 22 February 2006 by R Finlayson

Material: charcoal: Salicaceae, single fragment (R Gale 2006)

Initial comment: the charcoal in this sample was recovered from material 22036, surrounding heat shattered cobbles at the base of circular cut feature 22039 towards the northern end of trench 22. While the heat shattered cobbles suggested *in situ* burning, there was no indication of the effects of heat, or scorching to the surrounding area. The feature lay within the area encompassed by the large enclosure and also within an area defined by a curving ditch. No other dating evidence was recovered from its fill. The feature was stratigraphically below 22004, and above 22027 and 22037 (from which other samples were also submitted).

Objectives: to establish the date of this material for which no other dating evidence is available. The pit is a feature which may relate to the use of the enclosure, and it would be useful to know if the activity it represents relates to a datable period within the use of the large enclosure. One of four closely stratigraphically related samples.

Calibrated date: *1σ:* cal AD 670–780
                      2σ: cal AD 660–870

Final comment: P Marshall (13 October 2014), the pit features within the D-shaped enclosure, contexts (14002, 5001, and pit 22039) all produced Anglo-Saxon dates, apart from the lowest fill (context 22037) of pit 22039 (SUERC-10173). Chronological modelling of the dates (SUERC-10161–4 and SUERC-10170–2) provide an estimate for the start of Anglo-Saxon pit digging of *cal AD 640–775 (95% probability)* and probably in *cal AD 640–775 (68% probability)*.

Laboratory comment: English Heritage (6 July 2006), the two results from context 22036 are statistically consistent (T'=0.2; T'(5%)=3.8; ν=1; Ward and Wilson 1978) and reliably date the context as they could be of the same actual age.

References: Ward and Wilson 1978

SUERC–10162 1290 ±35 BP

$\delta^{13}C$: -27.0‰

Sample: Sample 103; context 22036 B, submitted on 22 February 2006 by R Finlayson

Material: charcoal: *Quercus* sp., sapwood, single fragment (R Gale 2006)

Initial comment: as SUERC-10161

Objectives: as SUERC-10161

Calibrated date: *1σ:* cal AD 660–770
                      2σ: cal AD 650–780

Final comment: see SUERC-10161

Laboratory comment: see SUERC-10161

SUERC–10163 1200 ±35 BP

$\delta^{13}C$: -24.5‰

Sample: Sample 104; context 5001 A, submitted on 22 February 2006 by R Finlayson

Material: charcoal: *Ulex/Cytisus* sp., single fragment (R Gale 2006)

Initial comment: the upper fill of a posthole or pit 5002, *c* 0.50m wide and *c* 0.76m deep. The feature was located close to the northern return of the enclosure ditch. The only dating evidence within this fill was a glass fragment (sf 757) of uncertain date. This upper fill is likely to provide a date by which this feature was no longer in use.

Objectives: to establish the date of this material for which limited other dating evidence is available. The pit is a feature which may relate to the use of the enclosure it would be useful to know if the activity it represents relates to a datable period within the use of the large enclosure or not.

Calibrated date: *1σ:* cal AD 770–890
                 2σ: cal AD 690–950

Final comment: see SUERC-10161

Laboratory comment: English Heritage (6 July 2006), the two results from context 5001 are statistically consistent (T′=1.0; T′(5%)=3.8; ν=1; Ward and Wilson 1978) and reliably date the context as they could be of the same date.

References: Ward and Wilson 1978

SUERC–10164 1250 ±40 BP

δ¹³C: -26.7‰

Sample: Sample 104; context 5001 B, submitted on 22 February 2006 by R Finlayson

Material: charcoal: *Corylus* sp., single fragment (R Gale 2006)

Initial comment: as SUERC-10163

Objectives: as SUERC-10163

Calibrated date: *1σ:* cal AD 680–780
                 2σ: cal AD 660–890

Final comment: see SUERC-10163

Laboratory comment: see SUERC-10163

SUERC–10168 3715 ±35 BP

δ¹³C: -23.0‰

Sample: Sample 105; context 19003 A, submitted on 22 February 2006 by R Finlayson

Material: charcoal: *Quercus* sp., roundwood; 5 growth rings, single fragment (R Gale 2006)

Initial comment: this charcoal sample was from the backfill 19003 of curving ditch 19004 (a possible barrow). The feature was *c* 0.50m from ground surface below the ploughsoil and subsoil. Three sherds of pottery recovered from the sandy backfill 19003 of this ditch, including a sherd of Cord Impressed Ware, suggests a Bronze Age date.

Objectives: to help establish the accuracy of the dating of the samples where there is other secure dating evidence for comparison.

Calibrated date: *1σ:* 2200–2030 cal BC
                 2σ: 2210–1980 cal BC

Final comment: P Marshall (13 October 2014), the fill [19003] of curving ditch [19004] contained ceramics that suggested a Bronze Age date. Although not at odds with the radiocarbon dates, the latest date SUERC-10169 only provides a *terminus post quem* for this context.

Laboratory comment: (6 July 2006), the two measurements from context 19003 are not statistically consistent (T′=60.4; T′(5%)=3.8; ν=1; Ward and Wilson 1978), and suggest that either residual or intrusive material was dated in this context. The latest date (SUERC-10169) should be used to provide a *terminus post quem* for this the context. This is the only Bronze Age feature dated by this means from this site.

References: Ward and Wilson 1978

SUERC–10169 3330 ±35 BP

δ¹³C: -24.0‰

Sample: Sample 105; context 19003 B, submitted on 22 February 2006 by R Finlayson

Material: charcoal: *Fraxinus* sp., roundwood; single fragment (R Gale 2006)

Initial comment: as SUERC-10168

Objectives: as SUERC-10168

Calibrated date: *1σ:* 1660–1540 cal BC
                 2σ: 1730–1510 cal BC

Final comment: see SUERC-10168

Laboratory comment: see SUERC-10168

SUERC–10170 1210 ±35 BP

δ¹³C: -29.3‰

Sample: Sample 102; context 14002 A, submitted on 22 February 2006 by R Finlayson

Material: charcoal: Pomoideae, single fragment (R Gale 2006)

Initial comment: located at the southern edge of trench 14, *c* 1m below ground, this sample came from the upper fill 14002 of a pebble-lined pit 14022 which lay within the area encompassed by a large enclosure, close to its entrance. The large enclosure may have been utilised from the Bronze Age to the Roman period. No other datable material was recovered from this pit.

Objectives: to establish the date of this pit backfill for which no other dating evidence is available. The pit is a feature which may relate to the use of the enclosure, and it would be useful to know if the activity it represents relates to a datable period within the use of the enclosure or not.

Calibrated date: *1σ:* cal AD 720–890
                 2σ: cal AD 680–940

Final comment: see SUERC-10161

Laboratory comment: English Heritage (6 July 2006), the two measurements on samples from context 14002 are statistically consistent (T′=1.0; T′(5%)=3.8; ν=1; Ward and Wilson 1978) and reliably date the context as they could be of the same actual age.

References: Ward and Wilson 1978

SUERC–10171 1260 ±35 BP

δ¹³C: -26.5‰

Sample: Sample 102; context 14002 B, submitted on 22 February 2006 by R Finlayson

Material: charcoal: Salicaceae, single fragment (R Gale 2006)

Initial comment: as SUERC 10170

Objectives: as SUERC 10170

Calibrated date: 1σ: cal AD 680–780
                       2σ: cal AD 660–880

Final comment: see SUERC 10170

Laboratory comment: see SUERC 10170

SUERC–10172 1270 ±35 BP

δ¹³C: -25.3‰

Sample: Sample 106; context 22004, submitted on 22 February 2006 by R Finlayson

Material: charcoal: Salicaceae, single fragment (R Gale 2006)

Initial comment: the charcoal in this sample was recovered from the upper fill 22004 of a cobble filled feature 22039. 22004 was a thin layer of fine grained sand strained dark by charcoal with occasional burnt bone and charcoal flecks. The lowest fill of the feature comprised heat-shattered cobbles with sand and charcoal in the voids 22036, and these cobbles appeared to have been shattered *in situ*, but there was no scorching or burning of surrounding sand suggesting that this could not have been the case. The cut widened towards the top and the cobbles 22005, although burnt and shattered had clearly been re-deposited with no suggestion of *in situ* burning. The cobbles may have acted as packing, although there was no void representing a removed timber. This feature may have acted as a posthole or post-pad or served another unknown function. The feature lay within the area encompassed by the large enclosure and also within an area defined by a curving ditch. No other dating evidence was recovered from its fill. It was stratigraphically above 22036, 2202, and 22037 (from which other samples were also submitted).

Objectives: to establish the date of this material for which no other dating evidence is available. The pit is a feature which may relate to the use of the enclosure, and it would be useful to know if the activity it represents relates to a datable period within the use of the large enclosure or not.

Calibrated date: 1σ: cal AD 670–780
                       2σ: cal AD 660–870

Final comment: see SUERC-10161

SUERC–10173 2185 ±30 BP

δ¹³C: -27.2‰

Sample: Sample 108; context 22037, submitted on 22 February 2006 by R Finlayson

Material: charcoal: cf *Corylus* sp., single fragment (R Gale 2006)

Initial comment: at the eastern end of trench 22, stratigraphically above a silted up ditch or stream bed 22044, was a layer very slightly silty, fine-grained sand 22037, which appeared to be an accumulation or build up of material incorporating debris including small lenses of burnt sand, occasional charcoal, and burnt cobbles. Embedded in the top of this material was a scatter of burnt and heat shattered cobbles 22038. This horizon extended over an area *c* 14m square, extending beyond the northern and eastern edges of excavation. These cobbles appeared to have been dumped rather than placed with the intention of serving any structural purpose, although it is conceivable that, located close to the edge of the former stream bed, they could have been intended to mitigate against what may have been a waterlogged or area of 'soft' ground. The slightly silty sand 22027/22041 above the horizon cobbles 22038 was similar to the build-up 22037 and each of these layers was *c* 0.1–0.15m deep. 22027 extended further south than 22037, beyond the edge of excavation. The context was stratigraphically below 22027, 22036, and 22004 (from which samples were also submitted).

Objectives: to compare with other samples closely stratigraphically related.

Calibrated date: 1σ: 360–190 cal BC
                       2σ: 370–160 cal BC

Final comment: P Marshall (13 October 2014), the Iron Age date from the lowest fill of pit [22039] suggests that either residual material was incorporated when the cobbles were placed at the base or that it was brought in through unidentified animal burrowing.

SUERC–10511 1095 ±35 BP

δ¹³C: -21.4‰
δ¹⁵N (diet): +10.1‰
C/N ratio: 4.2

Sample: Sample 101; context 11017, SK1 B, submitted on 22 February 2006 by R Finlayson

Material: human bone (femur) (K Tucker 2005)

Initial comment: located in the south-east quadrant of trench 11, *c* 0.7m below ground. Inhumation 11017, where the upper body had been truncated by a modern cut 11010, but the lower part of the body appeared to be undisturbed and was articulated. No grave cut was discernable. A fragment of the back of a skull lay to the north of 11017 and may have represented the remains of another inhumation, 11018.

Objectives: to establish the date of this inhumation, for which no other dating evidence is available.

Calibrated date: 1σ: cal AD 890–1000
                       2σ: cal AD 880–1020

Final comment: P Marshall (13 October 2006), the inhumation, although Anglo-Saxon in date, has a *98% probability* of being older than the estimated date for the end of pit digging.

Laboratory comment: English Heritage (6 July 2006), as the measurements on two distinct bones from the burial were statistically consistent (T'=0.5; T'(5%)=3.8; ν=1; Ward and Wilson 1978), these measurements were combined by taking a weighted mean (1078 ±25 BP) prior to calibration, providing an estimate for the death of the individual of cal AD 890–1020 at 95% confidence (Reimer *et al* 2004).

References: Reimer *et al* 2004
                   Ward and Wilson 1978

SUERC–10642 1060 ±35 BP

$\delta^{13}C$: -20.6‰
$\delta^{15}N$ *(diet):* +11.6‰
C/N ratio: 3.4

Sample: Sample 101; context 11017, SK1 A, submitted on 22 February 2006 by R Finlayson

Material: human bone (part of acetabulum oscoxa) (K Tucker 2005)

Initial comment: a replicate of SUERC-10511.

Objectives: as SUERC-10511

Calibrated date: 1σ: cal AD 970–1020
                        2σ: cal AD 890–1030

Final comment: see SUERC-10511

Laboratory comment: see SUERC-10511

Anglo-Saxon graves and grave goods (female graves, Bedfordshire, Buckinghamshire, Cambridgeshire, Gloucestershire, Humberside, Kent, Lincolnshire, North Yorkshire, Oxfordshire, Suffolk and Sussex

Location: *see* individual sites
Project manager: J Hines and F G McCormac
                   (Cardiff University and Queen's
                   University, Belfast), 1998–2013

Description: typological analysis, seriation by correspondence analysis, high-precision radiocarbon dating, and Bayesian chronological modelling of the corpus of furnished Anglo-Saxon female graves covering the later part of the early Anglo-Saxon period in England. The final seriation included 300 individual grave-assemblages and 81 different artefact-types. Fifty-two radiocarbon dated graves could be included in the chronological models of this seriation.

Objectives: to provide a relative and absolute chronology for furnished female graves of the later part of the early Anglo-Saxon period in England.

Final comment: A Bayliss (27 September 2013), a relative chronology for the female graves was successfully obtained through correspondence analysis (Bayliss *et al* 2013b, fig 7.62). This series was partitioned on the basis of leading artefact-types (Bayliss *et al* 2013b, e-fig 7.3) and on the basis of the two-dimensional map of the correspondence analysis producing the same phases (Bayliss *et al* 2013b, fig 7.62a). A chronological model combining these identical partitions with the associated radiocarbon dates is presented by Bayliss *et al* 2013b fig 7.65). All analyses for this project have been undertaken using the project-specific calibration data of McCormac *et al* (2004; 2008). Full details of the analysis of the female graves is provided by Bayliss *et al* (2013b, chapter 7).

Laboratory comment: English Heritage (27 September 2013), the entries below list the radiocarbon dates from female graves that were commissioned as part of this project. Relevant radiocarbon dates from other graves were also incorporated in the statistical analyses. These are BloodH12 (UB-4914, 1397 ±18 BP) and BloodH22 (UB-4910, 1365 ±15 BP) (Lucy *et al* 2009, 322–9) ; But1674 (UB-4042, 1407 ±20 BP) and But4275 (UB-4077, 1476 ±24 BP) (Scull 2009, 261–7); BuD391a (UB-4959, 1420 ±20 BP) and BuD391B (UB-4960, 1611 ±181 BP) (Parfitt and Anderson 2012, 360–6); Ber008 (OxA-18188, 1635 ±26 BP; Ber035 (OxA-18189, 1583 ±25 BP), and Ber073 (OxA-18214, 1569 ±29 BP) (Hills and O'Connell 2009); and WHes176 (HAR-8242, 1510 ±40 BP) and WHes177 (HAR-8243, 1610 ±40 BP) (Haughton and Powlesland 1999, 308–12).

References: Bayliss *et al* 2013b
                   Haughton and Powlesland 1999
                   Hills and O'Connell 2009
                   Lucy *et al* 2009
                   McCormac *et al* 2004
                   McCormac *et al* 2008
                   Parfitt and Anderson 2012
                   Scull 2009

Anglo-Saxon graves and grave goods (female graves): Apple Down, Compton, West Sussex

Location: SU 79431509
                   Lat. 50.55.44 N; Long. 00.52.12 W
Project manager: A Down (Chichester Museum), 1982–7
Archival body: Chichester District Museum

Description: an Anglo-Saxon cemetery of the fifth to seventh centuries.

Objectives: to test and refine the artefact dating of furnished female burials of the late-sixth and seventh centuries AD.

Final comment: A Bayliss (27 September 2013), three female graves have been dated from this site (ApD107, ApD117, and ApD134), all dating to the mid-sixth century. One replicate analysis failed.

References: Bayliss *et al* 2013b
                   Down and Welch 1990

UB–4965 1475 ±21 BP

$\delta^{13}C$: -20.9 ±0.5‰
$\delta^{13}C$ *(diet):* -20.4 ±0.2‰
$\delta^{15}N$ *(diet):* +8.1 ±0.3‰
C/N ratio: 3.2 %C: 42.1 %N: 15.3

Hydroxyproline	Aspartic	Glutamic	Proline	Glycine	Alanine	Arginine
80.0	56.0	89.0	129.0	318.0	112.0	47.0

Sample: Grave 117, submitted on 11 August 2003 by C Scull

Material: human bone (396g) (femurs and right humerus) (S Mays 2003)

Initial comment: an inhumation of a moderately preserved skeleton of an elderly female buried with bead-types BE1–Koch49/50, BE1–Koch20Ye, BE1–Koch34Wh, BE1–Koch34Bl, and BE1–CylRound.

Objectives: to test the integrity of the early seventh-century 'gap' apparent for female burials in the interim seriation.

Calibrated date: *1σ:* cal AD 565–615
                   2σ: cal AD 545–645

Final comment: A Bayliss (26 September 2013), this grave falls into the first phase of the female seriation (AS-FB), and is dated to *cal AD 545–580 (95% probability; UB-4965 (ApD117)*; Bayliss *et al* 2013b, fig 7.65).

Laboratory comment: English Heritage (26 September 2013), a second sample from this burial failed to produce sufficient carbon for radiocarbon dating, although a duplicate stable isotope and amino acid analyses were undertaken.

Laboratory comment: Rafter Radiocarbon Laboratory (26 September 2013), duplicate stable isotope analyses were run for carbon and nitrogen, in both cases the measurements are statistically consistent and weighted means are provided here (*see* Beavan *et al* 2011; tables 1–2 for full details). The replicate amino acid analyses on this skeleton show greater variability than the ±5% quoted error for hydroxyproline, proline, and glycine. This skeleton has a low Gly/Asp ratio (5.7), but the C:N ratio (3.3) suggests that protein preservation was adequate for accurate dating. Further details on the replicate amino acid analyses can be found in Beavan *et al* (2011, tables 3–4).

References: Beavan *et al* 2011

UB–5208 1481 ±20 BP

δ¹³C: -20.6 ±0.5‰
δ¹³C (diet): -20.3 ±0.4‰
δ¹⁵N (diet): +7.9 ±0.3‰
C/N ratio: 3.3 *%C:* 42.3 *%N:* 15.0

Hydroxyproline	Aspartic	Glutamic	Proline	Glycine	Alanine	Arginine
74.0	58.0	94.0	103.0	348.0	116.0	47.0

Sample: Grave 107; female B1, submitted on 12 September 2005 by C Scull

Material: human bone (201g) (femurs) (S Mays 2005)

Initial comment: inhumation of a moderately preserved adult female skeleton accompanied by bead-types BE1–DotReg, BE1–Koch20Ye, BE1–Koch20Wh, and BE1–Koch34Wh.

Objectives: to test and refine the date and integrity of preliminary phase B1 for the female graves in the interim seriation.

Calibrated date: *1σ:* cal AD 565–610
                   2σ: cal AD 545–640

Final comment: A Bayliss (26 September 2013), this grave falls into the first phase of the female seriation (AS-FB), and is dated to *cal AD 545–580 (95% probability; UB-5208 (ApD107)*; Bayliss *et al* 2013b, fig 7.65).

Laboratory comment: Rafter Radiocarbon Laboratory (26 September 2013), this skeleton had a slightly low Gly/Asp ratio (6.0), but the C:N ratio (3.3) suggests that protein preservation was adequate for accurate dating.

Anglo-Saxon graves and grave goods (female graves): Aston Clinton, Tring Hill, Buckinghamshire

Location: SP 905113
                Lat. 51.47.32 N; Long. 00.41.17 W

Project manager: R Masefield (RPS), 2001

Archival body: Buckinghamshire County Council

Description: multi-period activity excavated in advance of road construction.

Objectives: to test and refine the artefact dating of furnished female burials of the late sixth and seventh centuries AD.

Final comment: A Bayliss (27 September 2013), one grave was dated from this site falling in the mid-sixth century.

UB–4975 1517 ±19 BP

δ¹³C: -21.0 ±0.5‰
δ¹³C (diet): -20.8 ±0.3‰
δ¹⁵N (diet): +9.4 ±0.4‰
C/N ratio: 3.3 *%C:* 39.3 *%N:* 14.0

Hydroxyproline	Aspartic	Glutamic	Proline	Glycine	Alanine	Arginine
79.0	60.0	97.0	137.0	321.0	112.0	47.0

Sample: Grave 12, submitted on 2 September 2003 by C Scull

Material: human bone (300g) (left femur and tibia) (S Mays 2003)

Initial comment: an inhumation of a moderately well-preserved skeleton of an elderly female buried with bead-types BE1–MelonBl, BE1–DotReg, BE1–Koch20Ye, BE1–Koch34Ye, and a brooch of type BR2–a which is a skeuomorph of a BR2–b3.

Objectives: to test the integrity and refine the dating of phase B3 in the preliminary seriation.

Calibrated date: *1σ:* cal AD 540–575
                   2σ: cal AD 435–600

Final comment: A Bayliss (26 September 2013), this grave falls into the first phase of the female seriation (AS-FB), and is dated to *cal AD 540–575 (95% probability; UB-4975 (AstCli12)*; Bayliss *et al* 2013b, fig 7.65).

Laboratory comment: Rafter Radiocarbon Laboratory (26 September 2013), this skeleton has a slightly low Gly/Asp ratio (5.3), but the C:N ratio (3.3) suggests that protein preservation was adequate for accurate dating.

Anglo-Saxon graves and grave goods (female graves): Buckland Dover, Kent

Location: TR 30874291
                Lat. 51.88.17 N; Long. 01.19.59 E

Project manager: K Parfitt and T Anderson (Canterbury
                    Archaeological Trust), 1994

Archival body: Dover Museum

Description: almost 250 early Anglo-Saxon burials recovered in the southern part of the Buckland Dover cemetery to the south of the railway cutting.

Objectives: to test and refine the artefact dating of furnished female burials of the late sixth and seventh centuries AD.

Final comment: A Bayliss (27 September 2013), three female graves were dated as part of this project, but a further two relevant results are available from previous research (BuD391A and BuD391B). Four of these graves date to the mid-sixth century and one to the mid-seventh century.

References: Parfitt and Anderson 2012

UB–6472 1550 ±19 BP

δ¹³C: -20.1 ±0.5‰
δ¹³C (diet): -19.7 ±0.2‰
δ¹⁵N (diet): +9.7 ±0.2‰
C/N ratio: 3.2 %C: 41.0 %N: 15.1

Hydroxyproline	Aspartic	Glutamic	Proline	Glycine	Alanine	Arginine
77.0	53.0	95.0	105.0	339.0	117.0	50.0

Sample: Grave 222, submitted on 3 May 2005 by C Scull

Material: human bone (296g) (femurs, tibiae, left humerus and ulna, skull vault) (S Mays 2005)

Initial comment: an inhumation of a poorly preserved skeleton of an elderly female, buried with buckle-types BU3 and BU4–e, brooch-type BR2–b2, and bead-types BE1–MelonBl, BE1–Koch49/50, BE1–CylPen, BE1–Koch34Ye, and BE1–Koch34Wh.

Objectives: to test and refine the dating and integrity of phase B1 in the preliminary female seriation.

Calibrated date: 1σ: cal AD 430–550
2σ: cal AD 425–565

Final comment: A Bayliss (26 September 2013), this grave falls into the first phase of the female seriation (AS-FB), and is dated to *cal AD 530–570 (95% probability; UB-6472 (Bud222)*; Bayliss *et al* 2013b, fig 7.65).

Laboratory comment: Rafter Radiocarbon Laboratory (26 September 2013), duplicate stable isotope analyses were run for carbon and nitrogen, in both cases the measurements are statistically consistent and weighted means are provided here (*see* Beavan *et al* 2011, tables 1–2 for full details).

References: Beavan *et al* 2011

UB–6473 1572 ±22 BP

δ¹³C: -20.1 ±0.5‰
δ¹³C (diet): -19.8 ±0.4‰
δ¹⁵N (diet): +10.4 ±0.3‰
C/N ratio: 3.2 %C: 39.5 %N: 14.4

Hydroxyproline	Aspartic	Glutamic	Proline	Glycine	Alanine	Arginine
75.0	56.0	99.0	98.0	340.0	114.0	50.0

Sample: Grave 250, submitted on 3 May 2005 by C Scull

Material: human bone (292g) (left tibia and fibula, right femur) (S Mays 2005)

Initial comment: an inhumation of a moderately preserved adult female skeleton buried with bead-types BE1–Reticella, BE1–CylPen, and BE1–CylRound, BE1–Mosaic, wire ring-type WR3, pi-type PI1–f, and pendant-types PE2–e and PE3–b.

Objectives: to test and refine the integrity and dating of phase B2b in the interim female seriation.

Calibrated date: 1σ: cal AD 425–540
2σ: cal AD 415–550

Final comment: A Bayliss (26 September 2013), this is a diverse grave-asemblage that caused recurrent problems when incorporated in the correspondence analysis. It appears to have atypical early incidences of wire ring-type WR3 and pin-type PI1–f, artefact-types which have not been included in the final seriation for this reason.

This grave falls into the first phase of the female seriation (AS-FB), and is dated to *cal AD 525–565 (95% probability; UB-6473 (BuD250)*; Bayliss *et al* 2013b, fig 7.65).

Laboratory comment: Rafter Radiocarbon Laboratory (26 September 2013), this skeleton has a slightly low Gly/Asp ratio (6.0), but the C:N ratio (3.2) suggests that protein preservation was adequate for accurate dating.

UB–6476 1592 ±17 BP

δ¹³C: -20.0 ±0.5‰
δ¹³C (diet): -19.7 ±0.4‰
δ¹⁵N (diet): +8.9 ±0.3‰
C/N ratio: 3.1 %C: 41.9 %N: 15.6

Hydroxyproline	Aspartic	Glutamic	Proline	Glycine	Alanine	Arginine
75.0	49.0	88.0	104.0	354.0	115.0	51.0

Sample: Grave 339, submitted on 3 May 2005 by C Scull

Material: human bone (369g) (femurs, tibiae, and humeri) (S Mays 2005)

Initial comment: an inhumation of a poorly preserved adult female skeleton buried with buckle-types BU2–d and BU2–h, pendant-type PE2–a, and bead-types BE1–Reticalla, BE1–MelonY-G, BE1–MelonBl, BE1–Koch49/50, BE1–CylPen, BE1–Mosaic, and BE1–CylRound.

Objectives: to test and refine the dating and integrity of phase B2b in the interim female seriation.

Calibrated date: 1σ: cal AD 420–535
2σ: cal AD 410–540

Final comment: A Bayliss (26 September 2013), this grave falls into the first phase of the female seriation (AS-FB), and is dated to *cal AD 525–550 (95% probability; UB-6476 (BuD339)*; Bayliss *et al* 2013b, fig 7.65).

Anglo-Saxon graves and grave goods (female graves): Castledyke South, Barton-on-Humber, Humberside

Location:	TA 031217
	Lat. 53.40.53 N; Long. 00.26.21 W
Project manager:	M Foreman (Humberside Archaeology), 1982–90
Archival body:	Scunthorpe Museum

Description: an extensive Anglo-Saxon cemetery, mostly of inhumations dated by grave goods to the sixth and seventh centuries. The site lies 300m south-west of the late Saxon St Peter's Church and its associated cemetery.

Objectives: to test and refine the artefact dating of furnished female burials of the late sixth and seventh centuries AD.

Final comment: A Bayliss (27 September 2013), seven female graves, and one stratigraphically-related male grave, have been dated from this site (CasD120, CasD013, CasD096, CasD134, CasD182, CasD183, CasD053, and CasD088) spanning the entire period of the female seriation.

References: Drinkall and Foreman 1998

UB–6034 1502 ±17 BP

$\delta^{13}C$: -21.2 ±0.5‰
$\delta^{13}C$ *(diet):* -20.9 ±0.3‰
$\delta^{15}N$ *(diet):* +9.7 ±0.4‰
C/N ratio: 3.6   *%C:* 45.4   *%N:* 14.7

Sample: Grave 120, submitted on 15 December 2003 by J Hines

Material: human bone (280g) (left femur) (S Mays 2003)

Initial comment: an inhumation of a moderately well-preserved teenage ?female skeleton buried with brooch-type BR3–d.

Objectives: to test and refine the dating of burials assigned to phase C in the interim female seriation.

Calibrated date: *1σ:* cal AD 545–595
                        2σ: cal AD 540–610

Final comment: A Bayliss, this grave was assigned to phases AS-FC-E in the female seriation on the basis of the occurrence of brooch-type BR3–d which is confined to these phases. It dates to *cal AD 560–610 (95% probability; UB-6034 (CasD120)*; Bayliss *et al* 2013b, fig 7.65).

UB–6035 1517 ±15 BP

$\delta^{13}C$: -21.6 ±0.5‰
$\delta^{13}C$ *(diet):* -21.3 ±0.3‰
$\delta^{15}N$ *(diet):* +11.0 ±0.4‰
C/N ratio: 2.8   *%C:* 43.1   *%N:* 18.1

Sample: Grave 96, submitted on 15 December 2003 by J Hines

Material: human bone (310g) (right femur and left tibia) (S Mays 2003)

Initial comment: an inhumation of a moderately preserved adult female skeleton buried with bead-types BE1–Koch34Wh and BE2–c, and brooch-type BR3–d.

Objectives: to test and refine the dating of graves asigned to phase B2 in the interim female seriation.

Calibrated date: *1σ:* cal AD 540–575
                        2σ: cal AD 535–600

Final comment: A Bayliss (26 September 2013), this grave falls into the second phase of the female seriation (AS-FC), and is dated to *cal AD 560–600 (95% probability; UB-6035 (CasD096)*; Bayliss *et al* 2013b, fig 7.65).

UB–6036 1421 ±17 BP

$\delta^{13}C$: -21.2 ±0.5‰
$\delta^{13}C$ *(diet):* -20.9 ±0.3‰
$\delta^{15}N$ *(diet):* +11.5 ±0.4‰
C/N ratio: 2.7   *%C:* 42.2   *%N:* 18.3

Sample: Grave 13, submitted on 15 December 2003 by J Hines

Material: human bone (280g) (left femur) (S Mays 2003)

Initial comment: an inhumation of a moderately preserved skeleton of a young adult female buried with brooch-tpe BR3–c and BR3–d.

Objectives: to test and refine the dating of graves of phase C in the interim female seriation.

Calibrated date: *1σ:* cal AD 615–650
                        2σ: cal AD 600–655

Final comment: A Bayliss (27 September 2013), this grave falls into the second phase of the female seriation (AS-FC), and is dated to *cal AD 570–630 (95% probability; UB-6036 (CasD013)*; Bayliss *et al* 2013b, fig 7.65).

UB–6037 1544 ±14 BP

$\delta^{13}C$: -20.9 ±0.5‰
$\delta^{13}C$ *(diet):* -20.6 ±0.3‰
$\delta^{15}N$ *(diet):* +10.0 ±0.4‰
C/N ratio: 3.1   *%C:* 48.5   *%N:* 18.5

Sample: Grave 134, submitted on 15 December 2003 by C Scull

Material: human bone (230g) (left femur) (S Mays 2003)

Initial comment: an inhumation of a moderately preserved teenage ?female skeleton buried with pendant-type PE2–a, bead-type BE1–Reticella, and brooch-type BR3–c.

Objectives: to test and refine the dating of burials allocated to phase B2 in the interim seriation.

Calibrated date: *1σ:* cal AD 435–550
                        2σ: cal AD 425–565

Final comment: A Bayliss (26 September 2013), this grave falls into the first phase of the female seriation (AS-FB), and is dated to *cal AD 535–570 (95% probability; UB-6037 (CasD134)*; Bayliss *et al* 2013b, fig 7.65).

UB–6038 1449 ±14 BP

$\delta^{13}C$: -21.3 ±0.5‰
$\delta^{13}C$ *(diet):* -21.0 ±0.3‰
$\delta^{15}N$ *(diet):* +9.9 ±0.4‰
C/N ratio: 3.1   *%C:* 44.7   *%N:* 16.7

Sample: Grave 183, submitted on 15 December 2003 by J Hines

Material: human bone (260g) (left femur and tibia) (S Mays 2003)

Initial comment: an inhumation of a moderately preserved skeleton of a young adult female buried with bead-types BE1–Koch34Bl and BE1–Dghnt, accessory-type wooden cup, Ae mts, comb, HT, Wooden box, and WBX, buckle-type BU9–a, and pin-type PI2–a. This grave cuts unfurnished male grave CasD182 (UB-6041).

Objectives: to test and refine the dating of phase C in the interim female seriation.

Calibrated date: 1σ: cal AD 600–640
2σ: cal AD 580–650

Final comment: A Bayliss (27 September 2013), this grave falls into the penultimate phase of the female seriation (AS-FD), and is dated to *cal AD 600–650 (95% probability; UB-6038 (CasD183);* Bayliss *et al* 2013b, fig 7.65).

UB–6040 1535 ±15 BP

$\delta^{13}C$: -21.4 ±0.5‰
$\delta^{13}C$ *(diet):* -21.1 ±0.3‰
$\delta^{15}N$ *(diet):* +10.2 ±0.4‰
C/N ratio: 2.9 *%C:* 57.3 *%N:* 23.4

Sample: Grave 53, submitted on 15 December 2003 by J Hines

Material: human bone (380g) (left femur) (S Mays 2003)

Initial comment: an inhumation of a moderatel well-preserved adult female skeleton buried with brooch-type BR3–c and accessory-type Wooden cup, Ae mts.

Objectives: to test and refine the dating of burials assigned to phase B1 in the interim female seriation.

Calibrated date: 1σ: cal AD 535–560
2σ: cal AD 430–575

Final comment: A Bayliss (26 September 2013), this grave falls into the first phase of the female seriation (AS-FB), and is dated to *cal AD 535–570 (95% probability; UB-6040 (CasD053);* Bayliss *et al* 2013b, fig 7.65).

UB–6041 1515 ±15 BP

$\delta^{13}C$: -20.8 ±0.5‰
$\delta^{13}C$ *(diet):* -20.5 ±0.3‰
$\delta^{15}N$ *(diet):* +9.9 ±0.4‰
C/N ratio: 3.1 *%C:* 44.7 *%N:* 16.7

Sample: Grave 182, submitted on 15 December 2003 by J Hines

Material: human bone (240g) (radii, ulnae, right clavicle and humerus) (S Mays 2003)

Initial comment: an inhumation of a modeately well-preserved skeleton of a young adult man. This grave was unfurnished, but cut by CasD183 (UB-6038), which is included in the female seriation and removed the pelvis and legs of this skeleton.

Objectives: to constrain the date of CasD183 (UB-6038) which is included in the female seriation.

Calibrated date: 1σ: cal AD 540–575
2σ: cal AD 535–600

Final comment: A Bayliss (27 September 2013), this grave dates to the middle of the sixth century and does not usefully constrain the date of CasD183 which dates to the early seventh century.

UB–6042 1323 ±13 BP

$\delta^{13}C$: -20.6 ±0.5‰
$\delta^{13}C$ *(diet):* -20.3 ±0.3‰
$\delta^{15}N$ *(diet):* +10.1 ±0.4‰
C/N ratio: 4.2 *%C:* 50.8 *%N:* 14.0

Sample: Grave 88, submitted on 15 December 2003 by J Hines

Material: human bone (290g) (left femur) (S Mays 2003)

Initial comment: an inhumation of a moderately well-preserved skeleton of a young adult female buried with brooch-type BR3–d and wire ring-type WR4.

Objectives: to test and refine the dating of burials assigned to phase C in the interim female seriation.

Calibrated date: 1σ: cal AD 660–680
2σ: cal AD 655–690

Final comment: A Bayliss (27 September 2013), this grave falls into the last phase of the female seriation (AS-FE), and is dated to *cal AD 655–675 (95% probability; UB-6042 (CasD088);* Bayliss *et al* 2013b, fig 7.65). This radiocarbon date has good individual agreement (A:117; Bronk Ramsey 1995, 429) with its position in the seriation, and so despite the suspected presence of exogenous carbon contamination suggested by the C:N ratio, this measurement appears to be accurate.

Laboratory comment: Rafter Radiocarbon Laboratory (27 September 2013), this C:N ratio is the only significant outlier in the Anglo-Saxon dataset (Grubbs test, Z-score 6.985, P<0.05). This result with 50.8%C and 14%N suggests the possible presence of an exogenous carbon contaminant.

Anglo-Saxon graves and grave goods (female graves): Coddenham, Suffolk

Location:	TM 120538 Lat. 52.08.28 N; Long. 01.05.52 E
Project manager:	S Anderson (Suffolk County Council), 1999
Archival body:	Suffolk County Council

Description: a multi-period (Iron Age-Saxon) site excavated in advance of gravel extraction.

Objectives: to test and refine the artefact dating of furnished female burials of the late sixth and seventh centuries AD.

Final comment: A Bayliss (27 September 2013), a single grave has been dated from this site, and falls in the middle decades of the seventh century.

UB–4964 1417 ±16 BP

$\delta^{13}C$: -20.7 ±0.5‰
$\delta^{13}C$ *(diet):* -20.4 ±0.3‰
$\delta^{15}N$ *(diet):* +10.3 ±0.4‰
C/N ratio: 3.1 *%C:* 45.9 *%N:* 17.1

Hydroxyproline	Aspartic	Glutamic	Proline	Glycine	Alanine	Arginine
79.0	54.0	90.0	130.0	321.0	117.0	48.0

Sample: Grave 0308, submitted on 11 August 2003 by C Scull

Material: human bone (332g) (left femur) (S Mays 2003)

Initial comment: an inhumation of a poorly preserved adult female skeleton interred on a bed buried with bead-types BE2–c, BE1–Orange, BE1–Amethyst, BE1–AnnTw, BE2–a, BE1–WoundSp, bucket, Fe frame, pendant-types PE2–c, PE11, and PE7–a, a gold looped tremissis accessory-type of Dagobert I (King of the Merovingian Franks, reigned AD 629–39), a silver sheet impression of a primary sceatt, and wire ring-type WR2.

Objectives: to test and refine the dating of the latest phase of furnished burial in the interim female seriation.

Calibrated date: 1σ: cal AD 630–650
                     2σ: cal AD 605–655

Final comment: A Bayliss (26 September 2013), this grave falls into the last phase of the female seriation (AS-FE), and is dated to *cal AD 635–660 (95% probability; UB-4964 (Cod30)*; Bayliss *et al* 2013b, fig 7.65).

Laboratory comment: Rafter Radiocarbon Laboratory (26 September 2013), this skeleton has slightly low Gly/Ala and Gly/Asp ratios (2.7 and 5.9 respectively), but the C:N ratio (3.1) suggests that protein preservation was adequate for accurate dating.

Anglo-Saxon graves and grave goods (female graves): Lechlade, Butler's Field, Gloucestershire

Location:	SU 213995 Lat. 51.41.35 N; Long. 01.41.31 W
Project manager:	D Miles and S Palmer (Oxford Archaeological Unit), 1985
Archival body:	Corinium Museum

Description: an Anglo-Saxon cemetery. Objectives: to test and refine the artefact dating of furnished female burials of the late-sixth and seventh centuries AD.

Final comment: A Bayliss (27 September 2013), seven female graves were dated from this site (Lec014, Lec138, Lec148, Lec179, Lec172/2, Lec187, and Lec018), one of which dated to the mid-sixth century, the rest of which dated to the mid-seventh century. One further sample Lec136 failed.

References:	Boyle *et al* 1998 Boyle *et al* 2011

UB–4984 1507 ±20 BP

$\delta^{13}C$: -20.7 ±0.5‰

$\delta^{13}C$ *(diet)*: -20.5 ±0.2‰

$\delta^{15}N$ *(diet)*: +8.7 ±0.3‰

C/N ratio: 3.3 *%C:* 38.2 *%N:* 13.7

Hydroxyproline	Aspartic	Glutamic	Proline	Glycine	Alanine	Arginine
76.0	61.0	95.0	128.0	306.0	112.0	47.0

Sample: Grave 18, submitted in September 2003 by C Scull

Material: human bone (362g) (left femur and part of proximal left tibia) (S Mays 2003)

Initial comment: an inhumation of a moderately preserved adult female skeleton accompanied by brooch-types BR1-b and BR2-a, pendant-types PE10-b and PE2-c, wire ring-type W3, and accessory-type Wooden cup, Ae mts.

Objectives: to test and refine the dating of phase B2 in the preliminary seriation.

Calibrated date: 1σ: cal AD 640–595
                     2σ: cal AD 535–610

Final comment: A Bayliss (26 September 2013), this grave assemblage was persistently problematic in the seriation and seems to have included a number of atypical or curated artefact types. The example of pendant-type PE2-c from this grave has a perforated centre and it maybe that this detail is sufficient to identify it as a different type, similarly the example of wire ring-type WR3 is unusual and also seems not to be a true representative of this artefact-type. Consequently, these incidences have been excluded from the seriation matrix. The example of pendant-type PE10-b appears to be an atypical early occurrence of an artefact-type which is predominantly found in later burials, and so again this occurrence has been suppressed. With these incidences suppressed from the seriation, this grave falls into the first phase of the female seriation (AS-FB), and is dated to *cal AD 540–575 (95% probability; UB-4984 (Lech018)*; Bayliss *et al* 2013b, fig 7.65).

Laboratory comment: Rafter Radiocarbon Laboratory (26 September 2013), this skeleton has a low Gly/Ala and Gly/Asp ratios (2.7 and 5.1 respectively), but the C:N ratio (3.3) suggests that protein preservation was adequate for accurate dating. Duplicate stable isotope analyses were run for carbon and nitrogen, in the case of carbon the measurements are statistically consistent (at 95% confidence) but for nitrogen the measurements are only statistically consistent at 99% confidence. Weighted means are provided here (*see* Beavan *et al* 2011; tables 1–2 for full details).

References:	Beavan *et al* 2011

Anglo-Saxon graves and grave goods (female graves): St Peter's Tip, Broadstairs, Kent

Location:	TR 375693 Lat. 51.22.20 N; Long. 01.24.43 E
Project manager:	A Hogarth (Chatham House School Archaeology Society), 1969–71
Archival body:	British Museum

Description: an inhumation cemetery of the late sixth and seventh centuries AD.

Objectives: to test and refine the artefact dating of furnished female burials of the late sixth and seventh centuries AD.

Final comment: A Bayliss (27 September 2013), two female graves have been dated from this site (SPTip208 and SPTip073A), one dating to the decades around AD 600, and the other to the mid-seventh century.

UB–4963 1432 ±21 BP

$\delta^{13}C$: -20.1 ±0.5‰
$\delta^{13}C$ (diet): -19.8 ±0.3‰
$\delta^{15}N$ (diet): +9.5 ±0.4‰
C/N ratio: 3.3 %C: 43.4 %N: 15.4

Hydroxyproline	Aspartic	Glutamic	Proline	Glycine	Alanine	Arginine
82.0	69.0	97.0	137.0	327.0	123.0	48.0

Sample: Grave 208, submitted on 12 August 2003 by C Scull

Material: human bone (278g) (fragments of clavicles, humeri, radii, ulnae, femora, and skull) (S Mays 2003)

Initial comment: an inhumation of a poorly preserved skeleton of an elderly female buried with bead-types BE1–Koch20Ye, BE1–Koch20Wh, BE1–CylPen, BE1–Dot34, BE1–Koch34Ye, BE1–CylRound, BE1–SegGlob, BE1–Orange, and BE1–Cowrie.

Objectives: to test the integrity and refine the dating of phase B3 in the preliminary female seriation.

Calibrated date: 1σ: cal AD 605–650
2σ: cal AD 590–655

Final comment: A Bayliss (26 September 2013), this grave falls into the second phase of the female seriation (AS-FC), and is dated to *cal AD 570–630 (95% probability; UB-4963 (SPTip208)*; Bayliss *et al* 2013b, fig 7.65).

Laboratory comment: Rafter Radiocarbon Laboratory (26 September 2013), this skeleton has a slightly low Gly/Ala and Gly/Asp ratios (2.7 and 4.7 respectively), but the C:N ratio (3.3) suggests that protein preservation was adequate for accurate dating.

UB–6032 1422 ±17 BP

$\delta^{13}C$: -21.2 ±0.5‰
$\delta^{13}C$ (diet): -20.9 ±0.3‰
$\delta^{15}N$ (diet): +8.5 ±0.4‰
C/N ratio: 3.2 %C: 58.8 %N: 21.2

Hydroxyproline	Aspartic	Glutamic	Proline	Glycine	Alanine	Arginine
99.0	51.0	90.0	102.0	328.0	121.0	50.0

Sample: Grave 73A, submitted on 15 December 2003 by J Hines

Material: human bone (260g) (scapulae, humeri, tibiae, femora, fibulae, right ulna, left radius, right clavicle) (S Mays 2003)

Initial comment: an inhumation of a poorly preserved teenage ?female skeleton buried with bead-type BE1–Cowrie, wire ring-type WR1–a, and pendant-type PE4.

Objectives: to test and clarify the dating of graves assigned to phase C2 in the initial female seriation.

Calibrated date: 1σ: cal AD 615–650
2σ: cal AD 600–655

Final comment: A Bayliss (27 September 2013), this grave falls into the last phase of the female seriation (AS-FE), and is dated to *cal AD 635–660 (95% probability; UB-6032 (SPTip073)*; Bayliss *et al* 2013b, fig 7.65).

Laboratory comment: Rafter Radiocarbon Laboratory (27 September 2013), this skeleton has a low Gly/Ala ratio (2.7), but the C:N ratio (3.2) suggests that protein preservation was adequate for accurate dating.

Anglo-Saxon graves and grave goods (male graves) Cambridgeshire, Gloucestershire, Humberside, Kent, Oxfordshire, Suffolk, Surrey and Wiltshire

Location: see individual sites

Project manager: J Hines, F G McCormac, and C Scull (Cardiff University and Queen's University, Belfast), 1998–2013

Description: typological analysis, seriation by correspondence analysis, high-precision radiocarbon dating, and Bayesian chronological modelling of the corpus of furnished Anglo-Saxon male graves covering the later part of the early Anglo-Saxon period in England. The final seriation included 272 individual grave-assemblages and 78 different artefact-types. Thirty-eight radiocarbon dated graves could be included in the chronological models of this seriation.

Objectives: to provide a relative and absolute chronology for furnished male graves of the later part of the early Anglo-Saxon period in England.

Final comment: A Bayliss (25 September 2013), a relative chronology for the male graves was successfully obtained through correspondence analysis (Bayliss *et al* 2013b, fig 6.49). This series was partitioned in two alternative ways, on the basis of leading artefact-types (Bayliss *et al* 2013b, e-fig 6.6) and on the basis of the two-dimensional map of grave-assemblages (Bayliss *et al* 2013b, fig 6.49a). These alternative partitions are very similar. Chronological models combining these partitions with the associated radiocarbon dates are presented by Bayliss *et al* 2013b (figs 6.52–3). All analyses for this project have been undertaken using the project specific calibration data of McCormac *et al* (2004; 2008). Full details of the analysis of the male graves is provided by Bayliss *et al* (2013b, chapter 6).

Laboratory comment: English Heritage (25 September 2013), the entries below list the radiocarbon dates from male graves that were commissioned as part of this project. Relevant radiocarbon dates from other graves were also incorporated in the statistical analyses. These are UB-4958 (BuD375) 1493 ±18 BP (Parfitt and Anderson 2012, 360–6); UB-4074 (BuT2297), 1419 ±23 BP; UB-4039 (BuT3971), 1441 ±20 BP (Scull 2009, 261–7); BM-640, 1427 ±45 BP (SutH01), and UB-4423, 1420 ±28 BP (SutH17) (Carver 2005, 54–5); UB-4880 (BOB91), 1318 ±18 BP (Cowie and Blackmore 2012, 307–12); UB-4642 (WHes72), 1487 ±19 BP, and HAR-8242 (WHes176) 1510 ±40 BP (Haughton and Powlesland 1999, 107–9; 308–9).

References: Bayliss *et al* 2013b
Carver 2005
Cowie and Blackmore 2012
Haughton and Powlesland 1999
McCormac *et al* 2004
McCormac *et al* 2008
Parfitt and Anderson 2012
Scull 2009

Anglo-Saxon graves and grave goods (male graves): Buckland Dover, Kent

Location: TR 30874291
Lat. 51.08.17 N; Long. 01.17.59 E

Project manager: K Parfitt and T Anderson (Canterbury Archaeological Trust), 1994

Archival body: Dover Museum

Description: almost 250 early Anglo-Saxon burials recovered in the southern part of the Buckland Dover cemetery to the south of the railway cutting.

Objectives: to test and refine the artefact dating of furnished male burials of the late sixth and seventh centuries AD.

Final comment: A Bayliss (25 September 2013), four male graves from Dover Buckland were included in this study (BuD264, BuD323, BuD375, and BuD414). All the graves selected for dating fall in the first two phases of the male seriation and date to the third quarter of the sixth century AD.

References: Evison 1987
Parfitt and Anderson 2012

UB–6474 1528 ±17 BP

$\delta^{13}C$: -20.0 ±0.5‰
$\delta^{13}C$ *(diet):* -19.9 ±0.2‰
$\delta^{15}N$ *(diet):* +9.4 ±0.2‰
C/N ratio: 3.2 *%C:* 40.9 *%N:* 15.2

Hydroxyproline	Aspartic	Glutamic	Proline	Glycine	Alanine	Arginine
75.0	49.0	89.0	106.0	347.0	116.0	51.0

Sample: Grave 264, submitted on 3 May 2005 by C Scull

Material: human bone (661g) (left and right femurs and tibia) (S Mays 2005)

Initial comment: inhumation of a poorly preserved adult male skeleton accompanied by shield boss-type SB3–b3+4, sword-type SW3–a, and spearhead-type SP1–a4. This grave cuts grave BuD347.

Objectives: to test and refine the dating and integrity of preliminary male phase A2c.

Calibrated date: *1σ:* cal AD 535–565
2σ: cal AD 430–585

Final comment: A Bayliss (25 September 2013), this grave falls in the first phase of the male seriation (AS-MB/AS-Mp), and is dated to *cal AD 530–560 (95% probability; UB-6477(BuD414);* Bayliss *et al* 2013b, figs 6.52–3).

Laboratory comment: Rafter Radiocarbon Laboratory (25 September 2013), duplicate stable isotope analyses were run for carbon and nitrogen, in both cases the measurements are statistically consistent and weighted means are provided here (*see* Beavan *et al* 2011, tables 1–2 for full details).

References: Beavan *et al* 2011

UB–6475 1491 ±18 BP

$\delta^{13}C$: -20.1 ±0.5‰
$\delta^{13}C$ *(diet):* -19.8 ±0.4‰
$\delta^{15}N$ *(diet):* +9.8 ±0.3‰
C/N ratio: 3.2 *%C:* 37.4 *%N:* 13.8

Hydroxyproline	Aspartic	Glutamic	Proline	Glycine	Alanine	Arginine
75.0	53.0	95.0	101.0	322.0	116.0	52.0

Sample: Grave 323, submitted on 3 May 2005 by C Scull

Material: human bone (301g) (left tibia and humeri) (S Mays 2005)

Initial comment: inhumation of a poorly preserved adult male skeleton buried with shield boss-type SB3–b3+4 and spearhead-type SP3–b. This grave cuts grave BuD324.

Objectives: to test and refine the integrity of preliminary male phase A2b.

Calibrated date: *1σ:* cal AD 555–605
2σ: cal AD 540–620

Final comment: A Bayliss (25 September 2013), this grave falls in the first phase of the male seriation (AS-Mp) based on the two-dimensional map of correspondence analysis, but can only be allocated to either phase AS-MB or AS-MC in the partition based on leading artefact-types on the basis of the occurrence of shield boss-type SB3–b3+4 which is diagnostic of those phases. Consequently, this grave is dated to *cal AD 545–585 (95% probability; UB-6475 (BuD323);* Bayliss *et al* 2013b, fig 6.52) in the partition based on leading artefact-types, or to *cal AD 545–565 (95% probability; UB-6475 (BuD323);* Bayliss *et al* 2013b, fig 6.53) in the partition based on the two-dimensional map of grave-assemblages.

Laboratory comment: Rafter Radiocarbon Laboratory (25 September 2013), this skeleton had slightly low Gly/Asp ratio (6.0), but the C:N ratio (3.2) suggests that protein preservation was adequate for accurate dating.

UB–6477 1570 ±20 BP

$\delta^{13}C$: -20.1 ±0.5‰
$\delta^{13}C$ *(diet):* -19.8 ±0.4‰
$\delta^{15}N$ *(diet):* +9.4 ±0.3‰
C/N ratio: 3.2 *%C:* 40.0 *%N:* 14.8

Hydroxyproline	Aspartic	Glutamic	Proline	Glycine	Alanine	Arginine
75.0	52.0	94.0	105.0	333.0	117.0	51.0

Sample: Grave 414, submitted on 3 May 2005 by C Scull

Material: human bone (330g) (left femur and left tibia) (S Mays 2005)

Initial comment: inhumation of a poorly preserved adult male skeleton buried with shield boss-type SB3–b3+4, sword-type SW4, and spearhead-type SP1–a4.

Objectives: to test and refine the integrity of preliminary male phase A2c.

Calibrated date: *1σ:* cal AD 425–540
2σ: cal AD 415–550

Final comment: A Bayliss (25 September 2013), this grave falls in the first phase of the male seriation (AS-MB/AS-Mp), and is dated to *cal AD 530–560 (95% probability; UB-6477 (BuD414);* Bayliss *et al* 2013b, figs 6.52–3).

Anglo-Saxon graves and grave goods (male graves): Bury St Edmunds, Westgarth Gardens, Suffolk

Location:	TL 843634
	Lat. 52.14.14 N; Long. 00.41.56 E
Project manager:	S West (Suffolk County Council), 1972
Archival body:	Moyses Hall Museum, Bury St Edmunds

Description: a late-fifth to early seventh-century Anglo-Saxon inhumation cemetery.

Objectives: to test and refine the artefact dating of furnished male burials of the late sixth and seventh centuries AD.

Final comment: A Bayliss (26 September 2013), two male graves have been dated from this site (WG11 and WG66), both dating to the mid sixth century.

References:	Bayliss *et al* 2013b
	West 1988

UB–4985 1528 ±18 BP

$\delta^{13}C$: -21.1 ±0.5‰
$\delta^{13}C$ *(diet):* -20.8 ±0.3‰
$\delta^{15}N$ *(diet):* +8.4 ±0.4‰
C/N ratio: 3.3 *%C:* 42.9 *%N:* 15.3

Hydroxyproline	Aspartic	Glutamic	Proline	Glycine	Alanine	Arginine
79.0	56.0	93.0	131.0	321.0	116.0	448.0

Sample: Grave 11; male, Siegmund 5, submitted in September 2003 by C Scull

Material: human bone (355g) (right femur and tibia) (S Mays 2003)

Initial comment: inhumation of a poorly preserved articulated skeleton of an adult male, buried in a grave accompanied by shield boss-type SB3–b2 and spearhead-types SP2–a2c and SP2–b1a3.

Objectives: to test and refine the dating of male burials assigned to Siegmund 1998 phase NR5.

Calibrated date:	*1σ:* cal AD 535–565
	2σ: cal AD 430–590

Final comment: A Bayliss (23 September 2013), this grave falls in the first phase of the male seriation (AS-MB/AS-Mp), and is dated to *cal AD 535–560 (95% probability; UB-4985 (WG11)*; Bayliss *et al* 2013b, figs 6.52) in the partition based on leading artefact-types, and *cal AD 540–565 (95% probability; UB-4985 (WG11)*; Bayliss *et al* 2013b, figs 6.53) in the partition based on the two-dimensional map of grave-assemblages.

Laboratory comment: Rafter Radiocarbon Laboratory (23 September 2013), this skeleton had a slightly low Gly/Asp ratio (5.7), but the C:N ratio (3.3) suggests that protein preservation was adequate for accurate dating.

References:	Siegmund 1998

Anglo-Saxon graves and grave goods (male graves): Butler's Field, Lechlade, Gloucestershire

Location:	SU 213995
	Lat. 51.41.35 N; Long. 01.41.31 W
Project manager:	D Miles and S Palmer (Oxford Archaeological Unit), 1985
Archival body:	Corinium Museum

Description: an Anglo-Saxon cemetery.

Objectives: to test and refine the artefact dating of furnished male burials of the late sixth and seventh centuries AD.

Final comment: A Bayliss (25 September 2013), four male graves (Lec040, Lec115, Lec172/1, and Lec183) were dated from this site. Grave 183 dated to the end of the sixth century cal AD in the early part of the seriation, but the other three graves all dated to the third quarter of the seventh century AD and belonged to the last phase of the seriation.

References:	Bayliss *et al* 2013b
	Boyle *et al* 1998
	Boyle *et al* 2011
	Siegmund 1998

UB–4981 1469 ±18 BP

$\delta^{13}C$: -20.6 ±0.5‰
$\delta^{13}C$ *(diet):* -20.3 ±0.3‰
$\delta^{15}N$ *(diet):* +8.0 ±0.4‰
C/N ratio: 3.3 *%C:* 39.8 *%N:* 14.1

Hydroxyproline	Aspartic	Glutamic	Proline	Glycine	Alanine	Arginine
80.0	58.0	93.0	131.0	311.0	115.0	48.0

Sample: Grave 183, submitted in September 2003 by C Scull

Material: human bone (350g) (right femur) (S Mays 2003)

Initial comment: inhumation of a moderately preserved skeleton of an elderly male accompanied by seax-type SX1–b. This grave was cut by grave 181 (undated).

Objectives: to test and refine the dating of the 'latest' category of male burials.

Calibrated date:	*1σ:* cal AD 575–615
	2σ: cal AD 555–645

Final comment: A Bayliss (24 September 2013), this grave assemblage is allocated to the second or third phases of the male seriation (AS-MC or AS-MD/AS-Mq or AS-Mr), on the basis of the occurrence of seax-type SX1–b which is diagnostic of those phases. The burial is dated to *cal AD 560–605 (95% probability; UB-4981 (Lec183)*; Bayliss *et al* 2013b, fig 6.52) in the partition based on leading artefact-types, and *cal AD 565–605 (95% probability; UB-4981 (Lec183*; Bayliss *et al* 2013b, fig 6.53) in the partition based on the two-dimensional map of grave-assemblages.

Laboratory comment: Rafter Radiocarbon Laboratory (24 September 2013), this skeleton had slightly low Gly/Ala and Gly/Asp ratios (2.7 and 5.4 respectively), but the C:N ratio (3.3) suggests that protein preservation was adequate for accurate dating.

UB–4982 1361 ±17 BP

$\delta^{13}C$: -21.0 ±0.5‰

$\delta^{13}C$ (diet): -20.7 ±0.3‰

$\delta^{15}N$ (diet): +9.7 ±0.4‰

C/N ratio: 3.3 %C: 44.8 %N: 15.7

Hydroxyproline	Aspartic	Glutamic	Proline	Glycine	Alanine	Arginine
82.0	38.0	95.0	136.0	322.0	114.0	48.0

Sample: Grave 155, submitted in September 2003 by C Scull

Material: human bone (375g) (right femur) (S Mays 2003)

Initial comment: inhumation of a well-preserved skeleton of a young adult male associated with a spearhead-type SP1–a2, buckle-type BU6, and seax-types SX2–a and SX4–d.

Objectives: to test and refine the chronology of male burials of Siegmund (1998) phases NR 9–10 in the English seriation.

Calibrated date: 1σ: cal AD 650–665
     2σ: cal AD 645–675

Final comment: A Bayliss (24 September 2013), since the incidence of spearhead-type SP1–a2 appears to be an anachronistic survival in this later grave-assemblage and artefact types BU6 and SX4–d do not appear in the final seriation this grave is allocated to the last phase in the seriation (AS-MF/AS-Mt) on the basis of seax-type SX2–a which is diagnostic of that phase. This grave can thus be dated to *cal AD 645–670 (95% probability; UB-4683 (Lec155);* Bayliss *et al* 2013b, figs 6.52–3).

References: Siegmund 1998

Anglo-Saxon graves and grave goods (male graves): Castledyke South, Barton-on-Humber, Humberside

Location: TA 031217
     Lat. 53.40.53 N; Long. 00.26.21 W

Project manager: M Foreman (Humberside Archaeology), 1982–90

Archival body: Scunthorpe Museum

Description: an extensive Anglo-Saxon cemetery, mostly of inhumations dated by grave goods to the sixth and seventh centuries. The site lies 300m south-west of the late Saxon St Peter's Church and its associated cemetery.

Objectives: to test and refine the artefact dating of furnished male burials of the late sixth and seventh centuries AD.

Final comment: A Bayliss (25 September 2013), only a single male grave (CasD094) was dated from this site. Both of the artefact types which appeared in the final seriation of male graves seemed to be anachronistic late survivals in this grave, which cannot therefore be placed in the seriation.

References: Drinkall and Foreman 1998

UB–6039 1412 ±14 BP

$\delta^{13}C$: -20.8 ±0.5‰

$\delta^{13}C$ (diet): -20.5 ±0.3‰

$\delta^{15}N$ (diet): +10.1 ±0.4‰

C/N ratio: 3.0 %C: 46.5 %N: 17.8

Sample: Grave 94, SK1452, submitted on 15 September 2003 by J Hines

Material: human bone (375g) (right femur) (S Mays 2003)

Initial comment: inhumation of a poorly preserved, articulated, adult male skeleton buried with buckle-type BU3–g and seax-type SX1–a.

Objectives: to test annd refine the dating of male burials attributed to Siegmund (1998) NR phase 7 in the English seriation.

Calibrated date: 1σ: cal AD 635–655
     2σ: cal AD 610–660

Final comment: A Bayliss (25 September 2013), the radiocarbon date from this grave is much later than would be expected from the two artefact-types which appear in the seriation. The example of SX1–a in this grave is the smallest specimen of this type in the typological survey, and could therefore lie on an ambiguous typological boundary between large knives and small seaxes. It also has the first example of buckle type BU-g to appear in the seriation, a type otherwise found in phase AS-MC/AS-Mq. It appears likely that both these artefacts were some decades old when buried in this grave. For this reason this burial has not been included in the final seriation.

Laboratory comment: Rafter Radiocarbon Laboratory (25 June 2013), this C:N ratio is a significant outlier (Grubbs test, Z-score 3.39, P<0.05), but is within the acceptable range (DeNiro 1985) and values of 46.5%C and 17.8%N also suggest adequate protein preservation.

References: DeNiro 1985
     Siegmund 1998

Anglo-Saxon graves and grave goods (male graves): Edix Hill (Barrington A), Cambridgeshire

Location: TL 375495
     Lat. 52.07.33 N; Long. 00.00.30 E

Project manager: T Malim (Cambridgeshire County Council Archaeological Field Unit), 1989–91

Archival body: Cambridgeshire County Council

Description: the remains of 149 individuals from 119 graves were recovered from a late-fifth to early seventh-century Anglo-Saxon inhumation cemetery.

Objectives: to test and refine the artefact dating of furnished male burials of the late sixth and seventh centuries AD.

Final comment: A Bayliss (25 September 2013), four male graves (EH007, EH012, EH033, and EH048) were included in the final male seriation spanning the period from the middle decades of the sixth century to the early decades of the seventh century AD.

References: Bayliss *et al* 2013b
     Malim and Hines 1998

UB–4922 1508 ±19 BP

$\delta^{13}C$: -20.9 ±0.5‰
$\delta^{13}C$ (diet): -20.6 ±0.3‰
$\delta^{15}N$ (diet): +9.5 ±0.4‰
C/N ratio: 3.3 %C: 37.4 %N: 13.4

Hydroxyproline	Aspartic	Glutamic	Proline	Glycine	Alanine	Arginine
73.0	57.0	89.0	110.0	316.0	126.0	50.0

Sample: Skeleton 148; Grave 48, submitted on 4 August 2003 by C Scull

Material: human bone (373g) (left femur) (S Mays 2003)

Initial comment: as UB-4510

Objectives: as UB-4510

Calibrated date: *1σ:* cal AD 540–585
              2σ: cal AD 535–605

Final comment: see UB-4510

Laboratory comment: see UB-4510

Laboratory comment: Rafter Radiocarbon Laboratory (24 September 2013), *see* UB-4510. This skeleton had slightly low Gly/Ala and Gly/Asp ratios (2.5 and 5.5 respectively), but the C:N ratio (3.3) suggests that protein preservation was adequate for accurate dating.

UB–4923 1572 ±20 BP

$\delta^{13}C$: -20.6 ±0.5‰
$\delta^{13}C$ (diet): -20.4 ±0.2‰
$\delta^{15}N$ (diet): +10.6 ±0.3‰
C/N ratio: 3.2 %C: 38.9 %N: 14.2

Hydroxyproline	Aspartic	Glutamic	Proline	Glycine	Alanine	Arginine
76.0	44.0	83.0	106.0	364.0	114.0	51.0

Sample: Grave 7; Skeleton 11, submitted on 4 June 2003 by C Scull

Material: human bone (333g) (left femur) (S Mays 2003)

Initial comment: inhumation of a well-preserved adult male skeleton in a grave accompanied by shield boss-type SB3–b3+4 and spearhead-type SP2–a1a2.

Objectives: to test and refine the dating of male burials attributed to Siegmund (1998) NR phases 9–10.

Calibrated date: *1σ:* cal AD 425–540
              2σ: cal AD 415–550

Final comment: A Bayliss (24 September 2013), this grave falls in the first phase of the male seriation (AS-MB/AS-Mp), and is dated to *cal AD 530–560 (95% probability; UB-4923 (EH007)*; Bayliss *et al* 2013b, fig 6.52–3).

Laboratory comment: Rafter Radiocarbon Laboratory (24 September 2013), duplicate stable isotope analyses were run for carbon and nitrogen, in both cases the measurements are statistically consisent and weighted means are provided here (*see* Beavan *et al* 2011, tables 1–2 for full details).

References: Beavan *et al* 2011
              Siegmund 1998

Anglo-Saxon graves and grave goods (male graves): Eriswell, Lakenheath, Suffolk

Location: TL 72958038
              Lat. 52.23.36 N; Long. 00.32.30 E

Project manager: J Caruth (Suffolk County Council), 1998

Archival body: Suffolk County Council

Description: an Anglo-Saxon cemetery.

Objectives: to test for the possibility of dietary offsets in the radiocarbon age of human bone in the Saxon period.

Final comment: A Bayliss (26 September 2013), sampling of the contemporary horse and rider from grave 323 suggests that dietary offsets in this period amount to a few decades at most.

UB–6347 1604 ±20 BP

$\delta^{13}C$: -20.1 ±0.5‰
$\delta^{13}C$ (diet): -19.8 ±0.4‰
$\delta^{15}N$ (diet): +9.2 ±0.3‰
C/N ratio: 3.4 %C: 38.8 %N: 13.5

Hydroxyproline	Aspartic	Glutamic	Proline	Glycine	Alanine	Arginine
100.0	51.0	95.0	100.0	312.0	119.0	51.0

Sample: 104 4222, submitted on 15 October 2004 by C Scull

Material: human bone (270g) (right tibia and fibula) (S Mays 2004)

Initial comment: inhumation of an articulated adult male skeleton interred within a coffin accompanied by a rich grave assemblage, including an articulated horse skeleton, sword type SW1–a, shield boss-type SB1–b, and spearhead-type SP2–B1a2.

Objectives: this weapon burial is earlier than the assemblage of male graves included in this project, but has been dated so that the radiocarbon age on a purely terrestrial (horse) sample can be compared to the result on contemporary human bone.

Calibrated date: *1σ:* cal AD 415–530
              2σ: cal AD 395–540

Final comment: A Bayliss (25 September 2013), the result on the horse and man buried together in grave 323 are statistically consistent (T'=1.1; T'(5%)=3.8; ν=1; Ward and Wilson 1978), suggesting that there was no significant marine or freshwater offset in the diet of the dated human being.

Taking into account the presence of grave goods decorated in Style I (Salin 1904), this burial can be dated to *cal AD 460–470 (5% probability; Eris104*; Bayliss *et al* 2013b, fig 6.1), or *cal AD 490–535 (90% probability)*.

Laboratory comment: Rafter Radiocarbon Laboratory (25 September 2013), this skeleton had slightly low Gly/Ala and Gly/Asp ratios (2.6 and 6.1 respectively), but the C:N ratio (3.4) suggests that protein preservation was adequate for accurate dating.

References: Salin 1904
              Ward and Wilson 1978

UB–6348 1611 ±20 BP

δ¹³C: -22.9 ±0.5‰

Sample: ERL 104 4206, submitted on 15 October 2004 by C Scull

Material: animal bone (horse): *Equus* sp., right humerus (495g) (P Baker 2004)

Initial comment: as UB-6347

Objectives: as UB-6347

Calibrated date: *1σ:* cal AD 410–505
                   2σ: cal AD 395–535

Final comment: see UB-6347

Anglo-Saxon graves and grave goods (male graves): Ford, Laverstock, Wiltshire

Location: SU 172332
                     Lat. 51.05.49 N; Long. 01.45.16 W

Project manager: J Musty (Ministry of Public Buildings and Works), 1964

Archival body: Salisbury and South Wiltshire Museum

Description: an Anglo-Saxon barrow burial.

Objectives: to test and refine the artefact dating of furnished male burials of the late sixth and seventh centuries AD.

Final comment: A Bayliss (25 September 2013), only one male skeleton was dated from this barrow (Fo2) which dated to the third quarter of the seventh century AD and fell in the last phase of the seriation.

References: Musty 1969

UB–4976 1464 ±16 BP

δ¹³C: -20.6 ±0.5‰
δ¹³C (diet): -20.4 ±0.3‰
δ¹⁵N (diet): +9.0 ±0.4‰
C/N ratio: 3.3 *%C:* 40.4 *%N:* 14.4

Hydroxyproline	Aspartic	Glutamic	Proline	Glycine	Alanine	Arginine
77.0	53.0	93.0	135.0	308.0	117.0	48.0

Sample: Barrow 2 [18/1964], submitted on 5 September 2003 by C Scull

Material: human bone (418g) (right femur) (S Mays 2003)

Initial comment: primary inhumation of a moderately preserved adult male skeleton under barrow 2 accompanied by seax-type SX1–c, buckle-type BU7, shield boss-type SB5–b+c, and spearhead-type SP2–a1b1.

Objectives: to test and refine the emergent seriation of male graves by dating a late example.

Calibrated date: *1σ:* cal AD 580–620
                   2σ: cal AD 565–645

Final comment: A Bayliss (24 September 2013), the incidence of seax-type SX1–c appears to be an anomalously late burial of a much earlier type and so has been suppressed in the final seriation. This places the grave in the latest phase of the male seriation (AS-MF/AS-Mt), and is dated to *cal AD 620–650 (95% probability; UB-4976 (Fo2)*; Bayliss *et al* 2013b, fig 6.52–3).

Laboratory comment: English Heritage (24 September 2013), this measurement is probably a slightly early outlier since it is consistently slightly earlier than expected from its position in the male seriation (Bayliss *et al* 2013b, chapter 6.4).

Laboratory comment: Rafter Radiocarbon Laboratory (24 September 2013), this skeleton had slightly low Gly/Ala and Gly/Asp ratios (2.6 and 5.8 respectively), but the C:N ratio (3.3) suggests that protein preservation was adequate for accurate dating.

Anglo-Saxon graves and grave goods (male graves): Gally Hills, Banstead, Surrey

Location: TQ 250607
                     Lat. 51.19.51 N; Long. 00.12.22 W

Project manager: J Barfoot (Surrey Archaeological Society) D Price-Williams (Nonsuch Antiquarian Society), 1972

Archival body: Bourne Hall Museum, Ewell

Description: an Anglo-Saxon barrow burial.

Objectives: to test and refine the artefact dating of furnished male burials of the late sixth and seventh centuries AD.

Final comment: A Bayliss (26 September 2013), the single skeleton (GaH) sampled for dating from this site has been treated in PVA. The repeat measurement (UB-4727) is probably the more accurate estimate of the radiocarbon age of the primary burial, although in the light of the contamination problems, the radiocarbon date of this individual must be interpreted with some caution.

References: Barfoot and Price-Williams 1976, 59–76

UB–4920 1419 ±18 BP

δ¹³C: -20.5 ±0.5‰
δ¹³C (diet): -20.2 ±0.3‰
δ¹⁵N (diet): +10.4 ±0.4‰
C/N ratio: 3.3 *%C:* 42.5 *%N:* 15.0

Hydroxyproline	Aspartic	Glutamic	Proline	Glycine	Alanine	Arginine
77.0	52.0	95.0	101.0	362.0	114.0	42.0

Sample: Primary Burial, submitted on 4 June 2003 by C Scull

Material: human bone (481g) (left femur) (S Mays 2003)

Initial comment: primary inhumation of an adult male skeleton of a moderately preserved adult male skeleton in a barrow accompanied by spearhead-type SP2–a1b1 and shield boss-type SB5–b+c.

Objectives: as UB-4727; as the analysis developed this date would also validate the attribution of this English burial to Siegmund (1998) phase NR 9–10. This replicate sample was also run to test the validity of the anomalously early result obtained from UB-4727 which may have been due to an unrecognised consolidant.

Calibrated date: *1σ:* cal AD 615–650
                         2σ: cal AD 600–660

Final comment: A Bayliss (24 September 2013), this grave falls in the last phase of the male seriation (AS-MF/AS-Mt), and is dated to *cal AD 620–660 (95% probability; UB-4920 (GaH)*; Bayliss *et al* 2013b, figs 6.52–3).

Laboratory comment: English Heritage (24 September 2013), this skeleton appears to have been treated with some form of consolidant between its excavation in 1972 and the retrieval of the sample for dating in 2001. No conservation records or other information about the treatment was found in the site archive, although Fourier Transform Infra Red Spectroscopy (FTIRS) demonstrated the presence of Polyvinyl Acetate (PVA; McCormac *et al* 2011, fig 21). A replicate measurement subsequently undertaken (see UB-4920) is statistically significantly younger than this measurement (T′=8.0; T′(5%)=3.8; *v*=1; Ward and Wilson 1978). It seems probable therefore that this measurement is anomalously old because of the incomplete removal of PVA during the pretreatment process.

Laboratory comment: University of Belfast (24 September 2013), this sample was processed using a soxhlet extraction system to remove contaminants by continual rinsing with circulating clean solvents (Bruhn *et al* 2001). The sequence of solvents used in this case was: tetrahydrofuran, chloroform, petroleum ether, acetone, methanol, and water. Each solvent was applied for two cycles. A further test for contamination using FTIRS confirmed the removal of a significant amount, although traces may have remained (McCormac *et al* 2011, fig 21). In the light of the contamination problems encountered with this bone, the radiocarbon date of this individual must be interpreted with some caution.

Laboratory comment: Rafter Radiocarbon Laboratory (24 September 2013), the replicate amino acid analyses on this skeleton show greater variability than the ±5% quoted error for glycine. Difficulties were reported with the first analysis and so the values reported for UB-4920 more probably reflect the protein content of the dated bone. The C:N ratio (3.3) suggests that protein preservation was adequate for accurate dating.

References: Bruhn *et al* 2001
               McCormac *et al* 2011
               Siegmund 1998

Anglo-Saxon graves and grave goods (male graves): Melbourn, Water Lane, Cambridgeshire

Location: TL 383439
                   Lat. 52.04.31 N; Long. 00.01.04 E

Project manager: H Duncan (Albion Archaeology), 2000

Archival body: Albion Archaeology

Description: an Anglo-Saxon cemetery.

Objectives: to test and refine the artefact dating of furnished male burials of the late sixth and seventh centuries AD.

Final comment: A Bayliss (26 September 2013), a series of three intercutting male graves (MelSG077, MelSG079, and MelSG080) have been dated along with a fourth intercutting female grave (MelSG078; UB-4885). This sequence of graves spans the early and mid-seventh century AD. The radiocarbon dates are fully compatible with the stratigraphic sequence.

References: Bayliss *et al* 2013b
               Duncan *et al* 2003

UB–6345 1516 ±23 BP

$δ^{13}C$: -19.8 ±0.5‰
$δ^{13}C$ *(diet):* -19.5 ±0.4‰
$δ^{15}N$ *(diet):* +8.7 ±0.3‰
C/N ratio: 3.2   %C: 38.3   %N: 13.8

Hydroxyproline	Aspartic	Glutamic	Proline	Glycine	Alanine	Arginine
98.0	50.0	92.0	99.0	324.0	116.0	51.0

Sample: SK 1204 SG77, submitted on 15 October 2004 by C Scull

Material: human bone (305g) (left femur and tibia) (S Mays 2004)

Initial comment: a replicate of UB-4886. An inhumation of a well-preserved skeleton of a young adult male buried in a grave containing spearhead-type SP4 and buckle-type BU7. This grave lay at the bottom of a sequence of four intercutting graves and is stratigraphically earlier than grave SG078 (UB-4885).

Objectives: at an early stage of the analysis to refine the dating of the male sequence in an apparent gap in the early seventh century, and to constrain the date for grave SG078 which is stratigraphically later.

Calibrated date: *1σ:* cal AD 540–580
                         2σ: cal AD 430–605

Final comment: A Bayliss (24 September 2013), this grave falls in the penultimate phase of the male seriation (ASME/AS-Ms), and is dated to *cal AD 585–620 (95% probability; MelSG077*; Bayliss *et al* 2013b, fig 6.52) in the partition based on leading artefact-types, and *cal AD 590–625 (95% probability; MelSG077*; Bayliss *et al* 2013b, fig 6.53) in the partition based on the two-dimensional map of grave-assemblages.

Laboratory comment: English Heritage (24 September 2013), the two measurements on this skeleton (UB-4886 (1458 ± 20BP) and UB-6345) are statistically consistent (T′=3.6; T′(5%)=3.8; *v*=1; Ward and Wilson 1978).

Laboratory comment: Rafter Radiocarbon Laboratory (24 September 2013), the replicate amino acid analyses on this skeleton show greater variability than the ±5% quoted error for hydroxyproline and proline. Difficulties were reported with the first analysis and so the values reported for UB-6345 more probably reflect the protein content of the dated bone. The C:N ratio (3.3) suggests that protein preservation was adequate for accurate dating. Duplicate stable isotope analyses were run for carbon and nitrogen, in both cases the measurements are statistically consistent and weighted means are quoted here (for full details *see* Beavan *et al* 2011, tables 1–2).

References: Beavan *et al* 2011
               Ward and Wilson 1978

Anglo-Saxon graves and grave goods (male graves): Mill Hill, Deal, Kent

Location:	TR 36315074
	Lat. 51.12.22 N; Long. 01.22.57 E
Project manager:	K Parfitt (Dover Archaeological Group), 1986–9
Archival body:	Dover Museum

Description: an Anglo-Saxon cemetery.

Objectives: to test and refine the artefact dating of furnished male burials of the late sixth and seventh centuries AD.

Final comment: A Bayliss (26 September 2013), four male graves were dated from Mill Hill (MH040, MH079, MH081, and MH093), all falling into the first or second phase of the seriation and dating to the mid-sixth century AD.

References:	Bayliss *et al* 2013b
	Parfitt and Brugmann 1997

UB–4921 1560 ±16 BP

$\delta^{13}C$: -20.6 ±0.5‰
$\delta^{13}C$ *(diet)*: -20.3 ±0.3‰
$\delta^{15}N$ *(diet)*: +9.3 ±0.4‰
C/N ratio: 3.3 %C: 21.4 %N: 7.7

Hydroxyproline	Aspartic	Glutamic	Proline	Glycine	Alanine	Arginine
73.0	51.0	91.0	115.0	322.0	127.0	47.0

Sample: Grave 81, submitted on 4 June 2003 by C Scull

Material: human bone (378.50g) (left femur and left tibia) (S Mays 2003)

Initial comment: from an articulated inhumation of an adult male buried with buckle-types BU2–d and BU2–h, sword-type SW4, and shield boss-type SB3–c.

Objectives: to refine and test the dating of male burials attributed to Siegmund (1998) phase SNR5.

Calibrated date:	*1σ:* cal AD 430–545
	2σ: cal AD 425–555

Final comment: A Bayliss (23 September 2013), this grave falls in the first phase of the male seriation (AS-MB/AS-Mp), and is dated to *cal AD 535–560 (95% probability; UB-4921 (MH081)*; Bayliss *et al* 2013b, figs 6.52–3).

Laboratory comment: Rafter Radiocarbon Laboratory (23 September 2013), this skeleton had a slightly low Gly/Ala ratio (2.8), but the C:N ratio (3.3) suggests that protein preservation was adequate for accurate dating.

References:	Siegmund 1998

UB–6479 1555 ±22 BP

$\delta^{13}C$: -19.8 ±0.5‰
$\delta^{13}C$ *(diet)*: -19.5 ±0.4‰
$\delta^{15}N$ *(diet)*: +8.8 ±0.3‰
C/N ratio: 3.2 %C: 39.9 %N: 14.7

Hydroxyproline	Aspartic	Glutamic	Proline	Glycine	Alanine	Arginine
72.0	55.0	90.0	114.0	325.0	129.0	49.0

Sample: Grave 40, submitted on 17 June 2005 by C Scull

Material: human bone (382g) (left and right tibia) (S Mays 2005)

Initial comment: inhumation of a poorly preserved adult male skeleton accompanied by sword-type SW4, shield boss-type SB3–c, and spearhead-types SP1–a3 and SP4.

Objectives: to test and refine the dating of preliminary male phase A2c.

Calibrated date:	*1σ:* cal AD 430–545
	2σ: cal AD 420–565

Final comment: A Bayliss (25 September 2013), this grave lies in a liminal area of the seriation and belongs to phase AS-MC in the partition based on leading artefact-types, where it has poor individual agreement (A:45), and to phase AS-Mp in the partition based on the two-dimensional map of grave assemblages in the correspondence analysis, where it has good individual agreement (A:104). The date of this burial is estimated to be *cal AD 550–575 (95% probability; UB-6479 (MH040)*; Bayliss *et al* 2013b, fig 6.52) in the partition based on artefact-types, or *cal AD 535–565 (95% probability; UB-6479 (MH040)*; Bayliss *et al* 2013b, fig 6.53) in the partition based on the two-dimensional map of grave assemblages.

Laboratory comment: Rafter Radiocarbon Laboratory (25 September 2013), this skeleton had slightly low Gly/Ala and Gly/Asp ratios (2.5 and 5.9 respectively), but the C:N ratio (3.2) suggests that protein preservation was adequate for accurate dating.

Anglo-Saxon graves and grave goods (male graves): St Peter's Tip, Broadstairs, Kent

Location:	TR 375693
	Lat. 51.22.20 N; Long. 01.24.43 E
Project manager:	A Hogarth (Chatham House School Archaeology Society), 1969–71
Archival body:	British Museum

Description: an inhumation cemetery of the late sixth and seventh centuries AD.

Objectives: to test and refine the artefact dating of furnished male burials of the late sixth and seventh centuries AD.

Final comment: A Bayliss (26 September 2013), ten male graves (SPTip008, SPTip042, SPTip068, SPTip113, SPTip194, SPTip196, SPTip212, SPTip250, SPTip263, and SPTip318) were selected for dating from this cemetery representing all but the first phase of the male seriation.

References:	Bayliss *et al* 2013b

UB–4924 1261 ±16 BP

$\delta^{13}C$: -20.0 ±0.5‰
$\delta^{13}C$ *(diet)*: -19.7 ±0.3‰
$\delta^{15}N$ *(diet)*: +8.3 ±0.4‰
C/N ratio: 3.2 %C: 41.4 %N: 15.0

Hydroxyproline	Aspartic	Glutamic	Proline	Glycine	Alanine	Arginine
71.0	46.0	98.0	120.0	344.0	128.0	51.0

Sample: Grave 113, submitted on 4 August 2003 by C Scull

Material: human bone (321g) (right femur, left and right tibia) (S Mays 2003)

Initial comment: inhumation of a poorly preserved skeleton of a young adult male buried with buckle-type BU7 and seax-type SX3–a.

Objectives: to test and refine the dating of male burials of Siegmund (1998) phase NR 8–9 in the English series.

Calibrated date: 1σ: cal AD 690–770
    2σ: cal AD 680–775

Final comment: A Bayliss (24 September 2013), this grave falls in the last phase of the male seriation (AS-MF/AS-Mt), and is dated to *cal AD 655–680 (95% probability; (SPTip113);* Bayliss *et al* 2013b, figs 6.52–3).

Laboratory comment: English Heritage (24 September 2013), the two radiocarbon results (UB-4924 and UB-6534) on this skeleton are not statistically consistent (T′=4.3; T′(5%)=3.8; v=1; Ward and Wilson 1978). Although these results are statistically different at 95% confidence they are consistent at 99% confidence. This suggests that one of these measurements is simply a slight statistical outlier and so a weighted mean has been taken before further analysis.

Laboratory comment: Rafter Radiocarbon Laboratory (24 September 2013), the replicate amino acid analyses on this skeleton show greater variability than the ±5% quoted error for proline and glycine. Difficulties were reported with the first analysis and so the values reported for UB-6534 more probably reflect the protein content of the dated bone. The C:N ratio (3.2) suggests that protein preservation was adequate for accurate dating. Duplicate stable isotope analyses, however, produced statistically consistent results for both carbon and nitrogen (*see* UB-6534 and Beavan *et al* 2011, tables 1–2 for full details).

References: Beavan *et al* 2011
    Siegmund 1998
    Ward and Wilson 1978

UB–4925 1466 ±16 BP

δ¹³C: -19.7 ±0.5‰
δ¹³C (diet): -19.4 ±0.2‰
δ¹⁵N (diet): +10.3 ±0.3‰
C/N ratio: 3.2 *%C:* 40.0 *%N:* 14.8

Hydroxyproline	Aspartic	Glutamic	Proline	Glycine	Alanine	Arginine
60.0	51.0	98.0	118.0	346.0	125.0	52.0

Sample: Grave 68, submitted on 4 June 2003 by C Scull

Material: human bone (342g) (right femur and tibia) (S Mays 2003)

Initial comment: inhumation of a poorly preserved, articulated, adult male skeleton buried with spearhead-type SP4, shield boss-type SB4–b1+2, and buckle-type BU4–b.

Objectives: to test and refine the dating of male burials attributed to Siegmund (1998) phases NR 7–8.

Calibrated date: 1σ: cal AD 580–620
    2σ: cal AD 560–645

Final comment: A Bayliss (23 September 2013), this grave falls in the middle phase of the male seriation (AS-MD/AS-Mr), and is dated to *cal AD 570–605 (95% probability;*

UB-4925 (SPTip068); Bayliss *et al* 2013b, figs 6.52) in the partition based on leading artefact-types, and *cal AD 575–610 (95% probability; UB-4925 (SPTip068);* Bayliss *et al* 2013b, figs 6.53) in the partition based on the two-dimensional map of grave-assemblages.

Laboratory comment: Rafter Radiocarbon Laboratory (24 September 2013), duplicate stable isotope analyses were run for carbon and nitrogen, in both cases the measurements are statistically consistent and weighted means are provided here (*see* Beavan *et al* 2011, tables 1–2 for full details).

References: Beavan *et al* 2011
    Siegmund 1998

UB–4926 1537 ±18 BP

δ¹³C: -20.3 ±0.5‰
δ¹³C (diet): -20.0 ±0.3‰
δ¹⁵N (diet): +10.4 ±0.4‰
C/N ratio: 3.3 *%C:* 37.7 *%N:* 13.5

Hydroxyproline	Aspartic	Glutamic	Proline	Glycine	Alanine	Arginine
70.0	49.0	98.0	118.0	341.0	126.0	53.0

Sample: Grave 212, submitted on 4 June 2003 by C Scull

Material: human bone (362g) (left femur and tibiae) (S Mays 2003)

Initial comment: inhumation of a poorly preserved adult male articulated skeleton in a grave accompanied by spearhead-type SP2–a2c, seax-type SX1–b, and buckle-type BU7.

Objectives: to test and refine the dating of male burials attributed to Siegmund (1998) phase NR 6–7 in the English sequence.

Calibrated date: 1σ: cal AD 475–560
    2σ: cal AD 425–575

Final comment: A Bayliss (23 September 2013), this grave falls in the second phase of the male seriation (AS-MC/AS-Mq), and is dated to *cal AD 550–580 (95% probability; UB-4926 (SPTip212);* Bayliss *et al* 2013b, figs 6.52) in the partition based on leading artefact-types, and *cal AD 555–580 (95% probability; UB-4926 (SPTip212);* Bayliss *et al* 2013b, figs 6.53) in the partition based on the two-dimensional map of grave-assemblages.

Laboratory comment: Rafter Radiocarbon Laboratory (23 September 2013), this skeleton had a slightly low Gly/Ala ratio (2.7), but the C:N ratio (3.3) suggests that protein preservation was adequate for accurate dating.

References: Siegmund 1998

UB–4928 1458 ±18 BP

δ¹³C: -20.4 ±0.5‰
δ¹³C (diet): -20.1 ±0.3‰
δ¹⁵N (diet): +9.5 ±0.4‰
C/N ratio: 3.2 *%C:* 39.8 *%N:* 14.5

Hydroxyproline	Aspartic	Glutamic	Proline	Glycine	Alanine	Arginine
69.0	53.0	99.0	119.0	334.0	124.0	52.0

Sample: Grave 250, submitted on 4 June 2003 by C Scull

Material: human bone (334g) (femurs) (S Mays 2003)

Initial comment: inhumation of an articulated, poorly preserved skeleton of an elderly male, buried with shield boss-type SB4–b1+2 and buckle-type BU3–a.

Objectives: to test and refine the dating of male burials attributed to Siegmund (1998) phases 7–8.

Calibrated date: *1σ:* cal AD 590–640
                      2σ: cal AD 565–645

Final comment: A Bayliss (23 September 2013), this grave falls in the middle phase of the male seriation (AS-MD/AS-Mr), and is dated to *cal AD 575–605 (95% probability; UB-4928 (SPTip250);* Bayliss *et al* 2013b, figs 6.52) in the partition based on leading artefact-types, and *cal AD 575–610 (95% probability; UB-4428 (SPTip250);* Bayliss *et al* 2013b, figs 6.53) in the partition based on the two-dimensional map of grave-assemblages.

Laboratory comment: Rafter Radiocarbon Laboratory (23 September 2013), this skeleton had a slightly low Gly/Ala ratio (2.7), but the C:N ratio (3.2) suggests that protein preservation was adequate for accurate dating.

References: Siegmund 1998

UB–4929 1485 ±18 BP

$δ^{13}C$: -20.3 ±0.5‰
$δ^{13}C$ *(diet):* -20.0 ±0.3‰
$δ^{15}N$ *(diet):* +9.9 ±0.4‰
C/N ratio: 3.2 *%C:* 41.2 *%N:* 15.0

Hydroxyproline	Aspartic	Glutamic	Proline	Glycine	Alanine	Arginine
70.0	50.0	97.0	116.0	345.0	125.0	52.0

Sample: Grave 194, submitted on 4 June 2003 by C Scull

Material: human bone (346g) (right femur) (S Mays 2003)

Initial comment: inhumation of a poorly preserved adult male skeleton in a grave accompanied by spearhead-type SP2–a3 and buckle-type BU7.

Objectives: to test and refine the dating of male burials attributed to Siegmund (1998) phases NR 8–9 in the English sequence.

Calibrated date: *1σ:* cal AD 560–605
                      2σ: cal AD 545–625

Final comment: A Bayliss (24 September 2013), this grave falls in the penultimate phase of the male seriation (AS-ME/AS-Ms), and is dated to *cal AD 585–630 (95% probability; UB-4929 (SPTip194);* Bayliss *et al* 2013b, fig 6.52) in the partition based on leading artefact-types, and *cal AD 590–630 (95% probability; UB-4929 (SPTip194);* Bayliss *et al* 2013b, fig 6.53) in the partition based on the two-dimensional map of grave-assemblages.

References: Siegmund 1998

UB–4930 1414 ±19 BP

$δ^{13}C$: -19.7 ±0.5‰
$δ^{13}C$ *(diet):* -19.4 ±0.3‰
$δ^{15}N$ *(diet):* +8.7 ±0.4‰
C/N ratio: 3.2 *%C:* 44.5 *%N:* 16.1

Hydroxyproline	Aspartic	Glutamic	Proline	Glycine	Alanine	Arginine
69.0	53.0	97.0	119.0	339.0	126.0	51.0

Sample: Grave 42, submitted on 4 June 2003 by C Scull

Material: human bone (406g) (left and right femurs) (S Mays 2003)

Initial comment: inhumation of a poorly preserved, articulated skeleton of a young adult male buried with buckle-type BU3–h and spearhead-type SP4.

Objectives: to test and refine the dating of male burials attributed to Siegmund (1998) phases NR 7–8 in the English sequence.

Calibrated date: *1σ:* cal AD 635–655
                      2σ: cal AD 600–660

Final comment: A Bayliss (23 September 2013), this grave-assemblage can only be allocated to phases AS-MC-AS-ME, or AS-Mp-AS-Ms, on the basis of the occurrence of spearhead-type SP4, which is diagnostic of those phases.

This grave is dated to *cal AD 585–640 (95% probability; SPTip042;* Bayliss *et al* 2013b, figs 6.52–3).

Laboratory comment: English Heritage (23 September 2013), the two measurements on this burial (UB-4930 and UB-6346) are statistically consistent (T'=0.7; T' (5%)=3.8; ν=1; Ward and Wilson 1978).

Laboratory comment: Rafter Radiocarbon Laboratory (23 September 2013), the replicate amino acid analyses on this skeleton show greater variability than the ±5% quoted error for hydroxyproline, proline, and glycine. Difficulties were reported with the first analysis and so the values reported for UB-6346 more probably reflect the protein content of the dated bone. The statistical consistency of the radiocarbon ages for this burial suggest that the protein preservation was adequate for accurate dating. Duplicate stable isoptope analyses, however, for carbon and nitrogen produced statistically consistent values (*see* UB-6346 and Beavan *et al* 2011, tables 1–2 for full details).

References: Beavan *et al* 2011
                      Siegmund 1998
                      Ward and Wilson 1978

UB–4931 1498 ±21 BP

$δ^{13}C$: -20.3 ±0.5‰
$δ^{13}C$ *(diet):* -19.4 ±0.4‰
$δ^{15}N$ *(diet):* +8.8 ±0.3‰
C/N ratio: 3.2 *%C:* 39.8 *%N:* 14.4

Hydroxyproline	Aspartic	Glutamic	Proline	Glycine	Alanine	Arginine
69.0	53.0	98.0	120.0	340.0	124.0	49.0

Sample: Grave 318, submitted on 4 June 2003 by C Scull

Material: human bone (348g) (left and right femurs and left tibia) (S Mays 2003)

Initial comment: inhumation of a poorly preserved adult male skeleton buried with spearhead-type SP1–a2 and seax-type SX1–b.

Objectives: to test and refine the dating of male burials attributed to Siegmund (1998) phases NR 6–7 in the English sequence.

Calibrated date: *1σ:* cal AD 545–600
                      2σ: cal AD 535–615

Final comment: A Bayliss (23 September 2013), this grave falls in the second phase of the male seriation (AS-MC/AS-Mq), and is dated to *cal AD 550–585 (95% probability;*

UB–4931 (SPTip318); Bayliss *et al* 2013b, figs 6.52) in the partition based on leading artefact-types, or *cal AD 555–585 (95% probability; UB-4931 (SPTip318)*; Bayliss *et al* 2013b, figs 6.53) in the partition based on the two-dimensional map of grave-assemblages.

Laboratory comment: Rafter Radiocarbon Laboratory (23 September 2013), this skeleton had a slightly low Gly/Ala ratio (2.7), but the C:N ratio (3.2) suggests that protein preservation was adequate for accurate dating.

References: Siegmund 1998

UB–4961 *1447 ±17 BP*

δ¹³C: -19.8 ±0.5‰
δ¹³C (diet): -19.5 ±0.3‰
δ¹⁵N (diet): +9.5 ±0.4‰
C/N ratio: 3.2 *%C:* 45.5 *%N:* 16.5

Hydroxyproline	Aspartic	Glutamic	Proline	Glycine	Alanine	Arginine
83.0	58.0	93.0	130.0	314.0	115.0	45.0

Sample: Grave 8, submitted on 12 August 2003 by C Scull

Material: human bone (396g) (left and right femurs) (S Mays 2003)

Initial comment: inhumation of a pooly preserved skeleton of an elderly male buried with spearhead-type SP1–a4 and buckle-type BU7.

Objectives: to test the integrity and refine the dating of the grave-assemblages assigned to Siegmund (1998) phase NR 8 in the English sequence.

Calibrated date: *1σ:* cal AD 600–645
                       2σ: cal AD 575–650

Final comment: A Bayliss (24 September 2013), this grave falls in the middle phase of the male seriation (AS-MD/AS-Mr), and is dated to *cal AD 575–605 (95% probability; UB-4961 (SPTip008)*; Bayliss *et al* 2013b, fig 6.52) in the partition based on leading artefact-types, and *cal AD 575–610 (95% probability; 4961 (SPTip008)*; Bayliss *et al* 2013b, fig 6.53) in the partition based on the two-dimensional map of grave-assemblages.

Laboratory comment: Rafter Radiocarbon Laboratory (24 September 2013), this skeleton had slightly low Gly/Ala and Gly/Asp ratios (2.7 and 5.5 respectively), but the C:N ratio (3.2) suggests that protein preservation was adequate for accurate dating.

References: Siegmund 1998

UB–4962 *1445 ±16 BP*

δ¹³C: -20.2 ±0.5‰
δ¹³C (diet): -19.9 ±0.3‰
δ¹⁵N (diet): +9.1 ±0.4‰
C/N ratio: 3.2 *%C:* 44.9 *%N:* 16.3

Hydroxyproline	Aspartic	Glutamic	Proline	Glycine	Alanine	Arginine
77.0	56.0	92.0	135.0	311.0	120.0	47.0

Sample: Grave 196, submitted on 12 August 2003 by C Scull

Material: human bone (420g) (right femur and tibia) (S Mays 2003)

Initial comment: inhumation of a poorly preserved adult male skeleton in a grave accompanied by shield boss-type SB4–b1+2, spearhead-type SP1–a3, and buckle-type BU7.

Objectives: to test the integrity and refine the dating of grave-assemblages assigned to Siegmund (1998) phase NR 8 in the English sequence.

Calibrated date: *1σ:* cal AD 600–645
                       2σ: cal AD 580–650

Final comment: A Bayliss (24 September 213), this grave falls in the penultimate phase of the male seriation (AS-ME/AS-Ms), and is dated to *cal AD 590–635 (95% probability; UB-4962 (SPTip196)*; Bayliss *et al* 2013b, fig 6.52) in the partition based on leading artefact-types, and *cal AD 595–635 (95% probability; UB-4962 (SPTip196)*; Bayliss *et al* 2013b, fig 6.53) in the partition based on the two-dimensional map of grave-assemblages.

Laboratory comment: Rafter Radiocarbon Laboratory (24 September 2013), this skeleton had slightly low Gly/Ala and Gly/Asp ratios (2.6 and 5.6 respectively), but the C:N ratio (3.2) suggests that protein preservation was adequate for accurate dating.

References: Siegmund 1998

UB–6346 *1435 ±16 BP*

δ¹³C: -19.3 ±0.5‰
δ¹³C (diet): -19.0 ±0.4‰
δ¹⁵N (diet): +8.1 ±0.3‰
C/N ratio: 3.2 *%C:* 41.3 *%N:* 15.0

Hydroxyproline	Aspartic	Glutamic	Proline	Glycine	Alanine	Arginine
119.0	53.0	112.0	113.0	218.0	141.0	56.0

Sample: Grave 42, submitted on 15 October 2004 by C Scull

Material: human bone (285g) (left and right tibiae, radii, humeri, and ulnae) (S Mays 2004)

Initial comment: a replicate of UB-4930.

Objectives: as UB-4930

Calibrated date: *1σ:* cal AD 605–645
                       2σ: cal AD 595–655

Final comment: see UB-4930

Laboratory comment: see UB-4930

Laboratory comment: Rafter Radiocarbon Laboratory (24 September 2013), *see* UB-4930. This skeleton had a low Gly/Ala and Gly/Asp ratios (1.6 and 4.1 respectively). The C:N ratio (4.2) is also out of the expected range, but the consistency of this radiocarbon age with UB-4930 suggests that the collagen was sufficiently well-preserved for accurate dating.

UB–6478 *1414 ±16 BP*

δ¹³C: -20.4 ±0.5‰
δ¹³C (diet): -20.1 ±0.4‰
δ¹⁵N (diet): +10.6 ±0.3‰
C/N ratio: 3.2 *%C:* 40.2 *%N:* 14.5

Hydroxyproline	Aspartic	Glutamic	Proline	Glycine	Alanine	Arginine
67.0	55.0	98.0	122.0	336.0	125.0	51.0

Sample: Grave 360; male C, submitted on 3 May 2005 by C Scull

Material: human bone (190g) (left and right humeri, radii, and ulnae) (S Mays 2005)

Initial comment: an articulated adult male inhumation from a discrete grave pit associated with buckle-type BU7 and seax-type SX3–a.

Objectives: to test and refine the dating of preliminary male phase C.

Calibrated date: 1σ: cal AD 635–655
2σ: cal AD 605–660

Final comment: A Bayliss (23 September 2013), this burial falls in the latest phase of the Anglo-Saxon seriation of male graves (AS-MB/AS-Mp), and is estimated to date to *cal AD 620–660 (95% probability; UB-6478 (SPTip360);* Bayliss *et al* 2013b, figs 6.52–3).

Laboratory comment: Rafter Radiocarbon Laboratory (23 September 2013), this skeleton had slightly low Gly/Alaand Gly/Asp ratios (2.7 and 6.1 respectively), but the C:N ratio (3.2) suggests that protein preservation was adequate for accurate dating.

UB–6534 1311 ±18 BP

$\delta^{13}C$: -19.7 ±0.5‰
$\delta^{13}C$ (diet): -19.4 ±0.4‰
$\delta^{15}N$ (diet): +8.8 ±0.3‰
C/N ratio: 3.2 %C: 39.8 %N: 14.4

Hydroxyproline	Aspartic	Glutamic	Proline	Glycine	Alanine	Arginine
75.0	47.0	93.0	103.0	358.0	119.0	50.0

Sample: Grave 113, submitted on 3 June 2005 by C Scull

Material: human bone (320g) (left femur, left and right humeri) (S Mays 2005)

Initial comment: as UB-4924

Objectives: as UB-4924

Calibrated date: 1σ: cal AD 665–690
2σ: cal AD 660–770

Final comment: see UB-4924

Laboratory comment: see UB-4924

Laboratory comment: Rafter Radiocarbon Laboratory (24 September 2013), see UB-4924. The amino acid and C:N ratios for this sample are within the expected range for well preserved bone collagen.

Antler Maceheads Project, Cumbria, Derbyshire, Greater London, Nottinghamshire and Yorkshire (East Riding)

Location: see individual sites

Project manager: see individual sites

Description: a number of antler maceheads were identified for dating throughout England. Antler maceheads comprise the lower sections of red deer antler beams, and are perforated

for hafting. A total of 58 have been identified the majority of which (41) come from the Thames valley (Simpson 1996).

Objectives: first, to establish the temporal range of maceheads and the degree of contemporaneity between the Thames Valley and northern burial series; second, to establish the Neolithic/Mesolithic credentials of the Thames Valley Maceheads; third, to establish the potential for lattice decorated maceheads as prototypes for the stone Maesmawr series; and fourth, to establish by means of AMS dating, the temporal horizon for the introduction of spiral decoration into southern England as witnessed by the Garboldisham macehead.

Final comment: R Loveday (2007), the results demonstrate that both the middle Thames specimens, and those from northern Britain, date to the second half of the fourth millennium cal BC. This suggests a degree of contemporaneity between riverine activity in the south and 'prestige' burial in the north, although the possibility that this is a function of the radiocarbon calibration curve cannot be discounted. The possibility that lattice decorated maceheads can be regarded as prototypes for the Maesmore series of fine stone maceheads is considered, but the failure of two out of three decorated examples to produce radiocarbon determinations means that the debate cannot yet be settled (Loveday *et al* 2007, 381).

Laboratory comment: English Heritage (2007), samples from six antler maceheads were submitted for dating in 2002. Of these samples, one (Mortlake; Simpson no. 40) failed due to a poor collagen yield following pretreatment, while the remaining five produced results. However, following the identification of a problem with the ultrafiltration procedures undertaken as part of bone pretreatment at the Oxford Radioacrbon Accelerator Unit in October 2002 (Bronk Ramsey *et al* 2004a), the results could not be regarded as reliable and consequently all five were withdrawn.

One of the maceheads from Windmill Lane (MoL 01154C) was subequently resampled, and four additional samples were submitted in 2004. These samples were processed according to the new pretreatment ultrafiltration stage outlined in Bronk Ramsey *et al* (2004a). In addition, collagen from the original extraction procedures undertaken on the samples from Duggleby Howe (Hull and East Riding Museum; Simpson 1996, no. 4), Attenborough (Nottingham University Museum; Simpson 1996, no. 1), and Windmill Lane (MoL 01154D; Simpson 1996, no. 45), was also subjected to the new ultrafiltration procedures and dated. Unfortunately, it was not possible to re-date the macehead from Liffs Low due to a lack of collagen surviving from the original sample pretreatment.

Of the eight samples, six produced results, and two failed due to poor collagen yields (Burwell Fen; Cambridge University 241981; Simpson 1996, no. 55 and Hammersmith; MoL A13687; Simpson 1996, no. 32). A replicate sample from Burwell Fen was also submitted to the radiocarbon dating laboratory at the University of Groningen.

References: Bronk Ramsey *et al* 2004a
Loveday *et al* 2007
Simpson 1996

Antler Maceheads Project: Thames Valley, Greater London

Location: *see* individual results

Project manager: R Loveday (University of Leicester) and A Gibson (University of Bradford)

Archival body: Museum of London and Cambridge University

Description: a selection of crown antler maceheads from the Thames Valley and a comparative example from Burwell Fen, Cambridgeshire.

Objectives: first, to establish the Mesolithic/Neolithic credentials of the Thames Valley maceheads; second, to establish the temporal range of maceheads and the degree of contemporaneity between the Thames Valley and the northern burial series; and third, to establish the potential of the lattice decorated examples to be regarded as prototypes for the Maesmawr series as represented by the Knowth specimen.

Final comment: A Bayliss and P Marshall (2007), the four maceheads from the Thames Valley (OxA-13207 (4611 ±37 BP), -13440 (4684 ±37 BP), -14192, and -14193) are all Neolithic in date and give calibrated date ranges that span the second half of the fourth millennium cal BC.

The two maceheads from Windmill Lane (OxA-13207 and -13440) have produced statistically consistent results (T'=1.9; T'(5%)=3.8; v=1; Ward and Wilson 1978), and could therefore be of the same actual age.

Laboratory comment: English Heritage (24 June 2014), two further samples from this series were dated before 2003 (OxA-13207 and -13440) and are published in Bayliss *et al* 2016, 23).

References: Loveday *et al* 2007
               Simpson 1996
               Ward and Wilson 1978

GrA–27417 3920 ±60 BP

δ¹³C: -25.9‰

Sample: Burwell Fen 241981, submitted on 7 October 2004 by A Gibson

Material: antler: *Cervus elaphus*, undecorated macehead (D D A Simpson 1996)

Initial comment: a chance find by peat cutters at approximately TL 5768, Burwell Fen, Cambridgeshire.

Objectives: to establish the degree of contemporaneity between the Northern Burial and the Thames macehead series; to provide a geographical link between the Northern Burial and the Thames series; and to establish the Mesolithic/Neolithic credentials of maceheads.

Calibrated date: *1σ:* 2480–2290 cal BC
                        2σ: 2580–2200 cal BC

Final comment: A Bayliss and P Marshall (2007), a replicate sample from the Burwell macehead failed to produce a result at the Oxford laboratory due to a poor collagen yield. Additionally, the δ13C value of the sample is significantly different from the others obtained in this study and from typical values. The value suggests contamination, most probably from humic acids, and so is not a reliable measurement of the actual age of the sample and should thus be treated with caution.

Laboratory comment: Rijksuniversitat Groningen (AMS) (26 January 2005), the 'collagen' carbon yield is only 4.5%, so the quality of the sample is very poor. The date is not very reliable.

Ascott under Wychwood: long barrow, Oxfordshire

Location: SP 299175
                  Lat. 51.51.16 N; Long. 01.33.57 W

Project manager: D Benson (Oxford City and County Museum), 1965–9

Archival body: Oxford City and County Museum; The Natural History Museum

Description: Ascott-under-Wychwood Long Barrow, Oxfordshire, was a Cotswold-Severn type Neolithic monument. This site was a limestone and earthen trapezoidal long barrow and was excavated between 1965 and 1969. There were two sets of two lateral chambers constructed from limestone orthstats within the matrix of the barrow. The long axis of the barrow was orientated east-west, the barrow was 45m in length and 15m in width, and there was evidence for a forecourt constructed at its eastern end with northern and southern horn-works. Four quarry pits were excavated to the north-west of the monument. The barrow itself was composed of bay divisions, which had then been enclosed by internal and external revetment walls, these had been constructed from courses of limestone plaques. Both sets of chambered areas revealed extensive deposits of human bone. Two timber structures were located under the stonework to the east of the chamber areas. There was an oblique zone of midden material across this area of the site. The timber structure and the midden were Neolithic, although they were considered as evidence for 'pre-mound' activities. There were pits, hearths, and a tree-throw within the buried soil that were evidence of earlier activity. These earlier features in the same area as three distinct distributions of Mesolithic microlithic worked flint.

Objectives: to identify features associated with the early/late Mesolithic worked flint assemblages; to determine whether there was a gap between this activity and the Neolithic occupation (including the midden) sealed beneath the barrow, and if so to determine the duration of this gap; to establish the absolute date and duration of the Neolithic occupation sealed beneath the barrow; to determine whether there was a hiatus between the end of the Neolithic occupation and the construction of the barrow; to date the initial construction of the monument (including the stone cists); to investigate the duration of the infilling of the quarry ditches; to determine when the barrow was extended to the east; and to establish the absolute dates and duration of the Neolithic burial activity within the cists.

Final comment: A Bayliss (2007), chronological modelling of the 44 radiocarbon measurements now available from the barrow were presented in Bayliss *et al* (2007).Three alternative models were proposed, the favoured one suggests the following sequence: pre-barrow occupation including small timber structures and a midden was followed by a gap long enough to allow a turfline to form. Cists and the primary barrow were then initiated and the first human remains inserted into the cists; there was subsequently a secondary extension to the barrow. In this model, occupation goes back to the fortieth century cal BC, the midden being quite short-lived in the latter part of the fortieth or first part of the thirty-ninth century cal BC. The gap was probably not less than 50 years long, in the latter part of the thirty-ninth century cal BC and the first half of the thirty-eighth century cal BC. The barrow was begun between *3760–3695 cal BC (95% probability; primary_construction; Bayliss* et al *2007e, fig 6)* and extended in *3745–3690 cal BC (95% probability; secondary_construction; Bayliss* et al *2007e, fig 6)* , probably within a generation. The first bodies were inserted in *3755–3690 cal BC (95% probability; first_ bodies; Bayliss* et al *2007e, fig 7)* , contemporaneously with the primary barrow, and the last remains were probably deposited in the *3645–3590 cal BC (95% probability; last_body; Bayliss* et al *2007e, fig 7)* cal BC. The use of the monument probably did not exceed three to five generations.

References: Bayliss *et al* 2007e
Benson and Clegg 1978
Benson and Whittle 2007
Evans 1971
Selkirk 1971
T C Darvill 1982b

GrA–23828 4940 ±50 BP

δ¹³C: -22.2‰

Sample: AB 5 (774), submitted on 16 June 2003 by L McFadyen

Material: animal bone: *Bos* sp., skull (J Mulville 2003)

Initial comment: from the surface of buried soil, directly underneath mound material.

Objectives: to establish sequence of pre-mound activity and how this connects to mound construction (cattle skulls were laid out to mark the long axis of the Beckhampton Road Long Mound, Wiltshire).

Calibrated date: 1σ: 3780–3650 cal BC
2σ: 3910–3640 cal BC

Final comment: A Bayliss (2007), one of six results that date the material used in the construction of the primary barrow (GrA-23828–9, -25295, OxA-13315, and previous dates BM-832–3; 4942 ±74 BP and 5020 ±92 BP), estimated to have been in *3760–3695 cal BC (95% probability; primary construction;* fig 6; Bayliss *et al* 2007e).

Laboratory comment: English Heritage (2007), chronological modelling provides a posterior density estimate for this bone of *3790–3710 cal BC (95% probability;* table 1; Bayliss *et al* 2007e).

GrA–23829 5050 ±50 BP

δ¹³C: -21.6‰

Sample: AB 8 (1044), submitted on 16 June 2003 by L McFadyen

Material: antler (tine) (J Mulville 2003)

Initial comment: a broken off tip of antler tine in primary fill of a quarry pit 2.98m from the surface.

Objectives: to provide a *terminus ante quem* for quarry 3.

Calibrated date: 1σ: 3960–3770 cal BC
2σ: 3970–3700 cal BC

Final comment: see GrA-23828

Laboratory comment: English Heritage (2007), chronological modelling provides a posterior density estimate for this sample of *3800–3710 cal BC (95% probability;* table 1; Bayliss *et al* 2007e).

GrA–23831 4700 ±50 BP

δ¹³C: -23.2‰

Sample: AB11 (1041), submitted on 16 June 2003 by L McFadyen

Material: antler (pick (shed antler)) (J Mulville 2003)

Initial comment: antler pick from the primary fill of the quarry pit, 0.16m from the base.

Objectives: to date the *terminus ante quem* for quarry 4.

Calibrated date: 1σ: 3630–3370 cal BC
2σ: 3640–3360 cal BC

Final comment: A Bayliss (2007), this result is surprisingly late and does not appear to relate to the primary construction of the monument as initially anticipated. Quarry 4 therefore appears not to be part of initial construction, and could conceivably be connected with closure or cessation of activity at the site. It was excluded from the chronological modelling.

GrA–23927 4970 ±45 BP

δ¹³C: -22.5‰

Sample: AB 4 (234), submitted on 16 June 2003 by L McFadyen

Material: animal bone: *Cervus elaphus*, pelvis with butchery marks (J Mulville 2003)

Initial comment: from hearth (F50) within a buried soil, associated with a distribution of microliths.

Objectives: to establish the phase of Mesolithic activity.

Calibrated date: 1σ: 3800–3690 cal BC
2σ: 3940–3650 cal BC

Final comment: A Bayliss (2007), this result is considered to be a *terminus ante quem* for the start of the pre-barrow phase of Neolithic occupation, and also for the construction of the secondary barrow. These results (OxA-12679, -12680, and GrA-23927) date the linear spread of hearths F50, and probably fall in the thirty-eighth century cal BC. It may be contemporary with either the construction of the primary barrow or its initial use. BM-492 (4735 ±70 BP) on the same deposit is now thought to be inaccurate.

Laboratory comment: English Heritage (2007), chronological modelling provides a posterior density estimate for this bone of *3935–3860 cal BC (11% probability)* or *38415–3690 cal BC (84% probability;* table 1; Bayliss *et al* 2007e).

GrA–23933 5105 ±45 BP

$\delta^{13}C$: -20.4‰

Sample: AB 1 (511), submitted on 16 June 2003 by L McFadyen

Material: animal bone: *Sus* sp., epiphyses has been glued back on, evidence for articulation (J Mulville 2003)

Initial comment: from the lower fill of pit F7 associated with hearth F48. The pit cuts through the upper levels of the buried soil. This bone is partly scorched and is evidence for a joint of meat having been held over a fire to roast (Mulville pers comm). Located 0.40m deep in the fill of the pit, this cuts through the upper levels of buried soil.

Objectives: to establish the sequence of pre-mound activity.

Calibrated date: 1σ: 3970–3800 cal BC
2σ: 3990–3780 cal BC

Final comment: A Bayliss (2007), the best estimate for the date of pit F7 is provided by this result (a previous result (BM-491b; 4893 ±70 BP) is now considered to be inaccurate).

Laboratory comment: English Heritage (2007), chronological modelling provides a posterior density estimate for this bone of *3980–3820 cal BC (95% probability;* table 1; Bayliss *et al* 2007e).

GrA–25292 4880 ±40 BP

$\delta^{13}C$: -21.9‰
$\delta^{15}N$ (diet): +8.5‰

Sample: HB2 391/137 (A2), submitted in June 2003 by A Whittle

Material: human bone (right ulna) (D Galer 2003)

Initial comment: individual A2 was part of deposit A of human bone from chamber 3. This individual remains are relatively isolated to the north-west corner of the chamber, and were excavated at the same stage as individual A3 (which overlies A1). The skeleton was partially articulated. A replicate of BM-1976R (4930 ±100 BP).

Objectives: to establish the sequence of a Neolithic collective bone deposit.

Calibrated date: 1σ: 3700–3640 cal BC
2σ: 3720–3630 cal BC

Final comment: A Bayliss (2007), on the assumption that the cists were constructed at the same time as the primary barrow, all the results on human bone from the cists are regarded as later than the primary construction (especially this is true for seven largely complete or partially articulated individuals dated by GrA-25292, -25294, -25304, and OxA-

13319–20, -13400–1, and previous dates BM-1974R (4680 ±160 BP) and BM-1976R (4930 ±100 BP), and is a stronger assumption for the five disarticulated examples (GrA-25305–6, OxA-13402–3, and previously dated BM-1975R (3870 ±100 BP).

Laboratory comment: English Heritage (2007), chronological modelling provides a posterior density estimate for this bone of *3715–3635 cal BC (95% probability;* table 1; Bayliss *et al* 2007e).

GrA–25294 4840 ±40 BP

$\delta^{13}C$: -21.7‰

Sample: HB 4 S46/154 (D1), submitted in June 2003 by A Whittle

Material: human bone (right tibia) (D Galer 2003)

Initial comment: individual D1 was part of a deposit D of human bone in chamber 5. The main deposit of human bone was in a matrix of dark brown stony loam.

Objectives: to establish the sequence of a Neolithic collective bone deposit.

Calibrated date: 1σ: 3660–3630 cal BC
2σ: 3700–3530 cal BC

Final comment: see GrA-25292

Laboratory comment: English Heritage (2007), chronological modelling provides a posterior density estimate for this bone of *3700–3625 cal BC (95% probability;* table 1; Bayliss *et al* 2007e).

Laboratory comment: Rijksuniversitat Groningen (AMS) (23 April 2004), the amount of collagen of the sample was not enough to also measure $\delta^{15}N$.

GrA–25295 4940 ±45 BP

$\delta^{13}C$: -22.2‰
$\delta^{15}N$ (diet): +5.4‰

Sample: AB 47 735, submitted on 21 January 2004 by L McFadyen

Material: animal bone: *Bos* sp., tibia with articulating astragulus (J Mulville 2004)

Initial comment: the sample comes from cutting DVIII, at the eastern end of the primary barrow. In yellow-brown clayey loam deposit of the upcast mound and sealed between layers of limestone rubble and further clayey loams.

Objectives: to establish the phase and sequence of the primary barrow construction.

Calibrated date: 1σ: 3780–3650 cal BC
2σ: 3900–3640 cal BC

Final comment: see GrA-23828

Laboratory comment: English Heritage (2007), chronological modelling provides a posterior density estimate for this bone of *3790–3710 cal BC (95% probability;* table 1; Bayliss *et al* 2007e).

GrA–25296 4965 ±40 BP

$\delta^{13}C$: -22.0‰
$\delta^{15}N$ (diet): +4.4‰

Sample: AB66 786, submitted on 21 January 2004 by L McFadyen

Material: animal bone: *Bos* sp., distal tibia - unfused but has epiphysis (J Mulville 2004)

Initial comment: the sample comes from a brown loam deposit in an upcast mound sealed between layers of limestone rubble deposits. In the eastern end of the secondary barrow, found within cutting DXII, in a deposit that spanned through cuttings DXI-DXII, and had preponderance of skulls and of left-side limb elements of animal bone making the deposit unique within the mound architecture. Bone rots down in about six months to one year and so the epiphysis attached to the unfused tibia is significant.

Objectives: to establish the phase and sequence of secondary barrow.

Calibrated date: 1σ: 3790–3690 cal BC
                       2σ: 3910–3650 cal BC

Final comment: A Bayliss (2007), one of four samples (GrA-25296, OxA-12675–6, and OxA-13318) from within the secondary barrow, estimated to have been constructed in *3745–3670 cal BC (95% probability; secondary construction;* fig 6; Bayliss *et al* 2007e).

Laboratory comment: English Heritage (2007), chronological modelling provides a posterior density estimate for this bone of *3725–3650 cal BC (95% probability;* table 1; Bayliss *et al* 2007e).

GrA–25304 4890 ±40 BP

$\delta^{13}C$: -22.3‰
$\delta^{15}N$ (diet): +8.4‰

Sample: HB7 (530/125) (B2), submitted in February 2004 by A Whittle

Material: human bone (left ulna) (D Galer 2003)

Initial comment: individual B2 was not separated from the adult material during laboratory analysis - bones of a discreet individual could be directly identified on the archaeological plans. Individual B2 was found beneath individual B1 (under stone packing) and was lying diagonally across the chamber with one leg flexed so that the foot was tucked beneath the pelvis. Many elements were articulated and the majority of the skeleton is in the correct anatomical alignment.

Objectives: to establish the sequence of a Neolithic collective bone deposit.

Calibrated date: 1σ: 3710–3640 cal BC
                       2σ: 3760–3630 cal BC

Final comment: see GrA-25292

Laboratory comment: English Heritage (2007), chronological modelling provides a posterior density estimate for this bone of *3715–3635 cal BC (95% probability;* table 1; Bayliss *et al* 2007e).

GrA–25305 4820 ±40 BP

$\delta^{13}C$: -21.9‰
$\delta^{15}N$ (diet): +8.5‰

Sample: HB9 (330/116) (C1/C2/C3/C4/C5), submitted in February 2004 by A Whittle

Material: human bone (left ulna) (D Galer 2003)

Initial comment: it was not possible to identify individuals or parts of individuals from deposit C, either through laboratory analysis, or from the archaeological plans. The deposit represents a complete jumble of remains in no apparent arrangement or order. No elements were found in articulation. The bones in the deposit were interleaved and intermixed throughout with stone, some which was of a type used in the outer face of the cairn. The deposit also contained stone. In places within the deposit there was loose brown soil and smaller stones, some soil showing evidence and smaller stones, some soil showing evidence of worm activity. Rodent-gnawed human bones were present in various parts of the deposit.

Objectives: to establish the sequence of a Neolithic collective bone deposit.

Calibrated date: 1σ: 3650–3530 cal BC
                       2σ: 3660–3520 cal BC

Final comment: see GrA-25292

Laboratory comment: English Heritage (2007), chronological modelling provides a posterior density estimate for this bone of *3700–3620 cal BC (95% probability;* table 1; Bayliss *et al* 2007e).

GrA–25306 4805 ±40 BP

$\delta^{13}C$: -21.2‰
$\delta^{15}N$ (diet): +8.4‰

Sample: HB10 (330/65) (C1/C2/C3/C4/C5), submitted in February 2004 by A Whittle

Material: human bone (left ulna) (D Galer 2003)

Initial comment: as GrA-25305

Objectives: as GrA-25305

Calibrated date: 1σ: 3650–3530 cal BC
                       2σ: 3660–3510 cal BC

Final comment: see GrA-25292

Laboratory comment: English Heritage (2007), chronological modelling provides a posterior density estimate for this bone of *3700–3610 cal BC (95% probability;* table 1; Bayliss *et al* 2007e).

GrA–27093 5100 ±45 BP

$\delta^{13}C$: -21.8‰

Sample: (m 25, 657) PBM 1, submitted in September 2004 by A Whittle

Material: animal bone (sheep/goat, upper molar 3) (J Mulville 2003)

Initial comment: from pre-barrow midden square m25, sealed by more than 1m of the overlying long barrow on calcareous limestone subsoil.

Objectives: dating so far shows that there is material under the barrow of both early fourth millenium BC and late fifth millenium BC date. Sheep/goat is by general agreement Neolithic in age in southern Britain. This tooth from the midden should therefore be Neolithic, and help to date the formation of the midden at some point before the construction of the barrow in the thirty-eighth century BC (current estimate).

Calibrated date: *1σ:* 3970–3800 cal BC
                 2σ: 3990–3780 cal BC

Final comment: A Bayliss (2007), one of six results from the midden (GrA-27093-4, -27096, -27100, -27102, and OxA-13135) which suggest the midden represents a relatively short period of activity, conceivably a single event or a short series of closely connected events. The midden was formed during the fortieth century cal BC or the thirty-ninth century cal BC.

Laboratory comment: English Heritage (2007), chronological modelling provides a posterior density estimate for this bone of *3960–3800 cal BC (95% probability;* table 1; Bayliss *et al* 2007e).

GrA–27094 5095 ±45 BP

δ¹³C: -21.7‰

Sample: (m 39, 886) PBM 2, submitted in September 2004 by A Whittle

Material: animal bone (sheep/goat teeth, lower molar 3) (J Mulville 2003)

Initial comment: from square m30, as GrA-27093.

Objectives: as GrA-27093

Calibrated date: *1σ:* 3970–3800 cal BC
                 2σ: 3980–3780 cal BC

Final comment: see GrA-27093

Laboratory comment: English Heritage (2007), chronological modelling provides a posterior density estimate for this bone of *3960–3800 cal BC (95% probability;* table 1; Bayliss *et al* 2007e).

GrA–27096 5095 ±45 BP

δ¹³C: -21.7‰

Sample: (o 25, 868) PBM 3, submitted in September 2004 by A Whittle

Material: animal bone (sheep/goat tooth molar 1 or 2) (J Mulville 2003)

Initial comment: from square o25, as GrA-27093.

Objectives: as GrA-27093

Calibrated date: *1σ:* 3970–3800 cal BC
                 2σ: 3980–3780 cal BC

Final comment: see GrA-27093

Laboratory comment: English Heritage (2007), chronological modelling provides a posterior density estimate for this bone of *3960–3800 cal BC (95% probability;* table 1; Bayliss *et al* 2007e).

GrA–27098 6180 ±45 BP

δ¹³C: -24.4‰

Sample: (m 23, 709) PBM 4, submitted in September 2004 by A Whittle

Material: animal bone: *Capreolus capreolus,* radius (J Mulville 2003)

Initial comment: from square m23, as GrA-27093.

Objectives: from the pre-barrow midden; a red/roe deer bone and polished flint axe fragments are closely associated in spatial terms. Polished axes are by general agreement Neolithic in date. This bone from the midden should therefore also be Neolithic, or help to date the formation of the midden at some point before the construction of the barrow.

Calibrated date: *1σ:* 5220–5050 cal BC
                 2σ: 5300–4990 cal BC

Final comment: A Bayliss (2007), the date on this bone is unexpectedly early (as is GrA-27099). The date was excluded from the chronological modelling.

GrA–27099 6000 ±45 BP

δ¹³C: -24.2‰

Sample: (k 26, 474) PBM 5, submitted in September 2004 by A Whittle

Material: animal bone: *Capreolus capreolus,* 2nd phalanx (J Mulville 2003)

Initial comment: from square k26, as GrA-27093.

Objectives: as GrA-27098

Calibrated date: *1σ:* 4950–4800 cal BC
                 2σ: 5010–4780 cal BC

Final comment: A Bayliss (2007), the date on this bone is unexpectedly early (as is GrA-27098). The date was excluded from the chronological modelling.

GrA–27100 5010 ±45 BP

δ¹³C: -23.3‰

Sample: (m 24, 667) PBM 6, submitted in September 2004 by A Whittle

Material: antler: *Cervus elaphus,* fragment (J Mulville 2003)

Initial comment: from square m24, as GrA-27093.

Objectives: as GrA-27098

Calibrated date: *1σ:* 3940–3710 cal BC
                 2σ: 3960–3670 cal BC

Final comment: see GrA-27093

Laboratory comment: English Heritage (2007), chronological modelling provides a posterior density estimate for this sample of *3950–3790 cal BC (95% probability;* table 1; Bayliss *et al* 2007e).

GrA–27102 5130 ±45 BP

δ¹³C: -23.2‰

Sample: (o 24, 936) PBM 7, submitted in September 2004 by A Whittle

Material: antler: *Cervus elaphus*, fragment (J Mulville 2003)

Initial comment: from square o24, as GrA-27093.

Objectives: as GrA-27098

Calibrated date: 1σ: 3980–3820 cal BC
                      2σ: 4040–3790 cal BC

Final comment: see GrA-27093

Laboratory comment: English Heritage (2007), chronological modelling provides a posterior density estimate for this sample of *3965–3895 cal BC (34% probability)* or *3965–3895 cal BC; 61% probability;* table 1; Bayliss *et al* 2007e).

OxA–12675 5050 ±33 BP

δ¹³C: -26.2‰

Sample: CH7 (256, A9), submitted on 16 June 2003 by L McFadyen

Material: charcoal: Pomoideae, single fragment (G Campbell 2003)

Initial comment: A9 cuts through the buried soil and underlies the secondary phase of mound construction. 1.42m from the top of the mound, localised subsoil was sandy with degraded limestone.

Objectives: to date the phase of secondary mound construction.

Calibrated date: 1σ: 3950–3790 cal BC
                      2σ: 3960–3710 cal BC

Final comment: A Bayliss (2007), *see* GrA-25296. This sample is probably residual and was excluded from the chronological model.

OxA–12676 4992 ±33 BP

δ¹³C: -24.4‰

Sample: CH 8 (260, A10), submitted on 16 June 2003 by L McFadyen

Material: charcoal: *Corylus* sp., one large piece (G Campbell 2003)

Initial comment: A10 cuts through the buried soil and underlies the secondary phase of mound construction. 1.42m from the top of the mound.

Objectives: as OxA-12676

Calibrated date: 1σ: 3800–3700 cal BC
                      2σ: 3940–3690 cal BC

Final comment: see GrA-25296

Laboratory comment: English Heritage (2007), chronological modelling provides a posterior density estimate for this sample of *3730–3655 cal BC (95% probability;* table 1; Bayliss *et al* 2007e).

OxA–12677 5353 ±32 BP

δ¹³C: -25.1‰

Sample: CH9 (F.16) A, submitted on 18 July 2003 by L McFadyen

Material: charcoal: *Fagus* sp., single fragment (G Campbell 2003)

Initial comment: posthole (F16) cuts the buried soil and is part of a timber structure associated with pre-mound activity.

Objectives: to establish the phase of pre-mound activity.

Calibrated date: 1σ: 4260–4070 cal BC
                      2σ: 4330–4050 cal BC

Final comment: A Bayliss (2007), the posthole produced two measurements both of which provide evidence of further activity on the site in the last quarter of the fifth millennium cal BC. This date was excluded from the chronological modelling.

OxA–12678 5246 ±32 BP

δ¹³C: -25.3‰

Sample: CH9 (F.16) B, submitted on 18 July 2003 by L McFadyen

Material: charcoal: *Fagus* sp., single fragment (G Campbell 2003)

Initial comment: as OxA-12677

Objectives: as OxA-12677

Calibrated date: 1σ: 4060–3990 cal BC
                      2σ: 4230–3970 cal BC

Final comment: A Bayliss (2007), *see* OxA-12677. This date was excluded from the chronological modelling.

OxA–12679 4930 ±31 BP

δ¹³C: -26.4‰

Sample: CH6 (538) A, submitted on 16 June 2003 by L McFadyen

Material: charcoal: *Corylus avellana*, single fragment (G Campbell 2003)

Initial comment: charcoal associated with hearth setting F50 on buried soil. The underlying upcast mound material from a possible phase of secondary mound construction, 0.80m from the top of the mound.

Objectives: to date the *terminus post quem* beneath a possible phase of secondary mound construction.

Calibrated date: 1σ: 3720–3650 cal BC
                      2σ: 3780–3640 cal BC

Final comment: see GrA-23927

Laboratory comment: English Heritage (2007), chronological modelling provides a posterior density estimate for this sample of *3790–3690 cal BC (95% probability;* table 1; Bayliss *et al* 2007e).

OxA–12680 4989 ±31 BP

δ¹³C: -25.1‰

Sample: CH6 (538) B, submitted on 16 June 2003 by L McFadyen

Material: charcoal: Pomoideae, single fragment (G Campbell 2003)

Initial comment: as OxA-12679

Objectives: as OxA-12679

Calibrated date: 1σ: 3800–3700 cal BC
                       2σ: 3940–3690 cal BC

Final comment: see GrA-23927

Laboratory comment: English Heritage (2007), chronological modelling provides a posterior density estimate for this bone of *3925–3875 cal BC (9% probability)* or *3810–3700 cal BC; 86% probability;* table 1; Bayliss *et al* 2007e).

OxA–13135 4950 ±100 BP

δ¹³C: -30.6‰

Sample: PT1 (167) P15, submitted on 16 June 2003 by L McFadyen

Material: carbonised residue

Initial comment: located on the buried soil, directly west of the area of midden material. On the surface of the buried soil.

Objectives: to establish the phase of earlier Neolithic activity.

Calibrated date: 1σ: 3920–3640 cal BC
                       2σ: 3970–3520 cal BC

Final comment: see GrA-27093

Laboratory comment: English Heritage (2007), chronological modelling provides a posterior density estimate for this vessel of *3955–3790 cal BC; 95% probability;* table 1; Bayliss *et al* 2007e).

OxA–13315 4962 ±30 BP

δ¹³C: -21.4‰

Sample: AB68, 400, submitted on 21 January 2004 by L McFadyen

Material: antler: *Cervus elaphus,* tip (J Mulville 2004)

Initial comment: the sample comes from cutting DIX, in the limestone rubble of the primary barrow at the eastern end. In limestone rubble in the upcast mound and sealed between layers of loam and limestone sandy rubble.

Objectives: to establish the phase and sequence of the primary barrow architecture.

Calibrated date: 1σ: 3780–3700 cal BC
                       2σ: 3800–3650 cal BC

Final comment: see GrA-23828

Laboratory comment: English Heritage (2007), chronological modelling provides a posterior density estimate for this sample of *3785–3710 cal BC; 95% probability;* table 1; Bayliss *et al* 2007e).

OxA–13316 1130 ±24 BP

δ¹³C: -20.3‰

Sample: AB67 368, submitted on 21 January 2004 by L McFadyen

Material: animal bone: *Cervus elaphus,* red deer, proximal tibia unfused but has kept epiphysis (J Mulville 2004)

Initial comment: sample from brown-yellow clayey loam deposit, in cutting DVII, of primary barrow at the western end. Sealed between layers of limestone rubble and yellow sand and rubble deposits. Bone rots down in about six months to a year and so the epiphysis attached to unfused tibia is significant.

Objectives: to establish the phase and sequence of the primary barrow architecture.

Calibrated date: 1σ: cal AD 885–970
                       2σ: cal AD 780–985

Final comment: A Bayliss (2007), the two measurements (OxA-13316-7) on this deer from the primary barrow are statistically consistent (T'=0.5; T'(5%)=3.8; ν=1; Ward and Wilson 1978) and proved to date from 870–970 cal AD (95% confidence; Reimer *et al* 2004), and derive from an area of mound that had some later disturbance. These dates were excluded from the chronological model.

References: Reimer *et al* 2004
                  Ward and Wilson 1978

OxA–13317 1153 ±24 BP

δ¹³C: -20.2‰

Sample: AB 67 368, submitted on 21 January 2006 by L McFadyen

Material: animal bone: *Cervus elaphus,* proximal tibia unfused but has kept epiphysis (J Mulville 2004)

Initial comment: a replicate of OxA-13316.

Objectives: as OxA-13316

Calibrated date: 1σ: cal AD 780–945
                       2σ: cal AD 775–975

Final comment: see OxA-13316

OxA–13318 5222 ±31 BP

δ¹³C: -19.8‰

Sample: AB12 450, submitted on 21 January 2004 by L McFadyen

Material: animal bone: *Canis* sp., left and right mandible (J Mulville 2004)

Initial comment: in the brown loam deposit of the mound, in cutting DX, in a deposit that spanned through cuttings DX, DXI, and DXII, in the western end of the secondary barrow. It was sealed between layers of limestone rubble deposits. It had preponderance of skulls and of left-side limb elements of animal bone, making the deposit unique within the mound architecture.

Objectives: to establish the phase and sequence of the secondary barrow construction.

Calibrated date: 1σ: 4050–3970 cal BC
2σ: 4220–3960 cal BC

Final comment: A Bayliss (2007), *see* GrA-25296. This sample is probably residual and was excluded from the chronological model.

OxA–13319 4984 ±29 BP

δ¹³C: -20.7‰
δ¹⁵N (diet): +9.5‰
C/N ratio: 3.1

Sample: HB 1 391/31 (A1), submitted in June 2003 by A Whittle

Material: human bone (left tibia) (D Galer 2003)

Initial comment: individual A1 was part of a deposit A of human bone from chamber 3. A1 was the earliest individual in the sequence. Individual A3 was overlying A1. The skeleton was partially articulated. The main deposit of human bone was found in a matrix of dark brown stony loam.

Objectives: to establish the sequence of a Neolithic collective bone deposit.

Calibrated date: 1σ: 3800–3700 cal BC
2σ: 3910–3690 cal BC

Final comment: see GrA-25292

Laboratory comment: English Heritage (2007), chronological modelling provides a posterior density estimate for this bone of *3750–3690 cal BC (92% probability)* or *3685–3670 cal BC; 3% probability;* table 1; Bayliss *et al* 2007e).

OxA–13320 4974 ±29 BP

δ¹³C: -20.6‰
δ¹⁵N (diet): +10.1‰
C/N ratio: 3.0

Sample: HB3 391/28 (A3), submitted in June 2003 by A Whittle

Material: human bone (right ulna (articulating)) (D Galer 2003)

Initial comment: individual A3 was part of a deposit A of human bone from chamber 3. A3 was later than A1 in terms of deposition; the majority of the remains overlie individual A1. The main deposit of human bone was found in a matrix of dark brown stony loam.

Objectives: to establish the sequence of a Neolithic collective bone deposit.

Calibrated date: 1σ: 3790–3700 cal BC
2σ: 3900–3660 cal BC

Final comment: see GrA-25292

Laboratory comment: English Heritage (2007), chronological modelling provides a posterior density estimate for this bone of *3735–3655 cal BC; 95% probability;* table 1; Bayliss *et al* 2007e).

OxA–13400 4876 ±33 BP

δ¹³C: -20.6‰
δ¹⁵N (diet): +9.4‰
C/N ratio: 3.3

Sample: HB5 (534/36) (E1), submitted in February 2004 by A Whittle

Material: human bone (right humerus) (D Galer 2003)

Initial comment: individual E1 is an isolated burial, found at the southern end of the monument, (adjacent to the empty cist). The remains represent one individual only. It was relatively articulated, with the legs tightly flexed. Burial deposit E had been placed on stones already introduced as part of the passage filling. The uppermost bones lay in a matrix on crumbly brown soil; bones which had slipped down into crevices in the filling were in a darker soil matrix, assumed to have been the result of earthworm activity.

Objectives: to establish the sequence of a Neolithic collective bone deposit.

Calibrated date: 1σ: 3700–3640 cal BC
2σ: 3710–3630 cal BC

Final comment: see GrA-25292

Laboratory comment: English Heritage (2007), chronological modelling provides a posterior density estimate for this bone of *3700–3635 cal BC; 95% probability;* table 1; Bayliss *et al* 2007e).

OxA–13401 4765 ±31 BP

δ¹³C: -20.3‰
δ¹⁵N (diet): +9.5‰
C/N ratio: 3.3

Sample: HB6 (530/154) (B1), submitted in February 2004 by A Whittle

Material: human bone (left femur) (D Galer 2003)

Initial comment: individual B1 found overlying individual B2 (separated by stony packing). The remains are developmentally consistent with a juvenile aged 7–8 years. It was relatively complete, with no evidence that the individual was articulated when deposited. The main deposit of human bone was in a matrix of dark brown stony loam.

Objectives: as OxA-13400

Calibrated date: 1σ: 3640–3520 cal BC
2σ: 3640–3380 cal BC

Final comment: see GrA-25292

Laboratory comment: English Heritage (2007), chronological modelling provides a posterior density estimate for this bone of *3650–3600 cal BC; 95% probability;* table 1; Bayliss *et al* 2007e).

OxA–13402 4964 ±32 BP

$\delta^{13}C$: -20.7‰
$\delta^{15}N$ (diet): +10.2‰
C/N ratio: 3.3

Sample: HB8 (530/346) (B3/B4/B5), submitted in February 2004 by A Whittle

Material: human bone (left humerus) (D Galer 2003)

Initial comment: most of the adult bones (not including adult B2) were in a pile in the north-west corner of the chambered area containing deposit B. Very few bones were recovered considering the MNI indicated a further three individuals. These bones were very mixed, although several vertebrae and a left arm were in articulation. The main deposit of human bone was in a matrix of dark brown stony loam.

Objectives: as OxA-13400

Calibrated date: 1σ: 3780–3700 cal BC
                              2σ: 3900–3650 cal BC

Final comment: see GrA-25292

Laboratory comment: English Heritage (2007), chronological modelling provides a posterior density estimate for this bone of *3740–3655 cal BC; 95% probability;* table 1; Bayliss *et al* 2007e).

OxA–13403 4816 ±31 BP

$\delta^{13}C$: -20.5‰
$\delta^{15}N$ (diet): +9.7‰
C/N ratio: 3.3

Sample: HB11 (330/7) (C1/C2/C3/C4/C5), submitted in February 2004 by A Whittle

Material: human bone (left ulna) (D Galer 2003)

Initial comment: it was not possible to identify individuals or parts of individuals from deposit C, either through laboratory analysis or from archaeological plans. The deposit represents a complete jumble of remains in no apparent arrangement or order. No elements were found in articulation. The bones in the deposit were interleaved and intermixed throughout with stone, some which was of a type used in the outer face of the cairn. The deposit also contained stone. In places within the deposit there was loose brown soil and smaller stones, some soil showing evidence of worm activity. Rodent-gnawed human bones were present in various parts of the deposit.

Objectives: to establish the sequence of a Neolithic collective bone deposit.

Calibrated date: 1σ: 3650–3530 cal BC
                              2σ: 3660–3520 cal BC

Final comment: see GrA-25292

Laboratory comment: English Heritage (2007), chronological modelling provides a posterior density estimate for this bone of *3695–3680 cal BC (2% probability)* or *3665–3620 cal BC; 93% probability;* table 1; Bayliss *et al* 2007e).

OxA–13404 4945 ±32 BP

$\delta^{13}C$: -20.1‰
$\delta^{15}N$ (diet): +10.8‰
C/N ratio: 3.2

Sample: HB12 (546/132) (D3/D4), submitted in February 2004 by A Whittle

Material: human bone (left scapula) (D Galer 2003)

Initial comment: the mixed remains of two individuals (D3 and D4), both adult, were found clustered to the south-west corner of the chambered area containing deposit D, with no evidence of articulation, and relatively incomplete individuals. The main deposit of human bone was in a matrix of dark brown stony loam.

Objectives: to establish the sequence of a Neolithic collective bone deposit.

Calibrated date: 1σ: 3770–3660 cal BC
                              2σ: 3790–3650 cal BC

Final comment: see GrA-25292

Laboratory comment: English Heritage (2007), chronological modelling provides a posterior density estimate for this bone of *3740–3655 cal BC; 95% probability;* table 1; Bayliss *et al* 2007e).

Aveley Marshes, Essex

Location: TQ 543790
                              Lat. 51.29.22 N; Long. 00.13.18 E

Project manager: R Batchelor (Royal Holloway College), December 2004

Archival body: Royal Holloway College

Description: the site is situated on reclaimed marshland, located on the north bank of the River Thames, approximately 200m from the river itself. The site is now owned by the RSPB and is also a Site of Special Scientific Interest (SSSI). A series of cores were taken from Aveley Marsh reaching a depth of 8.93m below the surface. The sequence traverses two peat horizons contained within three clay-rich layers. The base of the sequence is marked by a large piece of wood that the drilling equipment was unable to wholly penetrate. Some, however, was extracted though it is unknown whether it is preserved *in situ*.

Objectives: these dates will act as range finder dates for the whole of the Aveley Marsh sequence, allowing an estimate of the rate of sedimentation and an age-depth model to be calculated. A pollen diagram for the site is currently being constructed; the age-depth model will allow dating of the changes in pollen stratigraphy that occur.

Final comment: R Batchelor (28 November 2005), dating of the Aveley Marsh sequence has proved to be very successful. A total of eight dates were taken from the 9m sequence; six from the main peat unit, and a further two from the smaller peat unit located below. The dates will lead to the creation of an age-depth model and allow the estimation of pollen stratigraphic changes - early evidence has already suggested that the species *Taxus* cf *baccata* colonised Aveley Marsh prior to Hornchurch Marshes and Beckton (located further west along the Thames valley); and changes in the rate of relative sea-level rise.

The first and older peat unit was forming around *c* 6000 cal BP (dates were only ascertained from the centre of this smaller unit). The main peat unit formed from *c* 5500 to 2600 cal BP (its initiation and cessation controlled by changes in relative sea-level rise). The date ascertained for the top of the peat was a little later than expected (late Iron/early Bronze Age) suggesting that inundation occurred later at Aveley Marsh than other local sites. Also of interest are pronounced changes in peat accumulation rates.

References: Branch *et al* 2012
               Meddens 1996
               Sidell *et al* 2000
               Thomas and Rackham 1996
               Wilkinson *et al* 2000

SUERC–7354 2580 ±35 BP

δ¹³C: -27.5‰

Sample: AMR 186 187, submitted on 27 May 2005 by R Batchelor

Material: peat (humic acid)

Initial comment: this sample is located 1.86m below the surface. It marks the top of the peat layer and is capped by inorganic estuarine clay deposited by an increase in the rate of sea-level rise.

Objectives: to establish a date for the top of the peat horizon and thus the inundation of the peat by an increase in the rate of sea-level rise. This date will also act as the final range finder date through the peat sequence and will allow estimates of sedimentation rates as well as an age-depth model to be constructed.

Calibrated date: 1σ: 800–770 cal BC
                    2σ: 810–670 cal BC

Final comment: R Batchelor (28 November 2005), SUERC-7354 together with SUERC-7355 have provided a successful, though later than expected date for the top of peat sequence. Despite the two radiocarbon dates being statistically different, they both suggest a late Iron Age/early Bronze Age date for peat inundation, which is later than other local sites. Further analysis needs to be carried out before the most realistic date can be ascertained.

Laboratory comment: English Heritage (20 September 2005), the two measurements (SUERC-7354 and SUERC-7355) are not statistically consistent (T'=5.9; ν=1; T'(5%)=3.8; Ward and Wilson 1978), and therefore the two dates cannot be combined. It is not possible with the present information to determine which of these two dates may be too old or too young.

References: Ward and Wilson 1978

SUERC–7355 2700 ±35 BP

δ¹³C: -28.4‰

Sample: AMR 186 187, submitted on 27 April 2005 by R Batchelor

Material: peat (humin)

Initial comment: as SUERC-7354

Objectives: as SUERC-7354

Calibrated date: 1σ: 900–810 cal BC
                    2σ: 920–800 cal BC

Final comment: see SUERC-7354

Laboratory comment: see SUERC-7354

SUERC–7356 3650 ±35 BP

δ¹³C: -28.6‰

Sample: AMR 274 275, submitted on 27 April 2005 by R Batchelor

Material: waterlogged plant macrofossil: *Alnus glutinosa* (R Gale 2005)

Initial comment: this sample is located 2.73m below the surface. It is *c* 1m from the top of the peat and is capped by inorganic estuarine clay deposited by an increase in the rate of sea level rise.

Objectives: this date will act as the third range finder date through the peat sequence and will allow estimates of sedimentation rates, as well as an age-depth model to be constructed. A pollen diagram for this site is currently being constructed; the age-depth model will allow dating of the changes in pollen stratigraphy that occur.

Calibrated date: 1σ: 2120–1950 cal BC
                    2σ: 2140–1920 cal BC

Final comment: R Batchelor (28 November 2005), this date has been very successful. It suggests a Bronze Age or Neolithic date mid way through the peat sequence as expected and will contribute towards the age-depth model currently under construction.

SUERC–7360 3980 ±40 BP

δ¹³C: -27.4‰

Sample: AMR 350 351, submitted on 27 April 2005 by R Batchelor

Material: waterlogged plant macrofossil: *Alnus glutinosa*, single fragment (R Gale 2005)

Initial comment: this sample is located 3.5m below the surface. It is *c* 1.7m from the top of the peat and is capped by inorganic estuarine clay deposited by an increase in the rate of sea level rise.

Objectives: this date will act as the second range finder date through the peat sequence and will allow estimates of sedimentation rates, as well as an age-depth model to be constructed. A pollen diagram for the site is currently being constructed; the age-depth model will allow dating of the changes in pollen stratigraphy that occur.

Calibrated date: 1σ: 2570–2460 cal BC
                    2σ: 2580–2350 cal BC

Final comment: R Batchelor (28 November 2005), this date falls within the Neolithic period as estimated. Interestingly, it also reveals a rapid period of peat growth between SUERC-7356 and SUERC-7360 of *c* 75cm in 500–600 years. Before and after this period accumulation appears to be far slower.

SUERC–7361 4665 ±35 BP

δ¹³C: -27.5‰

Sample: AMR 428 429, submitted on 27 April 2005 by
R Batchelor

Material: waterlogged plant macrofossil: *Alnus glutinosa*,
roundwood; >10mm diameter (R Gale 2005)

Initial comment: this sample is located 4.28m below the
surface. It is 9cm above the base of the peat and *c* 2.5m
from the top of the peat and is capped by inorganic estuarine
clay deposited by an increase in the rate of sea-level rise.

Objectives: to establish a date for the base of the peat horizon
and thus the inundation of the peat by an increase in the
rate of sea-level rise. This date will also act as the range
finder date through the peat sequence and will allow
estimates of sedimentation rates as well as an age-depth
model to be constructed.

Calibrated date: 1σ: 3520–3360 cal BC
                     2σ: 3630–3360 cal BC

Final comment: R Batchelor (28 November 2005), a second
radiocarbon date was carried out at this depth to provide a
reliable date from the base date of the sequence. SUERC-
7361 and SUERC-7364 are near identical and thus provide
a very firm estimation for the initiation of peat growth. In
addition, these dates, as expected fall within the estimated
Neolithic/Mesolithic period.

SUERC–7362 5250 ±35 BP

δ¹³C: -26.0‰

Sample: AMR 572 573A, submitted on 27 April 2005 by
R Batchelor

Material: waterlogged plant macrofossil: *Alnus glutinosa*
(R Gale 2005)

Initial comment: this sample is located 5.72m below the
surface. It marks the centre of a small later of peat located
between 5.55m and 5.77m below the surface. It was
deposited on organic clay due to a slowing in the rate of sea-
level rise allowing peat to grow. The peat is capped by a
second by a second organic clay deposited by an increase in
the rate of sea-level rise.

Objectives: to establish a date for the second layer of peat and
thus provide an approximate date for changes in the rate of
sea level rise. This date will also act as the range finder date
for the lower part of the Aveley sequence. It will allow
estimates of sedimentation rates, as well as an age-depth
model to be constructed.

Calibrated date: 1σ: 4150–3990 cal BC
                     2σ: 4230–3970 cal BC

Final comment: R Batchelor (28 November 2005), the two
dates, SUERC-7362 and SUERC-7363, were taken from
different *Alnus* plant macrofossils were submitted and have
provided near identical dates for the centre of the second
peat unit in the Aveley sequence. They indicate a very
reliable date that falls within the estimated Neolithic-
Mesolithic period, thus indicating fluctuations in sea level
allowed the formation of the older peat unit on the
floodplain surface around this time.

Laboratory comment: (20 September 2005), the samples
(SUERC-7362 and SUERC-7363) from this level are
statistically consistent (T'=0.5; ν=1; T'(5%)=3.8; Ward and
Wilson 1978) and suggest that no residual or intrusive
material was dated from this level. In this case, the latest
date (SUERC-7362, AMR 572 573 A) most reliably dates
this level.

References: Ward and Wilson 1978

SUERC–7363 5285 ±35 BP

δ¹³C: -25.4‰

Sample: AMR 572 573B, submitted on 27 April 2005 by
R Batchelor

Material: waterlogged plant macrofossil: *Alnus glutinosa*
(R Gale 2005)

Initial comment: as SUERC-7362

Objectives: as SUERC-7362

Calibrated date: 1σ: 4230–4040 cal BC
                     2σ: 4240–3990 cal BC

Final comment: see SUERC-7362

Laboratory comment: see SUERC-7362

SUERC–7364 4675 ±35 BP

δ¹³C: -28.2‰

Sample: AMR 429 430, submitted on 27 April 2005 by
R Batchelor

Material: waterlogged plant macrofossil: *Alnus glutinosa*
(R Gale 2005)

Initial comment: as SUERC-7361

Objectives: as SUERC-7361

Calibrated date: 1σ: 3520–3370 cal BC
                     2σ: 3630–3360 cal BC

Final comment: see SUERC-7361

Baguley Hall: north wing, Greater Manchester

Location: SJ 81628874
                    Lat. 53.23.41 N; Long. 02.16.35 W

Project manager: D Hamilton (English Heritage), 2005

Archival body: English Heritage

Description: Baguley Hall is an exemplary medieval timber-
framed building. The hall and north wing were
dendrochronologically dated (Nayling 2003; 2005), although
the date of the north wing rests on only two tree-ring dates.
As nearly all the timbers are fast-grown in the north wing,
only two are being wiggle-matched (BAG-H core 1 and core
3) and the remainder are having their outer ten rings
measured. All of the samples end in the heartwood/sapwood
boundary. Two wiggle-matches (core 1: 7 blocks, and core
3: 3 blocks) and 12 outer 10–ring samples on various pieces
of 'primary' phase timber were submitted for dating.

Objectives: to provide independent evidence for the construction date of this range and relate it to the chronology of the main hall which has been securely dated with tree-ring dates on 30 timbers. Detail of the timber framing between the hall and the north wing suggests that the latter may be earlier. The hall was constructed shortly after AD 1398/9. this research aimed to determine whether the north wing was really mid-fifteenth century in date, and thus the structural evidence misleading.

Final comment: D Hamilton (22 October 2014), the radiocarbon dates and subsequent wiggle-matching and Bayesian modelling were successful and provide solid support for the tree-ring date of AD 1398/9. By dating multiple timbers from the primary construction of the north wing it was possible to demonstrate that the felling date produced was not on a non-primary timber.

Laboratory comment: English Heritage (3 September 2007), the wiggle-match of core 1 produced a probability estimate of *cal AD 1425–1440 (95% probability; heartSap; Hamilton* et al *2007, fig 3)* for the heartwood/sapwood boundary of that timber. The two separate cores from this same timber, dated with dendrochronology (Nayling 2003; 2005) have heartwood/sapwood boundaries of AD 1433 (sample 8) and AD 1432 (sample 55).

The combination of AMS radiocarbon dating and Bayesian modelling has also shown that a substantial number of timbers in the north wing of Baguley Hall do, in fact, post-date the main hall range by slightly over a half century. So while it may be difficult to imagine pegs being driven into place with the south-wall frame *in situ*, what is plausible is that the south-wall frame had been pegged *ex situ* and then moved into position prior to constructing the remainder of the north wing. The fact that the north wing could have, at least in part, been pre-fabricated might also explain empty motices that were never pegged, as the building design might have altered between the initial design and the raising of the wall.

References: Hamilton *et al* 2007
                   Nayling 2003
                   Nayling 2005

OxA–14586 460 ±29 BP

$\delta^{13}C$: -24.8‰

Sample: BAG-H Core 3 Part 1, submitted on 17 March 2005 by D Hamilton

Material: wood: *Quercus* sp. (R Howard 2005)

Initial comment: core 3 is from the third bridging beam from the east. It is the outer 10 rings (1–10) starting from the heartwood/sapwood boundary.

Objectives: to date this bridging beam and provide independent evidence for the date of the wing.

Calibrated date: 1σ: cal AD 1430–1450
                      2σ: cal AD 1410–1460

Final comment: D Hamilton (22 October 2014), the result is in good agreement with the sequence of dates from this core. The overall agreement of the wiggle-match for core 3 was A=105.4%(An=40.8%; Hamilton *et al* 2007, fig 4).

Laboratory comment: English Heritage (22 October 2014), the two measurements (OxA-14586-7) on this sample are statistically consistent (T'=0.1; v=1; T'(5%)=3.8: Ward and Wilson 1978).

References: Ward and Wilson 1978

OxA–14587 472 ±28 BP

$\delta^{13}C$: -25.1‰

Sample: BAG-H Core 3 Part 1, submitted on 17 March 2005 by D Hamilton

Material: wood: *Quercus* sp. (R Howard 2005)

Initial comment: as OxA-14586. Auto-replicate.

Objectives: as OxA-14586

Calibrated date: 1σ: cal AD 1420–1450
                      2σ: cal AD 1410–1460

Final comment: see OxA-14586

Laboratory comment: see OxA-14586

OxA–14588 503 ±28 BP

$\delta^{13}C$: -25.3‰

Sample: BAG-H Core 3 Part 2, submitted on 17 March 2005 by D Hamilton

Material: wood: *Quercus* sp. (R Howard 2005)

Initial comment: core 3 is from the third bridging beam from the east. It is the middle 10 rings (11–20) starting from the heartwood/sapwood boundary.

Objectives: to date this bridging beam and provide independent evidence for the date of the wing.

Calibrated date: 1σ: cal AD 1410–1440
                      2σ: cal AD 1400–1450

Final comment: D Hamilton (22 October 2014), the result is in good agreement with the sequence of dates from this core. The overall agreement of the wiggle-match for core 3 was A=105.4% (An=40.8%; Hamilton *et al* 2007, fig 4).

OxA–14594 562 ±27 BP

$\delta^{13}C$: -25.6‰

Sample: BAG-H Core 3 part 3, submitted on 17 March 2005 by D Hamilton

Material: wood: *Quercus* sp. (R Howard 2005)

Initial comment: core 3 is from the third bridging beam from the east. It is the inner 10 rings (21–30) starting from the heartwood/sapwood boundary.

Objectives: to date this bridging beam and provide independent evidence for the date of the wing.

Calibrated date: 1σ: cal AD 1320–1420
                      2σ: cal AD 1310–1430

Final comment: D Hamilton (22 October 2014), the result is in good agreement with the sequence of dates from this core. The overall agreement of the wiggle-match for Core 3 was A=105.4% (An=40.8%; Hamilton *et al* 2007, fig 4).

OxA–14595 482 ±26 BP

δ¹³C: -25.1‰

Sample: BAG-H Core 7 part 1, submitted on 17 March 2005 by D Hamilton

Material: wood: *Quercus* sp. (R Howard 2005)

Initial comment: core 7 is from the first floor joist 9 from the south, west side of the spine beam. It is the outer 10 rings, ending in the heartwood/sapwood boundary.

Objectives: to date this floor joist and provide independent evidence for the date of the wing.

Calibrated date: 1σ: cal AD 1420–1450
2σ: cal AD 1410–1450

Final comment: D Hamilton (22 October 2014), the result fits well within a Bayesian model for the construction of the north wing (Hamilton *et al* 2007, figs 3–9).

OxA–14596 489 ±27 BP

δ¹³C: -25.7‰

Sample: BAG-H Core 8 Part 1, submitted on 17 March 2005 by D Hamilton

Material: wood: *Quercus* sp. (R Howard 2005)

Initial comment: core 8 is from the first floor joist 5 from the south, east side of the spine beam, in the West Room. It is the outer 10 rings, ending in the heartwood/sapwood boundary

Objectives: to date this floor joist and provide independent evidence for the date of the wing.

Calibrated date: 1σ: cal AD 1410–1450
2σ: cal AD 1400–1450

Final comment: see OxA-14595

OxA–14597 475 ±27 BP

δ¹³C: -25.6‰

Sample: BAG-H Core 9 part 1, submitted on 17 March 2005 by D Hamilton

Material: wood: *Quercus* sp. (R Howard 2005)

Initial comment: core 9 is from the first floor joist 6 from the south, east side of the spine beam, in the West Room. It is the outer 10 rings, ending in the heartwood/sapwood boundary.

Objectives: to date this floor joist and provide independent evidence for the date of the wing.

Calibrated date: 1σ: cal AD 1420–1450
2σ: cal AD 1410–1460

Final comment: see OxA-14595

OxA–14598 387 ±28 BP

δ¹³C: -24.7‰

Sample: BAG-H Core 10 Part 1, submitted on 17 March 2005 by D Hamilton

Material: wood: *Quercus* sp. (R Howard 2005)

Initial comment: core 10 is from the first floor joist 7 from the south, east side of the spine beam, in the West Room. It is the outer ten rings, ending in the heartwood/sapwood boundary.

Objectives: to date this floor joist and provide independent evidence for the date of the wing.

Calibrated date: 1σ: cal AD 1450–1620
2σ: cal AD 1440–1630

Final comment: see OxA-14595

OxA–14599 487 ±26 BP

δ¹³C: -25.6‰

Sample: BAG-H Core 11 part 1, submitted on 17 March 2005 by D Hamilton

Material: wood: *Quercus* sp. (R Howard 2005)

Initial comment: core 11 is from the first floor joist 6 from the south, west side of the partition wall, in the Middle Room. it is the outer ten rings, ending in the heartwood/sapwood boundary.

Objectives: to date this floor joist and to provide independent evidence for the date of the wing.

Calibrated date: 1σ: cal AD 1410–1450
2σ: cal AD 1410–1450

Final comment: see OxA-14595

OxA–14600 455 ±27 BP

δ¹³C: -24.1‰

Sample: BAG-H Core 14 Part 1, submitted on 17 March 2005 by D Hamilton

Material: wood: *Quercus* sp. (R Howard 2005)

Initial comment: core 14 is from the first floor, joist 9, from the south, west side of the partition wall, in the Middle Room. It is the outer 10 rings, ending in the heartwood/sapwood boundary.

Objectives: to date this floor joist and provide independent evidence for the date of the wing.

Calibrated date: 1σ: cal AD 1430–1450
2σ: cal AD 1420–1460

Final comment: see OxA-14595

OxA–14601 458 ±26 BP

δ¹³C: -23.7‰

Sample: BAG-H Core 13 Part 1, submitted on 17 March 2005 by D Hamilton

Material: wood: *Quercus* sp. (R Howard 2005)

Initial comment: core 13 is from the first floor, joist 8 from the south, west side of the partition wall, in the Middle Room. It is the outer ten rings, ending in the heartwood/sapwood boundary.

Objectives: to date this floor joist and provide independent evidence for the date of the wing.

Calibrated date: *1σ:* cal AD 1430–1450
                          2σ: cal AD 1410–1460

Final comment: see OxA-14595

OxA–14602 476 ±27 BP

δ¹³C: -24.5‰

Sample: BAG-H Core 12 Part 1, submitted on 17 March 2005 by D Hamilton

Material: wood: *Quercus* sp. (R Howard 2005)

Initial comment: core 12 is from the first floor, joist 7 from the south, on the west side of the partition wall, in the Middle Room. It is the outer 10 rings, ending in the heartwood/sapwood boundary.

Objectives: to date this floor joist and provide independent evidence for the date of the wing.

Calibrated date: *1σ:* cal AD 1420–1450
                          2σ: cal AD 1410–1460

Final comment: see OxA-14595

SUERC–6552 485 ±20 BP

δ¹³C: -24.3‰

Sample: BAG-H Core 1 Part 1, submitted on 17 March 2005 by D Hamilton

Material: wood: *Quercus* sp. (R Howard 2005)

Initial comment: core 1 is the main post (identical to sample 09 in Nayling 2005), in the Main Wing. It ends in heartwood/sapwood boundary. The sample is the outer ten rings.

Objectives: to wiggle-match date this post and provide independent evidence for the date of the wing.

Calibrated date: *1σ:* cal AD 1420–1440
                          2σ: cal AD 1410–1450

Final comment: D Hamilton (22 October 2014), the result is in good agreement with the sequence of dates from this core. The overall agreement of the wiggle-match for core 1 was A=105.4% (An=26.7%; Hamilton *et al* 2007, fig 3).

SUERC–6568 550 ±20 BP

δ¹³C: -23.5‰

Sample: BAG-H Core 1 Part 2, submitted on 17 March 2005 by D Hamilton

Material: wood: *Quercus* sp. (R Howard 2005)

Initial comment: core 1 is the main post (identical to sample 09 in Nayling 2005), from the North Wing. It ends in the heartwood/sapwood boundary. This sample is the second decadal block (rings 11–20) starting from the boundary.

Objectives: to wiggle-match date this post and provide independent evidence for the date of the wing.

Calibrated date: *1σ:* cal AD 1395–1420
                          2σ: cal AD 1320–1425

Final comment: see SUERC-6552

SUERC–6572 545 ±20 BP

δ¹³C: -24.3‰

Sample: BAG-H Core 1 Part 3, submitted on 17 March 2005 by D Hamilton

Material: wood: *Quercus* sp. (R Howard 2005)

Initial comment: core 1 is the main post (identical to sample 09 in Nayling 2005), from the North Wing. It ends in the heartwood/sapwood boundary. This sample is the third decadal block (rings 21–30).

Objectives: to wiggle-match date this post and provide independent evidence for the date of the wing.

Calibrated date: *1σ:* cal AD 1400–1420
                          2σ: cal AD 1320–1430

Final comment: see SUERC-6552

SUERC–6573 570 ±20 BP

δ¹³C: -24.2‰

Sample: BAG-H Core 1 Part 4, submitted on 17 March 2005 by D Hamilton

Material: wood: *Quercus* sp. (R Howard 2005)

Initial comment: core 1 is the main post (identical to sample 09 in Nayling 2005), from the North Wing. It ends in the heartwood/sapwood boundary. This sample is the fourth decadal block (rings 31–40).

Objectives: to wiggle-match date this post and provide independent evidence for the date of the wing.

Calibrated date: *1σ:* cal AD 1320–1410
                          2σ: cal AD 1310–1420

SUERC–6574 635 ±20 BP

δ¹³C: -24.7‰

Sample: BAG-H Core 1 Part 5, submitted on 17 March 2005 by D Hamilton

Material: wood: *Quercus* sp. (R Howard 2005)

Initial comment: core 1 is the main post (identical to sample 09 in Nayling 2005), from the North Wing. It ends in the heartwood/sapwood boundary. This sample is from the fifth decadal block (rings 41–50).

Objectives: to wiggle-match date this post and provide independent evidence for the date of the wing.

Calibrated date: *1σ:* cal AD 1295–1390
                          2σ: cal AD 1285–1395

Final comment: see SUERC-6552

SUERC–6575 645 ±20 BP

$\delta^{13}C$: -23.9‰

Sample: BAG-H Core 1 Part 6, submitted on 17 March 2005 by D Hamilton

Material: wood: *Quercus* sp. (R Howard 2005)

Initial comment: core 1 is the main post (identical to sample 09 in Nayling 2005, from the North Wing. It ends in the heartwood/sapwood boundary. This sample is the sixth decadal block (rings 51–60).

Objectives: to wiggle-match date this post and provide independent evidence for the date of the wing.

Calibrated date: *1σ:* cal AD 1290–1390
                       2σ: cal AD 1285–1395

Final comment: see SUERC-6552

SUERC–6579 665 ±20 BP

$\delta^{13}C$: -24.7‰

Sample: BAG-H Core 1 Part 7, submitted on 17 March 2005 by D Hamilton

Material: wood: *Quercus* sp. (R Howard 2005)

Initial comment: core 1 is the main post (identical to sample 09 in Nayling 2005), form the North Wing. It ends in the heartwood/sapwood boundary. This sample is the seventh decadal block (rings 61–70).

Objectives: to wiggle-match date this post and provide independent evidence for the date of the wing.

Calibrated date: *1σ:* cal AD 1285–1385
                       2σ: cal AD 1280–1390

Final comment: see SUERC-6552

SUERC–6580 460 ±20 BP

$\delta^{13}C$: -24.6‰

Sample: BAG-H Core 15 Part 1, submitted on 17 March 2005 by D Hamilton

Material: wood: *Quercus* sp. (R Howard 2005)

Initial comment: core 15 is from the first floor, joist 10 from the south, west side of the partition wall, in the Middle Room. It is the outer 10 rings, ending in the heartwood/sapwood boundary.

Objectives: to date this floor joist and provide independent evidence for the date of the wing.

Calibrated date: *1σ:* cal AD 1430–1450
                       2σ: cal AD 1420–1455

Final comment: D Hamilton (22 October 2014), the result fits well within a Bayesian model for the construction of the North Wing of the Hall (Hamilton *et al* 2007, figs 3–9).

SUERC–6581 460 ±20 BP

$\delta^{13}C$: -23.5‰

Sample: BAG-H Core 4 Part 1, submitted on 17 March 2005 by D Hamilton

Material: wood: *Quercus* sp. (R Howard 2005)

Initial comment: core 4 is from the second bridging beam from the east. It is the outer 10 rings ending in the heartwood/sapwwod boundary.

Objectives: to date this bridging beam and provide independent evidence for the date of the wing.

Calibrated date: *1σ:* cal AD 1430–1450
                       2σ: cal AD 1420–1455

Final comment: see SUERC-6580

SUERC–6582 460 ±20 BP

$\delta^{13}C$: -24.0‰

Sample: BAG-H Core 5 Part 1, submitted on 17 March 2005 by D Hamilton

Material: wood: *Quercus* sp. (R Howard 2005)

Initial comment: core 5 is from the first bridging beam from the east. It is the outer 10 rings, ending in the heartwood/sapwood boundary.

Objectives: to date this bridging beam and provide independent evidence for the date of the wing.

Calibrated date: *1σ:* cal AD 1430–1450
                       2σ: cal AD 1420–1455

Final comment: see SUERC-6580

SUERC–6583 470 ±20 BP

$\delta^{13}C$: -24.4‰

Sample: BAG-H Core 6 Part 1, submitted on 17 March 2005 by D Hamilton

Material: wood: *Quercus* sp. (R Howard 2005)

Initial comment: core 6 is from the first floor joist 3W from the south in the West Room. It is the outer 10 rings, ending in the heartwood/sapwood boundary.

Objectives: to date this floor joist and provide independent evidence for the date of the wing.

Calibrated date: *1σ:* cal AD 1425–1445
                       2σ: cal AD 1415–1450

Final comment: see SUERC-6580

Binchester, Durham

Location: NZ 210313
                     Lat. 54.40.34 N; Long. 01.40.27 W

Project manager: (University of Birmingham), 1976–81, 1986–91

Description: a Roman fort of the first to fourth/fifth centuries AD. There was also subsequent sub-Roman and Anglo-Saxon activity, and medieval settlement.

Objectives: to understand the stratigraphic sequence that runs from the mid-fourth to the mid-sixth centuries AD, and in order to understand the transitional Roman to sub-Roman to Saxon use of the site.

Final comment: P Marshall (12 October 2012), the pattern of activity at Binchester appears to be one of a gradual decline into the fifth century, but with some residual importance retained. The recognition of a cemetery dating to the middle Saxon period by radiocarbon dating underlines the possibility of a redefinition of the relationship between power and status after the end of Roman occupation. The radiocarbon results have demonstrated that the technique does have a place in understanding the chronology of the Roman and post-Roman period in England.

References: Ferris 2010

Binchester: butchery, Durham

Location: NZ 210313
                     Lat. 54.40.36 N; Long. 01.40.28 E

Project manager: I Ferris (University of Birmingham), 1976–81 and 1986–91

Archival body: Bowes Museum and Barnard Castle

Description: one room of a former commandant's house was turned into a slaughterhouse, with structures for penning animals, and an associated cess deposit. Some articulated cattle bones were found in this horizon (A60). Nearby were dumps of bone and cess-like material in and around the former western *praefurnium*.

Objectives: to date the butchery in the sub-Roman sequence and also the dumping.

Laboratory comment: English Heritage (24 June 2014), 11 further samples from this site were dated before 2003 (OxA-8705–14, -8781, and -8792; Bayliss *et al* 2016, 58-60)

OxA–12370 1714 ±26 BP

δ¹³C: -20.8‰

Sample: A1884(b), submitted in 2005 by I Ferris

Material: animal bone: *Bos* sp., articulated first and second phalanges (S Davis)

Initial comment: a replicate of OxA-8706. From a cess-like deposit (A1884) dumped in the backfill of the western *praefurnium* and containing butchery waste, some of which was articulated. This deposit post-dated the last firing of the flue (A1591), and pre-dated an horizon of bone and antler working (A5068).

Objectives: to provide absolute dating for the butchery activity in the disused *praefurnium* of the bath house. In particular, to confirm that this epsisode is post-Roman in date.

Calibrated date: *1σ:* cal AD 250–390
                     2σ: cal AD 240–400

Final comment: P Marshall (12 October 2012), this result, when combined with others from the butchery phase in a chronological model, provides an estimate for a short-lived phase of activity after the end of use of the bath house furnace in the late-fourth century cal AD.

Laboratory comment: English Heritage (12 October 2012), the duplicate measurements on A1884(b) (OxA-8706, 1600 ±40 BP and OxA-12370) are statistically inconsistent (T'=5.7; T'(5%)=3.8; ν=1; Ward and Wilson 1978).

References: Ward and Wilson 1978

OxA–12371 1723 ±27 BP

δ¹³C: -21.2‰

Sample: A1884(d), submitted in 2005 by I Ferris

Material: animal bone: *Bos* sp., articulated first and second phalanges (S Davis)

Initial comment: a replicate of OxA-8707. As OxA-12370.

Objectives: as OxA-12370

Calibrated date: *1σ:* cal AD 250–390
                     2σ: cal AD 240–400

Final comment: P Marshall (12 October 2012), the duplicate measurements on A1884(d) (OxA-8707, 1610 ±40 BP and OxA-12371) are statistically inconsistent (T'=5.5; T'(5%)=3.8; ν=1; Ward and Wilson 1978).

Laboratory comment: see OxA-8707

References: Ward and Wilson 1978

OxA–12372 1761 ±30 BP

δ¹³C: -21.1‰

Sample: A1821(d), submitted in 2005 by I Ferris

Material: animal bone: *Bos* sp., articulated first, second, and third phalanges (S Davis)

Initial comment: a replicate of OxA-8711. As OxA-12370.

Objectives: as OxA-12370

Calibrated date: *1σ:* cal AD 230–340
                     2σ: cal AD 210–380

Final comment: see OxA-12370

Laboratory comment: English Heritage (12 October 2012), the duplicate measurements on A1821(d) (OxA-8711, 1735 ±35 BP and OxA-12372) are statistically consistent (T'=0.3; T'(5%)=3.8; ν=1; Ward and Wilson 1978), and therefore a weighted mean can be calculated before calibration (1750 ±23 BP; cal AD 230–380 at 95% confidence; Reimer *et al* 2004).

References: Reimer *et al* 2004
                     Ward and Wilson 1978

Binchester: Saxon burial A1584, Durham

Location:	NZ 210313
	Lat. 54.40.36 N; Long. 01.40.28 E
Project manager:	I Ferris (University of Birmingham), 1976–81 and 1986–91
Archival body:	Bowes Museum and Barnard Castle

Description: a burial of a female in a shallow grave, accompanied by grave goods, including a brooch of mid sixth-century type. Other grave goods included objects of bone and antler.

Objectives: to provide tighter dating of the burial which closes the extended sequence of activity that starts in the mid-fourth century AD.

Final comment: P Marshall (12 October 2012), the Anglo-Saxon burial was found just outside the now-disused Roman bath house cut into a layer of rubble (from the collapsed roof of the main bath house building) that sealed the butchery deposits in the former west *praefurnium*. The burial is spatially and culturally separate from the other burials lying to the north.

Although the grave post-dates the butchery phase, there is no demonstrable stratigraphic relationship between A1584 and A5068.

Three samples were dated: SF1861, a human bone (OxA-9058), and two grave goods: SF1908, an antler object (OxA-9059 and OxA-14991), and SF1907, a bone artefact (OxA-9232). It is not thought that the grave goods were curated artefacts (ie significantly older than the burial itself).

The two measurements on SF1908 are not statistically consistent (T'=26.8; T'(5%)=6.0; v=2; Ward and Wilson 1978). As OxA-9059 is considerably younger than the articulated skeleton it was buried with we have chosen to exclude this measurement from the published analysis.

The dating of this burial to the early Anglo-Saxon period is significant for understanding the post-Roman north of England.

Laboratory comment: English Heritage (24 June 2014), three further samples were dated prior to 2003 (OxA-9058–9 and OxA-9232), and are published in Bayliss *et al* 2016, 62).

References:	Ward and Wilson 1978

OxA–14991 1637 ±29 BP

$\delta^{13}C$: 22‰

Material: antler

Initial comment: a replicate of OxA-9059. From a furnished female inhumation, cut into a rubble surface formed from stone and tufa collapse from the bath house building. This collapse post-dates the dumping of the cess-like butchery waste and the more general levelling of this area (A1821 and A1884).

Objectives: this deposit is stratigraphically later than the disuse of the praefurnium and butchery deposits. Dating of this grave will constrain the calibration of the dates from the sub-Roman activity and provide absolute dating for the artefact assemblage associated with the burial.

Calibrated date:	*1σ:* cal AD 390–430
	2σ: cal AD 340–540

Final comment: P Marshall (12/10/2012) the two measurements on SF1908 (OxA-9059 and OxA-14991) are not statistically consistent (T'=26.8; T'(5%)=6.0; v=2; Ward and Wilson 1978). As OxA-9059 (1380 ±40 BP) would seem to be inaccurate, OxA-14991 provides the best estimate for the date of the artefact.

References:	Ward and Wilson 1978

Boden Vean, Cornwall

Location:	SW 66852405
	Lat. 50.04.15 N; Long. 05.15.34 W
Project manager:	C Johns (Cornwall County Council), October 2003

Description: the site is situated 1km south of the village of Manaccan. The field containing the fogou was until 2003 regularly ploughed and used for growing corn.

The fogou, was surrounded by a rectangular ditched enclosure. Evaluation revealed the fogou to be a partially stone-walled feature with both underground and 'above ground' elements. Associated pottery has been provisionally dated to the fifth century BC. The enclosure ditch was *c* 3m wide and excavated to a depth of 2.5m without being bottomed, it contained fragments of pottery provisionally dated to the Iron Age.

Outside the enclosure ditch a Bronze Age structure was found, the floor of which was covered with broken sherds representing five vessels. Most of the sherds belonged to one very large Trevisker style vessel with unusual decoration.

Objectives: to provide dates for the Bronze Age structure, deposition and last use of the vessel; to provide a chronology for the round and occupation of the site; to provide a date for the fogou and creep; and to provide secure dates for the Gwithian ceramic style.

Final comment: J Gossip (30 January 2013), samples were taken from the Bronze Age structure, the fogou, the enclosure ditch, and features within the enclosure. Three principal phases of activity have been confirmed, taking place in the middle-late Bronze Age (probably *1400–1190 cal BC (95% probability; SUERC-6170)*, the middle to later Iron Age (*530–400 cal BC; 95% probability; start fogou*, to *370–160 cal BC; 95% probability; end enclosure*), and during the post-Romano-British periods (*cal AD 610–660; 95% probability; OxA-14560*).

These dates are consistent with the ranges of pottery found from each period and confirm that the large vessel T1 from the Bronze Age structure (deposit 107) fits into the accepted Trevisker Ware sequence.

References:	Gossip 2004
	Gossip and Johns undated
	Gossip 2013
	Linford 1998
	Linford 2004
	Rose and Preston-Jones 1991

Boden Vean: Bronze Age structure, Cornwall

Location: SW 66852405
                Lat. 50.04.15 N; Long. 05.15.34 W

Project manager: C Johns (Cornwall County Council),
                October 2003

Archival body: Royal Cornwall Museum

Description: trench 1 was positioned to investigate a linear geophysical anomaly that appeared to be physically linked with the round as well as a large amorphous un-numbered anomaly to its south. The linear anomaly proved to be a narrow, shallow ditch of uncertain date [113] (1.6m wide and 0.35m deep), and a second linear ditch feature [109] 1.6m wide and 0.52m deep) on the same alignment was located 2.6m to the south.

The larger amorphous anomaly to the south, when cleaned appeared to be a backfilled curved-edged hollow cut into the shillet. This feature is probably the remains of a Bronze Age structure, approximately 8m in diameter. The trench, positioned in the centre of the anomaly, cut through the western half of the house; the geophysical survey therefore suggests there is another similar feature to the west of the trench.

The top of the unfilled hollow was revealed in plan at depth of approximately 0.6m below the present surface of the field. Excavation of the upper stony fills (105) and (106) revealed the remains of some collapsed stone walling (118), perhaps serving as a partial stone kerb around the edge of the hollow. These upper fills suggested deliberate infilling of the hollow. Further investigation was restricted to the south-western quadrant of the hollow, the floor of which was covered by sherds of pottery from a very large decorated Trevisker vessel (T1), many sherds of which were placed with decoration (incised and cord impressed chevrons and lines) upward. These sherds were lying within a charcoal-rich silt clay deposit (107) 0.12m deep that was bulk sampled for environmental and dating analysis. Much pottery was recovered but it is likely that more remains buried beneath the adjacent baulk. Although the quadrant was excavated to its apparent base, no structural features such as post holes were revealed. A large sherd from a different decorated Trevisker vessel was also recovered from this deposit.

Objectives: it is important to obtain absolute dates for the structure and the vessel. It is clearly not a standard Trevisker Ware vessel, and therefore cannot be assumed to fall neatly into that fairly well-dated middle Bronze Age date. The unusual form and decoration make its closest comparanda vessels associated with features of earlier date (ie funerary contexts) and it may have been curated.

Final comment: J Gossip (30 January 2013), samples taken from charcoal and residue on pottery from basal deposit (107) produced radiocarbon dates (OxA-14517, OxA-14567, SUERC-6169, and SUERC-6170) consistent with activity in the middle to late Bronze Age, demonstrating that the large vessel T1 fits into the accepted Trevisker Ware sequence. Recent excavation by Cornwall Archaeological Society have revealed the structure to be a roundhouse deliberately backfilled following abandonment.

Laboratory comment: English Heritage (24 February 2010), three of the four measurements from this deposit are statistically consistent (T'=3.0; T'(5%)=6.0; v=2; Ward and Wilson 1978), which suggests that this deposit dates to between *1750–1260 cal BC (95% probability; start_Bronze Age)* and *1400–850 cal BC (95% probability; end_Bronze Age)*, probably in the thirteenth or fourteenth century cal BC. The Trevisker-ware vessel (T1) was probably last used in *1400–1190 cal BC (95% probability; SUERC-6170)*.

References: Gossip 2013
                Ward and Wilson 1978

OxA–14517 3085 ±30 BP

$\delta^{13}C$: -25.7‰

Sample: context (107) <1019>, submitted on 16 March 2005 by C Johns

Material: charcoal: *Ulex/Cytisus* sp., single fragment (R Gale 2004)

Initial comment: from a dark grey/black silt layer in the Bronze Age roundhouse containing frequent pottery and charcoal.

Objectives: it is very important to get dates for this deposit. It is clearly not a standard Trevisker Ware vessel, and therefore cannot be assumed to fall neatly into the fairly well-dated middle Bronze Age currency of this pottery form. The unusual form and decoration make its closest comparanda vessels associated with features of earlier date (ie funerary contexts). Is it a particularly early form of Bronze Age house of this type (for Cornwall); or is it not a domestic structure; or has the vessel(s) been curated for a long period of time (generations)?

Calibrated date: 1σ: 1410–1290 cal BC
                2σ: 1430–1260 cal BC

Final comment: J Gossip (30 January 2013), the radiocarbon date ranges are consistent with activity in the middle to late Bronze Age, demonstrating that the large vessel T1 fits into the accepted Trevisker Ware sequence.

OxA–14567 2277 ±33 BP

$\delta^{13}C$: -23.4‰

Sample: context (107) 3/73, submitted on 16 March 2005 by C Johns

Material: carbonised residue

Initial comment: from a large pottery vessel in dark grey/black slit layer 107 in the Bronze Age roundhouse containing frequent pottery sherds and charcoal.

Objectives: as OxA-14517

Calibrated date: 1σ: 400–360 cal BC
                2σ: 410–210 cal BC

Final comment: (24 February 2010), the two radiocarbon measurements on the carbonised residues from vessel T1 are not statistically consistent (T'=229.5; T'(5%)=3.8; v=1; Ward and Wilson 1978). This may be because OxA-14567 had an extremely low carbon content and not all the surrounding soil contaminants were removed.

SUERC–6169 3055 ±35 BP

$\delta^{13}C$: -25.7‰

Sample: context (107) <1036>, submitted on 16 March 2005 by C Johns

Material: charcoal: *Corylus* sp., roundwood; single fragment (R Gale 2004)

Initial comment: from a dark grey/black silt layer in the Bronze Age roundhouse containing frequent pottery and charcoal.

Objectives: as OxA-14517

Calibrated date: *1σ:* 1400–1260 cal BC
                     2σ: 1420–1210 cal BC

Final comment: see OxA-14517

SUERC–6170 3005 ±35 BP

$\delta^{13}C$: -22.7‰

Sample: context (107) 3/26, submitted on 16 March 2005 by C Johns

Material: carbonised residue

Initial comment: a very large pottery vessel from dark grey/black silt layer (107) in the Bronze Age roundhouse containing frequent pottery and charcoal. A replicate of OxA-14567.

Objectives: as OxA-14517

Calibrated date: *1σ:* 1290–1200 cal BC
                     2σ: 1390–1120 cal BC

Final comment: see OxA-14567

Boden Vean: Iron Age activity, Cornwall

Location: SW 66852405
                     Lat. 50.04.15 N; Long. 05.15.34 W

Project manager: C Johns (Cornwall County Council), October 2003

Archival body: Royal Cornwall Museum

Description: trench 2 investigated the ditch forming the northern side of the rectilinear round containing the fogou. The 3m-wide ditch [202] was excavated to a depth of approximately 2.5m, at which point the edges showed little sign of narrowing so that, unless the ditch is flat-bottomed, it is likely to be considerably deeper. Larger sherds of pottery, provisionally dated to the Iron Age, were recovered from the ditch fills. Fourteen distinct fills were recorded, suggesting erosion of the shillet edges following construction, gradual silting, and refuse dumping and deliberate backfilling.

Trench 3 was positioned to investigate a possible entrance through the western side of the round. The ditch 315 was approximately 4m wide and no break was identified.

Trenches 4, 5, 6, 7, 8, and 9 investigated the fogou itself. Two parallel lines of stones were found on either side of a deep vertical-sided cut into the shillet representing the linear anomaly approaching the stone-walled fogou passage. This trench had been backfilled with a number of loose stony deposits, some of which yielded pottery provisionally identified as Iron Age or Romano-British in date. The purpose of this feature is not yet fully understood. It is possible that the stones may have been placed to mark the line of the fogou approach after it had been infilled.

To the west of this linear cut another diagonal cut [412] extended into the shillet. Within the backfill of this feature a stone 'box' [425] had been constructed from a number of small orthostatic stones. The fill of this feature (426) was sampled in its entirety.

A curvilinear anomaly ('creep') was investigated in Trench 4 as context [431] and in Trench 6 as [609]. This feature proved to be a rock-cut ditch approximately 1.5m wide by 1m deep. Two sections were excavated through the 'ditch', which terminated just to the west of the anomaly leading south from the fogou [612]. The two fills of the feature ((606) and (610)) were silty and charcoal rich, and contained large amounts of fragmented burnt bone. Finds included two ceramic beads and a fragment of copper alloy brooch. To the west of the feature was a spread of small stones (608), which may have been placed on an *ad hoc* basis to form a series of drainage channels, or alternatively comprise the rubble from a structure. The linear anomaly [612] leading south from the fogou was only partially excavated, but had a more gradual profile than in Trench 4. The backfilled ditch had been superseded by a small rubble-filled pit cut into the fill of the ditch. Iron Age/Romano-British pottery was recovered from the ditch fill.

Trenches 8 and 9 were excavated in order to elucidate the nature, extent, and preservation of the fogou structure, revealing coursed stonework (801/901). Two large orthostats on either side of the fogou defined a point at which the tunnel narrowed. The coursed stone walling could be seen to become deeper, curving to the north-east in the direction of the existing open tunnel. The walling was corbelled and there were no *in situ* roof stones, although there were stones in the fill of the fogou that could have spanned the roof. The interior of the fogou was mostly filled by a homogeneous deposit of shillet and clay, 1.1m deep (804/900), which appeared to indicate deliberate backfilling. Above the floor of the fogou were stony silty clay deposits 805, 807, and 904, and 806 from which pottery of provisional Iron Age date was recovered. Each of these deposits was around 0.1m in thickness.

Objectives: to add to the range of dates for the enclosure, and possibly provide one of the earlier dates closer to the construction. Determinations from context 806 would provide secure dates for the ?later early Iron Age pottery from that context and a *terminus post quem* for use of the fogou.

A date from the curvilinear ditch/creep would help to define this enigmatic feature and clarify the chronological relationship with the fogou and enclosure. Determinations from the 'stone-box sequence' will provide a good date for activity within the enclosure perhaps associated with the fogou.

Final comment: J Gossip (30 January 2013), samples were submitted from charcoal and residue on pottery from (806), a dark silt above the base of the fogou. Dates suggest that the main period of use of the fogou took place in the fourth century BC, broadly consistent with other recently excavated fogous in Cornwall.

Laboratory comment: English Heritage (24 February 2010), chronological modelling indicates that the deposit which provides a *terminus ante quem* for the fogou's construction of *530–400 cal BC (95% probability; start_Fogou)*, and probably by *460–410 cal BC (68% probability)*. The fogou's primary use is likely to be confined to the fourth century cal BC. The chronological model for activity in the enclosure suggests that activity within began in *580–410 cal BC (95% probability; start_enclosure)*, probably in *490–420 cal BC (68% probability)*. Activity in the enclosure ended in *370–160 cal BC (95% probability; end_enclosure)*, probably either in the middle decades of the fourth century cal BC (*360–330 cal BC at 11% probability*) or in the third century cal BC (*280–200 cal BC at 58% probability*). The enclosure was probably used for *60–390 years (95% probability)*, however the distribution of this span is strongly bi-modal. The enclosure probably was either used for *80–100 years (11% probability)*, or was used for two or three centuries (*175–310 years; 57% probability*). There is, unfortunately, an insufficient number of radiocarbon dates to determine between these two possibilities, although the consistency of the measurements suggests that the latter possibility may be an artefact of statistical scatter.

References: Gossip 2013

OxA–14486 2205 ±37 BP

$\delta^{13}C$: -25.3‰

Sample: context (806) <1018>c, submitted on 16 March 2005 by C Johns

Material: carbonised residue

Initial comment: from a very dark silt layer (806) 100mm thick overlying the fogou floor. Also contained pottery provisionally dated to the fifth century BC. Overlain by layer (805), 100mm thick and above that the homogeneous backfill of the fogou chamber approximately 2.0m deep.

Objectives: a secure date for this deposit would give a likely *terminus post quem* for use of the fogou and be useful for comparison for other dates obtained for the round and other features associated with the fogou. It is also important to have secure dates for this pottery style.

Calibrated date: 1σ: 370–200 cal BC
                          2σ: 390–170 cal BC

Final comment: J Gossip (30 January 2013), the range of ages from these samples suggests that the deposit comprised trampled or disturbed silt which accumulated over several centuries spanning the middle to late Iron Age, with the main focus of activity likely to be during the fourth century BC. Dates are consistent with the pottery excavated from the deposit.

Laboratory comment: English Heritage (24 February 2010), five samples were dated from this context. The measurements are not statistically consistent (T'=75.1;

T'(5%=11.1; ν=5; Ward and Wilson 1978), suggesting that the deposit contains material of a range of actual ages.

References: Ward and Wilson 1978

OxA–14487 2144 ±36 BP

$\delta^{13}C$: -26.4‰

Sample: context (806) <1018>a, submitted on 16 March 2005 by C Johns

Material: carbonised residue

Initial comment: as OxA-14486

Objectives: as OxA-14486

Calibrated date: 1σ: 350–110 cal BC
                          2σ: 360–50 cal BC

Final comment: see OxA-14486

Laboratory comment: see OxA-14486

OxA–14514 2261 ±28 BP

$\delta^{13}C$: -25.7‰

Sample: context (806) <1018>g, submitted on 16 March 2005 by C Johns

Material: charcoal: *Corylus* sp., single fragment (R Gale 2004)

Initial comment: as OxA-14486

Objectives: a secure date for this deposit would give a likely *terminus post quem* for use of the fogou and be useful for comparison for other dates obtained for the round and other features associated with the fogou.

Calibrated date: 1σ: 390–250 cal BC
                          2σ: 400–200 cal BC

Final comment: see OxA-14486

Laboratory comment: see OxA-14486

OxA–14515 2425 ±29 BP

$\delta^{13}C$: -24.3‰

Sample: context (806) <1018>f, submitted on 16 March 2005 by C Johns

Material: charcoal: *Quercus* sp., roundwood; single fragment (R Gale 2004)

Initial comment: as OxA-14486

Objectives: as OxA-14514

Calibrated date: 1σ: 730–400 cal BC
                          2σ: 750–400 cal BC

Final comment: see OxA-14486

Laboratory comment: English Heritage (24 February 2010), the two measurements on this single fragment of charcoal are statistically consistent (T'=0.8; T'(5%)=3.8; ν=1; Ward and Wilson 1978). The results were combined before calibration (2444 ±21 BP; 750–405 cal BC at 95% confidence; Reimer *et al* 2004).

References: Reimer *et al* 2004
                    Ward and Wilson 1978

OxA–14516 2462 ±29 BP

δ¹³C: -24.8‰

Sample: context (806) <1018>f, submitted on 6 March 2005 by C Johns

Material: charcoal: *Quercus* sp., roundwood; single fragment (R Gale 2004)

Initial comment: as OxA-14486. A replicate of OxA-14515.

Objectives: as OxA-14514

Calibrated date: *1σ:* 760–510 cal BC
                              2σ: 770–410 cal BC

Final comment: see OxA-14486

Laboratory comment: see OxA-14515

OxA–14518 2463 ±28 BP

δ¹³C: -22.8‰

Sample: context (610) <1015>a, submitted on 16 March 2005 by C Johns

Material: charcoal: *Quercus* sp., roundwood; single fragment (R Gale 2004)

Initial comment: from the primary dark silty fill 150mm thick in curvilinear ditch/creep [609] (same as [431]).

Objectives: a date from this would help to define this enigmatic feature, in particular to clarify the chronological relationship with the fogou and enclosure.

Calibrated date: *1σ:* 760–510 cal BC
                              2σ: 770–410 cal BC

Final comment: J Gossip (30 January 2013), these date ranges suggest activity within the enclosure of a possible domestic nature at the same time as the fogou, probably during the fourth century BC.

Laboratory comment: English Heritage (24 February 2010), the two samples from this context provided statistically inconsistent measurements (T'=19.4; T'(5%)=3.8; *v*=1; Ward and Wilson 1978). They were also inconsistent with the measurement from context 432 (T'=23.5; T'(5%)=6.0; *v*=2; Ward and Wilson 1978). Deposits 610 and 432 appear to be equivalent basal deposits of the same feature. The best estimate for the construction of the ditch is probably provided by the latest dated material in the ditch (*SUERC-6171*), with date ranges of either *410–340 cal BC (59% probability)* or *330–220 cal BC (36% probability)*.

References: Ward and Wilson 1978

OxA–14520 2459 ±28 BP

δ¹³C: -25.1‰

Sample: context (432) <1009>a, submitted on 16 March 2005 by C Johns

Material: charcoal: *Alnus* sp., single fragment (R Gale 2004)

Initial comment: from the primary dark silty fill 150mm thick in curvilinear ditch/creep [431] (same as [609]). Unlikely to be intrusive residual or disturbed. 1.5m below the ground surface and 0.42m below the upper layer of fill in the ditch cut.

Objectives: a date from this would help to define this enigmatic feature, in particular to clarify the chronological relationship with the fogou and enclosure and compare with date from sample <1015>.

Calibrated date: *1σ:* 750–500 cal BC
                              2σ: 770–410 cal BC

Final comment: see OxA-14518

Laboratory comment: see OxA-14518

OxA–14521 2272 ±28 BP

δ¹³C: -25.3‰

Sample: context (201) <1035>a, submitted on 16 March 2005 by C Johns

Material: charcoal: Pomoideae, single fragment (R Gale 2004)

Initial comment: from a dark charcoal-rich layer of fill (201) in round ditch [202]. The deposit was well stratified. Layer of fill in the side of the Iron Age enclosure ditch, 1m from ground surface and 0.5 m from the excavated extent of the ditch, which was not bottomed.

Objectives: this deposit was fairly well stratified in enclosure ditch, but is obviously a discarded deposit - not burnt *in situ*. The deposit that contained probable later early Iron Age pottery, so it would add to a range of dates for the enclosure, and possibly provide one of the earlier dates - closer to construction of the round?

Calibrated date: *1σ:* 400–260 cal BC
                              2σ: 400–210 cal BC

Final comment: J Gossip (30 January 2013), the fourth century BC radiocarbon dates from the enclosure suggest origins contemporary with the fogou and internal features such as the curvilinear structure [431]/[609].

Laboratory comment: English Heritage (24 February 2010), the two measurements on separate fragments of charcoal in ditch fill 201 are statistically consistent (T'=3.0; T'(5%)=3.8; Ward and Wilson 1978). They provide a *terminus ante quem* for the digging of the enclosure ditch at around 400 cal BC.

References: Ward and Wilson 1978

OxA–14522 2240 ±28 BP

δ¹³C: -24.5‰

Sample: context (411) <1024>a, submitted on 16 March 2005 by C Johns

Material: charcoal: *Quercus* sp., roundwood; single fragment (R Gale 2004)

Initial comment: from a stony dark brown silt, the primary fill in [412].

Objectives: this would provide a good date for activity within the enclosure perhaps associated with the fogou. Could provide an interesting comparison for dates from the fogou floor (initial disuse?) and with dates from samples <1006> and <1021>.

Calibrated date: 1σ: 380–210 cal BC
                       2σ: 400–200 cal BC

Final comment: see OxA-14518

OxA–14523 2253 ±28 BP

δ¹³C: -23.6‰

Sample: context (426) <1006>a, submitted on 16 March 2005 by C Johns

Material: charcoal: *Ulex* sp., single fragment (R Gale 2004)

Initial comment: from the fill 150mm thick of stone box [425].

Objectives: this would provide a good date for activity within the enclosure perhaps associated with the fogou. Could provide an interesting comparison for dates from the fogou floor (initial disuse?).

Calibrated date: 1σ: 390–230 cal BC
                       2σ: 400–200 cal BC

Final comment: see OxA-14518

SUERC–6168 2335 ±35 BP

δ¹³C: -25.3‰

Sample: context (806) <1018>e, submitted on 16 March 2005 by C Johns

Material: charcoal: *Corylus avellana*, single fragment (R Gale 2004)

Initial comment: as OxA-14486

Objectives: as OxA-14514

Calibrated date: 1σ: 410–390 cal BC
                       2σ: 420–360 cal BC

Final comment: see OxA-14486

SUERC–6171 2265 ±35 BP

δ¹³C: -25.3‰

Sample: context (610) <1015>b, submitted on 16 March 2005 by C Johns

Material: charcoal: *Quercus* sp., single fragment (R Gale 2004)

Initial comment: as OxA-14518

Objectives: as OxA-14518

Calibrated date: 1σ: 400–230 cal BC
                       2σ: 400–200 cal BC

Final comment: see OxA-14518

Laboratory comment: see OxA-14518

SUERC–6172 2315 ±35 BP

δ¹³C: -25.4‰

Sample: context (414) <1021>b, submitted on 16 March 2005 by C Johns

Material: charcoal: *Corylus avellana*, roundwood; single fragment (R Gale 2004)

Initial comment: as OxA-14519

Objectives: as OxA-14519

Calibrated date: 1σ: 410–380 cal BC
                       2σ: 410–260 cal BC

Final comment: J Gossip (30 January 2013), *see* OxA-14518. These dates are consistent with the earlier phases of activity within the enclosure during the middle Iron Age and suggest domestic occupation at this time.

Laboratory comment: English Heritage (24 February 2010), the chronological model suggests that cut [416] dates to *420–350 cal BC (95% probability; SUERC-6172).* Cut [412] probably also dates to the fourth century cal BC, and the stone 'box' probably dates to the third century cal BC.

SUERC–6173 2350 ±35 BP

δ¹³C: -26.9‰

Sample: context (201) <1035>b, submitted on 16 March 2005 by C Johns

Material: charcoal: Pomoideae, single fragment (R Gale 2004)

Initial comment: as OxA-14621

Objectives: as OxA-14621

Calibrated date: 1σ: 410–390 cal BC
                       2σ: 510–380 cal BC

Final comment: see OxA-14621

Laboratory comment: see OxA-14621

SUERC–6177 2190 ±35 BP

δ¹³C: -24.7‰

Sample: context (411) <1024>b, submitted on 16 March 2005 by C Johns

Material: charcoal: *Corylus avellana*, single fragment (R Gale 2004)

Initial comment: as OxA-14522

Objectives: as OxA-14522

Calibrated date: 1σ: 360–190 cal BC
                       2σ: 380–160 cal BC

Final comment: J Gossip (30 January 2013), these dates are consistent with the middle to later Iron Age activity within the enclosure, suggesting the continuation of domestic occupation. Also *see* OxA-14518.

SUERC–6178 2255 ±35 BP

δ¹³C: -25.9‰

Sample: context (426) <1006>b, submitted on 16 March 2005 by C Johns

Material: charcoal: *Corylus avellana*, single fragment (R Gale 2004)

Initial comment: as OxA-14523

Objectives: as OxA-14523

Calibrated date: 1σ: 390–230 cal BC
2σ: 400–200 cal BC

Final comment: see SUERC-6177

Boden Vean: post-Roman activity, Cornwall

Location: SW 66852405
Lat. 50.04.15 N; Long. 05.15.34 W

Project manager: C Johns (Cornwall County Council),
October 2003

Archival body: Royal Cornwall Museum

Description: the presence of this Style was first indicated by the distinctive low-walled platters with sanded, not grass-marked bases, in contexts 314, an upper ditch fill near the possible entrance 430, upper fill of curvilinear ditch or creep 431, and unstratified elsewhere in trench 4. These platters at Gwithian form part of an assemblage in which necked jars with slightly concave rims otherwise predominate and which are currently dated, on the basis of association with post-Roman import wares, to the fifth and sixth centuries AD. The Gwithian material has recently been assessed (Thomas *et al* 2004) and it is expected that analysis will provide more details of the Style that appears to succeed the late Roman Gabbroic forms in west Cornwall. There are problems here because at Trethurgy the Style does not occur and late-Roman forms appear to continue into the sixth century AD (Quinnell 2004). There is the possibility that the date for the Style given for Gwithian is too early. The Style has not previously been identified for assemblages that have been published but a review of the literature shows that the platters are present at Goldherring (Guthrie 1969) and at Carngoon Bank, at the latter site with possible other forms of similar date (McAvoy *et al* 1980).

Objectives: to obtain an absolute date for this pottery style, currently dated by association to the early post-Roman period, and facilitate comparison with dates from the Gwithian assemblage.

Final comment: J Gossip (30 January 2013), the date range from the submitted samples indicate continued use of the enclosure as a domestic site into the sixth and seventh centuries AD, consistent with dates recently obtained for Gwithian-style platters from Gwithian.

Laboratory comment: English Heritage (24 February 2010), the late fill 314 from the enclosure ditch provided two statistically inconsistent radiocarbon measurements (T'=456.8; T'(5%)=3.8; v=1; Ward and Wilson 1978). The earlier of these is undoubtedly residual from the initial use of the enclosure. The other, a carbonised residue on the Gwithian-style platter dates to the post-Roman period.

References: Gossip 2013
Guthrie 1969
McAvoy *et al* 1980
Quinnell 2004
Thomas *et al* 2004

OxA–14519 2269 ±27 BP

$\delta^{13}C$: -24.6‰

Sample: context (414) <1021>a, submitted on 16 March 2005 by C Johns

Material: charcoal: *Corylus* sp., roundwood; single fragment (R Gale 2004)

Initial comment: from a stony dark brown silt, primary fill in cut [416].

Objectives: this would provide a good date for activity within the enclosure perhaps associated with the fogou. Could provide an interesting comparison for dates from the fogou floor (initial disuse?) and with dates from samples <1006> and <1024>.

Calibrated date: 1σ: 400–260 cal BC
2σ: 400–210 cal BC

Final comment: J Gossip (30 January 2013), this sample appears to date residual Iron Age material associated with the early phases of the site.

OxA–14560 1417 ±29 BP

$\delta^{13}C$: -25.6‰

Sample: context (314) pottery, submitted on 18 March 2005 by C Johns

Material: carbonised residue (H Quinnell 2005)

Initial comment: pottery from a very dark silt layer (314) containing charcoal and burnt bone, an upper layer in a possible terminal of the round enclosure ditch - context [315] in this trench (trench 3).

Objectives: to obtain an absolute date for this pottery style, currently dated by association to the early post-Roman period, and facilitate comparison with dates from the Gwithian assemblage.

Calibrated date: 1σ: cal AD 610–660
2σ: cal AD 590–670

Final comment: J Gossip (30 January 2013), the residue has provided a clear seventh-century AD date for the use of Gwithian-style ceramics and illustrates the continued use of the site in the post-Roman period.

Bouldnor Cliff: BCII and BCIV, Isle of Wight

Location: SK 605101
Lat. 52.41.05 N; Long. 01.06.18 W

Project manager: G Momber (Hampshire and Wight Trust for Maritime Archaeology), 2003

Archival body: Hampshire and Wight Trust for Maritime Archaeology

Description: the site is a submerged landscape 12m to 4m below OD consisting of three separate peat units intercalated within mineral sediments. The sediment series is laterally consistent, running over 1km. The causal mechanism for these sedimentary units was controlled by a progressive positive eustatic sea-level rise. Sampling for radiocarbon dating and chronological modelling took place, in addition to the radiocarbon wiggle-match. Context 7 contains archaeological material of Mesolithic age.

Objectives: the primary objectives of the dating programme were to establish a chronology for the defined palaeoenvironmental sequence at the site; and to determine the chronological relationship between the human occupation and the palaeoenvironmental sequence. More specific aims included: providing a date estimate for peat formation with rising sea level; providing a date estimate for human occupation and its relationship to the archaeological contexts; dating the final marine inundation of the site; and providing a date estimate for the floating tree-ring chronology.

Final comment: G Member (25 October 2014), the maximum span of years between the top of context 2007 (within which the majority of Mesolithic archaeological artefacts were found) and the bottom of the peaty matrix that forms context 2003 and related to the accumulation of the gravelly alluvium comprising context 2006 is estimated at *1–50 years* (95% probability; *alluviation;* Member *et al* 2011, fig 3.25), and more likely at *1–30 years (68% probability).* Finally, the model estimates that marine inundation was complete by *5990–5900 cal BC* (95% probability; fig 2.12, *Marine inundation;* Member *et al* 2011, fig 3.23).

References: Member *et al* 2011
                     Member 2000

SUERC–7560 7105 ±35 BP

δ¹³C: -29.3‰

Sample: MS08 08 BCII, submitted on 3 August 2005 by J Gillespie

Material: wood (waterlogged): *Alnus glutinosa,* roundwood; single fragment (R Gale 2005)

Initial comment: from the basal few centimetres of context 3, above context 7. Compact humified peat heterogeneous colour, black but with patches of very dark reddish brown. Roundwood pieces embedded within the peat. Increasing mineral content. Strongly oxidised and desiccated peat suggestive of significant period of lowered groundwater levels.

Objectives: there is no cultural material associated with this context but macrofossils will provide a date for the end of the gravel deposition and the onset of peat formation. This will help to answer the questions of sea-level change and the timing of human occupation and its relationship to the archaeological contexts.

Calibrated date: *1σ: 6020–5930 cal BC*
                       2σ: 6050–5910 cal BC

Final comment: G Member (25 October 2014), this sample provides a date for the end of the gravel deposition and the onset of peat formation associated with sea-level rise.

SUERC–7561 7175 ±40 BP

δ¹³C: -29.3‰

Sample: MS07 01 BCII, submitted on 3 August 2005 by J Gillespie

Material: wood (waterlogged): *Alnus glutinosa,* roundwood; single fragment (R Gale 2005)

Initial comment: from the basal few centimetres of context 7. Organic sand. Medium sandy clay, slightly calcareous. Roundwood fragments and highly humified organic fragments (dark brown lenses). Heterogeneous colour comprising dominantly very dark grey/grey. Occasional sub-rounded flint granules and pebbles. *In situ* freshly knapped flints.

Objectives: radiocarbon dating of selected deposits is necessary in order to answer questions of sea level change and the timing of human occupation and its relationship to the archaeological contexts. Context 7 contains the archaeological material and the bottom few centimetres of this deposit marks the end of this event.

Calibrated date: *1σ: 6070–6000 cal BC*
                       2σ: 6100–5980 cal BC

Final comment: G Member (25 October 2014), this is the base of the archaeological context and is organic sand that contains many unabraded knapped flints. This base layer is at an absolute depth of 11.24m (± 0.1m) the top of this layer being at 11.04m (±0.1m).

SUERC–7562 7130 ±35 BP

δ¹³C: -28.5‰

Sample: MS20 03 BCII, submitted on 3 August 2005 by J Gillespie

Material: wood (waterlogged): *Alnus glutinosa,* roundwood; single fragment (R Gale 2005)

Initial comment: from the upper few centimetres of context 7, below context 6. Organic sand. Medium sandy clay, slightly calcareous. Roundwood fragments and highly humified organic fragments (dark brown lenses). Heterogeneous colour comprising dominantly very dark grey/grey. Occasional sub-rounded flint granules and pebbles. *In situ* freshly knapped flints.

Objectives: dating of the top of this deposit will provide a *terminus post quem* for the deposition of the overlying gravels, which is necessary in order to answer questions of sea-level change and the timing of human occupation and its relationship to the archaeological contexts.

Calibrated date: *1σ: 6030–5980 cal BC*
                       2σ: 6070–5920 cal BC

Final comment: G Member (25 October 2014), the top of the archaeological layer at an absolute depth of 11.06m (± 0.1m) An estimate of the minimum time span for the accumulation of this layer of sand and Mesolithic flints is a minimum duration of *1–250 years* (95% probability; *span of context 2007;* Member *et al* 2011, fig 3.24), but the skewed shape of the probability distribution model suggests that *1–130 years (68% probability)* is a more likely representation.

SUERC–7579 6925 ±35 BP

$\delta^{13}C$: -26.8‰

Sample: MS06 12 BCII, submitted on 3 August 2005 by J Gillespie

Material: waterlogged plant macrofossils: monocot, leaf (R Gale 2005)

Initial comment: from the top few centimetres of context 3. Compact humified peat. No plant macrofossils, heterogeneous colour, black but with patches of very dark reddish brown. Roundwood pieces embedded within the peat. Increasing mineral content. Small fragments of *Quercus* sp. Strongly oxidised and desiccated peat suggestive of significant period of lowered groundwater levels.

Objectives: context 3 underlies the alluvium of context 2, this will provide a date for the cessation of peat formation.

Calibrated date: *1σ:* 5850–5740 cal BC
                       2σ: 5900–5720 cal BC

Final comment: G Member (25 October 2014), the top of the compacted peat section is at an absolute depth of 10.68m (± 0.1m) providing and provides an index date for the caseation of the peat formation.

SUERC–7580 7115 ±35 BP

$\delta^{13}C$: -22.7‰

Sample: SO31 10 BCII, submitted on 3 August 2005 by J Gillespie

Material: wood (waterlogged): *Alnus glutinosa*, roundwood; single fragment (R Gale 2005)

Initial comment: from the interface between context 3 and context 6. Context 6 consists of gravel containing decayed roots, which interfaces with the compact humified peat of context 3 above it. Strongly oxidised and desiccated peat suggestive of significant period of lowered groundwater levels, which is followed by a high energy process such as alluvial fan deposition.

Objectives: macrofossils will provide a date for the end of the gravel deposition and the onset of peat formation in this area, which overlies that of the gravel formation. This will help to answer the questions of sea-level change and the timing of human occupation and its relationship to the archaeological contexts.

Calibrated date: *1σ:* 6020–5980 cal BC
                       2σ: 6060–5910 cal BC

Final comment: G Member (25 October 2014), the interface between the gravel and peat is at an absolute depth of 10.83m (± 0.1m), this gives an index date for the onset of peat formation and marks the end of human occupation on this land surface.

SUERC–8157 7110 ±40 BP

$\delta^{13}C$: -27.7‰

Sample: MS08 05 BCII, submitted on 3 August 2005 by J Gillespie

Material: wood (waterlogged): *Alnus glutinosa*, roundwood; single fragment (R Gale 2005)

Initial comment: context 7 (top few cms) below context 6. Organic sand. Medium sandy clay, slightly calcareous. Roundwood fragments and highly humified organic fragments (dark brown lenses). Heterogeous colour comprising dominantly very dark grey/grey. Occasional sub-rounded flint granules and pebbles. *In situ* freshly knapped flints.

Objectives: it is possible that some/more context 7 material was eroded in this area by the same process that deposited context 6, therefore this material is from a different monolith than the previous.

Calibrated date: *1σ:* 6020–5930 cal BC
                       2σ: 6060–5900 cal BC

Final comment: G Member (25 October 2014), the evidence at this level suggests that the deposit was formed in shallow water, the absolute depth of this sample was 11.19m (± 0.1m).

Bouldnor Cliff: BCII, BCIV, and BCV, Isle of Wight

Location: SK 605101
                     Lat. 52.41.05 N; Long. 01.06.18 W

Project manager: G Member (Hampshire and Wight Trust for Maritime Archaeology), 2003

Archival body: Hampshire and Wight Trust for Maritime Archaeology

Description: a submerged Mesolithic landscape in 12m of water situated some 200m offshore of Bouldnor beach in the western Solent. Soft Holocene silts run parallel with the foreshore intercalated with three outcrops of peat. Excavations have revealed Mesolithic lithics from a deposit immediately below the submerged vegetated landscape dated to 6640–6390 cal BP (Beta-140104; 7640 ±70 BP). The causal mechanism for these sedimentary units was controlled by a progressive positive eustatic sea level rise. Context 7 contains archaeological material of Mesolithic age. The sedimentary series consists of: grey alluvium silts (unit 1) overlaying silty alluvium with dark organic staining (unit 2), this overlays a laminated peat deposit inlaid with timber (unit 3). Beneath this is a gravel deposit with timber inclusions and rolled worked flint (unit 6), this covers fine grey sandy/silt with timber inclusions and knapped flint of Mesolithic age (unit 7). The deposit rises from 12m below OD to 4m below OD. The sediment series is laterally consistent running over 1km.

Objectives: radiocarbon dating of selected deposits is necessary in order to answer questions of sea-level change and the timing of human occupation and its relationship to the archaeological contexts, and to define the pollen zones.

Final comment: G Member (25 October 2014), the maximum span of years between the top of context 2007 (within which the majority of Mesolithic archaeological artefacts were found) and the bottom of the peaty matrix that forms context 2003 and related to the accumulation of the gravelly alluvium comprising context 2006 is estimated at *1–50 years (95% probability; alluviation;* Member *et al* 2011, fig 3.25), and more likely at *1–30 years (68%*

probability). Finally, the model estimates that marine inundation was complete by *5990–5900 cal BC* (*95% probability*; *Marine inundation*; Momber *et al* 2011, fig 3.23).

The samples from BC-IV provide the shallowest and therefore the youngest palaeo-vegetation data for the basal peats along the submerged land-surfaces of Bouldnor Cliff. This represents an old land surface between 6.46m and 6.64m below OD with associated Mesolithic archaeology. The land-surface is subject to increasing wetness resulting in inception of peat and ultimately brackish/marine transgression. Radiocarbon measurements estimate the top of the peat/organic unit to date to *(5330–5210 cal BC; 95% probability; OxA-15717;* Momber *et al* 2011, fig 5.4).

Laboratory comment: English Heritage (23 March 2015), three further samples from this series were subsequently dated (SUERC-11284–6).

References: Momber *et al* 2011
                     Momber 2000

OxA–15695 6369 ±34 BP

δ¹³C: -27.6‰

Sample: MS10 30–33 BCIV 03, submitted in March 2006 by J Gillespie

Material: wood (waterlogged): unidentified, twigs (G Campbell 2006)

Initial comment: from context 2. A mineral deposit with 1mm lenses of horizontally bedded, highly-humified, dark brown organics, with piddock burrows common. Becoming organic mud with a stiff silty clay consistency, non-calcareous, with highly fragmented, unidentifiable humified and fresh plant macros. Rare sub-rounded pebbles. Heterogeneous with paddock infestation. The boundary with context 3 is sharp and wavy.

Objectives: this important profile demonstrates the local establishment of lime (*Tilia*) woodland, which is now considered to have been the dominant, or at least co-dominant, wood element throughout the middle and early part of the late-Holocene of southern and eastern England. Goodwin's pollen zone scheme suggests arrival during the late Boreal period, but this has rarely been evidenced, and might only be expected in the coastal zone. There is a need to define the pollen zones within this profile.

Calibrated date: *1σ:* 5380–5310 cal BC
                      2σ: 5470–5300 cal BC

Final comment: G Momber (25 October 2014), this sample was from the base of an organic horizon that was influenced by a rising sea level that resulted in the formation of peat. The macrofossil was recovered from the interface between the dry land surface peat and the transitional peat.

OxA–15696 7013 ±36 BP

δ¹³C: -24.4‰

Sample: MS08 14 BCII 03, submitted in March 2006 by J Gillespie

Material: wood (waterlogged): *Alnus glutinosa*, twig (G Campbell 2006)

Initial comment: from context 3. A compact, humified peat. Heterogeneous colour, black but with patches of very dark reddish brown. Roundwood pieces embedded within the peat. Increasing mineral content.

Objectives: to define the pollen zones within context 3.

Calibrated date: *1σ:* 5980–5840 cal BC
                      2σ: 5990–5800 cal BC

Final comment: G Momber (25 October 2014), this date assists in defining the pollen zones at an absolute depth of 10.96m (± 0.1m) as *Pinus* declines and *Quercus* and *Ulmus* become more important.

OxA–15697 7110 ±34 BP

δ¹³C: -26.9‰

Sample: MS07 22 BCII 03, submitted in March 2006 by J Gillespie

Material: wood (waterlogged): cf *Betula* sp., of ?7–years' growth (G Campbell 2006)

Initial comment: from context 7. An organic sand. Medium sandy clay, slightly calcareous. Roundwood fragments and highly-humified organic fragments (dark brown lenses). Heterogeneous colour comprising dominantly very dark grey/grey. Occasional sub-rounded flint granules and pebbles. *In situ* freshly knapped flints.

Objectives: context 7 contains the archaeological material and definition of pollen zones is needed within the context.

Calibrated date: *1σ:* 6020–5980 cal BC
                      2σ: 6050–5910 cal BC

Final comment: G Momber (25 October 2014), an absolute depth of 11.58m (± 0.1m) at the base of context 7 gives an index level from which to define pollen zones within this layer.

OxA–15698 6956 ±35 BP

δ¹³C: -26.0‰

Sample: MS05 16 BCII, submitted in March 2006 by J Gillespie

Material: wood (waterlogged): *Betula* sp., of 10–years' growth (<5g) (G Campbell 2006)

Initial comment: from context 3. A stiff, non-calcareous clay, highly fragmented, increasingly organic to the base. Beneath are the lower coarse sands and it is capped by salt marsh sediments.

Objectives: this sample comes from a transition zone, and dating is necessary to establish whether differences are temporal or spatial in the habitat and age of peat formation also for the definition of pollen zones.

Calibrated date: *1σ:* 5890–5770 cal BC
                      2σ: 5980–5730 cal BC

Final comment: G Momber (25 October 2014), the absolute depth of this sample is 10.94m (±0.1m) coming from a predominantly peat environment with *Quercus* and *Corylus Pinus* declining.

Laboratory comment: English Heritage (25 November 2014), the two results on this fragment of roundwood (OxA-15698 and OxA-15721) are statistically consistent (T′=0.6; T′(5%)=3.8; *ν*=1; Ward and Wilson 1978), and so a weighted mean can be taken (6938 ±26 BP), which calibrates to 5890–5730 cal BC (95% confidence; Reimer *et al* 2004).

References: Reimer *et al* 2004
            Ward and Wilson 1978

OxA-15699 7203 ±36 BP

$\delta^{13}C$: -27.4‰

Sample: MS17 48 BCV, submitted in March 2006 by J Gillespie

Material: wood (waterlogged): *Alnus glutinosa*, twig (G Campbell 2006)

Initial comment: from a pit or hearth. The site is submerged in some 12m of seawater. The sample is from the base of the fibrous peat. The stratigraphy comprises timber capping fibrous peat. This was underlain by a less fibrous humic peat which enveloped clay nodules and lithic material (burnt flint), this was sitting on soft grey clay with organic inclusions (possibly plant roots).

Objectives: to date the hearth.

Calibrated date: 1σ: 6080–6020 cal BC
            2σ: 6210–6000 cal BC

Final comment: G Member (25 October 2014), coming from the base of the pit this gives a date for when the pit was in use.

OxA-15716 6335 ±40 BP

$\delta^{13}C$: -28.1‰

Sample: MS10 27–30 BCIV 03, submitted in March 2006 by J Gillespie

Material: wood (waterlogged): unidentified, knot (G Campbell 2006)

Initial comment: from context 2. A mineral deposit with 1mm lenses of horizontally bedded, highly-humified, dark brown organics, with paddock burrows common. Becoming organic mud with a stiff silty clay consistency, non-calcareous, with highly fragmented, unidentifiable humified and fresh plant macros. Rare sub-rounded pebbles. Heterogeneous with piddock infestation. The boundary with context 3 is sharp and wavy.

Objectives: this important profile demonstrates the local establishment of lime (*Tilia*) woodland, which is now considered to have been dominant, or at least co-dominant, wood element throughout the middle and early part of the late-Holocene of southern and eastern England. Goodwin's pollen zone scheme suggests arrival during the late Boreal period, but this has rarely been evidenced, and might only be expected in the coastal zone. There is a need to define the pollen zones within this profile.

Calibrated date: 1σ: 5360–5300 cal BC
            2σ: 5470–5210 cal BC

Final comment: G Member (25 October 2014), the dated sample came from wood within a humic horizon that was formed under the influence of a rising sea-level. The wood sample came from the transitional peat just prior to inundation. The samples from the transition peat into the soft silty estuarine mudflat clay deposit provide index points for the sea-level transgression *c* 5200 cal BC.

OxA-15717 6300 ±40 BP

$\delta^{13}C$: -26.5‰

Sample: MS10 23–26 BCIV 03, submitted in March 2006 by J Gillespie

Material: wood (waterlogged): unidentified, twig (G Campbell 2006)

Initial comment: from context 2. A mineral deposit with 1mm lenses of horizontally bedded, highly-humified, dark brown organics, with paddock burrows common. Becoming organic mud with a stiff silty clay consistency, non-calcareous, with highly fragmented, unidentifiable humified and fresh plant macros. Rare sub-rounded pebbles. Heterogeneous with paddock infestation. The boundary with context 3 is sharp and wavy.

Objectives: this important profile demonstrates the local establishment of lime (*Tilia*) woodland, which is now considered to have been dominant, or at least co-dominant, wood element throughout the middle and early part of the late-Holocene of southern and eastern England. Goodwin's pollen zone scheme suggests arrival during the late Boreal period, but this has rarely been evidenced, and might only be expected in the coastal zone. There is a need to define the pollen zones within this profile.

Calibrated date: 1σ: 5320–5220 cal BC
            2σ: 5370–5210 cal BC

Final comment: see OxA-15716

OxA-15718 7175 ±45 BP

$\delta^{13}C$: -27.2‰

Sample: MS08 40 BCII 03, submitted in March 2006 by J Gillespie

Material: wood (waterlogged): *Corylus avellana*, roundwood; 15mm diameter (G Campbell 2006)

Initial comment: from context 7. Medium coarse sand, slightly clayey (clay content increases slightly with depth).

Objectives: to define the pollen zones within contexts 3 and 7.

Calibrated date: 1σ: 6070–6000 cal BC
            2σ: 6100–5980 cal BC

Final comment: G Member (25 October 2014), sample recovered from a sample silt context with humic horizons and un-abraided Mesolithic flint flakes and tools.

OxA–15719 6320 ±40 BP

δ¹³C: -25.0‰

Sample: MS10 20–23 BCIV 03, submitted in March 2006 by J Gillespie

Material: wood (waterlogged): *Quercus* sp., twig, ?7–years' growth (G Campbell 2006)

Initial comment: from context 2. A mineral deposit with 1mm lenses of horizontally bedded, highly-humified, dark brown organics, with paddock burrows common. Becoming organic mud with a stiff silty clay consistency, non-calcareous, with highly fragmented, unidentifiable humified and fresh plant macros. Rare sub-rounded pebbles. Heterogeneous with paddock infestation. The boundary with context 3 is sharp and wavy.

Objectives: this important profile demonstrates the local establishment of lime (*Tilia*) woodland, which is now considered to have been dominant, or at least co-dominant, wood element throughout the middle and early part of the late-Holocene of southern and eastern England. Goodwin's pollen zone scheme suggests arrival during the late Boreal period, but this has rarely been evidenced, and might only be expected in the coastal zone. There is a need to define the pollen zones within this profile.

Calibrated date: 1σ: 5330–5230 cal BC
                    2σ: 5380–5210 cal BC

Final comment: G Momber (25 October 2014), the dated sample came from a twig from the soft silty estuarine mudflat clay (located above OxA-15717) that was formed with the onset of marine conditions. The samples from the transition peat into the mud flat deposit provide index points for the sea-level transgression *c* 5200 cal BC.

OxA–15720 7125 ±45 BP

δ¹³C: -24.5‰

Sample: MS07 10–12 BCII 03, submitted in March 2006 by J Gillespie

Material: wood (waterlogged): *Alnus glutinosa* (G Campbell 2006)

Initial comment: from context 7. An organic sand. Medium sandy clay, slightly calcareous. Roundwood fragments and highly-humified organic fragments (dark brown lenses). Heterogeneous colour comprising dominantly very dark grey/grey. Occasional sub-rounded flint granules and pebbles. *In situ* freshly knapped flints.

Objectives: radiocarbon dating of selected deposits is necessary in order to answer questions of sea-level change and the timing of human occupation and its relationship to the archaeological contexts. Context 7 contains the archaeological material and definition of pollen zones is needed within the context.

Calibrated date: 1σ: 6030–5980 cal BC
                    2σ: 6070–5910 cal BC

Final comment: G Momber (25 October 2014), the result is comparable to pollen zone 1c which indicates that it was a time of predominantly peat accretion. The absolute depth is 11.20m (± 0.1m).

OxA–15721 6915 ±40 BP

δ¹³C: -26.2‰

Sample: MS05 16 BCII, submitted in March 2006 by J Gillespie

Material: wood (waterlogged): *Betula* sp., of ?10 years' growth (G Campbell 2006)

Initial comment: a replicate of OxA-15698.

Objectives: as OxA-15698

Calibrated date: 1σ: 5850–5730 cal BC
                    2σ: 5900–5710 cal BC

Final comment: see OxA-15698

Laboratory comment: see OxA-15698

OxA–15722 7230 ±45 BP

δ¹³C: -28.3‰

Sample: MS17 31 BCV, submitted in March 2006 by J Gillespie

Material: wood (waterlogged): unidentified, ?root (G Campbell 2006)

Initial comment: from a pit or hearth. The site is submerged in some 12m of seawater. The sample is from the base of the fibrous peat. The stratigraphy comprises timber capping fibrous peat. This was underlain by a less fibrous humic peat which enveloped clay nodules and lithic material (burnt flint), this was sitting on soft grey clay with organic inclusions (possibly plant roots).

Objectives: to date the base of the vegetation that developed following abandonment of the hearth.

Calibrated date: 1σ: 6100–6040 cal BC
                    2σ: 6220–6010 cal BC

Final comment: G Momber (25 October 2014), further up the sequence this sample indicates that the pit was no longer in use and vegetation had begun to develop over the site. The chronological model estimates that the pit was no longer in use by *6080–5990 cal BC (95% probability; OxA-15722; Momber et al* 2011, fig 4.11).

OxA–15723 7170 ±45 BP

δ¹³C: -27.2‰

Sample: MS17 38.5 BCV, submitted in March 2006 by J Gillespie

Material: wood (waterlogged): *Alnus glutinosa* (G Campbell 2006)

Initial comment: as OxA-15722

Objectives: as OxA-15722

Calibrated date: 1σ: 6070–6000 cal BC
                    2σ: 6100–5980 cal BC

Final comment: G Momber (25 October 2014), see OxA-15722, this layer of material was deposited after the feature fell into disuse.

Bouldnor Cliff: wiggle-matching, Isle of Wight

Location: SK 605101
Lat. 52.41.05 N; Long. 01.06.18 W

Project manager: G Momber (Hampshire and Wight Trust for Maritime Archaeology), 2003

Archival body: Hampshire and Wight Trust for Maritime Archaeology

Description: 11 oak samples from Bouldnor Cliff cross-matched to form a 285–year ring-width site mean tree-ring chronology (Bouldnor T11). This cross-matches with similar site chronologies from Goldcliff, Redwick, and Avonmouth in the Severn Estuary (Momber *et al* 2011, table 6.2), although none of these chronologies is currently dated by dendrochronology.

Objectives: a series of six sequential decadal blocks of samples from tree D606/12 were consequently submitted for radiocarbon dating. The last ring of the wiggle-match sequence is ring 258 of the 285–year site mean (Bouldnot T11).

Final comment: G Momber (25 October 2014), the results of wiggle-matching the radiocarbon results on a sequence of six decadal blocks from a tree-ring sequence. Despite two of the six measurements having low individual indices of agreement (UB-6862 [A=13.5%]; UB-6859 [A=42.0%], A'=60.0%), the model has a good overall index of agreement ($A_{overall}$ =32.4%, A_n =28.9%) and suggests that the results are consistent with their relative order. The model estimates the date of the last ring of the sequence to be *6030–5990 cal BC (95% probability; last ring;* Momber *et al* 2011, fig 6.1), and probably *6020–6000 cal BC (68% probability).*

References: Momber *et al* 2011

UB–6858 7168 ±42 BP

$\delta^{13}C$: -26.0±0.2‰

Sample: Q10745A, submitted on 13 January 2006 by N Nayling

Material: wood (waterlogged): *Quercus* sp. (N Nayling 2006)

Initial comment: rings 161–170 from tree DS06/12.

Objectives: to provide calendar dating for a series of cross-matching, but floating, tree-ring master chronologies.

Calibrated date: 1σ: 6070–6000 cal BC
2σ: 6090–5980 cal BC

Final comment: G Momber (25 October 2014), this radiocarbon date has good agreement with its position in the wiggle-match sequence.

UB–6859 7115 ±42 BP

$\delta^{13}C$: -27.0±0.2‰

Sample: Q10745B, submitted on 13 January 2006 by N Nayling

Material: wood (waterlogged): *Quercus* sp. (N Nayling 2006)

Initial comment: rings 171–180 from tree DS06/12.

Objectives: as UB-6858

Calibrated date: 1σ: 6030–5980 cal BC
2σ: 6070–5900 cal BC

Final comment: G Momber (25 October 2014), this radiocarbon date has slightly poor individual agreement (A:42) with its position in the wiggle-match sequence.

UB–6860 7127 ±40 BP

$\delta^{13}C$: -25.0±0.2‰

Sample: 10745C, submitted on 13 January 2006 by N Nayling

Material: wood (waterlogged): *Quercus* sp. (N Nayling 2006)

Initial comment: rings 181–190 from tree DS06/12.

Objectives: as UB-6858

Calibrated date: 1σ: 6030–5980 cal BC
2σ: 6070–5910 cal BC

Final comment: see UB-6858

UB–6861 7191 ±41 BP

$\delta^{13}C$: -25.0±0.2‰

Sample: Q10745D, submitted on 13 January 2006 by N Nayling

Material: wood (waterlogged): *Quercus* sp. (N Nayling 2006)

Initial comment: rings 191–200 from tree DS06/12.

Objectives: as UB-6858

Calibrated date: 1σ: 6080–6010 cal BC
2σ: 6200–5990 cal BC

Final comment: see UB-6858

UB–6862 7259 ±41 BP

$\delta^{13}C$: -25.0±0.2‰

Sample: Q10745E, submitted on 13 January 2006 by N Nayling

Material: wood (waterlogged): *Quercus* sp. (N Nayling 2006)

Initial comment: rings 201–210 from tree DS06/12.

Objectives: as UB-6858

Calibrated date: 1σ: 6220–6060 cal BC
2σ: 6230–6020 cal BC

Final comment: G Momber (25 October 2014), this radiocarbon date has poor individual agreement (A:14) with its position in the wiggle-match sequence.

UB–6863 7156 ±41 BP

$\delta^{13}C$: -27.0±0.2‰

Sample: Q10745F, submitted on 13 January 2006 by N Nayling

Material: wood (waterlogged): *Quercus* sp., single fragment (N Nayling 2006)

Initial comment: rings 211–220 from tree DS06/12.

Objectives: as UB-6858

Calibrated date: *1σ:* 6060–6000 cal BC
                2σ: 6080–5930 cal BC

Final comment: see UB-6858

Brandon: Staunch Meadow, Suffolk

Location: TL 778864
             Lat. 52.26.46 N; Long. 00.36.59 E

Project manager: A Tester (Suffolk Archaeological Unit), 1981–2 and 1984–5

Description: a middle Saxon settlement including buildings, an industrial area, church, and attendant cemeteries all concentrated within a readily defined island. The occupation of the bulk of the site is restricted to the middle Saxon period. The settlement sits beside a 1km wide arm of the Fenland which follows the valley of the Little Ouse river *c* 6km inland from Hockwold Fen; Brandon was probably the lowest crossing point of the River Ouse until recent times. The site occupies a sand ridge surrounded by peat, and stands as an island in time of flood. The river is some 50m north of the 'island' while the southern margin of the peat deposits (ie the edge of the flood plain) is *c* 80m to the south. The island is *c* 350m east-west by 150m north-south at its widest point with an area of some 4.75ha; of this *c* 1.5ha at the west end appears to have been unoccupied and a further *c* 1.25ha at the east end of the island has been scheduled as an Ancient Monument.

Objectives: the scientific dating programme was designed to achieve the following objectives: to provide a chronological framework for interpreting palaeoecological results, a precise date for the wooden causeway and bridge, a date for the use of cemetery 2, dates for the buildings, a precise estimate for the period of use of the cemetery located south of the church, and to provide dates for the waterfront activity.

References: Carr *et al* 1988
              Tester *et al* 2014

Brandon: Staunch Meadow, cemetery 2, Suffolk

Location: TL 77908656
             Lat. 52.26.51 N; Long. 00.37.05 E

Project manager: S Anderson (Suffolk County Council Archaeological Service), 1984

Archival body: Suffolk County Council Archaeological Service

Description: cemetery 2 located to the north of the church, was presumed to represent the second phase of burial in the settlement. It was only partially excavated with 31 individuals recovered. Associated with these burials was a clay pad 2m ∞ 3m that is interpreted as being a mortuary structure. The clay pad was cut by some burials and its surface then reinstated, suggesting it had significance to the whole cemetery rather than individual burials.

Objectives: these samples are being submitted in order to determine the date of cemetery 2, an incompletely excavated cemetery to the north-east of the excavated area. This cemetery may be a late development in the middle Saxon settlement, or it may be of medieval date. These samples were chosen for dating as part of the assessment stage of the project. Bone condition on the site is generally poor, and the submission of these bones will provide an opportunity to assess their suitability for radiocarbon dating.

Final comment: P Marshall (14 November 2013), the radiocarbon dates have confirmed the late Saxon date for the cemetery. Unfortunately, the poor collagen preservation meant that it was not possible to obtain sufficient determinations to provide a precise date for the cemetery's use.

Laboratory comment: English Heritage (14 November 2013), given the three samples from cemetery 2 submitted in 2000 all had sufficient amounts of well-preserved collagen to produce radiocarbon measurements, an additional eight samples were selected in 2006 to provide estimates for the length of use of the cemetery, and the date of construction and subsequent modification of the clay pad mortuary structure. Unfortunately, only four of these produced sufficient carbon for dating.

Laboratory comment: English Heritage (24 June 2013), two further samples from this series were dated before 2003 (GU-5817–8). Four further samples were also subsequently dated after 2006 (SUERC-11287–8 and -11292–3; Bayliss *et al* forthcoming).

References: Carr 1992
              Tester *et al* 2014

GU–6050 1290 ±60 BP

$δ^{13}C$: -20.1‰
$δ^{15}N$ *(diet):* +10.5‰
C/N ratio: 3.3

Sample: BRD018 4587, submitted in April 2004 by S Anderson

Material: human bone (285g) (left femur of possible female) (S Anderson 2004)

Initial comment: from burial 4587, which was part of cemetery group 2. The burial was from the middle line of the excavated area of the cemetery, it cuts building 4531 and surface 4607. The burial was articulated and from a grave 0.75–1m deep cut into natural sand.

Objectives: to establish the period of use of Cemetery 2 at the northern end of the island, and compare this skeleton with those previously dated from either end of the excavated area (4584, GU-5817; and 4842, GU-5818; Bayliss *et al* 2016, 66).

Calibrated date: *1σ:* cal AD 660–780
                 2σ: cal AD 640–890

Final comment: see series comment

Brandon: Staunch Meadow, waterfront activity, Suffolk

Location:	TL 778864 Lat. 52.26.46 N; Long. 00.36.59 E
Project manager:	R Carr (Suffolk Archaeological Unit), 1982 and 1985
Archival body:	Suffolk County Council Archaeological Service

Description: on the riverward face of the island evidence for intense industrial activity was found, together with fence lines and three artificial mounds or 'islands' constructed by the dumping of sand. All these waterfront contexts with dated material are stratigraphically later than (6188) which contained two coins, type Q II and type Q or R II that date from AD 720–740. They therefore provide a *terminus post quem* for the overlying sequence.

Objectives: to date the industrial activity.

Final comment: P Marshall (14 November 2013), the dates contribute to understand the chronology of the middle Saxon settlement.

References:	Carr *et al* 1988 Tester *et al* 2014

OxA–14569 1317 ±34 BP

$\delta^{13}C$: -26.3‰

Sample: 9774 B, submitted on 21 March 2005 by A Tester

Material: wood (waterlogged): Pomoideae, hawthorn/*Sorbus* group, roundwood, diameter 18mm, *c* 12 growth rings (R Gale 2004)

Initial comment: partially charred, waterlogged wood. From layer 5764, which is associated with wooden structure 5736. 5764 is above layer 61889 (group number). And below layers 5763 which is below layers 5755 and 5756. Layer accumulated in peat, with peat growth above.

Objectives: to establish the date of the industrial process. Stratigraphically later than the context with two coins AD 710 and AD 720. Sealed by deposits containing Ipswich Ware pottery, (end date *c* AD 850). The objective is to establish the time span for this activity.

Calibrated date:	*1σ:* cal AD 660–770 *2σ:* cal AD 650–770

Final comment: P Marshall (14 November 2013), the date provides an estimate for the age of the industrial activity.

Laboratory comment: English Heritage (14 November 2013), two fragments of charcoal from layer (5764) are statistically consistent (T'=0.0; *v*=1; T'(5%)=3.8; Ward and Wilson 1978).

References:	Ward and Wilson 1978

OxA–14593 1313 ±30 BP

$\delta^{13}C$: -27.9‰

Sample: 9774, submitted on 21 March 2005 by A Tester

Material: wood (waterlogged): *Corylus avellana*, roundwood, bark *in situ*, diameter 14mm, 8 growth rings (R Gale 2004)

Initial comment: as OxA-14569

Objectives: as OxA-14569

Calibrated date:	*1σ:* cal AD 660–770 *2σ:* cal AD 650–770

Final comment: see OxA-14569

Laboratory comment: see OxA-14569

OxA–14603 1298 ±29 BP

$\delta^{13}C$: -26.4‰

Sample: 6422, submitted on 21 March 2005 by A Tester

Material: wood (waterlogged): *Corylus avellana*, roundwood 40mm. Very degraded, remnants of bark retained (R Gale 2004)

Initial comment: a post from a wattle fence (context 6394) dug into peat and buried soon after use in wet sand and peat. The sample is from a structure associated with a waterfront textile working industry suggested to be after AD 720 but before AD 850. Buried by sand and peat soon after use.

Objectives: to establish the date of this structure. Stratigraphically later than context 6188 (group number) with two coins AD 710 and AD 720.

Calibrated date:	*1σ:* cal AD 660–770 *2σ:* cal AD 650–780

Final comment: P Marshall (14 November 2013), provides an estimate for the date of the wattle fence.

Laboratory comment: English Heritage (14 November 2013), the three measurements from posts of wattle fence line [6394] are not statistically consistent (T'=11.4; *v*=2; T'(5%)=6.0; Ward and Wilson 1978).

References:	Ward and Wilson 1978

OxA–14604 1263 ±29 BP

$\delta^{13}C$: -24.4‰

Sample: 6362, submitted on 21 March 2005 by A Tester

Material: wood (waterlogged): *Salix/Populus sp.*, wide roundwood, very degraded. Outer surface removed for dating (R Gale 2004)

Initial comment: the timber was part of a wooden structure with wattles that was built onto peat. It was sealed by dumped sand after use (context 5751).

Objectives: sample is from a structure associated with a waterfront textile working industry suggested to be after AD 720 but before AD 850. The objective is to establish the general time span for this activity.

Calibrated date: 1σ: cal AD 680–780
2σ: cal AD 660–860

Final comment: P Marshall (14 November 2013), provides an estimate for the date of the waterfront textile working industry.

Laboratory comment: English Heritage (14 November 2013), OxA-14604 was replicated by the laboratory (OxA-14607) and the two measurements were combined (1246 ±20 BP; T'=0.7; ν=1; T'(5%)=3.8; Ward and Wilson 1978).

References: Ward and Wilson 1978

OxA–14605 1184 ±28 BP

δ¹³C: -27.2‰

Sample: 6408, submitted on 21 March 2005 by A Tester

Material: wood (waterlogged): *Corylus avellana*, roundwood; diameter 45mm. Very degraded, outer surface removed for dating (R Gale 2004)

Initial comment: a post from wattle fence (context 6394) dug into peat and buried soon after use in wet sand and peat.

Objectives: the sample is from a structure associated with a waterfront textile working industry suggested to be after AD 720 but before AD 850. The objective is to establish the general time span for this activity.

Calibrated date: 1σ: cal AD 770–890
2σ: cal AD 720–950

Final comment: see OxA-14603

Laboratory comment: see OxA-14603

OxA–14606 1178 ±27 BP

δ¹³C: -25.1‰

Sample: 6437, submitted on 21 March 2005 by A Tester

Material: wood (waterlogged): *Corylus/Alnus* sp. (R Gale 2004)

Initial comment: a post from wattle fence (context 6394) dug into peat and buried soon after use in wet sand and peat.

Objectives: to establish the date of this structure. Stratigraphically later than context with two coins AD 710 and AD 720.

Calibrated date: 1σ: cal AD 770–900
2σ: cal AD 770–950

Final comment: see OxA-14603

Laboratory comment: see OxA-14603

OxA–14607 1230 ±28 BP

δ¹³C: -24.8‰

Sample: 6362, submitted on 21 March 2005 by A Tester

Material: wood (waterlogged): *Salix/Populus sp.*, wide roundwood, very degraded. Outer surface removed for dating (R Gale 2004)

Initial comment: a replicate of OxA-14604.

Objectives: as OxA-14604

Calibrated date: 1σ: cal AD 710–860
2σ: cal AD 680–890

Final comment: see OxA-14604

Laboratory comment: see OxA-14604

Bridgwater Bay, Somerset

Location: ST 270480
Lat. 51.13.32 N; Long. 03.02.43 W

Project manager: R Brunning (Somerset County Council), September 2003

Archival body: Somerset Heritage Centre

Description: a group of dispersed wooden fishing structures in the intertidal mudflats south of the River Parrett in Bridgwater Bay. The site comprised a fishing structure made of wooden stakes driven into stiff clay and sand deposits in the intertidal area of Bridgwater Bay. The tops of the stakes protrude above the clay to varying lengths between 5cm and 50cm. It is uncertain over how long a period the fish weir was used and repaired.

Objectives: until the recent fieldwork there was absolutely no evidence for the date of the numerous wooden fishing structures known to exist on Stert Flats. The initial dating and species identification data showed that two of the individual V-shaped weirs are likely to be tenth century in date (dendrochronological *terminus post quos* as no sapwood on samples) while two more substantial structures are partially composed of species imported within the last 400 years.

Final comment: R Brunning (31 May 2005), this series of dates has provided first unequivocal dating for this group of fishing structures. They prove that fish weirs were being constructed in that area from the late Saxon period. The dates have confirmed the *terminus post quem* dendrochronological dated (no sapwood) for two of the weirs. The series of dates have allowed us to link different types of weir to particular periods and begin to develop a typology for such structures.

References: Groves *et al* 2004

GU–6002 1090 ±50 BP

δ¹³C: -27.7‰

Sample: Site 307; 307/1, submitted on 15 January 2004 by R Brunning

Material: wood (waterlogged): *Corylus avellana*, roundwood, *c* 23 rings (110g) (R Gale 2004)

Initial comment: as the tops of the wooden piles protrude above the inter-tidal deposits they may be affected by any contamination in the water. However, the Severn Estuary has an enormous tidal range that effectively flushes the system out twice a day, helping to prevent the build up of any contamination that may be present. No contaminants or vegetation were visible on the intertidal surface.

At low tide the water table is at the ground surface. The tops of the stakes are therefore above the water table for *c* 3 hours at every tide. Attack by marine borers appears to be minimal although the top of the piles display visible decay and erosion damage. The samples were taken as low down as possible on the stakes to obtain the best preserved and least damaged areas of wood. The visual condition of the portion of the stakes that is buried in clay is excellent and toolmarks survive very well.

Objectives: the samples from this structure will help to determine if certain types of weir are datable to particular periods and establish over what periods that fishing ground was in use. The radiocarbon measurement will significantly contribute to an assessment of significance of the structures and the potential and need for further work.

Calibrated date: *1σ:* cal AD 890–1020
                          2σ: cal AD 770–1030

Final comment: R Brunning (31 January 2005), this date confirms that this post row is part of a larger complex of late Saxon fish weirs.

Laboratory comment: English Heritage (11 January 2005), the measurements on samples from site 307 [GU-6002 and 6003; 307/1 and 2] are statistically consistent (T'=0.7; ν=1; T'(5%)=3.8; Ward and Wilson 1978) suggesting that these timbers are also from a single construction episode that dates to the mid-late Saxon period.

References: Ward and Wilson 1978

GU–6003 1150 ±50 BP

δ¹³C: -29.4‰

Sample: Site 307; 307/2, submitted on 15 January 2004 by R Brunning

Material: wood (waterlogged): *Alnus glutinosa*, roundwood and bark (7 rings) (230g) (R Gale 2004)

Initial comment: as GU-6002

Objectives: as GU-6002

Calibrated date: *1σ:* cal AD 770–980
                          2σ: cal AD 720–1000

Final comment: see GU-6002

Laboratory comment: see GU-6002

GU–6004 1150 ±60 BP

δ¹³C: -29.5‰

Sample: Site 309; 309/1, submitted on 15 January 2004 by R Brunning

Material: wood (waterlogged): *Betula* sp., roundwood (*c* 10 rings) (100g) (R Brunning 2004)

Initial comment: as GU-6002

Objectives: as GU-6002

Calibrated date: *1σ:* cal AD 770–980
                          2σ: cal AD 690–1020

Final comment: R Brunning (31 January 2005), this provides a late Saxon date for V-shaped fish weir, previously undated.

Laboratory comment: English Heritage (11 January 2005), the measurements on samples from site 309 [GU-6004 and 6005; 309/1 and 3] are statistically consistent (T'=0.1; ν=1, T'(5%)=3.8; Ward and Wilson 1978) suggesting that these timbers are from a single construction episode that dates to the mid-late Saxon period.

References: Ward and Wilson 1978

GU–6005 1170 ±50 BP

δ¹³C: -25.3‰

Sample: Site 309; 309/3, submitted on 15 January 2004 by R Brunning

Material: wood (waterlogged): *Quercus* sp., roundwood (including heartwood and sapwood) (110g) (R Gale 2004)

Initial comment: as GU-6002

Objectives: as GU-6002

Calibrated date: *1σ:* cal AD 770–950
                          2σ: cal AD 690–990

Final comment: see GU-6004

Laboratory comment: see GU-6005

GU–6006 430 ±50 BP

δ¹³C: -27.4‰

Sample: Site 202; 202/1, submitted on 15 January 2004 by R Brunning

Material: wood (waterlogged): *Quercus* sp., roundwood (8 rings) (250g) (R Gale 2004)

Initial comment: as GU-6002

Objectives: as GU-6002

Calibrated date: *1σ:* cal AD 1430–1480
                          2σ: cal AD 1410–1630

Final comment: R Brunning (31 January 2005), this provides dating for the putt rank previously undated, and helps to provide the earliest known date for such structures in this area.

Laboratory comment: English Heritage (11 January 2005), the measurements on samples from site 202 [GU-6006 and 6007; 202/1 and 2] are statistically consistent (T'=1.6; ν=1, T'(5%)=3.8; Ward and Wilson 1978) suggesting that these timbers are from a single construction episode that dates to the fifteenth or sixteenth century.

References: Ward and Wilson 1978

GU–6007 340 ±50 BP

δ¹³C: -28.0‰

Sample: Site 202; 202/2, submitted on 15 January 2004 by R Brunning

Material: wood (waterlogged): *Fraxinus excelsior*, roundwood (8 rings) (530g) (R Gale 2004)

Initial comment: as GU-6002

Objectives: as GU-6002

Calibrated date: *1σ:* cal AD 1460–1650
                     2σ: cal AD 1440–1660

Final comment: see GU-6006

Laboratory comment: see GU-6006

GU–6008 960 ±50 BP

δ¹³C: -28.7‰

Sample: Site 306; 306/1, submitted on 15 January 2004 by R Brunning

Material: wood (waterlogged): *Alnus glutinosa*, roundwood (27mm diameter) (240g) (R Gale 2004)

Initial comment: as GU-6002

Objectives: as GU-6002

Calibrated date: *1σ:* cal AD 1020–1160
                     2σ: cal AD 980–1210

Final comment: R Brunning (1 February 2005), the dates confirm that this horizontal group of poles is associated with the fish weir they adjoin. They probably therefore represent a branch trackway alongside the weir to allow access to the baskets at the end of the weir.

Laboratory comment: English Heritage (11 January 2005), the measurements on samples from site 306 (GU-6008 and 6009; 306/1 and 2) are statistically consistent ($T'=1.6$; $v=1$; $T'(5\%)=3.8$; Ward and Wilson 1978) suggesting that these timbers are from a single construction episode that dates to the late Saxon or medieval period.

References: Ward and Wilson 1978

GU–6009 1050 ±50 BP

δ¹³C: -31.0‰

Sample: Site 306; 306/2, submitted on 15 January 2004 by R Brunning

Material: wood (waterlogged): *Alnus glutinosa*, roundwood (40mm diameter) (440g) (R Gale 2004)

Initial comment: as GU-6002

Objectives: as GU-6002

Calibrated date: *1σ:* cal AD 960–1030
                     2σ: cal AD 880–1120

Final comment: see GU-6008

Laboratory comment: see GU-6008

GU–6010 1060 ±50 BP

δ¹³C: -29.7‰

Sample: Site 204; 204/12, submitted on 15 January 2004 by R Brunning

Material: wood (waterlogged): *Alnus glutinosa*, fast-grown roundwood (60mm diameter) (200g) (R Gale 2004)

Initial comment: as GU-6002

Objectives: as GU-6002

Calibrated date: *1σ:* cal AD 900–1030
                     2σ: cal AD 880–1040

Final comment: R Brunning (1 February 2005), this provides a late Saxon date for this V-shaped weir showing that this is an early type. This fits in with the dates of three other V-shaped weirs from the area, and also suggests that weir 204 and the adjoining wooden structure 306 are contemporary.

Laboratory comment: English Heritage (11 January 2005), the measurements on samples from site 204 (GU-6010 and 6011; 204/12 and 18) are statistically consistent ($T'=1.4$; $v=1$, $T'(5\%)=3.8$; Ward and Wilson 1978) suggesting that these timbers are from a single construction episode that dates to the late Saxon or medieval period.

References: Ward and Wilson 1978

GU–6011 1160 ±70 BP

δ¹³C: -26.0‰

Sample: Site 204; 204/18, submitted on 15 January 2004 by R Brunning

Material: wood (waterlogged): *Quercus* sp., with sapwood (100g) (R Gale 2004)

Initial comment: as GU-6002

Objectives: as GU-6002

Calibrated date: *1σ:* cal AD 770–980
                     2σ: cal AD 680–1020

Final comment: see GU-6010

Laboratory comment: see GU-6010

GU–6038 1050 ±50 BP

δ¹³C: -26.2‰

Sample: Site 205; 205/5, submitted on 15 January 2004 by R Brunning

Material: wood (waterlogged): *Corylus avellana*, roundwood (>100g) (R Gale 2004)

Initial comment: as GU-6002

Objectives: as GU-6002

Calibrated date: *1σ:* cal AD 960–1030
                     2σ: cal AD 880–1120

Final comment: R Brunning (1 February 2005), this provides evidence that this V-shaped weir is late Saxon in date and that this sample is from a stake of probable original build with GU-6039 from late repair.

Laboratory comment: English Heritage (11 January 2005), the measurements on samples from site 205 (GU-6038 and 6039; 2054/5 and 6) are statistically consistent ($T'=2.4$; $v=1$, $T'(5\%)=3.8$; Ward and Wilson 1978) suggesting that these timbers are from a single construction episode that dates to the late Saxon or medieval period.

References: Ward and Wilson 1978

GU–6039 940 ±50 BP

δ¹³C: -28.3‰

Sample: Site 205; 205/6, submitted on 15 January 2004 by R Brunning

Material: wood (waterlogged): *Alnus glutinosa*, roundwood (25mm) (>100g) (R Gale 2004)

Initial comment: as GU-6002

Objectives: as GU-6002

Calibrated date: *1σ:* cal AD 1020–1170
     2σ: cal AD 990–1220

Final comment: R Brunning (1 February 2005), this suggests repairs made to the late Saxon weir in the early Norman period.

Laboratory comment: see GU-6038

Callington: St Sampson's Church, Cornwall

Location:	SX 358696 Lat. 50.30.14 N; Long. 04.18.59 W
Project manager:	J Gossip (Cornwall County Council), September 2000
Archival body:	Royal Cornwall Museum, Parochial Church Council (human remains reburied)

Description: an archaeological watching brief during drainage works at St Sampson's Church, South Hill, Callington. The hand excavation of a trench around the church (for insertion of trench drain) by archaeologist and contractors (average 0.8m deep, 0.5–1.5m wide) revealed the buried remains of fifteen individuals, four within stone cist graves, others within earth-cut grave pits. The church dates to the fourteenth century AD but has confirmed Norman origins, and there is a possibility that an early medieval site pre-dates this.

Objectives: to date different burial types (cist graves and earth-cut graves). Also to identify time depth in the cemetery and establish whether different cist types are of different date.

Final comment: J Gossip (30 January 2013), measurements were also taken on burials 20, 25, and 27 providing a date range cal AD 1020–1280 (all at 95% confidence; Reimer *et al* 2004). All radiocarbon ages are consistent with the expected date of these burials but with none pre-dating the Norman church. The dates suggest that the cist and earth-cut grave traditions were contemporaneous.

Laboratory comment: English Heritage (8 September 2005), the four measurements on burial 39 are statistically consistent (T′=0.6; T′(5%)=7.8; *ν*=3; Ward and Wilson 1978). The best estimate of its radiocarbon age is provided by the pooled mean of these measurements (878 ±16 BP), which calibrates to cal AD 1050–1220 at 95% confidence (Reimer *et al* 2004).

References: Reimer *et al* 2004
     Ward and Wilson 1978

OxA–14584 888 ±27 BP

δ¹³C: -19.9‰
δ¹⁵N (diet): +9.9‰
C/N ratio: 3.2

Sample: burial 39B1, submitted in March 2005 by J Gossip

Material: human bone (right femur) (H Gestsdottir 2001)

Initial comment: the skeleton was laid in an extended supine position on the base of a cist grave and orientated west-north-west to east-south-east. This comprised large orthostatic slate slabs set within a vertical grave cut. The cist was mostly intact and covered with similar slabs to its sides - but as its eastern end the cist was disturbed, having been truncated by the construction of a corner buttress to the tower. The cist was filled (loosely) with a silty clay deposit. The capstones were sealed below a minimum of 0.15m and maximum of 0.9m of soil.

Objectives: this cist, whilst being the best preserved, has its axis on a different alignment to that of the other cists and earth-cut graves recorded during the watching brief. It is possible that this cist relates to burial activity pre-dating the earliest 'known' elements of the surviving church (possibly twelfth century, but more probably fourteenth century), and may indicate an early medieval Christian site (already suggested by the presence of the inscribed stone and possible 'Iann'). Dating therefore helps the define periods of cemetery use, and adds to the body of evidence now growing to help in the problematic chronology of cist graves. Comparison of dates for this grave, other cist graves in the churchyard, and the earth-cut graves might help to establish the time-depth of the churchyard.

Calibrated date: *1σ:* cal AD 1050–1210
     2σ: cal AD 1040–1220

Final comment: J Gossip (30 January 2013), the date is taken from a bone belonging to a burial within a cist. The radiocarbon age confirms that it relates to the Norman phase of the church and not an earlier structure and is statistically consistent with the other measurements from burial 39.

OxA–14808 882 ±31 BP

δ¹³C: -19.7‰
δ¹⁵N (diet): +11.1‰

Sample: burial 39B2, submitted in March 2005 by J Gossip

Material: human bone (right femur) (H Gestsdottir 2001)

Initial comment: a replicate measurement of OxA-14584.

Objectives: as OxA-14584

Calibrated date: *1σ:* cal AD 1050–1220
     2σ: cal AD 1040–1230

Final comment: see OxA-14584

SUERC–6137 805 ±35 BP

δ¹³C: -20.4‰
δ¹⁵N (diet): +9.3‰
C/N ratio: 3.4

Sample: burial 20, submitted in March 2005 by J Gossip

Material: human bone (right humerus) (H Gestsdottir 2001)

Initial comment: the skeleton was laid in an extended supine position close to the modern paved surface adjacent to the church and orientated west-north-west to east-south-east. The lower half of the skeleton had been truncated by the foundations of the tower buttress. The skeleton was laid on the base of a grave-cut up to 0.15m deep, which had been partially lined with orthostatic slabs. The original level from which the grave was cut is unknown. Despite truncation by the tower buttress the bone samples can be confidently described as *in situ.*

Objectives: this skeleton appeared to lie within an earth-cut grave containing a partial stone lining. It is unlikely that this represents a truncated or disturbed cist, and is therefore an unusual grave type identified by the watching brief. A date would add to the potential dating range from the churchyard and help to answer questions relating to the period of use of this burial type, and whether there is chronological cross-over with true cist graves and/or earth-cut graves pits.

Calibrated date: 1σ: cal AD 1210–1270
             2σ: cal AD 1160–1280

Final comment: J Gossip (30 January 2013), the date is taken from a bone belonging to an earth-cut grave pit. The radiocarbon age confirms that the burial relates to the Norman phase of the church is consistent with the dates of other burials, confirming both earth-cut and cist grave traditions were occurring simultaneously.

SUERC–6138 960 ±35 BP

$\delta^{13}C$: -20.3‰
$\delta^{15}N$ *(diet):* +9.1‰
C/N ratio: 3.2

Sample: burial 25, submitted in March 2005 by J Gossip

Material: human bone (right tibia) (H Gestsdottir 2001)

Initial comment: the partial remains of an inhumation lay in an extended supine position, orientated west-east, within a disturbed cist. The cist and skeleton had been disturbed by earlier drainage improvements and by the insertion of burial 27 (grave cut 34). Whilst only fragments of the lower limbs survived, these were clearly contained within a cist formed by orthostatic slate slabs, and displaced cap stones confirmed that the bones were *in situ.* The cut of the cist extended to a depth of 0.24m, but the original top level of this cut is unknown. The grave had been cut into the natural subsoil.

Objectives: despite the poor preservation, burial 25 was clearly contained within a cist of similar type to burial 39, but on a closer alignment to that of the church (ie west-east). A date for this burial will provide a dating comparison between cists, as well as establishing a chronology for changing burial rite (the grave is truncated by burial 27).

Calibrated date: 1σ: cal AD 1020–1160
             2σ: cal AD 1010–1170

Final comment: J Gossip (30 January 2013), this cist grave burial is broadly consistent with all other dated burials (with a slightly earlier start date of cal AD 1010), and is associated with the Norman church.

SUERC–6139 910 ±35 BP

$\delta^{13}C$: -20.2‰
$\delta^{15}N$ *(diet):* +9.6‰
C/N ratio: 3.3

Sample: burial 27, submitted in March 2005 by J Gossip

Material: human bone (left femur) (H Gestsdottir 2001)

Initial comment: a well-defined earth-cut grave below 0.87m of churchyard soils with its base at 1.26m below surface. The skeleton had been laid in an extended supine position (west-east) on the base of the grave cut. There was an indication (tentative) of possible stone lining on the southern edge of the grave cut. The grave cut clearly truncated cist grave 26 containing burial 25. Preservation was poor, and there was some possibility of disturbance in antiquity. The long bones were articulated. The grave had been cut into the natural subsoil.

Objectives: a well-defined earth-cut grave appears to be a later grave type whilst some possibility remains that partial stone lining was used in order to 'formalise' the grave cut. The stratigraphic relationship with cist grave 26 (burial 25) allows a date for this burial to provide a *terminus ante quem* for the cist grave and assist the date range for the earth-cut grave type.

Calibrated date: 1σ: cal AD 1040–1170
             2σ: cal AD 1020–1220

Final comment: J Gossip (30 January 2013), the date comes from an earth-cut grave associated with the Norman church, proving that both earth-cut and cist grave traditions were occurring simultaneously.

SUERC–6865 880 ±35 BP

$\delta^{13}C$: -20.8‰

Sample: burial 39A1, submitted in March 2005 by J Gossip

Material: human bone (right femur) (H Gestsdottir 2001)

Initial comment: a replicate of OxA-14584.

Objectives: as OxA-14584

Calibrated date: 1σ: cal AD 1050–1220
             2σ: cal AD 1030–1250

Final comment: see OxA-14584

SUERC–6932 855 ±35 BP

$\delta^{13}C$: -20.4‰

Sample: burial 39A2, submitted in March 2005 by J Gossip

Material: human bone (right femur) (H Gestsdottir 2001)

Initial comment: a replicate of OxA-14584.

Objectives: as OxA-14584

Calibrated date: 1σ: cal AD 1160–1230
             2σ: cal AD 1040–1270

Final comment: see OxA-14584

Carlton Colville: Bloodmoor Hill, Suffolk

Location: TM 52089002
                    Lat. 52.26.58 N; Long. 01.42.34 E

Project manager: A Dickens (Cambridge Archaeological Unit), 1998

Description: an early Anglo-Saxon settlement and cemetery was excavated at Bloodmoor Hill, near Lowestoft, Suffolk. The area excavated covered just over 3ha, and produced material from two main periods: first- and second-century Romano-British, associated with a ditched field and track system; and sixth–eighth-century Anglo-Saxon associated with dense settlement remains including *Grubenhäuser*, post-buildings, pits, a midden, a cemetery, and evidence of industrial activity. Twenty-nine inhumation burials also were excavated. Twenty-six were buried in a formal but unenclosed cemetery within the central area of the settlement. Three graves were located 50m to the east, but still within the settlement area and contemporary with the main cemetery. The Anglo-Saxon features both overlay, and were in parts contained by, the Romano-British system. There were also small prehistoric and medieval/post-medieval elements. The features were relatively well-preserved, though with some truncation as a result of medieval and later ploughing.

The primary focus for both the initial and subsequent excavations was the early Anglo-Saxon settlement. The northern boundary of the settlement lies under 1980s and 1990s housing development, while western, eastern, and southern limits have now been defined.

The area is situated on the south-eastern slope of a broad, shallow valley, with hills to both sides rising gently to a height of *c* 18m. The valley is a little over 2km wide, and runs south-west to north-east, flowing into the eastern end of Lake Lothing, and then to the sea at Lowestoft. The site lies on sand at 8–10m AOD.

Objectives: the radiocarbon programme was designed to achieve the following: to provide overall estimates of the start, end, and duration of the settlement activity; to refine the dating of the cemetery, and to understand its development; to date the isolated group of burials to the east of the site; to understand the process of filling disused *Grubenhäuser*; and to provide absolute dating for the ceramic typologies.

References: Lucy *et al* 2009

Carlton Colville: Bloodmoor Hill, settlement and pottery, Suffolk

Location: TM 52089002
                    Lat. 52.26.58 N; Long. 01.42.34 E

Project manager: A Dickens (Cambridge Archaeological Unit), 1998

Archival body: Suffolk Museums

Description: 49 radiocarbon age determinations were obtained on samples of animal bone and charred residues on the interior of pottery sherds from contexts across the settlement.

Objectives: to investigate the chronology of the settlement, to determine the absolute date of the settlement and its duration, and to investigate its development, particularly in areas with stratigraphic sequences (the cemetery and the midden). It was hoped to confirm when the settlement started, how long it was used for while establishing the chronological development of the buildings, how it relates to the date of the cemetery, and when the settlement was abandoned. This was to be achieved by dating residues on sherds from sealed contexts across the site which are part of primary disposals, ie several sherds from the same vessel, deposited together, and which it is believed were deposited directly/very soon after vessel breakage; and also a number of animal articulations and an animal cremation from sealed contexts.

A secondary aim relates to *Grubenhäuser* and the formation of their fills. By dating several primary disposals from the fills of the same *Grubenhaus*, it is hoped that the duration of time it took for the pits to fill in can be examined. The dating of single sherds from the fill of these pits, primarily sampled to examine the ceramic sequence will also enable an examination of the nature of their fills, ie whether or not the material has been in a midden first.

A further aim was to refine the dating of the Anglo-Saxon ceramic sequence, which is otherwise difficult to date very accurately; to examine the dating of fabric types, in particular to determine whether biotite-tempered and grog-tempered pottery is early, ie sixth century, and also determine whether chaff-tempered pottery is late, ie later sixth- and seventh-century; and, to determine the date of decorated pottery, to see if bossed, stamped, and incised pottery is (early) sixth-century in date, and also to determine the date of combed decoration, and to see if there is a chronological sequence between different decoration types.

Final comment: P Marshall (25 October 2013), the radiocarbon dating programme has demonstrated that the Anglo-Saxon settlement at Bloodmoor Hill was probably in use from sometime after cal AD 500 until shortly after cal AD 700. In the mid-seventh century cal AD human burial is seen for the first time, with the start of an inhumation cemetery that eventually becomes the core of the settlement as buildings are constructed to its south and west. Although the dating of the ceramic typologies was not completely successful, it demonstrated the potential for providing precise chronologies for ceramics in this period.

Laboratory comment: English Heritage (25 October 2006), the radiocarbon dating programme has been successful in achieving most of the aims as outlined above. The results provide an estimate for the start of settlement activity at Bloodmoor Hill of *cal AD 480–540 (95% probability)* with it being more probable *(85%)* that activity started after *cal AD 500*. The end of activity is estimated at *cal AD 660–710 (95% probability; start;* Lucy *et al* 2009, fig 6.4). The site was in use for a relatively short period of time, estimated *at 140–210 years (95% probability; settlement use;* Lucy *et al* 2009, fig 6.6).

The results do not conclusively refute any of the main processes by which the infilling of *Grubenhäuser* are thought to take place.

The absolute dating of the ceramic typologies remains the one original aim that has not been completely fulfilled. The reasons for this are threefold; firstly inaccurate measurements on carbonised residues are still apparent in the results described below with results that are both too old and too young for the pottery types from which they come. This suggests that we do not still have an adequate understanding of the chemistry of dating from carbonised residues. This is not a site specific problem at Bloodmoor Hill but a methodological problem inherent in the dating of carbonised residues from any site/period. Secondly, our archaeological understanding of chronological changes in fabric types might be flawed, although this might in part be due to a paucity of excavated sites with large assemblages from the early-middle Saxon phase in this part of the country. Thirdly, at Bloodmoor Hill we probably did not have a large enough sample of carbonised residues on different fabric types to adequately test the hypothesis provided by the ceramic analysis.

References: Lucy *et al* 2009

GrA–25563 1375 ±35 BP

δ¹³C: -30.0‰

Sample: 00480AA, Structure 35, submitted on 18 March 2004 by J Tipper

Material: carbonised residue (internal)

Initial comment: from fill [03268] of a *Grubenhaus* F333 (Structure 35), located in the north-east part of the site. [003268] is the upper fill of the *Grubenhaus* pit. The sample is from the south-east quadrant of the pit. The *Grubenhaus* cut a Roman feature (F204), but was not itself cut by any later features.

Objectives: the sherd (now broken into four) is a chaff-tempered fabric, which may indicate a seventh-century date. This will provide a date for the use of this fabric type. It will also provide a date for this part of the settlement.

Calibrated date: *1σ:* cal AD 640–670
                 2σ: cal AD 610–690

Final comment: P Marshall (25 October 2013), the radiocarbon date is in agreement with others obtained from organic tempered pottery on the site that suggest its currency dates from at least the mid-sixth century onwards - in accordance with the date range suggested by Hammerow *et al* (1994).

Laboratory comment: English Heritage (25 October 2006), the two measurements from carbonised residue on separate fragments of a single sherd of organic tempered pottery 00480AA (GrA-25563 and OxA-13755) are statistically consistent (T′=2.7; *v*=1; T′(5%)=3.8; Ward and Wilson 1978).

References: Hammerow *et al* 1994
                 Ward and Wilson 1978

GrA–25589 1385 ±35 BP

δ¹³C: -28.1‰

Sample: 11990AD, submitted on 18 March 2004 by J Tipper

Material: carbonised residue (internal)

Initial comment: from upper fill [04603] of a *Grubenhaus* F514, located in the central-north part of the site. The sample is from spit B of the north-west quadrant of the pit. The *Grubenhaus* cut an earlier Roman ditch, and was cut by pit F517. The *Grubenhaus* and pit were sealed by midden F1.

Objectives: the sherd is from a primary disposal. There were two sherds (77g) of this vessel from this *Grubenhaus*. It is hoped that the date will provide a *terminus post quem* for the backfilling of this pit, and by implication the use/disuse of the original building. The dating of this residue will also provide a date for chaff-tempered fabric, believed to be of seventh-century date. The date can also be compared to other samples from the fill of the same *Grubenhaus*. The interval between the samples will provide an estimate for the duration of the infilling It will also provide a date for this part of the settlement.

Calibrated date: *1σ:* cal AD 640–670
                 2σ: cal AD 600–680

Final comment: see GrA-25563

Laboratory comment: English Heritage (25 October 2006), 11990AE (OxA-13726) and 11990AD (GrA-25589) come from a possible primary disposal V128 (organic tempered). The two measurements are not statistically consistent (T′=7.8; *v*=1; T′(5%)=3.8; Ward and Wilson 1978).

References: Ward and Wilson 1978

GrA–25590 1425 ±35 BP

δ¹³C: -29.2‰

Sample: 02234AC, Pit group I, submitted on 18 March 2004 by J Tipper

Material: carbonised residue (internal)

Initial comment: from upper fill [03521] of pit cut [03520], within pit complex F372, located in the central part of the site. The pit was immediately west of a similar sized and shaped pit [03525], but there was no clear stratigraphic relationship between them. It was not cut by any later features.

Objectives: sherds 02234AB and 02234C are two decorated sherds from the same vessel. Both sherds are decorated with hollow vertical bosses. Dating will provide a date for this type of decoration, which is of probable late fifth- or sixth-century date. The sherds are from probable primary disposal, and there were three sherds (44g) from this fill [03521]. A date will also provide a *terminus post quem* for the backfilling of the pit and for this part of the settlement.

Calibrated date: *1σ:* cal AD 600–660
                 2σ: cal AD 560–670

Final comment: P Marshall (25 October 2013), the Roman radiocarbon date for this vessel (OxA-14019) is clearly too old given the vessel is Anglo-Saxon. GrA-25590 is therefore preferred as the more accurate estimate for the age of this vessel.

Laboratory comment: English Heritage (25 October 2006), samples 02234AB (OxA-14019) and 02234AC (GrA-25590) came from carbonised residues on two sherds (sandstone/sand tempered, decorated with bosses) of primary disposal V89. The two measurements are not statistically consistent (T′=9.9; ν=1; T′(5%)=3.8; Ward and Wilson 1978) indicating that material of two different ages was dated.

References: Ward and Wilson 1978

GrA–25592 1440 ±35 BP

δ¹³C: -29.6‰

Sample: 11976AD, submitted on 18 March 2004 by J Tipper

Material: carbonised residue (external)

Initial comment: from upper fill [04603] of *Grubenhaus* F514, located in the central-north part of the site. The fill [04603A] was possibly part of midden F1 which sealed the *Grubenhaus* (or was a mixed interface between the midden and the *Grubenhaus*). Fill [04603B] was the upper fill of the *Grubenhaus*. It is not known whether sherd 11976AD was from either [04603] A or B. The sherd was from the north-east quadrant of the pit. The *Grubenhaus* truncated a Roman ditch, and was cut by pit F517.

Objectives: sherd 11976AD is from a primary disposal. There were 15 sherds of this vessel (292g), 13 of which (260g) were from the fill of *Grubenhaus* F514. The dating of the external residue on this sherd will establish a *terminus post quem* for the backfilling of the pit, and by implication, a *terminus post quem* for the use of the original building. This date can also be compared to other dates on sherds from the same *Grubenhaus*. The intervals from between the samples will provide an estimate for the duration of the infilling of the pit. It will also provide a date for this part of the settlement. In addition, the vessel is decorated with comb marks, and this decoration may indicate a fifth- or sixth-century date. The dating of this residue will provide a date for the use of this type of decoration.

Calibrated date: 1σ: cal AD 590–650
                      2σ: cal AD 550–660

Final comment: P Marshall (25 October 2013), the result provides an archaeologically acceptable date for the age of the vessel.

Laboratory comment: English Heritage (25 October 2006), 11983AA (OxA-13966) and 11976AD (GrA-25592) come from primary disposal V125 (calcitic tempered). The two measurements are statistically consistent (T′=0.1; ν=1; T′(5%)=3.8; Ward and Wilson 1978).

References: Ward and Wilson 1978

GrA–25923 1400 ±35 BP

δ¹³C: -27.3‰

Sample: 00211AA, Structure 12, submitted on 18 March 2004 by J Tipper

Material: carbonised residue (internal)

Initial comment: from posthole fill [00237] of a *Grubenhaus* F5 (Structure 12), located within the central part of the site. [00237] is the fill of the eastern (central) gable posthole of the pit. The posthole was 0.4m in depth below the base of the *Grubenhaus* pit. It contained several different fills (not separately numbered). The exact relationship between the posthole and the lower fills of the *Grubenhaus* was not recorded, ie it was not certain whether or not the post rotted *in situ*. Nine sherds (134g) from the same vessel were from fill [00205], located beneath upper fill [00204]; and four sherds (39g) from lower fill [00216]. Only sherd 00211A was from the fill of posthole [00237]. The *Grubenhaus* cut two Roman ditches, but was not itself cut by any later features.

Objectives: the sherd (now broken in two) is from a primary disposal. There were 14 sherds of this vessel from this *Grubenhaus*. It is hoped that the date will provide a *terminus post quem* for the backfilling of this pit, and by implication the use/disuse of the original building. This can be compared to sample 00182 AA ((OxA-14004), also a primary disposal from upper fill [00204]. The interval between the two samples will provide an estimate for the duration of the infilling and the earliest of the two will provide a *terminus ante quem* for the construction of the building. It will also provide a date for this part of the settlement.

Calibrated date: 1σ: cal AD 630–660
                      2σ: cal AD 590–670

Final comment: P Marshall (25 October 2013), the result provides an archaeologically acceptable date for the age of the vessel.

GrA–25925 1305 ±40 BP

δ¹³C: -30.3‰

Sample: 00490AA, Structure 25, submitted on 18 March 2004 by J Tipper

Material: carbonised residue (internal)

Initial comment: from fill [03428] of a *Grubenhaus* F335 (Structure 25), located in the central part of the site. [003428] is the upper fill of the *Grubenhaus* pit. The sample is from the south-east quadrant of the pit. The *Grubenhaus* does not cut any earlier features, and was also not cut by any later features.

Objectives: the sherd is from a primary disposal. There were two sherds (176g) of this vessel from this *Grubenhaus*. It is hoped that the date will provide a *terminus post quem* for the backfilling of the pit, and by implication the use/disuse of the original building. The dating of this residue will also provide a date for chaff-tempered fabric, believed to be of seventh-century date. The fill also contained an imported wheel-made vessel which dating can therefore, by association, provide a *terminus post quem* for its deposition. It will also provide a date for this part of the settlement.

Calibrated date: 1σ: cal AD 660–770

2σ: cal AD 650–780

Final comment: see GrA-25563

Laboratory comment: English Heritage (25 October 2006), the measurements on the two samples 00490AB (OxA-13710) and 00490AA (GrA-25925), carbonised residues on sherds of primary disposal V28 (organic tempered), are statistically consistent (T'=0.1; ν=1; T'(5%)=3.8; Ward and Wilson 1978).

References: Ward and Wilson 1978

GrA–25926 1505 ±40 BP

δ¹³C: -27.7‰

Sample: 13153AA, Structure 24, submitted on 18 March 2004 by J Tipper

Material: carbonised residue (internal)

Initial comment: from upper fill [05115] of a *Grubenhaus* F626 (structure 24), located in the southern part of the site. The sample is from the south-east quadrant of the pit. The *Grubenhaus* cut Roman ditch F620, and was not cut by any later features.

Objectives: the sherd is tempered with grog and quartz, which may indicate a sixth-century date. Dating will provide a date for the use of this fabric type, and can be compared to dates on samples from the same fill. It will also provide a date for this part of the settlement.

Calibrated date: 1σ: cal AD 530–610

2σ: cal AD 420–650

Final comment: P Marshall (25 October 2013), the result provides an archaeologically acceptable date for the age of the vessel.

Laboratory comment: English Heritage (25 October 2006), the two samples, from the upper fill [05115] comprised 13141AA (GrA-25950) from primary disposal V143 (sandstone sand tempered) and 13153AA (GrA-25926) from an organic tempered sherd. These two measurements are not statistically consistent (T'=10.3; ν=1; T'(5%)=3.8; Ward and Wilson 1978).

References: Ward and Wilson 1978

GrA–25927 1610 ±40 BP

δ¹³C: -27.8‰

Sample: 13198AA, Structure 24, submitted on 18 March 2004 by J Tipper

Material: carbonised residue (internal)

Initial comment: from lower fill [05116] of a *Grubenhaus* F626 (Structure 24), located in the southern part of the site. The sample is from the south-west quadrant of the pit, and is of probable primary disposal. The *Grubenhaus* cut Roman ditch F620, and was not cut by any later features.

Objectives: the sherd is tempered with biotite and quartz, which may indicate a sixth-century date. Dating will provide a date for the use of this fabric type, and can be compared to dates on samples from the same fill. There are four sherds of

this vessel (104g). Dating will help establish a *terminus post quem* for the backfilling of the pit, and by implication the use of the original building. This sample can be compared to other sherds from the fill of *Grubenhaus* F626. The interval between the earliest and latest samples will provide an estimate for the duration of the infilling of the pit. It will also provide a date for this part of the settlement.

Calibrated date: 1σ: cal AD 390–540

2σ: cal AD 350–550

Final comment: P Marshall (25 October 2013), the result provides an archaeologically acceptable date for the age of the vessel.

Laboratory comment: English Heritage (25 October 2006), the fill of the pit was characterised by two fills; the lower fill [5116] of which provided three samples from carbonised residues; 13187AB (OxA-13752) from V99 (quartz-tempered) a probable primary disposal; 13198AA (GrA-25927) from primary disposal V352 (biotite), and 13187AF (OxA-14018) from a biotite-tempered sherd. The three measurements are not statistically consistent (T'=9.7; ν=2; T'(5%)=6.0; Ward and Wilson 1978).

References: Ward and Wilson 1978

GrA–25929 1505 ±40 BP

δ¹³C: -27.5‰

Sample: 11963AB, submitted on 18 March 2004 by J Tipper

Material: carbonised residue (external)

Initial comment: from upper fill [04603] of a *Grubenhaus* F514, located in the central-north part of the site. The sample is from the south-east quadrant of the pit, and is of probable primary disposal. The fill [04603A] was possibly part of midden F1 which sealed the *Grubenhaus* (or was a mixed interface between the midden and the *Grubenhaus*). Fill [04603B] was the upper fill of the *Grubenhaus*. It is not known whether sherd 11963AB was from either [04603] A or B. The *Grubenhaus* cut a Roman ditch, and was cut by pit F517. A replicate of GrA-25937 from a sherd of the same vessel.

Objectives: the sherd is from a possible primary disposal. There are 21 sherds of this vessel (439g), five of which are from the fill of *Grubenhaus* F514. The dating will establish a *terminus post quem* for the backfilling of this pit, and by implication the use of the original building. This can be compared to dates from primary disposals from a fill of the same *Grubenhaus*. The interval between the earliest and latest samples will provide an estimate for the duration of the infilling of the pit. It will also provide a date for this part of the settlement. In addition, the sherds are tempered with biotite and quartz, which may indicate a sixth-century date. The dating will provide a date for the use of this fabric type.

Calibrated date: 1σ: cal AD 530–610

2σ: cal AD 420–650

Final comment: P Marshall (25 October 2013), the result provides an archaeologically acceptable date for the age of the vessel.

Laboratory comment: English Heritage (25 October 2006), biotite/granite fabric can clearly be differentiated as early and definitely sixth-century in date, although four measurements from primary disposal V178 (T'=2.4; *v*=3; T'(5%)=7.8; Ward and Wilson 1978) (OxA-13882, OxA-13754, GrA-25937 and GrA-25929) make up 40% of the sample from this fabric group.

References: Ward and Wilson 1978

GrA–25931 1530 ±40 BP

δ¹³C: -28.5‰

Sample: 02175AC, Pit group J, submitted on 18 March 2004 by J Tipper

Material: carbonised residue (internal)

Initial comment: from initial cleaning deposit [03363] within pit complex F350, located in the north-east part of the site. The area comprised to pits: [03364] cut by pit [03440]. Context [03363] is probably the same deposit as [03365], the upper fill of pit [03340]. The pit complex cut Roman ditch F204, but was not cut by any later features.

Objectives: the sherd is from a primary disposal, of which there are three sherds of this vessel from the same fill (66g). A date will provide a *terminus post quem* for the backfilling of the pit and for this part of the settlement. The vessel is also chaff-tempered, which may indicate a seventh-century date. Dating will therefore provide a date for the use of this fabric type.

Calibrated date: *1σ:* cal AD 430–580
                        2σ: cal AD 420–620

Final comment: see GrA-25563

GrA–25935 1505 ±40 BP

δ¹³C: -29.6‰

Sample: 01826AA, Structure 17, submitted on 18 March 2004 by J Tipper

Material: carbonised residue (internal)

Initial comment: from upper fill [01764] of *Grubenhaus* F212 (structure 17), located in the central-north part of the site. The fill was indistinguishable from surface spread F275 overlying the *Grubenhaus* and other features in this area. The sample is from the north-east quadrant of the pit. The *Grubenhaus* cut two Roman ditches, and was not cut by any later features.

Objectives: the sherd is from a primary disposal. There were three sherds (101g) of this vessel from fill [01764]. It is hoped that the date will provide a *terminus post quem* for the backfilling of this pit, and by implication the use/disuse of the original building. The dating of this residue will also provide a date for the biotite and quartz-tempered fabric, believed to be of sixth-century date. This date will also be compared to 00425AA, a single decorated sherd from the same fill. The interval between the two primary disposals will provide an estimate for the duration of the infilling. It will also provide a date for this part of the settlement.

Calibrated date: *1σ:* cal AD 530–610
                        2σ: cal AD 420–650

Final comment: P Marshall (25 October 2013), the result provides an archaeologically acceptable date for the age of the vessel.

Laboratory comment: English Heritage (25 October 2006), three samples (all carbonised residues on pottery sherds) came from the upper fill [01764] of the structure; 09528AA (OxA-13708) from primary disposal V106 (quartz-tempered); sample 01826AA (GrA-25935) from primary disposal V282 (biotite-tempered); and sample 00425AA (OxA-13728) from a quartz-tempered, stamped and incised sherd. The three measurements are not statistically consistent (T'=6.5; *v*=2; T'(5%)=6.0; Ward and Wilson 1978).

References: Ward and Wilson 1978

GrA–25936 1480 ±40 BP

δ¹³C: -29.6‰

Sample: 02167AA, Structure 30, submitted on 18 March 2004 by J Tipper

Material: carbonised residue (internal)

Initial comment: from the western gable posthole [03339] within *Grubenhaus* F341 (structure 30), located in the central-east part of the site. The sample was located at the interface of the two fills within the posthole. There was no evidence to suggest the post had decayed *in situ*, which means it may have been removed prior to the infilling of the pit. The *Grubenhaus* cut the edge of surface deposit F342; it was not cut by any later features.

Objectives: the large rim sherd is from a probable primary disposal as it is large and in a good state of preservation. It is hoped that the date will provide a *terminus post quem* for the backfilling of this pit, and by implication the use/disuse of the original building. This date will also be compared to other samples from the same *Grubenhaus*. The interval between the primary disposals will provide an estimate for the duration of the infilling of the pit. It will also provide a date for this part of the settlement.

Calibrated date: *1σ:* cal AD 540–640
                        2σ: cal AD 430–650

Final comment: P Marshall (25 October 2013), the result provides an archaeologically acceptable date for the age of the vessel.

GrA–25937 1490 ±40 BP

δ¹³C: -29.7‰

Sample: 10509AA, submitted on 18 March 2004 by J Tipper

Material: carbonised residue (internal)

Initial comment: from a 0.1m lower spit B within a one metre text square; context [04040] through midden deposit F1, in the central-north part of the site. A replicate of GrA-25929 on a sherd from the same vessel.

Objectives: the sherd is tempered with biotite and quartz, which may indicate a sixth-century date. The dating will provide a date for the use of this fabric type. In addition, the sherds are from a possible primary disposal spread across several different adjacent 1m squares through the midden, and also within the fill of *Grubenhaus* F514. There are 21 sherds of this vessel (238g). It is hoped that the dates will help to establish a closer chronology for the accumulation of midden deposit F1, and the infilling of *Grubenhaus* F514 in this part of the site.

Calibrated date: 1σ: cal AD 540–620
                                   2σ: cal AD 430–650

Final comment: P Marshall (25 October 2013), the result provides an archaeologically acceptable date for the age of the vessel.

Laboratory comment: see GrA-25929

GrA–25949 1455 ±45 BP

δ¹³C: -28.7‰

Sample: 11843AA, Structure 10, submitted on 18 March 2004 by J Tipper

Material: carbonised residue (internal)

Initial comment: from the upper fill [04588] of *Grubenhaus* F512 (structure 10), located in the central-north part of the site. The sample was from the south-east quadrant of the pit. The *Grubenhaus* cut two Roman ditches [5152] and [5154]; it was not cut by any later features, although there was extensive animal burrowing.

Objectives: the sherd (now broken into three) is tempered with biotite and quartz, which may indicate a sixth-century date. It is hoped that dating will provide a date for the use of this fabric type. It will also provide a date for this part of the settlement.

Calibrated date: 1σ: cal AD 560–650
                                   2σ: cal AD 530–660

Final comment: P Marshall (25 October 2013), the result provides an archaeologically acceptable date for the age of the vessel.

GrA–25950 1710 ±50 BP

δ¹³C: -28.5‰

Sample: 13141AA, Structure 24, submitted on 18 March 2004 by J Tipper

Material: carbonised residue (internal)

Initial comment: from upper fill [05115] of a *Grubenhaus* F626 (structure 24), located in the southern part of the site. The sample is from the north-east quadrant of the pit. The *Grubenhaus* cut Roman ditch F620, and was not cut by any later features.

Objectives: the sherd is from a primary disposal, and there were three sherds of this vessel (126g). It is hoped that the date will establish a *terminus post quem* for the backfilling of this pit, and by implication the use/disuse of the original building. This can be compared to other dates on primary disposals from the lower fill of the *Grubenhaus*. The interval between the earliest and latest samples will provide an estimate for the duration of the infilling. It will also provide a date for this part of the settlement.

Calibrated date: 1σ: cal AD 250–400
                                   2σ: cal AD 220–430

Final comment: P Marshall (25 October 2013), given the sherd is from an Anglo-Saxon vessel the radiocarbon date clearly does not provide an accurate age for the vessel.

Laboratory comment: English Heritage (25 October 2006), OxA-14006, OxA-14017, and GrA-25950 were measurements made on carbonised residues adhering to the inside of sherds of identifiable Anglo-Saxon pottery, although they gave Roman ages. The carbonaceous fractions extracted physically and chemically from the inside of the sherds are assumed to represent organic-rich food remains, and thus should date the last use of the vessel in question. However, in this case the residue is clearly much too old for the pottery type in question. It has been suggested that clays may contain appreciable amounts of carbon which may remain in pottery even after firing (Nakamura *et al* 2001). Such a mechanism as this would introduce 'old' carbon and may therefore provide an explanation for the apparent erroneous measurements.

References: Nakamura *et al* 2001

GrA–26355 1805 ±35 BP

δ¹³C: -21.7‰

Sample: 07250, Structure 44, submitted on 18 March 2004 by J Tipper

Material: animal bone: *Ovis* sp. (J Tipper 2004)

Initial comment: this sample is from the fill [01924] of a posthole F246, located in the central part of the site. The posthole is located within, and associated with, a posthole building that is stratigraphically earlier than the cemetery.

Objectives: 07250 is an articulated sheep skeleton packed within the fill of the posthole, which is part of a building. It is hoped that this sample will provide a *terminus post quem* for the building. This will help to establish the chronological time-span of this part of the site. This sample can be compared with 01894AA - pottery residue from posthole [01950] nearby, and also to sample 13251 - a cremated sheep burial, cut by a posthole from the same building.

Calibrated date: 1σ: cal AD 130–250
                                   2σ: cal AD 120–340

Final comment: P Marshall (25 October 2013), the Roman date for the sheep was wholly unexpected, given the sample came from an animal disposal in the fill of an Anglo-Saxon postholed building. The building therefore seems to have cut through earlier Roman deposits.

Laboratory comment: English Heritage (25 October 2006), sample 01894AA (OxA-14005) a carbonised residue on a quartz-tempered sherd and 7250 an articulated sheep packed within the fill of a posthole (OxA-14044, GrA-26355, and UB-6185) (weighted mean 1804 ±13 BP; T'=4.4; v=2; T'(5%)=6.0; Ward and Wilson 1978) come from 2 of the 26 postholes that make up the building.

The articulated sheep, however, clearly relates to an earlier Roman episode of activity. The sample comes from a posthole within structure 44 and it seems to have been fortuitous this structure was located over an earlier building. Such an explanation also explains the Roman date of OxA-13892, part of a neonatal sheep burial within a small posthole/pit that was cut by a post-hole [01903] of structure 44. The measurements from both samples from both postholes with sheep packed in them are in fact statistically consistent (T'=6.0; ν=3; T'(5%)=7.8; Ward and Wilson 1978) and so could be of the same actual age.

References: Ward and Wilson 1978

GrA–26357 1500 ±35 BP

$\delta^{13}C$: -21.8‰

Sample: 13429, Pit Group C, submitted on 18 March 2004 by J Tipper

Material: animal bone: *Ovis* sp., metatarsal (A Grieve 2004)

Initial comment: this sample is from the fill [04642] of pit F521, located in the central-north part of the site. The fill of this pit was excavated as a single context; however, three fills were distinguished in section (A, B, and C). The pit cut two earlier pits F500a and F500b; all three pits were sealed by midden F1.

Objectives: it is hoped that the dating of the animal bone will establish a *terminus post quem* for the backfilling of the pit. This can be compared with a sample on residue from a pottery sherd from the same fill, and also to the animal bone sample from pit complex F500. The interval between these samples will provide an estimate for the duration of the infilling of the pit, and also a *terminus ante quem* for the cutting of this pit. It will also provide a date for this part of the settlement.

Calibrated date: 1σ: cal AD 540–610
2σ: cal AD 430–650

Final comment: P Marshall (25 October 2013), the Roman date for the sheep was wholly unexpected, given the sample came from an animal disposal in the fill of an Anglo-Saxon post-holed building. The building therefore seems to have cut through earlier Roman deposits.

Laboratory comment: English Heritage (25 October 2006), two samples from pit F521 comprised 12147AC (OxA-14017) a carbonised residue from V298, a probable primary disposal (organic-tempered) from the upper fill [4642], and 13429 (OxA-13757 and GrA-26357) from an articulated sheep. The two measurements from the sheep (13429) are statistically consistent (T'=0.1; ν=1; T'(5%)=3.8; Ward and Wilson 1978) and so allow a weighted mean to be calculated (1506 ±21 BP). However, all three measurements from the upper fill [4642] are not statistically consistent (T'=32.9; ν=2; T'(5%)=6.0; Ward and Wilson 1978).

References: Ward and Wilson 1978

OxA–13707 1398 ±25 BP

$\delta^{13}C$: -28.8‰

Sample: 02130AC, Structure 30, submitted on 18 March 2004 by J Tipper

Material: carbonised residue (internal)

Initial comment: this sample is from the upper fill [03311] of *Grubenhaus* F341 (structure 30), located in the central-east part of the site. The sample comes from the north-west

quadrant of the *Grubenhaus*, which cut the surface deposit F342; it was not cut by any later features. There was some animal burrowing.

Objectives: this sherd is from a primary disposal. There are ten sherds in total (169g) from this vessel from the fill of this *Grubenhaus*. It is hoped that the dating of the residue will establish a *terminus post quem* for the backfilling of the pit, and by implication, the disuse/use of the original building. This can be compared to another date from the western gable posthole [03339] and from other fills from the *Grubenhaus*. The interval between the earliest and latest sample should provide an estimate for the duration of the infilling of the pit. It will also provide an absolute date for this part of the settlement.

Calibrated date: 1σ: cal AD 640–660
2σ: cal AD 600–670

Final comment: P Marshall (25 October 2013), the date provides an archaeologically acceptable age for the vessel.

Laboratory comment: English Heritage (25 October 2006), two samples from the upper fill of the pit comprised 2130AC (OxA-13707) from primary disposal V77 (calcitic-tempered) and sample 2131AH (OxA-14008) from a grog-tempered sherd. The two measurements are not statistically consistent (T'=9.9; ν=1; T'(5%)=3.8; Ward and Wilson 1978).

References: Ward and Wilson 1978

OxA–13708 1481 ±26 BP

$\delta^{13}C$: -27.6‰

Sample: 09528AA, Structure 17, submitted on 18 March 2004 by J Tipper

Material: carbonised residue (internal)

Initial comment: from upper fill [01764] of *Grubenhaus* F212 (structure 17), located in the central-north part of the site. The fill was indistinguishable from surface spread F275 overlying the *Grubenhaus* and other features in this area. The sample is from the north-east quadrant of the pit, spit 1, square G. The *Grubenhaus* cut two Roman ditches, and was not cut by any later features.

Objectives: the sherd is from a primary disposal. There were three sherds (30g) of this vessel from fill [01764]. It is hoped that the date will provide a *terminus post quem* for the backfilling of this pit, and by implication the use/disuse of the original building. The dating of this residue will also provide a date for use of the vessel decoration which comprises stamp impressions (Briscoe stamp motif A 5avii) and incised lines, which form a two-line pendant-triangle, believed to be of sixth-century date. This date will also be compared to other primary disposal sherds from the same fill. The interval between the two primary disposals will provide an estimate for the duration of the infilling, and the earliest of the two will provide a *terminus ante quem* for the construction of the building. It will also provide a date for this part of the settlement.

Calibrated date: 1σ: cal AD 560–620
2σ: cal AD 540–650

Final comment: P Marshall (25 October 2013), the date provides an archaeologically acceptable age for the vessel.

Laboratory comment: see GrA-25935

OxA–13709 1459 ±29 BP

δ¹³C: -28.0‰

Sample: 10449AA, submitted on 18 March 2004 by J Tipper

Material: carbonised residue (internal)

Initial comment: this sample is from context [04035], a one metre test square through the midden surface deposit F1, in the central-north part of the site. The test square was located immediately above, and sealed, *Grubenhaus* F514 (layer [04603]); the sherd derives from a 01.m spit B, the lower spit excavated in the midden. The upper fill of the *Grubenhaus* and the midden deposit were indistinguishable on excavation.

Objectives: the sherd is from a primary disposal. There were two large sherds (238g) of this vessel from fill [04035]. It is hoped that the date will provide a *terminus post quem* for the backfilling of this pit, and the accumulation of the midden deposit F1 in this part of the site, in combination with the other samples from the fill of *Grubenhaus* F514. The dating of this residue will also provide a date this part of the settlement where there is a complex stratigraphic sequence. In addition, the sherd is tempered with biotite and quartz, which may indicate a sixth-century date. It is hoped that dating will provide a date for the use of this fabric type.

Calibrated date: *1σ:* cal AD 570–640
                       2σ: cal AD 540–660

Final comment: P Marshall (25 October 2013), the date provides an archaeologically acceptable age for the vessel.

OxA–13710 1316 ±25 BP

δ¹³C: -30.4‰

Sample: 00490AB, Structure 25, submitted on 18 March 2004 by J Tipper

Material: carbonised residue (internal)

Initial comment: from fill [03428] of a *Grubenhaus* F335 (structure 25), located in the central part of the site. [003428] is the upper fill of the *Grubenhaus* pit. The sample is from the south-east quadrant of the pit. The *Grubenhaus* does not cut any earlier features, and was also not cut by any later features. A replicate of GrA-25925 from a sherd frm the same vessel.

Objectives: the sherd is from a primary disposal. There were two sherds (176g) of this vessel from this *Grubenhaus*. It is hoped that the date will provide a *terminus post quem* for the backfilling of this pit, and by implication the use/disuse of the original building. The dating of this residue will also provide a date for chaff-tempered fabric, believed to be of seventh-century date. The fill also contained an imported wheel-made vessel dating can therefore, by association, provide a *terminus post quem* for its deposition. It will also provide a date for this part of the settlement.

Calibrated date: *1σ:* cal AD 660–690
                       2σ: cal AD 650–770

Final comment: see GrA-25563

Laboratory comment: see GrA-25925

OxA–13711 1515 ±26 BP

δ¹³C: -27.6‰

Sample: 00457AT, Structure 38, submitted on 18 March 2004 by J Tipper

Material: carbonised residue (internal)

Initial comment: from upper fill [03001] of a *Grubenhaus* F286 (structure 38), located on the eastern edge of the site. The sample is from the north-east quadrant of the pit. The *Grubenhaus* does not cut any earlier features, and was also not cut by any later features.

Objectives: the sherd is tempered with biotite and quartz, which may indicate a sixth-century date. It is hoped that dating of the residue will provide a date for the use of this fabric type. This can then be compared with sample 00457AC, also from this context. It will also provide a date for this part of the settlement.

Calibrated date: *1σ:* cal AD 540–590
                       2σ: cal AD 430–610

Final comment: P Marshall (25 October 2013), the date provides an archaeologically acceptable age for the vessel.

Laboratory comment: English Heritage (25 October 2006), the measurements on the two samples from the upper fill: 00457AC (OxA-13883) from a quartz-tempered, stamped, and incised decorated sherd, and sample 00457AT (OxA-13711) from a biotite-tempered sherd, are statistically consistent (T'=1.4; ν=1; T'(5%)=3.8; Ward and Wilson 1978).

References: Ward and Wilson 1978

OxA–13726 1509 ±27 BP

δ¹³C: -27.8‰

Sample: 11990AE, submitted on 18 March 2004 by J Tipper

Material: carbonised residue (external)

Initial comment: from upper fill [04603] of a *Grubenhaus* F514, located in the central-north part of the site. The sample is from spit B of the north-west quadrant of the pit. The *Grubenhaus* cut an earlier Roman ditch, and was cut by pit F517. The *Grubenhaus* and pit were sealed by midden F1. A replicate of GrA-25589 from a sherd from the same vessel.

Objectives: the sherd is from a primary disposal. There were two sherds (77g) of this vessel from this *Grubenhaus*. It is hoped that the date will provide a *terminus post quem* for the backfilling of this pit, and by implication the use/disuse of the original building. The dating of this residue will also provide a date for chaff-tempered fabric, believed to be of seventh-century date. The date can also be compared to other samples from the fill of the same *Grubenhaus*.

The interval between the samples will provide an estimate for the duration of the infilling It will also provide a date for this part of the settlement.

Calibrated date: 1σ: cal AD 540–600
                2σ: cal AD 430–620

Final comment: see GrA-25563

Laboratory comment: see GrA-25589

OxA–13727 1474 ±29 BP

δ¹³C: -27.8‰

Sample: 12060AA, Structure 9, submitted on 18 March 2004 by J Tipper

Material: carbonised residue (external)

Initial comment: from middle fill [04618] of a *Grubenhaus* F514 (structure 9), located in the central-north part of the site. The sample is from the south-east quadrant of the pit. The *Grubenhaus* cut an earlier Roman ditch, and was cut by pit F517. The *Grubenhaus* and pit were sealed by midden F1.

Objectives: the sherd (now broken into five pieces) is from a primary disposal. There were four sherds (61g) of this vessel from this *Grubenhaus*. It is hoped that the date will provide a *terminus post quem* for the backfilling of this pit, and by implication the use/disuse of the original building. The date can also be compared to other (stratigraphically later) samples from the same *Grubenhaus*. The interval between the samples will provide an estimate for the duration of the infilling of the pit. It will also provide a date for this part of the settlement.

Calibrated date: 1σ: cal AD 560–620
                2σ: cal AD 540–650

Final comment: P Marshall (25 October 2013), the date provides an archaeologically acceptable age for the vessel.

OxA–13728 1579 ±29 BP

δ¹³C: -27.6‰

Sample: 00425AA, Structure 17, submitted on 18 March 2004 by J Tipper

Material: carbonised residue (internal)

Initial comment: the sample is from upper fill [01764] of *Grubenhaus* F212, located within the central-northern part of the site. It was located in spit 1, square D, of the north-west quadrant. The upper fill of the *Grubenhaus* and the surface deposit were indistinguishable. The *Grubenhaus* F212 cut two Roman ditches, but was not itself cut by any later features.

Objectives: the sherd is decorated with stamp impressions (Briscoe stamp motif A 5bii) and incised lines, which may indicate a sixth-century date. It is hoped that dating of the residue will provide a date for the use of this decoration type. This can then be compared with other samples from this context. It will also provide a date for this part of the settlement.

Calibrated date: 1σ: cal AD 420–540
                2σ: cal AD 400–560

Final comment: P Marshall (25 October 213), the date provides an archaeologically acceptable age for the vessel.

Laboratory comment: see GrA-25935

OxA–13752 1502 ±27 BP

δ¹³C: -27.0‰

Sample: 13187AB, Structure 24, submitted on 18 March 2004 by J Tipper

Material: carbonised residue (internal)

Initial comment: from lower fill [05116] of a *Grubenhaus* F626 (Structure 24), located in the southern part of the site. The sample is from the north-west quadrant of the pit. The *Grubenhaus* cut Roman ditch F620, and was not cut by any later features.

Objectives: the sherd is from a primary disposal, and there were five sherds of this vessel (111g). It is hoped that the date will establish a *terminus post quem* for the backfilling of this pit, and by implication the use/disuse of the original building. This can be compared to other dates on primary disposals from the fills of the *Grubenhaus*. The interval between the earliest and latest samples will provide an estimate for the duration of the infilling. It will also provide a date for this part of the settlement.

Calibrated date: 1σ: cal AD 540–600
                2σ: cal AD 430–620

Final comment: P Marshall (25 October 2013), the date provides an archaeologically acceptable age for the vessel.

Laboratory comment: see GrA-25927

OxA–13753 1501 ±28 BP

δ¹³C: -26.3‰

Sample: 08302AA, Structure 18, submitted on 18 March 2004 by J Tipper

Material: carbonised residue (external)

Initial comment: this sample is from the lowest fill [01670] of *Grubenhaus* F178, structure 18, located in the central part of the site. The sample is from square A of the north-west quadrant, and recovered from environmental sample <202>. The *Grubenhaus* cut through Roman ditch F421, and was truncated by post-medieval ditch F202.

Objectives: the sherd is from a primary disposal, and there were 12 sherds of this vessel (214g). It is hoped that the date will establish a *terminus post quem* for the backfilling of this pit, and by implication the use/disuse of the original building. The vessel is chaff-tempered, which may indicate a seventh-century date. Dating of this sherd may provide a date for the use of this fabric type. It will also provide a date for this part of the settlement.

Calibrated date: 1σ: cal AD 540–600
                2σ: cal AD 430–630

Final comment: see GrA-25563

OxA–13754 1530 ±26 BP

$\delta^{13}C$: -27.0‰

Sample: 10481AA, submitted on 18 March 2004 by J Tipper

Material: carbonised residue

Initial comment: this sample is from context [04038], lower spit B in a one metre test square through midden deposit F1, which seals *Grubenhaus* F514. A replicate of GrA-25939 from a sherd of the same vessel.

Objectives: the sherd is from a primary disposal, and there were 21 sherds of this vessel (238g). The vessel is tempered with biotite and quartz, which may indicate a sixth-century date. Dating of this sherd may provide a date for the use of this fabric type. It is hoped that the dates from several sherds from the midden deposit will help establish a closer chronology for the accumulation of the deposit, and the infilling of *Grubenhaus* F514. It will also provide a date for this part of the settlement.

Calibrated date: *1σ:* cal AD 470–570
                                 2σ: cal AD 420–600

Final comment: P Marshall (25 October 2013), the date provides an archaeologically acceptable age for the vessel.

Laboratory comment: see GrA-25929

OxA–13755 1449 ±28 BP

$\delta^{13}C$: -29.0‰

Sample: 00480AA, Structure 35, submitted on 18 March 2004 by J Tipper

Material: carbonised residue (internal)

Initial comment: from fill [03268] of a *Grubenhaus* F333 (structure 35), located in the north-east part of the site. [003268] is the upper fill of the *Grubenhaus* pit. The sample is from the south-east quadrant of the pit. The *Grubenhaus* cut a Roman (F204}, but was not itself cut by any later features. A replicate of GrA-25563.

Objectives: the sherd (now broken into four) is a chaff-tempered fabric, which may indicate a seventh-century date. This will provide a date for the use of this fabric type. It will also provide a date for this part of the settlement.

Calibrated date: *1σ:* cal AD 590–650
                                 2σ: cal AD 550–660

Final comment: see GrA-25563

Laboratory comment: see GrA-25563

OxA–13756 1490 ±26 BP

$\delta^{13}C$: -20.1‰

Sample: 13416, Pit Group C, submitted on 18 March 2004 by J Tipper

Material: animal bone: *Sus* sp., articulated neonate pig; scapulae, femur, and humerus (J Tipper 2004)

Initial comment: the sample is from lower fill [04719] of pit F500b, located in the central-north part of the site. It was recovered from environmental sample <695>. The pit was truncated by later pit F521, although its relationship to pit F500a was not established, this had been destroyed by pit F521. All the pits were sealed by midden F1.

Objectives: it is hoped that this sample will provide a *terminus post quem* for the backfilling of the pit. This can also be compared with animal bone sample 13429, and pottery residue sample 12060AA from the fill of pit F521. It will also provide a date for this part of the settlement.

Calibrated date: *1σ:* cal AD 550–610
                                 2σ: cal AD 530–640

Final comment: P Marshall (25 October 2013), the date has provided an archaeologically acceptable *terminus post quem* for the backfilling of the pit.

OxA–13757 1510 ±26 BP

$\delta^{13}C$: -21.1‰

Sample: 13429, Pit Group C, submitted on 18 March 2004 by J Tipper

Material: animal bone: *Ovis* sp., sheep fibula (A Grieve 2004)

Initial comment: a replicate of GrA-26357.

Objectives: as GrA-26357

Calibrated date: *1σ:* cal AD 540–600
                                 2σ: cal AD 430–620

Final comment: see GrA-26357

Laboratory comment: see GrA-26357

OxA–13882 1559 ±29 BP

$\delta^{13}C$: -26.4‰

Sample: 12045AC, Structure 9, submitted on 18 March 2004 by J Tipper

Material: carbonised residue (external)

Initial comment: from upper fill [04616] of a *Grubenhaus* F514 (structure 9), located in the central-north part of the site. The sample is from the north-east quadrant of the pit. The *Grubenhaus* cut an earlier Roman ditch, and was cut by pit F517. The *Grubenhaus* and pit were sealed by midden F1. A replicate of GrA-25929 on a sherd from the same vessel.

Objectives: the sherd (now broken into five pieces) is from a primary disposal. There were five sherds (126g) of this vessel from this *Grubenhaus*. The sherd is tempered with biotite and quartz, which may indicate a sixth-century date, and it is hoped dating will provide a date for the use of this type of fabric. It is also hoped that the date will provide a *terminus post quem* for the backfilling of this pit, and by implication the use/disuse of the original building. The date can also be compared to other samples from the same *Grubenhaus*. The interval between the samples will provide an estimate for the duration of the infilling of the pit. It will also provide a date for this part of the settlement.

Calibrated date: *1σ:* cal AD 420–550
                                 2σ: cal AD 410–580

Final comment: P Marshall (25 October 2013), the date provides an archaeologically acceptable age for the date of the vessel.

Laboratory comment: see GrA-25929

OxA–13883 1559 ±26 BP

$\delta^{13}C$: -28.5‰

Sample: 00457AC, Structure 38, submitted on 18 March 2004 by J Tipper

Material: carbonised residue (internal)

Initial comment: as OxA-13711

Objectives: the sherd is tempered with biotite and quartz; decorated with stamp impressions (Briscoe stamp motifs A 5aviii and A 6ai) and horizontal and diagonal incised lines forming a probable chevron pattern, all of which may indicate a sixth-century date. It is hoped that dating of the residue will provide a date for the use of this fabric type and decoration. This can then be compared with sample 00457AT, also from this context. It will also provide a date for this part of the settlement.

Calibrated date: 1σ: cal AD 420–550
2σ: cal AD 420–570

Final comment: P Marshall (25 October 2013), the date provides an archaeologically acceptable age for the date of the vessel.

Laboratory comment: see OxA-13711

OxA–13892 1842 ±26 BP

$\delta^{13}C$: -21.4‰

Sample: 13251, Structure 44, submitted on 18 March 2004 by J Tipper

Material: animal bone: *Ovis* sp., neonate humerus and femur (1.40g) (J Tipper 2004)

Initial comment: this sample is from the fill of a posthole/pit, located in the central part of the site. It contained a large quantity of cremated sheep/goat bones, and a partial neonatal sheep skeleton; it was uncertain whether there were one or more of the cremated individuals. The feature was cut by posthole F243, part of a building earlier than the cemetery.

Objectives: it is hoped that this sample will provide a *terminus ante quem* for the building. This will help to establish the chronological time-span of this part of the site. This sample can be compared with 01894AA - pottery residue from posthole [01950] nearby, and also to sample 07250 - a sheep burial, both relating to the same building.

Calibrated date: 1σ: cal AD 120–230
2σ: cal AD 80–250

Final comment: P Marshall (25 October 2013), the Roman date for the sheep was wholly unexpected, given the sample came from an animal disposal in the fill of an Anglo-Saxon post-holed building. The building therefore seems to have cut through earlier Roman deposits.

Laboratory comment: see GrA-26355

OxA–13966 1425 ±27 BP

$\delta^{13}C$: -28.3‰

Sample: 11983AA, submitted on 18 March 2004 by J Tipper

Material: carbonised residue (external)

Initial comment: from upper fill [04603] of *Grubenhaus* F514, located in the central-north part of the site. The fill [04603A] was possibly part of midden F1 which sealed the *Grubenhaus* (or was a mixed interface between the midden and the *Grubenhaus*). Fill [04603B] was the upper fill of the *Grubenhaus*. Sherd 11983AA was from Spit B, from the north-west quadrant of the pit. The *Grubenhaus* truncated a Roman ditch, and was cut by pit F517. A replicate of GrA-25592 on a sherd from the same vessel.

Objectives: the sherd is from a primary disposal. There were 13 sherds of this vessel (260g) from the fill of *Grubenhaus* F514. The dating of the external residue on this sherd will establish a *terminus post quem* for the backfilling of the pit, and by implication, a *terminus post quem* for the use of the original building. This date can also be compared to other dates on sherds from the same *Grubenhaus*. The intervals from between the samples will provide an estimate for the duration of the infilling of the pit. It will also provide a date for this part of the settlement. In addition, the vessel is decorated with comb marks, and this decoration may indicate a fifth- or sixth-century date. The dating of this residue will provide a date for the use of this type of decoration.

Calibrated date: 1σ: cal AD 600–660
2σ: cal AD 580–660

Final comment: P Marshall (25 October 2013), the date provides an archaeologically acceptable age for the date of the vessel.

Laboratory comment: see GrA-25592

OxA–13967 1510 ±26 BP

$\delta^{13}C$: -28.2‰

Sample: 01692AB, Pit F10, submitted on 18 March 2004 by J Tipper

Material: carbonised residue (internal)

Initial comment: this sample is from fill [01215] of Pit F10, located in the north-west part of the site. Pit F10 contained three fills, although it is not certain from which fill this sherd derived. The pit cuts ditch F48, and gully F59; it is not cut by any later features.

Objectives: the sherd is from a primary disposal. There were five sherds of this vessel (122g) from Pit F10. The dating of the internal residue on this sherd will establish a *terminus post quem* for the backfilling of the pit. In addition, the vessel is chaff-tempered, which may indicate a seventh-century date. The dating of this residue will provide a date for the use of this type of fabric.

Calibrated date: 1σ: cal AD 540–600
2σ: cal AD 430–620

Final comment: see GrA-25563

OxA–14004 1166 ±38 BP

$\delta^{13}C$: -28.5‰

Sample: 00182AA, Structure 12, submitted on 18 March 2004 by J Tipper

Material: carbonised residue (external)

Initial comment: this sample is from upper fill [00204] of *Grubenhaus* F5 (structure 12), located within the central part of the site. The *Grubenhaus* cut two earlier Roman ditches, but was not itself cut by any later features.

Objectives: the sherd (now broken into five pieces) is from a primary disposal. There were three joining sherds of this vessel (59g) from the same fill. The dating of the residue on this sherd will establish a *terminus post quem* for the backfilling of the pit, and by implication the use/disuse of the original building. This can be compared to sample 00211AA from the same *Grubenhaus*. The interval between the two samples will provide an estimate for the duration of the infilling, and the earliest of the two will provide a *terminus ante quem* for the construction of the building, and a date for this part of the settlement. In addition, the vessel is chaff-tempered, which may indicate a seventh-century date. The dating of this residue will provide a date for the use of this type of fabric.

Calibrated date: *1σ:* cal AD 770–950
                    2σ: cal AD 720–980

Final comment: P Marshall (25 October 2013), the date is at odds with the currency of organic-tempered vessels from the site and suggests it does not provide an accurate date for the vessel.

Laboratory comment: English Heritage (25 October 2006), OxA-14004 has produced a late Saxon date from a residue on what appears to be a mid-Saxon sherd. In this case, the contamination may be due to absorption of humic or fluvic acids or lipids from the surrounding soil (Hedges *et al* 1992a).

References: Hedges *et al* 1992a

OxA–14005 1515 ±40 BP

$\delta^{13}C$: -26.5‰

Sample: 01894AA, Structure 44, submitted on 18 March 2004 by J Tipper

Material: carbonised residue (internal)

Initial comment: this sample is from the fill of a posthole [01950], Structure 44, located in the central part of the site. It was one of a line of postholes that appeared to form the eastern wall of a building, which was stratigraphically earlier than the cemetery.

Objectives: it is hoped that this sample will provide a *terminus post quem* for the building. This will help to establish the chronological time-span of this part of the site. This can be compared with sample 07250 - a sheep burial from posthole [01924], and to sample 13251, a cremated sheep burial cut by posthole [01903], part of the western wall of the same building.

Calibrated date: *1σ:* cal AD 470–600
                    2σ: cal AD 420–640

Final comment: P Marshall (25 October 2013), the date provides an archaeologically acceptable age for the date of the vessel.

OxA–14006 2015 ±60 BP

$\delta^{13}C$: -28.3‰

Sample: 01860AB, Pit group G, submitted on 18 March 2004 by J Tipper

Material: carbonised residue (internal)

Initial comment: this sample is from the upper fill [01831] of pit F232, Pit Group G, located in the central part of the site. The pit cut another pit F235, and was itself cut by grave F440. The two pits and grave were truncated by post-medieval ditch F156.

Objectives: it is hoped that this sample will provide a *terminus ante quem* for the grave. This will help to establish the chronological time-span of this part of the site.

Calibrated date: *1σ:* 90 cal BC–cal AD 60
                    2σ: 180 cal BC–cal AD 130

Final comment: P Marshall (25 October 2013), the calibrated radiocarbon date, Roman, is clearly at odds with the Anglo-Saxon character of the sherd from which the sample was taken.

Laboratory comment: see GrA-25950

OxA–14007 1614 ±33 BP

$\delta^{13}C$: -26.5‰

Sample: 11471AC, submitted on 18 March 2004 by J Tipper

Material: carbonised residue (internal)

Initial comment: this sample from a one metre test square through surface deposit F503/F11, within the remains of a Roman trackway in the north-central part of the site. It was not cut by any later features.

Objectives: the sherd is decorated with stamp impressions (Briscoe stamp motif A 4ai) and an incised line, which may indicate a sixth-century date. It is hoped that a date on this sherd will provide a date for this type of decoration. It will also provide a date for this part of the settlement.

Calibrated date: *1σ:* cal AD 400–540
                    2σ: cal AD 380–550

Final comment: P Marshall (25 October 2013), the date provides an archaeologically acceptable age for the vessel.

OxA–14008 1615 ±65 BP

$\delta^{13}C$: -27.9‰

Sample: 02131AH, Structure 30, submitted on 18 March 2004 by J Tipper

Material: carbonised residue (internal)

Initial comment: from the upper fill [033111] of *Grubenhaus* F341 (structure 30), located in the central-east part of the site. The sample from the north-west quadrant.

The *Grubenhaus* cut the edge of surface deposit F342; it was not cut by any later features, although a large animal burrow was present.

Objectives: the sherd is from a primary disposal, with a total of eight sherds (103g) recovered from the fill of this *Grubenhaus*. It is hoped that the date will provide a *terminus post quem* for the backfilling of this pit, and by implication the use/disuse of the original building. This date will also be compared to other samples from the same *Grubenhaus*. The interval between the primary disposals will provide an estimate for the duration of the infilling of the pit. It will also provide a date for this part of the settlement. In addition, the sherds are tempered with grog and quartz, which may indicate a sixth-century date. It is hoped that a date on this sample will provide a date for the use of this fabric type.

Calibrated date: 1σ: cal AD 380–540
2σ: cal AD 250–590

Final comment: P Marshall (25 October 2013), the radiocarbon date is too early compared with the currency of biotite/granite fabric vessels from the site.

Laboratory comment: English Heritage (25 October 2006), two samples comprised 2130AC (OxA-13707) from primary disposal V77 (calcitic-tempered) and sample 2131AH (OxA-14008) from a grog-tempered sherd. The two measurements are not statistically consistent (T'=9.9; ν=1; T'(5%)=3.8; Ward and Wilson 1978). OxA-14008 could be a statistical outlier, although, given the extremely small sample size (reflected in the error on the measurement), it could also be inaccurate.

References: Ward and Wilson 1978

OxA–14016 1489 ±24 BP

δ¹³C: -27.1‰

Sample: 02267AA, Structure 26, submitted on 18 March 2004 by J Tipper

Material: carbonised residue (internal)

Initial comment: from the fill [03643] of possible southern gable posthole of *Grubenhaus* F379 (structure 26), located in the central part of the site. The sample was located on the surface of the infilled posthole, so may in fact belong to the lowest fill of the *Grubenhaus* [03290], or to a possible hearth/oven base also within the hollow of the *Grubenhaus*. The *Grubenhaus* did not cut, and was not cut by, any other features.

Objectives: the sherd is from a primary disposal, with a total of five sherds (82g) recovered from the fill of this posthole. It is hoped that the date will provide a *terminus post quem* for the backfilling of this pit, and by implication the use/disuse of the original building. This date will also be compared to other samples from the same *Grubenhaus*. It will also provide a date for this part of the settlement. In addition, the sherds are chaff-tempered which may indicate a seventh-century date. It is hoped that a date on this sample will provide a date for the use of this fabric type.

Calibrated date: 1σ: cal AD 555–605
2σ: cal AD 540–640

Final comment: see GrA-25563

OxA–14017 1697 ±26 BP

δ¹³C: -28.5‰

Sample: 12147, Pit Group C, submitted on 18 March 2004 by J Tipper

Material: carbonised residue (external)

Initial comment: this sample is from the fill [04642] of pit F521, located in the central-north part of the site. The fill of this pit was excavated as a single context; however, three fills were distinguished in section (A, B, and C). The pit cut two earlier pits F500a and F500b; all three pits were sealed by midden F1. The sherd is from the south-east quadrant.

Objectives: the sherd (now broken in four) is from a probable primary disposal; there were seven sherds of this vessel (210g). It is hoped that the dating of the residue will establish a *terminus post quem* for the backfilling of the pit. This can be compared with animal bone from the same fill, and also to the animal bone sample from pit complex F500. The interval between these samples will provide an estimate for the duration of the infilling of the pit, and also a date for this part of the settlement. In addition, the vessel is chaff-tempered which may indicate a seventh-century date, and a date on this residue will provide a date for the use of this fabric type.

Calibrated date: 1σ: cal AD 260–400
2σ: cal AD 250–410

Final comment: P Marshall (25 October 2013), the date is not agreement with the currency of organic-tempered ceramics from the site and the date does not provide an accurate estimate for the age of the vessel, that is Anglo-Saxon, and not Roman as the radiocarbon date suggests.

Laboratory comment: see GrA-26357 and GrA-25950

OxA–14018 1635 ±40 BP

δ¹³C: -27.5‰

Sample: 13187AF, Structure 24, submitted on 18 March 2004 by J Tipper

Material: carbonised residue (internal)

Initial comment: from lower fill [05116] of a *Grubenhaus* F626 (Structure 24), located in the southern part of the site. The sample is from the north-west quadrant of the pit. The *Grubenhaus* cut Roman ditch F620, and was not cut by any later features.

Objectives: the sherd is tempered with biotite and quartz, which may indicate a sixth-century date. It is hoped that a date on this residue will provide a date for the use of this fabric type. This sample can also be compared to other samples from this *Grubenhaus*. It will also provide a date for this part of the settlement.

Calibrated date: 1σ: cal AD 380–430
2σ: cal AD 330–540

Final comment: see OxA-14017

OxA–14019 1559 ±24 BP

δ¹³C: -27.4‰

Sample: 02234AB, Pit group I, submitted on 18 March 2004 by J Tipper

Material: carbonised residue (internal)

Initial comment: a replicate of GrA-25590.

Objectives: as GrA-25590

Calibrated date: *1σ:* cal AD 425–545
                     2σ: cal AD 420–565

Final comment: see GrA-25590

Laboratory comment: see GrA-25590

OxA–14044 1851 ±28 BP

δ¹³C: -21.7‰

Sample: 07250, Structure 44, submitted on 18 March 2004 by J Tipper

Material: animal bone: *Ovis* sp., (J Tipper 2004)

Initial comment: this sample is from posthole [01924] (F246), located in the central part of the site. A replicate of GrA-26355.

Objectives: this sample is from the fill of a posthole which is part of a post-built building. The building is stratigraphically earlier than the cemetery, and it is hoped that this sample will provide a *terminus post quem* for the building, and help to establish the chronological time-span on this part of the site. This sample can be compared with those from other postholes from the same building.

Calibrated date: *1σ:* cal AD 120–230
                     2σ: cal AD 70–240

Final comment: see GrA-26355

Laboratory comment: see GrA-26355

OxA–14232 1477 ±27 BP

δ¹³C: -28.6‰

Sample: 02100AH, Pit group L, submitted on 18 March 2004 by J Tipper

Material: carbonised residue (internal)

Initial comment: the sample is from fill [03087] of pit group L (F321), located in the central-east part of the site. The fill sealed three or four intercutting pits, and merged imperceptibly with their upper fills. The fill could be the remains of a later surface deposit formed in the hollow created by the earlier pits. The pits were not cut by any later features.

Objectives: the sherd is decorated with a small hollow vertical boss. It is hoped that the dating of the residue will help refine the date for this type of decoration, which is of likely sixth-century date. It will also provide a date for this part of the settlement.

Calibrated date: *1σ:* cal AD 560–620
                     2σ: cal AD 540–650

Final comment: P Marshall (25 October 2013), the date provides an archaeologically acceptable age for the vessel.

OxA–14244 1598 ±30 BP

δ¹³C: -27.7‰

Sample: 02071AB, Structure 38, submitted on 18 March 2004 by J Tipper

Material: carbonised residue (internal)

Initial comment: the sherd is from the middle fill [03002] of *Grubenhaus* F286 (Structure 38), located on the eastern edge of the site. The *Grubenhaus* does not cut, and is not cut by, any other features.

Objectives: the sherd is decorated with stamp impressions (Briscoe stamp motif N 1ai), which may indicate a sixth-century date. It is hoped that dating of the residue will provide a date for this decoration. This can then be compared with other samples from this structure. It will also provide a date for this part of the settlement.

Calibrated date: *1σ:* cal AD 410–540
                     2σ: cal AD 390–550

Final comment: P Marshall (25 October 2013), the date provides an archaeologically acceptable age for the vessel.

UB–6185 1779 ±20 BP

δ¹³C: -21.7 ±0.2‰

Sample: 07250, Structure 44, submitted on 18 March 2004 by J Tipper

Material: animal bone: *Ovis* sp., articulated sheep (250g) (J Tipper 2004)

Initial comment: a replicate of GrA-26355.

Objectives: as GrA-26455

Calibrated date: *1σ:* cal AD 230–320
                     2σ: cal AD 180–330

Final comment: see GrA-26355

Laboratory comment: see GrA-26355

Causewayed Enclosures: Beech Court Farm, Vale of Glamorgan

Location: SS 904766
                 Lat. 51.28.37 N; Long. 03.34.41 W

Project manager: R Lewis (Glamorgan-Gwent
                 Archaeological Trust), 1999–2002

Archival body: National Museums and Galleries of Wales

Description: an enclosure identified from air photographs. Excavation of approximately on third of the total area was undertaken in advance of quarrying. It revealed a single causewayed ditch of ovoid plan divided into five segments. There were almost no finds from the enclosure ditch. Those from internal features and superficial contexts were predominantly late Neolithic/early Bronze Age.

Objectives: two samples were submitted to determine whether the enclosure ditch was Neolithic. Three other samples were submitted to date interior features.

Final comment: F Healy (2012), the dating of the ditch makes an addition to the repertoire of local Iron Age enclosures and prompts caution in the interpretation of air photographs of enclosures with discontinuous ditches. First millennium cal BC dates for the enclosure ditch and for one of two postholes associated with an entrance show that, despite its causewayed plan, the enclosure was of Iron Age date. The remaining two samples from internal features reflect activity on the site in the second millennium cal BC (Whittle *et al* 2011, chapter 11).

References:	Glamorgan-Gwent Archaeol Trust 2000b
	Glamorgan-Gwent Archaeol Trust 2001
	Graves-Brown 1998
	Pearson 2003
	Whittle *et al* 2011
	Yates 2000a
	Yates 2002

GrA–27318 2230 ±40 BP

$\delta^{13}C$: -25.5‰

Sample: BCF sample 2016/A, submitted on 27 August 2004 by F Healy

Material: charcoal: *Prunus* sp., roundwood; single fragment (R Gale 2004)

Initial comment: ditch terminal 2072, lower-middle part of context 2085. In apparent dump of charcoal-rich material made in butt of segment after some silt had accumulated over the backfill covering BCF find 1578. *Prunus* is short-lived, as are the other species represented in the deposit, which seems to derive from a single episode of burning. *Prunus* roundwood is unlikely to be more than a few years old.

Objectives: to date the deposit and hence to help define the date of the enclosure.

Calibrated date:	*1σ:* 380–200 cal BC
	2σ: 400–190 cal BC

Final comment: F Healy (2012), this sample, from a context above the ditch floor, is so much older than that dated by OxA-14142 from a context on the ditch floor, that the dated fragment must have been reworked from an earlier context.

OxA–14142 2099 ±26 BP

$\delta^{13}C$: -25.0‰

Sample: BCF sample 2019, submitted on 27 August 2004 by F Healy

Material: charcoal: *Prunus* sp., roundwood; single fragment (R Gale 2004)

Initial comment: ditch terminal 2072, context 2098. One of several charcoal fragments scattered in a layer (probably backfill) covering the base of the ditch on which BCF find 1578 lay in the north-west terminal of a ditch segment, beside an exceptionally large causeway, probably an entrance. *Prunus* is short-lived and the roundwood of the sample is likely to have been only a few years old. The sample provides a *terminus post quem* for the deposition of apparent backfill on the ditch bottom, and comparison of its date and that of BCF find 1578 will provide an indication of whether one or the other was older than its context.

Objectives: to provide a *terminus post quem* for the layer and an indication of whether the cattle teeth beneath it are older than their context.

Calibrated date:	*1σ:* 180–50 cal BC
	2σ: 200–40 cal BC

Final comment: F Healy (2012), BCF find 1578 (a cattle molar) failed to date. The date of this fragment should be close in age to the digging and initial infilling of the ditch.

Causewayed Enclosures: Bury Hill, West Sussex

Location:	TQ 00231203
	Lat. 50.54.13 N; Long. 00.34.24 W
Project manager:	O Bedwin (Sussex Archaeological Field Unit), 1979
Archival body:	Arundel Castle

Description: an enclosure defined by single continuous pit-dug ditch with western entrance, containing early Neolithic artefacts.

Objectives: two samples were submitted during the project to refine the two existing fourth millennium cal BC measurements on an antler pick and an animal bone fragment from the ditch floor (HAR-3595, 4570±80 BP and HAR-3596, 4680 ±80 BP; Jordan *et al* 1994, 81–2).

Final comment: F Healy (2012), OxA-14175 and GrA-27320, measured on an articulating and a fitting bone sample from the primary silt, are statistically consistent (T

References:	Bedwin 1981a
	Jordan *et al* 1994
	Whittle *et al* 2011

GrA–27320 4890 ±45 BP

$\delta^{13}C$: -22.1‰

Sample: BH79/46/A, submitted on 2 June 2004 by F Healy

Material: animal bone: *Bos* sp., proximal end of right metatarsal with articulating tarsal (A Powell 2004)

Initial comment: area J, layer 46. This was the primary silt in a cutting roughly opposite the entrance. The recovery of the two articulating bones from the same layer in the confined space of a cutting *c* 2m wide by 4m long indicates that they were still in articulation when buried, so that the animal from which that came was recently dead and close in age to its context. From same context as sample for OxA-14175.

Objectives: to refine the chronology of the construction of the monument.

Calibrated date:	*1σ:* 3710–3640 cal BC
	2σ: 3770–3630 cal BC

Final comment: F Healy (2012), GrA-27320 and OxA-14175 together indicate a construction date for the site.

OxA–14175 4933 ±32 BP

δ¹³C: -20.0‰

Sample: BH79/46/B, submitted on 2 June 2004 by F Healy

Material: animal bone: *Sus* sp., distal end of right femur with unfused epiphysis (A Powell 2004)

Initial comment: area J, layer 46. This was the primary silt in a cutting roughly opposite the entrance. The recovery of the shaft end and its unfused epiphysis from the same layer in the confined space of a cutting *c* 2m wide ∞ 4m long indicates that the two were still conjoined when buried, so that the animal from which that came was recently dead and close in age to its context. From same context as sample for GrA-27320.

Objectives: to refine the chronology of the construction of the monument.

Calibrated date: 1σ: 3750–3650 cal BC
                        2σ: 3790–3640 cal BC

Final comment: see GrA-27320

Causewayed Enclosures: Chalk Hill, Kent

Location: TR 3635065350
                      Lat. 51.19.00 N; Long. 00.01.23 E

Project manager: G Shand, P Clark, and J Weekes
                      (Canterbury Archaeological Trust),
                      1997–8

Description: a causewayed enclosure *c* 150m across with three circuits. The inner ditches are fairly insubstantial; the outer ditch wider, deeper and richer in finds, formed from series of pits, extensively recut. Two parallel, fairly straight discontinuous ditches cutting middle and outer circuits, also Neolithic, cut in turn by possible cursus. The chalk into which the enclosure was cut is overlain by Brickearths. The ditch fills are thus less chalky than those of some downland enclosures and the variable preservation of the bone samples almost certainly reflects the proportion of chalk in the fills from which they came.

Objectives: samples were submitted during the project to establish the sequence of construction and absolute chronology of the three circuits; to establish the duration of use of the enclosure, especially for the repeated reworkings of the outer ditch; and to relate the site to other fourth millennium activity in the region.

Final comment: F Healy (2012), the aims were frustrated by a dearth of suitable samples from contexts other than the outer ditch. Although the evidence is heavily biassed by this one circuit, it was possible to estimate a start date for the site of *3780–3680 cal BC (95% probability*; Whittle *et al* 2011, fig 7.21: *start Chalk Hill*),) and a primary use of *45–165 years (95% probability*; Whittle *et al* 2011, fig 7.22: *use Chalk Hill*). Further samples from the site were subsequently submitted for dating by the Canterbury Archaeological Trust, leading to revised estimates which are very similar to these (Bayliss *et al* 2019). The parallel discontinuous ditches and the possible cursus have been assigned to later periods in the course of post-excavation analysis (J Weekes pers.

comm.). The numbering of ditch segments has changed since the samples were submitted.

References: Bayliss *et al* 2008b
                       Bayliss *et al* 2019
                       Clark *et al* 2019
                       Dyson *et al* 2000
                       Oswald *et al* 2001
                       Shand 1998
                       Shand 2001
                       Whittle *et al* 2011

Causewayed Enclosures: Chalk Hill, outer ditch, Kent

Location: TR 3635065350
                      Lat. 51.19.00 N; Long. 00.01.23 E

Project manager: G Shand, P Clark, and J Weekes
                      (Canterbury Archaeological Trust),
                      1997–8

Archival body: Canterbury Archaeological Trust

Description: a circuit up to 170m in diameter, deeper and wider than the inner and middle ditches. Nine segments were investigated within the excavated area. Complex recuts were found throughout. Finds were particularly abundant in the east.

Objectives: to date the circuit as a step towards dating the complex.

Final comment: F Healy (2012), modelling of sequences of short-life samples through complex series of recuts in two segments has provided an estimated construction date of *3760–3675 cal BC (95% probability*; Whittle *et al* 2011, fig 7.21: *build outer Chalk Hill*). NB segments were re-numbered during post-excavation analysis so that samples originally submitted from segments 7 and 8 are published in Gathering Time as from segments 2 and 3 and will be published in the Chalk Hill monograph as from segments 3 and 5.

References: Whittle *et al* 2011

GrA–30880 4730 ±40 BP

δ¹³C: -22.4‰

Sample: Articulation 36, submitted on 25 November 2005 by F Healy

Material: animal bone: *Ovis* sp., left humerus from among numerous bones from two animals (R Bendrey 2002)

Initial comment: segment 7, F1683 context 1473. Lowest fill of one of phase 3D recuts of segment, stratigraphically later than Articulation 10 and sherd group 98. The presence of most of the bones from two sheep skeletons in a single context indicates that the animals were either still articulated or recently butchered when deposited.

Objectives: to date the context as a step towards dating the enclosure.

Calibrated date: 1σ: 3640–3380 cal BC
                        2σ: 3640–3370 cal BC

Final comment: see GrA-30882

GrA–30882 4885 ±40 BP

δ¹³C: -20.6‰

Sample: Articulation 10/A, submitted on 25 November 2005 by F Healy

Material: animal bone: *Sus* sp., proximal phalanx, of identical size and development stage to another from the same context, probably from the same foot, retaining unfused epiphysis (R Bendrey 2002)

Initial comment: segment 7, F1574, context 1586. Fill of one of three phase 3A pits which were later joined into a single segment. Partly overlying pit base, partly overlying initial silts. It would have been deposited soon after the pit was dug. The presence of the two phalanges and the fitting epiphysis of one in the same context indicates that they were still or recently joined by soft tissue when buried and that the animal from which they came was not long dead.

Objectives: to date the context as a step towards dating the enclosure.

Calibrated date: *1σ:* 3700–3640 cal BC
                       2σ: 3750–3630 cal BC

Final comment: F Healy (2012), the result was in agreement with the stratigraphic position and with the model.

GrA–30884 4885 ±40 BP

δ¹³C: -22.0‰

Sample: Articulation 6, submitted on 25 November 2005 by F Healy

Material: animal bone: *Bos* sp., right humerus articulating with radius and ulna (R Bendrey 2002)

Initial comment: segment 8, F1671 context 1256. Fill of a phase 3D recut of segment. The articulation of the bones indicates that they were still or recently joined by soft tissue when buried.

Objectives: to date the context as a step towards dating the enclosure.

Calibrated date: *1σ:* 3700–3640 cal BC
                       2σ: 3750–3630 cal BC

Final comment: see GrA-30882

GrA–30885 4910 ±40 BP

δ¹³C: -22.4‰

Sample: Articulation 22, submitted on 25 November 2005 by F Healy

Material: animal bone: *Bos* sp., right ulna articulating with radius; radius ?heated (R Bendrey 2002)

Initial comment: segment 8, F44=F1672 context 59. Fill of a phase 3B recut, stratigraphically later than context of articulation 23.

Objectives: to date the context as a step towards dating the enclosure.

Calibrated date: *1σ:* 3710–3640 cal BC
                       2σ: 3780–3630 cal BC

Final comment: see GrA-30882

GrA–30886 4935 ±40 BP

δ¹³C: -22.3‰

Sample: Articulation 20, submitted on 25 November 2005 by F Healy

Material: animal bone: *Bos* sp., right radius articulating with ulna, small patch of burning on ulna (R Bendrey 2002)

Initial comment: segment 8, F44=F1672 context 59. Fill of a phase 3B recut, stratigraphically later than context of articulation 23. The presence in this context of seven sets of articulating cattle bone from at least three animals indicates that the deposit was made in the immediate aftermath of butchery and/or consumption, and that the bone, which is fresh and well-preserved, had not passed through intermediate contexts.

Objectives: to date the context as a step towards dating the enclosure.

Calibrated date: *1σ:* 3770–3650 cal BC
                       2σ: 3800–3640 cal BC

Final comment: see GrA-30882

GrA–30888 4825 ±50 BP

δ¹³C: -30.9‰

Sample: Sherd group 265/A, submitted on 25 November 2005 by F Healy

Material: carbonised residue (one sherd out of >15 from same Plain Bowl, four of them with fresh, well-preserved residue) (A Gibson 2005)

Initial comment: a replicate of OxA-15509 and OxA-17122. Segment 8, F1672=F44, context 1505=72. The fact that so much of the pot was present in one place indicates that it was deposited soon after its last use and breakage.

Objectives: to date the context as a step towards dating the enclosure.

Calibrated date: *1σ:* 3660–3530 cal BC
                       2σ: 3710–3510 cal BC

Final comment: see OxA-17122

OxA–15390 4874 ±33 BP

δ¹³C: -27.1‰

Sample: Sherd group 98, submitted on 25 November 2005 by F Healy

Material: carbonised residue (1 large body sherd among >10 from a single Neolithic Bowl. Looks as if further residue has been scraped from the others) (A Gibson 2005)

Initial comment: group 18. Segment 7, F1358, context 1272. Lowest fill of phase 3B recut of segment. The presence of several sherds from the same pot in one place indicates that they were deposited soon after the vessel's last use and breakage, before the sherds had become dispersed.

Objectives: to date the context as a step towards dating the enclosure.

Calibrated date: *1σ:* 3700–3640 cal BC
                       2σ: 3710–3630 cal BC

Final comment: see GrA-30882

OxA–15447 4750 ±32 BP

δ¹³C: -20.9‰

Sample: Articulation 37, submitted on 25 November 2005 by F Healy

Material: animal bone: *Ovis* sp., left humerus from among numerous bones from two sheep (R Bendrey 2002)

Initial comment: segment 7, F1683 context 1473. Lowest fill of one of phase 3D recuts of segment, stratigraphically later than articulation 10 and sherd group 98. The presence of most of the bones from two sheep skeletons in a single context indicates that the animals were either still articulated or recently butchered when deposited.

Objectives: to date the context as a step towards dating the enclosure.

Calibrated date: *1σ:* 3640–3510 cal BC
                      2σ: 3640–3370 cal BC

Final comment: see GrA-30882

OxA–15448 4952 ±33 BP

δ¹³C: -21.6‰

Sample: Articulation 23, submitted on 25 November 2005 by F Healy

Material: animal bone: *Bos* sp., left astragalus, articulating with tarsal (R Bendrey 2002)

Initial comment: segment 8. F56=F1667 context 55=60=1236=1445. Fill of one of the phase 3A primary pits eventually joined to form segment. The presence of the articulating bones in the same context suggests that they were still or recently held together by soft tissue when buried. This layer, in which there were almost no finds, was separated by *c* 0.40m of chalk rubble fill from the phase 3B recuts. The stratigraphic and probably temporal interval between it and a large amount of fresh, well-preserved cattle bone in context 59, much of it articulating, makes it most unlikely that articulation 23 came from any of the same animals as the samples from that context, since the context 59 bone seems to have been deposited soon after butchery and/or consumption, without passing through intermediate contexts.

Objectives: to date the context as a step towards dating the enclosure.

Calibrated date: *1σ:* 3780–3670 cal BC
                      2σ: 3800–3650 cal BC

Final comment: see GrA-30882

OxA–15449 4949 ±33 BP

δ¹³C: -21.8‰

Sample: Articulation 9, submitted on 25 November 2005 by F Healy

Material: animal bone: *Bos* sp., right radius articulating with ulna (R Bendrey 2002)

Initial comment: segment 8, F1304, context 1259. Upper fill of a phase 3B recut of segment. The articulation of the bones indicates that they were still or recently joined by soft tissue when buried.

Objectives: to date the context as a step towards dating the enclosure.

Calibrated date: *1σ:* 3770–3660 cal BC
                      2σ: 3800–3650 cal BC

Final comment: see GrA-30882

OxA–15509 4867 ±36 BP

δ¹³C: -27.3‰

Sample: Sherd group 265/B, submitted on 25 November 2005 by F Healy

Material: carbonised residue (one sherd out of >15 from same Plain Bowl, four of them with fresh, well-preserved residue) (A Gibson 2005)

Initial comment: a replicate of OxA-17122 and GrA-30888. Segment 8, F1672=F44, context 1505=72. The fact that so much of the pot was present in one place indicates that it was deposited soon after its last use and breakage.

Objectives: to date the context as a step towards dating the enclosure.

Calibrated date: *1σ:* 3660–3630 cal BC
                      2σ: 3710–3540 cal BC

Final comment: see OxA-17122

OxA–15543 4912 ±31 BP

δ¹³C: -21.5‰

Sample: Articulation 39, submitted on 25 November 2005 by F Healy

Material: animal bone: *Bos* sp., one of three fragments of right radius, articulating with ulna (R Bendrey 2002)

Initial comment: segment 8, F1665, context 1489. Lowest fill of phase 3C recut of segment. The articulation of the bones indicates that they were still or recently joined by soft tissue when buried.

Objectives: to date the context as a step towards dating the enclosure.

Calibrated date: *1σ:* 3710–3650 cal BC
                      2σ: 3770–3640 cal BC

Final comment: see GrA-30882

OxA–15544 4911 ±31 BP

δ¹³C: -20.5‰

Sample: Articulation 19, submitted on 25 November 2005 by F Healy

Material: animal bone: *Bos* sp., right radius articulating with ulna (R Bendrey 2002)

Initial comment: segment 8, F44=F1672 context 59. Fill of a phase 3B recut, stratigraphically later than context of articulation 23. The presence in this context of seven sets of articulating cattle bone from at least three animals indicates that the deposit was made in the immediate aftermath of butchery and/or consumption, and that the bone, which is fresh and well-preserved, had not passed through intermediate contexts.

Objectives: to date the context as a step towards dating the enclosure.

Calibrated date: 1σ: 3710–3650 cal BC
                        2σ: 3770–3640 cal BC

Final comment: see GrA-30882

OxA–17122 4839 ±31 BP

δ¹³C: -27.5‰

Sample: Sherd group 265/B, submitted in March 2007 by F Healy

Material: carbonised residue (one sherd out of >15 from same Plain Bowl, four of them with fresh, well-preserved residue) (A Gibson 2005)

Initial comment: a replicate of OxA-15509 and GrA-30888. Segment 8, F1672=F44, context 1505=72. The fact that so much of the pot was present in one place indicates that it was deposited soon after its last use and breakage.

Objectives: to elucidate confusion as to which sample was dated by OxA-15391 and which by OxA-15509.

Calibrated date: 1σ: 3650–3630 cal BC
                        2σ: 3670–3530 cal BC

Final comment: F Healy (2012), the statistical consistency of OxA-17122 with GrA-30888 and OxA-15509 (T'=0.6; T' (5%)=6.0; ν=2; Ward and Wilson 1978) confirms that OxA-15509 was measured on residue from sherd group 265. Confusion arose because the sample was marked in Oxford with the laboratory reference P17781 on the bag but entered in the laboratory database as P17783, so that it was unclear whether OxA-15509 dated sherd group 265 from the outer ditch or sherd group 10 from the inner ditch. The date for sherd group 10 (OxA-15391) is inconsistent with the mean of these three measurements (T'=9.8; T'(5%)=3.8; ν=1; Ward and Wilson 1978).

References: Ward and Wilson 1978

Causewayed Enclosures: Court Hill, West Sussex

Location: SU 89771375
                      Lat. 50.54.55 N; Long. 00.43.24 W

Project manager: O Bedwin (Sussex Archaeological Field
                      Unit), 1982

Archival body: Chichester District Museum

Description: a causewayed enclosure with a single circuit up to 175m in diameter. Three trenches cut across the ditch, one across the crescentic earthwork beyond it. Early Neolithic artefacts were recovered.

Objectives: two samples were submitted during the project to refine the dating, which rested on a single bulk animal bone sample submitted after the 1982 excavations (I-12893; 5420 ±180 BP). Very few potential samples were available; hence the choice of large, well-preserved disarticulated bones rather than articulated ones.

Final comment: F Healy (2012), the enclosure indeed appears to have been built more recently than I-12893 would suggest. The two new measurements are statistically consistent (T'=0.1; T'(5%)=3.8; ν=1; Ward and Wilson 1978) and both are later than the pre-existing measurement on a bulk sample from an overlying layer. Either I-12893 is inaccurate because the pretreatment protocol used in the early 1980s did not remove all contaminants, or the sample for it must have included already old bone. If the recently dated samples were freshly buried, a construction date of *3650-3530 cal BC (95% probability)* can be estimated for the enclosure (Whittle *et al* 2011, fig. 5.28: *build Court Hill*).

References: Bedwin 1984
                  Holden 1951
                  Oswald *et al* 2001
                  Ward and Wilson 1978
                  Whittle *et al* 2011

GrA–27321 4790 ±45 BP

δ¹³C: -22.3‰

Sample: Chichester District Museum A20095/A, submitted on 2 June 2004 by F Healy

Material: animal bone (cattle or deer. Two mandible fragments joining along a recent break) (A Powell 2004)

Initial comment: layer 6. No cutting is named on the bag. 6 was the lowest layer of the enclosure ditch in trenches C and D and the second lowest in trench A (Bedwin 1984, fig 3). There was also a layer 6 in trench B, a section across a different earthwork, but this is described as yielding 'only nine flint flakes', while the bone bagged from this unlocated layer 6 includes 3 animal teeth and 8 small bone fragments. The sample provides a *terminus post quem* for the construction of the monument.

Objectives: to check the reliability of a surprisingly early date of 5420 ±180 BP (I-12893) already obtained on a bulk sample of animal bone from layer 5 of the enclosure ditch in trench D.

Calibrated date: 1σ: 3640–3520 cal BC
                        2σ: 3660–3380 cal BC

Final comment: F Healy (2012), this measurement, from the lowest or possibly the second lowest layer in the enclosure ditch is more recent than I-12893, which was made on a bulk bone sample from layer 5 in cutting D (the second lowest layer). It therefore appears that, if I-11283 was accurate, one or more of the fragments which made up the sample for it was older than its context.

OxA–14176 4776 ±33 BP

δ¹³C: -21.6‰

Sample: Chichester District Museum A20095/B, submitted on 2 June 2004 by F Healy

Material: animal bone (cattle or deer. Mandible fragment) (A Powell 2004)

Initial comment: trench D, layer 6, 'from ditch floor' (bag). Layer 6 was the lowest layer (Bedwin 1984, fig 3: last section), and stratified below layer 5, the context of the bulk bone sample for I-12893. The sample provides a *terminus post quem* for the construction of the monument.

Objectives: as GrA-27321

Calibrated date: 1σ: 3640–3520 cal BC
                             2σ: 3650–3380 cal BC

Final comment: see GrA-27321

Causewayed Enclosures: Crickley Hill, Gloucestershire

Location: SO 9265016100
                          Lat. 51.30.31 N; Long. 02.06.24 W

Project manager: P Dixon (University of Nottingham), 1969–95

Description: Crickley Hill is a limestone spur projecting from the Cotswold scarp into the Severn valley. Excavation of an Iron Age hillfort there was initiated by Dr Philip Dixon in 1969 and led to the identification in 1971 of an extended Neolithic sequence. The first enclosures on the hill were two roughly concentric causewayed circuits. Postholes, pits, and artefacts were densest on the central knoll of the hill, within the inner enclosure. The two causewayed circuits were succeeded by a single, almost continuous ditch cut approximately on the line of the inner causewayed circuit for parts of its length. One of its entrances was linked by a bifurcating fenced track both to the central knoll of the hill and to a wooden structure, interpreted as a shrine, on a stone-built platform further west. Track, platform, and shrine were sealed by a long earthen mound with a stone circle on the site of the shrine. The successive enclosures are placed in the early-to-mid fourth millennium BC by their associated artefacts and by analogy with similar structures. The chronology of the later stages of the sequence is less clear.

Objectives: samples were submitted during the Dating Causewayed Enclosures project in an attempt to answer many questions, among them: What was the date of the earliest, pre-enclosure, structures on the hill? Were the inner and outer causewayed circuits built at the same time or successively? What was the interval between the two causewayed circuits and the continuous circuit? For how long was each enclosure in use? Was it possible to provide a date for the 'battle of Crickley', marked by concentrations of arrowheads at entrances in the continuous circuit? Could the chronology of the complex sequence in the long mound valley be elucidated? Could the post-built structures be shown by radiocarbon dating to be Neolithic? What was the relationship between Crickley Hill and Peak Camp, approximately 1km to the south-west?

Final comment: F Healy (2012), dates were obtained for only one pre-enclosure structure, the 'banana barrow' underlying the bank of the inner causewayed circuit; for which there are *termini post quos* in the late fifth/early fourth millennium cal BC. The outer causewayed circuit is poorly dated because of a scarcity of suitable samples; the inner has an estimated construction date in the late 37th century cal BC and was succeeded by the continuous circuit in the mid 36th century. The 'battle of Crickley' would have occurred in the 35th

century. Overall estimates for the Neolithic complex are of *3705–3600 cal BC (95% probability*; Whittle *et al* 2011, fig 9.7: *start Crickley Hill*) for its initiation and *3495–3395 cal BC (95% probability*; Whittle *et al* 2011, fig 9.7: *end Crickley Hill*), for its end, with a use-life of *125–285 years (95% probability*, Whittle *et al* 2011, fig 9.15: *use Crickley Hill*). The outer circuit of Peak Camp was probably built after the causewayed circuits on Crickley Hill and before the continuous circuit (Whittle *et al* 2011, fig 9.20). The complex and potentially long-lived long mound valley sequence remains problematic, mainly because of a shortage of suitable samples from its earlier stages. Some samples from beneath the latest, south-eastern, part of the long mound are of first millennium cal BC date (Whittle *et al* 2011, chapter 9).

References: Dixon 1971
                           Dixon 1972a
                           Dixon 1972b
                           Dixon 1979
                           Dixon 1981
                           Dixon 1988a
                           Dixon 1988b
                           Dixon 1994
                           Dixon 2005
                           Dixon and Borne 1977
                           Gale 1986
                           Hanson-James 1993
                           Hollos 1999
                           Oswald *et al* 2001
                           Palmer 1976
                           Savage 1988
                           Snashall 1997
                           Snashall 1998
                           Whittle *et al* 2011

Causewayed Enclosures: Crickley Hill, continuous ditch, Gloucestershire

Location: SO 9265016100
                          Lat. 51.50.33 N; Long. 02.06.24 W

Project manager: P Dixon (University of Nottingham), 1971–95

Archival body: Corinium Museum

Description: a single, almost continuous ditch, with three known entrances, succeeding two causewayed ditches. The ditch was backed by a low stone platform of cellular construction which supported a palisade towards its rear edge. More than 400 leaf-shaped arrowheads clustered in two entrances in the east of this circuit are evidence of a violent encounter.

Objectives: to date the construction and use of the circuit as a step to establishing the overall chronology of the complex.

Final comment: F Healy (2012), Bayesian modelling of the dates has provided estimates of *3580–3525 cal BC (95% probability*; Whittle *et al* 2011, fig 9.10: *build continuous*) for construction and *3535–3485 cal BC (95% probability*; Whittle *et al* 2011, fig 9.10: *modify continuous*) for recutting and modification.

References: Whittle *et al* 2011

GrA–27911 4780 ±40 BP

$\delta^{13}C$: -26.4‰

Sample: CH72 F603(6), submitted on 13 December 2004 by F Healy

Material: charcoal: *Corylus avellana*, single fragment (R Gale 2004)

Initial comment: from same find as samples OxA-14321 and OxA-14428. F603(6). Primary silt of continuous ditch. The quantity of charcoal, and the presence among it of substantial fragments indicate that it was freshly deposited. If short-life samples can be extracted they should be close in age to their context. 'Pink dust wash and silting' in base of limestone-cut ditch on well-drained limestone hill.

Objectives: to establish the date of this context as a step towards defining the chronology.

Calibrated date: 1σ: 3640–3520 cal BC
                      2σ: 3650–3380 cal BC

Final comment: F Healy (2012), the result is in agreement with the stratigraphic position and with the model.

GrA–27914 4660 ±40 BP

$\delta^{13}C$: -24.3‰

Sample: CH77 F2657 'C14 sample' A, submitted on 13 December 2004 by F Healy

Material: charcoal: *Fraxinus excelsior*, sapwood; single fragment (R Gale 2004)

Initial comment: from same charcoal find as sample for OxA-14322. F2657. One of several charcoal finds from palisade trench in surface of bank of continuous ditch. The trench survived as a band of burnt stone. From a belt of small/medium burnt stones in a limestone-built bank on a well-drained limestone hill.

Objectives: to establish the date of this context as a step towards defining the chronology of the complex.

Calibrated date: 1σ: 3520–3360 cal BC
                      2σ: 3630–3350 cal BC

Final comment: see GrA-27911 and OxA-14322

GrA–31105 4675 ±40 BP

$\delta^{13}C$: -24.4‰

Sample: CH77 2657/A, submitted on 19 December 2005 by F Healy

Material: charcoal: *Corylus* sp., single fragment (R Gale 2005)

Initial comment: N4: F2657(2). Palisade trench F2657. One of several charcoal finds from palisade trench in surface of bank of continuous ditch. The trench survived as a band of burnt stone, where the palisade had burnt down. If short life charcoal can be extracted from the sample it will provide a date close in time to the construction of the palisade.

Objectives: to refine the date of the context as a step to refining the chronology of the complex.

Calibrated date: 1σ: 3520–3370 cal BC
                      2σ: 3630–3360 cal BC

Final comment: see GrA-27911 and OxA-14322

GrA–31106 4710 ±40 BP

$\delta^{13}C$: -24.7‰

Sample: CH77 2657/B, submitted on 19 December 2005 by F Healy

Material: charcoal: *Corylus* sp., single fragment (R Gale 2005)

Initial comment: N4: F2657(2). Palisade trench F2657. One of several charcoal finds from palisade trench in surface of bank of continuous ditch. The trench survived as a band of burnt stone, where the palisade had burnt down. If short life charcoal can be extracted from the sample it will provide a date close in time to the construction of the palisade.

Objectives: to refine the date of the context as a step to refining the chronology of the complex.

Calibrated date: 1σ: 3630–3370 cal BC
                      2σ: 3640–3360 cal BC

Final comment: see GrA-27911 and OxA-14322

OxA–14321 4891 ±31 BP

$\delta^{13}C$: -25.9‰

Sample: CH72 F603(6) un-numbered B, submitted on 13 December 2004 by F Healy

Material: charcoal: *Corylus avellana*, single fragment (R Gale 2004)

Initial comment: a replicate of OxA-14428. F603(6). Primary silt of continuous ditch. The quantity of charcoal, and the presence among it of substantial fragments indicate that it was freshly deposited. If short-life samples can be extracted they should be close in age to their context. 'Pink dust wash and silting' in base of limestone-cut ditch on well-drained limestone hill.

Objectives: to establish the date of this context as a step towards defining the chronology of the complex.

Calibrated date: 1σ: 3700–3640 cal BC
                      2σ: 3710–3630 cal BC

Final comment: F Healy (2012), the result is statistically consistent with OxA-14428 (T'=0.2; T'=(5%)=3.8; ν=1; Ward and Wilson 1978). The sample for both, however, was probably redeposited, since they are older than GrA-27911, measured on another fragment from the same charcoal find.

References: Ward and Wilson 1978

OxA–14322 4567 ±33 BP

$\delta^{13}C$: -24.8‰

Sample: CH77 F2657 'C14 sample' B, submitted on 13 December 2004 by F Healy

Material: charcoal: *Fraxinus excelsior*, sapwood; single fragment (R Gale 2004)

Initial comment: from same find as sample for GrA-27914. F2657. One of several charcoal finds from palisade trench in surface of bank of continuous ditch. The trench survived as a band of burnt stone, where the palisade had burnt down. If short life charcoal can be extracted from the sample it will provide a date close in time to the construction of the palisade. From a belt of small/medium burnt stones in a limestone-built bank on a well-drained limestone hill.

Objectives: to establish the date of this context as a step towards defining the chronology of the complex.

Calibrated date: 1σ: 3370–3340 cal BC
2σ: 3490–3110 cal BC

Final comment: F Healy (2012), the latest of four dates for charcoal fragments from the palisade, the others being GrA-27914, -31105, and -31106.

OxA–14416 4890 ±32 BP

δ¹³C: -20.3‰

Sample: CH72 4807, submitted on 13 December 2004 by F Healy

Material: animal bone: *Sus* sp., proximal end of left femur, fitting unfused epiphysis (J Mulville and A Powell 2004)

Initial comment: a replicate of OxA-14417. F602(3). Topmost layer of the inner causewayed ditch, beneath the bank of the continuous ditch. The recovery of the fitting epiphysis and shaft from the same context indicates that they were not long separated when buried, so that this would have occurred soon after the death of the animal from which they came. The sample should be close in age to the construction of the continuous ditch. 'Major tumble of clean stone with white and pink limestone dust'.

Objectives: to establish the date of this context as a step towards defining the chronology of the complex.

Calibrated date: 1σ: 3700–3640 cal BC
2σ: 3710–3630 cal BC

Final comment: F Healy (2012), the result is statistically consistent with OxA-14417 (T′=2.2; T′(5%)=3.8; ν=1; Ward and Wilson 1978). Both, however, are older than dates for samples from lower in the inner ditch fills. The sample may have been moved from the bank of the inner ditch during levelling prior to the construction of the bank of the continuous ditch.

References: Ward and Wilson 1978

OxA–14417 4823 ±32 BP

δ¹³C: -20.4‰

Sample: CH72 4807, submitted on 13 December 2004 by F Healy

Material: animal bone: *Sus* sp., proximal end of left femur, fitting unfused epiphysis (J Mulville and A Powell 2004)

Initial comment: a replicate of OxA-14416.

Objectives: as OxA-14416

Calibrated date: 1σ: 3650–3530 cal BC
2σ: 3660–3520 cal BC

Final comment: see OxA-14416

OxA–14428 4913 ±34 BP

δ¹³C: -25.9‰

Sample: CH72 F603(6) un-numbered B, submitted on 13 December 2004 by F Healy

Material: charcoal: *Corylus avellana*, single fragment (R Gale 2004)

Initial comment: a replicate of OxA-14321.

Objectives: as OxA-14321

Calibrated date: 1σ: 3710–3650 cal BC
2σ: 3770–3640 cal BC

Final comment: see OxA-14321

UB–6394 4619 ±27 BP

δ¹³C: -22.2 ±0.5‰

Sample: CH71 F370, submitted on 13 January 2004 by F Healy

Material: animal bone: *Bos* sp., articulating right tibia, astragalus, and calcaneum from 2–3 year-old animal (J Mulville and A Powell 2004)

Initial comment: F370. From section across tail of bank of continuous ditch. The articulation of the sample shows that the bones were still joined by soft tissues when buried and that the animal from which it came would have been recently dead. It should be close in age to the construction of the bank.

Objectives: to establish the date of this context as a step towards defining the chronology of the complex.

Calibrated date: 1σ: 3500–3360 cal BC
2σ: 3500–3350 cal BC

Final comment: see GrA-27911

UB–6395 4803 ±22 BP

δ¹³C: -22.4 ±0.5‰

Sample: CH74 3396, submitted on 13 December 2004 by F Healy

Material: animal bone: *Bos* sp., articulating left ulna and radius, cut marks near proximal ends of both (J Mulville and A Powell 2004)

Initial comment: F1802, base. Found together at base of ditch but at N side of entrance. The articulation of the bones indicates that the animal from which they came had died shortly before they were buried; their exceptionally large size and good preservation are further arguments against redeposition.

Objectives: to establish the date of this context as a step towards defining the chronology of the complex.

Calibrated date: 1σ: 3640–3535 cal BC
2σ: 3645–3525 cal BC

Final comment: see GrA-27911

UB–6396 4681 ±20 BP

δ¹³C: -22.4 ±0.5‰

Sample: CH77 4517, submitted on 13 December 2004 by F Healy

Material: animal bone: *Bos* sp., complete right tibia (J Mulville and A Powell 2004)

Initial comment: F2688, bedrock, at a depth of 0.88–1.1m. From a recut in F2659 which extended down to a step in the limestone, overlain by layers 2659(3) and 2359(4). Stratified above CH77 4582. The completeness and exceptional preservation of the bone, which shows no sign of gnawing or weathering, suggest that it was fresh when buried and that the animal from which it came was recently dead. It should be close in age to the recut on the base of which it was found.

Objectives: to establish the date of this context as a step towards defining the chronology of the complex.

Calibrated date: 1σ: 3520–3370 cal BC
                     2σ: 3620–3370 cal BC

Final comment: see GrA-27911

UB–6397 4769 ±22 BP

δ¹³C: -22.6 ±0.5‰

Sample: CH77 4582, submitted on 13 December 2004 by F Healy

Material: animal bone: *Bos* sp., central of 4 lumbar vertebrae, 4 thoracic vertebrae, 2–3 indeterminate vertebrae. Also numerous vertebra fragments and some rib fragments. The size of the centra is such that they could all have come from the same vertebral column (J Mulville and A Powell 2004)

Initial comment: F2659, bedrock, at a depth of 1.60m. On the base of the continuous ditch. The fact that at least eight adjacent vertebrae were found together indicates that they were buried when still articulated, although their preservation is such that it is not possible to verify this. They are thus likely to have come from a recently dead animal and to be close in age to the original digging of the ditch.

Objectives: to establish the date of this context as a step towards defining the chronology of the complex.

Calibrated date: 1σ: 3635–3520 cal BC
                     2σ: 3640–3515 cal BC

Final comment: see GrA-27911

Causewayed Enclosures: Crickley Hill, inner causewayed ditch and 'banana barrow', Gloucestershire

Location: SO 9265016100
                   Lat. 51.50.33 N; Long. 02.06.24 W

Project manager: P Dixon (University of Nottingham), 1971–95

Archival body: Corinium Museum

Description: a causewayed ditch surrounding the central knoll of the hill, backed by a low stone platform with a timber palisade. The circuit had a complex history of re-cutting and reworking, with several original entrances blocked and several lengths of the ditch reddened by fire. There were wooden gates in the wider causeways. The alignment of an entrance in this ditch with another in the outer ditch which dated from after the construction of that ditch suggests that the inner ditch was built after the outer. Under the bank of the inner circuit of the causewayed enclosure lay a series of small pits surrounding an area measuring approximately 3m by 8m. This enigmatic feature is known as the 'banana barrow'.

Objectives: to date the construction and use of the circuit as a step to establishing the overall chronology of the complex, and to determine the date of the 'banana barrow'.

Final comment: F Healy (2012), Bayesian modelling of the available dates provides an estimated construction date of *3660–3595 cal BC (95% probability;* Whittle *et al* 2011, fig 9.9: *build inner*), and an estimated construction date for the 'banana barrow' of *4640–3970 cal BC (95% probability; build banana barrow;* Whittle *et al* 2011, fig 9.8).

References: Whittle *et al* 2011

GrA–27814 5270 ±40 BP

δ¹³C: -23.2‰

Sample: CH75 1347, submitted on 13 December 2004 by F Healy

Material: animal bone: unidentifiable, single fragment (J Mulville and A Powell 2004)

Initial comment: F2035. From one of a group of short lengths of ditch surrounding an ovoid area (known as the banana barrow). Beneath 2005/3, the remnant of the bank of the inner causewayed enclosure which inturn underlay the bank of the continuous enclosure. The sample provides a *terminus post quem* for the construction of the 'banana barrow' and for the sequence at the base of which it lies.

Objectives: to establish the date of this context as a step towards defining the chronology of the complex.

Calibrated date: 1σ: 4230–3990 cal BC
                     2σ: 4240–3970 cal BC

Final comment: F Healy (2012), statistically consistent with GrA-27815, OxA-14315 (T'=3.2; T'(5%)=6.0; ν=2; Ward and Wilson 1978). Together, these dates for comminuted and badly preserved fragments can be modelled to indicate a construction date for the 'banana barrow' and a *terminus post quem* for the construction of the inner circuit of *4640–3970 cal BC (95% probability;* Whittle *et al* 2011; fig 9.8).

References: Ward and Wilson 1978

GrA–27815 5215 ±40 BP

δ¹³C: -23.5‰

Sample: CH75 1945/B, submitted on 13 December 2004 by F Healy

Material: animal bone: unidentifiable, 1 of 40 unidentifiable bone fragments (J Mulville and A Powell 2004)

Initial comment: F2046(2). From the lower layer of one of a group of short lengths of ditch surrounding an ovoid area (known as the banana barrow). Beneath 2005/3, the remnant of the bank of the inner causewayed enclosure, which in turn underlay the bank of the continuous enclosure. The sample provides a *terminus post quem* for the construction of the 'banana barrow' and for the sequence at the base of which it lay. 'Medium rubble in darker clayey soil', in limestone-cut feature on well-drained limestone hill.

Objectives: to establish the date of this context as a step towards defining the chronology of the complex.

Calibrated date: *1σ:* 4050–3970 cal BC
      2σ: 4230–3950 cal BC

Final comment: see GrA-27814

GrA–27816 4885 ±45 BP

δ¹³C: -28.7‰

Sample: CH78 4275, submitted on 13 December 2004 by F Healy

Material: carbonised residue (internal; two recently broken fragments of a Neolithic Bowl sherd with residue under limey crust. External spalling suggests that it has been burnt. Sherd and ancient breaks under limey crust are fresh and unabraded) (F Healy 2004)

Initial comment: F3085(5). An upper layer of inner causewayed ditch segment F3085, overlying layers 4, 7, and 10. All the fills of F3085 were overlain by the bank of the continuous ditch. The freshness of the sherd and the good preservation suggest that it was buried soon after the final use and breakage of the pot from which it came. It is likely to be close in age to its context. 'Medium rubble in light brown soil' in limestone-cut ditch on well-drained limestone hill.

Objectives: to establish the date of this context as a step towards defining the chronology of the complex.

Calibrated date: *1σ:* 3710–3640 cal BC
      2σ: 3770–3540 cal BC

Final comment: F Healy (2012), the sherd was probably redeposited because GrA-27816 has poor agreement when constrained to be later than GrA-31100 and OxA-15575 from the initial silts of the same segment (A=11.8%).

GrA–27818 4770 ±40 BP

δ¹³C: -22.1‰

Sample: CH75 2920, submitted on 13 December 2004 by F Healy

Material: animal bone: *Bos* sp., cattle, proximal radius and ulna fragments, probably articulating (J Mulville and A Powell 2004)

Initial comment: F2033(2). Found with other animal bone in layer overlying ditch bottom at outer edge and overlying layer 1 at ditch centre. Layer 1 overlay ditch side and bottom at inner edge. Layer 2 was thus immediately above primary silt. The possible articulation of the bones, together with their large size and good preservation indicates that the animal from which they came was recently dead when they

were buried and that they are likely to be close in age to their context. 'Sticky limestone with large stones' in limestone-cut ditch on well-drained limestone hill.

Objectives: to establish the date of this context as a step towards defining the chronology of the complex.

Calibrated date: *1σ:* 3640–3520 cal BC
      2σ: 3650–3370 cal BC

Final comment: F Healy (2012), the result is in agreement with its stratigraphic position in the model.

GrA–27820 4770 ±40 BP

δ¹³C: -21.9‰

Sample: CH77 4721/B, submitted on 13 December 2004 by F Healy

Material: animal bone: *Bos* sp., right metacarpal articulating with two associated carpals (J Mulville and A Powell 2004)

Initial comment: a replicate of OxA-14414. F2615(3). A post-primary fill of segment F2615 of inner causewayed ditch, possibly the fill of a recut in turn cut by definite recut, in turn overlain by bank of continuous ditch (2601). The articulation of the bones indicates that they were still joined by softer tissue when buried and that the animal from which they came was recently dead. They should be close in age to their context. 'Red-brown wash with small limestone pieces' in limestone-cut ditch on well-drained limestone hill.

Objectives: to establish the date of this context as a step towards defining the chronology of the complex.

Calibrated date: *1σ:* 3640–3520 cal BC
      2σ: 3650–3370 cal BC

Final comment: F Healy (2012), the result is statistically consistent with OxA-14414 (T'=1.9; T'(5%)=3.8; v=1; Ward and Wilson 1978) and is in agreement with its stratigraphic position in the model.

References: Ward and Wilson 1978

GrA–27821 4815 ±40 BP

δ¹³C: -27.7‰

Sample: CH72 4856, 4980, submitted on 13 December 2004 by F Healy

Material: animal bone: *Sus* sp., two joining fragments of proximal end of pig femur (CH72 4856) fitting unfused epiphysis (CH72 4980). The presence of young pig mandible fragments and loose teeth in the same context suggests that it may be the same animal as OxA-14415 (J Mulville and A Powell 2004)

Initial comment: F853(3). From the rubble fill of the inner causewayed ditch, overlying a thin skin of initial silt on the ditch base and cut by a recut filled by 853(2). The refitting of the shaft and epiphysis indicates that they were not long separated when buried so that the animal from which they came would have been close in age to the context.

Objectives: to establish the date of this context as a step towards defining the chronology of the complex.

Calibrated date: *1σ:* 3650–3530 cal BC
      2σ: 3660–3520 cal BC

Final comment: F Healy (2012), the sample was probably redeposited because it is older than OxA-15574 from thin skin of silt on the base of the ditch.

Laboratory comment: English Heritage (26 November 2014), GrA-27821 and OxA-14415 are statistically consistent measurements (T'=2.7; T'(5%)=3.8; v=1; Ward and Wilson 1978), so it is possible that the samples were from the same piglet.

References: Ward and Wilson 1978

GrA–27828 4850 ±40 BP

$\delta^{13}C$: -28.4‰

Sample: CH78 4544, submitted on 13 December 2004 by F Healy

Material: carbonised residue (internal; Neolithic Bowl sherd with residue under limey encrustation) (F Healy 2004)

Initial comment: F3138(4). Layer of primary silt in bottom of segment F3138 of inner causewayed ditch. The good survival of the residue suggests that the sherd was buried soon after the final use and breakage of the pot from which it came and that it should be close in age to the original excavation of the ditch.

Objectives: to establish the date of this context as a step towards defining the chronology of the complex.

Calibrated date: 1σ: 3660–3630 cal BC
                          2σ: 3710–3530 cal BC

Final comment: see GrA-27818

GrA–31100 4710 ±40 BP

$\delta^{13}C$: -21.6‰

Sample: CH78 4421, submitted on 19 December 2005 by F Healy

Material: animal bone: *Bos* sp., distal humerus fragment probably from larger animal than metatarsal fragment CH78 4432 (A Powell 2005)

Initial comment: O6: F3085(10). Bottom layer of segment F3085 of the inner causewayed ditch, below 3085(9). The good condition of the bone suggests that it was freshly deposited. 'Small/medium rubble in yellow limestone dust', in limestone-cut ditch on well-drained limestone hill.

Objectives: to refine the date of the context as a step to refining the chronology of the complex.

Calibrated date: 1σ: 3630–3370 cal BC
                          2σ: 3640–3360 cal BC

Final comment: see GrA-27818

GrA–31101 4705 ±40 BP

$\delta^{13}C$: -22.0‰

Sample: CH78 4549, submitted on 19 December 2005 by F Healy

Material: animal bone: *Bos* sp., substantial distal femur fragment in good condition, probably from larger animal than CH78 4551 (A Powell 2005)

Initial comment: O6: F3138(4). Layer of primary silt in bottom of segment F3138 of inner causewayed ditch. The good condition of the bone suggests that it was freshly deposited.

Objectives: to refine the date of the context as a step to refining the chronology of the complex.

Calibrated date: 1σ: 3630–3370 cal BC
                          2σ: 3640–3360 cal BC

Final comment: see GrA-27818

OxA–14314 7288 ±36 BP

$\delta^{13}C$: -20.8‰

Sample: CH75 2490, submitted on 13 December 2004 by F Healy

Material: animal bone: unidentifiable, fragment (J Mulville and A Powell 2004)

Initial comment: F2039(2). From the lower layer of one of a group of short lengths of ditch surrounding an ovoid area (known as the banana barrow). Beneath 2005/3, the remnant of the bank of the inner causewayed enclosure, which in turn underlay the bank of the continuous enclosure. The sample provides a *terminus post quem* for the construction of the 'banana barrow' and for the sequence at the base of which it lies. In 'medium rubble with pink soil and grey leaching' in limestone-cut ditch on well-drained limestone hill.

Objectives: to establish the date of this context as a step towards defining the chronology of the complex.

Calibrated date: 1σ: 6220–6070 cal BC
                          2σ: 6240–6050 cal BC

Final comment: F Healy (2012), this sample was redeposited because it is much older than the other three dates on bone fragments from ditches of the 'banana barrow' (GrA-27814, -27815, and OxA-14315).

OxA–14315 5178 ±32 BP

$\delta^{13}C$: -22.8‰

Sample: CH75 1945/A, submitted on 13 December 2004 by F Healy

Material: animal bone: unidentifiable, long bone fragment, 1 of 40 unidentifiable bone fragments (J Mulville and A Powell 2004)

Initial comment: F2046(2). From the lower layer of one of a group of short lengths of ditch surrounding an ovoid area (known as the banana barrow). Beneath 2005/3, the remnant of the bank of the inner causewayed enclosure, which in turn underlay the bank of the continuous enclosure. The sample provides a *terminus post quem* for the construction of the 'banana barrow' and for the sequence at the base of which it lies. 'Medium rubble in darker clayey soil', in limestone-cut feature on well-drained limestone hill.

Objectives: to establish the date of this context as a step towards defining the chronology of the complex.

Calibrated date: 1σ: 4040–3960 cal BC
                          2σ: 4050–3950 cal BC

Final comment: see GrA-27814

OxA–14354 4746 ±33 BP

δ¹³C: -29.6‰

Sample: CH78 4433, 4437, submitted on 13 December 2004 by F Healy

Material: carbonised residue (internal; Neolithic Bowl body sherds joining along recent break to form large sherd approx 70mm ∞ 50mm, in fresh condition) (F Healy 2004)

Initial comment: F3085(10). Bottom layer of segment F3085 of the inner causewayed ditch, below 3085(9). The large size and good preservation of the sherd suggest that it was buried soon after the final use and breakage of the pot from which it came. It should be close in age to its context. 'Small/medium rubble in yellow limestone dust', in the limestone-cut ditch.

Objectives: to establish the date of this context as a step towards defining the chronology of the complex.

Calibrated date: 1σ: 3640–3380 cal BC
2σ: 3640–3370 cal BC

Final comment: see GrA-27818

OxA–14413 4786 ±32 BP

δ¹³C: -20.5‰

Sample: CH78 4053, submitted on 13 December 2004 by F Healy

Material: animal bone: *Bos* sp., articulating left metacarpal and magnum (A Powell 2004)

Initial comment: F3077. From a recut in inner causewayed ditch segment 3085, cutting layer 2 of that segment and overlain by layer 1, which was in turn overlain by the bank of the continuous ditch. The articulation of the bones indicates that they were still joined by softer tissue when buried, and that the animal from which they came was not long dead. They are thus likely to be close in age to their context. 'Very small stones in pink soil matrix' in limestone-cut ditch on well-drained limestone hill.

Objectives: to establish the date of this context as a step towards defining the chronology of the complex.

Calibrated date: 1σ: 3640–3520 cal BC
2σ: 3650–3510 cal BC

Final comment: see GrA-27818

OxA–14414 4696 ±35 BP

δ¹³C: -21.0‰

Sample: CH77 4721/A, submitted on 13 December 2004 by F Healy

Material: animal bone: *Bos* sp., right metacarpal articulating with two associated carpals (J Mulville and A Powell 2004)

Initial comment: a replicate of GrA-27820.

Objectives: as GrA-27820

Calibrated date: 1σ: 3630–3370 cal BC
2σ: 3640–3360 cal BC

Final comment: see GrA-27820

OxA–14415 4900 ±32 BP

δ¹³C: -20.7‰

Sample: CH72 4856, submitted on 13 December 2004 by F Healy

Material: animal bone: *Sus* sp., metapodial fragment fitting unfused epiphysis. The presence of young pig mandible fragments and loose teeth in the same context suggests that it may be the same animal as GrA-27821 (J Mulville and A Powell 2004)

Initial comment: F853(3). From rubble fill of inner causewayed ditch, overlying thin skin of initial silt on ditch base and cut by recut filled by 853(2). The refitting of the shaft and epiphysis indicates that they were not long separated when buried so that the animal from which they came would have been recently dead. The sample should be close in age to the infilling of the inner causewayed ditch. 'Large slabs of bedrock . . . Little fill between stones, the latter being loosely packed'.

Objectives: to establish the date of this context as a step towards defining the chronology of the complex.

Calibrated date: 1σ: 3710–3640 cal BC
2σ: 3750–3630 cal BC

Final comment: see GrA-27818

Laboratory comment: see GrA-27821

OxA–15574 4725 ±34 BP

δ¹³C: -22.3‰

Sample: CH75 323, submitted on 19 December 2005 by F Healy

Material: antler: *Capreolus capreolus*, shed; complete (A Powell 2004)

Initial comment: CIVa: F853(4). In thin skin of primary silt at bottom of ditch, under 853(3). The completeness and good condition of the small, delicate antler indicate that it was freshly shed when buried and should be close in age to its context.

Objectives: to refine the date of the context as a step to refining the chronology of the complex.

Calibrated date: 1σ: 3630–3380 cal BC
2σ: 3640–3370 cal BC

Final comment: see GrA-27818

OxA–15575 4698 ±35 BP

δ¹³C: -21.2‰

Sample: CH78 4432, submitted on 19 December 2005 by F Healy

Material: animal bone: *Bos* sp., distal metatarsal fragment probably from smaller animal than humerus fragment CH78 4421 (A Powell 2005)

Initial comment: O6: F3085(10). Bottom layer of segment F3085 of the inner causewayed ditch, below 3085(9). The good condition of the bone suggests that it was freshly deposited. 'Small/medium rubble in yellow limestone dust', in limestone-cut ditch on well-drained limestone hill.

Objectives: to refine the date of the context as a step to refining the chronology of the complex.

Calibrated date: 1σ: 3630–3370 cal BC
                     2σ: 3640–3360 cal BC

Final comment: see GrA-27818

Causewayed Enclosures: Crickley Hill, long mound valley sequence, Gloucestershire

Location: SO 9265016100
                     Lat. 51.50.33 N; Long. 02.06.24 W

Project manager: P Dixon (University of Nottingham), 1971–95

Archival body: Corinium Museum

Description: the long mound valley is a narrow natural gully in the hilltop, to the south-west of the central knoll of the hill. An entranceway common to both causewayed circuits and to the continuous ditch leads into the south-east end of the valley, and a track bounded by successive fences runs from that entrance along the valley to a low stone platform supporting a timber structure interpreted as a shrine. A second fenced track forked from this one to the central knoll of the hill, where there were rectangular post-built structures. In the long mound valley, further rectangular post-built structures lay near the track and parallel to it. Shrine, structures, and track were sealed by a long earthen mound, built in at least three stages, from north-east to south-west. The earliest, north-west, part of the mound, overlay the edge of the platform supporting the shrine and incorporated a stone cairn. A small stone circle, the slot for which cut through the end of the mound and cairn, was subsequently built over the platform and shrine. The central part of the mound incorporated a further cairn and the latest, south-east, part of the mound overlay a row of three postholes which would have held free-standing posts blocking that end of the valley near the earlier entrances. The dating of this whole sequence is uncertain.

Objectives: to date the sequence as a step to establishing the overall chronology of the complex.

Final comment: F Healy (2012), towards the start of the sequence, statistically consistent (T'=1.4; T'(5%)=7.8; v=3; Ward and Wilson 1978) fifth millennium cal BC dates for charcoal fragments from one layer of the stone platform (GrA-31110, -31111, -31113, and -31114) are difficult to reconcile with its being linked by fences to an entrance in the continuous enclosure built in the mid-36th century cal BC. The charcoal may have been gathered up with the stone from adjacent, probably Mesolithic, pre-enclosure occupation. These dates provide *termini post quos* for the north-west end of the long mound, which overlay the platform. Statistically consistent (T'=2.1; T'5%)=3.8; v=1; Ward and Wilson 1978) mid fourth millennium cal BC dates (GrA-27809, OxA-14311) for charred hazelnut shell fragments for an overlying layer, post-dating the north-west end of the long mound, may have been redeposited from a Neolithic context, given the minutely comminuted condition of the shells, although they were from a concentration of charred material. There were no suitable samples relating to

the central part of the mound, which was built onto the north-west section. Of the six samples relating to the south-east section of the mound, the last to be built, four were of animal bone from postholes of structures underlying it. Three of these (GrA-27806, -27810; OxA-14312) were of fourth millennium cal BC date and one (OxA-14313) of first millennium cal BC. Of two samples from animal bone apparently deliberately placed beneath stone slabs set into the mound in this area, one (OxA-14497) was of fourth millennium cal BC date and one (GrA-27808) of first millennium cal BC date. The two first millennium dates, both measured on disarticulated animal bone, provide *termini post quos* for at least one post-built structure, the deposits under the slabs, and the construction of the final, south-east section of the mound.

References: Ward and Wilson 1978
                     Whittle *et al* 2011

GrA–27806 4750 ±45 BP

δ¹³C: -22.8‰

Sample: CH78 2901, 3454, submitted on 13 December 2004 by F Healy

Material: animal bone: *Bos* sp., first, second and third phalanges, articulating (A Powell 2004)

Initial comment: F2806. Posthole of structure 1 (slightly trapezoid, *c* 6m ∞ 3.4m to 4.7m, formed of postholes F2798, F2802, F2806, F2811, F2826, F2833, and F2835). Beside and parallel to a stake-lined track leading from an entrance in the continuous enclosure to the platform and shrine. Both the track and at least some postholes of the structure, including F2806, underlay a palaeosol overlain by the third and final extension of the long mound (long mound c). The symmetry of the plan suggests that all the postholes belonged to the same structure and were contemporary (Snashall 1998, 16–20, and figs 13–18). The fact that so much of the mandible survives and that it has a single find number suggests that it was substantially complete and little degraded when buried. The sample provides a *terminus post quem* for the structure and for the third and final extension of the long mound. It was stratified below samples CH78 423 and CH78 2250. Sherds from F2806 include two fragments of grog- and ?flint-tempered protruding base angle which appear to be of second or even first millennium BC date (CH77 4660). There are also heavily gritted fragments including a base angle and body sherd with incised chevron. In 'dark grey soil/dust' in feature cut into naturally redeposited dissolved limestone in a fault (or gull).

Objectives: to establish the date of this context as a step towards defining the chronology of the monument.

Calibrated date: 1σ: 3640–3380 cal BC
                     2σ: 3650–3370 cal BC

Final comment: F Healy (2012), if F2835 and F2806 were both indeed postholes of structure 1, then the sample for this date and for OxA-14312 must have been redeposited, since a sample from F2806 dates to the first millennium cal BC (OxA-14313), and since and since F2806 and another posthole of the structure both contained post-Neolithic pottery.

References: Snashall 1998

GrA–27808 2265 ±35 BP

δ¹³C: -22.3‰

Sample: CH78 2250, submitted on 13 December 2004 by F Healy

Material: animal bone: *Sus* sp., left mandible, 2 joining fragments (recently broken) (J Mulville and A Powell 2004)

Initial comment: F2701/6. Under stone slab 2717 in body of third and final extension of the long mound. (Hollos 1999, 39–40, 327, and 343). The placing of the bone under the slab suggests that it relates to the construction of this part of the mound. It provides a *terminus post quem* for the third and final extension of the long mound and is stratigraphically later than sample CH78 2901+3454.

Objectives: to establish the date of this context as a step towards defining the chronology of the complex.

Calibrated date: *1σ:* 400–230 cal BC
                        2σ: 400–200 cal BC

Final comment: F Healy (2012), this result provides a *terminus post quem* for the construction of the south-east part of the long mound, indicating that the sample for OxA-14497, buried beneath a similar slab, was already old when buried.

References: Hollos 1999

GrA–27809 4630 ±40 BP

δ¹³C: -24.4‰

Sample: CH84 sample 730/A, submitted on 13 December 2004 by F Healy

Material: carbonised plant macrofossil (hazelnut shell fragment) (R Gale 2004)

Initial comment: X96: F5674. From the surface of cobbling forming the second floor of small stone circle cutting the north-west end of the long mound. It overlay the first floor of the circle. The sample is one of 64 fragments of charred hazelnut shell recorded from 21 findspots in this context. A local concentration of charred plant remains (as well as of charcoal and burnt bone fragments) suggests that the burnt material derived from the use of the circle. If so, the sample should be close in age to the second floor of the circle, and later than the construction of the first phase of the long mound.

Objectives: to establish the date of this context as a step towards defining the chronology of the complex.

Calibrated date: *1σ:* 3500–3360 cal BC
                        2σ: 3520–3340 cal BC

Final comment: see OxA-14311

GrA–27810 4735 ±45 BP

δ¹³C: -21.8‰

Sample: CH78 2766B, submitted on 13 December 2004 by F Healy

Material: animal bone: unidentifiable (J Mulville and A Powell 2004)

Initial comment: F2830. One of a row of three postholes across the south-east end of the long mound valley. Beyond the end of the second extension of long mound, covered by third extension (Hollos 1999, 137–9). Extracted from a find of bone fragments. The bone provides a *terminus post quem* for the row. In a posthole cut into naturally redeposited dissolved limestone in a fault (or gull) in solid limestone, on a well-drained limestone hill.

Objectives: to establish the date of this context as a step towards defining the chronology of the complex.

Calibrated date: *1σ:* 3640–3380 cal BC
                        2σ: 3640–3370 cal BC

Final comment: F Healy (2012), this sample was probably redeposited in posthole because another posthole in the row contained a post-Neolithic rim sherd.

References: Hollos 1999

GrA–31110 5435 ±40 BP

δ¹³C: -27.9‰

Sample: CH85 150/A, submitted on 19 December 2005 by F Healy

Material: charcoal: *Corylus* sp., single fragment (R Gale 2005)

Initial comment: W96: F6379. Incorporated into the south-east part of the limestone platform, underlying the cairn built beneath the north-west end of the long mound, and in turn underlying the stone circle and its successive floors (Snashall 1997). The large size and good preservation of the charcoal, by the standards of the site, indicates that the wood was freshly charred when buried. If short-life material can be extracted from it, it will provide a *terminus post quem* for the platform.

Objectives: to refine the date of the context as a step to refining the chronology of the complex.

Calibrated date: *1σ:* 4340–4250 cal BC
                        2σ: 4360–4230 cal BC

Final comment: F Healy (2012), the result is statistically consistent with GrA-31111, -31113, and -31114 from the same context (T'=1.4; T'(5%)=7.8; ν=3; Ward and Wilson 1978). Their fifth millennium cal BC date is, however, difficult to reconcile with the linkage of the platform by fences to an entrance of the continuous enclosure built in the mid 36th century cal BC. The charcoal may have been gathered up with the stone from adjacent, probably Mesolithic, pre-enclosure occupation.

References: Snashall 1998
                       Ward and Wilson 1978

GrA–31111 5465 ±45 BP

δ¹³C: -27.5‰

Sample: CH85 150/B, submitted on 19 December 2005 by F Healy

Material: charcoal: *Corylus* sp., single fragment (R Gale 2005)

Initial comment: as GrA-31110

Objectives: to refine the date of the context as a step to refining the chronology of the complex.

Calibrated date: *1σ:* 4350–4260 cal BC
                        2σ: 4370–4240 cal BC

Final comment: see GrA-31110

GrA–31113 5420 ±40 BP

δ¹³C: -27.4‰

Sample: CH85 189/A, submitted on 19 December 2005 by F Healy

Material: charcoal: *Corylus* sp., single fragment (R Gale 2005)

Initial comment: as GrA-31110

Objectives: to refine the date of the context as a step to refining the chronology of the complex.

Calibrated date: *1σ:* 4340–4240 cal BC
                        2σ: 4350–4170 cal BC

Final comment: see GrA-31110

GrA–31114 5480 ±40 BP

δ¹³C: -27.8‰

Sample: CH85 189/B, submitted on 19 December 2005 by F Healy

Material: charcoal: *Corylus* sp., single fragment (R Gale 2005)

Initial comment: as GrA-31110

Objectives: to refine the date of the context as a step to refining the chronology of the complex.

Calibrated date: *1σ:* 4360–4320 cal BC
                        2σ: 4370–4250 cal BC

Final comment: see GrA-31110

OxA–14311 4707 ±35 BP

δ¹³C: -24.3‰

Sample: CH84 sample 730/B, submitted on 13 December 2004 by F Healy

Material: carbonised plant macrofossil (hazelnut shell fragment) (R Gale 2004)

Initial comment: as GrA-27809

Objectives: to establish the date of this context as a step towards defining the chronology of the complex.

Calibrated date: *1σ:* 3630–3370 cal BC
                        2σ: 3640–3370 cal BC

Final comment: F Healy (2012), the result is statistically consistent with GrA-27809 from the same context (T'=2.1; T'(5%)=3.8; ν=1; Ward and Wilson 1978). The extreme comminution of the nut shell fragments may suggest that the burnt material had been redeposited from a fourth millennium context.

References: Ward and Wilson 1978

OxA–14312 4702 ±30 BP

δ¹³C: -22.3‰

Sample: CH78 2901, submitted on 13 December 2004 by F Healy

Material: animal bone: *Bos* sp., unfused femur head (A Powell 2004)

Initial comment: F2835. Posthole of structure 1 (slightly trapezoid, *c* 6m ∞ 3.4m to 4.7m, formed of postholes F2798, F2802, F2806, F2811, F2826, F2833, and F2835). Beside and parallel to a stake-lined track leading from an entrance in the continuous enclosure to the platform and shrine. Both the track and at least some postholes of the structure underlay a palaeosol overlain by the third and final extension of the long mound (long mound c). The symmetry of the plan suggests that all the postholes belonged to the same structure and were contemporary (Snashall 1998, 16–20, and figs 13–18). Sherds from two other postholes of the structure appear to be of second or even first millennium BC date. The sample provides a *terminus post quem* for the structure and for the third extension of the long mound and was stratified below samples CH78 423 and CH78 2250.

Objectives: to establish the date of this context as a step towards defining the chronology of the complex.

Calibrated date: *1σ:* 3630–3370 cal BC
                        2σ: 3630–3370 cal BC

Final comment: see GrA-27806

References: Snashall 1998

OxA–14313 2344 ±27 BP

δ¹³C: -21.2‰

Sample: CH77 4889, submitted on 13 December 2004 by F Healy

Material: animal bone: *Sus* sp., mandible fragments and 4 teeth, extracted from larger find of animal bones (A Powell 2004)

Initial comment: as GrA-27806

Objectives: to establish the date of this context as a step towards defining the chronology of the monument.

Calibrated date: *1σ:* 410–390 cal BC
                        2σ: 420–380 cal BC

Final comment: F Healy (2012), this sample provides a *terminus post quem* for structure and for the south-east part of long mound.

OxA–14418 2978 ±37 BP

δ¹³C: -21.3‰

Sample: CH77 2079, submitted on 13 December 2004 by F Healy

Material: antler: *Cervus elaphus*, weathered end of red deer antler pick; beam and brow tine present; bez tine seems never to have grown (J Mulville and A Powell 2004)

Initial comment: F2734(1). On track through the east entrance in the continuous ditch, between the bank terminals. Overlain by 2702(2) (wash, probably from the third and final extension of the long mound).

Objectives: to establish the date of this context as a step towards defining the chronology of the complex.

Calibrated date: 1σ: 1270–1120 cal BC
2σ: 1380–1050 cal BC

Final comment: F Healy (2012), this sample must relate to second millennium cal BC activity on the site.

OxA–14497 4480 ±33 BP

δ¹³C: -22.4‰

Sample: CH78 423, submitted on 13 December 2004 by F Healy

Material: animal bone: *Bos* sp., right astragalus (J Mulville and A Powell 2004)

Initial comment: F2740(3). Placed under stone slab 2792 in body of third and final extension of the long mound, at its south-east end, where the mound was delimited by cobbling (Hollos 1999, 39–40, 327, and 343). Side-by-side with proximal fragment of second phalanx, and fragments, some joining, of unidentified long bone, all eroded. Breaks appear to have occurred when the bone was already old. The placing of the bones under the slab suggests that they relate to the construction of this part of the mound. They provide a *terminus post quem* for the third and final extension of the long mound and are stratigraphically later than sample CH78 2901+3454.

Objectives: to establish the date of this context as a step towards defining the chronology of the complex.

Calibrated date: 1σ: 3340–3090 cal BC
2σ: 3350–3020 cal BC

Final comment: F Healy (2012), the sample must have been already old when placed under slab, since the sample for GrA-27808, placed under a similar slab, dates to the first millennium cal BC.

References: Hollos 1999

Causewayed Enclosures: Crickley Hill, outer causewayed ditch, Gloucestershire

Location: SO 9265016100
Lat. 51.50.33 N; Long. 02.06.24 W

Project manager: P Dixon (University of Nottingham), 1971–95

Archival body: Corinium Museum

Description: a causewayed ditch considerably slighter than the inner ditch and lying 30–50m outside it. Like the other Neolithic ditches, this was backed by a low stone platform and a palisade. This ditch saw some reworking, including the blocking and making of entrances. The alignment of an entrance dating from after the construction of this ditch with one in the inner ditch suggests that the outer ditch was built before the inner.

Objectives: to date the construction and use of the circuit as a step to establishing the overall chronology of the complex.

Final comment: F Healy (2012), the scarcity of samples makes it possible only to estimate a *terminus ante quem* of 3685–3595 cal BC (95% probability; Whittle et al 2011, fig 9.9: *taq outer*) for the construction of the circuit.

References: Whittle et al 2011

GrA–27813 4830 ±170 BP

δ¹³C: -21.8‰

Sample: CH72 4227, submitted on 13 December 2004 by F Healy

Material: animal bone: *Sus* sp., articulating first and second phalanges (J Mulville and A Powell 2004)

Initial comment: a replicate of GrA-30368. CIV: F674(1). Found with other scraps of bone in a probable treehole 240cm wide and 18cm deep partly overlain by F611(3a), the base of the bank of the outer causewayed ditch.

Objectives: to establish the date of this context as a step towards defining the chronology of the complex.

Calibrated date: 1σ: 3790–3370 cal BC
2σ: 3980–3110 cal BC

Final comment: F Healy (2012), the result is statistically consistent with GrA-30368 (T'=1.4; T'(5%)=3.8; ν=1; Ward and Wilson 1978), but more recent than OxA-14386, which should post-date them if the context was sealed by the bank. It seems probable that the phalanges were introduced into the hollow after the building of the earthwork.

References: Ward and Wilson 1978

GrA–30368 4625 ±40 BP

δ¹³C: -21.7‰

Sample: CH72 4227, submitted on 19 December 2005 by F Healy

Material: animal bone: *Sus* sp., articulating first and second phalanges (J Mulville and A Powell 2004)

Initial comment: a replicate of GrA-27813. CIV: F674(1). Found with other scraps of bone in a probable treehole 240cm wide and 18cm deep partly overlain by F611(3a), the base of the bank of the outer causewayed ditch.

Objectives: as GrA-27813

Calibrated date: 1σ: 3500–3360 cal BC
2σ: 3520–3340 cal BC

Final comment: see GrA-27813

GrA–31103 4450 ±45 BP

δ¹³C: -24.4‰

Sample: CH80 4652/A, submitted on 19 December 2005 by F Healy

Material: carbonised residue (internal; some of >40 shell-tempered Neolithic Bowl sherds, most of them from 1 pot, some fresh and well-preserved, many with internal residue.

They are not from the same pot as CH80 4674 (OxA-14386), because their temper is considerably less dense) (F Healy 2004)

Initial comment: a replicate of OxA-15704. M8: F4295=F4299(2). Second layer of outer ditch in trench M8, overlying F4929(3).

Objectives: to refine the date of the context as a step to refining the chronology of the complex.

Calibrated date: 1σ: 3330–3020 cal BC
2σ: 3350–2920 cal BC

Final comment: F Healy (2012), the result is statistically consistent with OxA-15704 (T'=1.6; T'(5%)= 3.8; ν=1; Ward and Wilson 1978). They provide a *terminus ante quem* for a late stage in the silting of the ditch.

References: Ward and Wilson 1978

OxA–14386 4736 ±30 BP

δ¹³C: -27.2‰

Sample: CH80 4674, submitted on 13 December 2004 by F Healy

Material: carbonised residue (internal; four small shell-tempered Neolithic Bowl body sherds, possibly from the same pot, with internal residue. They are from a different pot to CH80 4652 because their temper is considerably denser) (F Healy 2004)

Initial comment: F4299(3). In the layer immediately above the primary silt (layer 4) in segment F4299 of the outer causewayed ditch. The survival of the residue and the relative freshness of the sherds suggests that they were buried soon after the final use and breakage of the pot(s) from which they came. They should be close in age to their context.

Objectives: to establish the date of this context as a step towards defining the chronology of the complex.

Calibrated date: 1σ: 3630–3380 cal BC
2σ: 3640–3370 cal BC

Final comment: F Healy (2012), since four sherds were present, the pot may have been freshly broken when buried. If so, the date should provide a *terminus ante quem* for the building of the earthwork.

OxA–15704 4530 ±45 BP

δ¹³C: -29.4‰

Sample: CH80 4652/B, submitted on 19 December 2005 by F Healy

Material: carbonised residue (internal; some of >40 shell-tempererd Neolithic Bowl sherds, most of them from 1 pot, some fresh and well-preserved, many with internal residue. They are not from the same pot as CH80 4674 (OxA-14386), because their temper is considerably less dense) (F Healy 2004)

Initial comment: a relpicate of GrA-31103. M8: F4295=F4299(2). Second layer of outer ditch in trench M8, overlying F4929(3).

Objectives: as GrA-31103

Calibrated date: 1σ: 3360–3100 cal BC
2σ: 3370–3090 cal BC

Final comment: see GrA-31103

Laboratory comment: Oxford Radiocarbon Accelerator Unit (16 May 2006), a very small sample, 0.319mgs C was produced in the combustion of 2.4mgs of pretreated material from the sherd.

Causewayed Enclosures: Etton Woodgate, ditch, Cambridgeshire

Location: TF 13650735
Lat. 52.38.54 N; Long. 00.19.09 W

Project manager: F Pryor (Fenland Archaeological Trust), 1983

Archival body: British Museum and Natural History Museum

Description: two segments of ditch, almost certainly not part of a complete enclosure, contained early Neolithic artefacts and were close to contemporary pits and postholes. Beaker period activity was represented by further pits and a midden. The site lay on the south-east edge of Maxey 'island' on the lower Welland, 80m west of the Etton causewayed enclosure and separated from it by a palaeochannel. It was discovered in the course of a watching brief during overburden-stripping in advance of gravel quarrying and was the subject of rescue excavation. The ditches were F100, over 50m long, and F132, 75m long and waterlogged at its north end, from which the samples came.

Objectives: samples were submitted to date the construction of the site and establish its relation to the nearby Etton causewayed enclosure.

Final comment: F Healy (2012), sampling was restricted by limited excavation and by the fact that the ditches had silted naturally and relatively cleanly. Two roundwood samples from the ditch base provide an estimated construction date of *3645–3525 cal BC (95% probability;* Whittle *et al* 2011, fig 6.36: *start Woodgate),* after the Etton causewayed enclosure (Whittle *et al* 2011, fig 6.45).

References: French and Pryor 2005
Taylor 1998
Whittle *et al* 2011

GrA–29362 4740 ±40 BP

δ¹³C: -27.6‰

Sample: EW83 C14 (13)-A, submitted on 23 June 2005 by F Healy

Material: wood (waterlogged): unidentified, roundwood fragment; 15mm diameter, too degraded to identify (R Gale 2005)

Initial comment: F132 [1–2] (4) 6.50 m OD 3710/7420. Near the northern butt of segment F132, in a layer lying on the ditch base containing axed woodchips, rods and unmodified roundwood (Taylor 1998, 159). The fragments must result from nearby woodworking soon after the ditch was dug, and the small diameter of the roundwood means that it was not more than a few years old.

Objectives: to date the original excavation of the ditches in order to better define their relation to the Etton causewayed enclosure and adjacent monuments.

Calibrated date: 1σ: 3640–3380 cal BC
                     2σ: 3640–3370 cal BC

Final comment: F Healy (2012), GrA-29362 and OxA-15034 are statistically consistent (T'=1.3; T

References: Ward and Wilson 1978

OxA–14996 4985 ±34 BP

δ¹³C: -24.9‰

Sample: EW83 P806, submitted on 23 June 2005 by F Healy

Material: carbonised residue (internal; Neolithic Bowl body sherd with sooty residue) (F Healy 2005)

Initial comment: F132 [1–2] (4) 6.15 m OD 3705/7410. Found with other sherds near the northern butt of segment F132, in a layer lying on the ditch base containing axed woodchips, rods, and unmodified roundwood. The good preservation of the residue indicates that the sherd was buried soon after the final use and breakage of the pot.

Objectives: to date the original excavation of the ditches in order to better define their relation to the Etton causewayed enclosure and adjacent monuments.

Calibrated date: 1σ: 3800–3700 cal BC
                     2σ: 3940–3660 cal BC

Final comment: F Healy (2012), the residue is older than the two roundwood fragments from the same context (GrA-29362 and OxA-15034) and was probably redeposited in the ditch.

OxA–15034 4800 ±33 BP

δ¹³C: -29.5‰

Sample: EW83 C14 (13)-B, submitted on 23 June 2005 by F Healy

Material: wood (waterlogged): *Alnus glutinosa*, roundwood; one radial segment (R Gale 2005)

Initial comment: as GrA-29362

Objectives: to date the original excavation of the ditches in order to better define their relation to the Etton causewayed enclosure and adjacent monuments.

Calibrated date: 1σ: 3640–3530 cal BC
                     2σ: 3650–3520 cal BC

Final comment: see GrA-29362

Causewayed Enclosures: Etton, ditch, Cambridgeshire

Location: TF 13850735
              Lat. 52.38.54 N; Long. 00.19.46 W

Project manager: F Pryor (Fenland Archaeological Trust), 1981–7

Archival body: British Museum and Natural History Museum

Description: a single-circuit causewayed enclosure of ovoid plan with a maximum dimension of 187m, sited in a meander of a then-active channel on a gravel island in the lower Welland valley. The lower parts of the segments in the west of the circuit had remained waterlogged from construction to excavation. Both the ditch segments and the numerous features in the interior reflected a distinct depositional practices between the east and west of parts of the monument, corresponding to north-south fences dividing the interior.

Objectives: samples were submitted to determine whether the construction date and duration of the enclosure, base on dating achieved following the excavations (Ambers 1998) could be refined. Given the proximity of the Etton enclosure to two other partly excavated examples, at Etton Woodgate and Northborough, it was also sought to establish how the three sites related to each other.

Final comment: F Healy (2012), an estimated construction date of *3710–3645 cal BC (95% probability*; Whittle *et al* 2011, fig 6.33: *build Etton)* was followed by a sequence of events which mirrored the different character of the east and west sides of the enclosure. The waterlogged western arc seems to have seen a short history of activity, apart from being cut by a late fourth/early third millennium cal BC pit (dated by BM-2899, 4370 ±50 BP). The segments of the eastern arc saw an often complex sequence of recutting and deposition, continuing to an estimated date of *3330–3095 cal BC (95% probability*; Whittle *et al* 2011, fig 6.33: *end Etton)*, the infilling of the segments (which would have been rapid in a gravel-cut ditch if it were left to silt naturally), being extended by repeated intervention.

Laboratory comment: English Heritage (25 November 2014), eight bone and antler samples and one carbonised residue sample failed to date, in the case of the bone and antler because of inadequate collagen preservation (Brock *et al* 2007).

References: Ambers and Bowman 1998
               Beadsmore *et al* 2010
               Brock *et al* 2007
               Harris 2006
               Hedges *et al* 1996
               Oswald *et al* 2001
               Pryor *et al* 1985
               Pryor 1987
               Pryor 1988
               Pryor 1998
               Pryor and Kinnes 1982
               Whittle *et al* 2011

GrA–29353 4410 ±40 BP

δ¹³C: -29.4‰

Sample: P2391B, submitted on 23 June 2003 by 27/05/2008

Material: carbonised residue (thick, fresh internal residue from 1 of 2 fresh, well-preserved sherds from an Ebbsfleet Ware vessel)

Initial comment: segment 10 [0–205] (2). From the lower part of a rich deposit of cultural material, in one of the pits of F994, a complex deposit in the upper part of the segment (Pryor 1998, 41, fig 44:E).

Objectives: to establish the date of the context as a step towards defining the chronology of the enclosure.

Calibrated date: 1σ: 3100–2920 cal BC
                              2σ: 3330–2910 cal BC

Final comment: F Healy (2012), because this sample formed part of a major deposit of cultural material, it clearly derives from a late episode in the use of the enclosure.

GrA–29354 4560 ±45 BP

δ¹³C: -29.0‰

Sample: P3750, submitted on 23 June 2005 by F Healy

Material: carbonised residue (internal; 2 shell-tempered body sherds, almost certainly not from the same pot, with fresh, well-preserved residue, indicating they were buried soon after their final use and breakage) (F Healy 2005)

Initial comment: segment 12 [217–221] (3). Middle fill of phase 1C recut (Pryor 1998, 19, fig 74: B).

Objectives: to establish the date of this context as a step towards defining the chronology of the enclosure.

Calibrated date: 1σ: 3370–3130 cal BC
                              2σ: 3500–3090 cal BC

Final comment: F Healy (2012), this date has good agreement with the stratigraphic sequence in the model.

GrA–29355 4225 ±40 BP

δ¹³C: -28.8‰

Sample: P3821B, submitted on 23 June 2005 by F Healy

Material: carbonised residue (internal; large shell-tempered Neolithic body sherd with exceptionally fresh and thick residue) (F Healy 2005)

Initial comment: segment 12 [227–0] (2). Layer overlying phase 1C recut in the southern butt of segment (Pryor 1998, fig 74: C).

Objectives: to establish the date of this context as a step towards defining the chronology of the enclosure.

Calibrated date: 1σ: 2900–2780 cal BC
                              2σ: 2910–2690 cal BC

Final comment: F Healy (2012), GrA-29355 and OxA-14972 are statistically consistent (T′=1.9; T′(5%)=3.8; ν=1; Ward and Wilson 1978). Layer 2, in which the sherd lay, was separated by a skin of iron pan from the top of the

underlying phase 1C recut of the enclosure segment (Pryor 1998, fig 74:C), suggesting that the surface of the fill had consolidated before layer 2 was deposited and hence that some time had elapsed. The sample probably belongs with the later rather than the primary use of the enclosure.

References: Ward and Wilson 1978

GrA–29357 4875 ±40 BP

δ¹³C: -25.9‰

Sample: C14 (1), submitted on 23 June 2005 by F Healy

Material: wood (waterlogged): bark, single fragment; 12mm thick (R Gale 2005)

Initial comment: segment 1 [2–4] (8) *c* 6.15m OD, 85cm deep, 8784/7293. Extracted from roundwood twigs associated with wooden axe handle (W409) in basal layer (Pryor 1998, 21–24, 53–4, 148–9, and fig 59: B). These fragments seem to include bark. If recently grown material can be extracted from them it will provide samples close in age to their context.

Objectives: to establish the date of this context as a step towards defining the chronology of the enclosure.

Calibrated date: 1σ: 3700–3630 cal BC
                              2σ: 3710–3540 cal BC

Final comment: see GrA-29354

GrA–29358 4765 ±40 BP

δ¹³C: -26.5‰

Sample: C14 (4), submitted on 23 June 2005 by F Healy

Material: wood (waterlogged): unidentified, roundwood fragment; from undated remainder from sample for BM-2890, which included a twig fragment 6mm in diameter, and another slightly larger, but was mainly comminuted (R Gale 2005)

Initial comment: segment 1 [5–6] (3), *c* 6.20m OD 3780/7297. Near base of ditch, overlying layer 4+5 which was equivalent to layer 8 in [2–4] (Pryor 1998, 45–46, 63–64, and fig 60). The small diameter of this piece of roundwood indicates that it was at most a few years old when it entered the ditch.

Objectives: to establish the date of this context as a step towards defining the chronology of the enclosure.

Calibrated date: 1σ: 3640–3510 cal BC
                              2σ: 3650–3370 cal BC

Final comment: F Healy (2012), GrA-29358 and BM-2890 (4820 ±45 BP) are statistically consistent (T′=0.8; T′(5%)=3.8; ν=1; Ward and Wilson 1978) and in agreement with the stratigraphic position of the sample in the model.

References: Ward and Wilson 1978

GrA–29367 4665 ±40 BP

δ¹³C: -28.0‰

Sample: P3596, submitted on 23 June 2005 by F Healy

Material: carbonised residue (internal; large, well-preserved shell-tempered Neolithic Bowl body sherd, 1 of at least 10 probably from the same pot, some also with residue) (F Healy 2005)

Initial comment: segment 13 [239–0] (6) at 85cm deep. In S butt (Pryor 1998, fig 56: H). The size and freshness of the sherd, the freshness of the residue and the proximity of other sherds from the same vessel indicate that it was buried soon after its final use and breakage. NB The notations used here are from the bag. In the figure caption the layer is 4.

Objectives: to establish the date of this context as a step towards defining the chronology of the enclosure.

Calibrated date: 1σ: 3520–3360 cal BC
2σ: 3630–3360 cal BC

Final comment: see GrA-29354

GrA–29368 4865 ±40 BP

δ¹³C: -22.5‰

Sample: B479, submitted on 23 June 2005 by F Healy

Material: animal bone: *Bos* sp., one of a bundle of 20 ribs from both sides of the body of a juvenile, which had died in the first year of life. Cut marks on some indicate detachment from the sternum and vertebral column (M Armour-Chelu 1987)

Initial comment: segment 1 [5–6] (3). Found as a bundle Near base of ditch segment, overlying layer 4+5 which was equivalent to layer 8 in [2–4] (Pryor 1998, fig 60: A, B). The heaping of dismembered ribs from a single animal is strongly indicative of the immediate aftermath of consumption, and would have thus have occurred very soon after the death of the animal.

Objectives: to establish the date of this context as a step towards defining the chronology of the enclosure.

Calibrated date: 1σ: 3700–3630 cal BC
2σ: 3710–3530 cal BC

Final comment: see GrA-29354

GrA–29369 4810 ±40 BP

δ¹³C: -21.8‰

Sample: B5339, submitted on 23 June 2005 by F Healy

Material: animal bone (sheep/goat; left humerus articulating with ulna (B5342)) (A Powell 2005)

Initial comment: segment 3 [35–0] (3) 6.48m OD 3766/7338. In lowest fill of ditch (Pryor 1998, 25, fig 64: A). Heaped in segment butt with 25 other bones from a single 3–4–year-old sheep, some of them butchered (Armour-Chelu 1998, 278–9). The piling up of some many elements from the same animal suggests the immediate aftermath of slaughter and consumption.

Objectives: to establish the date of this context as a step towards defining the chronology of the enclosure.

Calibrated date: 1σ: 3650–3530 cal BC
2σ: 3660–3520 cal BC

Final comment: see GrA-29354

References: Armour-Chelu 1998

GrA–29372 4740 ±40 BP

δ¹³C: -22.2‰

Sample: B15481, B15500, submitted on 23 June 2005 by F Healy

Material: animal bone: *Bos* sp., 2 articulating phalanges. Size is such that they could be from the same animal as B13826 from layer 4 (J Powell 2005)

Initial comment: segment 13 [228–0] (5) at 105 cm deep. Layer overlying layer 7 and underlying layer 3 in the northern butt (Pryor 1998, fig 75: A). Found at same depth and same grid reference. This strongly suggests that they were still linked by soft tissue when buried and that the animal from which they came was not long dead.

Objectives: to establish the date of this context as a step towards defining the chronology of the enclosure.

Calibrated date: 1σ: 3640–3380 cal BC
2σ: 3640–3370 cal BC

Final comment: see GrA-29354

OxA–14883 4878 ±35 BP

δ¹³C: -28.5‰

Sample: P94, submitted on 23 June 2005 by F Healy

Material: carbonised residue (internal; Neolithic Bowl body sherd, from the same pot as a second sherd from context P93) (F Healy 2005)

Initial comment: segment 1 [0–1] (8). In the south-east butt of the segment, with complete pot on birch-bark mat (Pryor 1998, 21). Bag of P94 marked 'associated with P90 - whole pot on mat'. Layer 8 was on base of ditch (Pryor 1998, fig 59: A), but this may not have been the first cut of the segment (Pryor 1998, 21). The survival of the residue, as well as the presence of more than one sherd of the same pot, suggests that the vessel was freshly broken when buried

Objectives: to establish the date of this context as a step towards defining the chronology of the enclosure.

Calibrated date: 1σ: 3700–3640 cal BC
2σ: 3710–3630 cal BC

Final comment: see GrA-29354

OxA–14969 4809 ±36 BP

δ¹³C: -20.8‰

Sample: C14 (22)/2, submitted on 23 June 2005 by F Healy

Material: animal bone: *Sus* sp., unfused proximal tibia epiphysis from bag which formerly contained pig tibia which was sample for BM-2765 (A Powell 2005)

Initial comment: a replicate of BM-2765. Segment 1 [2–4] (8). In basal layer with waterlogged wood and axe haft (Pryor 1998, 21–4, 53–4, 148–9, and fig 59: B). The lifting and bagging of the shaft and epiphysis as a single find suggests that they were together in the ground, joined by soft tissue when buried, and hence that the animal from which they came was not long dead when the bone was buried.

Objectives: to obtain a replicate for BM-2765 to refine the date of the context as a step towards defining the chronology of the enclosure.

Calibrated date: 1σ: 3650–3530 cal BC
                       2σ: 3660–3520 cal BC

Final comment: F Healy (2012), OxA-14969 and BM-2765 (4960 ±90 BP) are statistically consistent (T'=2.5; T'(5%)=3.8; ν=1; Ward and Wilson 1978). Their mean is also statistically consistent with the results for the other two samples from this context (OxA-14883, GrA-29357; (T'=0.0; T'(5%)=6.0; ν=2; Ward and Wilson 1978).

References: Ward and Wilson 1978

OxA–14970 4751 ±34 BP

$\delta^{13}C$: -21.0‰

Sample: B505, B506, submitted on 23 June 2005 by F Healy

Material: animal bone (sheep/pig; unfused vertebra and vertebral column) (J Hamilton 2005)

Initial comment: segment 1 [15–16] (2). In an upper layer. The recovery of the two unfused elements from immediately adjacent findspots shows that they were still linked by soft tissue when buried and hence that the animal from which they came was recently dead.

Objectives: to establish the date of this context as a step towards defining the chronology of the enclosure.

Calibrated date: 1σ: 3640–3510 cal BC
                       2σ: 3640–3370 cal BC

Final comment: see GrA-29354

OxA–14971 3528 ±31 BP

$\delta^{13}C$: -21.9‰

Sample: B9700, submitted on 23 June 2005 by F Healy

Material: animal bone: *Capra* sp., 1 of 3 lumbar vertebrae (the others being B9701 and B9709), articulated with each other and sacrum (B9710). Recent fusion of epiphyses indicates that the animal was almost mature (J Hamilton 2005)

Initial comment: segment 6 [172–176] (6), grid ref 3877 7428, 6.33mOD. The vertebrae are almost certainly from F953, a phase 2 pit containing Peterborough Ware and alderwood bowls, which extended to both sides of section 176 and cut through the bottom of the ditch. Its limits, however, were not clearly defined (Pryor 1998, 29–30, and fig 71: B). The articulation of the vertebrae shows that they were still connected by soft tissue when buried and cannot have been redeposited from another context.

Objectives: to obtain a further date for F953 (BM-2891 is too late for the Peterborough Ware sherds) as a step to defining the chronology of the enclosure.

Calibrated date: 1σ: 1910–1770 cal BC
                       2σ: 1950–1750 cal BC

Final comment: F Healy (2012), OxA-14971 is even more recent than BM-2891 (3680 ±35 BP), measured on roundwood from the same context. Together they indicate that the pit and the bowls may post-date the apparently associated Peterborough Ware sherds.

OxA–14972 4300 ±36 BP

$\delta^{13}C$: -28.1‰

Sample: P3821A, submitted on 23 June 2005 by F Healy

Material: carbonised residue (internal; large shell-tempered Neolithic body sherd with exceptionally fresh and thick residue) (F Healy 2005)

Initial comment: segment 12 [227–0] (2). Layer overlying phase 1C recut in the southern butt of the segment (Pryor 1998, fig 74: C). A replicate of GrA-29355.

Objectives: to establish the date of this context as a step towards defining the chronology of the enclosure.

Calibrated date: 1σ: 2920–2890 cal BC
                       2σ: 3010–2880 cal BC

Final comment: see GrA-29355

OxA–14973 4836 ±36 BP

$\delta^{13}C$: -21.9‰

Sample: B13914, submitted on 23 June 2005 by F Healy

Material: animal bone: *Capra* sp., mandible from a sheep aged 18 months to two years (Armour-Chelu 1998, 280–1) (M Armour-Chelu 1998)

Initial comment: segment 12 [227–0] (6). Found together in a group of six bones, including vertyebrae, probably from the same sheep, in a lens of turf within gravel fills on the base of the southern ditch butt (Pryor 1998, fig 74: C).

Objectives: to establish the date of this context as a step towards defining the chronology of the enclosure.

Calibrated date: 1σ: 3660–3630 cal BC
                       2σ: 3700–3530 cal BC

Final comment: see GrA-29354

OxA–14995 4645 ±38 BP

$\delta^{13}C$: -27.7‰

Sample: P387, P393, P394, P399, P400, P501, P502, P503, submitted on 23 June 2005 by F Healy

Material: carbonised residue (7 joining Neolithic Bowl sherds, some with internal residue, forming a single fragment *c* 110mm ∞ 150mm) (F Healy 2005)

Initial comment: segment 1 [13–14] (2). In an upper layer. The fact that so many joining sherds were found together, even if they were a single large sherd when first buried, indicates that the pot from which they came was freshly broken.

Objectives: to establish the date of this context as a step towards defining the chronology of the enclosure.

Calibrated date: 1σ: 3500–3360 cal BC
                      2σ: 3620–3350 cal BC

Final comment: see GrA-29354

OxA–15033 4673 ±33 BP

δ¹³C: -26.3‰

Sample: C14 (7), submitted on 23 June 2005 by F Healy

Material: wood (waterlogged): *Quercus* sp., roundwood; single fragment, too degraded for maturity to be assessed (R Gale 2005)

Initial comment: segment 3 [35–0] (3) 6.48m OD 3766/7338. In lowest fill of ditch (Pryor 1998, 25, and fig 64: A). The small diameter of the twigs shows that they were only a few years old when buried.

Objectives: to establish the date of this context as a step towards defining the chronology of the enclosure.

Calibrated date: 1σ: 3520–3370 cal BC
                      2σ: 3630–3360 cal BC

Final comment: see GrA-29354

OxA–15039 4785 ±30 BP

δ¹³C: -21.2‰

Sample: B5356, submitted on 23 June 2005 by F Healy

Material: animal bone: *Sus* sp., cervical vertebra with unfused epiphysis and cut marks (J Hamilton 2005)

Initial comment: segment 3 [40–0] (3) 6.46–6.53m OD. Layer on base of ditch (Pryor 1998, fig 64: C).

Objectives: to establish the date of this context as a step towards defining the chronology of the enclosure.

Calibrated date: 1σ: 3640–3520 cal BC
                      2σ: 3650–3510 cal BC

Final comment: see GrA-29354

Causewayed Enclosures: Etton, internal features, Cambridgeshire

Location: TF 13850735
                   Lat. 52.38.54 N; Long. 00.19.46 W

Project manager: F Pryor (Fenland Archaeological Trust), 1981–7

Archival body: British Museum and Natural History Museum

Description: features inside the Etton causewayed enclosure, including both pits and postholes and linear features which served to divide the interior.

Objectives: to date those features which relate to the changing form and organisation of the enclosure.

Final comment: F Healy (2012), new samples were confined to pit F505, a feature near the north entrance cut by ditch F313 which was seen as part of a modification of the enclosure and the entrance following the waterlogging of the western arc. Dates for roundwood from the base of a recut of the pit are thus *termini post quos* for this reworking of the monument.

References: Whittle *et al* 2011

GrA-29359 4480 ±40 BP

δ¹³C: -26.4‰

Sample: C14 (18) A, submitted on 23 June 2005 by F Healy

Material: wood (waterlogged): unidentified, roundwood fragment; structurally collapsed and degraded, but possibly *Alnus glutinosa*, *Corylus avellana*, *Salix* sp., or *Populus* sp. (R Gale 2005)

Initial comment: F505 (6) 3844/7429 55–60cm. Extracted from sample of wood from basal layer of large pit just inside the north apex of the site (Pryor 1998, 98–99, and fig 104), cut by ditch F313 which probably took the place of segment 5 and others to the south as they became progressively wetter (Pryor 1998, 106, and fig 115). The smaller twigs, which can represent only one or two year's growth, provide a *terminus post quem* for the original excavation of F313 and for the construction of the Etton cursus which cuts F313.

Objectives: to provide a *terminus post quem* for the construction of F313 as a step to refining the chronology of the complex.

Calibrated date: 1σ: 3340–3090 cal BC
                      2σ: 3360–3010 cal BC

Final comment: F Healy (2012), OxA-14974 and GrA-29359 are statistically consistent (T'=1.2; T'(5%)=3.8; ν=1; Ward and Wilson 1978). They provide a *terminus post quem* for the cutting of F313.

References: Ward and Wilson 1978

OxA–14974 4539 ±37 BP

δ¹³C: -27.6‰

Sample: C14 (18)-B, submitted on 23 June 2005 by F Healy

Material: wood (waterlogged): unidentified, roundwood fragment; collapsed and with insufficient structure for identification (R Gale 2005)

Initial comment: F505 (6) 3844/7429 55–60cm. Extracted from sample of wood from basal layer of large pit just inside the north apex of the site (Pryor 1998, 98–99, and fig 104), cut by ditch F313 which probably took the place of segment 5 and others to the south as they became progressively wetter (Pryor 1998, 106, and fig 115). The smaller twigs, which can represent only one or two year's growth, provide a *terminus post quem* for the original excavation.

Objectives: to provide a *terminus post quem* for the construction of F313 as a step to refining the chronology of the complex.

Calibrated date: 1σ: 3360–3120 cal BC
                      2σ: 3370–3090 cal BC

Final comment: see GrA-29359

Causewayed Enclosures: Haddenham, Cambridgeshire

Location:	TL 4120073650 Lat. 52.20.31 N; Long. 00.04.21 E
Project manager:	C Evans and I Hodder (University of Cambridge), 1981–7
Archival body:	Cambridge Archaeological Unit

Description: a single-circuit causewayed enclosure with internal palisade, on the terrace of the Great Ouse. It is of irregular, sub-trapezoidal plan, and exceptionally large in size (8.75ha). It had multipple ditch recuts. Limited excavation of the interior revealed pits and postholes, a few of them contemporary with the enclosure. A dense Neolithic scatter lay 500m to the north. Radiocarbon dating was hindered by poor collagen preservation.

Objectives: to refine the dating of the enclosure and determine whether the palisade was Neolithic in origin.

Final comment: F Healy (2012), recovered from the ditches. This was exacerbated by the abysmal collagen preservation of the bone assemblage (Brock *et al* 2007), particularly unfortunate as both antler tools which may have been used to dig the ditches and articulated animal bone are still available. Four samples of animal bone were sent for AMS dating, but failed to produce sufficient protein for combustion. A very approximate construction date of *3960–3125 cal BC (95% probability*; Whittle *et al* 2011, fig 6.11: *start Haddenham*) can be estimated. No suitable new samples could be found from the palisade (Whittle *et al* 2011, chapter 6).

References:	Brock *et al* 2007 Evans and Hodder 2006 Oswald *et al* 2001 Whittle *et al* 2011

GrA–31184 4415 ±35 BP

δ¹³C: -25.3‰

Sample: C-14 CC (a), submitted on 3 January 2006 by F Healy

Material: charcoal: *Quercus* sp., sapwood; single fragment (R Gale 2006)

Initial comment: segment J, context 1866. 'Early recut of marl platform, marks extension of segment across entrance'. The material may have been primary, carried up as the marl 'erupted' through the ditch fills, or, more probably, may have lain in the base of a recut (Evans and Hodder 2006, 255–7).

Objectives: to refine the chronology of the enclosure.

Calibrated date:	1σ: 3100–2930 cal BC 2σ: 3320–2910 cal BC

Final comment: F Healy (2012), GrA-31184 is statistically consistent with GrA-31185 and with bulk charcoal sample HAR-8093 (4560 ±90 BP), both from the same context (T′=2.8; T′(5%)=6.0; ν=2; Ward and Wilson 1978). The taphonomy of the context being uncertain, the relation of the date to the building of the enclosure is equally so.

References:	Ward and Wilson 1978

GrA–31185 4400 ±35 BP

δ¹³C: -25.5‰

Sample: C-14 CC (b), submitted on 3 January 2006 by F Healy

Material: charcoal: *Corylus/Alnus* sp., single fragment (R Gale 2006)

Initial comment: segment J, context 1866. 'Early recut of marl platform, marks extension of segment across entrance'. The material may have been primary, carried up as the marl 'erupted' through the ditch fills, or, more probably, may have lain in the base of a recut (Evans and Hodder 2006, 255–7).

Objectives: to refine the chronology of the enclosure.

Calibrated date:	1σ: 3100–2920 cal BC 2σ: 3270–2910 cal BC

Final comment: see GrA-31184

Causewayed Enclosures: Hembury, Devon

Location:	ST 1125002980 Lat. 50.49.08 N; Long. 03.15.38 E
Project manager:	D Liddell and M Todd (Devon Archaeological Society and University of Exeter), 1930–5 and 1980–3

Description: an enclosure formed by the inner ditch cutting off the promontory. Fragments of another Neolithic ditch and a possible Neolithic ditch were found farther to the north. Neolithic pits were found both inside and outside the inner enclosure. Mesolithic presence was evidenced by at least 13 microliths (some late, some possibly early) and 6 microburins, mainly from cutting XI at the southern tip of the spur (Berridge 1986).Obscured by Iron Age, Romano-British, and later archaeology. Excavated first by Dorothy Liddell, when the causewayed enclosure was identified, and later by Malcolm Todd, whose 'Neolithic' ditch was stratigraphically pre-Iron Age but contained no artefacts.

Objectives: samples were submitted to estimate the date of the construction of the inner ditch, to date the construction of the outer ditch, to confirm whether the putatively Neolithic features north of the inner ditch were indeed of such a date, to determine the date of the two identified areas of Neolithic occupation, and to establish the duration of the complex as a whole.

Final comment: F Healy (2012), the results confirmed that the three late-fifth/early-fourth millennium cal BC dates obtained in the 1960s (BM-130, -136, and -138) are too old, probably because their samples included mature wood. Modelling of the newly obtained dates places the early Neolithic use of the site in the 37th to 35th centuries cal BC. It was not possible to date Todd's 'Neolithic' ditch, but a pit in the same area may reflect slight activity later in the fourth millennium cal BC (Whittle *et al* 2011).

References: Barker and Mackey 1968
Berridge 1986
Brown 1989
Fox 1963
Griffith 2001
Liddell 1930
Liddell 1931
Liddell 1932
Liddell 1935
Mercer 1999
Mercer 2003
Oswald *et al* 2001
Palmer 1976
Todd 1984a
Todd 1984b
Whittle *et al* 2011

Causewayed Enclosures: Hembury, discrete features, Devon

Location: ST 1125002980
Lat. 50.49.08 N; Long. 03.15.38 E

Project manager: D Liddell (Devon Archaeological Society)
M Todd (University of Exeter), 1930–5
and 1981

Archival body: Royal Albert Memorial Museum and Art
Gallery, Exeter

Description: early Neolithic pits and hearths, most extensively excavated at the southern tip of the spur, but also encountered immediately inside and outside the inner ditch and, in one case, some way to the north of it.

Objectives: to refine the dating of the period of Neolithic activity on the site.

Final comment: F Healy (2012), five dated pits on the southern tip of the spur and three dated pits in the area of the inner ditch all fall in the 37th or 36th century cal BC. One of the dated pits near the inner ditch lay within what has long been thought of as an early Neolithic house; two samples from a hearth within the same house, however, date to the first millennium cal BC (GrA-31548 and GrA-31550), raising doubt as to the date of the structure. Carbonised residue from the interior of a South-Western style pot from a pit to the north of the inner ditch has been dated to the last quarter of the fourth millennium cal BC (GrA-31544). If the date is accurate it points to some activity after the main use of the complex.

References: Whittle *et al* 2011

GrA–31094 4845 ±40 BP

$\delta^{13}C$: -25.5‰

Sample: CXI f cooking pit 10 /B, submitted on 20 January 2006 by F Healy

Material: carbonised plant macrofossil (charred hazelnut shell fragment; from a different nut to CXI f cooking pit 10 /A, from bag containing numerous fragments) (F Healy 2006)

Initial comment: C XI f, pit 10. One of numerous pits excavated on the southern tip of the spur, at least partly protected by the Iron Age rampart (Liddell 1931, 109, and fig 5; 1932, pl. XV). The date of the shells will help place this episode of activity in the sequence.

Objectives: to refine the chronology of the complex.

Calibrated date: 1σ: 3660–3630 cal BC
2σ: 3710–3530 cal BC

Final comment: F Healy (2012), the result is statistically consistent with GrA-31213, measured on a sample from the same pit (T'=0.4; T'(5%)=3.8; ν=1; Ward and Wilson 1978) is and in agreement with the model.

References: Liddell 1931
Liddell 1932
Ward and Wilson 1978

GrA–31201 4805 ±35 BP

$\delta^{13}C$: -24.2‰

Sample: CXVI c cooking pit/A, submitted on 20 January 2006 by F Healy

Material: carbonised plant macrofossil (fragment of charred hazelnut shell; from bag containing numerous fragments) (F Healy 2006)

Initial comment: CXVI c, cooking pit. One of several pits close to the inner ditch (Liddell 1932, 172, and pl. V). The date of the shells will help place this episode of activity in the sequence.

Objectives: to refine the chronology of the complex.

Calibrated date: 1σ: 3650–3530 cal BC
2σ: 3660–3520 cal BC

Final comment: F Healy (2012), the result is statistically consistent with GrA-31204, measured on a sample from the same pit (T'=0.2; T'(5%)=3.8; ν=1; Ward and Wilson 1978), and is in agreement with the model.

References: Liddell 1932
Ward and Wilson 1978

GrA–31204 4825 ±35 BP

$\delta^{13}C$: -24.1‰

Sample: CXVI c cooking pit/B, submitted on 20 January 2006 by F Healy

Material: carbonised plant macrofossil (charred hazelnut shell fragment; from bag containing numerous fragments) (F Healy 2006)

Initial comment: CXVI c, cooking pit. One of several pits close to the inner ditch (Liddell 1932, 172, and pl. V). The date of the shells will help place this episode of activity in the sequence.

Objectives: to refine the chronology of the complex.

Calibrated date: 1σ: 3650–3530 cal BC
2σ: 3660–3520 cal BC

Final comment: see GrA-31201

References: Liddell 1932

GrA–31205 4795 ±35 BP

δ¹³C: -25.8‰

Sample: CXI f cooking pit J /A, submitted on 20 January 2006 by F Healy

Material: carbonised plant macrofossil (charred hazelnut shell fragment; from a different nut to CXI f cooking pit J /B, from bag containing numerous fragments) (F Healy 2006)

Initial comment: CXI f, pit J. One of numerous pits excavated on the southern tip of the spur, at least partly protected by the Iron Age rampart (Liddell 1931, 109, and fig 5, pl. XV). The date of the shells will help place this episode of activity in the sequence.

Objectives: to refine the chronology of the complex.

Calibrated date: 1σ: 3640–3530 cal BC
                        2σ: 3650–3510 cal BC

Final comment: F Healy (2012), the result is statistically consistent with GrA-31206, measured on a sample from the same pit (T'=0.0; T'(5%)=3.8; *v*=1; Ward and Wilson 1978), and is in agreement with the model.

References: Liddell 1931
                        Ward and Wilson 1978

GrA–31206 4785 ±35 BP

δ¹³C: -28.8‰

Sample: CXI f cooking pit J /B, submitted on 20 January 2006 by F Healy

Material: carbonised plant macrofossil (charred hazelnut shell fragment; from a different nut to CXI f cooking pit J /A, from bag containing numerous fragments) (F Healy 2006)

Initial comment: CXI f, pit J. One of numerous pits excavated on the southern tip of the spur, at least partly protected by the Iron Age rampart (Liddell 1931, 109, fig 5, and pl. XV). The date of the shells will help place this episode of activity in the sequence.

Objectives: to refine the chronology of the complex.

Calibrated date: 1σ: 3640–3520 cal BC
                        2σ: 3650–3380 cal BC

Final comment: see GrA-31205

References: Liddell 1931

GrA–31207 4820 ±35 BP

δ¹³C: -28.8‰

Sample: CXI g cooking pit 11 (a), submitted on 20 January 2006 by F Healy

Material: carbonised plant macrofossil (charred hazelnut shell; 2 large fragments of charred hazelnut shell each from a different nut, from bag containing numerous fragments) (F Healy 2006)

Initial comment: CXI g, cooking pit 11. One of numerous pits excavated on the southern tip of the spur, at least partly protected by the Iron Age rampart (Liddell 1931,109, fig 5; 1932, and pl. XV). The date of the shells will help place this episode of activity in the sequence.

Objectives: to refine the chronology of the complex.

Calibrated date: 1σ: 3650–3530 cal BC
                        2σ: 3660–3520 cal BC

Final comment: F Healy (2012), the result is statistically slightly inconsistent with GrA-31209, measured on a sample from the same pit (T'=3.9; T'(5%)=3.8; *v*=1; Ward and Wilson 1978); but is in agreement with the model.

References: Liddell 1931
                        Liddell 1932
                        Ward and Wilson 1978

GrA–31209 4925 ±40 BP

δ¹³C: -29.9‰

Sample: CXI g cooking pit 11(b), submitted on 20 January 2006 by F Healy

Material: carbonised plant macrofossil (charred hazelnut shell; 2 large fragments of charred hazelnut shell each from a different nut, from bag containing numerous fragments) (F Healy 2006)

Initial comment: CXI g, cooking pit 11. One of numerous pits excavated on the southern tip of the spur, at least partly protected by the Iron Age rampart (Liddell 1931, 109, fig 5; 1932, and pl. XV). The date of the shells will help place this episode of activity in the sequence.

Objectives: to refine the chronology of the complex.

Calibrated date: 1σ: 3720–3650 cal BC
                        2σ: 3790–3640 cal BC

Final comment: see GrA-31207

References: Liddell 1931
                        Liddell 1932

GrA–31210 4855 ±40 BP

δ¹³C: -24.3‰

Sample: CXI k cooking pit H /B, submitted on 20 January 2006 by F Healy

Material: carbonised plant macrofossil (charred hazelnut shell; large fragments of charred hazelnut shell from a different nut to CXI k cooking pit H/A, from bag containing numerous fragments) (F Healy 2006)

Initial comment: CXI k, pit H. One of numerous pits excavated on the southern tip of the spur, at least partly protected by the Iron Age rampart (Liddell 1931,109, fig 5; 1932, and pl. XV). The date of the shells will help place this episode of activity in the sequence.

Objectives: to refine the chronology of the complex.

Calibrated date: 1σ: 3660–3630 cal BC
                        2σ: 3710–3530 cal BC

Final comment: F Healy (2012), the result is statistically consistent with GrA-31211, measured on a sample from the same pit (T′=0.0; T′(5%)=3.8; ν=1; Ward and Wilson 1978), and in agreement with the model.

References: Liddell 1931
                    Liddell 1932
                    Ward and Wilson 1978

GrA–31211 4855 ±40 BP

$\delta^{13}C$: -24.3‰

Sample: CXI k cooking pit H /B, submitted on 20 January 2006 by F Healy

Material: carbonised plant macrofossil (charred hazelnut shell; large fragments of charred hazelnut shell from a different nut to CXI k cooking pit H /A, from bag containing numerous fragments) (F Healy 2006)

Initial comment: CXI k, pit H. One of numerous pits excavated on the southern tip of the spur, at least partly protected by the Iron Age rampart (Liddell 1931,109, fig 5; 1932, and pl. XV). The date of the shells will help place this episode of activity in the sequence.

Objectives: to refine the chronology of the complex.

Calibrated date: *1σ:* 3660–3630 cal BC
                    2σ: 3710–3530 cal BC

Final comment: see GrA-31210

References: Liddell 1931
                    Liddell 1932

GrA–31213 4880 ±40 BP

$\delta^{13}C$: -24.7‰

Sample: CXI f cooking pit 10 /A, submitted on 20 January 2006 by F Healy

Material: carbonised plant macrofossil (charred hazelnut shell fragment; from a different nut to CXI f cooking pit 10 /B, from bag containing numerous fragments) (F Healy 2006)

Initial comment: C XI f, pit 10. One of numerous pits excavated on the southern tip of the spur, at least partly protected by the Iron Age rampart (Liddell 1931, 109, fig 5; 1932, and pl. XV). The date of the shells will help place this episode of activity in the sequence.

Objectives: to refine the chronology of the complex.

Calibrated date: *1σ:* 3700–3640 cal BC
                    2σ: 3720–3630 cal BC

Final comment: see GrA-31094

References: Liddell 1931
                    Liddell 1932

GrA–31463 4750 ±35 BP

$\delta^{13}C$: -22.3‰

Sample: CXXI extension charcoal & nuts (b), submitted on 20 January 2006 by F Healy

Material: carbonised plant macrofossil (*Corylus avellana* nutshell fragment) (R Gale 2006)

Initial comment: CXXI extension. 'Cooking hole'. One of several pits close to the inner ditch (Liddell 1935, 136). If short-life fragments can be extracted they will help place this episode of activity in the sequence.

Objectives: to refine the chronology of the complex.

Calibrated date: *1σ:* 3640–3380 cal BC
                    2σ: 3640–3370 cal BC

Final comment: F Healy (2012), the result is statistically consistent with GrA-31559, measured on a sample from the same pit (T′=3.7; T′(5%)=3.8; ν=1; Ward and Wilson 1978), and in agreement with the model.

References: Liddell 1935
                    Ward and Wilson 1978

GrA–31466 4770 ±35 BP

$\delta^{13}C$: -26.5‰

Sample: Ch 58. CXI d extn, pit 2 layer 2 charcoal (a), submitted on 20 January 2006 by F Healy

Material: carbonised plant macrofossil (*Corylus avellana* nutshell) (R Gale 2006)

Initial comment: CXI d extn, pit 2, layer 2. One of numerous pits excavated on the southern tip of the spur, at least partly protected by the Iron Age rampart (Liddell 1931, 111, fig 5, and pl. XXIV). If short-life fragments can be extracted they will help place this episode of activity in the sequence.

Objectives: to refine the chronology of the complex.

Calibrated date: *1σ:* 3640–3520 cal BC
                    2σ: 3650–3380 cal BC

Final comment: F Healy (2012), the result is statistically consistent with GrA-31467, measured on a sample from the same pit (T′=2.3; T′(5%)=3.6; ν=1; Ward and Wilson 1978), and in agreement with the model.

References: Liddell 1931
                    Ward and Wilson 1978

GrA–31467 4845 ±35 BP

$\delta^{13}C$: -24.0‰

Sample: Ch 58. CXI d extn, pit 2 layer 2 charcoal (b), submitted on 20 January 2006 by F Healy

Material: carbonised plant macrofossil (*Corylus avellana* nutshell fragment) (R Gale 2006)

Initial comment: CXI d extn, pit 2, layer 2. One of numerous pits excavated on the southern tip of the spur, at least partly protected by the Iron Age rampart (Liddell 1931, 111, fig 5, and pl. XXIV). If short-life fragments can be extracted they will help place this episode of activity in the sequence.

Objectives: to refine the chronology of the complex.

Calibrated date: *1σ:* 3660–3630 cal BC
                    2σ: 3700–3530 cal BC

Final comment: see GrA-31466

References: Liddell 1931

GrA–31544 4505 ±40 BP

δ¹³C: -29.6‰

Sample: H81 3 F41/A, submitted on 15 March 2006 by F Healy

Material: carbonised residue (from one sherd of many, probably from a single coarse, lugged vessel in the South-Western style. From same vessel as H81 3 F41/B) (F Healy 2006)

Initial comment: H81 3 F41. From area north of the inner ditch and west of the outer ditch, in a cutting adjacent to Liddell's CXXXIV. In the upper and mid fill of a shallow pit (F41), cut into backfilled deeper pit (F90; Todd 1984a, 257–8, and figs 5–6) containing numerous South-Western style sherds, probably from a single coarse, lugged pot, possibly broken *in situ*, since rim sherds were found at the top. This is the only Neolithic pit to have been excavated in this part of the site, and thus provides a glimpse of what must have taken place away from the concentrations of features from which the other samples have come.

Objectives: to refine the chronology of the complex.

Calibrated date: *1σ:* 3350–3090 cal BC
                 2σ: 3370–3020 cal BC

Final comment: F Healy (2012), if this measurement is included in the overall model for Neolithic activity at Hembury, it has poor agreement (A=1.2%). For this reason it has been excluded from the model. A replicate measurement from residue on another sherd of this vessel failed to produce a radiocarbon determination, and so it is possible that the material dated was poorly preserved. This date seems rather late for South-Western-style pottery.

References: Todd 1984a

GrA–31545 4770 ±35 BP

δ¹³C: -24.8‰

Sample: Ch 35. Cutting III c charcoal from pit (hearth) /A, submitted on 20 January 2006 by F Healy

Material: charcoal: *Corylus avellana*, single fragment (R Gale 2006)

Initial comment: CIII c layer 2. From pit (hearth) containing much burnt material, Neolithic Bowl pottery, and lithics including two arrowheads, under the northern butt of the Iron Age rampart at the southern side of the western entrance, beside a hearth, within a U-plan setting of postholes interpreted as Neolithic structure. Both features were at the edge of the 1931 cutting, so that part of each was excavated in 1931 and part in 1932. The excavation of more of the pit in 1932 indicates that the 1931 plan, which shows it complete outline, was subsequently revised. In 1932, it was described as a mound of white wood ash and charcoal 6ft in extent and 2ft deep, with two charred oak branches lying across it (Liddell 1931, 99, and figs 2–3: A; 1932, 170–2, and pls IV, VII). If short-life fragments can be extracted they will help place this episode of activity in the sequence.

Objectives: to refine the chronology of the complex.

Calibrated date: *1σ:* 3640–3520 cal BC
                 2σ: 3650–3380 cal BC

Final comment: F Healy (2012), this result is statistically consistent with GrA-31546, measured on a sample from the same pit (T′=0.5; T′(5%)=3.8; ν=1; Ward and Wilson 1978), and in agreement with the model.

References: Liddell 1931
                 Liddell 1932
                 Ward and Wilson 1978

GrA–31546 4735 ±35 BP

δ¹³C: -23.7‰

Sample: Ch 35. Cutting III c charcoal from pit (hearth) /B, submitted on 20 January 2006 by F Healy

Material: charcoal: *Corylus avellana*, single fragment (R Gale 2006)

Initial comment: as GrA-31545

Objectives: to refine the chronology of the complex.

Calibrated date: *1σ:* 3640–3380 cal BC
                 2σ: 3640–3370 cal BC

Final comment: see GrA-31545

GrA–31548 2485 ±35 BP

δ¹³C: -26.9‰

Sample: CIII e layer 2 charcoal & earth from ashes below hearth /A, submitted on 20 January 2006 by F Healy

Material: charcoal: *Betula* sp., single fragment (R Gale 2006)

Initial comment: CIII e layer 2. Charcoal and earth from ashes below hearth. 'Hearth' was a circle of stones laid flat and tightly packed around a mound of charcoal, ash and fire-cracked chert under the northern butt of the Iron Age rampart at the southern side of the west entrance, beside a 'cooking pit' within a U-plan setting of postholes interpreted as Neolithic structure. Both features were at the edge of the 1931 cutting, so that part of each was excavated in 1931 and part in 1932. (Liddell 1931, 99, and figs 2–3: C; 1932, 170–2, and pls IV and VIII). If short-life fragments can be extracted they will help place this episode of activity in the sequence.

Objectives: to refine the chronology of the complex.

Calibrated date: *1σ:* 770–530 cal BC
                 2σ: 790–410 cal BC

Final comment: F Healy (2012), this date and GrA-31550, measured on a sample from the same hearth, place the feature in the first millennium cal BC, raising the possibility that the structure in which it lay was also of this date.

References: Liddell 1931
                 Liddell 1932

GrA–31550 2635 ±35 BP

δ¹³C: -27.9‰

Sample: CIII e layer 2 charcoal & earth from ashes below hearth /B, submitted on 20 January 2006 by F Healy

Material: charcoal: *Betula* sp., single fragment (R Gale 2006)

Initial comment: as GrA-31548

Objectives: to refine the chronology of the complex.

Calibrated date: 1σ: 820–790 cal BC
2σ: 840–780 cal BC

Final comment: see GrA-31548

GrA–31559 4845 ±35 BP

δ¹³C: -25.7‰

Sample: CXXI extension charcoal nuts (a), submitted on 20 January 2006 by F Healy

Material: carbonised plant macrofossil (*Corylus* sp., nutshell fragment) (R Gale 2006)

Initial comment: CXXI extension. 'Cooking hole'. One of several pits close to the inner ditch (Liddell 1935, 136). If short-life fragments can be extracted they will help place this episode of activity in the sequence.

Objectives: to refine the chronology of the complex.

Calibrated date: 1σ: 3660–3630 cal BC
2σ: 3700–3530 cal BC

Final comment: see GrA-31463

References: Liddell 1935

Causewayed Enclosures: Kingsborough 1, Kent

Location: TQ 9770072000
Lat. 51.24.43 N; Long. 00.49.59 E

Project manager: S Stevens (Archaeology South-East), 1998

Archival body: Archaeology South-East

Description: a causewayed enclosure with an overall diameter of approximately 160m with three circuits. Much pottery and flint was found in the inner and middle circuits. An undated small central ditched enclosure with a four-post structure was also excavated. Kingsborough 1 lay approximately 100m south of the Kingsborough 2 enclosure. Excavation by Archaeology South-East in 1998 was followed by a further ditch section by Wessex Archaeology in 2004 to recover environmental samples.

Objectives: samples were submitted to determine whether Kingsborough 1 and 2 were in contemporary use, whether either or both were contemporary with the enclosure at Chalk Hill, Ramsgate, and for how long the Kingsborough enclosures were in use.

Final comment: F Healy (2012), animal bone preservation was poor and suitable samples were confined to the inner ditch. These make it possible to estimate a construction date of *3780–3520 cal BC (95% probability*; Whittle *et al* 2011: fig 7.15: start Kingsborough 1), probably after both Kingsborough 1 and Chalk Hill although in concurrent use with them (Whittle *et al* 2011, figs 7.23–4).

References: Allen *et al* 2008
Dyson *et al* 2000
Oswald *et al* 2001
Whittle *et al* 2011

GrA–29551 4770 ±40 BP

δ¹³C: -26.2‰

Sample: 46792–2530/1109/A, submitted on 4 May 2005 by F Healy

Material: carbonised plant macrofossil (charred hazelnut shell fragment) (C Stevens 2004)

Initial comment: inner ditch, feature 457. Context 2530, primary silt against inner edge of ditch. Stratified below context 2531. From a concentration of 28 hazelnut shell fragments apparently deposited in the course of a single event.

Objectives: to establish the date of this context as a step towards defining the chronology of the monument.

Calibrated date: 1σ: 3640–3520 cal BC
2σ: 3650–3370 cal BC

Final comment: F Healy (2012), the date is compatible with its stratigraphic position in the model.

GrA–29553 4760 ±40 BP

δ¹³C: -23.2‰

Sample: 46792–2531/1110/A, submitted on 4 May 2005 by F Healy

Material: carbonised plant macrofossil (charred hazelnut shell fragment) (C Stevens 2004)

Initial comment: inner ditch, feature 457. Context 2531.From a concentration of 30 charred hazelnut shell fragments, stratified above contexts 2530 and 2532 and below 2528 and 2529, the second of which which contained a semi-complete Whitehawk Style Bowl.

Objectives: to establish the date of this context as a step towards defining the chronology of the monument.

Calibrated date: 1σ: 3640–3510 cal BC
2σ: 3650–3370 cal BC

Final comment: see GrA-29551

GrA–29554 5110 ±40 BP

δ¹³C: -27.2‰

Sample: 46792–2532/1112/A, submitted on 4 May 2005 by F Healy

Material: carbonised plant macrofossil (charred hazelnut shell fragment) (C Stevens 2004)

Initial comment: inner ditch, feature 457. Context 2532. Primary silt on ditch base and outer side. From a concentration of 14 charred hazelnut fragments. Stratified below context 2531.

Objectives: to establish the date of this context as a step towards defining the chronology of the monument.

Calibrated date: 1σ: 3970–3810 cal BC
2σ: 3990–3790 cal BC

Final comment: F Healy (2012), this sample was probably redeposited because older than OxA-14768 from the same context. Since context 2532 was at the base of the ditch (Allen *et al* 2008, fig 4), the sample may derive from pre-construction activity in the area.

OxA–14766 4774 ±35 BP

$\delta^{13}C$: -24.1‰

Sample: 46792–2530/1109/B, submitted on 4 May 2005 by F Healy

Material: carbonised plant macrofossil (charred hazelnut shell fragment) (C Stevens 2004)

Initial comment: inner ditch, feature 457. Context 2530, primary silt against inner edge of ditch. Stratified below context 2531. From a concentration of 28 hazelnut shell fragments.

Objectives: to establish the date of this context as a step towards defining the chronology of the monument.

Calibrated date: 1σ: 3640–3520 cal BC
2σ: 3650–3380 cal BC

Final comment: see GrA-29551

OxA–14767 4719 ±34 BP

$\delta^{13}C$: -25.2‰

Sample: 46792–2531/1110/B, submitted on 4 May 2005 by F Healy

Material: carbonised plant macrofossil (charred hazelnut shell fragment) (C Stevens 2004)

Initial comment: inner ditch, feature 457. Context 2531. From a concentration of 30 charred hazelnut shell fragments, stratified above contexts 2530 and 2532 and below 2528 and 2529, the second of which which contained a semi-complete Whitehawk Style Bowl.

Objectives: to establish the date of this context as a step towards defining the chronology of the monument.

Calibrated date: 1σ: 3630–3370 cal BC
2σ: 3640–3370 cal BC

Final comment: see GrA-29551

OxA–14768 4704 ±35 BP

$\delta^{13}C$: -25.2‰

Sample: 46792–2532/1112/B, submitted on 4 May 2005 by F Healy

Material: carbonised plant macrofossil (charred hazelnut shell fragment) (C Stevens 2004)

Initial comment: inner ditch, feature 457. Context 2532. Primary silt on ditch base at outer side. From a concentration of 14 charred hazelnut fragments. Stratified below context 2531.

Objectives: to establish the date of this context as a step towards defining the chronology of the monument.

Calibrated date: 1σ: 3630–3370 cal BC
2σ: 3640–3370 cal BC

Final comment: see GrA-29551

Causewayed Enclosures: Kingsborough 2, Kent

Location: TQ 9770072350
Lat. 51.24.43 N; Long. 00.49.59 E

Project manager: R Greatorex (Wessex Archaeology), 2004

Archival body: Wessex Archaeology

Description: a causewayed enclosure formed of a single arc of segments possibly delimited to the north by a steepish slope rather than by earthworks, disappearing under newly built houses to south, and not yet traced to the west. Segments are of variable dimensions and seem to have silted naturally. Infilling of open cuttings suggests that primary silt could have accumulated in weeks. Sections suggest there was an external bank. The only hint of an extended history is a segment which almost completely occupies what was originally a very wide causeway (entrance?), cutting the terminal of an earlier segment. Finds are relatively few. The only features in the interior were one unaccompanied cremation and one narrow post-medieval ditch.

Objectives: samples were submitted to determine whether Kingsborough 1 and 2 were in contemporary use, whether either or both were contemporary with the Chalk Hill, Ramsgate, enclosure and for how long the Kingsborough enclosures were in use.

Final comment: F Healy (2012), samples were available only from the primary silts, because the circuit seems to have been left to infill naturally after its initial use. Its estimated construction date is *3790–3630 cal BC (76% probability)* or *3615–3535 cal BC (19% probability*; Whittle *et al* 2011, fig 7.17: *start Kingsborough 2*), before Kingsborough 1 and after Chalk Hill, although in concurrent use with them (Whittle *et al* 2011, fig 7.23–4).

References: Allen *et al* 2008
Whittle *et al* 2011

GrA–29555 4815 ±45 BP

$\delta^{13}C$: -25.5‰

Sample: 57170–M1248 @ 64 cm/A, submitted on 4 May 2005 by F Healy

Material: charcoal: Pomoideae, single fragment (C Chisham 2004)

Initial comment: segment 1, context 6132. From a discrete charcoal lens close to the base of the ditch, recovered from a monolith tin. Since the Pomoideae are short-lived, the charcoal should be close in age to the burning event and the construction of the earthwork.

Objectives: to establish the date of this context as a step towards defining the chronology of the monument.

Calibrated date: 1σ: 3650–3530 cal BC
2σ: 3700–3510 cal BC

Final comment: F Healy (2012), the result is compatible with its stratigraphic position in the model.

GrA–29557 4780 ±45 BP

δ¹³C: -23.9‰

Sample: 57170–6037 Sample 1207/A, submitted on 4 May 2005 by F Healy

Material: carbonised plant macrofossil (*Prunus spinosa* (sloe) stone) (C Stevens 2004)

Initial comment: segment 2, group 6037, context 6081. In a subconical heap of animal bone with a whetstone on the surface of the primary fill in the ditch butt. From the sample which also yielded two emmer glumes, one possible emmer grain, grass roots, and a Rosaceae thorn. The speed with which the sides of the excavated ditch segments collapsed shows that this deposit would have been put in place soon after the ditch was originally dug.

Objectives: to establish the date of this context as a step towards defining the chronology of the monument.

Calibrated date: 1σ: 3640–3520 cal BC
2σ: 3660–3380 cal BC

Final comment: see GrA-29555

OxA–14732 4858 ±34 BP

δ¹³C: -24.6‰

Sample: 57170–6213/B, submitted on 4 May 2005 by F Healy

Material: carbonised residue (internal; one of four well-preserved Neolithic Bowl body sherds from at least two vessels with residue, in this case fresh and abundant) (L Mepham 2004)

Initial comment: segment 2, group 6030, context 6213. Silt immediately overlying ditch base. Coarse, hand-made vessels are unlikely to have had a life of more than a few years. The sherds from the pots would have been incorporated in the ditch fills soon after the construction of the monument and, especially in the case of the largest sherd with the freshest residue, very soon after the last use of the vessel.

Objectives: to establish the date of this context as a step towards defining the chronology of the monument.

Calibrated date: 1σ: 3660–3630 cal BC
2σ: 3710–3530 cal BC

Final comment: see GrA-29555

OxA–14790 4779 ±36 BP

δ¹³C: -25.4‰

Sample: 57170–M1248 @ 64 cm/B, submitted on 4 May 2005 by F Healy

Material: charcoal: Pomoideae, single fragment (C Chisham 2004)

Initial comment: segment 1, context 6132. From a discrete charcoal lens close to the base of the ditch, recovered from a monolith tin. Since Pomoideae is short-lived, the charcoal should be close in age to the burning event and the construction of the earthwork.

Objectives: to establish the date of this context as a step towards defining the chronology of the monument.

Calibrated date: 1σ: 3640–3520 cal BC
2σ: 3650–3380 cal BC

Final comment: see GrA-29555

OxA–14791 4874 ±36 BP

δ¹³C: -25.8‰

Sample: 57170–6037 Sample 1207/B, submitted on 4 May 2005 by F Healy

Material: carbonised plant macrofossil (Rosaceae; cf *Prunus spinosa* (sloe) thorn) (C Stevens 2004)

Initial comment: as GrA-29557

Objectives: as GrA-29557

Calibrated date: 1σ: 3700–3630 cal BC
2σ: 3710–3630 cal BC

Final comment: see GrA-29555

Causewayed Enclosures: Knap Hill, Wiltshire

Location: SU 1210063650
Lat. 51.21.55 N; Long. 01.49.39 W

Project manager: M Cunnington (Independent) G Connah (University of Cambridge), 1908–9 and 1961

Archival body: Wiltshire Heritage Museum

Description: a single circuit causewayed enclosure conforming to the contours of the hill, possibly incomplete to the south (Oswald *et al* 2001, figs 2.6, 2.7, and 5.29). Following excavations in 1908, M E Cunnington recognised 'a method of defence hitherto unobserved in prehistoric fortifications in Britain . . . the entrenchment, instead of being continuous, . . . is broken up into short and irregular sections . . . It is believed that the camp is of early date, that it belongs to the bronze, or even to the late neolithic period.' (1909). Limited excavation by Connah in 1961 confirmed that it was a causewayed enclosure attributable to what was then known as the Windmill Hill Culture (1965).

Objectives: samples were to better define the construction date and duration of primary use, as well as the chronological relation of the site to Windmill Hill, some 7km to the north-west, and, as far as possible, to the local concentration of long barrows.

Final comment: F Healy (2012), Bayesian modelling of the dates provides an estimated construction date of *3530–3375 cal BC (91% probability*; Whittle *et al* 2011, fig 3.25: *start Knap Hill)* and an estimated end of primary use of *3525–3220 cal BC (92% probability*; Whittle *et al* 2011, fig 3.25:

end Knap Hill). Because there are relatively few measurements, the estimate for the duration of primary use is broad: *1–460 years (95% probability*; Whittle *et al* 2011, fig 3.26: *use Knap Hill).* The stratigraphy, however, suggests that this was actually short, since the ditch seems to have been left to silt naturally, without recutting, until an episode of Beaker occupation. The Knap Hill enclosure was probably built rather more than a century after the West Kennet long barrow and the Windmill Hill enclosure, although still during the primary use of the latter (Whittle *et al* 2011, fig 3.32).

References: Barker *et al* 1971
Connah 1965
Connah 1969
Cunnington 1909
Cunnington 1912
Oswald *et al* 2001
Palmer 1976
Whittle *et al* 2011

GrA–29808 4975 ±40 BP

$\delta^{13}C$: -21.4‰

Sample: K/I/(6)/A1, submitted on 2 September 2005 by F Healy

Material: antler: *Cervus elaphus,* fragments remaining from the measurement of BM-205, of which this is a replicate. Photograph in site album (Devizes Museum Library AA box 14) shows K/I/(6)/A.1 (the sample for BM-205) as a length of beam with one tine, projecting from section (F Healy 2005)

Initial comment: segment 3, cutting I, layer 6. Primary rubble of ditch, near base, sealed by 1–2m of undisturbed deposits at 46'4" ∞ 1'4" ∞ 5'4" (Connah 1965, fig 2: south-west face). The character and location of the sample indicate that it was an implement used to dig the ditch in which it was found.

Objectives: to confirm and refine BM-205, in order to date the context as a step towards dating the enclosure.

Calibrated date: 1σ: 3800–3700 cal BC
                  2σ: 3940–3650 cal BC

Final comment: F Healy (2012), the date is not statistically consistent with BM-205 (4710 ±115 BP; Barker *et al* 1971), of which it is a replicate (T'=4.6; T'(5%)=3.8; ν=1; Ward and Wilson 1978). Furthermore this date for an antler from the primary rubble in the ditch is earlier than those for three articulated samples sealed under the bank (OxA-15199, GrA-29810, OxA-15200, and GrA-29809). The antler may have been redeposited or the date may be an outlier.

References: Ward and Wilson 1978

GrA–29809 4755 ±40 BP

$\delta^{13}C$: -21.3‰

Sample: K/II/(8)/B107/B, submitted on 2 September 2005 by F Healy

Material: animal bone: *Bos* sp., proximal half of left metacarpal found articulated with two carpals (G Connah).

Could, on size, have come from same animal as K/II/(8)/B107/C, which was not dated (A Powell). Replicate of K/II/(8)/B107/A (G Connah) (G Connah 1961)

Initial comment: segment 5, cutting II, layer 8. Found articulated on old land surface under bank (Connah 1965, fig 3, pl. IIa). At 22'4" ∞ 3'4" ∞ 2'. The articulation of the three bones shows that they were still joined by soft tissue when buried, so that the animal from which they came was not long dead. The fact that they were undisturbed when the bank was built shows that there was only a short interval between their deposition and its construction. A replicate of OxA-15200.

Objectives: to date the context as a step towards dating the enclosure.

Calibrated date: 1σ: 3640–3380 cal BC
                  2σ: 3640–3370 cal BC

Final comment: see OxA-15200

GrA–29810 4775 ±40 BP

$\delta^{13}C$: -22.5‰

Sample: K/III/(8)/B18, submitted on 2 September 2005 by F Healy

Material: animal bone: *Bos* sp., right radius of mature individual found with fitting ulna (G Connah 1961)

Initial comment: segment 6, cutting III, layer 8. Under bank. 'Lying right on top of (8)' (findsbook). At 20' 10" ∞ 7'0" ∞ 2'1". Found together with ulna on old land surface beneath bank of enclosure (Connah 1965, fig 2). The two bones must have been deposited soon after the death of the animal from which they came, since they were still joined by soft tissue, and shortly before the bank was built, since they remained in place.

Objectives: to date the context as a step towards dating the enclosure.

Calibrated date: 1σ: 3640–3520 cal BC
                  2σ: 3650–3380 cal BC

Final comment: see OxA-15199

OxA–15199 4657 ±31 BP

$\delta^{13}C$: -22.4‰

Sample: K/I/(8)/B21, submitted on 2 September 2005 by F Healy

Material: animal bone: *Bos* sp., radius from immature individual, with articular ends missing, found with fragmentary fitting ulna (G Connah 1961)

Initial comment: segment 3, cutting I, layer 8. Under bank. At 21'9" ∞ 1'9" ∞ 1'10". Found together with ulna on old land surface beneath bank of enclosure (Connah 1965, fig 2). The two bones must have been deposited soon after the death of the animal from which they came, since they were still joined by soft tissue, and shortly before the bank was built.

Objectives: to date the context as a step towards dating the enclosure.

Calibrated date: 1σ: 3510–3360 cal BC
                      2σ: 3620–3360 cal BC

Final comment: F Healy (2012), this result, like the others for articulated samples sealed beneath the bank, is consistent with the sample's stratigraphic position and with the model.

OxA–15200 4699 ±37 BP

δ¹³C: -20.8‰

Sample: K/II/(8)/B107/A, submitted on 2 September 2005 by F Healy

Material: animal bone: *Bos* sp., proximal half of left metacarpal found articulated with two carpals (G Connah). Could, on size, have come from same animal as K/II/(8)/B107/C, which was not dated (A Powell). Replicate of K/II/(8)/B107/B (G Connah 1961)

Initial comment: segment 5, cutting II, layer 8. Found articulated on old land surface under bank (Connah 1965, fig 3, pl. IIa). At 22'4" ∞ 3'4" ∞ 2'. The articulation of the three bones shows that they were still joined by soft tissue when buried, so that the animal from which they came was not long dead. The fact that they were undisturbed when the bank was built shows that there was only a short interval between their deposition and its construction. A replicate of GrA-29809.

Objectives: to date the context as a step towards dating the enclosure.

Calibrated date: 1σ: 3630–3370 cal BC
                      2σ: 3640–3360 cal BC

Final comment: F Healy (2012), OxA-15200 is statistically with GrA-29809, its replicate measurement (T'=1.1; T'(5%)=3.8; ν=1; Ward and Wilson 1978). Otherwise as OxA-15199.

References: Ward and Wilson 1978

OxA–15305 4701 ±34 BP

δ¹³C: -27.7‰

Sample: K/III/(6)/P23, submitted on 2 September 2005 by F Healy

Material: carbonised residue (internal; Neolithic Bowl sherd (Connah 1965, 21). Sherd now formed of fragments glued together along recent breaks) (F Healy 2005)

Initial comment: segment 6, cutting III, layer 6. At 49'8" ∞ 1'4" ∞ 3' 6". At top of primary chalk rubbble (Connah 1965, fig 4). The freshness of the residue indicates that the sherd was buried soon after the last use of the vessel from which it came.

Objectives: to date the context as a step towards dating the enclosure.

Calibrated date: 1σ: 3630–3370 cal BC
                      2σ: 3640–3370 cal BC

Final comment: F Healy (2012), the result is consistent with the sample's stratigraphic position and with the model.

Causewayed Enclosures: Maiden Bower, Bedfordshire

Location: SP 9966022470
               Lat. 51.53.14 N; Long. 00.33.40 W

Project manager: W G Smith and C L Matthews (Antiquarian and Manshead Archaeological Society of Dunstable), late 1800s/early 1900s and mid-twentieth century

Description: a causewayed enclosure located on the Lower Chalk of Chiltern scarp. Five U-profiled ditch segments (Worthington Smith 1904b, 40; Oswald *et al* 2001, fig 2.18) and other features partly overlain by an Iron Age hillfort and extending to the west and north-west of it. Observed and salvaged during and in advance of quarrying by Worthington Smith and later by the Manshead Archaeological Society of Dunstable led by C L Matthews. Geophysical survey and fieldwalking was undertaken by M Hamilton and J Pollard, and earthwork survey and section drawing by RCHME. 'During the very extensive excavations for chalk by Messrs. Forder & Co. on the west side of Maiden Bower, numerous discoveries of shallow pits, filled with chalk rubble, broken bones, antlers of fallow-deer, broken and cut antlers of red-deer, flints, etc., have been found' (Smith 1915, 149). Decorated Bowl pottery has been recovered (Piggott 1931, fig 6; Matthews 1976, fig 4). Dense flint scatters have been found dating from both the early and later Neolithic, focussed on the earthwork. 'For a certain number of yards outside the camp the same abundance prevails, but beyond a given circuit both implements and flakes are rare (Smith 1904a, 160).

Objectives: samples were submitted once it was apparent that they could be assigned to contexts in the ditch segments or in an apparently internal pit in order to date what had almost certainly been a causewayed enclosure.

Final comment: F Healy (2012), three samples from Worthington Smith's collection were marked in ink as coming from Maiden Bower and corresponded to his descriptions of his finds from the probable ditch segments. Two further samples were clearly documented as coming from the pit excavated by the Manshead Society. Together, these point to a construction date of *3775–3380 cal BC (95% probability;* Whittle *et al* 2011, fig 6.4: *start Maiden Bower).*

References: Curwen 1930
                Davies 1956
                Dyer 1955
                Dyer 1961
                Matthews 1976
                Oswald *et al* 2001
                Palmer 1976
                Piggott 1931
                Piggott 1954
                Pollard and Hamilton 1994
                Smith 1894
                Smith 1904a
                Smith 1904b
                Smith 1915
                Whittle *et al* 2011

Causewayed Enclosures: Maiden Bower, ditch, Bedfordshire

Location: SP 9966022470
Lat. 51.53.14 N; Long. 00.33.40 W

Project manager: W Smith (Antiquarian) C Matthews (Manshead Archaeological Society of Dunstable), late 1800s/early1900s and mid-twentieth century

Archival body: Wardown Park Museum, Luton

Description: segments probably of a causewayed enclosure ditch exposed during chalk quarrying and further exposed by erosion of the abandoned quarry face. Smith observed five probable segments (1904b) and described two in greater detail: 'The two long grave-like excavations at 9 on fig 2 were remarkable. The smaller was 25ft [7.2m] long, 10ft [3.1m] wide, and 4ft [1.2m] deep, the larger 43ft [13.1m] long, 10ft [3.1m] wide, and 3ft [0.90m] deep. They were excavated in 1897 and found to contain many split, root-eaten bones and broken or cut antlers of red and roe-deer. The bones, skulls, and jaws were chiefly of *Bos longifrons*, red deer, horse, sheep or goat, pig, and large dog. Amongst the bones were parts of a broken up human skeleton which represented an aged person, with greatly worn-down teeth. Strange to say, there was also a part of a humerus of *Bos primigenius*, white and root-eaten, a notable find . . . The long bones represented hundreds of animals that had been killed and eaten; nearly all the bones were split or broken across the middle in pre-Roman times for the marrow. The bones were all rugose with age on the outer, inner, and split surfaces. On close examination of the long pieces of bone and antler I was enabled to rejoin some of them and partly rebuild the original bone or antler. In these examples none of the fractured surfaces was new and smooth, all were old and rugose.' (Smith 1915, 150).

Objectives: to establish the date of the segments, in order to confirm or disprove the impression that they were indeed part of a causewayed enclosure.

Final comment: F Healy (2012), of the three dates on samples probably from the ditch segments, GrA-29891–2 are statistically consistent (T'=0.1; T'(5%)=3.8; *v*=1; Ward and Wilson), but OxA-15079 is older (T'=15.8; T'=(5%)= 6.0; *v*=2; Ward and Wilson). This could reflect derivation from successive levels in the fills, in which case OxA-15079 would be closer to the construction date than the other two measurements. Alternatively, the implement dated by OxA-15079 could already have been old when buried, while the aurochs femur with unfused fitting epiphysis which was the sample for GrA-29892 was probably not. For this reason, GrA-29891–2 may provide a more reliable date for construction.

References: Smith 1904b
Smith 1915
Ward and Wilson 1978

GrA–29891 4710 ±40 BP

δ¹³C: -22.8‰

Sample: Wardown Park Museum 6/372/40/B, submitted on 29 July 2005 by F Healy

Material: antler: *Cervus elaphus*, antler base, shed, with fairly recently broken brow tine and beam. Highly unlikely to have come from the same animal as 6/372/40/A because it is much smaller. One of several antler fragments, some of them glued together (F Healy 2005)

Initial comment: the sole contextual evidence is in the words marked on the antler: 'Maiden Bower 1900. WGS'. The sample is highly likely to have come from a Neolithic feature, despite Iron Age and Roman activity on the hill, because Smith describes the apparent ditch segments as containing many broken or cut antlers, some of which he was able to rejoin.

Objectives: to determine whether the features observed by Smith were indeed of a date that would make them likely to have formed part of a causewayed enclosure.

Calibrated date: 1σ: 3630–3370 cal BC
2σ: 3640–3360 cal BC

Final comment: see GrA-29892

GrA–29892 4690 ±40 BP

δ¹³C: -21.8‰

Sample: Wardown Park Museum 2/372/40, submitted on 29 July 2005 by F Healy

Material: animal bone: *Bos primigenius*, distal end of right femur with fitting, unfused epiphysis (glued on) (W Smith, 1903 and J Mulville 2005)

Initial comment: the sole contextual evidence is in the words marked on the bone: 'Femur. *Bos primigens*. Maiden Bower 1.1903. WGS'. The recovery of fitting shaft and epiphysis, especially in salvage conditions, indicates that they were buried together and hence still joined by soft tissue, so that the animal from which they came would not have been long dead. The sample is highly likely to have come from a Neolithic feature, despite Iron Age and Roman activity on the hill, because the aurochs became extinct in Britain in the course of the second millennium cal BC.

Objectives: to determine whether the features observed by Smith were indeed of a date that would make them likely to have formed part of a causewayed enclosure.

Calibrated date: 1σ: 3630–3370 cal BC
2σ: 3640–3360 cal BC

Final comment: F Healy (2012), the result is consistent with the hypothesis.

OxA–15079 4866 ±31 BP

δ¹³C: -22.3‰

Sample: Wardown Park Museum 6/372/40/A, submitted on 29 July 2005 by F Healy

Material: antler: *Cervus elaphus*, antler base, shed, brow and bez tine and distal end of beam recently broken off. Highly unlikely to be from same animal as 6/372/40/B because it is much larger. One of several antler fragments, some of them glued together (F Healy 2005)

Initial comment: the sole contextual evidence is in the words marked on the antler: 'Maiden Bower 99 WGS'. The sample is highly likely to have come from a Neolithic feature, despite Iron Age and Roman activity on the hill.

Objectives: to determine whether the features observed by Smith were indeed of a date that would make them likely to have formed part of a causewayed enclosure.

Calibrated date: *1σ:* 3660–3630 cal BC
                    2σ: 3710–3630 cal BC

Final comment: F Healy (2012), the result is consistent with the hypothesis. The sample is, however, rather older than those for GrA-29891 and -29892.

Causewayed Enclosures: Maiden Bower, pit, Bedfordshire

Location:	SP 9966022470 Lat. 51.53.14 N; Long. 00.33.40 W
Project manager:	C Matthews (Manshead Archaeological Society of Dunstable), mid-twentieth century
Archival body:	Wardown Park Museum, Luton

Description: the pit measured 0.92m across and 0.46m deep, exposed in the eroding, abandoned quarry face at north side of Maiden Bower hillfort. The remaining part was excavated and found to contain 94 sherds from up to 8 Neolithic Bowls, some of them decorated (Matthews 1976, fig 4).

Objectives: to determine whether this pit and the apparent ditch segments from which Worthington Smith retrieved antler fragments and aurochs bones were contemporary.

Final comment: F Healy (2012), the two results are statistically consistent with each other (T′=0.6; T′(5%)=3.8; *v*=1; Ward and Wilson 1978) and with GrA-29891–2 (T′=1.0; T′(5%)=7.8; *v*=3; Ward and Wilson 1978), indicating that the pit and the ditch segments could indeed have been contemporary.

References: Matthews 1976
              Ward and Wilson 1978
              Whittle *et al* 2011

GrA–30026 4695 ±40 BP

δ¹³C: -22.5‰

Sample: Wardown Park Museum A248 M321 P248/A, submitted on 29 July 2005 by F Healy

Material: antler: *Cervus elaphus*, large tine with chipped, worn tip. Recently broken from beam close to ancient cut with parallel incomplete cutmarks (F Healy 2005)

Initial comment: pit 11. Exposed in eroding quarry face at north side of Maiden Bower hillfort, surviving to 0.92m across and 0.46m deep, containing Neolithic Bowl pottery, some of it decorated (Matthews 1976, 8–9). The modification of the antler and the wear on it indicate that it was part of an implement, which may have been used to dig the pit in which it was found.

Objectives: to date the pit and establish its relation to the probable circuits of causewayed enclosure nearby.

Calibrated date: *1σ:* 3630–3370 cal BC
                    2σ: 3640–3360 cal BC

Final comment: F Healy (2012), the pit and the probable ditch segments could indeed have been contemporary.

OxA–15098 4735 ±31 BP

δ¹³C: -22.1‰

Sample: Wardown Park Museum A248 M321 P248/B, submitted on 29 July 2005 by F Healy

Material: animal bone: *Bos* sp., articulating astragalus and distal end of tibia (F Healy 2005)

Initial comment: pit 11. Exposed in eroding quarry face at north side of Maiden Bower hillfort, surviving to 0.92m across and 0.46m deep, containing Neolithic Bowl pottery, some of it decorated (Matthews 1976, 8–9). The articulation of the bones shows that they were still joined by soft tissue when buried and hence came from an animal that was not long dead.

Objectives: to date the pit and establish its relation to the probable circuits of causewayed enclosure nearby.

Calibrated date: *1σ:* 3630–3380 cal BC
                    2σ: 3640–3370 cal BC

Final comment: see GrA-30026

Causewayed Enclosures: Maiden Castle, Dorset

Location:	SY 6693088480 Lat. 50.41.38 N; Long. 02.28.05 W
Project manager:	R E M Wheeler and N Sharples (Society of Antiquaries of London and Wessex Archaeology), 1934–7 and 1985–6

Description: a causewayed enclosure with two concentric circuits, approximately 15m apart, occupying the east knoll of the hill on which an Iron Age hillfort was later built, obscuring and sometimes destroying the Neolithic earthworks. There were Neolithic pits within the enclosure and beyond it to the east. The west side of the enclosure was overlain by the Long Mound, a 500m-long, east-west bank with flanking ditches, apparently built in three sections.

Objectives: to refine and expand the dating achieved following the 1985–6 excavations. In addition to questions of the chronology, sequence, and duration of the main Neolithic elements, specific questions included those of whether the central and eastern sections of the long mound could be distinguished, Richard Bradley having suggested (1983) that the central part was a long barrow onto which extended earthworks were built; and of whether the date of a bank between the two ditches, built after a turfline had formed over the silted inner ditch and covering a sherd of Peterborough Ware, could be more precisely defined.

Final comment: F Healy (2012), estimates from Bayesian analysis of the pre-existing and newly obtained dates indicate that the two circuits of the causewayed enclosure ditch were built in such rapid succession that their sequence could not be determined, with a construction date of *3580–3535 cal BC (95% probability;* Whittle *et al* 2011, fig 4.41: *start Maiden enclosure*). Both were infilled by *3555–3520 cal BC (95% probability;* Whittle *et al* 2011, fig 4.41: *end Maiden enclosure*), after a brief use-life of *1–50 years (95% probability;* Whittle *et al* 2011, fig 4.47: *use Maiden Castle*). The Long Mound was built in *3550–3500 cal BC (40% probability)* or *3480–3385 cal BC (55% probability;* Whittle *et al* 2011, fig 4.44: *start Maiden long mound*). There was no indication of any difference in date between the east and central parts of the mound. No new dates were obtained for the bank between the two enclosure ditches. Attempts to date internal carbonised residue on sherds from the site were not always successful, in some cases yielding results which were in poor agreement with their stratigraphic position.

References: Ambers *et al* 1989
Barclay and Bayliss 1999
Bradley 1983
Brothwell 1971
Cleal 2004
Hedges *et al* 2019b
Oswald *et al* 2001
Palmer 1976
Sharples 1986
Sharples 1987
Sharples 1991a
Sharples 1991b
Sharples n d
Smith 1966
Wainwright and Cunliffe 1985
Wheeler 1943
Whittle *et al* 2011

Causewayed Enclosures: Maiden Castle, inner ditch, Dorset

Location: SY 6693088480
Lat. 50.41.38 N; Long. 02.28.05 W

Project manager: R E M Wheeler (Society of Antiquaries of London) N M Sharples (Wessex Archaeology), 1934–7 and 1985–6

Archival body: Dorset County Museum, Dorset County Museum, Natural History Museum, Duckworth Laboratory, Society of Antiquaries of London

Description: this was the inner of two concentric causewayed circuits, and was considerably richer in finds than the outer, in large part due to a series of 'midden' layers, rich in charcoal, artefacts, and animal bone, which had entered the upper part of the ditch from the exterior. These were interleaved with loams derived from the interior which were equally finds-rich but lacked charcoal. Any accompanying bank had disappeared by the time the Long Mound was built over the by then infilled ditch.

Objectives: to refine the chronology of the circuit, as a step to building up the chronology of the complex as a whole.

Final comment: F Healy (2012), the dating of short-life and articulating samples from the chalk rubble fills of the ditch indicates that it was built later than the originally estimated 39th to 38th century cal BC construction date (Sharples 1991a, 104–5), Bayesian modelling of pre-existing and newly obtained dates provides an estimated construction date of *3575–3535 cal BC (95% probability;* Whittle *et al* 2011, fig 4.42: *dig Maiden inner*). The inner ditch was infilled by *3555–3525 cal BC (95% probability;* Whittle *et al* 2011, fig 4.42: *fill Maiden inner*).

References: Sharples 1991a
Whittle *et al* 2011

GrA–29107 4755 ±40 BP

$\delta^{13}C$: -22.1‰

Sample: 401 2205, submitted on 18 March 2005 by F Healy

Material: animal bone: *Bos* sp., distal right tibia fragment (fused) and astragalus, possibly articulating, although surface condition of two bones different (A Powell 2004)

Initial comment: trench I. Context 2205 (subdivision of 280). 'Midden' layer above 281 and below 2157, not extending into section (Sharples 1991a, fig 49). The articulation of the bones indicates that they were still connected by tendons when buried and that the animal from which they came was not long dead.

Objectives: to establish the date of the context as a step to establishing the chronology of the complex.

Calibrated date: 1σ: 3640–3380 cal BC
2σ: 3640–3370 cal BC

Final comment: see GrA-29109

GrA–29108 4915 ±40 BP

$\delta^{13}C$: -21.8‰

Sample: 401 14563/A, submitted on 18 March 2005 by F Healy

Material: animal bone: *Cervus* sp., large ungulate, probably *Cervus* sp., thoracic vertebra (M Armour-Chelu 1986)

Initial comment: a replicate of OxA-1144. Trench II/A. Context 554. Overlying primary silt 570 (Sharples 1991a, fig 59).

Objectives: to establish the date of the context as a step to establishing the chronology of the complex.

Calibrated date: 1σ: 3710–3650 cal BC
2σ: 3780–3630 cal BC

Final comment: F Healy (2012), this sample was dated to provide a check on its replicate, OxA-1144 (4550 ±80 BP), which seemed anomalously recent.

Laboratory comment: English Heritage (25 November 2014), this measurement is significantly older than the original determination on this sample (OxA-1144; T'=16.2; T'(5%)=3.8; ν=1; Ward and Wilson 1978).

References: Ward and Wilson 1978

GrA–29109 4860 ±40 BP

δ¹³C: -22.4‰

Sample: 401 136, submitted on 18 March 2005 by F Healy

Material: animal bone: *Bos* sp., unfused left distal metacarpal shaft end and fitting epiphyses (A Powell 1986)

Initial comment: trench I. Context 136 (subdivision of 130). Layer overlying 140 (Sharples 1991a, fig 51). The articulation of the bones shows that they were still connected by tendons when buried so that the animal from which they came was then not long dead.

Objectives: to establish the date of the context as a step to establishing the chronology of the complex.

Calibrated date: 1σ: 3660–3630 cal BC
                      2σ: 3710–3530 cal BC

Final comment: F Healy (2012), the date is compatible with its stratigrpahic position in the model.

GrA–29111 4815 ±40 BP

δ¹³C: -21.8‰

Sample: 401 567, submitted on 18 March 2005 by F Healy

Material: animal bone: *Bos* sp., proximal femur fragment and unfused epiphysis, probably fitting (A Powell 2004)

Initial comment: trench II/A. Context 567. 'Midden' layer over 553 and under 550 (Sharples 1991a, fig 59). From same context as sample for BM-2454. The proximity of the fitting shaft end epiphysis indicates that they were still joined by cartilage when buried, so that the animal from which they came was then not long dead.

Objectives: to establish the date of the context as a step to establishing the chronology of the complex.

Calibrated date: 1σ: 3650–3530 cal BC
                      2σ: 3660–3520 cal BC

Final comment: see GrA-29109

GrA–29112 4785 ±40 BP

δ¹³C: -21.8‰

Sample: 401 299/A, submitted on 18 March 2005 by F Healy

Material: animal bone: *Bos* sp., cattle-sized animal; rib fragment (A Powell 2004)

Initial comment: trench I. Context 299. One of the fills of feature 2233, which was cut by inner ditch 2235, above 2183 (Sharples 1991a, fig 49).

Objectives: to establish a *terminus post quem* for the cutting of the inner ditch as a step towards defining the chronology of the complex.

Calibrated date: 1σ: 3640–3520 cal BC
                      2σ: 3650–3380 cal BC

Final comment: F Healy (2012), together with OxA-14834, this provides a *terminus post quem* for the cutting of the inner ditch.

GrA–29143 4920 ±45 BP

δ¹³C: -28.8‰

Sample: 401 2180 vessel 4040, submitted on 18 March 2005 by F Healy

Material: carbonised residue (1 of a few sherds from a Neolithic Bowl, found together) (F Healy 2004)

Initial comment: trench I. Context 2180 (subdivision of 2206). Chalk rubble layer at equivalent level to 2169, above 2216 and below 298. It did not extend into section (Sharples 1991a, fig 49). The fresh condition of the residue and the fact that more than one sherd of the same vessel were found together suggest that the pot was freshly broken when the vessel was deposited.

Objectives: to establish the date of the context as a step to establishing the chronology of the complex.

Calibrated date: 1σ: 3720–3650 cal BC
                      2σ: 3790–3630 cal BC

Final comment: F Healy (2012), this result is older than dates for articulating and short-life samples from the same rubble fills (GrA-29743–4, OxA-14832, -14835, and -15097). The sherd may have been be redeposited or the measurement may be inaccurate.

GrA–29207 4935 ±45 BP

δ¹³C: -28.8‰

Sample: 401 553/A, submitted on 18 March 2005 by F Healy

Material: carbonised residue (1 of 2 Neolithic Bowl sherds with fresh residue) (F Healy 2004)

Initial comment: trench II/A. Context 553. Lowest charcoal-rich 'midden' layer over 568 and under 550 (Sharples 1991a, fig 59). The fresh condition of the residue suggests that the pot was freshly broken when the sherds were deposited.

Objectives: to establish the date of the context as a step to establishing the chronology of the complex.

Calibrated date: 1σ: 3770–3650 cal BC
                      2σ: 3800–3640 cal BC

Final comment: F Healy (2012), the result is statistically consistent with OxA-14734 (T′=3.6; T′(5%)=3.8; ν=1; Ward and Wilson 1978), but nonetheless older than articulating samples from equivalent contexts. The sherds may have been redeposited or the dates may be inaccurate.

References: Ward and Wilson 1978

GrA–29209 4910 ±45 BP

δ¹³C: -29.1‰

Sample: 401 284/A, submitted on 18 March 2005 by F Healy

Material: carbonised residue (3 sherds from same pot with fresh-looking residue) (F Healy 2004)

Initial comment: trench I. Context 284 (subdivision of 280). 'Midden' layer above 281 and below 2157, not extending into section (Sharples 1991a, fig 49). The freshness of the sherds and of the residue on them suggests that the pot was freshly broken when they were buried.

Objectives: to establish the date of the context as a step to establishing the chronology of the complex.

Calibrated date: 1σ: 3710–3640 cal BC
                       2σ: 3790–3630 cal BC

Final comment: F Healy (2012), this was measured on the same residue as OxA-14733, the two dates being statistically consistent (T'=1.6; T'(5%)=3.8; ν=1; Ward and Wilson 1978). They are also statistically consistent with two other residue dates from the same cotnext (GrA-29210, -29211; T'=3.0; T'(5%)=6.0; ν=2; Ward and Wilson 1978). All are, however, older than GrA-29107, measured on an articulating sample from the same context. The sherds may have been redeposited or the dates may be inaccurate.

References: Ward and Wilson 1978

GrA–29210 4975 ±40 BP

δ¹³C: -28.4‰

Sample: 401 2283, submitted on 18 March 2005 by F Healy

Material: carbonised residue (internal, well-preserved; Neolithic Bowl body sherd) (F Healy 2004)

Initial comment: trench I. Context 284 (subdivision of 280). 'Midden' layer above 281 and below 2157, not extending into section (Sharples 1991a, fig 49).

Objectives: to establish the date of the context as a step to establishing the chronology of the complex.

Calibrated date: 1σ: 3800–3700 cal BC
                       2σ: 3940–3650 cal BC

Final comment: F Healy (2012), this date is statistically consistent with other residue dates from the same context (GrA-29211, GrA-29209, and OxA-14733; T'=3.0; T'(5%)=6.0; ν=2; Ward and Wilson 1978). All are, however, older than GrA-29107, measured on an articulating sample from the same context. The sherds may have been redeposited or the dates may be inaccurate.

References: Ward and Wilson 1978

GrA–29211 4885 ±40 BP

δ¹³C: -28.3‰

Sample: 401 2284, submitted on 18 March 2005 by F Healy

Material: carbonised residue (internal, abundant and fresh; >5 conjoining sherds of a Neolithic Bowl) (F Healy 2004)

Initial comment: trench I. Context 284 (subdivision of 280). 'Midden' layer above 281 and below 2157, not extending into section (Sharples 1991a, fig 49). The presence of several conjoining sherds of the same pot and the thickness and freshness of the residue suggest that the pot was freshly broken when the sherds were buried.

Objectives: to establish the date of the context as a step to establishing the chronology of the complex.

Calibrated date: 1σ: 3700–3640 cal BC
                       2σ: 3750–3630 cal BC

Final comment: see GrA-29210

GrA–29743 4825 ±40 BP

δ¹³C: -26.3‰

Sample: 401 215 A, submitted on 18 March 2005 by F Healy

Material: charcoal: *Quercus* sp., roundwood; single fragment (R Gale 2005)

Initial comment: trench I. Context 215 (subdivision of 140). Layer immediately above initial silt (Sharples 1991a, fig 51). The quantity of charcoal in the chalk rubble fills in this cutting and the burning of some of the fills suggest an immediate event. If short life charcoal can be isolated it should provide a close *terminus post quem* for than event.

Objectives: to establish a *terminus post quem* for this context as a step towards defining the chronology of the complex.

Calibrated date: 1σ: 3650–3530 cal BC
                       2σ: 3700–3520 cal BC

Final comment: see GrA-29744

GrA–29744 4825 ±40 BP

δ¹³C: -24.5‰

Sample: 401 141 A, submitted on 18 March 2005 by F Healy

Material: charcoal: Pomoideae, single fragment (R Gale 2005)

Initial comment: trench I. Context 141 (subdivision of 140). Layer immediately above initial silt (Sharples 1991a, fig 51). If short-life material can be extracted from the sample it will provide a close *terminus post quem* for the layer.

Objectives: to establish a *terminus post quem* for this context as a step towards defining the chronology of the complex.

Calibrated date: 1σ: 3650–3530 cal BC
                       2σ: 3700–3520 cal BC

Final comment: F Healy (2012), the results is statistically consistent with five out of the six other dates on short-life or articulating samples from the rubble fills in this trench (T'=4.4; T'(5%)=11.1; ν=5; Ward and Wilson 1978).

References: Ward and Wilson 1978

OxA–14733 4980 ±32 BP

δ¹³C: -26.2‰

Sample: 401 284/B, submitted on 18 March 2005 by F Healy

Material: carbonised residue (3 sherds from same pot with fresh-looking residue) (F Healy 2004)

Initial comment: a replicate of GrA-29209.

Objectives: as GrA-29209

Calibrated date: *1σ:* 3800–3700 cal BC
                       2σ: 3910–3660 cal BC

Final comment: see GrA-29209

OxA–14734 4830 ±33 BP

δ¹³C: -26.2‰

Sample: 401 553/B, submitted on 18 March 2005 by
F Healy

Material: carbonised residue (1 of 2 Neolithic Bowl sherds
with fresh residue) (F Healy 2004)

Initial comment: a replicate of GrA-29207.

Objectives: as GrA-29207

Calibrated date: *1σ:* 3650–3540 cal BC
                       2σ: 3660–3530 cal BC

Final comment: see GrA-29207

OxA–14792 4922 ±39 BP

δ¹³C: -28.0‰

Sample: 401 109, submitted on 18 March 2005 by F Healy

Material: carbonised residue (internal; Neolithic Bowl body
sherd) (F Healy 2004)

Initial comment: trench I. Context 109 (subdivision of 98).
Lower horizon of pre-long mound soil in ditch top (Sharples
1991a, fig 51). The fresh condition of the sherd and the
residue suggest that it was incorporated in its context soon
after breakage. From the same context as the sample for
OxA-1147.

Objectives: to establish the date of the context as a step to
establishing the chronology of the complex.

Calibrated date: *1σ:* 3720–3650 cal BC
                       2σ: 3790–3640 cal BC

Final comment: F Healy (2012), although the context of this
sample overlay the 'midden' layers, this date is of similar age
to residue samples from them and is older than a
disarticulated bone sample from the same layer (OxA-1147,
4690 ±80 BP). The sherd may have been redeposited or the
date may be inaccurate.

OxA–14832 4886 ±35 BP

δ¹³C: -20.2‰
δ¹⁵N (diet): +10.9‰

Sample: 401 14577/A, submitted on 18 March 2005
by F Healy

Material: human bone (humerus; 3–4 year-old)
(J Henderson 1986)

Initial comment: a replicate of OxA-1148. Trench I. Context
215 (subdivision of 140). In top of rubble layer 140 which
immediately overlay the initial fine silts (Sharples 1991a, fig
51 - the skull of this skeleton is shown at the south-west
(left) end of the section, but is not labelled). The articulation
of the skeleton shows that the individual was recently dead
when buried.

Objectives: to establish the date of the context as a step to
establishing the chronology of the complex.

Calibrated date: *1σ:* 3700–3640 cal BC
                       2σ: 3710–3630 cal BC

Final comment: F Healy (2012), the result is statistically
consistent with OxA-1148 (4810 ±80 BP), confirming the
accuracy of the original measurement (T'=0.8; T'(5%)=3.8;
ν=1; Ward and Wilson 1978). This burial is also statistically
consistent with five of the other six dates on short-life or
articulating samples from the rubble fills in this trench
(T'=4.4; T'(5%)=11.1; ν=5; Ward and Wilson 1978).

References: Ward and Wilson 1978

OxA–14833 4804 ±37 BP

δ¹³C: -20.6‰

Sample: 401 291, submitted on 18 March 2005 by F Healy

Material: animal bone: *Sus* sp., lumbar vertebra with fitting
unfused epiphysis (A Powell 2004)

Initial comment: trench I. Context 291. Loam intercalated
with 'midden' layers, above 296 and beneath 293.

Objectives: to establish the date of the context as a step to
establishing the chronology of the complex.

Calibrated date: *1σ:* 3650–3530 cal BC
                       2σ: 3660–3520 cal BC

Final comment: see GrA-29109

OxA–14834 4734 ±35 BP

δ¹³C: -21.7‰

Sample: 401 299/B, submitted on 18 March 2005 by
F Healy

Material: animal bone (sheep/goat; right mandible fragment,
adult, with accessory mental foramen below P2) (A Powell
2004)

Initial comment: trench I. Context 299. One of fills of
feature 2233, which was cut by inner ditch 2235, above
2183 (Sharples 1991a, fig 49).

Objectives: to establish a *terminus post quem* for this context as
a step towards defining the chronology of the complex.

Calibrated date: *1σ:* 3640–3380 cal BC
                       2σ: 3640–3370 cal BC

Final comment: see GrA-29112

OxA–14835 4796 ±36 BP

δ¹³C: -22.2‰

Sample: 401 2180, submitted on 18 March 2005 by F Healy

Material: animal bone (sheep/goat; 2 possibly articulating
lumbar vertebrae) (A Powell 2004)

Initial comment: trench I. Context 2180 (subdivision of
2206). Chalk rubble layer at equivalent level to 2169, that
did not extend into the section (Sharples 1991a, fig 49).

The articulation of the bones indicates that they were still held together by tendons when buried, and that the animal from which they came was then not long dead.

Objectives: to establish the date of the context as a step to establishing the chronology of the complex.

Calibrated date: *1σ:* 3640–3530 cal BC
                       2σ: 3650–3510 cal BC

Final comment: see GrA–29744

OxA–15096 4303 ±30 BP

δ¹³C: -24.5‰

Sample: 401 141 B, submitted on 18 March 2005 by F Healy

Material: charcoal: *Corylus* sp., single fragment (R Gale 2005)

Initial comment: trench I. Context 141 (subdivision of 140). Layer immediately above initial silt (Sharples 1991a, fig 51). If short-life material can be extracted from the sample it will provide a close *terminus post quem* for the layer.

Objectives: to establish a *terminus post quem* for this context as a step towards defining the chronology of the complex.

Calibrated date: *1σ:* 2920–2890 cal BC
                       2σ: 3000–2880 cal BC

Final comment: F Healy (2012), this date is substantially more recent than the other five dates on short-life samples from the rubble fills in this trench and than dates for overlying articulating samples. The single fragment may have been intrusive or may have been an outlier.

OxA–15097 4868 ±33 BP

δ¹³C: -26.0‰

Sample: 401 215 B, submitted on 18 March 2005 by F Healy

Material: charcoal: *Quercus* sp., sapwood; single fragment (R Gale 2005)

Initial comment: trench I. Context 215 (subdivision of 140). Layer immediately above initial silt (Sharples 1991a, fig 51). The quantity of charcoal in the chalk rubble fills in this cutting and the burning of some of the fills suggest an immediate event. If short life charcoal can be isolated it should provide a close *terminus post quem* for than event.

Objectives: to establish a *terminus post quem* for this context as a step towards defining the chronology of the complex.

Calibrated date: *1σ:* 3660–3630 cal BC
                       2σ: 3710–3630 cal BC

Final comment: see GrA–29744

OxA-X–2135–46 4880 ±65 BP

δ¹³C: -27.5‰

Sample: 401 2336, submitted on 18 March 2005 by F Healy

Material: carbonised residue (substantial part of gabbroic vessel, with residue on 1 lower corner of 1 wall sherd). Pot formed of AOR 2336 (context 291) and rim sherd AOR 2321 (from later context 277, phase 3A) (Cleal 1991, fig 141:4, microfiche M9:C4–C5). The substantial representation of the pot and the large size of the sherds show that it, and the residue on it, were fresh when buried (R M J Cleal, Alexander Keiller Museum, Avebury; Cleal 1986–90)) (F Healy 2004)

Initial comment: trench I. Inner ditch, trench, context 291. Loam intercalated with 'midden' layers, not extending in to sections, above 296 and below 293.

Objectives: to establish the date of the context as a step to establishing the chronology of the complex.

Calibrated date: *1σ:* 3710–3630 cal BC
                       2σ: 3790–3520 cal BC

Final comment: F Healy (2012), this sample was submitted in an attempt to obtain a direct date for a gabbroic pot. It was a very small sample which produced a low yield after pre-treatment (2.85mg) and a small target. This gave a low current in the AMS and so the result has been reported as an experimental measurement and should be used with caution.

Causewayed Enclosures: Maiden Castle, long mound, Dorset

Location: SY 6693088480
                     Lat. 50.41.38 N; Long. 02.28.05 W

Project manager: R E M Wheeler (Society of Antiquaries of London) N M Sharples (Wessex Archaeology), 1934–7 and 1985–6

Archival body: Dorset County Museum, Dorset County Museum, Natural History Museum, Duckworth Laboratory, Society of Antiquaries of London

Description: an earth and chalk mound approximately 500m long and 22m wide, flanked by two ditches. Two major changes of height and direction suggest that it may have been built in three lengths: a western one well beyond the enclosure, a central one just outside the circuits and extending almost up to them, and an eastern one built over the inner and probably the outer circuit, running into the interior of the enclosure.

Objectives: to refine the chronology of the monument, as a step to building up the chronology of the complex as a whole.

Final comment: F Healy (October 2012), all seven antler samples from primary contexts in the long mound ditches (two measured after the 1985–6 excavations and five during the Dating Causewayed Enclosures project) are statistically consistent (T′=6.9; T′(5%)=12.6; ν=6; Ward and Wilson 1978), regardless of whether they came from its east or its central part, suggesting that the central part may not have been a pre-existing long barrow. Bayesian modelling of pre-existing and newly obtained dates provides an estimated construction date of *3545–3500 cal BC (40% probability)* or *3480–3385 cal BC (55% probability;* Whittle *et al* 2011, fig 4.44: *start Maiden long mound).*

References: Ward and Wilson 1978
                     Whittle *et al* 2011

GrA–29146 4710 ±45 BP

δ¹³C: -23.2‰

Sample: 401 1102, submitted on 18 March 2005 by F Healy

Material: antler: *Cervus elaphus*, trez tine and beam above trez tine. Chopped off above and below trez tine. Trez tine broken. Max dimension 1700mm (description from NHM box of Armour-Chelu archive) (M Armour-Chelu 1986)

Initial comment: trench III. Context 810. Initial fill, on base of ditch (Sharples 1991a, fig 57). The location, completeness and character of the implement suggest that it was used to dig the ditch.

Objectives: to establish the date of the context as a step to establishing the chronology of the complex.

Calibrated date: 1σ: 3630–3370 cal BC
                                  2σ: 3640–3360 cal BC

Final comment: see GrA-29336

GrA–29147 4740 ±45 BP

δ¹³C: -22.7‰

Sample: 401 1131, submitted on 18 March 2005 by F Healy

Material: antler: *Cervus elaphus*, shed antler with parts of bow tine, part of trez tine, max dimension 440mm. Brow tine detached. Antler chopped through above trez tine (description from box of Armour-Chelu archive in NHM). Possibly the same as 401 14591 (M Armour-Chelu 1986)

Initial comment: trench III. Context 991 (lowest fill of pit 2276). Pit cut into primary silt of the northern long mound ditch, west of the causeway, cutting clay layer 2262 which overlay the ditch bottom (Sharples 1991a, figs 56–7). Wear and areas of burning strongly suggest that it was a pick. If it was used to dig the pit it should be close in age to its context.

Objectives: to establish the date of the context as a step to establishing the chronology of the complex.

Calibrated date: 1σ: 3640–3380 cal BC
                                  2σ: 3640–3370 cal BC

Final comment: see GrA-29336

GrA–29336 4755 ±45 BP

δ¹³C: -21.2‰

Sample: ARC 1970 3054/B, submitted on 6 June 2005 by F Healy

Material: antler: *Cervus elaphus*, recently broken fragments of red deer antler beam and tine with worn tip (F Healy 2004)

Initial comment: object marked 'MC Q trench p 49 ext layer 6 black occ 15/10/37', ie site Q, trench p 49– extension, layer 6, dark 'occupation deposit'. Wheeler's site Q occupied the south side of the east end of the long mound (Wheeler 1943, pl. IV). Trench p 49 was in the east terminal of the south ditch of the mound (Sharples 1991a, fig 147). The marking on the object corresponds to the 'Black hearth-layer immediately over the rapid silt' in that terminal, which also contained cattle skulls (Wheeler 1943, 88), ie to an

equivalent stratigraphic position to the sample for OxA-1146. The worn tip of the tine suggests that the fragments formed part of a pick. Their position close to the base of the ditch suggests that the pick may have been used to dig that ditch and was thus close in age to its context. Dark 'hearth-like' deposit.

Objectives: to establish the date of the context as a step to establishing the chronology of the complex.

Calibrated date: 1σ: 3640–3380 cal BC
                                  2σ: 3650–3370 cal BC

Final comment: F Healy (2004), the date is statistically consistent with those for all six other antler samples from on or close to the base of the long mound ditch (GrA-29146, -29147, OxA-1145 (4660 ±80 BP), OxA-1349 (4660 ±80 BP), OxA-14831, and -14838; T'=6.9; T'(5%)=12.6; ν=6; Ward and Wilson 1978).

References: Cleal 1991
                              Ward and Wilson 1978
                              Wheeler 1943

OxA–14831 4783 ±35 BP

δ¹³C: -20.8‰

Sample: ARC 1970 3054/A, submitted on 6 June 2005 by F Healy

Material: antler: *Cervus elaphus*, large red deer antler tine with worn tip. Fresh break at proximal end (F Healy 2004)

Initial comment: as GrA-29336

Objectives: to establish the date of the context as a step to establishing the chronology of the complex.

Calibrated date: 1σ: 3640–3520 cal BC
                                  2σ: 3650–3380 cal BC

Final comment: see GrA-29336

OxA–14838 4674 ±35 BP

δ¹³C: -20.8‰

Sample: 401 1133, submitted on 18 March 2005 by F Healy

Material: antler (crown 'rake' with three worn tines. Cut-marks around stump of beam) (F Healy 2004)

Initial comment: trench III. Context 991 (lowest fill of pit 2276). Pit cut into primary fills of the northern long mound ditch, west of the causeway, cutting clay layer 2262 which overlay the ditch bottom (Sharples 1991a figs 56–7). Among antler fragments on floor of pit (Sharples n.d., 33). Wear and location suggest that it may have been used to dig the pit.

Objectives: to establish the date of the context as a step to establishing the chronology of the complex.

Calibrated date: 1σ: 3520–3370 cal BC
                                  2σ: 3630–3360 cal BC

Final comment: see GrA-29336

OxA–14881 2300 ±28 BP

δ¹³C: -26.8‰

Sample: 401 828, submitted on 18 March 2005 by F Healy

Material: carbonised residue (3 small sherds, 2 with fresh-looking residue) (F Healy 2004)

Initial comment: trench III. Context 828. Above primary fill 810 in central part of the north ditch of the long mound (Sharples 1991a, fig 57). The fresh condition of the residue suggests that the sherds were not long broken when buried. Fill with high chalk content derived from ditch sides.

Objectives: to establish the date of the context as a step to establishing the chronology of the complex.

Calibrated date: *1σ:* 400–370 cal BC
                2σ: 410–260 cal BC

Final comment: F Healy (2012), this sample was submitted in the belief that it came from context 828, which immediately overlay the primary fill of the long mound ditch. Its first millennium cal BC date suggests that it in fact probably came from Iron Age ditch 806, which cut into context 828.

Causewayed Enclosures: Maiden Castle, outer ditch, Dorset

Location: SY 6693088480
              Lat. 50.41.38 N; Long. 02.28.05 W

Project manager: R E M Wheeler and (Wheeler Society of Antiquaries of London) N Sharples (Wessex Archaeology), 1934–7 and 1985–6

Archival body: Dorset County Museum, Dorset County Museum, Natural History Museum, Duckworth Laboratory, and the MacDonald Institute

Description: this was the outer of two concentric causewayed circuits, and was considerably poorer in finds than the inner. Its particularly steep, unweathered profile and relatively clean chalk rubble fills suggest that it was backfilled soon after construction. Its stratigraphic relationship to the Long Mound remains undefined.

Objectives: to refine the chronology of the circuit, as a step to building up the chronology of the complex as a whole.

Final comment: F Healy (2012), no articulating or fitting animal bone samples could be found. Instead samples were dated from among substantial, well-preserved, disarticulated bone fragments in the basal layer, some of which were dated following the 1985–6 excavations. All six such samples from Sharples' trench II were statistically consistent (T′=10.4; T′(5%)=11.1; ν=5; Ward and Wilson 1978), as were all three from his trench V (T′=3.3; T′(5%)=6.0; ν=2; Ward and Wilson 1978), suggesting that none was redeposited. The human bone samples from trench II (BM-2451 (4860 ±70 BP), OxA-1338 (4930 ±90 BP), and OxA-14837) are not significantly earlier than the animal bones, contra Sharples (1991a) and Cleal (2004), so that no question of curation need arise. Since all the samples were

disarticulated, the estimated construction date of *3580–3525 cal BC (95% probability)* is based on the most recent of them (Whittle *et al* 2011, fig 4.43: *dig Maiden outer*).

References: Cleal 2004
               Sharples 1991a
               Whittle *et al* 2011

GrA–29113 4775 ±40 BP

δ¹³C: -21.6‰

Sample: 401 2030, submitted on 18 March 2005 by F Healy

Material: animal bone: *Ovis* sp., base of right horncore with attached skull fragment (M Armour-Chelu 1986)

Initial comment: trench II/A. Context 324. Basal fill of truncated ditch. Associated with disarticulated human bone and other disarticulated animal bone (Sharples 1991a, figs 50 and 54). In the absence of articulated samples from this context, the youngest of dates on several samples, including this one, will help to establish a *terminus post quem* for the excavation of the ditch. Mixed sand and clay, derived form natural bands of sand and clay through which ditch cut.

Objectives: to establish a *terminus post quem* for this context as a step towards defining the chronology of the complex.

Calibrated date: *1σ:* 3640–3520 cal BC
                2σ: 3650–3380 cal BC

Final comment: see OxA-14794

GrA–29120 4795 ±40 BP

δ¹³C: -22.0‰

Sample: 401 7014 18/19, submitted on 18 March 2005 by F Healy

Material: animal bone: *Bos* sp., left mandible. Comparison with the residue of OxA-1340 shows that this cannot be part of the same left mandible. The residue is from a larger jaw (M Armour-Chelu 1986)

Initial comment: trench V/F. Context 7014. Silt overlying ditch floor (Sharples 1991a, fig 53), stratified below context 7013.

Objectives: to establish a *terminus post quem* for this context as a step towards defining the chronology of the complex.

Calibrated date: *1σ:* 3640–3520 cal BC
                2σ: 3660–3380 cal BC

Final comment: F Healy (2012), the result is statistically consistent with the two other dates from this context (OxA-1339 (4740 ±80 BP), OxA-1340 (4650 ±70 BP; T′=3.3; T′(5%)=6.0; ν=2; Ward and Wilson 1978).

References: Ward and Wilson 1978

GrA–29145 4905 ±45 BP

δ¹³C: -22.0‰

Sample: 401 7012 1/30, submitted on 18 March 2005 by F Healy

Material: animal bone: *Bos* sp., 2 rib fragments with longitudinal cut-marks from filleting (M Armour-Chelu 1987)

Initial comment: trench V/F. Context 7012. Apparent backfill of unweathered ditch, overlying 7013. The speed with which the unweathered ditch was backfilled, and the completeness and good preservation of this bone suggest that the ditch was only recently dug when this layer was deposited. Very loose chalk rubble with some flint in chalk-cut ditch on well-drained chalk down.

Objectives: to establish a *terminus post quem* for this context as a step towards defining the chronology of the complex.

Calibrated date: 1σ: 3710–3640 cal BC
                   2σ: 3780–3630 cal BC

Final comment: F Healy (2012), the result is older than animal bone dates from underlying layers, and is therefore probably redeposited.

GrA–29213 4605 ±40 BP

$\delta^{13}C$: -29.5‰

Sample: 401 7850/A, submitted on 18 March 2005 by F Healy

Material: carbonised residue (5 sherds, extracted from larger group all from same Neolithic Bowl. Recorded in field as single sherd. The recovery of so much of a pot at one spot, the freshness of the sherds and the survival of the residue all suggest that the pot was freshly broken when buried (R M J Cleal, Alexander Keiller Museum, Avebury 1986–90)) (F Healy 2004)

Initial comment: trench V/F. Context 7013. Large chalk blocks overlying 7014, apparently deliberately laid (Sharples 1991a, fig 53). The size of the sherd (as it was when excavated) and the survival of the residue indicate that it was fairly fresh when deposited. Large chalk blocks, several of them scored by sharp object.

Objectives: to establish the date of this context as a step towards defining the chronology of the complex.

Calibrated date: 1σ: 3500–3350 cal BC
                   2σ: 3510–3130 cal BC

Final comment: F Healy (2012), the result is statistically inconsistent with its replicate, OxA-14793 (T'=17.2; T'(5%)= 3.8; v=1; Ward and Wilson 1978). It is impossible to tell which date, if either, is accurate.

References: Ward and Wilson 1978

OxA–14793 4870 ±50 BP

$\delta^{13}C$: -28.6‰

Sample: 401 7850/B, submitted on 18 March 2005 by F Healy

Material: carbonised residue (5 sherds, extracted from larger group all from same Neolithic Bowl. Recorded in field as single sherd. The recovery of so much of a pot at one spot, the freshness of the sherds and the survival of the residue all

suggest that the pot was freshly broken when buried (R M J Cleal, Alexander Keiller Museum, Avebury 1986–90)) (F Healy 2004)

Initial comment: a replicate of GrA-29213.

Objectives: as GrA-295213

Calibrated date: 1σ: 3700–3630 cal BC
                   2σ: 3760–3530 cal BC

Final comment: see GrA-29213

OxA–14794 4806 ±36 BP

$\delta^{13}C$: -25.5‰

Sample: 401 1580, submitted on 18 March 2005 by F Healy

Material: carbonised residue (internal; small Neolithic Bowl body sherd (in 2 recently broken fragments) (R M J Cleal, Alexander Keiller Museum, Avebury, 1986–90)) (F Healy 2004)

Initial comment: trench II/A. Context 324. Basal fill of truncated ditch. Associated with disarticulated human bone and other disarticulated animal bone (Sharples 1991a, figs 50 and 54). In the absence of articulated samples from this context, the youngest of dates on several samples, including this one, will help to establish a *terminus post quem* for the excavation of the ditch. Mixed sand and clay derived from natural bands of sand and clay through which ditch cut.

Objectives: to establish a *terminus post quem* for this context as a step towards defining the chronology of the complex.

Calibrated date: 1σ: 3650–3530 cal BC
                   2σ: 3660–3520 cal BC

Final comment: F Healy (2012), the result is statistically consistent with the other five samples from this context (BM-2451, -2452; GrA-29113; OxA-1338, and -14837; T'=10.4; T'(5%)=11.1; v=5; Ward and Wilson 1978).

References: Ward and Wilson 1978

OxA–14836 4819 ±34 BP

$\delta^{13}C$: -22.0‰

Sample: 401 7012 2/30, submitted on 18 March 2005 by F Healy

Material: animal bone: *Bos* sp., cervical vertebra, cut-marked (A Powell 2005)

Initial comment: trench V/F. Context 7012. Apparent backfill of unweathered ditch, overlying 7013. The speed with which the unweathered ditch was backfilled, and the good preservation of this bone suggest that the ditch was only recently dug when this layer was deposited.

Objectives: to establish a *terminus post quem* for this context as a step towards defining the chronology of the complex.

Calibrated date: 1σ: 3650–3530 cal BC
                   2σ: 3660–3520 cal BC

Final comment: F Healy (2012), the result is comparable to animal bone dates from the underlying layers.

OxA–14837 4794 ±38 BP

δ¹³C: -20.6‰

δ¹⁵N (diet): +9.2‰

Sample: 401 2026, submitted on 18 March 2005 by F Healy

Material: human bone (mandible; 3–5 year-old)
(J Henderson 1986)

Initial comment: trench II/A. Context 324. Basal fill of truncated ditch. Associated with disarticulated human bone and other disarticulated animal bone (Sharples 1991a, figs 50 and 54).

Objectives: to establish a *terminus post quem* for this context as a step towards defining the chronology of the complex.

Calibrated date: 1σ: 3640–3520 cal BC
                       2σ: 3660–3510 cal BC

Final comment: see OxA-14794

Causewayed Enclosures: Northborough, Cambridgeshire

Location: TF 1557008450
                     Lat. 52.39.39 N; Long. 00.17.27 W

Project manager: Time Team, 2004

Description: a causewayed enclosure formed of two concentric pairs of closely-spaced ovoid circuits, with a trace of a possible fifth circuit, lying at 5mOD, on a gravel 'island' in the floodplain of the Welland, within a cluster of enclosures which includes Etton and Etton Woodgate (Oswald *et al* 2001, fig 1.1; 109–10, fig 6.3). Some pit-like features were located in the interior. Geophysical survey and limited excavation were undertaken in 2004 by the Time Team and Wessex Archaeology. The pottery assemblage from the site as a whole comprises ten small, heavily abraded sherds in friable fabrics, most of which appear to be shelly. One decorated rim sherd is present (burnt fill 228 of inner circuit recut) and three decorated body sherds (context 104). The rim sherd can be identified as early Neolithic, and comparable forms are known from the large Mildenhall style assemblage from Etton (Kinnes 1998). The decorated body sherds could also be of early Neolithic date.

Objectives: to define the chronology of the monument and establish its relation to Etton and Etton Woodgate.

Final comment: F Healy (2012), samples were dated from only two circuits out of four or five, and only in the innermost ditch were there samples from the base of the sequence. On this slender foundation, a construction date of *3700–3550 cal BC (95% probability)* may be estimated (Whittle *et al* 2011, fig 6.39: *start Northborough*).

References: GSB Prospection 2004
                     Kinnes 1998
                     Lewis 2005
                     Oswald *et al* 2001
                     Time Team 2005
                     Whittle *et al* 2011

Causewayed Enclosures: Northborough, inner ditch, Cambridgeshire

Location: TF 1557008450
                     Lat. 52.39.39 N; Long. 00.17.27 W

Project manager: Time Team and Wessex Archaeology, 2004

Archival body: Wessex Archaeology

Description: the innermost ditch of a causewayed enclosure with four to five concentric circuits. Samples were submitted from two separate sections.

Objectives: to confirm the apparent date and character of the enclosure.

Final comment: F Healy (2012), the four results make it possible to estimate a construction date for the circuit of *3640–3555 cal BC (95% probability*; Whittle *et al* 2011, fig 6.39: *build Northborough innermost)* and a date of *3610–3525 cal BC (95% probability)* for its recutting (Whittle *et al* 2011, fig 6.39: *recut Northborough innermost*).

References: Whittle *et al* 2011

GrA–29141 4710 ±50 BP

δ¹³C: -27.4‰

Sample: B 107/14, submitted on 18 February 2005 by F Healy

Material: charcoal: *Alnus glutinosa*, roundwood, not from the same twig as C 107/14; single fragment (C Chisham 2005)

Initial comment: context 107. In ditch 108, part of the innermost of four or five concentric circuits. Small, compact deposit of charcoal and degraded sherds on the centre of the ditch floor. Stratified below samples from context 207 but in a different trench. Roundwood cannot have been more than a few years old when cut, and the coherence of the deposit indicates that it was freshly made when buried. From same bulk sample as C 107/14.

Objectives: to date the deposit as a step towards dating the circuit.

Calibrated date: 1σ: 3630–3370 cal BC
                      2σ: 3640–3360 cal BC

Final comment: F Healy (2012), the date is compatible with its stratigraphic position in the model.

GrA–29142 4800 ±45 BP

δ¹³C: -26.1‰

Sample: D 207/2, submitted on 18 February 2005 by F Healy

Material: charcoal: *Alnus glutinosa*, roundwood; single fragment (C Chisham 2005)

Initial comment: context 207. In ditch 210, part of the innermost of four to five concentric circuits. Discrete deposit of charcoal and reddish-brown burnt clay in middle fill of recut made in largely silted ditch, sealed by infilled bank material. Stratified above samples from context 107 but in a different trench. The coherence of the deposit and the short life of the sample indicate that it is close in age to its context. From same bulk sample as D 207/2.

Objectives: to date the deposit as a step towards dating the circuit.

Calibrated date: 1σ: 3650–3520 cal BC
2σ: 3660–3380 cal BC

Final comment: see GrA-29141

OxA–14469 4743 ±37 BP

δ¹³C: -26.8‰

Sample: C 107/14, submitted on 18 February 2005 by F Healy

Material: charcoal: *Alnus glutinosa*, single fragment; knotted twig (not the same as D 107/14 (C Chisham 2005)

Initial comment: context 107. In ditch 108, part of the innermost of four or five concentric circuits. Small, compact deposit of charcoal and degraded sherds on the centre of ditch floor. Stratified below samples from context 207 but in a different trench. Roundwood cannot have been more than a few years old when cut, and the coherence of the deposit indicates that it was freshly made when buried. From same bulk sample as B 107/14.

Objectives: to date the deposit as a step towards dating the circuit.

Calibrated date: 1σ: 3640–3380 cal BC
2σ: 3640–3370 cal BC

Final comment: see GrA-29141

OxA–14470 4795 ±38 BP

δ¹³C: -27.4‰

Sample: A 207/2, submitted on 18 February 2005 by F Healy

Material: charcoal: Pomoideae, small roundwood, *c* 7 years, bark removed before submission; single fragment (C Chisham 2005)

Initial comment: context 207. In ditch 210, part of the innermost of four or five concentric circuits. Discrete deposit of charcoal and reddish-brown burnt clay in middle fill of recut made in largely silted ditch, sealed by infilled bank material. Stratified above samples from context 107 but in a different trench. The coherence of the deposit and the short life of the sample indicate that it is close in age to its context. From same bulk sample as D 207/2.

Objectives: to date the deposit as a step towards dating the circuit.

Calibrated date: 1σ: 3640–3520 cal BC
2σ: 3660–3510 cal BC

Final comment: see GrA-29141

Causewayed Enclosures: Northborough, outer ditch, Cambridgeshire

Location: TF 1557008450
Lat. 52.39.39 N; Long. 00.17.27 W

Project manager: Time Team and Wessex Archaeology, 2004

Archival body: Wessex Archaeology

Description: the outermost ditch of a causewayed enclosure with four to five concentric circuits.

Objectives: to confirm the apparent date and character of the enclosure.

Final comment: F Healy (2012), the only available samples were from a concentration of charred hazelnut shells in a recut in a single section. An aurochs femur on the ditch base failed to date. The hazelnut shell dates provide a *terminus ante quem* for the construction of the circuit of *3640–3530 cal BC* (95% probability; Whittle *et al* 2011, fig 6.39: *taq Northborough outermost*).

References: Whittle *et al* 2011

GrA–30076 4710 ±40 BP

δ¹³C: -22.5‰

Sample: 720 (722) <7> B, submitted in 2005 by M Allen

Material: carbonised plant macrofossil (single hazelnut shell fragment (*Corylus avellana*)) (C Chisham 2005)

Initial comment: context 722(7). In ditch 720, part of the outermost of four to five concentric circuits. From concentration of 30+ charred hazelnut fragments and occasional charred grain in secondary fill, in uniform condition and stratified above the aurochs femur.

Objectives: to provide a *terminus ante quem* for the construction of the circuit.

Calibrated date: 1σ: 3630–3370 cal BC
2σ: 3640–3360 cal BC

Final comment: F Healy (2012), OxA-15325 and GrA-30076 are statistically consistent (T′=2.0; T′(5%)=3.8; *v*=1; Ward and Wilson 1978) and provide a *terminus ante quem* for the construction of the circuit.

References: Ward and Wilson 1978

OxA–15325 4784 ±33 BP

δ¹³C: -26.2‰

Sample: 720 (722) <7> A, submitted in 2005 by F Healy

Material: carbonised plant macrofossil (single hazelnut shell fragment (*Corylus avellana*)) (C Chisham 2005)

Initial comment: context 722(7). In ditch 720, part of the outermost of four to five concentric circuits. From a concentration of 30+ charred hazelnut fragments and occasional charred grain in secondary fill, in uniform condition and stratified above the aurochs femur.

Objectives: as GrA-30076

Calibrated date: *1σ:* 3640–3520 cal BC
                       2σ: 3650–3510 cal BC

Final comment: see GrA-30076

Causewayed Enclosures: Offham Hill, East Sussex

Location: TQ 39881175
                     Lat. 50.53.14 N; Long. 00.00.42 W

Project manager: P Drewett (Sussex Archaeological Field
                     Unit), 1976

Archival body: Barbican House Museum

Description: a causewayed enclosure composed of two circuits. The eastern part of the site has been quarried away, and the rest badly ploughed-down. Relatively shallow ditches (surviving to a maximum 0.80m), were naturally silted. More material was found in the outer ditch than the inner. Most artefacts and animal bone were abraded and could have been on the surface for some time. Few or no Neolithic features were found in the interior. Molluscs evidence was interpreted as suggesting that the outer ditch was dug later than inner because the population from under the outer bank was very like that of layer 3 (secondary fill) in the inner ditch, both indicative of a clearance episode.

Objectives: to refine the dating achieved following the 1976 excavations, which rested on dates measured on two bulk samples of oak charcoal from the inner ditch (BM-1414, 4925 ±80 BP; BM-1415, 474 ±60 BP) and on the differential vegetation history inferred from molluscan analysis. The dearth of appropriate material was such that the only suitable sample was from an articulated burial in the base of the outer ditch.

Final comment: F Healy (2012), the burial provides an estimated construction date for the outer circuit of *3635–3555 cal BC (66% probability)* or *3540–3490 cal BC (23% probability)* or *3435–3380 cal BC (6% probability;* Whittle *et al* 2011, fig 5.14: *burial 1).* BM-1414 and 1415 provide *termini post quos* for the inner ditch. It was not possible to determine the sequence of the ditches.

References: Drewett 1977
                    Oswald *et al* 2001
                    Whittle *et al* 2011

GrA–27322 4685 ±45 BP

δ¹³C: -20.9‰
δ¹⁵N (diet): +10.5‰

Sample: Burial 1. Barbican House Museum, Lewes 77.23, submitted on 2 June 2004 by F Healy

Material: human bone (proximal end of left femur of 20–25 year-old male) (T O'Connor 1976)

Initial comment: a replicate of OxA-14177. Outer ditch, segment 4, bottom. Buried articulated in a pit cut into the base of the outer ditch. The articulation of the skeleton shows that the individual was recently dead when buried and hence close in age to the construction of the circuit.

Objectives: to establish the date of this context as a step towards defining the chronology of the monument.

Calibrated date: *1σ:* 3630–3370 cal BC
                       2σ: 3640–3360 cal BC

Final comment: F Healy (2012), OxA-14177 and GrA-27322 are statistically consistent (T'=0.4; T'(5%)=3.8; ν=1; Ward and Wilson 1978) and provide the best available estimate of a construction date for the outer ditch.

References: Ward and Wilson 1978

OxA–14177 4722 ±32 BP

δ¹³C: -20.5‰

Sample: Burial 1. Barbican House Museum, Lewes 77.23, submitted on 2 June 2004 by F Healy

Material: human bone (proximal end of left femur of 20–25 year-old male) (T O'Connor 1976)

Initial comment: a replicate of GrA-27322.

Objectives: as GrA-27322

Calibrated date: *1σ:* 3630–3380 cal BC
                       2σ: 3640–3370 cal BC

Final comment: see GrA-27322

Causewayed Enclosures: Peak Camp, Gloucestershire

Location: SO 9243015020
                     Lat. 51.49.59 N; Long. 02.06.35 W

Project manager: T Darvill (University of Southampton),
                     1980–1

Archival body: Corinium Museum

Description: Peak Camp (called Birdlip Camp by Oswald *et al* (2001)) is an enclosure on a triangular promontory from the Cotswold scarp approximately 1km south of Crickley Hill. Its original area, now diminished by quarrying, would have been more than 1.14ha. Two leaf arrowheads were found near the tip of the promontory in 1919. Two earthworks across the spur, about 90m apart, have vestigial banks, but there is some doubt as to whether the inner circuit is an earthwork. Two sections were cut by Timothy Darvill in 1980–1. Area I was a trench through the outer earthwork. There were only 0.30m of rubble in bank and no old land surface beneath it. The ditch had at least three main cuts and a fourth best described as a gully. It had shifted east with each recutting. Neolithic pottery and flintwork were found in all ditch phases, as was animal bone. Site II (a cutting near west limit of preserved hilltop) revealed an east-west gully along the axis of the hill containing Abingdon Ware, a blade-based industry, much animal bone, at least six leaf arrowheads, a flake from a greenstone axe, and a shale object. It was partly overlain by a 0.25m thick limestone platform, and partly by burnt stones (the relationship between the two earthworks is uncertain).

Objectives: eight samples from the ditch on site I were submitted to refine estimates of the date of earthwork construction derived from measurements already obtained by the excavator, to permit comparison with the results from Crickley Hill, and to determine the duration of Neolithic use of the spur.

Final comment: F Healy (2012), the results, modelled with existing dates from the site, make it possible to estimate that the site I ditch was built in *3650–3550 cal BC (95% probability;* Whittle *et al* 2011, fig 9.19: *build outer Peak Camp).* The site II ditch yielded no suitable samples for further dating.

References:	Bayliss *et al* 2011
	Darvill 1981
	Darvill 1986
	Darvill 2011
	Gillespie *et al* 1985
	Hedges *et al* 1989b
	Oswald *et al* 2001
	T Darvill 1982a
	Whittle *et al* 2011

GrA–30028 5060 ±45 BP

δ¹³C: -25.1‰

Sample: I 10 626 (a), submitted on 26 September 2005 by F Healy

Material: charcoal: *Quercus* sp., roundwood; from same charcoal find as I 10 626 (b); single fragment (R Gale 2005)

Initial comment: area I, phase I, north face of trench, layer 10. Thin lens of burnt material overlying rubble fill in first of three ditch cuts. Stratigraphically equivalent to, though not continuous with, Layer 19 in the south face. If short-life charcoal can be extracted from the sample it will provide a date or dates for the episode of burning and hence for this stage of the infilling of the ditch.

Objectives: to date the context as a step to refining the chronology of the ditch.

Calibrated date: 1σ: 3960–3780 cal BC
  2σ: 3970–3710 cal BC

Final comment: F Healy (2012), both GrA-30028 and OxA-15250, from the same charcoal find, are older than samples from underlying layers and are therefore *termini post quos* for the deposit.

GrA–30029 4825 ±40 BP

δ¹³C: -26.3‰

Sample: I 19 782 (a), submitted on 26 September 2005 by F Healy

Material: charcoal: *Corylus avellana*, from same charcoal find as I 19 782 (b); single fragment (R Gale 2005)

Initial comment: area I, phase I, south face of trench, layer 19. Thin layer of burnt material overlying rubble fill in first of three ditch cuts. Stratigraphically equivalent to, though not continuous with, layer 10 in the north face. If short-life

charcoal can be extracted from the samples it will provide a date or dates for the dump of burnt material and hence for this stage of the infilling of the ditch.

Objectives: to date the context as a step to refining the chronology of the ditch.

Calibrated date: 1σ: 3650–3530 cal BC
  2σ: 3700–3520 cal BC

Final comment: F Healy (2012), this date and OxA-15251, from the same charcoal find, are older than some samples from underlying layers and are thus *termini post quos* for the deposit.

GrA–30030 4760 ±40 BP

δ¹³C: -22.7‰

Sample: I 22 924 (a), submitted on 26 September 2005 by F Healy

Material: animal bone: *Bos* sp., part of neural spine recently broken from young adult thoracic vertebra (926) from same context, which also contained 2 fragments of unfused epiphysis (936, 964), fitting the vertebra (E Hambleton 2005)

Initial comment: a replicate of I 22 924 (b) (OxA-15284). Area I, phase I, layer 22. Initial silt on ditch base, underlying all other area I samples. The presence of a virtually complete vertebra with fragments of its unfused epiphysis in a circumscribed area of the same layer suggests that the bones were still joined by soft tissue when buried, and that the animal from which they came was not long dead.

Objectives: to date the context, providing a *terminus post quem* for dates from the overlying layers and hence refining the date of the ditch.

Calibrated date: 1σ: 3640–3510 cal BC
  2σ: 3650–3370 cal BC

Final comment: F Healy (2012), this result is statistically consistent with OxA-15284 (T'=0.2; T'(5%)=3.8; v=1; Ward and Wilson 1978), and is also consistent with stratigraphic position and model.

References: Ward and Wilson 1978

GrA–30031 4790 ±40 BP

δ¹³C: -21.4‰

Sample: I 22 941, submitted on 26 September 2005 by F Healy

Material: animal bone: *Sus* sp., radius fragment (E Hambleton 2005)

Initial comment: area I, phase I, layer 22. Initial silt on ditch base, underlying all other area I samples.

Objectives: to provide a *terminus post quem* for dates from overlying layers and hence to refine the dating of the ditch.

Calibrated date: 1σ: 3640–3520 cal BC
  2σ: 3660–3380 cal BC

Final comment: F Healy (2012), the result is consistent with the stratigraphic position and with the model.

OxA–15249 4776 ±29 BP

δ¹³C: -21.4‰

Sample: I 22 953, submitted on 26 September 2005 by F Healy

Material: animal bone (sheep/goat tooth) (E Hambleton 2005)

Initial comment: area I, phase I, layer 22. Initial silt on ditch base, underlying all other area I samples.

Objectives: to provide a *terminus post quem* for dates from overlying layers and hence to refine the dating of the ditch.

Calibrated date: *1σ:* 3640–3520 cal BC
                     2σ: 3640–3510 cal BC

Final comment: F Healy (2012), the result is consistent with the stratigraphic position and with the model.

OxA–15250 5060 ±29 BP

δ¹³C: -25.2‰

Sample: I 10 626 (b), submitted on 26 September 2005 by F Healy

Material: charcoal: *Quercus* sp., sapwood; from same charcoal find as I 10 626 (a); single fragment (R Gale 2005)

Initial comment: area I, phase I, north face of trench, layer 10. Thin lens of burnt material overlying rubble fill in first of three ditch cuts. Stratigraphically equivalent to, though not continuous with, layer 19 in the south face. If short-life charcoal can be extracted from the sample it will provide a date or dates for the episode of burning and hence for this stage of the infilling of the ditch.

Objectives: to date the context as a step to refining the chronology of the ditch.

Calibrated date: *1σ:* 3950–3790 cal BC
                     2σ: 3960–3770 cal BC

Final comment: see GrA-30028

OxA–15251 4865 ±29 BP

δ¹³C: -26.1‰

Sample: I 19 782 (b), submitted on 26 September 2005 by F Healy

Material: charcoal: *Quercus* sp., sapwood; from same charcoal find as I 19 782 (a); single fragment (R Gale 2005)

Initial comment: area I, phase I, south face of trench, layer 19. Thin layer of burnt material overlying rubble fill in first of three ditch cuts. Stratigraphically equivalent to, though not continuous with, layer 10 in the north face. If short-life charcoal can be extracted from the samples it will provide a date or dates for the dump of burnt material and hence for this stage of the infilling of the ditch.

Objectives: to date the context as a step to refining the chronology of the ditch.

Calibrated date: *1σ:* 3660–3630 cal BC
                     2σ: 3710–3630 cal BC

Final comment: see GrA-30029

OxA–15284 4782 ±31 BP

δ¹³C: -21.7‰

Sample: I 22 924 (b), submitted on 26 September 2005 by F Healy

Material: animal bone: *Bos* sp., part of neural spine recently broken from young adult thoracic vertebra (926) from same context, which also contained 2 fragments of unfused epiphysis (936, 964), fitting the vertebra (E Hambleton 2005)

Initial comment: a replicate of GrA-30030 (I 22 924 (a)).

Objectives: as GrA-30030

Calibrated date: *1σ:* 3640–3520 cal BC
                     2σ: 3650–3510 cal BC

Final comment: see GrA-30030

Causewayed Enclosures: Robin Hood's Ball, Wiltshire

Location: SU 1011046040
                    Lat. 51.12.45 N; Long. 01.51.19 W

Project manager: N Thomas (Devizes Museum), 1956

Archival body: N Thomas and Salisbury and South Wiltshire Museum, Salisbury and South Wiltshire Museum

Description: the site is a causewayed enclosure still visible as an earthwork, with a double circuit. Limited excavation by Nicholas Thomas revealed trace of post-built structure beneath and/or in inner bank. Surface collection and small-scale excavation by Julian Richards outside the enclosure recorded a lithic and ceramic scatter and fourth millennium pits.

Objectives: to better define its place in the sequence of fourth millennium cal BC construction and other activity in the area in which Stonehenge was later to be built.

Final comment: F Healy (2012), sampling was severely restricted by the limited scale of the excavation, the original paucity of finds from the outer ditch, and failure to locate any of the faunal remains or charcoal. This was particularly disappointing since the excavator reports that 'the joints of several long bones were still in articulation when discovered' (N Thomas 1964, 20). The result is a tentatively estimated construction date of *3640–3500 cal BC (91% probability)* or *3430–3400 cal BC (4% probability;* Whittle *et al* 2011, fig 4.51: *build Robin Hood's Ball).*

References: McOmish *et al* 2002
                 Oswald *et al* 2001
                 Palmer 1976
                 Richards 1990
                 Thomas 1964
                 Whittle *et al* 2011

GrA–30038 4765 ±40 BP

δ¹³C: -29.9‰

Sample: RHB I (50), submitted in 2005 by F Healy

Material: carbonised residue (internal; the largest of three Neolithic Bowl sherds, 2 with residue) (F Healy 2005)

Initial comment: at the interface of layers K and G (Thomas 1964, fig 3). Stratified above RHBI (74). The fresh condition of the residue indicates that the sherd was buried soon after the final use of the vessel.

Objectives: to date the context as a step towards dating the circuit.

Calibrated date: 1σ: 3640–3510 cal BC
                    2σ: 3650–3370 cal BC

Final comment: F Healy (2012), the date is consistent with the pottery style and the context.

OxA–15254 4732 ±30 BP

δ¹³C: -27.0‰

Sample: RHB (I) 74, submitted in 2005 by F Healy

Material: carbonised residue (internal; Nelithic Bowl sherd from larger find) (F Healy 2005)

Initial comment: on the surface and in the very top of layer M, overlain by layers K and L (Thomas 1964, fig 3). The sample formed part of a spread of sherds, some joining, and of bone, on what would have been a temporary surface. Stratified below RHB I (50) and above RHB I (65). The number of sherds and the freshness of the residue indicate that the vessel (or vessels) was freshly broken.

Objectives: to date the context as a step towards dating the circuit.

Calibrated date: 1σ: 3630–3380 cal BC
                    2σ: 3640–3370 cal BC

Final comment: F Healy (2012), the date is consistent with the pottery style and the context.

OxA–15320 5199 ±35 BP

δ¹³C: -29.4‰

Sample: RHB I (65), submitted in 2005 by F Healy

Material: carbonised residue (internal; 1 of 2 Neolithic Bowl sherds, probably from the same vessel) (F Healy 2005)

Initial comment: in layer M, close to bottom of ditch (Thomas 1964, fig 3). Stratified below RBH I 74. The sample and nearby sherds would have been deposited very soon after the ditch was cut. The freshness of the residue indicates that the vessel was broken shortly before the sherd was buried.

Objectives: to date the context as a step towards dating the circuit.

Calibrated date: 1σ: 4040–3960 cal BC
                    2σ: 4050–3950 cal BC

Final comment: F Healy (2012), this result is to be treated with caution.

Laboratory comment: Oxford Radiocarbon Accelerator Unit (2005), OxA-15320 produced a yield of 0.75mg C from a burn-weight of 8.93mg. In addition, the δ¹³C value is more negative than usual. For these reasons, we give a health warning on the measurement.

Causewayed Enclosures: Staines, Surrey

Location: TQ 0241072610
             Lat. 51.26.33 N; Long. 00.31.35 W

Project manager: R Robertson-Mackay (Ministry of Works), 1961–3

Description: a causewayed enclosure formed of two circuits approximately 25m apart, eroded by stream on the flattened south-west side. The outer ditch was wider and deeper than the inner. Internal features of Neolithic, late Bronze Age/early Iron Age, Romano-British, and Saxon date.

Objectives: to date the two circuits; to establish their chronological relationship to each other and to local Neolithic activity; and to estimate the duration of the use of the enclosure.

Final comment: F Healy (2012), attempts to date bone samples from the site have repeatedly been defeated by poor collagen preservation, experienced by the National Physical Laboratory in the 1960s and AERE Harwell and the British Museum in the 1980s. In the early 2000s further bone samples were submitted with equal lack of success, for reasons made clearer by Brock *et al* (2007). No charcoal from Neolithic contexts could be found, and carbonised residues on pottery were scarce because of cleaning and restoration. Some residue samples either failed or yielded blatantly inaccurate results. As a result, only three reliable new dates were obtained. On this slender foundation the chronology of the site cannot be estimated with any precision (Whittle *et al* 2011). It should be noted that the extent to which features and other evidence for activity in the interior, such as quantities of burnt flint, are contemporary with the enclosure is uncertain, since charcoal samples from internal features, some of them submitted in previous decades in expectation of a Neolithic result, have yielded Saxon dates, some of them published by Hardiman *et al* (1992), prompting reconsideration of the conclusion of Robertson-Mackay *et al* that 'The Saxon and medieval finds . . . are therefore of considerable interest in that they . . . provide the what is probably the only archaeological, artefactual record of a Saxon and medieval landscape which has now been lost' (1981, 119).

References: Bradley 2004
              Brock *et al* 2007
              Hardiman *et al* 1992
              Oswald *et al* 2001
              Palmer 1976
              Robertson-Mackay *et al* 1981
              Robertson-Mackay 1962
              Robertson-Mackay 1965
              Robertson-Mackay 1987
              Whittle *et al* 2011

Causewayed Enclosures: Staines, inner ditch, Surrey

Location:	TQ 0241072610
	Lat. 51.26.33 N; Long. 00.31.35 W
Project manager:	R Robertson-Mackay (Ministry of Public Buildings and Works), 1961–3
Archival body:	English Heritage, Swindon and British Museum, Natural History Museum and British Museum

Description: the inner of two concentric causewayed circuits, approximately 50% of which was excavated.

Objectives: to date the construction and use of the circuit as a step to dating those of the monument.

Final comment: F Healy (2012), GrA-30035 and -30066 were from the same segment. GrA-30066, the lower of the two, provides a *terminus ante quem* for the construction of the circuit of *3625–3600 cal BC (6% probability)* or *3525–3380 cal BC (89% probability*; Whittle *et al* 2011, fig 8.3: GrA-30066).

References: Whittle *et al* 2011

GrA–30033 4535 ±40 BP

δ¹³C: -30.0‰

Sample: STA 2d (iii) (3), submitted on 10 October 2005 by F Healy

Material: carbonised residue (internal; 1 of numerous Neolithic Bowl body sherds, many large and well-preserved, 11 with some internal residue, from a thick, coarse, flint-tempered vessel) (F Healy 2005)

Initial comment: segment 32–36, trench 33, 2d (iii), L3. In a lower level of a ditch segment.

Objectives: to date the context as a step towards dating the monument.

Calibrated date:	*1σ:* 3360–3110 cal BC
	2σ: 3370–3090 cal BC

Final comment: see OxA-15252

GrA–30035 4705 ±40 BP

δ¹³C: -30.3‰

Sample: STA 2b (iv) (1) or (v) (1), submitted on 10 October 2005 by F Healy

Material: carbonised residue (internal; coarse Neolithic Bowl body sherd) (F Healy 2005)

Initial comment: segment 24–27, trench 26 or 27, 2b (iv), L1 or 2b (v), L1. Sherd is marked '2b (v) (1)', bag is marked '2b (iv) (1). In either case the sherd came from the topmost fill of the ditch in segment 24–27. The good preservation of the residue indicates that the pot from which the sherd came was freshly broken when the sherd was buried.

Objectives: to date the context as a step towards dating the monument.

Calibrated date:	*1σ:* 3630–3370 cal BC
	2σ: 3640–3360 cal BC

Final comment: F Healy (2012), the result is consistent with the stratigraphic position.

GrA–30036 3165 ±40 BP

δ¹³C: -26.1‰

Sample: STA 2c (xv) (3) [546] 187, submitted on 10 October 2005 by F Healy

Material: carbonised residue (internal; 1 of 44 sherds inc. 1 rim fragment from single ripple-burnished Neolithic Bowl (Robertson-Mackay 1987, fig 47: P134), all found together, several large and well-preserved, up to 10 sherds with internal residue) (R Robertson-Mackay 1964)

Initial comment: a replicate of STA 2c (xv) (3) [546] 97. Segment 13–19, trench 18, 2c (xv), L3. In a lower level of the northern butt of a ditch segment in a concentration of pottery next to a possible entrance (Robertson-Mackay 1987, figs 11 and 27. The fact that so many sherds of the same pot were found together indicates that they were deposited soon after the last use and breakage of the pot.

Objectives: to date the context as a step towards dating the monument.

Calibrated date:	*1σ:* 1500–1410 cal BC
	2σ: 1510–1310 cal BC

Final comment: F Healy (2012), GrA-30036 and OxA-15253, measured on parts of the same carbonised residue, are statistically inconsistent (T'=205.4; T'(5%)=3.8; ν=1; Ward and Wilson 1978) and provide anomalously late dates for the vessel. The fluctuating watertable at Staines may have, perhaps locally, introduced younger humic material to the charred residues which has not been removed by the laboratory protocols intended to remove such contaminants.

References: Robertson-Mackay 1987

     Ward and Wilson 1978

GrA–30066 4650 ±40 BP

δ¹³C: -26.4‰

Sample: STA 2b (v) (4), submitted on 10 October 2005 by F Healy

Material: carbonised residue (internal; Coarse, plain flint-tempered Neolithic Bowl body sherd extracted from among 8 sherds possibly from the same pot) (F Healy 2005)

Initial comment: segment 24–27, trench 26, 2b (v), L4. In a low layer, possibly the bottom layer, of a ditch segment. The exceptionally fresh condition of the residue indicates that the sherd was buried very soon after the final use and breakage of the vessel from which it came.

Objectives: to date the context as a step towards dating the monument.

Calibrated date:	*1σ:* 3510–3360 cal BC
	2σ: 3630–3350 cal BC

Final comment: F Healy (2012), the result is consistent with the stratigraphic position and provides a *terminus ante quem* for the construction of the circuit.

OxA–15252 4364 ±29 BP

δ¹³C: -26.5‰

Sample: STA 2c (xi) (3) [547], submitted on 10 October 2005 by F Healy

Material: carbonised residue (internal; 1 of numerous Neolithic Bowl body sherds, some large and well-preserved, found together, a few with internal residue, most from same pot) (F Healy 2005)

Initial comment: segment 13–19, trench 17, 2c (xi), L3. In a lower layer of a ditch segment. The presence of sherds from the same vessel in a single context indicates that it was deposited soon after breakage, in other words soon after its last use, so that the residue should be close in age to its context.

Objectives: to date the context as a step towards dating the monument.

Calibrated date: *1σ:* 3020–2910 cal BC
                         2σ: 3090–2900 cal BC

Final comment: F Healy (2012), the date is anomalously late for the vessel concerned. The fluctuating water-table at Staines may have locally, introduced younger humic material to the charred residues which has not been removed by the laboratory protocols intended to remove such contaminants.

OxA–15253 3869 ±27 BP

δ¹³C: -25.5‰

Sample: STA 2c (xv) (3) [546] 97, submitted on 10 October 2005 by F Healy

Material: carbonised residue (internal; 1 of 44 sherds from a ripple-burnished Neolithic Bowl (Robertson-Mackay 1987, fig 47: P134) all found together, several of them large and well-preserved, up to 10 sherds with internal residue) (R Robertson-Mackay 1964)

Initial comment: a replicate of GrA-30036 (STA 2c (xv) (3) [546] 187).

Objectives: as GrA-30036

Calibrated date: *1σ:* 2460–2290 cal BC
                         2σ: 2470–2200 cal BC

Final comment: see GrA-30036

References: Robertson-Mackay 1987

Causewayed Enclosures: Staines, outer ditch, Surrey

Location: TQ 0241072610
                     Lat. 51.26.33 N; Long. 00.31.35 W

Project manager: R Robertson-Mackay (Ministry of Public Buildings and Works), 1961–3

Archival body: English Heritage, Swindon and British Museum, Natural History Museum and British Museum

Description: the outer of two concentric causewayed circuits, approximately 40% of which was excavated.

Objectives: to date the construction and use of the circuit as a step to dating those of the monument.

Final comment: F Healy (2012), the only reliable date from the ditch, OxA-15319, provides a *terminus ante quem* of *3630–3555 cal BC (19% probability)* or *3540–3495 cal BC (21% probability)* or *3465–3375 cal BC (55% probability)* for the construction of the circuit (Whittle *et al* 2011, fig 8.3: *OxA-15319*).

References: Whittle *et al* 2011

GrA–30176 3530 ±60 BP

δ¹³C: -28.4‰

Sample: BI ext bone 9 on plan, submitted on 10 October 2005 by F Healy

Material: animal bone: *Bos* sp., proximal fragment of right radius from mature individual, found articulated with ulna (J Mulville and A Powell 2005)

Initial comment: segment 42–48, trench 44, B (i) ext, layer 6. Just above bottom of segment in layer immediately overlying first quick silting in a cluster of sherds and bone including a human skull ('skull A'; Robertson-Mackay 1987, 36, and fig 10). Planned side-by-side with ulna, both marked '[9]', on original plan of BI ext in NMR (ROBO2), photographed *in situ*, NMR neg. BB91/8600. The articulation of the bones shows that they were joined by soft tissue when buried and that the animal from which they came was not long dead.

Objectives: to date the context as a step towards dating the monument.

Calibrated date: *1σ:* 1950–1760 cal BC
                         2σ: 2030–1690 cal BC

Final comment: see GrA-30197

References: Robertson-Mackay 1987

GrA–30197 2660 ±45 BP

δ¹³C: -24.8‰

Sample: STA 789, 800, submitted on 10 October 2005 by F Healy

Material: animal bone: *Bos* sp., distal end of right metacarpal (789), with unfused epiphysis (800) glued on (J Mulville and A Powell 2005)

Initial comment: segment 16–21, trench 20, 1b (xi), L3. In a lower layer of a ditch segment. The presence of the metacarpal end and fitting epiphysis in the same context suggests that they were still joined, or had until recently been joined by soft tissue when buried, and that the animal from which they came was not long dead.

Objectives: to date the context as a step towards dating the monument.

Calibrated date: *1σ:* 840–790 cal BC
                         2σ: 910–780 cal BC

Final comment: F Healy (2012), the δ¹³C value from this sample is anomalously enriched, suggesting that the dated material was contaminated by more recent humic acids from the burial environment. Furthermore, the laboratory notes that 'The samples GrA-30176 and GrA-30197 were of a very poor quality bone. . . . [they] could be measured, but had a very low organic carbon content so the dates are questionable'.

OxA–15319 4735 ±32 BP

δ¹³C: -27.9‰

Sample: YEO61 M (x) (4), submitted on 10 October 2005 by F Healy

Material: carbonised residue (Neolithic Bowl body sherd with surviving residue, among more numerous sherds, including at least 1 other with residue) (F Healy 2005)

Initial comment: segment 6–10, trench 8, M (x), L4. In a lower layer of a ditch segment. The good preservation of the residue indicates that the sherd was buried soon after the breakage of the pot from which it came.

Objectives: to date the context as a step towards dating the monument.

Calibrated date: *1σ:* 3640–3380 cal BC
                     2σ: 3640–3370 cal BC

Final comment: F Healy (2012), the result is consistent with the stratigraphic position and provides a *terminus ante quem* for the construction of the circuit.

Causewayed Enclosures: The Trundle, West Sussex

Location:	SU 87741107
	Lat. 53.28.00 N; Long. 44.55.00 W
Project manager:	E Curwen, O Bedwin, and F Aldsworth
	(Independent and Sussex Archaeological
	Field Unit), 1928, 1930, and 1980

Description: a causewayed enclosure composed of three clearly defined circuits and further intersecting and overlapping ones, only parts of which can be detected on the surface. Outworks to the north and west may be of Neolithic or later date, perhaps relating to the Iron Age hillfort which occupies the same hiltop. In 1928 and 1930, segments were excavated in arbitrary spits approximately 10" (0.25m) deep.

Objectives: to answer questions of date and sequence and to test a potentially early date for ditch 2 inferred from I-11615 (5240 ±140 BP, measured on a disarticulated cattle femur; Drewett 2003, 40; Russell 1996, 58). Another question was whether the inner and second ditches were built together, given their proximity, similarity of plan, and probable concentricity.

Final comment: F Healy (2012), few suitable samples could be located, and some of the questions remain unanswered. Enough dates were, however, obtained to point to a start for construction in the first half of the fourth millennium cal BC (Whittle *et al* 2011).

References:	Bedwin 1981b
	Bradley 1969
	Curwen 1929
	Curwen 1931
	Down 1997
	Drewett *et al* 1988
	Drewett 2003
	Geophysical Surveys of Bradford 1989
	Institute of Archaeology 1987
	Kenny 1994
	Oswald *et al* 2001
	RCHME 1995a
	Russell 1996
	Thomas 1982
	Whittle *et al* 2011

Causewayed Enclosures: The Trundle, ditch 2, West Sussex

Location:	SU 87741107
	Lat. 53.28.00 N; Long. 44.55.00 W
Project manager:	E Curwen (Independent), 1928 and 1930
Archival body:	Barbican House Museum

Description: the second circuit of a complex causewayed enclosure, close to and concentric with the inner ditch.

Objectives: to date the construction of the circuit as a step to establishing the overall chronology of the complex.

Final comment: F Healy (2012), the only apparently suitable samples which could be found were residues from two sherds from the penultimate spit (4), one of which proved to date to the mid fourth millennium cal BC, the other to the first millennium, although the finds table records Iron Age pottery only in spit 1 (Curwen 1930, 39, and 79–80). The sherd may have been intrusive.

References:	Curwen 1930
	Whittle *et al* 2011

GrA–26817 2390 ±35 BP

δ¹³C: -30.0‰

Sample: Barbican House Museum, Lewes 59–17/K, submitted on 2 June 2004 by F Healy

Material: carbonised residue (1 of 2 coarse, plain Neolithic Bowl body sherds from different vessels in fresh condition with some internal residue) (F Healy 2004)

Initial comment: ditch 2. Cutting I. Spit 4. Spit 4 was the penultimate one, 24–36" (0.60m to 0.90m) from the surface. The ditch base was 54" (1.37m) from the surface (Curwen 1929, 39 and 80). The fresh, unweathered condition of the sherd and the survival of the internal residue indicate that it was buried soon after the pot from which it came was broken, in other words soon after the deposition of the residue. It is thus likely to be close in age to its context.

If it indeed came from chalk rubble fills, as is likely, these would have accumulated rapidly, so that the sherd would have been incorporated in the fill soon after the segment was cut. From same spit as sample for OxA-14024. No section was drawn of this segment, but, by analogy with other cuttings on the site, the fill at this level is highly likely to have been chalk rubble.

Objectives: to establish the date of this context as a step towards defining the chronology of the monument.

Calibrated date: 1σ: 510–400 cal BC
                      2σ: 740–390 cal BC

Final comment: F Healy (2012), the first millennium date is surprising, given the depth of the spit from which the sample came and the age of OxA-14024 from the same spit and I-11615 (5240 ±140 BP) and I-11616 (5040 ±170 BP) from the spit below. It suggests that an undetected Iron Age feature may have been cut into the ditch fills or that the relatively small sherd may have been otherwise intrusive.

References: Curwen 1929

OxA–14024 4792 ±28 BP

$\delta^{13}C$: -26.2‰

Sample: Barbican House Museum, Lewes 59–17/J, submitted on 2 June 2004 by F Healy

Material: carbonised residue (1 of 2 coarse, plain Neolithic Bowl body sherds from different vessels in fresh condition with some internal residue) (F Healy 2004)

Initial comment: as GrA-26817

Objectives: as GrA-26817

Calibrated date: 1σ: 3640–3530 cal BC
                      2σ: 3650–3520 cal BC

Final comment: F Healy (2012), since the sherd was not part of a well-represented pot, it could have been redeposited. Cautiously, then it should be seen as providing a *terminus post quem* for its context.

Causewayed Enclosures: The Trundle, inner ditch, West Sussex

Location: SU 87741107
                     Lat. 53.28.00 N; Long. 44.55.00 W

Project manager: E Curwen (Independent), 1928 and 1930

Archival body: Barbican House Museum

Description: the innermost circuit of a complex causewayed enclosure.

Objectives: to date the construction of the circuit as a step to establishing the overall chronology of the complex.

Final comment: F Healy (2012), the only suitable sample that could be found was carbonised residue from a sherd.

References: Whittle *et al* 2011

OxA–14009 5110 ±55 BP

$\delta^{13}C$: -30.8‰

Sample: Barbican House Museum, Lewes 59–17/L, submitted on 2 June 2004 by F Healy

Material: carbonised residue (one of two coarse, plain Neolithic Bowl body sherds in fresh condition, possibly from same pot, with vestigial internal residue) (F Healy 2004)

Initial comment: inner ditch. Cutting I. Spit 7. Spit 7 was the lowest in this cutting, immediately above the ditch floor, and was only 3" (0.08m) deep (Curwen 1929, 79, and pl III). The sherds would have been in primary fill on or just above the ditch floor. Their good preservation and the survival of residue on their surfaces indicate that they were buried soon after the pot from which they came was broken, in other words soon after the deposition of the residue. The proximity of two sherds from the same vessel is also consistent with this. They are thus likely to be close in age to their context and to date from close to the time at which the segment was cut.

Objectives: to establish the date of this context as a step towards defining the chronology of the monument.

Calibrated date: 1σ: 3970–3800 cal BC
                      2σ: 4040–3770 cal BC

Final comment: F Healy (2012), the result should be treated with caution. Note on certificate reads: 'this was a very small sample and we cannot be sure that the material dated was residue'.

References: Curwen 1929

Causewayed Enclosures: The Trundle, outer ditch, West Sussex

Location: SU 87741107
                     Lat. 53.28.00 N; Long. 44.55.00 W

Project manager: E Curwen (Independent), 1928

Archival body: Barbican House Museum

Description: the outermost excavated circuit of a complex causewayed enclosure, partly overlain by the Iron Age hillfort which occupied the same hilltop.

Objectives: to provide a *terminus ante quem* for the construction of the circuit as a step to establishing the overall chronology of the complex.

Final comment: F Healy (2012), a burial from the upper fill of this ditch proved to date to the first millennium cal BC. However it is interpreted, this surprising result provides a *terminus post quem* for the overlying counterscarp bank of the Iron Age hillfort.

References: Whittle *et al* 2011

GrA–26819 2135 ±30 BP

$\delta^{13}C$: -20.5‰
$\delta^{15}N$ *(diet):* +8.0‰

Sample: Barbican House Museum, Lewes/C, submitted on 2 June 2004 by F Healy

Material: human bone (rib fragment from skeleton of 25 to 30–year-old female) (F Parsons 1928)

Initial comment: outer ditch. Trial trench 2, at interface of coarse chalk rubble (effectively primary fill) and fine chalk rubble (effectively secondary fill), towards outer edge of ditch. Fine chalk rubble in turn overlain by counterscarp bank of Iron Age hillfort. Articulated, crouched, under small cairn of chalk blocks (Curwen 1929, pls VI and VII). The articulation of the skeleton shows that the burial was made soon after the death of the individual, so that the sample is close in age to its context. The burial would have been made after the rapid accumulation of chalk rubble in the base of the segment following its original cutting, and hence within relatively few years of the construction of the circuit. NB The photograph published by Oswald *et al* as of this burial (2001, fig 8.4) is in fact of skeleton II in ditch III at Whitehawk (compare Curwen 1929 pls VI and VII with Curwen 1934, pl. XVII: 2).

Objectives: to establish the date of this context as a step towards defining the chronology of the monument.

Calibrated date: *1σ:* 210–110 cal BC
               2σ: 350–50 cal BC

Final comment: F Healy (2012), OxA-13935 and GrA-26819 are statistically consistent (T′=0.1; T′(5%)=3.8; *v*=1; Ward and Wilson 1978). The first millennium date provides a *terminus post quem* for the construction of the Iron Age counterscarp bank which was built over the ditch in which the skeleton was found. The age of the skeleton is a surprise, since the ditch in which it was placed was dug in segments, like the indubitably Neolithic ditches on the hill (RCHME 1995, 12), although it yielded virtually no diagnostic finds (Curwen 1929, 46). There are three possible interpretations: the ditch was cleaned out down to the chalk rubble in the first millennium; the burial was in a grave which was not detected by Curwen; or the ditch was excavated in the first millennium, despite its causewayed plan.

References: Curwen 1929
                       Curwen 1934
                       Oswald *et al* 2001
                       RCHME 1995a
                       Ward and Wilson 1978

OxA–13935 2124 ±28 BP

δ¹³C: -19.6‰

Sample: Barbican House Museum, Lewes/B, submitted on 2 June 2004 by F Healy

Material: human bone (rib fragment from skeleton of 25 to 30-year-old female) (F Parsons 1928)

Initial comment: from the same skeleton as GrA-26819.

Objectives: as GrA-26819

Calibrated date: *1σ:* 200–100 cal BC
               2σ: 350–50 cal BC

Final comment: see GrA-26819

Causewayed Enclosures: Whitehawk, East Sussex

Location: TQ 33030477
                    Lat. 49.35.00 N; Long. 06.39.00 W

Project manager: R Ross Williamson, E Curwen, M Russell, and D Rudling (Brighton and Hove Archaeological Club, Sussex Archaeological Society, and Sussex Archaeological Field Unit), 1929, 1932–3, 1935, and 1991

Description: a causewayed enclosure composed of four clearly-defined circuits, with further intersecting and overlapping ones, only parts of which can be detected on the surface. Outer circuit joined by tangential earthworks, segmented in the north-east, continuous in the south-west. Two further possible ditches to the south were observed by Curwen during road building in 1935. Gate structures and discrete Neolithic features in the interior have been recorded. In 1929 and 1932–3, segments were excavated in arbitrary spits approximately 10in (0.25m) deep. In 1935 they were excavated stratigraphically.

Objectives: to clarify overall points of sequence and duration and to answer specific questions, including whether the contrast in plan and scale between the inner and outer pairs of circuits indicated diachronic construction, and the date of the south-west tangential ditch excavated in 1991.

Final comment: F Healy (2012), Bayesian modelling of the radiocarbon dates provided an estimated construction date of *3690–3635 cal BC (73%)* or *3620–3560 cal BC (22%;* Whittle *et al* 2011, fig 5.5: *build Whitehawk*). It was not possible to sequence the four main ditches, partly because of uneven sample availability, partly because of the relatively coarse resolution that came from working with samples largely recorded by spit rather than by stratigraphic unit. Their construction does not seem too have been separated by any great interval. The primary use of the site seems to have lasted *75–260 years (95% probability;* Whittle *et al* 2011, fig 5.11: *use Whitehawk*). The south-west tangential ditch was dug in the second millennium cal BC and was thus not part of the Neolithic monument. The date of the north-east tangential ditch, which, unlike the south-west one, was segmented, remains unknown (Whittle *et al* 2011).

References: Burstow 1942
                       Curwen 1934
                       Curwen 1935
                       Curwen 1936
                       Curwen 1954
                       Darvill and Fulton 1998
                       Drewett *et al* 1988
                       Geophysical Surveys of Bradford 1993
                       Oswald *et al* 2001
                       Palmer 1976
                       RCHME 1995b
                       Ross Williamson 1930
                       Russell and Rudling 1996
                       Thomas 1996
                       Underwood 1996
                       Whittle *et al* 2011

Causewayed Enclosures: Whitehawk Hill, ditch I, East Sussex

Location:	TQ 33030477 Lat. 49.35.00 N; Long. 06.39.00 W
Project manager:	R Ross Williamson (Brighton and Hove Archaeological Club) E Curwen (Sussex Archaeological Society), 1932–3 and 1935
Archival body:	Brighton and Hove Museum, Brighton and Hove Museum, Barbican House Museum, and Natural History Museum

Description: the innermost circuit of a complex causewayed enclosure.

Objectives: to date the construction of the circuit as a step to establishing the overall chronology of the complex.

Final comment: F Healy (2012), all the samples except GrA-32367 came from Ross Williamson's 1929 excavations and, while they came from different spits, all the spits but the lowest (6) crossed more than one stratigraphic context (Ross Williamson 1930, pl. III). It is possible that the samples from spits 5, 4, and 3 all came from the 'black mould', because 'it was in this black mould that practically all the finds occurred. Little else other than a few roughly worked flints were found in the chalk' (Ross Williamson 1930, 61). If the results for the samples from these spits are modelled as if they were stratigraphically successive the model has poor overall agreement; if they are modelled as if from a single phase, all but OxA-14157 show good agreement. The model provides an estimated construction date of *3635–3560 cal BC (95% probability;* Whittle *et al* 2011, fig 5.6: *dig Whitehawk I).*

References:	Ross Williamson 1930 Whittle *et al* 2011

GrA–26962 4715 ±35 BP

δ¹³C: -23.8‰

Sample: Brighton Museum R3162/169/N (1), submitted on 2 June 2004 by F Healy

Material: antler: *Cervus elaphus*, tine tip (J Mulville and A Powell 2004)

Initial comment: ditch I. Segment CI-CIII. Cutting II. Spit 6. This was the bottom spit (Ross Williamson 1930, pl. III: section C-D). At a lower level than sample Barbican House Museum, Lewes 29.46/D. Damage to the tine tip and an ancient break at tine base suggest that the sample it formed part of an antler pick used to dig the ditch and is thus close in age to its original excavation. In chalk-cut ditch on well-drained chalk down.

Objectives: to establish the date of this context as a step towards defining the chronology of the monument.

Calibrated date:	*1σ:* 3630–3370 cal BC *2σ:* 3640–3370 cal BC

Final comment: F Healy (2012), GrA-26962 and OxA-14126 are statistically consistent (T'=1.6; T'(5%)=3.8; ν=1; Ward and Wilson 1978) and consistent with an origin in the chalk rubble below the 'black band'.

References:	Ward and Wilson 1978

GrA–26963 4575 ±35 BP

δ¹³C: -29.5‰

Sample: Barbican House Museum, Lewes 36.37/E, submitted on 2 June 2004 by F Healy

Material: carbonised residue (internal; one of five Neolithic Bowl body sherds, from at least two separate vessels, in fairly fresh condition) (F Healy 2004)

Initial comment: ditch I. Segment CI-CIII. Cutting I. Spit 4. (Ross Williamson 1930, pl. III: section A-B). At a higher level than sample Barbican House Museum, Lewes 29.46/D. The condition of the sherds and the survival of residues indicate that they were freshly broken when deposited, and that the residue should be close in age to the context. The fills encountered in this spit are listed as 'mould and chalk' (Ross Williamson 1930, 88). His section A-B (ibid. pl. III) shows spit 4 cutting across chalk rubble and 'black mould'. The sherds are more likely to come from the latter, because it was richer in finds.

Objectives: to establish the date of this context as a step towards defining the chronology.

Calibrated date:	*1σ:* 3370–3340 cal BC *2σ:* 3500–3120 cal BC

Final comment: F Healy (2012), the result is consistent with an origin in the 'black mould'.

GrA–26965 4690 ±40 BP

δ¹³C: -29.1‰

Sample: Barbican House Museum, Lewes 36.37/H, submitted on 2 June 2004 by F Healy

Material: carbonised residue (internal; one of twenty-five Neolithic Bowl body sherds. From different vessel to samples Barbican House Museum, Lewes 36.37/G and /I) (F Healy 2004)

Initial comment: ditch I. Segment CVI, spit 3. There is no section of CVI; its average depth was 4 ft (Ross Williamson 1930, 61). Spit 3 lay at 20–30" (0.50–0.75m). At a higher level than any of the samples from segment CI-CII, but probably from the same stratigraphic horizon. On the evidence of the other segments, spit 3 would have been mainly in 'black mould' a conclusion supported by the volume of finds from it (ibid, 91).

Objectives: to establish the date of this context as a step towards defining the chronology of the monument.

Calibrated date:	*1σ:* 3630–3370 cal BC *2σ:* 3640–3360 cal BC

Final comment: see GrA-26963

GrA–32367 4805 ±35 BP

δ¹³C: -22.1‰

Sample: Brighton Museum R 4100/143/V, submitted on 23 August 2006 by F Healy

Material: animal bone: *Bos* sp., metatarsal fragment articulating with navicular-cuboid to which 2nd and 3rd tarsals fused (A Powell 2006)

Initial comment: site A, DI, layer 2. Typed on envelope: 'WHITEHAWK (East) EXCAVATIONS 26th Oct., 1935. Ditch 1. Level 2, black earth. OX, bones and teeth of' (Curwen 1936, 62–3, fig C). The fact that the two bones were recovered on the same day suggests that they were close together and hence still or recently joined by soft tissue when buried.

Objectives: to refine the chronology of the ditch and the complex.

Calibrated date: *1σ:* 3650–3530 cal BC
                       2σ: 3660–3520 cal BC

Final comment: F Healy (2012), the date is compatible with the stratigraphic position of this sample in the model.

References: Curwen 1936

OxA–14030 4809 ±29 BP

δ¹³C: -26.8‰

Sample: Barbican House Museum, Lewes 36.37/F, submitted on 2 June 2004 by F Healy

Material: carbonised residue (internal; 1 of 4 sherds, 3 possibly from same pot) (F Healy 2004)

Initial comment: ditch I. Segment CI-CIII, cutting II, spit 4 (Ross Williamson 1930, pl. III). At a higher level than sample Barbican House Museum, Lewes 29.46/D. The condition of the sherds and the survival of residues indicate that they were freshly broken when deposited, and that the residue should be close in age to the context. The fills encountered in this spit are listed as 'mould and chalk' (Ross Williamson 1930, 88). His section A-B (ibid. pl. III) shows spit 4 cutting across chalk rubble and 'black mould'. The sherds are more likely to come from the latter, because it was richer in finds.

Objectives: to establish the date of this context as a step towards defining the chronology of the monument.

Calibrated date: *1σ:* 3650–3530 cal BC
                       2σ: 3650–3520 cal BC

Final comment: see GrA-26963

OxA–14039 4602 ±39 BP

δ¹³C: -27.6‰

Sample: Barbican House Museum, Lewes 29.46/D, submitted on 2 June 2004 by F Healy

Material: carbonised residue (internal residue surviving on lowest part of large, well-preserved sherd of Carinated Neolithic Bowl, with faint channelling on lower body; joins with two others from DI CII 4) (F Healy 2004)

Initial comment: ditch I. Segment CI-CIII, cutting III, spit 5. Spit 5 was the penultimate one (Ross Williamson 1930, pl III: section A-B). At a higher level than sample Brighton Museum R3162.169/N and at a lower level than samples Barbican House Museum, Lewes 36.37/E and /F. The large size and good preservation of the sherd, the fact that other large, well-preserved sherds of the same vessel came from nearby, and the survival of superficial residue al indicate that the vessel was freshly broken and recently used when the

sherds were buried. Spit 5 is shown on Ross Williamson's pl III (section A-B) as cutting across chalk rubble and 'black mould'. It is listed as ' black mould, 36–45 in' (ibid. 90). This was a dark deposit rich in cultural material.

Objectives: to establish the date of this context as a step towards defining the chronology of the monument.

Calibrated date: *1σ:* 3490–3350 cal BC
                       2σ: 3500–3130 cal BC

Final comment: see OxA-14030

OxA–14040 4725 ±65 BP

δ¹³C: -27.2‰

Sample: Barbican House Museum, Lewes 36.37/I, submitted on 2 June 2004 by F Healy

Material: carbonised residue (internal; one of twenty-five Neolithic Bowl body sherds. From different vessel to samples Barbican House Museum, Lewes 36.37/G and /H) (F Healy 2004)

Initial comment: ditch I. Segment CVI, spit 3. There is no section of CVI; its average depth was 4 ft (Ross Williamson 1930, 61). Spit 3 lay at 20–30" (0.50–0.75m). At a higher level than any of the samples from segment CI-CII, but probably from the same stratigraphic horizon. On the evidence of the other segments, spit 3 would have been mainly in 'black mould' a conclusion supported by the volume of finds from it (ibid, 91).

Objectives: to establish the date of this context as a step towards defining the chronology of the monument.

Calibrated date: *1σ:* 3640–3370 cal BC
                       2σ: 3650–3360 cal BC

Final comment: see OxA-14030

OxA–14126 4774 ±31 BP

δ¹³C: -22.9‰

Sample: Brighton Museum R3162/169/N (2), submitted on 2 June 2004 by F Healy

Material: antler: *Cervus elaphus*, tine tip (J Mulville and A Powell 2004)

Initial comment: a replicate of GrA-26962.

Objectives: as GrA-26962

Calibrated date: *1σ:* 3640–3520 cal BC
                       2σ: 3650–3380 cal BC

Final comment: see GrA-26962

OxA–14157 4846 ±32 BP

δ¹³C: -27.6‰

Sample: Barbican House Museum, Lewes 36.37/G, submitted on 2 June 2004 by F Healy

Material: carbonised residue (internal; one of twenty-five Neolithic Bowl body sherds. From different vessel to samples Barbican House Museum, Lewes 36.37/H and /I) (F Healy 2004)

Initial comment: ditch I. Segment CVI, spit 3. There is no section of CVI; its average depth was 4 ft (Ross Williamson 1930, 61). Spit 3 lay at 20–30" (0.50–0.75m). At a higher level than any of the samples from segment CI-CII, but probably from the same stratigraphic horizon. On the evidence of the other segments, spit 3 would have been mainly in 'black mould' a conclusion supported by the volume of finds from it (ibid, 91).

Objectives: to establish the date of this context as a step towards defining the chronology of the monument.

Calibrated date: 1σ: 3660–3630 cal BC
                     2σ: 3700–3530 cal BC

Final comment: F Healy (2012), this is the only one of the samples from spits 3–5 in the 1929 excavation of this ditch to be inconsistent with an origin in the 'black mould'. The sherd may have been redeposited or the date may be inaccurate.

Causewayed Enclosures: Whitehawk Hill, ditch II, East Sussex

Location: TQ 33030477
                  Lat. 49.35.00 N; Long. 06.39.00 W

Project manager: R Ross Williamson (Brighton and Hove Archaeological Club) E Curwen (Sussex Archaeological Society), 1929, 1932–3, and 1935

Archival body: Brighton and Hove Museum, Brighton and Hove Museum, Barbican House Museum, and Natural History Museum

Description: the second circuit of a complex causewayed enclosure, close to and concentric with DI.

Objectives: to date the construction of the circuit as a step to establishing the overall chronology of the complex.

Final comment: F Healy (2012), Bayesian modelling of the radiocarbon dates provides an estimated construction date of *3675–3630 cal BC (72% probability)* or *3600–3545 cal BC (23% probability*; Whittle *et al* 2011, fig 5.7: *dig Whitehawk II*).

References: Whittle *et al* 2011

GrA–26966 4605 ±40 BP

$\delta^{13}C$: -21.2‰

Sample: Skeleton III. Brighton Museum R 4100/139 221788/U (1), submitted on 2 June 2004 by F Healy

Material: human bone (mandible from articulated skeleton of middle-aged male) (M Tildesley 1935)

Initial comment: site B, between Ditch I and Ditch II 'lying on the surface of the undisturbed chalk and covered only by a foot of topsoil'. Contracted, head to the east, face to the north, hands in front of face. Accompanied by three sherds of Neolithic pottery, land-molluscs, and two or three mussel shells near the head (Curwen 1936, 70). The skeleton was 3m from the inner lip of DII. If DII once had an internal bank, as well as the external one observed by RCHME, then

the skeleton would have lain in the area of that bank. Indeed, one of the few plausible explanations for the skeletons having survived intact is that it was protected by an overlying bank, at least up to the time of ploughing in the nineteenth century. In this case, it would pre-date the construction of ditch II and the deposition of sample Brighton Museum/A.

Objectives: to establish the date of this context as a step towards defining the chronology of the monument.

Calibrated date: 1σ: 3500–3350 cal BC
                     2σ: 3510–3130 cal BC

Final comment: F Healy (2012), OxA-14061 and GrA-26966 are slightly statistically inconsistent at 2σ (T'=6.2; T'(5%)=3.8; ν=1; Ward and Wilson 1978). Both are, however, more recent than OxA-14031, measured on a sample from layer 5 in an adjacent section across ditch II, and than samples from the fill of the ditch in other parts of the circuit. The skeleton was thus not buried under any bank internal to ditch II and may either have been placed in a grave cut entirely in the soil without reaching into the chalk or have been cut into an internal bank if one ever existed.

Laboratory comment: Rijksuniversitat Groningen (AMS) (11 January 2005), insufficient collagen was extracted to enable measurement of the $\delta^{15}N$.

References: Ward and Wilson 1978

OxA–14031 4897 ±29 BP

$\delta^{13}C$: -25.3‰

Sample: Brighton Museum/A, submitted on 2 June 2004 by F Healy

Material: carbonised residue (internal; Neolithic Bowl body sherd in fresh condition) (F Healy 2004)

Initial comment: ditch II. Site B, layer 5. Layer 5 does not figure in the published description or section of this ditch (Curwen 1936, 70–61 and fig E: first section). The fill is, however, described as consisting of 'four principal layers', which leaves open the possibility of others, and layer 5 is given as the context on the envelope in which the sherd was stored. Since layers were numbered from the top, it would have lain between layer 4 and the base of the ditch. This context must date to very shortly after the original digging of the segment. The fresh condition of the sherd and the survival of the residue indicate that it was broken and buried soon after use.

Objectives: to establish the date of this context as a step towards defining the chronology of the monument.

Calibrated date: 1σ: 3710–3640 cal BC
                     2σ: 3710–3630 cal BC

Final comment: F Healy (2012), the date is compatible with the sample's stratigraphic position.

References: Curwen 1936

OxA–14061 4739 ±36 BP

δ¹³C: -20.3‰
δ¹⁵N (diet): +10.0‰
C/N ratio: 3.4

Sample: Skeleton III. Brighton Museum R 4100/139 221788/U (2), submitted on 2 June 2004 by F Healy

Material: human bone (mandible from articulated skeleton of middle-aged   male) (M Tildesley 1935)

Initial comment: a replicate of GrA-26966.

Objectives: as GrA-26966

Calibrated date: *1σ:* 3640–3380 cal BC
    2σ: 3640–3370 cal BC

Final comment: see GrA-26966

Causewayed Enclosures: Whitehawk Hill, ditch III, East Sussex

Location: TQ 33030477
    Lat. 49.35.00 N; Long. 06.39.00 W

Project manager: R Ross Williamson (Brighton and Hove Archaeological Club) E Curwen (Sussex Archaeological Society), 1932–3 and 1935

Archival body: Brighton and Hove Museum, Brighton and Hove Museum, Barbican House Museum, and Natural History Museum

Description: one of perhaps several outer circuits of a complex causewayed enclosure, recut around at least part of its length.

Objectives: to date the construction of the circuit as a step to establishing the overall chronology of the complex.

Final comment: F Healy (2012), Bayesian modelling of the radiocarbon dates provides an estimated construction date of *3660–3560 cal BC (95% probability;* Whittle *et al* 2011, fig 5.8: *dig Whitehawk III).* This may, however, apply to the recut rather than to the original ditch.

References: Whittle *et al* 2011

GrA–26967 4545 ±35 BP

δ¹³C: -29.3‰

Sample: Brighton Museum R3688/71/D, submitted on 2 June 2004 by F Healy

Material: carbonised residue (internal; Neolithic Bowl body sherd) (F Healy 2004)

Initial comment: ditch III. Segment CVI-CVIII. Cutting VII. Spit 3. Spit 3 lay at 20–30" (0.50–0.75m) below the surface (Curwen 1934, fig 2: sections IV and V). At a higher level than sample Brighton Museum R. 3688/127/K. Good preservation of sherd and survival of residue indicate that it was freshly-broken and recently-used when buried.

Objectives: to establish the date of this context as a step towards defining the chronology of the monument.

Calibrated date: *1σ:* 3360–3120 cal BC
    2σ: 3370–3090 cal BC

Final comment: F Healy (2012), the most recent of the dates from the ditch. May reflect activity after the main use of the site.

References: Curwen 1934

GrA–26969 4660 ±35 BP

δ¹³C: -30.3‰

Sample: Brighton Museum R3688/74/G, submitted on 2 June 2004 by F Healy

Material: carbonised residue (internal; Plain Neolithic Bowl body sherd) (F Healy 2004)

Initial comment: ditch III. Segment CIII-CIV. Spit 5H (= 'spit 5 hearth'). Hearth with pottery and human and animal bone in 'occupation layer' (Curwen 1934, 111). Spit 5 lay 40–50" (1.0–1.27m) below the surface. Good preservation of sherd and survival of residue indicate that it was freshly-broken and recently-used when buried.

Objectives: to establish the date of this context as a step towards defining the chronology of the monument.

Calibrated date: *1σ:* 3520–3360 cal BC
    2σ: 3630–3360 cal BC

Final comment: F Healy (2012), the date is compatible with the stratigraphic position of the sample in the model.

References: Curwen 1934

GrA–26971 4795 ±40 BP

δ¹³C: -20.7‰
δ¹⁵N (diet): +10.0‰

Sample: Skeleton I. Brighton Museum R3688/128/S, submitted on 2 June 2004 by F Healy

Material: human bone (rib fragment from articulated skeleton of female, 25–30 years-old) (M Tildesley 1933)

Initial comment: ditch III. Segment CII, in 'occupation layer'. Articulated (Curwen 1934, fig 2: section I, marked 'S'). At a lower level than samples Brighton Museum R. 3688/122/H and NHM 1970.3068. The articulation of the skeleton ensures that the individual was recently dead when buried. In a dark, organic deposit with much cultural material.

Objectives: to establish the date of this context as a step towards defining the chronology of the monument.

Calibrated date: *1σ:* 3640–3520 cal BC
    2σ: 3660–3380 cal BC

Final comment: see GrA-26969

References: Curwen 1934

GrA–26975 4965 ±40 BP

δ¹³C: -27.5‰

Sample: Barbican House Museum, Lewes 36.37/M (Horniman 3941), submitted on 2 June 2004 by F Healy

Material: carbonised residue (internal residue from Plain Neolithic Bowl sherd, among several others) (F Healy 2004)

Initial comment: ditch III. Segment CIII-CV. Cutting V. Spit 3. Spit 3 lay 20–30" (0.50–0.75m) below the surface. Curwen's (1934) fig 2 (section III) suggests that spit 3 cut across secondary and tertiary silts. Good preservation of the sherd and survival of residue indicate that it was freshly-broken and recently-used when buried. Either in chalk rubble or in chalk rubble with dark material and many artefacts and food remains.

Objectives: to establish the date of this context as a step towards defining the chronology of the monument.

Calibrated date: *1σ:* 3790–3690 cal BC
                   2σ: 3910–3650 cal BC

Final comment: F Healy (2012), older than the underlying samples, probably redeposited.

References: Curwen 1934

GrA–26976 4710 ±45 BP

δ¹³C: -31.0‰

Sample: Brighton Museum R3688/73/E (1), submitted on 2 June 2004 by F Healy

Material: carbonised residue (internal; Neolithic Bowl body sherd) (F Healy 2004)

Initial comment: ditch III. Segment CIII-CV. Cutting CV. Spit 5. Spit 5 lay 40–50" (1.0–1.25m) below the surface. Good preservation of sherd and survival of residue indicate that it was freshly-broken and recently-used when buried.

Objectives: to establish the date of this context as a step towards defining the chronology of the monument.

Calibrated date: *1σ:* 3630–3370 cal BC
                   2σ: 3640–3360 cal BC

Final comment: F Healy (2012), the result is statistically consistent with its replicate, OxA-14041 (T'=0.6; T'(5%)=3.8; *v*=1; Ward and Wilson 1978) and compatible with the stratigraphic position of the sample in the model.

References: Ward and Wilson 1978

GrA–26977 4785 ±40 BP

δ¹³C: -21.1‰
δ¹⁵N (diet): +9.7‰

Sample: Skeleton IIa. Brighton Museum R3688/129/T (1), submitted on 2 June 2004 by F Healy

Material: human bone (rib fragment from articulated skeleton of female 20–25 years-old) (M Tildesley 1933)

Initial comment: ditch III. Segment CIII-CV. Cutting V, in 'lower part of occupation layer' (Curwen 1934, 108–10, pl XIV, fig 2: section III, pl. XVII: 2). Articulated, with articulated remains of infant, in elongated oval area surrounded by chalk blocks with 2 perforated chalk fragments, covered with soil to top of blocks. The articulation of the skeleton shows that the individual was recently dead when buried and that the sample is likely to be close in age to its context. NB The photograph published by

Oswald *et al* as of the burial excavated by Curwen in the outer ditch of the Trundle (2001, fig 8.4) is in fact of this skeleton (compare Curwen 1929 pls VI and VII with Curwen 1934, pl. XVII: 2).

Objectives: to establish the date of this context as a step towards defining the chronology of the monument.

Calibrated date: *1σ:* 3640–3520 cal BC
                   2σ: 3650–3380 cal BC

Final comment: F Healy (2012), GrA-26977 and OxA-14063 are statistically consistent (T'=0.0; T'(5%)=3.8; *v*=1; Ward and Wilson 1978) and compatible with the burial's stratigraphic position in the model.

References: Curwen 1929
                    Curwen 1934
                    Oswald *et al* 2001
                    Ward and Wilson 1978

GrA–27325 4770 ±45 BP

δ¹³C: -25.4‰

Sample: Brighton Museum R3688/125/J (2), submitted on 2 June 2004 by F Healy

Material: charcoal: Pomoideae, single fragment (R Gale 2004)

Initial comment: ditch III. Segment CIII-CV, cutting CV, spit 5. Spit 5 lay 40–50" (1–1.25m) below the surface. The large size and good preservation of the fragments means that they are likely to have burnt *in situ*. If short-life charcoal can be extracted from the sample, it should be close in age to its context. Either in chalk rubble or at base of chalk rubble mixed with darker deposit and much cultural material.

Objectives: to establish the date of this context as a step towards defining the chronology of the monument.

Calibrated date: *1σ:* 3640–3510 cal BC
                   2σ: 3650–3370 cal BC

Final comment: see GrA-26969

GrA–27326 4995 ±45 BP

δ¹³C: -24.0‰

Sample: Brighton Museum R3688/122/H (2), submitted on 2 June 2004 by F Healy

Material: charcoal: *Quercus* sp., sapwood; single fragment (R Gale 2004)

Initial comment: ditch III. Cutting II spit 4. Spit 4 lay 30–40" (0.75–1.0m) below the surface, above skeleton I (Curwen 1934, fig 2: section I). From same find as OxA-14143.

Objectives: to establish the date of this context as a step towards defining the chronology of the monument.

Calibrated date: *1σ:* 3910–3700 cal BC
                   2σ: 3950–3650 cal BC

Final comment: F Healy (2012), this sample was older than skeleton I, which underlay the sample. It was probably redeposited.

References: Curwen 1934

GrA–27327 4775 ±45 BP

δ¹³C: -25.5‰

Sample: Brighton Museum R3688/127/L (2), submitted on 2 June 2004 by F Healy

Material: charcoal: cf *Corylus Viburnum* sp.; single fragment (R Gale 2004)

Initial comment: ditch III. Segment CVI-CVIII Cutting VII spit 6. Spit 6 lay at 60–70 in (1.50–1.75 m) below the surface (Curwen 1934, fig 2: sections IV and V). At a lower level than sample Brighton Museum R. 3688/127/K. If short-life charcoal can be extracted from this sample it should be close in age to the context.

Objectives: to establish the date of this context as a step towards defining the chronology of the monument.

Calibrated date: *1σ:* 3640–3520 cal BC
                     2σ: 3650–3370 cal BC

Final comment: see GrA-26929

References: Curwen 1934

GrA–27328 4955 ±45 BP

δ¹³C: -24.7‰

Sample: Brighton Museum R3688/124/I (2), submitted on 2 June 2004 by F Healy

Material: charcoal: *Quercus* sp., probably sapwood; single fragment (R Gale 2004)

Initial comment: ditch III. Segment CIII-CV, cutting IV spit 5, hearth. Hearth with pottery and human and animal bone in 'occupation layer' (Curwen 1934, 111; pl XIV; fig 2: sections II and III). Spit 5 lay 40–50" (1.0–1.27m) below the surface. If short-life charcoal can be extracted from this sample, its *in situ* use as firewood will indicate that it was close in age to its context.

Objectives: to establish the date of this context as a step towards defining the chronology.

Calibrated date: *1σ:* 3790–3660 cal BC
                     2σ: 3910–3640 cal BC

Final comment: F Healy (2012), the result is older than other samples from the same hearth and from the wider 'occupation layer'. The sample may not in fact have been of sapwood.

References: Curwen 1934

GrA–27330 4760 ±45 BP

δ¹³C: -21.9‰

Sample: Brighton Museum/C (2), submitted on 2 June 2004 by F Healy

Material: animal bone: *Capra* sp., humerus articulating with radius from subadult individual. Animal bone from this spit includes elements from two goats, this one represented by at least 10 long bones and a scapula (J Mulville and and A Powell 2004)

Initial comment: ditch III. Cutting CI, spit 6. Spit 6 was the penultimate spit and lay at 50–60" (1.30–1.50m) from the surface and 18–28" (0.45–0.70m) above the uneven base of the ditch (Ross Williamson 1930, 96; pl. III: section GH). These deposits would have accumulated quickly, probably within a few years of the originally digging of the ditch. The articulation of the bones suggests that the animal from which they came was not long dead when its remains were buried. The number of bones recovered from a single individual makes it possible that the entire skeleton was present.

Objectives: to establish the date of this context as a step towards defining the chronology of the monument.

Calibrated date: *1σ:* 3640–3380 cal BC
                     2σ: 3650–3370 cal BC

Final comment: F Healy (2012), OxA-14178 and GrA-27330 are statistically consistent (T'=0.0; T' (5%)=3.8; ν=1; Ward and Wilson 1978), and compatible with their stratigraphic position.

References: Ross Williamson 1930
                     Ward and Wilson 1978

GrA–29363 4720 ±45 BP

δ¹³C: -20.9‰

Sample: NHM 1970.3068/2, submitted on 6 June 2004 by F Healy

Material: animal bone: *Bos* sp., proximal phalanx articulating with medial phalanx, which in turn might articulate with unprovenanced distal phalanx (J Mulville and A Powell 2004)

Initial comment: ditch III. Cutting II, spit 4. Spit 4 lay 30–40" (0.75–1.0m) below the surface, above skeleton I (Curwen 1934, fig 2: section I). The articulation of the phalanges suggests that they were still held together with tendons when buried, and that the animal from which they came was recently dead when they were buried, so that they are close in age to their context.

Objectives: to establish the date of this context as a step towards defining the chronology of the monument.

Calibrated date: *1σ:* 3630–3370 cal BC
                     2σ: 3640–3360 cal BC

Final comment: F Healy (2012), OxA-14062 and GrA-29363 are statistically consistent (T'=1.3; T' (5%)=3.8; ν=1; Ward and Wilson 1978) and compatible with the stratigraphic location of the sample.

References: Curwen 1934
                     Ward and Wilson 1978

OxA–14041 4820 ±130 BP

δ¹³C: -29.9‰

Sample: Brighton Museum R3688/73/E (2), submitted on 2 June 2004 by F Healy

Material: carbonised residue (internal; Neolithic Bowl body sherd) (F Healy 2004)

Initial comment: a replicate of GrA-26976.

Objectives: as GrA-26976

Calibrated date: 1σ: 3710–3380 cal BC
2σ: 3950–3350 cal BC

Final comment: see GrA-26976

OxA–14062 4785 ±35 BP

δ¹³C: -21.5‰

Sample: NHM 1970.3068, submitted on 2 June 2004 by F Healy

Material: animal bone: *Bos* sp., proximal phalanx articulating with medial phalanx, which in turn might articulate with unprovenanced distal phalanx (J Mulville and A Powell 2004)

Initial comment: a replicate of GrA-29363.

Objectives: as GrA-29363

Calibrated date: 1σ: 3640–3520 cal BC
2σ: 3650–3380 cal BC

Final comment: see GrA-29363

OxA–14063 4792 ±33 BP

δ¹³C: -20.6‰
δ¹⁵N (diet): +9.9‰
C/N ratio: 3.3

Sample: Skeleton IIa. Brighton Museum R3688/129/T (2), submitted on 2 June 2004 by F Healy

Material: human bone (rib fragment from articulated skeleton of female 20–25 years-old) (M Tildesley 1933)

Initial comment: a replicate of GrA-26977.

Objectives: as GrA-26977

Calibrated date: 1σ: 3640–3530 cal BC
2σ: 3650–3510 cal BC

Final comment: see GrA-26977

OxA–14143 4941 ±39 BP

δ¹³C: -23.9‰

Sample: Brighton Museum R3688/122/H (1), submitted on 2 June 2004 by F Healy

Material: charcoal: *Quercus* sp., sapwood; single fragment (R Gale 2004)

Initial comment: ditch III. Cutting II spit 4. Spit 4 lay 30–40" (0.75–1.0m) below the surface, above skeleton I (Curwen 1934, fig 2: section I). From same find as GrA-27326.

Objectives: to establish the date of this context as a step towards defining the chronology of the monument.

Calibrated date: 1σ: 3770–3650 cal BC
2σ: 3800–3640 cal BC

Final comment: see GrA-27326

References: Curwen 1934

OxA–14144 4835 ±33 BP

δ¹³C: -25.5‰

Sample: Brighton Museum R3688/127/L (1), submitted on 2 June 2004 by F Healy

Material: charcoal: *Corylus avellana*, single fragment (R Gale 2004)

Initial comment: ditch III. Segment CVI-CVIII Cutting VII spit 6. Spit 6 lay at 60–70" (1.50–1.75m) below the surface (Curwen 1934, fig 2: sections IV and V). At a lower level than sample Brighton Museum R. 3688/127/K. If short-life charcoal can be extracted from this sample it should be close in age to its context.

Objectives: to establish the date of this context as a step towards defining the chronology of the monument.

Calibrated date: 1σ: 3650–3630 cal BC
2σ: 3670–3530 cal BC

Final comment: see GrA-26969

References: Curwen 1934

OxA–14145 4844 ±34 BP

δ¹³C: -24.4‰

Sample: Brighton Museum R3688/124/I (1), submitted on 2 June 2004 by F Healy

Material: charcoal: *Quercus* sp., probably sapwood; single fragment (R Gale 2004)

Initial comment: ditch III. Segment CIII-CV, cutting IV spit 5, hearth. Hearth with pottery and human and animal bone in 'occupation layer' (Curwen 1934, 111; pl XIV; fig 2: sections II and III). Spit 5 lay 40–50" (1.0–1.27m) below the surface. If short-life charcoal can be extracted from this sample, its *in situ* use as firewood will indicate that it was close in age to its context. In a hearth within a dark deposit rich in cultural material and also including medium chalk rubble.

Objectives: to establish the date of this context as a step towards defining the chronology of the monument.

Calibrated date: 1σ: 3660–3630 cal BC
2σ: 3700–3530 cal BC

Final comment: see GrA-26969

References: Curwen 1934

OxA–14178 4755 ±32 BP

δ¹³C: -21.2‰

Sample: Brighton Museum/C (1), submitted on 2 June 2004 by F Healy

Material: animal bone: *Capra* sp., humerus articulating with radius from subadult individual. Animal bone from this spit includes elements from two goats, this one represented by at least 10 long bones and a scapula (J Mulville and A Powell 2004)

Initial comment: a replicate of GrA-27330.

Objectives: as GrA-27330

Calibrated date: 1σ: 3640–3510 cal BC
2σ: 3640–3370 cal BC

Final comment: see GrA-27330

OxA–14204 4729 ±32 BP

δ¹³C: -25.9‰

Sample: Brighton Museum R3688/125/J (1), submitted on 2 June 2004 by F Healy

Material: charcoal: *Corylus avellana*, single fragment (R Gale 2004)

Initial comment: ditch III. Segment CIII-CV, cutting CV, spit 5. Spit 5 lay 40–50" (1–1.25 m) below the surface. The large size and good preservation of the fragments means that they are likely to have burnt *in situ*. If short-life charcoal can be extracted from the sample, it should be close in age to its context. Either in chalk rubble or at the base of chalk rubble mixed with the darker deposit and much cultural material.

Objectives: to establish the date of this context as a step towards defining the chronology of the monument.

Calibrated date: 1σ: 3630–3380 cal BC
2σ: 3640–3370 cal BC

Final comment: see GrA-26976

Causewayed Enclosures: Whitehawk Hill, ditch IV, East Sussex

Location: TQ 33030477
Lat. 49.35.00 N; Long. 06.39.00 W

Project manager: E Curwen (Sussex Archaeological Society), 1932–3 and 1935

Archival body: Brighton and Hove Museum, Brighton and Hove Museum, Barbican House Museum, and Natural History Museum

Description: the outermost excavated circuit of a complex causewayed enclosure, recut at least in its southern part.

Objectives: to date the construction of the circuit as a step to establishing the overall chronology of the complex.

Final comment: F Healy (2012), Bayesian modelling of the dates provides an estimated construction date of *3650–3505 cal BC (95% probability*; Whittle *et al* 2011, fig 5.9: *dig Whitehawk IV*). This may, however, apply to the recut rather than to the original ditch.

References: Whittle *et al* 2011

GrA–26972 4830 ±40 BP

δ¹³C: -21.8‰

Sample: Brighton Museum R3688/138/B, submitted on 2 June 2004 by F Healy

Material: animal bone: *Bos* sp., 'Proximal end of right tibia of ox, found with skeleton of roe deer, Whitehawk, Jan 1933'. In weathered condition (J W Jackson 1933–4) (J Mulville and A Powell 2004)

Initial comment: ditch IV. Segment CV-CVI. Cutting V. Hole 5. Found with near-complete articulated skeleton of roe deer (missing April 2004) in pit cut into surface of fairly low causeway truncated by recutting of ditch IV. 'The south wall of the hole was partly broken away' suggests that it may have been truncated when the ditch was recut (Curwen 1934, pls XII, XV; fig 1: section IV). If so, the date of this bone will provide a *terminus post quem* for the recutting of the ditch. In chalk-cut pit under chalk rubble fill.

Objectives: to establish the date of this context as a step towards defining the chronology of the monument.

Calibrated date: 1σ: 3650–3530 cal BC
2σ: 3700–3520 cal BC

Final comment: F Healy (2012), the roe deer skeleton remains missing. The date provides a *terminus post quem* for its burial.

References: Curwen 1934

GrA–26973 4410 ±35 BP

δ¹³C: -23.5‰
δ¹⁵N (diet): +3.0‰

Sample: Brighton Museum R3688/139/M (1), submitted on 2 June 2004 by F Healy

Material: antler: *Cervus elaphus*, probable antler pick, weathered. Base and beam. Brow and bez tines broken off, brow ?recently, bez anciently. Numerous small antler fragments from the same spit suggest that the complete pick (even a second pick?) was present at the time of excavation (J W Jackson 1932)

Initial comment: ditch IV. Segment CV-CVI. Cutting V. Spit 7. (Curwen 1934, 102–4, fig 1: sections IV and V). Spit 7 was the penultimate one recorded, and lay 60–70 in (1.50–1.75m) below the surface. Given the unevenness of the ditch bottom (Curwen 1935, fig 1), it must sometimes have been the lowest spit. The sample would have lain close to the ditch bottom in rapidly accumulated fills and is highly likely to have been used to dig the ditch.

Objectives: to establish the date of this context as a step towards defining the chronology of the monument.

Calibrated date: 1σ: 3100–2930 cal BC
2σ: 3320–2910 cal BC

Final comment: F Healy (2012), GrA-26973 and OxA-14064 are statistically consistent (T'=0.2; T'(5%)=3.8; ν=1; Ward and Wilson 1978). The late fourth-/early third-millennium date of the antler suggests that it came from the deeper part of the cutting, where spit 7 would have been some way above the ditch bottom, and could even have come from secondary fills.

References: Curwen 1934
Curwen 1935
Ward and Wilson 1978

GrA–29364 4720 ±45 BP

δ¹³C: -20.9‰

Sample: Brighton Museum R4100/141/P/2, submitted on 10 June 2005 by F Healy

Material: antler: *Cervus elaphus*, tine tip (A Powell 2004)

Initial comment: ditch IV. Site A. DIV layer 4. In the lowest fill of the ditch (Curwen 1935, fig C), which is likely to have accumulated within a couple of years of its originally having been dug. The tine, anciently broken from the beam, is likely to have formed part of a pick with which the ditch was dug and hence to be close in age to that event.

Objectives: to establish the date of this context as a step towards defining the chronology of the monument.

Calibrated date: 1σ: 3630–3370 cal BC
                      2σ: 3640–3360 cal BC

Final comment: F Healy (2012), OxA-14065 and GrA-29364 are statistically consistent (T′=1.5; T′(5%)=3.8; *v*=1; Ward and Wilson 1978) and compatible with the stratigraphic position of the sample.

References: Curwen 1935
                 Ward and Wilson 1978

OxA–14064 4389 ±32 BP

$\delta^{13}C$: -23.1‰

Sample: Brighton Museum R3688/139/M (2), submitted on 2 June 2004 by F Healy

Material: antler: *Cervus elaphus*, single fragment (J Jackson 1932)

Initial comment: a replicate of GrA-26973.

Objectives: as GrA-26973

Calibrated date: 1σ: 3090–2920 cal BC
                      2σ: 3260–2910 cal BC

Final comment: see GrA-26973

OxA–14065 4650 ±35 BP

$\delta^{13}C$: -20.2‰

Sample: Brighton Museum R4100/141/P, submitted on 2 June 2004 by F Healy

Material: antler: *Cervus elaphus*, tine tip (A Powell 2004)

Initial comment: a replicate of GrA-29364.

Objectives: as GrA-29364

Calibrated date: 1σ: 3510–3360 cal BC
                      2σ: 3620–3350 cal BC

Final comment: see GrA-29364

Causewayed Enclosures: Whitesheet Hill, Wiltshire

Location: ST 8017035190
               Lat. 51.06.53 N; Long. 02.17.00 W

Project manager: J Stone, S Piggott, J Richards, and M Rawlings (Independents and Wessex Archaeology), 1951 and 1989–90

Description: a causewayed enclosure with single ovoid circuit within a larger earthwork complex including cross-dykes, round barrows, and a second, undated enclosure. Limited excavation by Stuart Piggott and J F S Stone in 1951 and by Julian Richards and Mick Rawlings in 1989–90.

Objectives: to refine the dating achieved following the 1989–90 excavations (Ambers and Bowman 1998).

Final comment: F Healy (2012), modelling of the results with pre-existing dates and the relevant stratigraphic information provides an estimated construction date of *3655–3630 cal BC (10% probability)* or *3610–3535 cal BC (85% probability*; Whittle *et al* 2011, fig 4.26: *use Whitesheet Hill*). The estimate for its primary use is *1–125 years (95% probability* (Whittle *et al* 2011, fig 4.27: *use Whitesheet Hill*). The dating applies to an assemblage of predominantly South-Western style Bowl pottery, including a single gabbroic vessel from the ditch.

References: Ambers and Bowman 1998
                 Geophysical Surveys of Bradford 1995
                 Maltby 2004
                 Oswald *et al* 2001
                 Palmer 1976
                 Piggott 1952
                 Rawlings *et al* 2004
                 Whittle *et al* 2011

Causewayed Enclosures: Whitesheet Hill, ditch, Wiltshire

Location: ST 8017035190
               Lat. 51.06.53 N; Long. 02.17.00 W

Project manager: J Stone and S Piggott (Independent)
                      J Richards and M Rawlings (Wessex Archaeology), 1951 and 1989–90

Archival body: Salisbury and South Wiltshire Museum, Salisbury and South Wiltshire Museum and Wiltshire Heritage Museum

Description: a single-circuit causewayed enclosure ditch, still visible as earthwork, exceptionally deep in the south-east, where a single recut was identified.

Objectives: to date the circuit as a step to dating the complex.

Final comment: F Healy (2012), four samples of antler and of articulating animal bone from the primary chalk rubble fill of the ditch yielded results statistically consistent with those for two bulk bone samples from the same context measured following the 1989–90 excavation (BM-2784, 4820 ±50 BP and BM-2785, 4800 ±70 BP), indicating that all the dated material from this context were freshly deposited (T′=7.9;T′(5%)=11.1; *v*=5; Ward and Wilson 1978).

References: Ward and Wilson 1978
                 Whittle *et al* 2011

GrA–30067 4695 ±40 BP

$\delta^{13}C$: -22.7‰

Sample: W347.1579, submitted on 7 November 2005 by F Healy

Material: antler: *Cervus elaphus*, beam with base of recently broken-off tine, found with many antler fragments. Almost certainly the remains of an antler pick (J Maltby 1990)

Initial comment: feature 1288, context 1354. Loose, unsorted chalk rubble with a few chalk nodules lying directly on base of ditch and up to 1.75 m deep. Antler found at 232.12 m OD, about 0.80 m above ditch base (Rawlings *et al* 2004, fig 5), at a similar level to sf 1584. Rubble could have accumulated quickly in the exceptionally deep and narrow ditch, so that the implement could have been used in the construction of the monument.

Objectives: to date the context as a step towards dating the monument.

Calibrated date: 1σ: 3630–3370 cal BC
                     2σ: 3640–3360 cal BC

Final comment: F Healy (2012), the date is compatible with the sample's stratigraphic position and with the chronological model.

References: Rawlings *et al* 2004

GrA–30068 4825 ±40 BP

δ¹³C: -23.0‰

Sample: W347.1584.2, submitted on 7 November 2005 by F Healy

Material: antler: *Cervus elaphus*, beam with one tine, and recent breaks. Found with many small antler fragments. Almost certainly the remains of an antler pick (F Healy 2005)

Initial comment: feature 1288, context 1354. Loose, unsorted chalk rubble with a few chalk nodules lying directly on base of ditch and up to 1.75m deep. Sample found at 231.97m OD, above middle of layer (Rawlings *et al* 2004, fig 5). Part of same small find as sample for BM-2784.

Objectives: to date the context as a step towards dating the monument.

Calibrated date: 1σ: 3650–3530 cal BC
                     2σ: 3700–3520 cal BC

Final comment: see GrA-30067

References: Rawlings *et al* 2004

OxA–15290 4822 ±32 BP

δ¹³C: -21.6‰

Sample: W347.1354.1, submitted on 7 November 2005 by F Healy

Material: animal bone: *Ovis* sp., 3 rib fragments from animal between 6 and 10 months old, represented by 49 bones. 'There is no evidence of butchery and it is assumed that this skeleton was dumped in an articulated state.' (M Maltby 2004)

Initial comment: most of the skeleton was recovered except the carpals, tarsals, and phalanges. The absence of these small bones may result from recovery bias or poor preservation and it is possible that the sheep was originally dumped as a complete carcass' (Maltby 2004). From feature 1288, context 1354. Loose, unsorted chalk rubble with a few chalk nodules lying directly on base of ditch and up to 1.75m deep. (Rawlings *et al* 2004, fig 5). There is no record of the depth at which the skeleton was found, although rubble could have accumulated quickly in the exceptionally deep and narrow ditch, so that, whatever its depth, the

skeleton could have been deposited soon after the construction of the monument.

Objectives: to date the context as a step towards dating the monument.

Calibrated date: 1σ: 3650–3530 cal BC
                     2σ: 3660–3520 cal BC

Final comment: F Healy (2012), the date is compatible with the sample's stratigraphic position and with the chronological model.

References: Maltby 2004
                       Rawlings *et al* 2004

OxA–15291 4768 ±33 BP

δ¹³C: -21.7‰

Sample: W347.1354.2/A, submitted on 7 November 2005 by F Healy

Material: animal bone: *Bos* sp., proximal phalanx, articulating with medial and distal phalanges (A Powell 2005)

Initial comment: feature 1288, context 1354. Loose, unsorted chalk rubble with a few flint nodules lying directly on base of ditch and up to 1.75m deep. (Rawlings *et al* 2004, fig 5). The bones were extracted from a bulk find and their precise position is unknown. Rubble could have accumulated quickly in the exceptionally deep and narrow ditch, so that, whatever its depth, the bones could have been deposited soon after the construction of the monument. Their articulation suggests that they were still or recently joined by soft tissue when buried and hence close in age to their context.

Objectives: to date the context as a step towards dating the monument.

Calibrated date: 1σ: 3640–3520 cal BC
                     2σ: 3640–3380 cal BC

Final comment: see OxA-30071

References: Rawlings *et al* 2004

GrA–30071 4800 ±45 BP

δ¹³C: -22.2‰

Sample: W347.1354.2/B, submitted on 7 November 2005 by F Healy

Material: animal bone: *Bos* sp., medial phalanx, articulating with proximal and distal phalanges (A Powell 2005)

Initial comment: a replicate of OxA-15291.

Objectives: as OxA-15291

Calibrated date: 1σ: 3650–3520 cal BC
                     2σ: 3660–3380 cal BC

Final comment: F Healy (2012), this and OxA-15291, measured on two articulating phalanges, are statistically consistent (T'=0.3; T'(5%)=3.8; ν=1; Ward and Wilson 1978). Their mean is compatible with the sample's stratigraphic position and with the chronological model.

References: Ward and Wilson 1978

Causewayed Enclosures: Whitesheet Hill, internal features, Wiltshire

Location: ST 8017035190
Lat. 51.06.53 N; Long. 02.17.00 W

Project manager: J Richards and M Rawlings (Wessex Archaeology), 1989–90

Archival body: Salisbury and South Wiltshire Museum, Salisbury and South Wiltshire Museum and Wiltshire Heritage Museum

Description: pits excavated along the route of a water pipeline which crossed the hill. They contained exceptional amounts of dumped burnt material, especially in the north-west of the route.

Objectives: to date the pits as a step towards dating the use of the complex.

Final comment: F Healy (2012), groups of short-life samples from three pits were each statistically consistent and, in two cases, were also statistically consistent with pre-existing dates on bulked animal bone (BM-3821; 4750 ±90 BP), or charred hazelnuts (BM-2822; 4790 ±50 BP and BM-2823; 4740 ±35 BP), indicating the all the contents could have been freshly deposited.

References: Rawlings *et al* 2004
Whittle *et al* 2011

GrA–30072 4765 ±40 BP

$\delta^{13}C$: -25.7‰

Sample: W347.1322.4, submitted on 7 November 2005 by F Healy

Material: carbonised plant macrofossil (hazelnut shell fragment) (P Hinton 1990)

Initial comment: feature 1295, context 1322. Extracted from basal fill of pit, underlying context 1321 (Rawlings *et al* 2004, fig 7). Single entity samples from this context will provide a check on the representativeness of existing dates on bulk samples from it.

Objectives: to date the context as a step towards dating the monument.

Calibrated date: 1σ: 3640–3510 cal BC
2σ: 3650–3370 cal BC

Final comment: F Healy (2012), the result is statistically consistent with the other three measurements from this context (OxA-15322, BM-2821, and BM-2823; T′=1.5; T′(5%)=7.8; ν=3; Ward and Wilson 1978).

References: Ward and Wilson 1978

GrA–30073 4845 ±40 BP

$\delta^{13}C$: -25.2‰

Sample: W347.1697.2, submitted on 7 November 2005 by F Healy

Material: carbonised plant macrofossil (hazelnut shell fragment) (P Hinton 1990)

Initial comment: feature 1303, context 1346. Basal fill of feature in interior of enclosure (Rawlings *et al* 2004, fig 6), associated with >100 Neolithic Bowl sherds. No sign of *in situ* burning, probably dumped burnt material. Fill with much charcoal and burnt flint in base of chalk-cut pit on a well-drained chalk down.

Objectives: to date the context as a step towards dating the monument.

Calibrated date: 1σ: 3660–3630 cal BC
2σ: 3710–3530 cal BC

Final comment: see OxA-15293

GrA–30074 4740 ±40 BP

$\delta^{13}C$: -22.0‰

Sample: W347 1699.2, submitted on 7 November 2005 by F Healy

Material: carbonised plant macrofossil (hazelnut shell fragment) (P Hinton 1990)

Initial comment: feature 1293, context 1350. Deposit of charcoal and charred plant remains up to 0.10m thick on pit base in lower part of some areas of 1323 (Rawlings *et al* 2004, fig 70). Charcoal deposit on base of a chalk-cut pit on a well-drained chalk down.

Objectives: to date the context as a step towards dating the monument.

Calibrated date: 1σ: 3640–3380 cal BC
2σ: 3640–3370 cal BC

Final comment: F Healy (2012), the result is statistically consistent with OxA-15324, measured on another charred hazelnut shell fragment from the same context (T′=0.4; T′(5%)=3.8; ν=1; Ward and Wilson 1978).

References: Ward and Wilson 1978

OxA–15292 4830 ±32 BP

$\delta^{13}C$: -19.4‰

Sample: W347.1342.1, submitted on 7 November 2005 by F Healy

Material: animal bone: *Sus* sp., right radius with both fitting unfused epiphyses. 'Most of the pig bones in these fills [of feature 1303] could have belonged to two immature animals' (M Maltby 1990–2, original report in archive, detail missing from published version) (A Powell 2005)

Initial comment: feature 1303, context 1342. Upper fill of the first of two successive pits, stratified above 1346 (Rawlings *et al* 2004, fig 6). The presence of a complete shaft with both fitting epiphyses suggests that the bone was deposited soon after the pig was butchered, while the epiphyses were still connected to the shaft by soft tissue.

Objectives: to date the context as a step towards dating the monument.

Calibrated date: 1σ: 3650–3540 cal BC
2σ: 3660–3530 cal BC

Final comment: F Healy (2012), the result is compatible with the sample's stratigraphic position, above four samples from context 1346.

OxA–15293 4823 ±33 BP

$\delta^{13}C$: -21.5‰

Sample: W347.1346.2, submitted on 7 November 2005 by F Healy

Material: animal bone: *Bos* sp., 1 of 2 consecutive thoracic vertebrae from same immature individual, a third probably consecutive vertebra coming from the same context (A Powell 2005)

Initial comment: feature 1303, context 1346. Basal fill of the first of two successive pits, stratified below 1342 (Rawlings *et al* 2004, fig 6). The recovery of the three vertebrae suggests that they were still in, or only shortly out of, articulation when buried. Fill with much charcoal and burnt flint in base of chalk-cut pit on a well-drained chalk down.

Objectives: to date the context as a step towards dating the monument.

Calibrated date: 1σ: 3650–3530 cal BC
                      2σ: 3660–3520 cal BC

Final comment: F Healy (2012), the result is statistically consistent with the other three samples from this context (OxA-15323, GrA-30073, BM-2822; (T′=6.4; T′(5%)=7.8; *v*=3; Ward and Wilson 1978).

References: Ward and Wilson 1978

OxA–15322 4797 ±33 BP

$\delta^{13}C$: -24.3‰

Sample: W347.1322.3, submitted on 7 November 2005 by F Healy

Material: carbonised plant macrofossil (hazelnut shell fragment) (P Hinton 1990)

Initial comment: feature 1295, context 1322. Extracted from basal fill of pit, underlying context 1321 (Rawlings *et al* 2004, fig 7). Single entity samples from this context will provide a check on the representativeness of existing dates on bulk samples from it.

Objectives: to date the context as a step towards dating the monument.

Calibrated date: 1σ: 3640–3530 cal BC
                      2σ: 3650–3520 cal BC

Final comment: see GrA-30072

OxA–15323 4726 ±34 BP

$\delta^{13}C$: -28.2‰

Sample: W347.1697.1, submitted on 7 November 2005 by F Healy

Material: carbonised plant macrofossil (hazelnut shell fragment) (P Hinton 1990)

Initial comment: feature 1303, context 1346. Basal fill of feature in interior of enclosure (Rawlings *et al* 2004, fig 6), associated with >100 Neolithic Bowl sherds. No sign of *in situ* burning, probably dumped burnt material. Fill with much charcoal and burnt flint in base of chalk-cut pit on a well-drained chalk down.

Objectives: to date the context as a step towards dating the monument.

Calibrated date: 1σ: 3630–3380 cal BC
                      2σ: 3640–3370 cal BC

Final comment: see OxA-15293

OxA–15324 4775 ±35 BP

$\delta^{13}C$: -26.5‰

Sample: W347 1699.1, submitted on 7 November 2005 by F Healy

Material: carbonised plant macrofossil (hazelnut shell fragment) (P Hinton 1990)

Initial comment: feature 1293, context 1350. Deposit of charcoal and charred plant remains up to 0.10m thick on pit base in lower part of some areas of 1323 (Rawlings *et al* 2004, fig 70). Charcoal deposit on base of a chalk-cut pit on a well-drained chalk down.

Objectives: to date the context as a step towards dating the monument.

Calibrated date: 1σ: 3640–3520 cal BC
                      2σ: 3650–3380 cal BC

Final comment: see GrA-30074

References: Ward and Wilson 1978

Causewayed Enclosures: Windmill Hill, Wiltshire

Location: SU 0867071440
                     Lat. 51.26.28 N; Long. 01.52.32 W

Project manager: H Kendall, A Keiller, I Smith, and A Whittle (Independent, Morvern Institute of Archaeological Research, and Cardiff University), 1922–3, 1925–9, 1957–8, and 1988

Description: a causewayed enclosure formed of three circuits (the inner, middle, and outer ditches), located on Middle Chalk off the summit of a down, facing north-west, rather than towards Avebury to the south-east. There are relatively few pits or other features in the interior, and some outside the enclosure, within an extensive flint scatter. So far (2004) all three circuits seem to be broadly contemporary, so that very distinctive depositional practices in each may be interpreted as reflecting differences in use rather than in date. The complex is one of the largest causewayed enclosures in England and was one of the first to be excavated extensively. Its assemblages have played a seminal role in demonstrating the extent of the long- and medium-distance transport of artefacts and materials during the early Neolithic, the character of contemporary animal husbandry, and the development of early and middle Neolithic pottery styles. Its location in the Avebury area places it near the beginnings of a unique monument complex, where later developments included Silbury Hill, the Avebury henge and avenues, and the West Kennet Farm timber settings.

Objectives: samples were submitted during the project to refine the date of construction of the enclosure as a whole, to determine whether its constituent circuits were built at the same time or successively, to determine the duration of primary use, and to relate the dating to that of other Neolithic monuments and activity in the area.

Final comment: F Healy (2012), estimates derived from Bayesian modelling of the results with previously obtained measurements indicate that work on the enclosure began in *3700–3640 cal BC (95% probability*; Whittle *et al* 2011, figs 3.8: *start Windmill Hill)*, the inner, outer, and middle circuits probably being built successively in the thirty-seventh century cal BC, over a period of *5–75 years (95% probability*; Whittle *et al* 2011, fig 3.16: *period construction)*. Its initial use continued for *290–390 years (94% probability*; Whittle *et al* 2011, fig 3.16: *use Windmill Hill)*, to *3365–3295 cal BC (94% probability*; Whittle *et al* 2011, figs 3.8: *end Windmill Hill)*, after which use of the enclosure seems to have diminished. The exercise showed the feasibility of working with the coarser resolution of samples largely recorded by spit rather than by stratigraphic unit. What had seemed a particularly full stratigraphic sequence in one segment of the outer ditch is now shown to be the result of a localised recut made at the turn of the fourth and third millennia cal BC (Whittle *et al* 2011).

References: Ambers *et al* 1991
Anon 1990
Barker and Mackey 1961
Brothwell 1965
Evans 1966
Grigson 1982
Grigson 1984
Grigson 1999
Hedges *et al* 1992b
Keiller 1934
Kendall 1923
Malone 1989
Oswald *et al* 2001
Palmer 1976
Pollard 1999
Rouse and Rowland 1999
Smith 1959
Smith 1965
Smith 1966
Smith 1971
Stukely 1743
Whittle *et al* 1999
Whittle *et al* 2011
Whittle and Pollard 1998
Zienkiewicz and Hamilton 1999

Causewayed Enclosures: Windmill Hill, inner ditch, Wiltshire

Location: SU 0867071440
Lat. 51.26.28 N; Long. 01.52.32 W

Project manager: A Keiller (Independent) I Smith (Independent) and A Whittle (Cardiff University), 1925–9, 1957–8, and 1998

Archival body: Alexander Keiller Museum and Wiltshire Heritage Museum

Description: the innermost and smallest circuit of the causewayed enclosure.

Objectives: to refine the chronology of the circuit as a step to building up the chronology of the complex as a whole.

Final comment: F Healy (2012), Bayesian modelling of the results and pre-existing measurements provides an estimated construction date of *3685–3635 cal BC (95% probability*; Whittle *et al* 2011, fig 3.9: *dig WH inner)*.

References: Whittle *et al* 2011

GrA–25379 4910 ±50 BP

δ¹³C: -24.5‰

Sample: charcoal ID VII bottom, submitted on 6 February 2004 by F Healy

Material: charcoal: *Corylus avellana*, single fragment (R Gale 2004)

Initial comment: inner ditch, segment ID VII, bottom. Part of a sample of comminuted chalk with charcoal. The segment is described by Pollard (1999, 53–6). The charcoal must have been deposited very soon after the ditch was dug. If short-life samples can be extracted from it, they should be close in age to the original excavation of the segment. Stratified beneath spit 5. From same find as OxA-13760.

Objectives: to establish the date of this context as a step towards defining the chronology of the monument.

Calibrated date: 1σ: 3710–3640 cal BC
2σ: 3790–3630 cal BC

Final comment: F Healy (2012), the hand-collection of the sample suggests that the charcoal patch was visible and coherent. The statistical consistency of GrA-25379 and OxA-13760 (T′=0.1; T′(5%)=3.8; ν=1; Ward and Wilson 1978) indicates that the find is unlikely to have included redeposited or reworked material.

References: Ward and Wilson 1978
Whittle *et al* 1999

GrA–25391 4360 ±50 BP

δ¹³C: -28.3‰

Sample: WH26 sherd 2896, submitted on 6 February 2004 by F Healy

Material: carbonised residue (internal; one of several large, well-preserved joining flint-tempered Neolithic Bowl sherds) (R Cleal 2004)

Initial comment: inner ditch, segment ID VII, spit 4 (joining sherds recorded at depths between 2.3ft and 3ft). Spit 4 was the penultimate spit and probably included parts of the primary and secondary fills (Pollard 1999, 53–6). The proximity of large, well-preserved joining sherds of the same vessel, with well-preserved internal residue, indicates that their last use and breakage preceded their burial by only a short interval, they should hence be close in age to their context. At a higher level than spit 5.

Objectives: as GrA-25379

Calibrated date: 1σ: 3080–2900 cal BC
2σ: 3270–2890 cal BC

Final comment: F Healy (2012), this date and OxA-13732, measured on residue from the same vessel, are statistically inconsistent (T'=21.4; T'(5%)=3.8; *v*=1; Ward and Wilson 1978). Since OxA-13732 is consistent with two dates on articulated animal bone from the same spit, this measurement was considered inaccurate and excluded from modelling.

References: Ward and Wilson 1978
    Whittle *et al* 1999

GrA–25558 4690 ±40 BP

δ¹³C: -20.9‰
δ¹⁵N (diet): +9.4‰

Sample: WH26 B22.a, submitted on 6 February 2004 by F Healy

Material: animal bone: *Canis* sp., mandible; found with skull fragments (C Grigson 1970)

Initial comment: inner ditch, segment ID VII, spit 4 (2.3–3.5ft). This was the penultimate spit and probably included parts of the primary and secondary fills (Pollard 1999, 53–56). The association of skull fragments and mandible indicates that the dog skull was complete when buried, and hence close in age to its context. Found with sheep/goat longbones B22.b, 22.c. At a higher level than spit 5. One sherd of Beaker and two sherds of early Bronze Age pottery were present as well as much Bowl. The few later sherds are likely to have been intrusive.

Objectives: as GrA-25379

Calibrated date: *1σ:* 3630–3370 cal BC
    2σ: 3640–3360 cal BC

Final comment: see OxA-13715

References: Whittle *et al* 1999

GrA–25560 4500 ±40 BP

δ¹³C: -22.1‰
δ¹⁵N (diet): +5.4‰

Sample: WH88 6419 (B1344), submitted on 6 February 2004 by F Healy

Material: animal bone: *Bos* sp., right proximal metatarsal fragment found in articulation with right navicular and posterior cuneiform (WH88 6420/ B1342, B1343) (C Grigson 1989)

Initial comment: inner ditch, trench F, discrete bone heap 630 on surface of context 610, at interface of primary and secondary fills. Metatarsal lay in articulation with navicular and cuneiform of same foot (Whittle *et al* 1999, fig 97: 9, 26). The articulation of the joint when buried indicates that it was close in age to its context. Stratified above context 613, probably stratigraphically equivalent to context 629. On the surface of the chalky silt with small, sub-angular chalk rubble (610), beneath fine grey silt with occasional chalk fragments (604).

Objectives: as GrA-25379

Calibrated date: *1σ:* 3350–3090 cal BC
    2σ: 3370–3020 cal BC

Final comment: F Healy (2012), the result is compatible with the sample's stratigraphic position in the ditch. It calls into question the inferred equivalence of contexts 630 and 629, from which a disarticulated cattle vertebra was already dated to 3260–2880 cal BC (95% confidence, Reimer *et al* 2004; 4370 ±50 BP; BM-2672). This may suggest that the surface on which bone group 630 was deposited remained stable for some time.

References: Reimer *et al* 2004
    Whittle *et al* 1999

GrA–29707 4725 ±35 BP

δ¹³C: -22.0‰

Sample: WH29 B322, submitted on 6 June 2005 by F Healy

Material: animal bone: *Bos* sp., complete right femur, articulating with right tibia (B340), also complete and in identical condition (J Hamilton 2004)

Initial comment: inner ditch, segment ID XVI, spit 3a. Spit 3 lay at 2ft-3ft and was the ante penultimate one (Pollard 1999, 53– 6, figs 49, 53, and 54). The pottery from the spit was mainly Bowl with five sherds of Peterborough Ware and four of indeterminate ?late Neolithic/early Bronze Age (Zienkiewicz and Hamilton 1999, table 166). Close to the south-west butt, a cattle pelvis, femur, tibia, and astragalus, all complete, lay close together in this layer (Pollard 1999, 56; Smith 1965, pl. Vb). The present sample almost certainly equates to the femur from this group. Whether or not it does, the articulation of the bones and their completeness, which is rare on the site, indicates that they were not long out of articulation when buried, and that the animal from which they came was not long dead.

Objectives: to establish the date of this context as a step towards defining the chronology of the monument.

Calibrated date: *1σ:* 3630–3380 cal BC
    2σ: 3640–3370 cal BC

Final comment: see OxA-14968

References: Smith 1965
    Whittle and Pollard 1998
    Zienkiewicz and Hamilton 1999

GrA–29708 4700 ±35 BP

δ¹³C: -22.9‰

Sample: WH26 B23, submitted on 6 June 2005 by F Healy

Material: antler: *Cervus elaphus*, tine with worn, battered tip, charred towards junction with beam (C Grigson 1970)

Initial comment: inner ditch, segment ID VII, spit 5 (3ft 5in to base). Excavation number 474. From same spit as 'fine deerhorn pick' (B24; not found 2003–5) and another pick (B25; dated by OxA-13813). The location and associations strongly suggest that it either formed part of an implement with which the ditch was dug or was removed in the course of making or modifying such an implement. In either case it would be close in age to the construction of the earthwork.

Objectives: as GrA-25379

Calibrated date: *1σ:* 3630–3370 cal BC
    2σ: 3640–3360 cal BC

Final comment: F Healy (2012), all four samples from this spit (GrA-29708, -29746; OxA-13815, -14975) are statistically consistent (T'=6.5; T'(5%)=7.8; ν=3; Ward and Wilson 1978), which would accord with their being freshly deposited.

References: Ward and Wilson 1978

GrA–29746 4685 ±40 BP

δ¹³C: -25.2‰

Sample: WH26 charcoal ID VII spit 5 B, submitted on 6 June 2005 by F Healy

Material: charcoal: *Corylus avellana*, single fragment (R Gale 2005)

Initial comment: inner ditch, segment ID VII, within spit 5 (3.5–5ft, the lowest spit), extracted from sample of chalk with charcoal fragments. The segment is described by Pollard (1999, 53–6, figs 50–2). The fact that the sample was taken shows that the find was perceived as a discrete deposit. At this depth it would have been in chalk rubble fills formed soon after the excavation of the ditch. If short-life samples can be extracted from it, they should be close in age to the original excavation of the segment. Stratified above GrA-25379 and OxA-13760.

Objectives: as GrA-25379

Calibrated date: 1σ: 3620–3370 cal BC
                       2σ: 3640–3360 cal BC

Final comment: see OxA-14975

References: Whittle and Pollard 1998

OxA–13715 4710 ±29 BP

δ¹³C: -21.0‰

Sample: WH26 B22.c, submitted on 6 February 2004 by F Healy

Material: animal bone (sheep/goat; left humerus, articulating with radius B22.b) (J Mulville and A Powell 2004)

Initial comment: inner ditch, segment ID VII, spit 4 (2.3–3.5ft). This was the penultimate spit and probably included parts of the primary and secondary fills (Pollard 1999, 53–56). The association of articulating radius and humerus indicates that the bones were articulated or only recently disarticulated when buried, and hence close in age to their context. Found with dog skull and mandible B22.a. At a higher level than spit 5.

Objectives: as GrA-25379

Calibrated date: 1σ: 3630–3370 cal BC
                       2σ: 3640–3370 cal BC

Final comment: F Healy (2012), this result is statistically consistent with GrA-25558, measured on another articulating sample from the same bone group (T'=0.2; T'(5%)=3.8; ν=1; Ward and Wilson 1978), and is also compatible with its stratigraphic relation to other samples.

References: Ward and Wilson 1978
                   Whittle and Pollard 1998

OxA–13732 4672 ±45 BP

δ¹³C: -12.8‰
δ¹⁵N (diet): +6.6‰
C/N ratio: 13.5

Sample: WH26 sherd 2896, submitted on 6 February 2004 by F Healy

Material: carbonised residue (internal; 1 of several large, well-preserved, joining flint- and sand -tempered Neolithic Bowl sherds) (R Cleal 2004)

Initial comment: a replicate of GrA-25391.

Objectives: as GrA-25379

Calibrated date: 1σ: 3520–3360 cal BC
                       2σ: 3630–3350 cal BC

Final comment: F Healy (2012), this date and GrA-25391, measured on residue from the same vessel, are statistically inconsistent (T'=21.4; T'(5%)=3.8; ν=1; Ward and Wilson 1978). Since OxA-13732 is consistent with two dates on articulated animal bone from the same spit (OxA-13715, GrA-25558; T'=0.5; T'(5%)=6.0; ν=2; Ward and Wilson 1978), this date was considered probably accurate and included in modelling.

References: Ward and Wilson 1978

OxA–13760 4891 ±31 BP

δ¹³C: -26.1‰

Sample: charcoal ID VII bottom, submitted on 6 February 2004 by F Healy

Material: charcoal: *Corylus avellana*, single fragment (R Gale 2004)

Initial comment: as GrA-25379

Objectives: as GrA-25379

Calibrated date: 1σ: 3700–3640 cal BC
                       2σ: 3710–3630 cal BC

Final comment: see GrA-25379

OxA–13815 4798 ±34 BP

δ¹³C: -22.6‰

Sample: WH26 B25, submitted on 6 February 2004 by F Healy

Material: antler: *Cervus elaphus*, beam with trez tine (C Grigson 1970)

Initial comment: inner ditch, segment ID VII, spit 5 at 4.5ft. Keiller catalogue annotated 'in chalk rubble at foot of ditch'. Spit 5 was the lowest spit and the antler at this depth would have been close to the base (Pollard 1999, 53–6). The antler's location suggests that it was used to dig the ditch and hence should date the construction.

Objectives: as GrA-25379

Calibrated date: 1σ: 3640–3530 cal BC
                       2σ: 3650–3520 cal BC

Final comment: see GrA-29708

References: Whittle and Pollard 1998

OxA–14968 4747 ±33 BP

δ¹³C: -20.0‰

Sample: WH29 B759, submitted on 6 June 2005 by F Healy

Material: animal bone: *Sus* sp., one of two fitting right metatarsals (C Grigson 1970)

Initial comment: inner ditch, segment ID XII, spit 2b. Spit 2 was 1–2ft below the surface and was the middle spit of three in a shallow segment. There is no record of the fills, although there were five distinct and substantial bone groups in spits 2 and 3 (Smith 1965 pl. Va; Pollard 1999, 61– 63, figs 49, 53, and 59–60). The pottery from spit 2 was mainly Bowl, with one sherd of Peterborough Ware (Zienkiewicz and Hamilton 1999, table 166).The articulation of the bones indicates that they were still connected by soft tissue when buried and that the animal from which they came was not long dead.

Objectives: as GrA-25379

Calibrated date: 1σ: 3640–3380 cal BC
                     2σ: 3640–3370 cal BC

Final comment: F Healy (2012), the date is compatible with the sample's position in the ditch.

References: Smith 1965
                     Whittle and Pollard 1998
                     Zienkiewicz and Hamilton 1999

OxA–14975 4703 ±36 BP

δ¹³C: -24.5‰

Sample: WH26 charcoal ID VII spit 5 A, submitted on 6 June 2005 by F Healy

Material: charcoal: *Corylus avellana*, single fragment (R Gale 2005)

Initial comment: inner ditch, segment ID VII, within spit 5 (3.5ft-5ft, the lowest spit), extracted from sample of chalk with charcoal fragments. The segment is described by Pollard (1999, 53–56, and figs 50–52). The fact that the sample was taken shows that the find was perceived as a discrete deposit. At this depth it would have been in chalk rubble fills formed soon after the excavation of the ditch. If short-life samples can be extracted from it, they should be close in age to the original excavation of the segment. Stratified above GrA-25379 and OxA-13760.

Objectives: as GrA-25379

Calibrated date: 1σ: 3630–3370 cal BC
                     2σ: 3640–3360 cal BC

Final comment: F Healy (2012), this date and GrA-29746, measured on a *Corylus avellana* fragment from the same find, are statistically consistent (T′=0.1; T′(5%)=3.8; ν=1; Ward and Wilson 1978), indicating that the find is unlikely to have included redeposited or reworked material. Otherwise as GrA-29708.

References: Ward and Wilson 1978
                     Whittle and Pollard 1998

Causewayed Enclosures: Windmill Hill, middle ditch, Wiltshire

Location: SU 0867071440
                     Lat. 51.26.28 N; Long. 01.52.32 W

Project manager: A Keiller (Independent) I Smith (Independent) A Whittle (Cardiff University), 1925–9, 1957–8, and 1988

Archival body: Alexander Keiller Museum and Alexander Keiller Museum

Description: the middle circuit of the causewayed enclosure.

Objectives: samples were submitted during the project to refine the chronology of the circuit, as a step to building up the chronology of the complex as a whole.

Final comment: F Healy (2012), Bayesian modelling of the results and pre-existing measurements provides an estimated construction date of *3655–3605 cal BC (95% probability*; Whittle *et al* 2011, fig 3.10: *dig WH middle*).

References: Whittle *et al* 2011

GrA–25368 3650 ±50 BP

δ¹³C: -21.1‰
δ¹⁵N (diet): +3.1‰

Sample: WH88 12371a, submitted on 6 February 2004 by F Healy

Material: animal bone (toad; bones of hind limbs from almost complete toad skeleton (Rouse and Rowland 1999, table 154). Identification by Rouse, sample segregated by Mulville and Powell. From same skeleton as sample for OxA-13730) (J Mulville and A Powell 2004)

Initial comment: middle ditch, trench E, segment MD XII, in low-density bone spread 527 within context 515 just above base of ditch. Stratified below contexts 525, 523, 510 (Whittle *et al* 1999, 100, and figs 89–90). A toad is unlikely to have burrowed to this depth (approximately 1.25m from the modern surface) to hibernate, so that the animal is likely to have fallen into the ditch as it began to silt, and to have remained undisturbed ever since. It is thus likely to have been close in age to its context.

Objectives: to establish the date of this context as a step towards defining the chronology of the monument.

Calibrated date: 1σ: 2130–1940 cal BC
                     2σ: 2200–1880 cal BC

Final comment: F Healy (2012), the late third/early second millennium cal BC date, confirmed by a second measurement (OxA-13730) on the same toad.

Laboratory comment: English Heritage (27 November 2014), the two measurements on this toad are not statistically consistent (T′=4.7; T′(5%)=3.8; ν=1; Ward and Wilson 1978).

References: Rouse and Rowland 1999
                     Ward and Wilson 1978

GrA–25554 4725 ±40 BP

δ¹³C: -21.8‰
δ¹⁵N (diet): +4.6‰

Sample: WH28 B114, submitted on 6 February 2004 by F Healy

Material: antler: *Cervus elaphus*, beam with trez tine; cut below trez, very smooth (C Grigson 1970)

Initial comment: middle ditch, segment MD IB, spit 5A (4ft-5ft). This was the lowest spit and would have been within the primary fills (Pollard 1999, 47–51, and fig 41). The location of the modified antler in rapidly accumulated fills close to the ditch bottom suggests that it was used to excavate the ditch and should be close in age to its context. At lower level than spit 4.

Objectives: to establish the date of this context as a step towards defining the chronology of the monument.

Calibrated date: *1σ:* 3630–3370 cal BC
                      2σ: 3640–3370 cal BC

Final comment: F Healy (2012), the date is compatible with the sample's position in the ditch fill.

References: Whittle and Pollard 1998

GrA–25555 4685 ±40 BP

δ¹³C: -23.8‰
δ¹⁵N (diet): +5.0‰

Sample: WH28 B369, submitted on 6 February 2004 by F Healy

Material: animal bone: *Bos* sp., right magnum from complete set of 5 right carpals, articulating (C Grigson 1970)

Initial comment: middle ditch. Segment MD IB, spit 4 (3ft-4ft). This was the penultimate spit and would probably have been mainly in the upper part of the primary fill (Pollard 1999, 47–50, and fig 42 bottom left). Articulation indicates that the bones were either articulated or recently disarticulated when buried and hence close in age to their context. Sample may have come from same deposit as the right cattle carpals from spit 3 in same segment, which immediately overlay spit 4. At a higher level than spit 5 and a lower level than spit 3.

Objectives: to establish the date of this context as a step towards defining the chronology of the monument.

Calibrated date: *1σ:* 3620–3370 cal BC
                      2σ: 3640–3360 cal BC

Final comment: see GrA-25559

References: Whittle and Pollard 1998

GrA–25556 4735 ±40 BP

δ¹³C: -23.2‰
δ¹⁵N (diet): +6.1‰

Sample: WH88 4225 (B1441), submitted on 6 February 2004 by F Healy

Material: animal bone (medium mammal; rib section, from different left rib to sample for OxA-13714) (J Mulville and A Powell 2004)

Initial comment: middle ditch, trench D, bone group 414, within context 411. Part of 'bundle' of rib fragments composed of WH88 4241 (B1442), 4225 (B1441), 4234 (B1459–64), 4235 (B1446), 4236 (B1456–7), 4241 (B1442), 4242 (B1449), 4243 (B1447), 4244 (B1444), 4245 (B1445), 4247 (B1448), 4238 (B1435), 4251 (B1452), 4255 (B1458), and 4256 (B1454–5) (Whittle *et al* 1999, figs 86–7). Discard of more than one rib suggests the immediate aftermath of consumption, so that the sample should be close in age to its context. Stratified above context 416 and below context 413.

Objectives: to establish the date of this context as a step towards defining the chronology of the monument.

Calibrated date: *1σ:* 3640–3380 cal BC
                      2σ: 3640–3370 cal BC

Final comment: F Healy (2012), the statistical consistency of this date with OxA-13714, measured on a different rib from the same 'bundle' (T′=0.0; T′(5%)=3.8; *ν*=1; Ward and Wilson 1978) accords with the impression that they were part of the debris of a single episode of consumption.

References: Ward and Wilson 1978
                      Whittle *et al* 1999

GrA–25559 4730 ±40 BP

δ¹³C: -22.9‰
δ¹⁵N (diet): +5.3‰

Sample: WH28 B374, submitted on 6 February 2004 by F Healy

Material: animal bone: *Bos* sp., right magnum, articulating with right scaphoid (also B374) (C Grigson 1970)

Initial comment: middle ditch. Segment MD IB, spit 4 (3ft-4ft). This was the penultimate spit and would probably have been mainly in the upper part of the primary fill (Pollard 1999, 47–50, and fig 42 bottom left). Articulation indicates that the bones were either articulated or recently disarticulated when buried and hence close in age to their context. Sample may have come from same deposit (or same animal) as the right cattle carpals from spit 3 in same segment, which immediately overlay spit 4. At a higher level than spit 5 and a lower level than spit 3.

Objectives: to establish the date of this context as a step towards defining the chronology of the monument.

Calibrated date: *1σ:* 3640–3380 cal BC
                      2σ: 3640–3370 cal BC

Final comment: F Healy (2012), the statistical inconsistency of dates on articulating samples from this spit (this date, GrA-25555, OxA-13812; T′=8.1; T′(5%)=6.0; *ν*=2; Ward and Wilson 1978) probably reflects origins in different stratigraphic contexts.

References: Ward and Wilson 1978
                      Whittle and Pollard 1998

GrA–25706 4740 ±45 BP

$\delta^{13}C$: -22.5‰
$\delta^{15}N$ (diet): +4.8‰

Sample: WH88 4330 (B1743), submitted on 6 February 2004 by F Healy

Material: animal bone: *Bos* sp., right radius, articulating with right ulna (4331/B1733) (C Grigson 1989)

Initial comment: middle ditch, trench D, context 416 (Whittle *et al* 1999, fig 86). Overlying ditch bottom and layer 417. Stratified below context 414. Articulation indicates that bones were either articulated or only just disarticulated when buried and hence close in age to their context.

Objectives: to establish the date of this context as a step towards defining the chronology of the monument.

Calibrated date: 1σ: 3640–3380 cal BC
                       2σ: 3640–3370 cal BC

Final comment: F Healy (2012), this date is statistically consistent with those for the remaining samples from context 416 (T'=14.1; T'(5%)=15.5; ν=8; Ward and Wilson 1978), indicating that all were freshly deposited, including a previously dated disarticulated cattle tibia (BM-2670; 4670 ±90 BP).

References: Ward and Wilson 1978
                          Whittle *et al* 1999

GrA–25707 4675 ±40 BP

$\delta^{13}C$: -23.1‰
$\delta^{15}N$ (diet): +5.3‰

Sample: WH88 12281 (B70), submitted on 6 February 2004 by F Healy

Material: animal bone: *Bos* sp., 6th lumbar vertebra found together with 5th lumbar vertebra and sacrum from same animal (C Grigson 1989)

Initial comment: middle ditch, trench E, segment MD XII, found together in upper part of bone deposit 525 in lower part of context 508 at top of primary fills (Whittle *et al* 1999, 99–101, and figs 89 and 92). Recovery from a single findspot indicates that the bones were still articulated when buried and are likely to be close in age to their context. Stratified above context 527.

Objectives: to establish the date of this context as a step towards defining the chronology of the monument.

Calibrated date: 1σ: 3520–3370 cal BC
                       2σ: 3630–3360 cal BC

Final comment: F Healy (2012), this and OxA-13713 are cmpatible with the samples' stratigraphic position in the ditch fills.

References: Whittle *et al* 1999

GrA–29706 4700 ±40 BP

$\delta^{13}C$: -21.3‰

Sample: WH88 4360 (B1425)/B, submitted on 6 June 2005 by F Healy

Material: antler: *Cervus elaphus*, base with brow tine; pick (C Grigson 1989)

Initial comment: middle ditch, trench D, bone deposit 418 in centre-base of layer 416, between a cattle skull, which overlay it, and the ditch base (Whittle *et al* 1999, fig 86). Stratified below most of context 416. Location in rapidly accumulated chalk rubble fill very close to base of ditch suggests that the implement was used to dig the ditch.

Objectives: to provide a replicate for UB-6186 in order to establish the date of this context as a step towards defining the chronology of the monument.

Calibrated date: 1σ: 3630–3370 cal BC
                       2σ: 3640–3360 cal BC

Final comment: F Healy (2012), the five replicate measurements made on this antler UB-6186, GrA-29706, OxA-15075, -15076, -15088) are statistically consistent (T'=5.2; T'(5%)=9.5; ν=4; Ward and Wilson 1978). They are also statistically consistent with those for the remaining samples from context 416 (T'=14.1; T'(5%)=15.5; ν=8; Ward and Wilson 1978), indicating that all were freshly deposited, including a previously dated disarticulated cattle tibia (BM-2670; 4670 ±90 BP).

References: Ward and Wilson 1978
                          Whittle *et al* 1999

OxA–13505 4649 ±30 BP

$\delta^{13}C$: -20.4‰
$\delta^{15}N$ (diet): +7.3‰
C/N ratio: 3.4

Sample: WH28 B106, submitted on 6 February 2004 by F Healy

Material: animal bone: *Canis* sp., four articulating right metacarpals from a substantial part of an articulated skeleton, if not a complete one. There are, for example, numerous articulating vertebrae (J Mulville and A Powell 1970)

Initial comment: middle ditch. Segment MD IB, spit 3 (2–3ft). This was the ante penultimate spit and would probably have been mainly in the secondary fills (Pollard 1999, 47–50, fig 42 bottom left). The presence of so many bones, including a vertebral column, indicates that the dog was articulated when buried.

Objectives: to establish the date of this context as a step towards defining the chronology of the monument.

Calibrated date: 1σ: 3500–3360 cal BC
                       2σ: 3520–3360 cal BC

Final comment: see OxA-13679

References: Whittle and Pollard 1998

OxA–13679 4839 ±32 BP

$\delta^{13}C$: -22.0‰
$\delta^{15}N$ (diet): +5.5‰
C/N ratio: 3.2

Sample: WH28 B372, submitted on 6 February 2004 by F Healy

Material: animal bone: *Bos* sp., right scaphoid, articulating with right magnum (also B372) (C Grigson 1970)

Initial comment: middle ditch. Segment MD IB, spit 3 (2–3ft). This was the ante penultimate spit and would probably have been mainly in the secondary fills (Pollard 1999, 47–50, and fig 42 bottom left). Articulation indicates that the bones were either articulated or recently disarticulated when buried and hence close in age to their context. Sample may have come from same deposit (or same animal) as right cattle fore and hind foot bones from spit 4 in same segment. At a higher level than spit 4.

Objectives: to establish the date of this context as a step towards defining the chronology of the monument.

Calibrated date: 1σ: 3660–3630 cal BC
                        2σ: 3700–3530 cal BC

Final comment: F Healy (2012), statistical inconsistency of dates on articulating samples from this spit (this date and OxA-13505; T'=18.8; T'(5%)=3.8; ν=1; Ward and Wilson 1978) probably reflects origins in different stratigraphic contexts.

References: Ward and Wilson 1978
                       Whittle and Pollard 1998

OxA–13680 4403 ±33 BP

δ¹³C: -21.0‰

Sample: WH25 B6, submitted on 6 February 2004 by F Healy

Material: antler: *Cervus elaphus*, crown; ?pecked off (C Grigson 1970)

Initial comment: middle ditch I (north part of segment, excavated 1925 by Gray). Catalogue entry annotated 'at 4 (Ex no 34')'. This places the antler at 4ft deep, near the junction of the primary chalk rubble and the secondary fills (Pollard 1999, fig 42, bottom left). It is difficult to relate the context of this sample to those of samples from the rest of the segment (middle ditch Ib) excavated by Keiller in 1928.

Objectives: to establish the date of this context as a step towards defining the chronology of the monument.

Calibrated date: 1σ: 3100–2920 cal BC
                        2σ: 3270–2910 cal BC

Final comment: F Healy (2012), the date suggests that the difficulties of correlating stratigraphy in the northern butt (MD I), from which the sample came, and the rest of the segment (MD Ib), from which the other MDI samples came, were underestimated. MD I was more than a foot deeper than MD Ib (Pollard 1999, fig 41) and the drawn section (ibid. fig 42) was not through its deepest part. The sample would have been up to 2.5 ft (0.50m) above the base of the ditch, probably near the top of the chalk rubble fills, and its relation to the sequence of samples from MDI b is difficult to establish. the date suggests that the sample in fact came from the secondary fills.

References: Whittle and Pollard 1998

OxA–13713 4695 ±38 BP

δ¹³C: -22.1‰
δ¹⁵N (diet): +5.3‰
C/N ratio: 3.1

Sample: WH88 12301 (B54), submitted on 6 February 2004 by F Healy

Material: animal bone: *Bos* sp., lunate from same forelimb as anterior cuneiform, hamatum (both 12291/B43, B44), and pisiform (12310/B55) (C Grigson 1989)

Initial comment: middle ditch, trench E, segment MD XII, lower part of bone deposit 525 in lower part of context 508 at top of primary fills (Whittle *et al* 1999, 99–101, figs 89, and 93). Hamatum and cuneiform close together, in same find. Lunate approx 0.10m away. Articulation and proximity indicate that bones were only recently disarticulated when buried and are likely to be close in age to their context. Stratified above context 527, stratigraphically equivalent to contexts 523, 510.

Objectives: to establish the date of this context as a step towards defining the chronology of the monument.

Calibrated date: 1σ: 3630–3370 cal BC
                        2σ: 3640–3360 cal BC

Final comment: F Healy (2012), this and GrA-25707 are compatible with the samples' stratigraphic position in the ditch fills.

References: Whittle *et al* 1999

OxA–13714 4746 ±32 BP

δ¹³C: -22.0‰

Sample: WH88 4255 (B1458), submitted on 6 February 2004 by F Healy

Material: animal bone (medium mammal; rib section, from different left rib to sample for GrA-25556) (J Mulville and A Powell 2004)

Initial comment: middle ditch, trench D, bone group 414, within context 411. Part of 'bundle' of rib fragments composed of WH88 4241 (B1442), 4225 (B1441), 4234 (B1459–64), 4235 (B1446), 4236 (B1456–7), 4241 (B1442), 4242 (B1449), 4243 (B1447), 4244 (B1444), 4245 (B1445), 4247 (B1448), 4238 (B1435), 4251 (B1452), 4255 (B1458), and 4256 (B1454–5) (Whittle *et al* 1999, figs 86–7). Discard of more than one rib suggests the immediate aftermath of consumption, so that the sample should be close in age to its context. Stratified above context 416 and below context 413.

Objectives: to establish the date of this context as a step towards defining the chronology of the monument.

Calibrated date: 1σ: 3640–3380 cal BC
                        2σ: 3640–3370 cal BC

Final comment: see GrA-25556

References: Whittle *et al* 1999

OxA–13730 3524 ±30 BP

δ¹³C: -20.0‰
δ¹⁵N (diet): +7.6‰
C/N ratio: 3.2

Sample: WH88 12371b, submitted on 6 February 2004 by F Healy

Material: animal bone (toad; bones of fore limbs from almost complete toad skeleton (Rouse and Rowland 1999, table 154). Identification by Rouse, sample segregated by Mulville and Powell. From same skeleton as sample for GrA-25368) (J Mulville and A Powell 2004)

Initial comment: as GrA-25368

Objectives: as GrA-25368

Calibrated date: 1σ: 1900–1770 cal BC
2σ: 1940–1750 cal BC

Final comment: see GrA-25368

OxA–13812 4826 ±33 BP

δ¹³C: -20.8‰

Sample: Toad MD Ib L4 at 3ft 6 in, submitted on 6 February 2004 by F Healy

Material: animal bone (toad; vertebrae and long bones from one toad (extracted from larger collection from all parts of body - no duplicates present)) (J Mulville and A Powell 2004)

Initial comment: middle ditch, segment MD IB, spit 4, at 3.6ft. This was the penultimate spit (the last was spit 5 (4ft to 5ft). At this depth, the find would probably have been near the top of the primary fills, an unlikely depth for a hibernation death.

Objectives: to establish the date of this context as a step towards defining the chronology of the monument.

Calibrated date: 1σ: 3650–3540 cal BC
2σ: 3660–3520 cal BC

Final comment: see GrA-25559

OxA–13813 4682 ±34 BP

δ¹³C: -21.8‰

Sample: WH88 4194 (B1600), submitted on 6 February 2004 by F Healy

Material: animal bone: *Bos* sp., part of 1 of 4 fragmentary dorsal vertebrae; the others are find 4188 (B1593–8) (C Grigson 1989)

Initial comment: middle ditch, trench D, bone deposit 413 in context 411. Recovery of fragments of more than one dorsal vertebrae in close proximity (Whittle *et al* 1999, fig 88: 11–12). suggests that they were articulated when buried and hence close in age to their context. Stratified above context 414.

Objectives: to establish the date of this context as a step towards defining the chronology of the monument.

Calibrated date: 1σ: 3520–3370 cal BC
2σ: 3630–3360 cal BC

Final comment: F Healy (2012), the date is compatible with the sample's position in the ditch fill.

References: Whittle *et al* 1999

OxA–13814 4807 ±32 BP

δ¹³C: -21.9‰

Sample: WH88 4328 (B1761), submitted on 6 February 2004 by F Healy

Material: animal bone: *Bos* sp., right radius articulating with ulna (4328/B1742) (C Grigson 1989)

Initial comment: middle ditch, trench D, context 416 (Whittle *et al* 1999, fig 86). Overlying initial silt 417 in angle of ditch base and wall, overlying ditch bottom elsewhere. Stratified below context 414. Articulation indicates that bones were either articulated or only just disarticulated when buried and hence close in age to their context.

Objectives: to establish the date of this context as a step towards defining the chronology of the monument.

Calibrated date: 1σ: 3650–3530 cal BC
2σ: 3660–3520 cal BC

Final comment: F Healy (2012), this date and OxA-14967, measured on the ulna articulating with this radius, are statistically consistent (T'=2.0; T'(5%)=3.8; ν=1; Ward and Wilson 1978). These measurements are also statistically consistent with those for the remaining samples from context 416 (T'=14.1; T'(5%)=15.5; ν=8; Ward and Wilson 1978), indicating that all were freshly deposited, including a previously dated disarticulated cattle tibia (BM-2670; 4670 ±90 BP).

References: Ward and Wilson 1978
Whittle *et al* 1999

OxA–14967 4729 ±33 BP

δ¹³C: -21.4‰

Sample: WH88 4328 (B1742), submitted on 6 June 2005 by F Healy

Material: animal bone: *Bos* sp., right ulna articulating with radius 4328 (B1761) which provided sample for OxA-13814 (C Grigson 1989)

Initial comment: a replicate of OxA-13814.

Objectives: as OxA-13814

Calibrated date: 1σ: 3630–3380 cal BC
2σ: 3640–3370 cal BC

Final comment: see OxA-13814

OxA–15075 4717 ±30 BP

δ¹³C: -20.6‰

Sample: WH88 4360 (B1425)/A, submitted on 6 June 2005 by F Healy

Material: antler: *Cervus elaphus*, base with brow tine; pick (C Grigson 1989)

Initial comment: a replicate of GrA-29706.

Objectives: as GrA-29706

Calibrated date: 1σ: 3630–3370 cal BC
                       2σ: 3640–3370 cal BC

Final comment: see GrA-29706

Laboratory comment: Oxford Radiocarbon Accelerator Unit (2005), this sample [P17176] has been dated three times. The first determination (OxA-15088) was an ultrafiltered AMS with a solvent extraction (acetone/methanol/chloroform/water). The pre-treatment yield was 42.9mg from 840mg starting weight with a CN of 3.2. A second sample of bone was then obtained (820 mg). The sample was again solvent washed and decalcified. It was then gelatinised, and split into two fractions; one third was freeze dried as the filtered gelatin fraction (OxA-15076) and the other two thirds was ultrafiltered, freeze-dried, and dated as OxA-15075 (with a yield of 30.9mg). For analytical reasons, we favour the ultrafiltered determinations because they more successfully remove small amounts of contaminating carbon from bone gelatin. OxA-15088 and OxA-15075, therefore, would be the most reliable determinations for this antler. These provide a weighted mean of 4740 ±22 BP (T′=1.41; T′(5%)= 3.84; ν=1; Ward and Wilson 1978), but in fact all three are indistinguishable as a group.

References: Ward and Wilson 1978

OxA–15076 4673 ±30 BP

δ¹³C: -20.8‰

Sample: WH88 4360 (B1425)/A, submitted on 6 June 2005 by F Healy

Material: antler: *Cervus elaphus*, base with brow tine; pick (C Grigson 1989)

Initial comment: a replicate of GrA-29706.

Objectives: as GrA-29706

Calibrated date: 1σ: 3520–3370 cal BC
                       2σ: 3630–3360 cal BC

Final comment: see GrA-29706

Laboratory comment: see OxA-15075

OxA–15088 4770 ±33 BP

δ¹³C: -20.7‰

Sample: WH88 4360 (B1425)/A, submitted on 6 June 2005 by F Healy

Material: antler: *Cervus elaphus*, base with brow tine (C Grigson 1989)

Initial comment: a replicate of GrA-29706.

Objectives: as GrA-29706

Calibrated date: 1σ: 3640–3520 cal BC
                       2σ: 3650–3380 cal BC

Final comment: see GrA-29706

OxA–15177 4686 ±33 BP

δ¹³C: -21.6‰

Sample: WH27 1924, submitted on 6 June 2005 by F Healy

Material: animal bone: *Bos* sp., left humerus articulating with scapula from partial cattle skeleton (C Grigson 1970)

Initial comment: middle ditch, segment IVB, spits 4 (2.3–3.5ft) and 5 (3.5ft-base). Mentioned in letter from Keiller to Childe 6/3/28: 'a skeleton, which has taken nearly eight months to reconstruct, of an almost complete ox including head, on the forehead of which are curious markings, apparently artificial, from the bottom two layers of cutting IV of the middle ditch of Windmill Hill'. No further surviving record (Pollard 1999, 42). 'Curious markings' on forehead are faint horizontal line crossed by several parallel oblique lines. The occurrence of the skeleton in two successive spits means that it extended from the lower spit into the upper. Its articulation indicates that it was not long dead when placed on the base of the ditch.. In chalk-cut ditch on well-drained chalk down. There is no surviving record of the matrix in which the skeleton lay. By analogy with other segments of the same ditch it is likely to have consisted of chalk rubble (eg Whittle *et al* 1999, figs 86 and 89).

Objectives: to establish the date of this context as a step towards defining the chronology of the monument.

Calibrated date: 1σ: 3520–3370 cal BC
                       2σ: 3630–3360 cal BC

Final comment: F Healy (2012), this sample was dated to extend coverage around the circuit. The result is consistent with others from primary levels in the ditch.

References: Whittle *et al* 1999
                 Whittle and Pollard 1998

UB–6186 4699 ±20 BP

δ¹³C: -21.2 ±0.5‰

Sample: WH88 4360 (B1425), submitted on 6 February 2004 by F Healy

Material: antler: *Cervus elaphus*, base with brow tine (C Grigson 1989)

Initial comment: a replicate of GrA-29706.

Objectives: as GrA-29706

Calibrated date: 1σ: 3520–3375 cal BC
                       2σ: 3630–3370 cal BC

Final comment: see GrA-29706

Causewayed Enclosures: Windmill Hill, outer ditch, Wiltshire

Location: SU 0867071440
                     Lat. 51.26.28 N; Long. 01.52.32 W

Project manager: H Kendall (Independent) A Keiller (Independent) I Smith (Independent) A Whittle (Cardiff University), 1922–3, 1925–9, 1957–8, and 1988

Archival body: Alexander Keiller Museum and Wiltshire Heritage Museum

Description: the outer and largest circuit of the causewayed enclosure.

Objectives: samples were submitted during the project to refine the chronology of the circuit, as a step to building up the chronology of the complex as a whole.

Final comment: F Healy (2012), Bayesian modelling of the results and pre-existing measurements provides an estimated construction date of *3685–3610 cal BC (95% probability*; Whittle *et al* 2011, fig 3.11: *dig Windmill Hill outer*). In trench B, immediately adjacent to Isobel Smith's trench OD V, a markedly lower proportion of chalk rubble fill than in any other section across the outer ditch and the presence of Ebbsfleet Ware close to the ditch bottom can both be interpreted as the result of a deep recut made in the later fourth or early third millennium cal BC, extending to the top of Whittle's context 228 (Whittle *et al* 1999, fig 77).

References: Whittle *et al* 1999
Whittle *et al* 2011

GrA–25367 3640 ±50 BP

δ¹³C: -21.9‰

Sample: WH57–58 B198, submitted on 6 February 2004 by F Healy

Material: human bone (femur from articulated infant skeleton, 7–7.5 months. Bones present were frontal, parietal, occipital, temporal, ribs, vertebrae, pelvis, humerus, radius, ulna, clavicle, femora, scapula, tibia, phalanges, and possibly tarsals, metatarsals, and carpals) (D Brothwell and J Weyman 1959)

Initial comment: outer ditch, trench OD V, segment V-VI. At interface of layers 4 and 3. Section drawing in Alexander Keiller Museum, Avebury (acc. no 78510392) shows the location approximately mid-way between the outer edge and the centre of the ditch, at the junction of the lower and upper secondary fills. Stratified above context 227 and below OD V layer 2. Articulated but partly disturbed by burrowing animal(s) (Smith 1965, 9). On slope of a thin run of fairly small chalk rubble which had succeeded a band of earthy silt. Burial directly overlain by localised dark brown patch of irregular profile, possibly mound of earth over skeleton, possibly fill of animal burrow, possibly mound in turn overlain by earthy silts.

Objectives: to establish the date of this context as a step towards defining the chronology of the monument.

Calibrated date: 1σ: 2130–1930 cal BC
2σ: 2200–1880 cal BC

Final comment: F Healy (2012), statistically consistent with OxA-13759, measured on the same skeleton (T′=1.8; T′(5%)=3.8; ν=1; Ward and Wilson 1978). The late date for a burial long thought to be early Neolithic accords with its insertion into a largely silted recut in the Neolithic ditch, inferred from GrA-25550 and OxA-13500. The burial may well relate to Beaker and early Bronze Age pottery in the upper fills (Whittle *et al* 1999, table 156).

References: Brothwell 1965
Smith 1965
Ward and Wilson 1978
Whittle *et al* 1999

GrA–25389 4050 ±150 BP

δ¹³C: -29.1‰

Sample: WH88 23250b, submitted on 6 February 2004 by F Healy

Material: carbonised residue (part of a substantial portion of a Plain Bowl with external and internal residue (Whittle *et al* 1999, fig 187: 516)) (L Zienkiewicz 1989)

Initial comment: outer ditch, trench B, segment V-VI, bone deposit 229, between layers 228 and 210, within a few centimetres of the ditch base (Whittle *et al* 1999, fig 78:33). The presence of so much of the vessel, often in large fragments, and the good preservation of the superficial residue on it both suggest that it was buried soon after it was last used, without being subjected to weathering or redeposition. The internal residue should be the remnant of food cooked shortly before this. From same context as WH88 23207 (B4600). Stratified above samples on ditch base and below context 210. In chalk-cut ditch on well-drained chalk down. On surface of chalk rubble made up of small to medium angular chalk fragments with silt lenses (228) which overlay fine silt on ditch base. Overlain by rubble with smaller, more rounded chalk fragments (210).

Objectives: to establish the date of this context as a step towards defining the chronology of the monument.

Calibrated date: 1σ: 2880–2410 cal BC
2σ: 2930–2140 cal BC

Final comment: see GrA-25821

Laboratory comment: Rijksuniversitat Groningen (AMS) (29 July 2004), the alkali-soluble fraction was dated.

References: Whittle *et al* 1999

GrA–25545 4780 ±40 BP

δ¹³C: -22.8‰
δ¹⁵N (diet): +4.2‰

Sample: WH28 B370, submitted on 6 February 2004 by F Healy

Material: animal bone: *Bos* sp., right magnum articulating with hamatum (both B370) and with metacarpal B441 from spit 6 (J Mulville and A Powell 2004)

Initial comment: outer ditch, segment OD IB, spit 7 (6ft-7ft). This was the penultimate spit, the lowest being spit 8 (7ft-8ft). Both would have been within the primary fills (Pollard 1999, fig 26: top left). The recovery of two such small articulating bones indicates that the carpals were articulated or only recently disarticulated when buried and hence close in age to their context. Articulation with a metacarpal in spit 6 may mean that all three bones were articulated when buried, since the spits were arbitrary. Alternatively, the metacarpal may have been subsequently displaced from the carpals. In the same spit as WH28 B671. At a lower level than spit 6.

Objectives: to establish the date of this context as a step towards defining the chronology of the monument.

Calibrated date: 1σ: 3640–3520 cal BC
2σ: 3650–3380 cal BC

Final comment: see OxA-13501

References: Whittle and Pollard 1998

GrA–25546 4765 ±40 BP

δ¹³C: -22.2‰
δ¹⁵N (diet): +4.1‰

Sample: WH88 1687 (B5338), submitted on 6 February 2004 by F Healy

Material: animal bone (large mammal; part of 1 of 3 interleaved proximal rib fragments (WH88 1688 (B5330), 1686 (B5337), and 1687 (B5338))) (C Grigson 1989)

Initial comment: outer ditch, trench A, bone group 115 in top of context 111. One of three interleaved large mammal rib fragments (Whittle *et al* 1999, 90), probably discarded together when still attached by ligaments. Stratified above context 117.

Objectives: to establish the date of this context as a step towards defining the chronology of the monument.

Calibrated date: 1σ: 3640–3510 cal BC
2σ: 3650–3370 cal BC

Final comment: F Healy (2012), statistical inconsistency with a measurement on another rib fragment from the same 'bundle' (OxA-13504; T'=8.2; T'(5%)=3.8; ν=1; Ward and Wilson 1978) suggests that they were not, as supposed, the debris of a single consumption event.

References: Ward and Wilson 1978

GrA–25549 4740 ±40 BP

δ¹³C: -27.6‰

Sample: WH57–58 86, submitted on 6 February 2004 by F Healy

Material: carbonised residue (Plain shell-tempered Neolithic Bowl body sherd with internal residue under chalky deposit. In fresh condition, including the ancient breaks, which are covered by the same skin of chalky deposit as the faces. No sign of weathering) (I Smith 1959)

Initial comment: outer ditch, trench OD V, segment V-VI. Bottom of ditch, beside WH57–8 85. Freshness of surfaces and fractured edges suggest that sherd buried rapidly, as does preservation of residue, since soft, shelly fabric and fine sooty residue would have degraded rapidly if exposed to weathering. The residue is likely to derive from the last use of the pot from which the sherd came, and to be close in age to its context. Stratigraphically earlier than context 229.

Objectives: to establish the date of this context as a step towards defining the chronology of the monument.

Calibrated date: 1σ: 3640–3380 cal BC
2σ: 3640–3370 cal BC

Final comment: F Healy (2012), the result is compatible with the position at the bottom of the stratigraphic sequence in the ditch.

GrA–25550 4300 ±40 BP

δ¹³C: -21.5‰
δ¹⁵N (diet): +6.1‰

Sample: WH88 23059 (B3783), WH88 23067 (B3817), submitted on 6 February 2004 by F Healy

Material: animal bone: *Sus* sp., left ilium from new-born piglet, many of whose bones were found together (hind legs, pelvis, some vertebrae, some ribs). Finds 23059 (B3783), 23067 (B3817), and ?23063 (B3792) (C Grigson 1989)

Initial comment: outer ditch, Trench B, segment OD V-VI. Bone deposit in context 210 (Whittle *et al* 1999, 86; Grigson 1999, 189). Presence of so much of the piglet suggests that it was articulated when buried and was hence close in age to its context. Stratified above context 229 and below context 227.

Objectives: to establish the date of this context as a step towards defining the chronology of the monument.

Calibrated date: 1σ: 2920–2890 cal BC
2σ: 3020–2870 cal BC

Final comment: F Healy (2012), the surprisingly late date of this sample, from close to the bottom of the ditch, and of those successively overlying it (OxA-13500 and the replicates GrA-25367 and OxA-13759) can best be interpreted by positing a later fourth or early third millennium cal BC recut reaching to within *c* 0.50m of the ditch bottom. *See* also series comments.

References: Zienkiewicz and Hamilton 1999

GrA–25553 4755 ±40 BP

δ¹³C: -22.5‰
δ¹⁵N (diet): +4.4‰

Sample: WH88 23207 (B4600), submitted on 6 February 2004 by F Healy

Material: animal bone: *Bos* sp., proximal phalanx from same foot as another from same context (WH88 23201/B4613) (C Grigson 1989)

Initial comment: outer ditch, trench B, segment V-VI, bone deposit 229, between layers 228 and 210, within a few cm of ditch base (Whittle *et al* 1999, fig 78: 2, 5), This sample and another proximal phalanx from the same foot were not found in articulation, but *c* 0.25m apart, lying one at either end of a cattle tibia shaft WH88 23200 (Whittle *et al* 1999, fig 78: 2, 5, and 6). Their proximity in the same context, however, suggests that they were not long out of articulation when buried and were hence close in age to their context. From same context as WH88 23250. Stratified above samples on ditch base and below context 210. In chalk-cut ditch on well-drained chalk down. On surface of chalk rubble made up of small to medium angular fragments with silt lenses (228) which overlay fine silt on ditch base. Overlain by rubble with smaller, more rounded chalk fragments (210).

Objectives: to establish the date of this context as a step towards defining the chronology of the monument.

Calibrated date: 1σ: 3640–3380 cal BC
2σ: 3640–3370 cal BC

Final comment: F Healy (2012), the date is compatible with the stratigraphic position just above the bottom of the ditch.

GrA-25821 3980 ±50 BP

δ¹³C: -29.9‰

Sample: WH88 23250b, submitted on 6 February 2004 by F Healy

Material: carbonised residue (part of a substantial portion of a Plain Bowl with external and internal residue (Whittle *et al* 1999, fig 187: 516)) (L Zienkiewicz 1989)

Initial comment: a replicate of GrA-25389.

Objectives: as GrA-25389

Calibrated date: 1σ: 2570–2460 cal BC
2σ: 2620–2340 cal BC

Final comment: F Healy (2012), the three dates on carbonised residue from this pot are statistically inconsistent (this date, GrA-25389, OxA-13561; T'=391.2; T'(5%)=6.0; ν=2; Ward and Wilson 1978) and all are more recent than the pot and its context would indicate. It appears that some younger chemical contaminant has not been completely removed and that the results are inaccurate.

Laboratory comment: Rijksuniversitat Groningen (AMS) (29 July 2004), the acid and alkali insoluble fraction was dated.

References: Ward and Wilson 1978

GrA-29711 4615 ±40 BP

δ¹³C: -21.7‰
δ¹⁵N (diet): +11.9‰

Sample: WH29 B209 b, submitted on 6 June 2005 by F Healy

Material: human bone (left ilium of articulated skeleton of child of 2–3 years (Smith 1965, pl. VIIIa)) (I Smith 1965)

Initial comment: outer ditch, segment OD IIIB, spit 5 (4ft-5ft). IIIb was the central part of the segment, which encompassed two sub-segments and a higher ridge between them. The skeleton lay on the base of the ditch in its shallowest part, against the inner side (Smith 1965, 9; Pollard 1999, 30–4). Its articulation shows that the child was recently dead when buried, and likely to be close in age to its context. The section shows the burial on the chalk floor of the ditch, covered by fine silt overlain by chalk rubble.

Objectives: to establish the date of this context as a step towards defining the chronology of the monument.

Calibrated date: 1σ: 3500–3350 cal BC
2σ: 3520–3340 cal BC

Final comment: F Healy (2012), statistically consistent with a second measurement on the same skeleton (OxA-14966; T'=3.1; T'(5%)=3.8; ν=1; Ward and Wilson 1978). The later fourth millennium cal BC date was a surprise, and

prompts a reinterpretation of the stratigraphy, on two possible lines: (1) two originally separate segments may subsequently linked by a relatively shallow cut on the base of which the burial was placed or (2) a recut in OD IIIB may have gone deeper than shown in the published section drawing, which was constructed not in the field, but from measurements (Pollard 1999, 30, and fig 26).

References: Smith 1965
Ward and Wilson 1978
Whittle and Pollard 1998

GrA-29712 4715 ±35 BP

δ¹³C: -23.2‰

Sample: WH88 10455 (B74), submitted on 6 June 2005 by F Healy

Material: animal bone: *Bos* sp., left metatarsal shaft with fitting unfused epiphyses (10454), articulating with navicular (10464), which articulates with posterior cuneiform (10477) (Whittle *et al* 1999, 12, 18, 19) (C Grigson 1989)

Initial comment: outer ditch, trench C, segment OD IV, context 321. Bone deposit within 320. Compact group almost entirely of cattle bones, many of them conjoining or articulating. In silt overlying primary rubble and silt fills (Whittle *et al* 1999, figs 83–4). Stratified above context 308 and below context 317.

Objectives: to establish the date of this context as a step towards defining the chronology of the monument.

Calibrated date: 1σ: 3630–3370 cal BC
2σ: 3640–3370 cal BC

Final comment: F Healy (2012), all four measurements on cattle bone from this context, whether articulated or disarticulated, are statistically consistent, confirming the impression that the deposit could represent a single event (OxA-2401, -2402, GrA-29712, and -29713; T'=1.8; T'(5%)=7.8; ν=3; Ward and Wilson 1978).

References: Ward and Wilson 1978

GrA-29713 4675 ±40 BP

δ¹³C: -22.7‰

Sample: WH88 10458 (B248), submitted on 6 June 2005 by F Healy

Material: animal bone: *Bos* sp., 1 of 3 articulating dorsal vertebrae, 2 with fitting unfused epiphyses (Whittle *et al* 1999, fig 84: 7) (C Grigson 1989)

Initial comment: outer ditch, trench C, segment OD IV, context 321. Bone deposit within 320. Compact group almost entirely of cattle bones, many of them conjoining or articulating. In silt overlying primary rubble and silt fills (Whittle *et al* 1999, figs 83–4). Stratified above context 308 and below context 317. The articulation of the vertebrae indicates that they were still connected by soft tissue when buried and that the animal from which they came was not long dead.

Objectives: to establish the date of this context as a step towards defining the chronology of the monument.

Calibrated date: *1σ:* 3520–3370 cal BC
  2σ: 3630–3360 cal BC

Final comment: see GrA-29712

GrA–29714 4120 ±35 BP

δ¹³C: -24.9‰

Sample: WH88 10414, submitted on 6 June 2005 by F Healy

Material: charcoal: *Corylus* sp., single fragment from among 11.9g recovered from the context (C Cartwright 1989)

Initial comment: outer ditch, trench C, segment OD IV, context 308. This was a silt layer derived from the interior, near the top of the chalk rubble fills (Whittle *et al* 1999, fig 83). The quantity of charcoal of a single species, which is exceptional for the site suggests that it derived from a single event. Stratified below context 321, the source of samples for OxA-2401, and -2402.

Objectives: to establish the date of this context as a step towards defining the chronology of the monument.

Calibrated date: *1σ:* 2860–2580 cal BC
  2σ: 2880–2570 cal BC

Final comment: F Healy (2012), this sample was submitted in the belief that it came from context 308, described above. The lateness of the date was a surprise, especially as it post-dated articulating samples from bone deposit 321, supposedly stratified above it (GrA-29712, -29713) as well as the disarticulated samples noted above. Furthermore, it is statistically consistent with OxA-14965, now found to come from context 305 (T′=0.4; T′(5%)=3.8; v=1; Ward and Wilson 1978). This prompted re-examination of its context: in this case, however, the supervisor's notebook for trench C records the sample as from context 308. It may be that a '5' and an '8' were confused at some stage. *See* OxA-14965.

References: Ward and Wilson 1978

OxA–13499 4728 ±32 BP

δ¹³C: -27.6‰

Sample: WH57–58 85, submitted on 6 February 2004 by F Healy

Material: carbonised residue (Plain shell-tempered Neolithic Bowl body sherd with internal residue under chalky deposit. In fresh condition, including the ancient breaks, which are covered by the same skin of chalky deposit as the faces. No sign of weathering) (I Smith 1959)

Initial comment: outer ditch, trench OD V, segment V-VI. Bottom of ditch, beside WH57–8 86. Freshness of surfaces and fractured edges suggest that sherd buried rapidly, as does preservation of residue, since soft, shelly fabric and fine sooty residue would have degraded rapidly if exposed to weathering. The residue is likely to derive from the last use of the pot from which the sherd came, and to be close in age to its context. Stratigraphically earlier than context 229.

Objectives: to establish the date of this context as a step towards defining the chronology of the monument.

Calibrated date: *1σ:* 3630–3380 cal BC
  2σ: 3640–3370 cal BC

Final comment: see GrA-25549

OxA–13500 4021 ±29 BP

δ¹³C: -21.0‰
δ¹⁵N (diet): +9.1‰
C/N ratio: 3.2

Sample: WH88 23113 (B429), submitted on 6 February 2004 by F Healy

Material: animal bone: *Canis* sp., metatarsal articulating with proximal phalanx 23107 (B560) (C Grigson 1989)

Initial comment: outer ditch, trench B, segment V-VI, bone deposit 227 on surface of 210 (Whittle *et al* 1999, 82–85, fig 79: 36; Grigson 1999, 189, 231). This and the other dog foot bones were not articulated but lay in an area approximately 0.40m across. Their proximity and good condition suggest that they had been disarticulated for only a short time and were close in age to their context. At interface of primary and lower secondary fills. Stratified above 210 and below layer 3/4 interface in OD V.

Objectives: to establish the date of this context as a step towards defining the chronology of the monument.

Calibrated date: *1σ:* 2580–2480 cal BC
  2σ: 2620–2470 cal BC

Final comment: see GrA-25550

References: Zienkiewicz and Hamilton 1999

OxA–13501 4860 ±31 BP

δ¹³C: -21.3‰
δ¹⁵N (diet): +4.7‰
C/N ratio: 3.2

Sample: WH28 B671, submitted on 6 February 2004 by F Healy

Material: animal bone: *Bos* sp., 1 of several caudal vertebrae, with unfused epiphyses (C Grigson 1970)

Initial comment: outer ditch, segment OD IB, spit 7 (6–7ft). This was the penultimate spit, the lowest being spit 8 (7ft-8ft). Both would have been within the primary fills (Pollard 1999, fig 26: top left). The recovery of several small, fragile vertebrae together with their unfused epiphyses shows that they were articulated when buried and hence close in age to their context. In same spit as WH28 B370. At a lower level than spit 6.

Objectives: to establish the date of this context as a step towards defining the chronology of the monument.

Calibrated date: *1σ:* 3660–3630 cal BC
  2σ: 3700–3540 cal BC

Final comment: F Healy (2012), statistical consistency with another date on an articulating sample from the same spit (GrA-25545; T′=2.3; T′(5%)=3.8; v=1; Ward and Wilson 1978) indicates that both samples were indeed fresh when buried.

References: Ward and Wilson 1978
  Whittle and Pollard 1998

OxA–13502 4164 ±35 BP

δ¹³C: -22.2‰

Sample: WH28 B145a, submitted on 6 February 2004
by F Healy

Material: animal bone: *Cervus elaphus*, one of two proximal
phalanges (lateral and medial), likely to have been side-by-
side in same foot because comparable in size and muscle
attachments (J Mulville and A Powell 2004).

Initial comment: outer ditch, segment OD IB, spit 3 (2ft-3ft).
The spit would probably have been within the primary fills
at the inner edge of the ditch and within the secondary fills
at the outer (Pollard 1999, fig 26: top left). The recovery of
adjacent phalanges from a species the skeletal remains of
which are rare on the site indicates that they were articulated
or only recently disarticulated when buried and hence close
in age to their context. At a higher level than spit 6.

Objectives: to establish the date of this context as a step
towards defining the chronology of the monument.

Calibrated date: 1σ: 2880–2670 cal BC
                        2σ: 2890–2620 cal BC

Final comment: F Healy (2012), the third millennium cal BC
date of the sample indicates that it came from secondary
rather than primary silts.

References: Whittle and Pollard 1998

OxA–13503 4825 ±32 BP

δ¹³C: -22.2‰
δ¹⁵N (diet): +4.8‰
C/N ratio: 3.2

Sample: WH88 1712 (B18), submitted on 6 February 2004
by F Healy

Material: animal bone: *Bos* sp., proximal metatarsal fragment
(B18), articulating with complete navicular (B19) and
complete cuneiform (B20) (C Grigson 1989)

Initial comment: outer ditch, trench A, bone deposit 117, in
the top of context 112. Lying *c* 0.35m from a cattle frontlet,
approximately 1m above ditch base (Whittle *et al* 1999, fig
81, fig 82: 8). Stratified beneath context 115. In the top of a
layer of relatively small chalk rubble (112) derived from the
inside of the ditch, overlying a succession of rubble layers
interleaved with silt.

Objectives: to establish the date of this context as a step
towards defining the chronology of the monument.

Calibrated date: 1σ: 3650–3540 cal BC
                        2σ: 3660–3530 cal BC

Final comment: F Healy (2012), the date is compatible with
the sample's stratigraphic position in the ditch fills.

OxA–13504 4620 ±31 BP

δ¹³C: -21.3‰
δ¹⁵N (diet): +4.7‰
C/N ratio: 3.2

Sample: WH88 1688 (B5330), submitted on 6 February
2004 by F Healy

Material: animal bone (large mammal; 1 of 3 interleaved
proximal rib fragments (WH88 1688 (B5330), 1686
(B5337), and 1687 (B5338)) (C Grigson 1989).

Initial comment: outer ditch, trench A, bone group 115 in top
of context 111. One of three interleaved large mammal rib
fragments (Whittle *et al* 1999, 90) probably discarded
together when still attached by ligaments. Stratified above
context 117.

Objectives: to establish the date of this context as a step
towards defining the chronology of the monument.

Calibrated date: 1σ: 3500–3360 cal BC
                        2σ: 3510–3350 cal BC

Final comment: see GrA-25546

OxA–13561 2770 ±40 BP

δ¹³C: -28.3‰
δ¹⁵N (diet): +3.2‰
C/N ratio: 9.4

Sample: WH88 23250a, submitted on 6 February 2004
by F Healy

Material: carbonised residue (part of a substantial portion
of a plain bowl with external and internal residue (Whittle
et al 1999, fig 187: 516)) (L Zienkiewicz 1989)

Initial comment: a replicate of GrA-25389.

Objectives: as GrA-25389

Calibrated date: 1σ: 980–840 cal BC
                        2σ: 1020–820 cal BC

Final comment: see GrA-25821

OxA–13759 3716 ±28 BP

δ¹³C: -20.5‰

Sample: WH57–58 B198, submitted on 6 February 2004 by
F Healy

Material: human bone (femur from articulated infant
skeleton, 7–7.5 months. Bones present were frontal, parietal,
occipital, temporal, ribs, vertebrae, pelvis, humerus, radius,
ulna, clavicle, femora, scapula, tibia, phalanges, and possibly
tarsals, metatarsals and carpals (Brothwell 1965)) (D
Brothwell and J Weyman 1959)

Initial comment: a replicate of GrA-25367.

Objectives: as GrA-25367

Calibrated date: 1σ: 2200–2030 cal BC
                        2σ: 2210–2020 cal BC

Final comment: see GrA-25367

References: Brothwell 1965

OxA–14965 4089 ±34 BP

δ¹³C: -24.4‰

Sample: WH88 10343, submitted on 6 June 2005 by F
Healy

Material: charcoal: *Corylus* sp., single fragment (1.30g) (C Cartwright 1989)

Initial comment: outer ditch, trench C, segment OD IV, context 308. This was a silt layer derived from the interior, near the top of the chalk rubble fills (Whittle *et al* 1999, fig 83). The quantity of charcoal of a single species, which is exceptional for the site suggests that it derived from a single event.

Objectives: to establish the date of this context as a step towards defining the chronology of the monument.

Calibrated date: 1σ: 2840–2570 cal BC
2σ: 2870–2490 cal BC

Final comment:, see GrA-29714

OxA–14966 4521 ±35 BP

δ¹³C: -21.1‰
δ¹⁵N (diet): +11.9‰

Sample: WH29 B209 a, submitted on 6 June 2005 by F Healy

Material: human bone (left ilium of articulated skeleton of child of 2–3 years (Smith 1965, pl. VIIIa)) (I Smith 1965)

Initial comment: a replicate of GrA-29711.

Objectives: to establish the date of this context as a step towards defining the chronology of the monument.

Calibrated date: 1σ: 3360–3100 cal BC
2σ: 3370–3090 cal BC

Final comment: see GrA-29711

References: Smith 1965

Chesham Bois House, Buckinghamshire

Location: SU 969998
Lat. 51.41.16 N; Long. 00.35.54 W

Project manager: Y Edwards (Institute of Archaeology, London), May and July 2005

Archival body: Chess Valley Archaeological and Historical Society and Buckinghamshire County Council Museum

Description: Chesham Bois House stands at the northern extremity of a plateau ridge overlooking the Chess Valley, which is 65m below. Documentary records indicate that the house was part of one of five manorial estates of middle Saxon Chesham and that its history is entwined with that of the adjacent church, which was established prior to AD 1203 and the lord of the manor's private chapel from AD 1213. Between AD 1433 and AD 1735 the Cheyne family developed the house and the gardens so that by the early eighteenth century the house was a large complex of buildings surrounded by extensive pleasure gardens. Early plans of the house indicate significant structural evolution, probably starting from a four-sided courtyard house. The present house was built in AD 1820, within the footprint of the old house, but about one sixth of the size. Resistivity

surveys and excavations have taken place in the gardens. There are five acres of land, including areas which were formal gardens, a bowling green, and the church.

The samples for dating were selected from bulk samples (10 litres) of two charcoal rich layers (contexts 022 and 023) which occupied most of pit 1 (021). The charcoal was associated with large 'bowl' shaped pieces of slag - identified as smithing slag.

Objectives: secure dates for these charcoal fragments will help to fix a point in time when the smithy was in operation. Furthermore, it will provide a *terminus post quem* for the overlying features: a pitch tile hearth and a chalk floor. Dating of the charcoal would also help to date the earliest occupation of this area at the Chesham Bois House site and thereby improve our understanding of the evolution of the Chesham Bois House settlement.

Final comment: Y Edwards (20 March 2013), this dating coincided with two peaks in the tree calibration curve. Immediately adjacent to the smithing pit from which the samples came a silver groat dated AD 1422–27, which favours of the earliest calibrated date for the pit.

Laboratory comment: (22 May 2006), it is highly unlikely that the infilling of the pit with smithy debris occurred over an extended period of time, and the four measurements all calibrate to cal AD 1440–1650 at 95% confidence (Reimer *et al* 2004). This provides the best estimate for the age of this feature.

References: Edwards *et al* 2010
Reimer *et al* 2004

SUERC–10182 350 ±35 BP

δ¹³C: -25.3‰

Sample: CBH-05:CHAR 023 A, submitted on 28 February 2006 by Y Edwards

Material: charcoal: *Quercus* sp., sapwood; single fragment (R Gale 2006)

Initial comment: pit 1 (021) lay at the north-east end of a pitch-tile hearth about 8 cm below the bottom of the tiles. A plan drawn prior to the discovery of the pit records that the top surface of the pit was largely sealed with pinkish-grey mortar and broken tiles, although there was a breakthrough of charcoal in two patches immediately beneath the hearth tiles. At the time of drawing the plan these patches were interpreted as possible burnt timbers. The hearth tiles did not extend across the surface of the pit but had fallen away onto the mortar surface at the north-east corner of the hearth due to some subsidence of the pit contents.

The charcoal samples CBH-05: CHAR 22 were taken from the uppermost charcoal-rich layer of Pit 1 and samples CBH-05: CHAR 23 from the less charcoal-rich layer beneath. These contexts showed no evidence of disturbance since disposition.

The trench in which pit 1 was found lies beneath the south lawn of Chesham Bois House. The ground generally slopes downwards from west to east and there are signs of levelling and terracing in the past. Pit 1 appears to have been dug into natural clay. The surface of the pit lay *c* 50cm below the present day surface.

Objectives: the charcoal submitted for radiocarbon dating derives from pit 1 (021), which was uncovered at the lowest level within trench CBH05–003 at Chesham Bois House. This trench contains part of a room with a large pitch-tile hearth set into a thick, beaten chalk floor (about 50cm below ground level) and part of a second room to the west with evidence of floor renewal on several occasions during its history.

The top of pit 1 (021) dug into natural clay, lay about 12cm below the hearth and 55cm below the present land surface. The presence of charcoal fragments in this smithying waste pit provides an excellent opportunity to date the earliest occupation of this area at the Chesham Bois House site and will prove a *terminus post quem* for the overlying pitch tile hearth and earliest chalk floor. At present we have no dates associated with any of the features in trench CBH05–003. The floors and hearth were remarkably clean and only a few fragments of unglazed pottery were recovered from the lowest levels. A single silver groat of the 'long cross' type was found on a level with the hearth but in an area where there was some evidence of reworking.

While we can safely surmise that the smithy was eventually replaced by a building with at least three rooms, one of which contained a large domestic hearth, none of these activities are as yet located in time. Obtaining a radiocarbon date for the wood from the charcoal pit would represent an important adjunct to understanding the evolution of the Chesham Bois House site by fixing a point in time when the smithy, perhaps associated with a farm or house, was in operation.

Calibrated date: 1σ: cal AD 1460–1640
                  2σ: cal AD 1440–1650

Final comment: see series comments

SUERC–10183 350 ±35 BP

$\delta^{13}C$: -22.7‰

Sample: CBH-05: CHAR 023 B, submitted on 28 February 2006 by Y Edwards

Material: charcoal: *Fagus* sp., roundwood; single fragment (R Gale 2006)

Initial comment: as SUERC-10182

Objectives: as SUERC-10182

Calibrated date: 1σ: cal AD 1460–1640
                  2σ: cal AD 1440–1650

Final comment: see series comments

SUERC–10184 355 ±35 BP

$\delta^{13}C$: -28.4‰

Sample: CBH-05: CHAR 022 A, submitted on 28 February 2006 by Y Edwards

Material: charcoal: *Fagus* sp., roundwood, 12 growth rings; single fragment (R Gale 2006)

Initial comment: as SUERC-10182

Objectives: as SUERC-10182

Calibrated date: 1σ: cal AD 1460–1640
                  2σ: cal AD 1440–1650

Final comment: see series comments

SUERC–10188 360 ±35 BP

$\delta^{13}C$: -24.5‰

Sample: CBH-05: CHAR 022 B, submitted on 28 February 2006 by Y Edwards

Material: charcoal: *Quercus* sp., sapwood; single fragment (R Gale 2006)

Initial comment: as SUERC-10182

Objectives: as SUERC-10182

Calibrated date: 1σ: cal AD 1450–1640
                  2σ: cal AD 1440–1650

Final comment: see series comments

Claydon Pike, Lechlade, Gloucestershire

Location: SU 191996
                  Lat. 51.41.43 N; Long. 01.43.28 W

Project manager: D Miles (Oxford Archaeological Unit), 1979–83

Archival body: Oxford Archaeology, Corinium Museum

Description: there were two main areas of occupation. To the north at Warrens Field was an un-nucleated middle Iron Age farmstead represented by a series of roundhouse gullies, ditches, and pits, spread over three gravel islands. It was probably representative of a shifting settlement. Further to the south at Longdoles Field, was an extensive area of occupation ranging in date from the late Iron Age to the late-/sub-Roman period. The dating is based upon pottery and coin evidence.

There are three main phases that have been identified at Longdoles Field: 1) early first-early second century AD, characterised by a nucleated settlement comprising enclosures, linear ditches, and gullies; 2) early second-early fourth century AD, characterised by a radically changed site composed of rectangular enclosures and aisled buildings (possibly the site of an official Roman depot); and 3) early fourth-late fourth/early fifth century AD, characterised by a small masonry villa and shrine.

The samples submitted for radiocarbon dating come from a small inhumation cemetery of eight bodies within and around a small double enclosure, and with a further two bodies located *c* 20m east.

Objectives: to determine which phase of the site the cemetery belongs.

Final comment: A Smith (18 March 2004), the results of the Longdoles Field radiocarbon dating have been extremely useful in fulfilling the original objective of dating the post-Roman activity on the site. Furthermore, it is of great significance in terms of understanding middle Saxon activity in the region.

Laboratory comment: English Heritage (21 November 2013), three measurements (HAR-5409, HAR-5410, and HAR-5411) were undertaken previously on samples from this site and are published in Bayliss *et al* 2012a, 68–70).

References: Bayliss *et al* 2012a, 68–70
                 Miles *et al* 2007
                 Miles and Palmer 1982
                 Miles and Palmer 1983
                 Miles and Palmer 1990

UB–4896 1233 ±60 BP

$\delta^{13}C$: -19.8 ±0.2‰
$\delta^{13}C$ *(diet)*: -19.5 ±0.3‰
$\delta^{15}N$ *(diet)*: +10.0 ±0.3‰
C/N ratio: 3.2

Sample: FCP A, submitted on 11 February 2003 by A Smith

Material: human bone (247g) (right femur) (A Witkin 2003)

Initial comment: inhumation 1971 was recovered fully articulated. The grave cutting was through wall 1587 and room 10 of the villa. The skeleton of a male, aged 35–45. The grave was cut through the walls and rubble of the building, which consists largely of limestone.

Objectives: to establish a date for the burial group and a date for post-villa activity.

Calibrated date: *1σ:* cal AD 680–890
                      2σ: cal AD 660–970

Final comment: A Smith (18 March 2004), the dating of the skeleton confirms a middle Saxon date for the post-Roman activity on the site.

UB–4897 1187 ±60 BP

$\delta^{13}C$: -20.1 ±0.2‰
$\delta^{13}C$ *(diet)*: -19.8 ±0.3‰
$\delta^{15}N$ *(diet)*: +9.8 ±0.3‰
C/N ratio: 3.3

Sample: FCP B, submitted on 11 February 2003 by A Smith

Material: human bone (258g) (right femur, tibia and fibula shafts) (A Witkin 2003)

Initial comment: inhumation 2105 is fully articulated. It was located in a grave cutting robber trench 2123 of the villa. Possibly the skeleton of a male, over age 50. The grave cuts through the rubble of the building, which consists primarily of limestone.

Objectives: as UB-4896

Calibrated date: *1σ:* cal AD 720–950
                      2σ: cal AD 670–990

Final comment: see UB-4896

UB–4898 1271 ±60 BP

$\delta^{13}C$: -19.9 ±0.2‰
$\delta^{13}C$ *(diet)*: -19.6 ±0.3‰
$\delta^{15}N$ *(diet)*: +8.4 ±0.3‰
C/N ratio: 3.3

Sample: FCP C, submitted on 11 February 2003 by A Smith

Material: human bone (261g) (right femur and fibula) (A Witkin 2003)

Initial comment: inhumation 2129 is fully articulated. It lies within a grave that cuts wall 1587 of the villa. The skeleton is a female, aged 35–45. The grave cuts through the wall and rubble of the building, consisting primarily of limestone.

Objectives: as UB-4896

Calibrated date: *1σ:* cal AD 660–780
                      2σ: cal AD 650–900

Final comment: see UB-4896

Dover Buckland, Kent

Location: TR 30904285
                     Lat. 51.08.15 N; Long. 01.18.01 E

Project manager: K Parfitt (Canterbury Archaeological
                     Trust), 1994

Archival body: British Musuem

Description: an early Anglo-Saxon cemetery forming an extension to that excavated in the 1950s. About 250 graves were excavated in 1994, making this one of the largest Kentish cemeteries. About half the graves contained grave goods, indicating that this part of the cemetery was in use between *c* AD 450 and AD 650. This series of radiocarbon dates came from the leg bones of skeletons 391A, 391B, and 365. Six samples from further burials were subsequently dated as part of the Anglo-Saxon chronology project (Bayliss *et al* 2013; SK 222 (UB-6472, 1550 ±19 BP), SK 250 (UB-6473, 1572 ±22 BP), SK 264 (UB-6474, 1528 ±17 BP), Sk 323 (UB-6475, 1491 ±18 BP), SK 339 (UB-6476, 1592 ±17 BP), and SK 414 (UB-6477, 1570 ±20 BP).

Objectives: to resolve the problems of dating. Skeletons 391A and 391B occur in the same grave, one (A) above the other (B). The grave goods with 'A', although stratigraphically later, are stylistically earlier than 'B'. Grave 375 seems to be one of the latest on the site - how late?

Final comment: A Bayliss (4 August 2011), both the seriation of Anglo-Saxon male graves by correspondence analysis (Bayliss *et al* 2012b, fig 6.49a and e-fig 6.6) and the radiocarbon date of the skeleton confirm the grave 375 is probably the latest of the male burials sampled for radiocarbon dating from this site, although the correspondence analysis assigns grave 251 from the more recently excavated burial area at Buckland to the same phase (AS-MC/AS-Mq). The radiocarbon dates of graves 391A and 391B, combined with phasing on the basis of seriation by correspondence analysis and Bayesian modelling of the radiocarbon dates, confirm that these two interments were very probably made around a century apart, if not even more. Whether the relationship between them was deliberate, accidental, or simply opportunistic, is therefore a matter for further consideration. Both burials are relatively well-furnished for their periods: grave 391A has amethyst beads and bulla pendants, characteristic of phase AS-FE of the middle or later seventh century, while grave 391B contained, *inter alia,* a belt-set of the early to mid-sixth century and amber beads that were most common at that time too.

Laboratory comment: Rafter Radiocarbon Laboratory (2012), all the samples fall in the optimum C:N ratio range for well-preserved bone protein of 2.9 to 3.6 established by DeNiro (1985). The %N values for these samples do not indicate contamination with exogenous carbon. The protein in these samples is therefore suitable for accurate radiocarbon dating.

References: Bayliss *et al* 2012b
             DeNiro 1985
             Evison 1987
             Parfitt and Anderson 2012

UB–4958 1456 ±17 BP

$\delta^{13}C$: -22.3 ±0.2‰
$\delta^{13}C$ *(diet):* -19.8 ±0.3‰
$\delta^{15}N$ *(diet):* +9.0 ±0.4‰
C/N ratio: 3.2   *%C:* 43.0   *%N:* 15.9

Sample: Grave 375, submitted on 7 August 2003 by K Parfitt

Material: human bone (325g) (right femur) (T Anderson 1995)

Initial comment: an isolated male inhumation buried with spearhead-type SP3–a, sword-type SW4, buckle-type BU9–a, and shield boss-type SB4–b2.

Objectives: to confirm the dating of the grave goods, and to help establish how long this part of the cemetery was in use.

Calibrated date: *1σ:* cal AD 595–640
               2σ: cal AD 565–645

Final comment: A Bayliss (3 October 2013), this grave assemblage falls in the second phase of the national male seriation (AS-MC/AS-Mq), dating to *cal AD 550–585 (95% probability; UB-4958 BuD375)*; Bayliss *et al* 2012b, fig 6.52) in the position based on leading artefact-types, or *cal AD 555–585 (95% probability; UB-4958 (BuD375)*; Bayliss *et al* 2012b, fig 6.53) in the partition based on the two-dimensional map of grave-assemblages.

UB–4959 1418 ±19 BP

$\delta^{13}C$: -20.7 ±0.2‰
$\delta^{13}C$ *(diet):* -20.3 ±0.3‰
$\delta^{15}N$ *(diet):* +9.8 ±0.4‰
C/N ratio: 3.2   *%C:* 28.5   *%N:* 10.3

Sample: Grave 391A, submitted on 7 August 2003 by K Parfitt

Material: human bone (235g) (right femur and tibia and left femur) (T Anderson 1995)

Initial comment: this is the upper of the two skeletons, one above the other. The excavators believe that these bodies were buried one immediately after the other. The grave goods in this grave are earlier, however. This female grave contained bead-type BE1–Amethyst, and pendant-type PE-8.

Objectives: skeleton 391A is later than 391B, but the associated grave goods are earlier. Can radiocarbon dating help resolve this issue? Also, grave 391A contains amethyst beads which are used as a date indicator. An associated radiocarbon date would help refine their period of use.

Calibrated date: *1σ:* cal AD 615–655
               2σ: cal AD 600–660

Final comment: A Bayliss (3 October 2013), this grave assemblage falls in the last phase of the national female seriation (AS-FE), dating to *cal AD 635–660 (95% probability; UB-4959 (BuD391A)*; Bayliss *et al* 2012b, fig 7.65). Amethyst beads occur in all but the first phase of this seriation (AS-FC-AS-FE) occurring from the second half of the sixth century AD.

UB–4960 1598 ±17 BP

$\delta^{13}C$: -20.1 ±0.2‰
$\delta^{13}C$ *(diet):* -20.0 ±0.3‰
$\delta^{15}N$ *(diet):* +9.2 ±0.4‰
C/N ratio: 3.3   *%C:* 48.3   *%N:* 17.3

Sample: Grave 391B, submitted on 7 August 2003 by K Parfitt

Material: human bone (200g) (right femur and tibia) (T Anderson 1995)

Initial comment: this is the lower (female) of the two skeletons, one above the other. The dating of the associated grave goods does not equate with the stratigraphic evidence. Grave 391A may have been cut into 391B, or they could be contemporary. This grave contained a bucket with a copper-alloy frame which falls in the early sixth century, as well as buckle-types BU2–d and BU2–h, bead-type BE1–Koch20Wh, and wire ring-type WR1–a.

Objectives: grave 391B is stratigraphically earlier than 391A, but the dating of the associated grave goods is ambiguous. Thus, there is a problem with the dating which could have implications for the dating of grave goods beyond this specific grave.

Calibrated date: *1σ:* cal AD 415–535
               2σ: cal AD 405–540

Final comment: A Bayliss (3 October 2013), this grave assemblage falls in the first phase of the national female seriation (AS-FB), dating to *cal AD 520–550 (95% probability; UB-4960 (BuD391B)*; Bayliss *et al* 2012b, fig 7.65).

Duxford: Hinxton Road, Cambridgeshire

Location: TL 48104580
           Lat. 52.05.27 N; Long. 00.09.37 W

Project manager: J Roberts (Archaeological Field Unit, Fulbourn Community Centre Site), 2002

Archival body: Cambridgeshire County Council

Description: the Hinxton Road site sites on a chalk slope west of and overlooking the River Cam. The site comprises features that span the Iron Age through the post-medieval period. The Iron Age remains include ritual features (possible shrine or temple), pits (one containing a crouched burial), inhumations, and cremations. Saxon settlement remains have been found in the southern part of the site and one burial was dated by pottery found in the grave fill to this period. A lime kiln and mortar mixer were also identified in the southern part of the site. Post-medieval structural and garden features have truncated some of the remains.

Objectives: to provide a date for the use of the lime kiln and by association the nearby mortar mixer and to determine the period of use of the cemetery.

Final comment: A Lyons (3 July 2014), the radiocarbon dating undertaken at this site allowed for the many features, including a complex Iron Age to Romano-British cemetery, lime kiln, and a mortar mixer to be successfully phased and interpreted. The lime kiln structure was subsequently reinterpreted as a drying building, in use in the late Roman/early Saxon period and possibly beyond.

Laboratory comment: English Heritage (2004), chronological modelling of the radiocarbon results showed good agreement between the results and the stratigraphic record ($A_{overall}$ =94.4%); however, GU-5930 has an individual index agreement only just within the acceptable limit (A=60.9%), and could be an earlier outlier. Without additional radiocarbon dates from the cemetery it is impossible to determine whether there are two discrete phases of use, or if GU-5930 (DUXHR02 4065) simply represents a slightly earlier burial at this site.

Laboratory comment: English Heritage (3 July 2014), ten further dates were previously funded from this site (GU-5919–20, and GU-5924–31; Bayliss *et al* 2016, 93-4).

References: Lyons 2011

GU–5999 2050 ±50 BP

$\delta^{13}C$: -19.7‰
$\delta^{13}C$ *(diet)*: -20.1‰
$\delta^{15}N$ *(diet)*: +8.4‰
C/N ratio: 3.5

Sample: 3609, submitted on 15 December 2003 by J Roberts

Material: human bone (200g) (left femur) (C Duhig 2003)

Initial comment: a fully articulated skeleton in a grave between 0.3–0.4m deep.

Objectives: to identify temporal patterning in the burial ground used over the Iron Age and Roman periods.

Calibrated date: *1σ:* 160 cal BC–cal AD 10
                        2σ: 200 cal BC–cal AD 70

Final comment: A Lyons (3 July 2014), this radiocarbon date confirms a late Iron or early Romano-British date for the burial.

Laboratory comment: English Heritage (2004), chronological modelling indicates this burial dates to *180 cal BC–cal AD 60 (at 95% probability*; Lyons 2011, fig App 1.1, 127).

GU–6000 1910 ±70 BP

$\delta^{13}C$: -19.8‰
$\delta^{13}C$ *(diet)*: -19.8‰
$\delta^{15}N$ *(diet)*: +7.5‰
C/N ratio: 3.2

Sample: 4106, submitted on 15 December 2003 by J Roberts

Material: human bone (350g) (left femur) (C Duhig 2003)

Initial comment: the western end of the grave was cut by a ditch, removing the feet. The skeleton was fully articulated. The grave was very shallow, 0.18m below surface, and the skeleton's face had been severely damaged/truncated. It was cut into the chalk and has been disturbed by a post-Roman ditch at the west end and by ground levelling in antiquity.

Objectives: to try to identify the temporal patterning in burial ground used over the Iron Age and Roman periods.

Calibrated date: *1σ:* cal AD 20–210
                        2σ: 50 cal BC–cal AD 250

Final comment: A Lyons (3 July 2014), this confirms a late Iron Age or Romano-British date for the burial.

Laboratory comment: English Heritage (2004), chronological modelling indicates this burial dates to *50 cal BC–cal AD 260 (at 95% probability*; Lyons 2011, fig App 1.1, 127).

GU–6001 1960 ±50 BP

$\delta^{13}C$: -19.9‰
$\delta^{13}C$ *(diet)*: -19.7‰
$\delta^{15}N$ *(diet)*: +9.3‰
C/N ratio: 3.2

Sample: 4139, submitted on 15 December 2003 by J Roberts

Material: human bone (241g) (right femur) (C Duhig 2003)

Initial comment: grave (4138) was disturbed at the western end by a modern tree.

Objectives: to more fully establish the period of use of the cemetery. Previous dates show use of this part of the site from the Iron Age and Roman periods.

Calibrated date: *1σ:* 20 cal BC–cal AD 90
                        2σ: 60 cal BC–cal AD 140

Final comment: A Lyons (3 July 2014), the modelled terminal Iron Age to early Romano-British date of this skeleton is particularly interesting as the bones display evidence for tuberculosis.

Laboratory comment: English Heritage (2004), chronological modelling indicates this burial dates to *100–70 cal BC (at 1% probability)* or *60 cal BC- cal AD 140 (at 93% probability)* or *cal AD 150–180 (at 1% probability*; Lyons 2011, fig App 1.1, 127). It is likely that the 93% probability range most accurately dates this specimen, placing it in the terminal Iron Age-early Roman period.

Exmoor Iron Project, Devon and Somerset

Location: see individual sites

Project manager: G Juleff (University of Exeter), 1996–2008

Description: archaeological survey and excavation on Exmoor during the late 1990s and early 2000s has, for the first time, recognised extensive evidence for iron-working on the upland since the late first millennium cal BC. Survey and excavations have focussed on field evidence for all stages or primary iron production, from ore extraction to smelting and initial metal refining. These surveys have demonstrated near-continuous exploitation of the upland for metallurgic products over at least the last 2000 years. Until now, explanations of landscape change and economic development of the upland have revolved around its use as an agricultural resource and a recognition of the role of this and other uplands as centres of extractive metal industries demands a rethink of a range of issues. These include reviewing existing ideas around the social and economic functioning of Exmoor in the wider region, and from an environmental perspective the relative importance of woodland management and clearance in relation to agrarian and industrial purposes.

Objectives: the investigation of the development of iron-working as a rural industry through time and the role it has played in shaping the wider physical and cultural landscapes that are now threatened by agricultural improvement and forestry. The development of a methodological approach to the investigation of pre-industrial metal-working. The project will explore ways in which the repertoire of techniques can be extended to help bridge the gap between archaeology and archaeometallurgy.

Exmoor Iron: Blacklake Wood, Somerset

Location: SS 90452870
                     Lat. 51.02.46 N; Long. 03.33.45 W

Project manager: G Juleff (University of Exeter), 2004

Archival body: University of Exeter

Description: the site is an iron smelting site comprising a substantial slag heap and associated working area. Radiocarbon dating of charcoal from the slag heap during survey in the late 1990s gave a date of cal AD 410–650 (Beta-132445; 1520 ±60 BP; Reimer *et al* 2004). Excavation revealed a complex sequence of slag dumping and a working area comprising charcoal and ore storage and preparation areas as well as one furnace base. The apparent early medieval dating and the well-preserved nature of the site make it of national significance. Different elements of the site were submitted for dating: a slag heap: three contexts representing separate episodes of slag and smelting debris dumping. These are from the top of the heap (207), the middle of the heap (217), and the bottom (233); a furnace base: two contexts that lie in the base of a small furnace that stratigraphically pre-dates the main accumulation of debris in the slag heap. Contexts (258) and (259) are the residues of the last smelt or smelts that were carried out in the furnace; the metal-working activities in trench 1: two discrete contexts that relate to specific metallurgical activities in the main working area of the site. The first (146) represents a discrete dump of smelting debris from a single smelt of smelting campaign. It is analogous with the debris contexts in the slag heap but has been excavated in plan. The second (142) is another discrete dump of smelting debris from a short-lived smelting campaign. This context also contained important finds, including evidence of a failed smelt and a small metal bloom.

Objectives: to date the chronological span of smelting activity that lead to the formation of the deepest part of the slag heap and probably represents the core activity on the site; to

date the earliest identified technological activity on the site associated with the furnace base; to indicate the chronology of activity in the main metal-working area of the site and to provide a valuable comparison with the dated sequence from the adjacent slag heap. The dating of (142) will give a date for the rare find of a possible early medieval bloom.

Final comment: P Marshall (11 November 2013), the eleven measurements are statistically consistent (T'=11.5; *v*=10; T'(5%)=18.3; Ward and Wilson 1978) and thus the samples could all be of the same actual age. However, it is possible that if all the furnaces were used over a relatively short period of time they could produce such a group of results. The dates indicate that a short but intensive period of iron smelting took place on the site at some point in the second half of the third or first half of the fourth centuries cal AD. The results indicate that the initial radiocarbon determination from the site (Beta-132445) is not related to the primary phase of metallurgical activity on the site, and that it is Roman in date and not early medieval.

References: Ward and Wilson 1978

OxA–14507 1756 ±27 BP

δ¹³C: -23.2‰

Sample: 259 A 23, submitted on 22 February 2005 by G Juleff

Material: charcoal: *Quercus* sp., sapwood; single fragment (R Gale 2005)

Initial comment: this context formed the basal fill of the furnace, and is firmly sealed by context 258 and later contexts. A date from this sample would provide a *terminus post quem* for the first use of the furnace.

Objectives: not only would a date from this sample provide a date for the first use of the furnace, but due to the furnace's location at the base of the stratigraphic sequence in trench 2, it would provide a date for the first activity in this part of the site. This activity pre-dates the start of the accumulation of slag in this part of the site.

Calibrated date: *1σ:* cal AD 240–340
                      2σ: cal AD 220–380

Final comment: see series comments

OxA–14508 1770 ±28 BP

δ¹³C: -24.7‰

Sample: 233 308, submitted on 15 December 2004 by G Juleff

Material: charcoal: *Quercus* sp., roundwood, *c* 20 rings; single fragment (R Gale 2004)

Initial comment: the context containing the sample was situated at the base of the first phase of slag deposition, and was sealed up to 2m below ground level. The context contained large amounts of slag and furnace-lining, and is interpreted as a waste dump.

Objectives: the sample will provide the first of a sequence of dates through the first phase of slag deposition.

Calibrated date: 1σ: cal AD 230–330
2σ: cal AD 170–350

Final comment: see series comments

OxA–14509 1742 ±27 BP

δ¹³C: -23.3‰

Sample: 207 58 B, submitted on 15 December 2004 by G Juleff

Material: charcoal: *Quercus* sp., roundwood, *c* 20 rings; single fragment (R Gale 2004)

Initial comment: the sample is derived from a context located towards the top of the first phase of slag deposition. It is sealed at a depth of approximately 0.5m. The context is located in a well-stratified sequence.

Objectives: to provide one of a sequence of dates through a well-stratified slag deposit.

Calibrated date: 1σ: cal AD 240–340
2σ: cal AD 230–390

Final comment: see series comments

OxA–14510 1718 ±27 BP

δ¹³C: -25.4‰

Sample: 142 282, submitted on 15 December 2004 by G Juleff

Material: charcoal: *Quercus* sp., roundwood, *c* 13 rings; single fragment (R Gale 2004)

Initial comment: the sample was sealed within context 142 which comprised large fragments of slag and furnace lining with metallic debris. This deposit is interpreted as representing a single deposition event of material derived from a failed smelt, and has possibly been deliberately deposited with a linear morphology in order to form a rough revetment to the slag heap. The location of the sample amongst large fragments of technological debris reduces the chance of intrusion, while the interpretation of the context would suggest it is not residual.

Objectives: the context containing the sample was located near the top of the stratigraphic sequence. A date would provide a *terminus ante quem* for the sequence in trench 1.

Calibrated date: 1σ: cal AD 250–390
2σ: cal AD 240–400

Final comment: see series comments

OxA–14511 1744 ±27 BP

δ¹³C: -25.5‰

Sample: 217 201, submitted on 15 December 2004 by G Juleff

Material: charcoal: *Betula* sp., roundwood, *c* 9 rings; single fragment (R Gale 2004)

Initial comment: the context containing the sample is part of a well-stratified sequence within the first phase of slag deposition. The sample was sealed at a depth of *c* 1m below the surface. Context 217 consisted of a discrete dump of charcoal within the slag deposit.

Objectives: to provide one of a sequence of dates through the slag deposit.

Calibrated date: 1σ: cal AD 240–340
2σ: cal AD 230–390

Final comment: see series comments

Laboratory comment: English Heritage (28 November 2014), the two measurements on this fragment of charcoal are statistically consistent (T'=0.9; T'(5%)=3.8; ν=1; Ward and Wilson 1978).

References: Ward and Wilson 1978

OxA–14512 1781 ±27 BP

δ¹³C: -25.7‰

Sample: 217 201, submitted on 15 December 2004 by G Juleff

Material: charcoal: *Betula* sp., roundwood, *c* 9 rings; single fragment (R Gale 2004)

Initial comment: a replicate of OxA-14511.

Objectives: as OxA-14511

Calibrated date: 1σ: cal AD 220–330
2σ: cal AD 130–340

Final comment: see series comments

Laboratory comment: see OxA-14511

SUERC–5817 1680 ±35 BP

δ¹³C: -24.9‰

Sample: 142 199, submitted on 15 December 2004 by G Juleff

Material: charcoal: *Quercus* sp., roundwood, 23mm diameter, 4 growth rings; single fragment (R Gale 2004)

Initial comment: as OxA-14510

Objectives: as OxA-14510

Calibrated date: 1σ: cal AD 330–410
2σ: cal AD 250–430

Final comment: see series comments

SUERC–5818 1715 ±35 BP

δ¹³C: -23.9‰

Sample: 207 58 A, submitted on 15 December 2004 by G Juleff

Material: charcoal: *Quercus* sp., roundwood, 23 growth rings; single fragment (R Gale 2004)

Initial comment: as OxA-14509

Objectives: as OxA14509

Calibrated date: 1σ: cal AD 250–390
2σ: cal AD 230–410

Final comment: see series comments

SUERC–5819 1710 ±35 BP

δ¹³C: -25.0‰

Sample: 217 159, submitted on 15 December 2004 by
G Juleff

Material: charcoal: *Quercus* sp., roundwood, 21 growth rings;
single fragment (R Gale 2004)

Initial comment: as OxA-14511

Objectives: as OxA-14511

Calibrated date: 1σ: cal AD 250–390
2σ: cal AD 240–410

Final comment: see series comments

SUERC–5820 1685 ±35 BP

δ¹³C: -25.1‰

Sample: 233 307, submitted on 15 December 2004 by
G Juleff

Material: charcoal: *Betula* sp., roundwood, 16 growth rings;
single fragment (R Gale 2004)

Initial comment: as OxA-14508

Objectives: as OxA-14508

Calibrated date: 1σ: cal AD 330–410
2σ: cal AD 250–430

Final comment: see series comments

SUERC–5821 1710 ±35 BP

δ¹³C: -25.2‰

Sample: 259 B 23, submitted on 22 February 2005 by
G Juleff

Material: charcoal: *Quercus* sp., sapwood/roundwood; single
fragment (R Gale 2005)

Initial comment: as OxA-14507

Objectives: as OxA-14507

Calibrated date: 1σ: cal AD 250–390
2σ: cal AD 240–410

Final comment: see series comments

Exmoor Iron: Roman Lode, Devon

Location: SS 753381
Lat. 51.07.40 N; Long. 03.46.54 W

Project manager: G Juleff (University of Exeter), 2002

Archival body: University of Exeter

Description: Roman Lode is a multi-period iron mining site,
consisting of a 600m long open work on Burcombe near
Simonsbath. At its eastern end is an area of earthworks
consisting of sub-circular depressions (0.5m high maximum).
Originally these were thought to be the features left by hand
sorting of ore and therefore possibly representative of early
activity on the site. The excavation, consisting of a 4m by 3m

trench across one depression and its surrounding hummocks
was intended to investigate this theory, and any possible link
with smelting at Sherracombe. In the event a cut feature,
possibly a prospecting/mining pit was found to underlie the
depression although no dating evidence was found.

Objectives: to provide a date for the mining activity.

Final comment: P Marshall (11 November 2013), the two
dates from the hearth provide *termini post quos* for the later
mining activity at Roman Lode. It remains a matter of
interpretation whether the extraction of ore took place in
the Bronze Age.

References: Juleff and Bray 2007

OxA–13871 3526 ±35 BP

δ¹³C: -25.7‰

Sample: 122 A, submitted in June 2003 by G Juleff

Material: charcoal: *Quercus* sp., sapwood; single fragment
(R Gale 2004)

Initial comment: the sample comes from burnt material from
an *in situ* hearth. Charcoal was lying within the bowl of heated
material.

Objectives: to establish the period for activity excavated at Roman
Lode, where no other dating evidence has been found.

Calibrated date: 1σ: 1910–1770 cal BC
2σ: 1950–1740 cal BC

Final comment: P Marshall (11 November 2013), the date
provides a *terminus post quem* for the mining activity.

Laboratory comment: English Heritage (11 November 2013),
the two measurements (OxA-13871 and OxA-13890) are
statistically consistent (T'=0.2; T'(5%)=3.8; ν=1, Ward and
Wilson 1978) and thus both could be of the same age.

References: Ward and Wilson 1978

OxA–13890 3508 ±29 BP

δ¹³C: -25.5‰

Sample: 122 B, submitted in June 2003 by G Juleff

Material: charcoal: *Quercus* sp., sapwood; single fragment
(R Gale 2004)

Initial comment: as OxA-13871

Objectives: as OxA-13871

Calibrated date: 1σ: 1890–1770 cal BC
2σ: 1930–1740 cal BC

Final comment: see OxA-13871

Laboratory comment: see OxA-13871

Exmoor Iron: Roman Lode
(peat sequence), Devon

Location: SS 75343815
Lat. 51.07.41 N; Long. 03.46.52 W

Project manager: R Fyfe (University of Plymouth), 2004

Archival body: University of Exeter

Description: a peat section *c* 100m to the south-east of Roman Lode was sampled in the field using 0.5m monolith tins in August 2004. The section represents the deepest blanket peat adjacent to the iron extraction site at Roman Lode.

Objectives: to provide an environmental context for iron extraction activities at the site. More specifically, to provide a long-term record of atmospheric deposition of pollutants, and determine local impacts on the landscape, associated with iron extractive processes.

Final comment: P Marshall (11 November 2013), there is no evidence for any hiatus in the sequence from the pollen analysis (Fyfe 2009), which suggests continuous peat growth at the site since the Iron Age. The radiocarbon results at first glance suggest they provide accurate age estimates for the layers from which the samples are taken, however, the offset with the tephra age estimates implies this is not true.

References: Fyfe 2009
                  Matthews 2008

OxA–15750 1.065 ±0.003 fM

$\delta^{13}C$: -26.8‰

Sample: RLBP 2006:1, submitted on 21 March 2006 by D E Robinson

Material: peat (humin) (D E Robinson 2006)

Initial comment: the sample is of blanket peat. It was taken from 10–11cm below the modern surface of the peat, just below the base of root mat of the living surface vegetation. The vertical extent of the sample (10mm) almost certainly represents several decades of peat accumulation. The sample was taken from peat *c* 40cm above the mineral soil. The peat is highly humified and penetrated by abundant root material. Microscope examination suggested that, although the sample was taken from just below the root mat of the living surface vegetation, it only contained dead and decayed herbaceous material. no living roots were seen.

Objectives: to provide a date for the uppermost blanket peat at the site, immediately below the living vegetation, for comparison with a date for peat inception provided by sample RLBP 2006:2.

Calibrated date: *1σ:* cal AD 1956–2006
                       2σ: 1 cal AD 1956–2006

Final comment: P Marshall (11 November 2013), the date provides an age for the top of the blanket peat just below the living vegetation. This is compatible with the recovery of tephra from the eruption of Hekla in AD 1947 from 12cm (Matthews 2008).

Laboratory comment: English Heritage (11 November 2013), the two measurements (OxA-15750 and OxA-15825) are statistically consistent (T′=0.1; T′(5%)=3.8; *v*=1, Ward and Wilson 1978) and thus both could be of the same age. The weighted mean calibrates to cal AD 1956–2006 (Bomb13NH1; Hua *et al* 2013).

Laboratory comment: English Heritage (28 November 2014), this measurement has been calibrated using the post-nuclear bomb atmospheric calibration curve for zone 1 of the northern hemisphere (Bomb13NH1; Hua *et al* 2013).

Laboratory comment: Oxford Radiocarbon Accelerator Unit (29 June 2006), the acid and alkali insoluble residue of this sample was dated.

References: Hua *et al* 2013
                  Matthews 2008
                  Ward and Wilson 1978

OxA–15825 1.066 ±0.003 fM

$\delta^{13}C$: -27.7‰

Sample: RLBP 2006:1, submitted on 21 March 2006 by D E Robinson

Material: peat (humic acid) (D E Robinson 2006)

Initial comment: as OxA-15750

Objectives: as OxA-15750

Calibrated date: *1σ:* cal AD 1956–2006
                       2σ: cal AD 1956–2006

Final comment: see OxA-15750

Laboratory comment: English Heritage (28 November 2014), this measurement has been calibrated using the post-nuclear bomb atmospheric calibration curve for zone 1 of the northern hemisphere (Bomb13NH1; Hua *et al* 2013).

Laboratory comment: Oxford Radiocarbon Accelerator Unit (29 June 2006), the alkali soluble fraction of this sample was dated.

OxA–15826 1.019 ±0.003 fM

$\delta^{13}C$: -27.6‰

Sample: RLBP 2006:1, submitted on 21 March 2006 by D E Robinson

Material: peat (humin) (D E Robinson 2006)

Initial comment: as OxA-15750

Objectives: as OxA-15750

Calibrated date: *1σ:* cal AD 1966–1957
                       2σ: cal AD 1955–1958

Final comment: see OxA-15750

Laboratory comment: English Heritage (11 November 2013), this repeat measurement on the humin fraction at this level is significantly older than OxA-15750 and OxA-15825 (T′=160.2; T′(5%)=6.0; *v*=2; Ward and Wilson 1978).

Laboratory comment: English Heritage (28 November 2014), this measurement has been calibrated using the post-nuclear bomb atmospheric calibration curve for zone 1 of the northern hemisphere (Bomb13NH1; Hua *et al* 2013).

Laboratory comment: Oxford Radiocarbon Accelerator Unit (29 June 2006), the acid and alkali insoluble residue of this sample was dated.

References: Ward and Wilson 1978

OxA–15827 2184 ±29 BP

δ¹³C: -28.2‰

Sample: RLBP 2006:2, submitted on 21 March 2006 by
D E Robinson

Material: peat (humin) (D E Robinson 2006)

Initial comment: the sample is of blanket peat. It was taken
from 52–53cm below the modern surface of the peat, just
above what appears to be a gleyed mineral soil. The vertical
extent of the sample (10mm) almost certainly represents
several decades of peat accumulation. The sample was taken
from peat formed over a 2cm thick layer of apparently gleyed
mineral topsoil. The peat is very highly humified with very
little preserved macro plant material. Under the gleyed
topsoil is a clay with stones subsoil, the upper part of which
(*c* 8cm) is humus-stained.

Objectives: to provide a date for peat inception at the site,
being the first peat formed over the gleyed mineral soil.

Calibrated date: *1σ:* 360–190 cal BC
                     2σ: 370–160 cal BC

Final comment: P Marshall (11 November 2013), the
radiocarbon samples are from the same stratigraphic level
that contains the OMH-185 tephra that has been dated to
c 2705–2630 cal BC (van den Bogaard *et al* 2002; Plunkett
et al 2004; Larsen and Eiríksson 2008). Given the age
of the tephra the peat from 52–3cm is at least 310 years too
young. Anamoulously young radiocarbon ages can be
produced in slow-growing blanket peats (like those found at
Roman Lode) through rootlet penetration, or through the
mobilisation, downward movement, and subsequent
absorption of humic acids (Head *et al* 2007). This second
factor is not thought to be a dominant influence at Roman
Lode, as the dated humic and humin fractions produced
age determinations that are statistically consistent (*see*
laboratory comments). Without the OMH-185 tephra layer
it would be difficult to test for or identify this offset, and
the age-depth model for Roman Lode would be inaccurate,
possibly leading to spurious interpretations of
palaeoenvironmental data.

Laboratory comment: English Heritage (11 November 2013),
the two measurements (OxA-15827 and OxA-15865) are
statistically consistent (T'=2.1; T'(5%)=3.8; ν =1, Ward
and Wilson 1978) and thus both could be of the same age.

Laboratory comment: Oxford Radiocarbon Accelerator Unit
(29 June 2006), the acid and alkali insoluble residue of this
sample was dated.

References: Head *et al* 2007
                     Larsen and Eiríksson 2008
                     Plunkett *et al* 2004
                     van den Bogaard and Schmincke 2002
                     Ward and Wilson 1978

OxA–15865 2127 ±26 BP

δ¹³C: -26.7‰

Sample: RLBP 2006:2, submitted on 21 March 2006 by
D E Robinson

Material: peat (humic acid) (D E Robinson 2006)

Initial comment: as OxA-15827

Objectives: as OxA-15827

Calibrated date: *1σ:* 200–110 cal BC
                     2σ: 350–50 cal BC

Final comment: see OxA-15827

Laboratory comment: see OxA-15827

Laboratory comment: Oxford Radiocarbon Accelerator
Unit (29 June 2006), the alkali soluble fraction of this
sample was dated.

Exmoor Iron: Sherracombe Ford, Devon

Location: SS 720366
                    Lat. 56.06.48 N; Long. 03.49.42 W

Project manager: G Juleff (University of Exeter), 2002

Archival body: University of Exeter

Description: iron-production site of probable Romano-British
date on the western side of Exmoor. It consists of several
platforms on the sides of a steep-sided combe opening
out to the west-south-west. Downslope of the platforms lie
slagheaps; some of these are being eroded by a stream that
runs down to the base of the combe. The whole site is
under grass with a little light scrub (generally grazed).
A spring-mire runs through part of the site (away from the
excavated area), probably developing after the iron-
production activity. The site excavated in 2002 consisted
of a rectangular trench on one platform (which revealed
remains of furnaces and a smithing floor). A narrow trench
was also dug into the hillslope above the platform and
through the slagheap below it.

Objectives: to date the production of material from the
slagheap and furnace.

Final comment: P Marshall (11 November 2013), the
statistical consistency of the dated material from the
individual contexts indicates that each deposit does relate
to a single 'dump' of slag and other waste from a discrete
episode of metallurgical activity. Chronological modelling
of all the radiocarbon dates shows good agreement with
the stratigraphy and suggests that the start of activity took
place very soon after the Roman invasion of Britain and
ended in the late third century cal AD.

Laboratory comment: English Heritage (24 June 2014), 12
further samples were dated from this site after 2003 (OxA-
13773–81, -13884–5, and -13891).

OxA–13773 1876 ±29 BP

δ¹³C: -24.7‰

Sample: 505 A, submitted in June 2003 by G Juleff

Material: charcoal: *Betula* sp., single fragment (R Gale 2004)

Initial comment: sample taken from a layer within a shallow
pit dug into redeposited natural shillet that makes up the
platform. A discrete dump of waste material, unlikely to
have travelled far.

Objectives: to provide an approximate date for the activity excavated on the platform that provided no other dating evidence.

Calibrated date: *1σ:* cal AD 80–140
                       2σ: cal AD 60–230

Final comment: see series comments

Laboratory comment: English Heritage (11 November 2013), the two measurements from context 505 (OxA-13773 and OxA-13774) are statistically consistent (T'=1.0; T'(5%)=3.8; ν=1, Ward and Wilson 1978) and thus both could be of the same age.

References: Ward and Wilson 1978

OxA–13774 1916 ±28 BP

δ¹³C: -24.4‰

Sample: 505 B, submitted in June 2003 by G Juleff

Material: charcoal: *Corylus* sp., single fragment (R Gale 2004)

Initial comment: as OxA-13773

Objectives: as OxA-13773

Calibrated date: *1σ:* cal AD 60–130
                       2σ: cal AD 20–140

Final comment: see series comments

OxA–13775 1698 ±29 BP

δ¹³C: -27.0‰

Sample: 248 A, submitted in June 2003 by G Juleff

Material: charcoal: *Corylus* sp., single fragment (R Gale 2004)

Initial comment: charcoal embedded in the lining of a furnace - the context being the innermost and final phase of furnace setting 327.

Objectives: to establish the last period of use of the furnace setting 327. Derived from the latest phase of this feature - use of each phase not long-lived, therefore the sample will provide a close date.

Calibrated date: *1σ:* cal AD 260–400
                       2σ: cal AD 250–420

Final comment: see series comments

Laboratory comment: English Heritage (11 November 2013), the two measurements from context 248 (OxA-13775 and OxA-13776) are statistically consistent (T'=0.7; T'(5%)=3.8; ν=1, Ward and Wilson 1978) and thus both could be of the same age.

References: Ward and Wilson 1978

OxA–13776 1731 ±27 BP

δ¹³C: -27.9‰

Sample: 248 B, submitted in June 2003 by G Juleff

Material: charcoal: *Alnus* sp., single fragment (R Gale 2004)

Initial comment: as OxA-13775

Objectives: as OxA-13775

Calibrated date: *1σ:* cal AD 250–380
                       2σ: cal AD 230–390

Final comment: see series comments

OxA–13777 1785 ±30 BP

δ¹³C: -26.5‰

Sample: 238 A, submitted in June 2003 by G Juleff

Material: charcoal: *Betula* sp., roundwood 25mm diameter; single fragment (R Gale 2004)

Initial comment: context 238 is the fill of 239, a circular feature within a smithing floor, thought to be produced by a post for an anvil. The floor partly built-up around the post; the anvil must have been out of use for context 238 to go into the feature; it is sealed over by layers of the smithing floor, therefore this context must have been deposited while the smithing floor was in use.

Objectives: to establish the period of use of the smithing floor; stratigraphically earlier than sample 222, which is also related to the floor, therefore the span of the floor's use may be established.

Calibrated date: *1σ:* cal AD 220–320
                       2σ: cal AD 130–340

Final comment: see series comments

Laboratory comment: English Heritage (11 November 2013), the two measurements from context 238 (OxA-13777 and OxA-13891) are statistically consistent (T'=0.4; T'(5%)=3.8; ν=1, Ward and Wilson 1978) and thus both could be of the same age.

References: Ward and Wilson 1978

OxA–13778 1825 ±30 BP

δ¹³C: -23.8‰

Sample: 706, submitted in June 2003 by G Juleff

Material: charcoal: *Quercus* sp., sapwood; single fragment (R Gale 2003)

Initial comment: within context in clear stratigraphy of section through slag heap B. The section was created by erosion from the stream but was cleaned back to remove the possibility of contamination.

Objectives: to establish a sequence of dates through the section in slag heap B.

Calibrated date: *1σ:* cal AD 130–240
                       2σ: cal AD 90–320

Final comment: see series comments

OxA–13779 1825 ±28 BP

δ¹³C: -22.8‰

Sample: 710, submitted in June 2003 by G Juleff

Material: charcoal: *Quercus* sp., sapwood; single fragment (R Gale 2004)

Initial comment: as OxA-13778

Objectives: as OxA-13778

Calibrated date: *1σ:* cal AD 130–240
                        2σ: cal AD 90–320

Final comment: see series comments

OxA–13780 1893 ±30 BP

δ¹³C: -24.1‰

Sample: 701, submitted in June 2003 by G Juleff

Material: charcoal: *Corylus* sp., single fragment (R Gale 2004)

Initial comment: the sample was retrieved from a sealed context in a cleaned river-cut section directly above the natural at the base of a well-stratified sequence, approximately 1.5m below ground level.

Objectives: dating of this sample would provide a *terminus post quem* for the start of slag accumulation in this part of slag heap B.

Calibrated date: *1σ:* cal AD 70–140
                        2σ: cal AD 50–220

Final comment: see series comments

OxA–13781 1818 ±28 BP

δ¹³C: -25.3‰

Sample: 222 A, submitted in June 2003 by G Juleff

Material: charcoal: *Corylus* sp., single fragment (R Gale 2004)

Initial comment: sample from the fill of hole 223. This hole is thought to be for a post upon which an anvil sat. A smithing floor has built up around the post (the floor being to hard to drive a post through) through use of the anvil. The fill that has been sampled must have gone into the hole after the anvil - and presumably the smithing floor - had gone out of use.

Objectives: to establish an end date for the use of the smithing floor; coming later in a sequence with context 238, a similar deposit entirely sealed within the smithing floor.

Calibrated date: *1σ:* cal AD 130–250
                        2σ: cal AD 120–320

Final comment: see series comments

Laboratory comment: English Heritage (11 November 2013), the two measurements from context 222 (OxA-13781 and OxA-13885) are statistically consistent (T′=1.3; T′(5%)=3.8; *ν*=1, Ward and Wilson 1978) and thus both could be of the same age.

References: Ward and Wilson 1978

OxA–13884 1851 ±26 BP

δ¹³C: -25.8‰

Sample: 343 A, submitted in June 2003 by G Juleff

Material: charcoal: *Quercus* sp., roundwood; single fragment (R Gale 2004)

Initial comment: the context was sealed ina well-stratified sequence below the smithing floor, approximately 0.6m below ground level. The deposit was charcoal rich.

Objectives: given the stratigraphic location of this context, dating the sample will provide a *terminus post quem* for the start of the build-up of the smithing floor.

Calibrated date: *1σ:* cal AD 120–230
                        2σ: cal AD 80–240

Final comment: see series comments

OxA–13885 1863 ±28 BP

δ¹³C: -26.6‰

Sample: 222 B, submitted in June 2003 by G Juleff

Material: charcoal: *Alnus* sp., single fragment (R Gale 2004)

Initial comment: from the fill of hole 223. This hole is thought to be for a post upon which an anvil has sat. A smithing floor has built up around the post (the floor being too hard to drive a post through), through use of the anvil. The fill that has been sampled must have gone into the hole after the anvil, and presumably the smithing floor, had gone out of use.

Objectives: to establish an end date for the use of the smithing floor; coming later in the sequence with 238, a similar deposit entirely sealed within the smithing floor.

Calibrated date: *1σ:* cal AD 80–220
                        2σ: cal AD 70–240

Final comment: see series comments

OxA–13891 1810 ±26 BP

δ¹³C: -25.3‰

Sample: 238 B, submitted in June 2003 by G Juleff

Material: charcoal: Pomoideae, single fragment (R Gale 2004)

Initial comment: as OxA-13777

Objectives: as OxA-13777

Calibrated date: *1σ:* cal AD 130–250
                        2σ: cal AD 120–320

Final comment: see series comments

Fiskerton: auger survey, Lincolnshire

Location:	TF 052716
	Lat. 53.13.50 N; Long. 00.25.27 W
Project manager:	J Rackham (The Environmental Archaeology Consultancy), 2003
Archival body:	Lincolnshire County Museum

Description: as part of the programme of survey work at Fiskerton in advance of the proposal for a countryside stewardship scheme, an auger survey of the post-glacial deposits was undertaken. This programme was designed to build on the auger transect carried out as part of the excavations in 2001 that investigated the Iron Age timber causeway at Fiskerton. The transect ran the length of the river bank. These showed major differences in the sub-surface topography ranging from deep channels up to 4m in depth, to sand banks that rise up above the surrounding modern ground surface. Furthermore, it was possible to identify in the dyke side a palaeochannel of much more recent date which appears to lie on the course of the present parish boundary. The channel is approximately 20m wide and may represent a medieval or earlier watercourse. The current programme of auguring was designed to complement this information on the northern side of the Witham valley.

Objectives: to develop a chronology for the sediments within the valley that can be attached to the deposit model, and date the periods when the valley first became waterlogged, when it was tidal, and when the low lying floodplain across which the causeway was constructed first became flooded/marshy. It is also hoped that a date for the uppermost surviving strata can be obtained to compare with the dendrochronological dates for the causeway.

Final comment: J Rackham (July 2004), the auger survey was conducted across an area of approximately 32ha along the northern floodplain of the River Witham. The survey identified the ancient topography of the site, its broad environmental history, those areas of potential archaeological importance and the problems of preservation. A sequence of dated deposits were retrieved ranging from the early Mesolithic, through to the Bronze and Iron Ages, and into the early medieval period. A sealed and preserved Mesolithic landscape may lie beneath later alluvial sediments, with peat and silt samples dated to *c* 9500 BP.

References: Jones 2003
                  Martin 2002
                  Rackham 2004
                  Rylatt *et al* 2011

OxA–12868 2669 ±30 BP

δ¹³C: -26.6‰

Sample: NS3/A1/40–45, submitted on 27 August 2003 by J Rackham

Material: wood (waterlogged): *Corylus/Alnus* sp., roundwood; diameter 5mm (5g) (R Gale 2003)

Initial comment: a sample taken from peat deposits in core NS3/A1, which are extremely humified with considerable recent and ancient root penetration. The layer lies below the plough depth and is seasonally dessicated.

Objectives: this core is being used to establish the broad chronology of the sediments in the infilled channel of the Old Witham. This sample represents the highest secure deposit in this core that can be safely dated without considerable problems of contamination, and is believed to be broadly contemporary with the Iron Age timber trackway at Fiskerton.

Calibrated date: *1σ:* 840–800 cal BC
                       2σ: 900–790 cal BC

Final comment: J Rackham (27 October 2014), this borehole lies some 50m east of the Iron Age timber trackway. The late Bronze Age date indicates that surviving well-preserved

and sealed peats contemporary with the Iron Age timber trackway (constructed 456 BC) probably no longer survive in the immediate neighbourhood of the trackway, but have been lost to the ploughsoil as a result of dessication and shrinkage of the peat (*see also* Rylatt *et al* 2011).

OxA–12869 2945 ±32 BP

δ¹³C: -27.7‰

Sample: NS3/A1/120–125, submitted on 27 August 2003 by J Rackham

Material: waterlogged plant macrofossil: monocot, herbaceous stem (R Gale 2003)

Initial comment: this sample derives from slightly humified waterlogged fibrous peats in core NS3/A1. There is some ancient root disturbance, but minimal modern rootlet intrusion.

Objectives: this core is being used to establish the broad chronology of the sediments in the infilled channel of the Old Witham. This sample represents a period of transition from reedy open water at the site into full thick reed beds.

Calibrated date: *1σ:* 1220–1110 cal BC
                       2σ: 1260–1040 cal BC

Final comment: J Rackham (27 October 2014), this sample was obtained from deposits 20.5cm below OxA-12868 at 0m OD and has yielded a date at the end of the second millennium cal BC, a few hundred years earlier than the sample above. It appears to mark a period when this location moves from being on the margins of the open waterway to a thick reed bed. This might reflect movement of the channel to the south and/or a falling sea level and a change from tidal to freshwater conditions.

OxA–12870 4018 ±30 BP

δ¹³C: -27.4‰

Sample: NS3/A1/262, submitted on 27 August 2003 by J Rackham

Material: wood (waterlogged): *Alnus glutinosa*, roundwood; diameter 4mm (4g) (R Gale 2003)

Initial comment: this sample derives from a well-preserved reed peat within core NS3/A1, with some ancient root disturbance, but no modern intrusion.

Objectives: this core is being used to establish the broad chronology of the sediments in the infilled channel of the Old Witham. This sample represents a period when the location changed from a reed bed into open water.

Calibrated date: *1σ:* 2580–2480 cal BC
                       2σ: 2620–2460 cal BC

Final comment: J Rackham (27 October 2014), this sample recovered from 1.395m below OxA-12869 marks the end of a period of reed beds at this location before the site became open water. The change is believed to mark a rising sea level and the expansion of tidal waters over earlier freshwater reed beds and alder carr in the later Neolithic.

OxA–12892 4955 ±50 BP

$\delta^{13}C$: -29.0‰

Sample: NS3/A1/367–372, submitted on 27 August 2003 by J Rackham

Material: wood (waterlogged): *Alnus glutinosa* (R Gale 2003)

Initial comment: a sample from core NS3/A1, from well-preserved organic waterlain silts with some ancient root disturbance, but no modern intrusion. It is permanently waterlogged at the time of sampling, and oxidises on exposure to the air.

Objectives: this core is being used to establish the broad chronology of the sediments in the infilled channel of the Old Witham. This sample represents the date of the transition on the site from reedy open water to a reedy alder carr environment - a lower woodland phase.

Calibrated date: *1σ:* 3790–3650 cal BC
                  2σ: 3940–3640 cal BC

Final comment: J Rackham (27 October 2014), a further 1.08m below OxA-12870 at -2.475m OD a sample from grey silts laid down in open watermark a period when this location lay in the main river channel or adjacent mere. It is overlain by organic silts possibly reflecting river margin sediments and indicates that the area formerly dry has been inundated by rising water levels in the early Neolithic as a result of sea-level rises downstream in the North Sea and the development of extensive marsh and peats deposits downstream.

OxA–12914 5102 ±37 BP

$\delta^{13}C$: -27.6‰

Sample: NS3/A1/400, submitted on 27 August 2003 by J Rackham

Material: wood (waterlogged): *Alnus glutinosa*, roundwood; diameter 5mm (R Gale 2003)

Initial comment: a sample from core NS3/A1, taken from a dark grey sandy silt with some ancient root disturbance. The deposit may be part of a 'buried soil', and was permanently waterlogged at the time of sampling.

Objectives: this core is being used to establish the broad chronology of the sediments in the infilled channel of the Old Witham. This sample represents the base of the sequence in this core, and should give an approximate date for the onset of waterlogged conditions at this point.

Calibrated date: *1σ:* 3970–3800 cal BC
                  2σ: 3980–3790 cal BC

Final comment: J Rackham (27 October 2014), the basal date in this borehole at -2.775m OD, the sample derives from an organic silt overlying a possible palaeosol and underlying sands at -3m OD. It marks the initial waterlogging of this area on the south side of the main river channel in the late Mesolithic and early Neolithic as a result of rising water levels in the valley, and rising sea levels in the North Sea.

OxA–12915 4058 ±38 BP

$\delta^{13}C$: -24.9‰

Sample: NS3/A2/334, submitted on 27 August 2003 by J Rackham

Material: wood (waterlogged): bark, unidentified from large wood (R Gale 2003)

Initial comment: this sample was taken from core NS3/A2, from the top of clean laminate tidal sediments. Specifically it was taken from a laminate grey sandy silt which represents an intertidal episode. There was some minor ancient root penetration, but no modern intrusions.

Objectives: some of the cores show an intertidal sequence of finely laminated silts and fine sands. This core has this sequence and samples have been taken from the top and bottom of the deposits in order to date this marine phase.

Calibrated date: *1σ:* 2830–2490 cal BC
                  2σ: 2850–2470 cal BC

Final comment: J Rackham (27 October 2014), this borehole lies in the main Neolithic inter-tidal channel, filled with laminated sediments and saltmarsh silts. The sample derives from grey clayey silts immediately above the upper laminated silts and suggests a change from intertidal channel and mudflat environments to saltmarsh conditions in the middle to later Neolithic.

OxA–12916 4124 ±31 BP

$\delta^{13}C$: -26.1‰

Sample: NS3/A2/425, submitted on 27 August 2003 by J Rackham

Material: wood (waterlogged): *Alnus glutinosa*, fragment of large wood (15g) (R Gale 2003)

Initial comment: this sample was taken from core NS3/A2, from the base of laminate tidal sediments. Specifically it was taken from the base of a laminate grey sandy silt which represents an intertidal episode.

Objectives: as OxA-12915

Calibrated date: *1σ:* 2860–2620 cal BC
                  2σ: 2880–2570 cal BC

Final comment: J Rackham (27 October 2014), 0.91m below OxA-12915 at -3.082m OD, and lying in the base of a cut 'tidal' channel beneath laminated intertidal sediments this alder trunk or branch dates a period of tidal creek down-cutting in the middle to later Neolithic. It is possible that the trunk reflects a left over from earlier carr woodland on the site or timber washed in by the new tidal channel. The relatively short period between the two dates in this borehole and a much earlier date for alder roundwood in the borehole 50m to the south (OxA-12914) at a slightly higher altitude (-2.775m OD) suggests it was more likely to have been washed in. The two dates from this borehole (OxA-12915 and 12916) suggest a rapid build-up of tidal sediments at this location in the mid-late Neolithic.

OxA–12917 9450 ±45 BP

$\delta^{13}C$: -28.1‰

Sample: NS5/A4/653–658, submitted on 27 August 2003 by J Rackham

Material: wood (waterlogged): bark, very small and degraded (R Gale 2003)

Initial comment: a sample from core AS5/A4, from waterlain grey brown sands with organics and shells, laid down in freshwater conditions over liassic clays at the base of the sequence.

Objectives: this series is designed to place the sediments in a chronology. The borehole represents the deepest sequence found on site, lying within approximately the centre of the ancient channel. These lowermost sediments represent, potentially, some of the earliest on site. They are deposited in a freshwater environment below peats, which are below the intertidal sediments noted in core NS3/A3. They lie immediately over liassic clays and reflect a channel where the river has scoured to the bedrock and removed all the glacial sands and gravels.

Calibrated date: *1σ:* 8800–8650 cal BC
                     2σ: 8840–8620 cal BC

Final comment: J Rackham (27 October 2014), this date effectively proves that the early Holocene course of the River Witham in Fiskerton lay along the northern boundary of the floodplain, where it was cutting into the valley side. The early Mesolithic date indicates that the palaeochannels of the River Witham probably hold an environmental history spanning from the early Mesolithic to the medieval period, with the early Holocene river down cutting to over -4.5m OD in the Fiskerton area.

OxA–12918 3409 ±34 BP

δ¹³C: -27.5‰

Sample: NS17/A3/134–139, submitted on 27 August 2003 by J Rackham

Material: wood (waterlogged): *Alnus glutinosa*, narrow roundwood; diameter 3mm (R Gale 2003)

Initial comment: a sample of twig and wood fragments extracted from a sandy palaeosol at the base of the post-glacial sediment sequence on the eastern floodplain. This sediment appears to be the old gravel surface prior to the area becoming inundated and waterlogged.

Objectives: as part of the assessment of the chronology of the sediments infilling the valley, two locations on the low lying floodplain, east and west of the causeway, have been selected to date the period when they first became inundated/marshy, and the land ceased to be available for dry pasture or arable. This sample is designed to give a date to the initial waterlogging of the eastern floodplain.

Calibrated date: *1σ:* 1750–1660 cal BC
                     2σ: 1870–1620 cal BC

Final comment: J Rackham (27 October 2014), the sample was intended to indicate when the floodplain of the River Witham, beyond the levees, became waterlogged. The early to middle Bronze Age date indicates the beginning of the waterlogging and build-up of peats on the floodplain, and also marks the end of the agricultural use of this landscape and the construction of burial mounds on it, although the latter may have been possible for a few hundred years on the higher sand levees boardering the late glacial river channel. This is appreciably earlier (4–500 years) than the date of peats forming over the slightly higher ground upstream (OxA-12919).

OxA–12919 3616 ±34 BP

δ¹³C: -27.6‰

Sample: WT20/A1/190–200, submitted on 27 August 2003 by J Rackham

Material: wood (waterlogged): *Alnus glutinosa*, fragment of large wood (R Gale 2003)

Initial comment: the sampled deposit forms the base of a channel sequence of organic silts that overlie a reed peat, over an alder carr or woody peat deposit. The sequence continues to a depth of 3.85m in core WT20/A1.

Objectives: this core is the only one at the western end of the site to pick up the edge of the old river channel as it turns north again. The wood sample is primarily designed to establish that this is a channel contemporary with that at the eastern end of the site, so that we can confidently plot the course of the old river without the complication of channels being of different ages.

Calibrated date: *1σ:* 2030–1920 cal BC
                     2σ: 2120–1880 cal BC

Final comment: J Rackham (27 October 2014), the sample was taken from deposits at -0.498m OD, a level lying between two of the samples in NS3/A1 (OxA-12892 and OxA-12914) which yielded late Neolithic and late Bronze Age dates. The early Bronze Age date for this peat below organic silts is consistent with these results and dates a period of peat and woody peat formation on the edge of the river channel between two periods of organic silt deposition reflecting a river marginal habitat. It identifies this channel as contemporary with the channel recorded downstream in the causeway field, although reflecting a more marginal habitat.

OxA–12920 3057 ±32 BP

δ¹³C: -28.1‰

Sample: WT15/A2/100–115, submitted on 24 September 2003 by J Rackham

Material: wood (waterlogged): *Alnus glutinosa*, roundwood (R Gale 2003)

Initial comment: the wood sample was taken from grey sands overlying buff/yellow natural sands at the base of the sequence of alluvial silts and peats, and below an horizon interpreted as a soil in core WT15/A2.

Objectives: this area of the site represents a plateau on the floodplain of the ancient river, and at some time in the past became flooded and permanently boggy with alder carr and reed beds. This sample is designed to date the time at which this flooding first became permanent and able to preserve wood and other organics in a waterlogged environment, making the floodplain unsuitable for occupation.

Calibrated date: *1σ:* 1400–1260 cal BC
                     2σ: 1420–1210 cal BC

Final comment: J Rackham (27 October 2014), the wood was identified as alder roundwood and therefore was suitable for dating. The result indicates that at this point on the river floodplain the area became waterlogged in the middle to late Bronze Age and was no longer tenable for occupation or agriculture. The sample was taken from deposits at a level of

0.83m OD some 0.326m higher than the sample in a similar stratigraphic position in auger transect NS17/A3 (OxA-12918) downstream. The difference of some 400 years reflects the timescale for the slightly higher floodplain a kilometre upstream of the Iron Age causeway to be impacted upon by the rising water levels in the valley. It also means that the sand levees formed by the glacial river must have remained as islands in the wetland for over half a millennium and may have continued to be important as either ritual locations (barrows are sited on some of the levees) or even occupation sites.

OxA–13080 1572 ±27 BP

δ¹³C: -27.0‰

Sample: NS5/A4/24–30, submitted on 27 August 2003 by J Rackham

Material: peat (humin)

Initial comment: the sample derived from a very dark brown well humified (woody) peat in core NS5/A4. It is the most recent deposit in the surviving organic sequence, and there was recent root penetration.

Objectives: this series is designed to place the sediments in a chronology. This core represents the deepest sequence found on the site, lying within approximately the centre of the ancient channel of the river. The upper sediments represented by the humified peats on top of the woody peat therefore represent, potentially, the most recent sediment in the whole sequence within this part of the valley.

Calibrated date: *1σ:* cal AD 420–550
                      2σ: cal AD 410–560

Final comment: J Rackham (27 October 2014), located near the northern valley side and overlying the deepest channel deposits one would expect this deposit to reflect perhaps the most recent surviving peats on the site. With a result placing the date in the immediate post-Roman period this has proved true. At a height of 1.731m OD it lies well above the surviving remains of the Iron Age causeway and appears to show that peats were forming across the floodplain in the immediate post-Roman period. Sea-level rises are known to be responsible for the inundation of late Roman agricultural sites on the Fens in the late or immediately post-Roman period and this formation of peat at Fiskerton in the post-Roman period may mark a renewed rise in groundwater after a period of lower water levels during the Roman period, when studies suggest the floodplain peats may have been utilised for grazing.

GR99 dating experiment, Hampshire

Location: SU 277418
                  Lat. 51.10.26 N; Long. 01.36.14 W

Project manager: G Campbell (English Heritage), 1999

Archival body: English Heritage

Description: the individual *Triticum spelta* L. grains dated come from a large destruction deposit consisting entirely of a low mound of charred grain 50–100cm thick lying on the floor of a third to fourth century AD corn-drying oven (oven 4) at Grateley South, Hampshire. This deposit was sealed by the roof and superstructure of the oven that had collapsed onto it as a result the building burning down whilst in use (Cunliffe and Poole 2008, 71–7; Campbell 2008). For each chemical, 10 grains of *Triticum spelta* were combined with 800ml of clay soil and soaked in 1 litre of water for three days, to which 2.5ml of the respective chemical had been added. The sample was then floated using a simple wash-over technique and the individual grains extracted from the resulting flots. In addition to the grain samples, the chemicals used in the experiment were also subject to radiocarbon measurement.

Objectives: to examine the effects that the various chemicals, which are used as a way of breaking up clay-rich samples to make processing to recover charred plant remains and other biological remains easier and faster, have on the resulting radiocarbon determinations. In particular, whether the use of chemicals containing carbonates contaminates the charred plant material with 'old' (geological carbon) and meaning that material returns dates older than their true age.

Final comment: G Campbell (2 October 2014), the results indicate that the use of various chemicals to break down clay-rich samples by soaking them in water to which a small amount of the chemical is added prior to floating for the recovery of charred macroscopic plant remains (grain in particular) does not significantly affect the date obtained from the grain. In particular, the use of chemicals containing carbon does not lead to the infiltration of 'old' carbon into the grain. The possibility that hydrogen peroxide, due to its oxidising properties (Pearsall 1989: Zhao and Pearsall 1998), may damage any surviving organic material within the grain requires further work.

Laboratory comment: English Heritage (15 September 2006), three samples (sodium hexametaphosphate, tetra sodium pyrophosphate decahydrate, and hydrogen peroxide) failed to produce a sample for dating prior to any combustion because they had an insufficient carbon content. As these were samples of chemicals which should not contain carbon this is unsurprising. All of the 'contaminated' results are statistically consistent with the control result. This would suggest that none of the chemicals has significantly affected the resulting measurement. A comparison of samples 3–7 shows that the results from Calgon and hydrogen peroxide (samples 3 and 7 respectively) are significantly inconsistent. These two samples represent the youngest and oldest radiocarbon age determinations, and while this may be the result of minor contamination, it is also possible to that it is the result of statistical scatter associated with the measurement process. If it is the result of minor contamination, this would likely only be an issue that would need addressing in situations where different chemicals are being used at different times during the floating process. More work would be required to reliably answer this question.

References: Campbell 2008
                 Cunliffe and Poole 2008
                 Pearsall 1989
                 Zhao and Pearsall 1998

OxA–15714 24440 ±140 BP

δ¹³C: -28.3‰

Sample: chemical, sample 3, submitted on 17 March 2006 by G Campbell

Material: organic matter (Calgon (EEC regulations: 5–15%, polycarboxylates 30% and more, Zeolites))

Initial comment: a sample of Calgon from the supermarket.

Objectives: to establish a date for Calgon for comparison with the control grain and a grain 'contaminated' with Calgon.

Final comment: G Campbell (2 October 2014), this confirms that commercial sources of Calgon contain carbon derived from geological sources and is as expected.

OxA–15715 27560 ±180 BP

δ¹³C: -7.6‰

Sample: chemical, sample 4, submitted on 17 March 2006 by G Campbell

Material: organic matter (sodium bicarbonate)

Initial comment: a sample from a bulk supplier.

Objectives: to establish a date for sodium bicarbonate for comparison with the control grain and a grain 'contaminated' with sodium bicarbonate.

Final comment: G Campbell (2 October 2014), this confirms that commercial sources of sodium bicarbonate contain carbon derived from geological sources and is as expected.

OxA–15828 1728 ±28 BP

δ¹³C: -22.6‰

Sample: 3227, sample 2, submitted on 17 March 2006 by G Campbell

Material: grain: *Triticum spelta* L., single carbonised grain (G Campbell 2006)

Initial comment: a sample from a destruction deposit consisting entirely of a low mound of charred grain 50–100mm thick lying on the floor of a third- to fourth-century corn-drier, and sealed by the roof of the corn drier that had collapsed onto it.

Objectives: to establish a date for the grain and a control date for the dating experiment.

Calibrated date: *1σ:* cal AD 250–380
                       2σ: cal AD 230–400

Final comment: G Campbell (2 October 2014), this date confirms the third- to fourth-century AD date for corn-drying oven 4 at Grateley South (Cunliffe and Poole 2008, 71) and provides a date for the uncontaminated grain.

Laboratory comment: English Heritage (15 September 2006), this control sample is statistically consistent with samples 3, 4, 5, 6, and 7 (T'=0.3, 0.0, 0.6, 0.7, and 2.9 respectively, where T'(5%)=3.8; *ν*=1; Ward and Wilson 1978).

References: Ward and Wilson 1978

OxA–15923 1708 ±25 BP

δ¹³C: -22.7‰

Sample: 3227, sample 3, submitted on 17 March 2006 by G Campbell

Material: grain: *Triticum spelta* L., single carbonised grain (G Campbell 2006)

Initial comment: as OxA-15828

Objectives: to establish a date for the grain and whether the addition of Calgon affects the final date of the sample as part of the dating experiment.

Calibrated date: *1σ:* cal AD 260–390
                       2σ: cal AD 250–400

Final comment: G Campbell (2 October 2014), the date obtained is consistent with the control sample which would indicate that soaking samples in water to which Calgon (which contains various carboxylates) has been added in order to break down the clay in the sample before processing, does not contaminate charred grain with 'old' carbon.

Laboratory comment: English Heritage (15 September 2006), this sample is statistically consistent with samples 4, 5, and 6 (T'=0.3, 1.9, and 2.0 respectively, where T'(5%)=3.8; *ν*=1; Ward and Wilson 1978). It is statistically inconsistent with sample 7 (T'=5.6; T'(5%)=3.8; *ν*=1; Ward and Wilson 1978).

References: Ward and Wilson 1978

OxA–15924 1728 ±26 BP

δ¹³C: -21.6‰

Sample: 3227, sample 4, submitted on 17 March 2006 by G Campbell

Material: grain: *Triticum spelta* L., single carbonised grain (G Campbell 2006)

Initial comment: as OxA-15828

Objectives: to establish a date for the grain and whether the addition of sodium bicarbonate affects the final date of the sample as part of the dating experiment.

Calibrated date: *1σ:* cal AD 250–380
                       2σ: cal AD 240–400

Final comment: G Campbell (2 October 2014), the date obtained is consistent with the control sample which would indicate that soaking samples in water to which sodium bicarbonate has been added in order to break down the clay in the sample before processing does not contaminate charred grain with 'old' carbon.

Laboratory comment: English Heritage (15 September 2006), this sample is statistically consistent with samples 5, 6, and 7 (T'=0.7, 0.7, and 3.1 respectively, where T'(5%)=3.8; *ν*=1; Ward and Wilson 1978).

References: Ward and Wilson 1978

OxA–15925 1758 ±26 BP

δ¹³C: -20.9‰

Sample: 3227, sample 5, submitted on 17 March 2006 by G Campbell

Material: grain: *Triticum spelta* L., single carbonised grain (G Campbell 2006)

Initial comment: as OxA-15828

Objectives: to establish a date for the grain and whether the addition of sodium hexametaphosphate affects the final date of the sample as part of the dating experiment.

Calibrated date: 1σ: cal AD 240–340
2σ: cal AD 220–380

Final comment: G Campbell (2 October 2014), the date obtained is consistent with the control sample which would indicate that soaking samples in water to which sodium hexametaphosphate has been added in order to break down the clay in the sample before processing does not affect the date of the charred grain. As this chemical does not contain carbon, this is unsurprising. However, there is some concern that due to its high pH this compound may have a detrimental effect on organic materials so should not be used (cf. Zhao and Pearsall 1998).

Laboratory comment: English Heritage (15 September 2006), this sample is statistically consistent with samples 6 and 7 (T'=0.0 and 0.9 respectively, where T'(5%)=3.8; *v*=1; Ward and Wilson 1978).

References: Ward and Wilson 1978

OxA–15926 1759 ±26 BP

δ¹³C: -21.2‰

Sample: 3227, sample 6, submitted on 17 March 2006 by G Campbell

Material: grain: *Triticum spelta* L., single carbonised grain (G Campbell 2006)

Initial comment: as OxA-15828

Objectives: to establish a date for the grain and whether the addition of tetrasodium pyrophosphate decahydrate affects the final date of the sample as part of the dating experiment.

Calibrated date: 1σ: cal AD 240–340
2σ: cal AD 220–380

Final comment: G Campbell (2 October 2014), the date obtained is consistent with the control sample which would indicate that soaking samples in water to which tetrasodium pyrophosphate decahydrate has been added in order to break down the clay in the sample before processing does not affect the date of the charred grain.

Laboratory comment: English Heritage (15 September 2006), this sample is statistically consistent with sample 7 (T'=0.9;T'(5%)=3.8; *v*=1; Ward and Wilson 1978).

References: Ward and Wilson 1978

OxA–15927 1793 ±26 BP

δ¹³C: -21.2‰

Sample: 3227, sample 7, submitted on 17 March 2006 by G Campbell

Material: grain: *Triticum spelta* L., single carbonised grain (G Campbell 2006)

Initial comment: as OxA-15828

Objectives: to establish a date for the grain and whether the addition of hydrogen peroxide affects the final date of the sample as part of the dating experiment.

Calibrated date: 1σ: cal AD 210–320
2σ: cal AD 130–330

Final comment: G Campbell (2 October 2014), the date obtained is consistent with the control sample which would indicate that soaking samples in water to which hydrogen peroxide has been added in order to break down the clay in the sample before processing does not affect the date of the charred grain. However given that the date obtained is slighter older than would be expected and inconsistent with sample 3 (*see* above) it is possible that the addition of hydrogen peroxide, an oxidation agent (Pearsall 1989, 85–6) has damaged elements of any original tissue that may have survived within the grain, or destroyed any more recent fungal material that may have been present within the grain.

Laboratory comment: see results above

Groundwell Ridge, Wiltshire

Location: SU 14088935
Lat. 51.36.07 N; Long. 01.47.48 W

Project manager: P Wilson (English Heritage), July 2003

Archival body: Swindon Museum and Art Gallery

Description: Groundwell Ridge is Roman-period villa complex with clearly defined religious elements, most notably a *nymphaeum* that was discovered during development in 1996. In 1996–7, limited salvage recording, and evaluation (trenches A-E), coupled with extensive geophysics and topographic survey, demonstrated the existence of a complex site that includes high-status buildings and significant quantities of high-quality material culture. Further excavations in 2003–5, explored one of the domestic buildings of the complex (building 2) with much of the work focussed on part of a bath suite. The excavations undertaken in 2003–5 focused on seven trenches assessing the different features located in different areas of the site. The site is located on a sloping hillside, upon which there are a series of terraces running east-west. The topography and layout of the site is influenced by the position of springs which previously fed the *nymphaeum* and also provided water to the bath suite and villa buildings. The main building complex was shown to be located at the bottom of the slope.

Objectives: the objectives of dating the ploughsoil were to establish its age relative to an adjacent spring deposit which contains Roman period material and to give an increased understanding of how the hillside was exploited in the past. The main dating programme focused on the Roman bath suite sequence and the succeeding post-built structure.

Final comment: P Wilson (10 March 2014), many dates from the series are consistent with the excavated evidence, but there are a number that seem surprisingly early, including: those from the phase 3 timber building (OxA-18645–8 and SUERC-18022–4, and -18028) which is believed to be very late or post-Roman, but could be explained by earlier Roman-period material being incorporated into the cut features. Less readily explained are the apparently early dates from articulated bone from the backfill of room 1 (OxA-18638 and SUERC-18020) and from the demolition/collapse material in room 4 ((SUERC-18221).

Laboratory comment: (2011), a total of 27 radiocarbon measurements were obtained from 25 samples, consisting of 20 fragments of carbonised plant remains and five articulated animal bones, in addition to the two samples from the ploughsoil.

Chronological modelling of the radiocarbon results validated the phasing derived from artefactual and stratigraphic evidence, which suggests that this chronology can be extended to areas of the site, which have been phased but did not produce taphonomically secure radiocarbon samples. It indicates that building 2 was occupied throughout most, if not all of the Roman period, and part was used as a bath suite over much of this time. The earlier eastern furnace was used in the second century cal AD, but was then replaced by the southern furnace, most probably in *cal AD 210–260 (68% probability)*. The second furnace was also used for about a century, but the building probably ceased to function as a bath suite by the middle of the fourth century cal AD.

Laboratory comment: English Heritage (17 November 2014), 20 further samples were dated after 2006 (Bayliss *et al* forthcoming).

References: Hamilton 2006
                   Linford 2010
                   Meadows *et al* 2012

OxA–13476 1848 ±27 BP

δ¹³C: -24.4‰

Sample: Trench 3, 4020, submitted on 8 December 2003 by G Ayala

Material: charcoal: *Corylus avellana*, single fragment (G Campbell 2003)

Initial comment: the samples come from a buried ploughsoil located in trench 3 of the excavation. Trench 3 was located across a spring line that had earlier been the site of a Thames Water trench. Within the spring there was a waterlogged deposit with associated Roman material. This archaeological spring deposit appeared to cut into what is being considered, in the preliminary analysis, a buried ploughsoil. This layer appeared in the field as a sealed A horizon with a very homogenous distribution of charcoal throughout. The charcoal samples were taken from this ploughsoil which in turn buried under *c* 25–30cm of modern hillwash, in order to increase out understanding of its relationship to the spring deposit.

Objectives: the dating of the samples will be fundamental in understanding the development and cultural significance of both the spring and the surrounding hillslopes. The

possibility to date the ploughsoil would allow us to identify the period of the introduction of charcoal into the soil reflecting cultivation practices. This would clarify ideas as to how the hillslope was being exploited during the time when Roman material was being incorporated into the spring deposit. Specifically, to aid in verifying if the spring was being used for cult purposes with the Roman material reflecting 'offerings'. If the hillside was being cultivated in that period, then the ploughsoil and the spring could have had some form of cultural relationship. Instead, if the dating results show that the buried layer was cultivated in subsequent periods, the incorporation of later hillwash encourage by tillage having moved archaeological material from upslope to accumulated in the depression of the spring.

Calibrated date: *1σ:* cal AD 120–230
                       2σ: cal AD 80–240

Final comment: P Wilson (10 March 2014), the wide date range demonstrates that the deposits encountered were not from the Roman period and are therefore unrelated to any ritual activity on the site, particular as the spring line in question lies below the level of the *nymphaeum*.

OxA–13609 3315 ±55 BP

δ¹³C: -26.0‰

Sample: Trench 3, 4014, submitted on 8 December 2003 by G Ayala

Material: charcoal: *Fraxinus excelsior*, sapwood; single fragment (0.07g) (R Gale 2003)

Initial comment: as OxA-13476

Objectives: as OxA-13476

Calibrated date: *1σ:* 1670–1510 cal BC
                       2σ: 1750–1450 cal BC

Final comment: see OxA-13476

Harehaugh Hillfort, Northumberland

Location: NY 96959980
                  Lat. 55.17.35 N; Long. 02.02.53 W

Project manager: R Carlton (Museum of Antiquities, University of Newcastle), 2002

Archival body: Tyne and Wear Archives and Museums

Description: a multi-vallate hillfort with single rampart across an undulating interior otherwise devoid of obvious earthwork features. Excavation revealed rampart construction of different forms in several phases, as well as sub-surface structure, burnt areas, etc in the interior.

Objectives: to indicate the date of burning activity associated with a spread of domestic hearth material including pottery and pot boilers, thereby indicating a date at which this part of the monument was in use; to indicate the date at which a feature interpreted as a fence was constructed, and when this part of the monument interior was in use; and to indicate the date of metalworking activities in a part of the monument interior.

Final comment: P Marshall (13 October 2014), the radiocarbon dates indicate that activity at the hillfort was taking place in the mid and late Iron Age.

Laboratory comment: English Heritage (25 August 2004), of the three pairs of measurements two are statistically consistent (OxA-13461 and OxA-13462; T′=1.3; *v*=1 (5%)=3.8, Ward and Wilson 1978) and (OxA-13465 and OxA-13466; T′=0.1; *v*=1 (5%)=3.8, Ward and Wilson 1978) and thus the samples from these two contexts could be of the same actual age. The two measurements on the samples from H2002/415 are not statistically consistent (OxA-13463 and OxA-13464; T′=10.5; *v*=1 (5%)=3.8, Ward and Wilson 1978) meaning they represent the results of two distinct phases of activity.

References: Carlton 2011
Ward and Wilson 1978

OxA–13461 2161 ±29 BP

δ¹³C: -26.5‰

Sample: H2002/603/A, submitted on 19 February 2004 by R Carlton

Material: charcoal: *Betula* sp., single fragment (R Gale 2004)

Initial comment: the sample came from a context of firm dark earth, clearly a feature used or associated with iron working activities.

Objectives: to indicate the date of activities responsible for the production of a burnt (hearth) area and significant quantities of metalworking slag, thereby indicating when this part of the monument was in use.

Calibrated date: 1σ: 350–170 cal BC
2σ: 360–110 cal BC

Final comment: P Marshall (13 October 2014), the radiocarbon dates indicate that activity at the hillfort was taking place in the mid and late Iron Age.

OxA–13462 2114 ±30 BP

δ¹³C: -24.6‰

Sample: H2002/603/B, submitted on 19 February 2004 by R Carlton

Material: charcoal: *Betula* sp., single fragment (R Gale 2004)

Initial comment: as OxA-13461

Objectives: as OxA-13461

Calibrated date: 1σ: 200–90 cal BC
2σ: 350–40 cal BC

Final comment: see OxA-13461

OxA–13463 2091 ±31 BP

δ¹³C: -27.1‰

Sample: H2002/415/A, submitted on 19 February 2004 by R Carlton

Material: charcoal: *Alnus glutinosa,* single fragment (R Gale 2004)

Initial comment: the sample came from a context of firm dark brown silty soil, clearly a surface used for open fires and textually very much distinct from looser material above, which also contained large stones/boulders interpreted as tumble from the adjacent rampart wall, and from material below which did not contain similar quantities of organic remains.

Objectives: to indicate the date of burning activity associated with a spread of domestic hearth material including pottery and pot boilers, thereby indicating a date at which this part of the monument was in use. This will also provide dating evidence for two distinct pottery fabrics which have also been recorded together elsewhere in the region.

Calibrated date: 1σ: 180–50 cal BC
2σ: 200–30 cal BC

Final comment: see OxA-13461

OxA–13464 2231 ±31 BP

δ¹³C: -25.8‰

Sample: H2002/415/B, submitted on 19 February 2004 by R Carlton

Material: charcoal: *Alnus glutinosa*, single fragment (R Gale 2004)

Initial comment: as OxA-13463

Objectives: as OxA-13463

Calibrated date: 1σ: 380–200 cal BC
2σ: 390–200 cal BC

Final comment: see OxA-13461

OxA–13465 2194 ±30 BP

δ¹³C: -25.2‰

Sample: H2002/560/A, submitted on 19 February 2004 by R Carlton

Material: charcoal: *Corylus avellana*, roundwood, diameter 15mm, 16 growth rings; single fragment (R Gale 2004)

Initial comment: the sample came from the fill of a probable construction trench, apparent as a dark band across the subsoil.

Objectives: to indicate the date of construction of this feature, interpreted as a fence, subsequently burnt *in situ.* No other dates have previously been derived for the monument interior, nor are other forms of secure dating evidence available for this part of the monument.

Calibrated date: 1σ: 360–190 cal BC
2σ: 370–170 cal BC

Final comment: see OxA-13461

OxA–13466 2178 ±29 BP

δ¹³C: -25.0‰

Sample: H2002/560/B, submitted on 19 February 2004 by R Carlton

Material: charcoal: *Corylus avellana*, roundwood, diameter 15mm, 16 growth rings; single fragment (R Gale 2004)

Initial comment: as OxA-13465

Objectives: as OxA-13465

Calibrated date: *1σ:* 360–190 cal BC
                       2σ: 360–160 cal BC

Final comment: see OxA-13461

Higham Ferrers: Kings Meadow Lane (Roman): burial group 1, Northamptonshire

Location: SP 95446921
                     Lat. 52.18.43 N; Long. 00.36.00 W

Project manager: S Lawrence and A Smith (Oxford Archaeology), 1993–2003

Archival body: Oxford Archaeology

Description: a series of archaeological investigations were undertaken on land around Kings Meadow Lane from 1993–2003. This work revealed occupation from the Mesolithic through to the medieval period, including part of a substantial Roman roadside settlement and shrine.

Burial group 1: (UB-5215–21): four stratigraphically related inhumations cut into ditch 12880 (groups 10970, 12685, 10950, and 12655), and cremation groups 1915, 10935, and 10945 which may pre-date the inhumations and are thought to be contemporary with ditch 12880.

Objectives: to provide a chronological framework that could be used to understand and interpret the burial record against the background of the phased site development. This included providing dates for individual burials and estimating the start and end dates of burial activity at the locations of two spatially distinct burial groups.

Laboratory comment: English Heritage (2009), chronological modelling of the results suggests activity in burial group 1 began in *10 cal BC–cal AD 210 (95% probability; fig 4.57; start_burial group 1)* and probably *cal AD 40–150 (68% probability)* and ended in *cal AD 250–440 (95% probability; fig 4.57; end_burial group 1)* and probably *cal AD 260–360 (68% probability)*. The opverall span of burial activity in group 1 is estimated at *60–400 years (95% probability; fig 4.58; span burial group 1)* and probably *150–310 years (68% probability;* Lawrence and Smith 2009).

Further analysis shows that there is a 98% probability that activity associated with burial in group 2 preceded the cessation of activity in group 1. Furthermore, there is a 69% probability that the order of events is *start_burial group 1> start_burial group 2>end_burial group 1>end_burial group 2.*

References: Lawrence and Smith 2009

UB–5215 1649 ±20 BP

$δ^{13}C$: -21.2 ±0.2‰
$δ^{15}N$ *(diet):* +11.7 ±0.15‰
$δ^{13}C$ *(diet):* -19.5 ±0.1‰
C/N ratio: 3.3

Sample: sk 10922, submitted on 19 January 2006 by S Lawrence

Material: human bone (208g) (right femur, ulna, humerus, radius, and ulna; left fibula) (A Witkin 2005)

Initial comment: skeleton 10922 belongs to grave group 10950 which truncated/overlay earlier grave group 10970 (skeleton 10954). 10970 is cut into (and therefore later than) the backfilled ditch 12880.

Objectives: this inhumation was the latest within the group cut into ditch 12880, and will therefore provide an end date (with grave group 12655, sk 12656) for use of the specific burial area. It will also aid the dating of earlier burial 10970. This forms part of the broader dating programme targeting the burial groups across the site, and will significantly aid the interpretation of these.

Calibrated date: *1σ:* cal AD 385–420
                       2σ: cal AD 345–425

Final comment: A Lawrence and A Smith (2009), burial group 13045 consisted of five adult inhumations (10950, 10970, 12655, 12685, and 12725) cut into the top of ditch 12880 and following its alignment. These must post-date the backfilling of the ditch in the late second century AD. A late-Roman date for the group seems likely, particulary given that two of the burials (10970 and 12655) were prone - a rite that only became common during the fourth century AD. This date, on skeleton 10922 from burial 10950, which stratigraphically post-dated prone burial 10970, therefore confirms this interpretation.

Laboratory comment: English Heritage (2009), chronological modelling of the results indicate that the posterior density estimate for this burial is *cal AD 260–280 (6% probability)* or *cal AD 335–425 (89% probability;* Lawrence and Smith 2009, table 4.6).

UB–5216 1719 ±20 BP

$δ^{13}C$: -20.8 ±0.2‰
$δ^{15}N$ *(diet):* +11.2 ±0.15‰
$δ^{13}C$ *(diet):* -19.9 ±0.1‰
C/N ratio: 3.2

Sample: sk 10938, submitted on 19 January 2006 by S Lawrence

Material: human bone (206g) (right tibia and fibula) (A Witkin 2005)

Initial comment: skeleton 10938 belongs to grave group 10965 which is later than grave group 10960 (skeleton 10951). 10965 truncates the north-eastern side of 10960, removing some of the bones (right leg and arm) from this burial. These were incorporated into the backfill of the later grave.

Objectives: this inhumation has a direct stratigraphic relationship with earlier grave group 10960, and as such will provide complimentary dating evidence to refine the date range within this burial group. This forms part of the broader dating programme targeting the burial groups across the site, and will significantly aid the overall site interpretation of these. A spatial relationship to the surrounding plot boundaries and phased development of this part of the site can also be inferred, suggesting a fourth-century AD date for this group of burials. With the series of stratigraphic relationships and pottery dates available, the objective is to accurately date the development of burial areas and practices across the site.

Calibrated date: 1σ: cal AD 255–380
                      2σ: cal AD 250–390

Final comment: S Lawrence and A Smith (2009), the practise of burial alongside the rear boundary ditch of the plot continued into phase 5 (late third to fourth centuries AD), with at least two inhumation burials belonging to this phase. One of these burials was decapitated (this skeleton), a rite most common in the fourth century AD and confirmed by this radiocarbon date.

Laboratory comment: English Heritage (2009), chronological modelling of the results indicate that the posterior density estimate for this burial is *cal AD 250–385 (95% probability;* Lawrence and Smith 2009, table 4.6).

UB–5217 1798 ±18 BP

$\delta^{13}C$: -20.6 ±0.2‰
$\delta^{15}N$ *(diet):* +12.3 ±0.15‰
$\delta^{13}C$ *(diet):* -19.1 ±0.1‰
C/N ratio: 3.2

Sample: sk 10951, submitted on 19 January 2006 by S Lawrence

Material: human bone (219g) (left femur and right ulna) (A Witkin 2005)

Initial comment: skeleton 10951 belongs to grave group 10960 which was earlier than grave group 10965 (skeleton 10938). 10965 truncated the north-eastern side of 10960, removing some of the bones (right leg and arm) from this burial.

Objectives: this inhumation has a direct stratigraphic relationship with later grave group 10965, and as such will provide complimentary dating evidence to refine the date range within this burial group. This forms part of the broader dating programme targeting burial groups across the site, and will significantly aid the overall site interpretation of these. A spatial relationship to the surrounding plot boundaries and phased devleopment of this part of the site can also be inferred, suggesting a fourth-century AD date for this group of burials.

Calibrated date: 1σ: cal AD 220–245
                      2σ: cal AD 135–320

Final comment: S Lawrence and A Smith (2009), this radiocarbon date confirms this burial dates to the late second to third centuries AD (phase 4). The interment of inhumation burials alongside ditch 11170 commenced at this time, and up to seven burials including this one may belong to this phase (10740, 10790, 10955, 10960, 12810, 12815, and 12820).

Laboratory comment: English Heritage (2009), chronological modelling of the results indicate that the posterior density estimate for this burial is *cal AD 135–260 (94% probability)* or *cal AD 305–315 (1% probability;* Lawrence and Smith 2009, table 4.6).

UB–5218 1784 ±17 BP

$\delta^{13}C$: -20.1 ±0.2‰
$\delta^{15}N$ *(diet):* +12.0 ±0.15‰
$\delta^{13}C$ *(diet):* -19.4 ±0.1‰
C/N ratio: 3.2

Sample: sk 10954, submitted on 19 January 2006 by S Lawrence

Material: human bone (219g) (left humerus and tibia; right fibula) (A Witkin 2006)

Initial comment: skeleton 10954 belongs to grave group 10970, which is cut into (later than) the backfilled ditch 12880. Inhumation group 10950 (skeleton 10922) truncated/overlay this grave.

Objectives: this inhumation provides the middle part of a stratigraphic sequence spanning ditch 12880 (this grave and grave 10950; skeleton 10922). As such, it will provide a necessary date to complete the sequence. This forms part of the broader dating programme targeting the burial groups across the site, and will significantly aid the interpretation of these.

Calibrated date: 1σ: cal AD 230–315
                      2σ: cal AD 175–330

Final comment: S Lawrence and A Smith (2009), this date was less helpful, producing relatively broad date ranges spanning the second to fourth centuries AD.

Laboratory comment: English Heritage (2009), chronological modelling of the results indicate that the posterior density estimate for this burial is *cal AD 140–200 (11% probability)* or *cal AD 205–265 (60% probability)* or *cal AD 280–330 (24% probability;* Lawrence and Smith 2009, table 4.6).

UB–5219 1701 ±21 BP

$\delta^{13}C$: -21.5 ±0.2‰
$\delta^{15}N$ *(diet):* +11.1 ±0.15‰
$\delta^{13}C$ *(diet):* -19.5 ±0.1‰
C/N ratio: 3.2

Sample: sk 12656, submitted on 19 January 2006 by S Lawrence

Material: human bone (203g) (left femur and right tibia) (A Witkin 2006)

Initial comment: skeleton 12656 belongs to grave group 12655 which truncates the southern end of inhumation grave group 12685 (sk 12686), removing the torso region of this skeleton. The earlier inhumation is cut into ditch 12880.

Objectives: this inhumation provides the final part of the stratigraphic sequence spanning ditch 12880 and graves 12655 and 12685 (skeletons 12656 and 12686). As such it will provide a necessary date for the complete sequence aiding the dating of these. This forms part of the broader dating targeting the burial groups across the site, and will significantly aid in the interpretation of these.

Calibrated date: 1σ: cal AD 265–390
                      2σ: cal AD 255–400

Final comment: S Lawrence and A Smith (2009), the late dating of this burial group is supported by this determination on prone burial 12655 (*see also* UB-5215).

Laboratory comment: English Heritage (2009), chronological modelling of the results indicate that the posterior density estimate for this burial is *cal AD 255–305 (29% probability)* or *cal AD 315–400 (66% probability;* Lawrence and Smith 2009, table 4.6).

UB–5220 1793 ±20 BP

$\delta^{13}C$: -19.9 ±0.2‰
$\delta^{15}N$ *(diet):* +12.0 ±0.15‰
$\delta^{13}C$ *(diet):* -19.3 ±0.1‰
C/N ratio: 3.2

Sample: sk 12686, submitted on 19 January 2006 by S Lawrence

Material: human bone (207g) (left fibula; right tibia and fibula) (A Witkin 2005)

Initial comment: skeleton 12686 belongs to grave group 12685 which is cut into (later than) the backfilled ditch 12880. Inhumation group 12655 (skeleton 12656) truncated this grave, and removed the torso area of the skeleton.

Objectives: this inhumation provides the middle part of the stratigraphic sequence spanning ditch 12880, this grave, and grave 12655 (skeleton 12656). As such, it will provide a necessary date to be able to compile a complete sequence. This forms part of the broader dating programme targeting the burial groups across the site, aimed at significantly aiding the interpretation of these in the context of the site and Roman burial practices.

Calibrated date: 1σ: cal AD 220–250
   2σ: cal AD 135–325

Final comment: see UB-5218

Laboratory comment: English Heritage (2009), chronological modelling of the results indicate that the posterior density estimate for this burial is *cal AD 130–260 (87% probability)* or *cal AD 295–320 (8% probability;* Lawrence and Smith 2009, table 4.6).

UB–5221 1885 ±18 BP

$\delta^{13}C$: -20.5 ±0.2‰
$\delta^{15}N$ *(diet):* +12.7 ±0.15‰
$\delta^{13}C$ *(diet):* -19.9 ±0.1‰
C/N ratio: 3.2

Sample: sk 12814, submitted on 19 January 2006 by S Lawrence

Material: human bone (245g) (right femur) (A Witkin 2005)

Initial comment: skeleton 12814 belongs to grave group 12810, which is earlier than grave group 12815 (skeleton 12816). 12815 truncated the southern end of 12810, removing the legs of skeleton 12814 from the knees.

Objectives: this inhumation has a direct stratigraphic relationship with later grave group 12815, and as such will provide complimentary dating evidence to refine the date range within this burial group. This forms part of the broader dating programme targeting the burial groups across the site, and will significantly aid the overall site interpretation of these. A spatial relationship to the surrounding plot boundaries and phased development of this part of the site can also be inferred, suggesting a fourth-century AD date for this group of burials.

Calibrated date: 1σ: cal AD 80–135
   2σ: cal AD 70–140

Final comment: see UB-5217

Laboratory comment: English Heritage (2009), chronological modelling of the results indicate that the posterior density estimate for this burial is *cal AD 80–220 (95% probability;* Lawrence and Smith 2009, table 4.6).

UB–5222 1774 ±0 BP

$\delta^{13}C$: -19.5 ±0.2‰
$\delta^{15}N$ *(diet):* +11.6 ±0.15‰
$\delta^{13}C$ *(diet):* -18.6 ±0.1‰
C/N ratio: 3.2

Sample: sk 12816, submitted on 19 January 2006 by S Lawrence

Material: human bone (296g) (left femur) (A Witkin 2005)

Initial comment: skeleton 12816 belongs to grave group 12815 that truncates graves 12810 and 12820 (skeletons 12814 and 12902). 12815 truncated the southern end of 12810, removing the legs of skeleton 12814 from the knees, and the eastern end of 12820, removing the skull and several torso bones of skeleton 12902. Skeleton 12816 was fully articulated and undisturbed having been buried in a coffin represented by numerous coffin nails. Several bones from the earlier disturbed burials were incorporated into the backfill of 12815, but were clearly identifiable during excavation.

Objectives: this inhumation has a direct stratigraphic relationship with earlier grave groups 12810 amd 12820, and as such will provide complimentary dating evidence to refine the date range within this burial group. This forms part of the broader dating programme targeting these burial groups across the site, and will significantly aid the overall site interpretation of these. A spatial relationship to the surrounding plot boundaries and phased development of this part of the site can also be inferred, suggesting a fourth-century AD date for this group of burials.

Calibrated date: 1σ: cal AD 240–315
   2σ: cal AD 235–325

Final comment: see UB-5217

Laboratory comment: English Heritage (2009), chronological modelling of the results indicate that the posterior density estimate for this burial is *cal AD 210–340 (95% probability;* Lawrence and Smith 2009, table 4.6).

UB–5223 1869 ±20 BP

$\delta^{13}C$: -19.4 ±0.2‰
$\delta^{15}N$ *(diet):* +11.9 ±0.15‰
$\delta^{13}C$ *(diet):* -18.7 ±0.1‰
C/N ratio: 3.2

Sample: sk 12902, submitted on 19 January 2006 by S Lawrence

Material: human bone (262g) (right tibia and humerus) (A Witkin 2005)

Initial comment: skeleton 12902 belongs to grave group 12820, which is earlier than grave group 12815 (skeleton 12816). 12815 truncated the eastern end of 12820, removing the skull and several torso area bones from skeleton 12902. these were incorporated in the backfill of the later grave, with the skull placed on top of the coffin in this grave.

Objectives: this inhumation has a direct stratigraphic relationship with the later grave group 12815, and as such will provide complimentary dating evidence to refine the date range within this burial group. This forms part of the broader dating programme tageting burial groups across the site, and will significantly aid the overall site interpretation of these. A spatial relationship to the surrounding plot boundaries and phased develepment of this part of the site can be inferred, suggesting a fourth-century date for ths group of burials.

Calibrated date: 1σ: cal AD 85–135
2σ: cal AD 75–225

Final comment: see UB-5217

Laboratory comment: English Heritage (2009), chronological modelling of the results indicate that the posterior density estimate for this burial is *cal AD 85–225 (95% probability;* Lawrence and Smith 2009, table 4.6).

Higham Ferrers: Kings Meadow Lane (Roman): burial group 2, Northamptonshire

Location: SP 95446921
Lat. 52.18.43 N; Long. 00.36.00 W

Project manager: S Lawrence and A Smith (Oxford Archaeology), 2002–3

Archival body: Oxford Archaeology

Description: burial group 2 comprised eight inhumations (including a single neonate), several of which had stratigraphic relationships (eg grave group 12815, 12810, 12820, 10965, and 10960. Other burials were arranged with spatial consideration to these, and the burial area was neatly defined by a rear ditch boundary and plot division boundaries to the rear of building 12900. The later ditches in the sequence appear to date from the third and fourth centuries AD.

Objectives: to provide a chronological framework that could be used to understand and interpret the burial record against the background of the phased site development. This included providing dates for individual burials and estimating the start and end dates of burial activity at the locations of the two spatially distinct burial groups.

Laboratory comment: English Heritage (2009), activity within burial group 2 is estimated to have begun in *10 cal BC–cal AD 320 (95% probability;* fig 4.57; *start_burial group 2)* and probably *cal AD 100–250 (68% probability)* and ended in *cal AD 350–570 (90% probability;* Fig 4.57; *end_burial group 2)* and probably *cal AD 380–490 (68% probability).* The overall

span of burial activity in group 2 is estimated at *80–500 years (95% probability;* fig 4.58; *span burial group 2)* and probably *160–370 years (68% probability).* The overall duration of burial activity within the two loci is estimated at *210–540 years (95% probability;* fig 4.59) and probably *270–440 years (68% probability;* Lawrence and Smith 2009). *See* also burial group 1 series comments.

References: Lawrence and Smith 2009

SUERC–9649 1810 ±35 BP

δ¹³C: -17.8‰

Sample: crem 10912, submitted on 19 January 2006 by S Lawrence

Material: calcined human bone (9g) (A Witkin 2005)

Initial comment: from an isolated feature (group 10915) with no direct stratigraphic relationship to other features. It was spatially related to cremation groups 10935 and 10945, together forming a localised burial area accompanied by a single neonatal inhumation (group 10925). This was located to the immediate south-west of the inhumation burials cut into ditch 12880. Burnt animal bone was also noted within the deposit, with the human cremated bone weighing 204g. The whole body is not therefore represented by this deposit, although some truncation had occurred.

Objectives: the date of this group of cremations, ditch 12880, and the inhumations, is not known. The style of the burial, its positioning adjacent to the ditch, and the date range of the site as a whole suggetsts that it may be be contemporary with the ditch in the second century AD (earlier than the inhumations). Radiocarbon dating would allow for interpretation of this burial area as a whole, and establishing an overall sequence is considered important for understanding and interpreting the boundaries between burial groups and the longevity of burial practices. This is especially so for burial areas that appear otherwise undefined by archaeological boundaries.

Calibrated date: 1σ: cal AD 130–250
2σ: cal AD 120–330

Final comment: S Lawrence and A Smith (2009), the date ranges provided by this radiocarbon date are fairly broad, although the rite of cremation is generally more typical of the earlier rather than the later Roman period, and so this burial has been assigned to phase 3 (second century AD).

Laboratory comment: English Heritage (2009), chronological modelling of the results indicate that the posterior density estimate for this burial is *cal AD 125–265 (86% probability)* or *cal AD 280–325 (9% probability;* Lawrence and Smith 2009, table 4.6).

SUERC–9650 1760 ±35 BP

δ¹³C: -15.6‰

Sample: crem 10931, submitted on 19 January 2006 by S Lawrence

Material: calcined human bone (7g) (A Witkin 2005)

Initial comment: as SUERC-9649. From group 10935, this cremation only weighed 27g and so does not represent the whole body.

Objectives: as SUERC-9649

Calibrated date: *1σ:* cal AD 230–340
              2σ: cal AD 140–390

Final comment: see SUERC-9649

Laboratory comment: English Heritage (2009), chronological modelling of the results indicate that the posterior density estimate for this burial is *cal AD 160–200 (5% probability)* or *cal AD 205–385 (90% probability;* Lawrence and Smith 2009, table 4.6).

SUERC–9651 1835 ±35 BP

δ¹³C: -20.1‰

Sample: crem 10933, submitted on 19 January 2006 by S Lawrence

Material: calcined human bone (5g) (A Witkin 2005)

Initial comment: as SUERC-9649. From group 10945, this weighed only 10g, so the whole body was not represented.

Objectives: as SUERC-9649

Calibrated date: *1σ:* cal AD 120–240
              2σ: cal AD 80–320

Final comment: see SUERC-9649

Laboratory comment: English Heritage (2009), chronological modelling of the results indicate that the posterior density estimate for this burial is *cal AD 90–255 (95% probability;* Lawrence and Smith 2009, table 4.6).

Holme-next-the-Sea, Norfolk

Location:	TF 7112545263
	Lat. 52.58.37 N; Long. 00.32.56 E
Project manager:	W Boismier (Norfolk Archaeological Unit), 1999, 2003, 2004, and 2008

Description: in 1998, a circle of timber posts surrounding an up-turned tree was found on Holme Beach. A subsequent programme of surveying, recording, and dating revealed that the structure was built in the spring or early summer of 2049 BC, during the early Bronze Age (Brennard and Taylor 2003). It has become known as the Holme I timber circle or as 'Seahenge'.

During 1999, a walkover survey was conducted within the immediate environs of Holme I. This revealed a possible trackway, two possible fish-traps, and two logs surrounded by a ring of posts and wattling/hurdling (Brennard and Taylor 2003). Subsequent erosion showed that two outer circles of oak posts surrounded the two logs and wattling/hurdling. The structure has become known as the Holme II timber circle. Samples were collected from all four structures for radiocarbon dating.

Between 1999 and 2003, members of the public reported two further monuments on Holme Beach. One of these was a V-shaped fish-trap; the other was a possible pit.

In 2003, a second walkover survey was carried out. It examined all of the areas of beach within the Holme Dunes Nature Reserve. In total, 13 timber structures, 5 collections of posts, 13 or 14 individual posts, 3 or 4 planks, and a possible pit/erosion scour were observed and recorded. The condition of each was assessed along with potential threats. Of the monuments observed, timber circle Holme II, a Saxon fishtrap, a V-shaped fishtrap, and the possible pit have been noted before. The timbers within all of the sampled monuments are embedded in one of two deposits. Timbers in five are embedded in exposures of silt, which are thought to have developed from about 5900–4850 cal BC in saltmarsh conditions (Funnell and Pearson 1984).

Timbers in the nine other monuments are embedded in peat beds which overlie the silt. Various peat beds have been exposed by tidal erosion on Holme Beach at least since the 1940s (when they were observed during aerial photography). They form a series of islands and are up to 0.3m thick. It is thought that the peat accumulated from *c* 1970–1740 cal BC; 3530 ±40 BP, OxA-10207; sample taken *c* 150m to the west of the structure; *see* above).

Objectives: to establish absolute dates for the individual monuments, to establish a chronology of land use, and to establish a chronology of environment change on Home Beach.

Final comment: D Robertson (9 May 2005), the 33 radiocarbon dates, together with six radiocarbon dates from core HDR1 (*see* Holme Dunes Reserve; Bayliss *et al* 2016, 140-1), and dendrochronology dates from the Holme I timber circle (NER 33771), established an absolute chronology for activity and monument construction at Holme Beach. They have also contributed to the understanding of past environmental changes.

The dates suggest three to four phases of monument construction on Holme Beach. Two to three phases have been identified during the Bronze Age, each of which was separated by a change in environmental conditions. There was one phase of activity in the Saxon period.

References:	Brennand and Taylor 2003
	Funnell and Pearson 1984

Holme-next-the-Sea: walk-over survey (2003), Norfolk

Location:	TF 6929244674to TF 7279244950
	Lat. 52.58.22 N; Long. 00.31.09 E, to
	Lat. 52.58.27 N; Long. to 00.34.19 E
Project manager:	D Robertson (Norfolk Archaeological Unit), 2003–4, and 2008
Archival body:	Norfolk Museums and Archaeology Service

Description: a walk-over survey was conducted within the intertidal zone of Holme Beach in February and March 2003. A 3.5km long stretch of beach was examined. As the average high and low water marks were taken as the northern and southern boundaries, the width of the survey area varied between 0.5km and 1km. A bi-monthly monitoring programme of the timber structures, collections of posts, and the possible pit/erosion scour began in July 2004. This is an ongoing project that will continue until 2008. Between September 2003 and February 2004, samples were collected for radiocarbon dating and timber species identification from twelve of the monuments. (Samples from two monuments were not suitable for radiocarbon dating).

Objectives: as tidal erosion has removed the deposits that once covered all of the 12 monuments, no firm stratigraphic information is available for any of them. As yet no artefactual material has been recovered from the peat and silt deposits directly adjacent to the monuments. In the absence of stratigraphic and artefactual information, radiocarbon dating of timbers from the monuments will provide absolute dates for them.

It is hoped that the absolute dates will allow the 12 monuments to be fitted into the known environmental sequence for Holme Beach. It is also hoped that they will allow the structures to be placed within the Holme Beach archaeological sequence. At present Bronze Age (timber circles Holme I and Holme II) and Saxon monuments (Holme fishtraps I and II) have been identified. Many of the structures have the potential to be similar dates and, if so, the dating of the samples would contribute to the understanding of the landscape around the previously dated monuments. If some or all are of earlier or later date, the work will provide details of how the local settlement and land-use evolved over time. It will also help structural and constructional comparisons to be drawn with other British and north-western European monuments of similar date.

Final comment: D Robertson (9 May 2005), the 24 radiocarbon dates, together with 9 submitted in 2000 (GU-5800–8), six radiocarbon dates from core HDR1 (OxA-9610, OxA-9611, OxA-9748, OxA-10207, OxA-10208, and OxA-10209), and dendrochronology dates from the Holme I timber circle (NER 33771), established an absolute chronology for activity and monument construction at Holme Beach. They have also contributed to the understanding of past environmental changes.

The dates built upon the chronology established by the sample series GU-5800–8. They suggest three to four phases of monument construction on Holme Beach. Two or three phases have been identified during the Bronze Age, each of which was separated by a change in environmental conditions. There was one phase of activity in the Saxon period.

Laboratory comment: English Heritage (17 November 2014), two further samples were dated after 2006 (GrN-31823–4). Nine dates were submitted in 2000 (GU-5800–8), and are published in Bayliss *et al* 2016 (145-7).

References: Bayliss *et al* 1999
               Brennand and Taylor 2003

GU–6012 1210 ±50 BP

$\delta^{13}C$: -28.0‰

Sample: 38042–3201–1, submitted on 2 March 2004 by D Robertson

Material: wood (waterlogged): *Salix/Populus sp.* (175g) (M Taylor 2004)

Initial comment: the sample comes from a post (context 1) that formed part of structure Norfolk HER 38042, a fishtrap (Holme fishtrap I; TF 7130645220) located on Holme

Beach. HER 38042 was first identified and recorded in 1999. At this time it comprised 102 individual posts placed in a V-shaped arrangement, with the open end of the V to the south. The longer of the two lines was located in the west and ran in a south-west to north-east direction for a distance of 29.5m. From the north-eastern tip of this line a second arrangement of posts ran southwards for 13.6m.

At the time of sampling only the southernmost 6m of the western line of posts survived, comprising 12 posts. The upper part of the southern upright post was collected from this surviving alignment (sample 3201, taken so to give a full cross-section through the post). It was made from roundwood, was slightly distorted, and was firmly embedded in a peat bed. As tidal and wave erosion had destroyed the stratigraphy above, at the time it was not possible to firmly establish its stratigraphic relationship with the peat. However, it is most likely that it was probably driven in from above the peat had finished forming. This is because between 1999 and 2001, the structure was dated to between the fifth and tenth centuries (cal AD 420–900, GU-5800, and GU-5801); it was thought then that it was used in a beach environment rather than an environment in which peat formed.

The monument was first observed in May 1999; it is not clear how long after its first exposure this was. As the timber was firmly embedded in the peat it is very likely that is was found in its original position and, other than it being eroded and smoothed by the sea, it is unlikely it had been disturbed from its original position.

Objectives: as tidal erosion has removed deposits that once covered the structure (HER 38042), no firm stratigraphic information is available. As yet no artefactual material has been recovered from the peat in the vicinity of the monument. In the absence of stratigraphic and artefactual information, between 1999 and 2001, two timbers were subjected to radiocarbon dating to provide an absolute date for the structure. Both the dates fell between the fifth and tenth centuries, but were widely spaced (cal AD 660–900, GU-5800 and cal AD 420–650, GU-5801). This wide spacing could suggest that the monument was used and repaired over a series of centuries or could be a reflection of problems in applying radiocarbon dating methods to material from the later first millennium AD.

Dating of the timber, in conjunction with a second sample from another timber (38042–3202–2), should help clarify the existence of wide spacing in the dates of the timbers from the structure and may help establish a reason for it. This will then provide useful information to enable the fitting of the monument in the known environmental sequence for Holme Beach. It will also allow the structure to be placed more firmly within the Holme Beach archaeological sequence. As the structure is Saxon in date and there is the potential that other nearby monuments may be similar date, the dating of the sample would contribute to the understanding of the development of contemporary monuments and the surrounding landscape. It will also help structural and constructional comparisons to be drawn with other British and north-western European monuments of similar date.

Calibrated date: *1σ:* cal AD 710–890
                     2σ: cal AD 670–970

Final comment: D Robertson (9 May 2005), this is statistically consistent with the date of the second sample, and similar to one of the samples dated in 2000 (GU–5800). The date is different to the second sample dated in 2000 (GU–5801). This anomaly could be the result of the monument being constructed in the early Saxon period and then maintained and modified over several centuries. Another possibility is that old timbers may have been re-used during construction in the middle Saxon period.

The dating of the sample confirmed the use of Holme Beach and the exploitation of saltwater resources during the early/middle Saxon period. It will also allow for comparisons to be made with other known early/middle Saxon fish-traps in the future.

Laboratory comment: English Heritage (November 2014), the four measurements on this structure (GU-6012–3 and GU-5800–01) are not statistically consistent (T′=22.8; T′(5%)=7.8; *v*=3; Ward and Wilson 1978).

GU–6013 1260 ±50 BP

$\delta^{13}C$: -28.1‰

Sample: 38042–3202–2, submitted on 2 March 2004 by D Robertson

Material: wood (waterlogged): *Betula* sp. (325g) (M Taylor 2004)

Initial comment: as GU-6012, except the sample comes from the upper part of the upright post (context 2; sample 3202).

Objectives: as GU-6012, in conjunction with second sample from another timber (38042–3201–1).

Calibrated date: 1σ: cal AD 670–780
                2σ: cal AD 650–890

Final comment: see GU-6012

Laboratory comment: see GU-6012

GU–6014 3530 ±50 BP

$\delta^{13}C$: -30.2‰

Sample: 38195–3402–2, submitted on 2 March 2004 by D Robertson

Material: wood (waterlogged): *Alnus glutinosa* (220g) (M Taylor 2004)

Initial comment: the sample comes from a timber (context 2) that formed part of structure Norfolk HER 38195 (TF 6965144842), located on Holme Beach. The structure comprised nine timbers laid horizontally which ranged in length between 0.2m and 2.13m and between 0.06m and 0.15m in diameter. Five roughly parallel were aligned north-west to south-east whilst, just to the north-west, were four pieces orientated north-east to south-west. This rectilinear arrangement suggested that they had been deliberately laid within the surrounding peat bed. As such, they may have been the remains of a trackway or a platform.

The sample (3402) was taken from a roughly north-east to south-west aligned horizontally laid timber (context 2) located in the north-west of the structure. The westernmost 0.15m of the timber was sawn off to provide a full cross-section. It was made from a flattened radially split piece comprising less than a quarter of a tree. It measured 65mm wide by 33mm thick and was firmly embedded in a peat bed. As tidal and wave erosion had destroyed the stratigraphy above the timber it is not possible to say whether it was originally laid on the surface of the peat after the peat had formed or whether it was placed within the peat whilst it was forming. The monument was first observed in March 2003; it is not clear how long after its first exposure this was. As the timber was firmly embedded in the peat it is very likely that it was found in its original position and, other than it being eroded and smoothed by the sea, it is unlikely it had been disturbed from its original position.

Objectives: as tidal erosion has removed deposits that once covered the structure (HER 38195), no stratigraphic information is available. As yet no artefactual material has been recovered from the peat in the vicinity of the monument. In the absence of stratigraphic and artefactual information, radiocarbon dating of timbers from the structure will provide an absolute date for the structure.

Dating of the timber, in conjunction with a second sample from another timber (38195–3403–3), will provide the absolute date. This will then provide useful information to enable the fitting of the monument in the known environmental sequence for Holme Beach. It will also allow the structure to be placed more firmly within the Holme Beach archaeological sequence. At present Bronze Age and Saxon monuments have been identified. The structure has the potential to be Bronze Age in date and if it is the dating of the sample would contribute to the understanding of the landscape around Holme I and Holme II timber circles. If it is of earlier or later date the work will provide details on how the local settlement and land-use evolved over time. It will also help structural and constructional comparisons to be drawn with other British and north-western European monuments of similar date.

Calibrated date: 1σ: 1940–1770 cal BC
                2σ: 2020–1690 cal BC

Final comment: D Robertson (9 May 2005), the sample successfully dated the possible platform to the early Bronze Age. It also demonstrated the use of Holme Beach in the few centuries after the construction of the Holme I timber circle (NHER 33771; dated by dendrochronology to 2049 BC). It will allow broader understandings of the local Bronze Age landscape to be developed and will enable comparisons to be made with other known Bronze Age structures.

Laboratory comment: English Heritage (November 2014), the two measurements from this structure (GU6014–5) are statistically consistent (T′=1.0; T′(5%)=3.8; *v*=1; Ward and Wilson 1978).

References: Ward and Wilson 1978

GU–6015 3600 ±50 BP

$\delta^{13}C$: -31.0‰

Sample: 38195–3403–3, submitted on 2 March 2004 by D Robertson

Material: wood (waterlogged): *Alnus glutinosa* (400g) (M Taylor 2004)

Initial comment: as GU-6014; except sample comes from context 3, sample 3403, taken from a horizontally laid timber (aligned north-west to south-east) located in the east of the structure. The northernmost 0.19m of the timber was sawn off to provide a full cross-section; this was cut into two sub-samples and one of these has been submitted. The timber was made from a roundwood piece, which overtime eroded and decayed; its original diameter would have been 0.12m plus but only 0.055m survived.

Objectives: as GU-6014

Calibrated date: 1σ: 2030–1890 cal BC
2σ: 2140–1770 cal BC

Final comment: see GU-6014

Laboratory comment: see GU-6014

GU–6016 3530 ±50 BP

δ¹³C: -29.1‰

Sample: 38197–3601–1, submitted on 2 March 2004 by D Robertson

Material: wood (waterlogged): *Alnus glutinosa* (225g) (M Taylor 2004)

Initial comment: the sample comes from one (context 1) of the three horizontally laid logs that formed structure Norfolk HER 38197, located on Holme Beach. All three were located close together and aligned north-east to south-west, parallel to one another. These relationships suggested that they were not naturally placed and had probably been deliberately located amongst the peat. Their purpose was not clear, although they may have formed part of a platform, trackway, or reclamation feature.

The sample (3601) was taken from the northernmost horizontally laid timber (1) which measured about 6m long by 0.062m by 0.04m. A section close to the north-eastern end of the timber was sawn out to provide a full cross-section; this was cut into two sub-samples and one of these has been submitted. The timber was made from a roundwood piece which overtime eroded and been squashed. It was firmly embedded in a peat bed. As tidal and wave erosion had destroyed the stratigraphy above the timber it is not possible to say whether it was originally laid on the surface of the peat after the peat had formed or whether it was placed within the peat whilst it was forming. The monument was first observed in March 2003; it is not clear how long after its first exposure this was. As the timber was firmly embedded in the peat it is very likely that it was found in its original position and, other than it being eroded and smoothed by the sea, it is unlikely it had been disturbed from its original position.

Objectives: as tidal erosion has removed deposits that once covered the structure (HER 38197), no stratigraphic information is available. As yet no artefactual material has been recovered from the peat in the vicinity of the monument. In the absence of stratigraphic and artefactual information, radiocarbon dating of timbers from the structure will provide an absolute date for the structure.

Dating of the timber, in conjunction with a second sample from another timber (38197–3602–2), will provide the absolute date. This will then provide useful information to enable the fitting of the monument in the known environmental sequence for Holme Beach. It will also allow the structure to be placed more firmly within the Holme Beach archaeological sequence. At present Bronze Age and Saxon monuments have been identified. The structure has the potential to be Bronze Age in date and if it is the dating of the sample would contribute to the understanding of the landscape around Holme I and Holme II timber circles. If it is of earlier or later date the work will provide details on how the local settlement and land-use evolved over time. It will also help structural and constructional comparisons to be drawn with other British and north-western European monuments of similar date.

Calibrated date: 1σ: 1940–1770 cal BC
2σ: 2020–1690 cal BC

Final comment: see GU-6014

GU–6017 3460 ±50 BP

δ¹³C: -29.2‰

Sample: 38197–3602–2, submitted on 2 March 2004 by D Robertson

Material: wood (waterlogged): *Alnus glutinosa* (225g) (M Taylor 2004)

Initial comment: as GU-6016, except sample comes from context 2, sample 3602, taken from the southernmost horizontally laid timber, which measured about 9.3m long by 0.09m by 0.05m. A section close to the centre of the timber was sawn out to provide a full cross-section; this was cut into two sub-samples and one of these has been submitted. The timber was made from a tangentially split piece from the outer part of a tree (only bark and sapwood).

Objectives: as GU-6016

Calibrated date: 1σ: 1880–1690 cal BC
2σ: 1920–1630 cal BC

Final comment: see GU-6014

GU–6018 3340 ±50 BP

δ¹³C: -27.7‰

Sample: 38198–2601–1, submitted on 2 March 2004 by D Robertson

Material: wood (waterlogged): *Alnus glutinosa* (700g) (M Taylor 2004)

Initial comment: the sample comes from a timber (context 1) structure Norfolk HER 38198, located on Holme Beach. The structure comprised two upright posts and a rectilinear arrangement of timbers. The posts were located in the south-west of the monument, were circular or oval in cross-section, had a diameter of 0.13m, and stood proud of the surrounding peat by 0.1m. To the north-east of them at least eight straight logs or branches were situated, aligned north-east to south-west. One piece of wood was orientated perpendicular to them with two orientated at near right angles to them. The largest timber was 7.4m long and had a maximum width of 0.35m.

The sample (2601) came from the southernmost upright post (the uppermost part was sawn off to provide a full cross-section of the timber). It was made from roundwood and was firmly embedded in a peat bed. As tidal and wave erosion had destroyed the stratigraphy above the timber it is not possible to say whether it was originally laid on the surface of the surface of the peat after the peat had formed or whether it was placed within the peat whilst it was forming. The monument was first observed in March 2003; it is not clear how long after its first exposure this was. As the timber was firmly embedded in the peat it is very likely that it was found in its original position and, other than it being eroded and smoothed by the sea, it is unlikely it had been disturbed from its original position.

Objectives: as tidal erosion has removed deposits that once covered the structure (HER 38198), no stratigraphic information is available. As yet no artefactual material has been recovered from the peat in the vicinity of the monument. In the absence of stratigraphic and artefactual information, radiocarbon dating of timbers from the structure will provide an absolute date for the structure.

Dating of the timber, in conjunction with a second sample from another timber (38198–3602–2), will provide the absolute date. This will then provide useful information to enable the fitting of the monument in the known environmental sequence for Holme Beach. It will also allow the structure to be placed more firmly within the Holme Beach archaeological sequence. At present Bronze Age and Saxon monuments have been identified. The structure has the potential to be Bronze Age in date and if it is the dating of the sample would contribute to the understanding of the landscape around Holme I and Holme II timber circles. If it is of earlier or later date the work will provide details on how the local settlement and land-use evolved over time. It will also help structural and constructional comparisons to be drawn with other British and north-western European monuments of similar date.

Calibrated date: 1σ: 1690–1530 cal BC
2σ: 1750–1500 cal BC

Final comment: D Robertson (9 May 2005), the sample successfully dated the possible platform to the early Bronze Age. Although the second sample (GU-6019) was also early Bronze Age in date, the two dates are not statistically consistent. This could mean that the monument was constructed between 2010–1690 cal BC and then repaired or refurbished between 1750–1510 cal BC. On the other hand, it may mean that an old timber was re-used during its construction. The collection and dating of further samples would test these hypotheses.

The dating of the two samples demonstrated the use of Holme Beach in the few centuries after the construction of the Holme I timber circle (NHER 33771; dated by dendrochronology to 2049 BC). It will allow broader understandings of the local Bronze Age landscape to be developed and will enable comparisons to be made with other known Bronze Age structures.

Laboratory comment: English Heritage (2014), the two measurements from this structure (GU-6018–9) are not statistically consistent (T′=6.5; T′(5%)=3.8; ν=1; Ward and Wilson 1978).

References: Ward and Wilson 1978

GU–6019 3520 ±50 BP

$\delta^{13}C$: -31.7‰

Sample: 38198–2602–2, submitted on 2 March 2004 by D Robertson

Material: wood (waterlogged): *Alnus glutinosa* (400g) (M Taylor 2004)

Initial comment: as GU-6018, except sample comes from context 2, sample 2602, taken from a horizontally laid timber (2) in the east of the structure. The easternmost 0.18m of the timber was sawn off to provide a full cross-section of the timber; this was separated into two 0.09m long sub-samples, one of which has been submitted. The timber was made from a heavily worn, roughly tangentially split piece and was firmly embedded in a peat bed.

Objectives: as GU-6018

Calibrated date: 1σ: 1920–1760 cal BC
2σ: 2010–1690 cal BC

Final comment: D Robertson (9 May 2005), the sample successfully dated the possible platform to the early Bronze Age. Although the second sample (GU-6018) was also early Bronze Age in date, the two dates are not statistically consistent. This could mean that the monument was constructed between 2010–1690 cal BC and then repaired or refurbished between 1750–1510 cal BC. On the other hand, it may mean that an old timber was re-used during its construction. The collection and dating of further samples would test these hypotheses.

The dating of the two samples demonstrated the use of Holme Beach in the few centuries after the construction of the Holme I timber circle (NHER 33771; dated by dendrochronology to 2049 BC). It will allow broader understandings of the local Bronze Age landscape to be developed and will enable comparisons to be made with other known Bronze Age structures.

Laboratory comment: see GU-6018

GU–6020 3550 ±50 BP

$\delta^{13}C$: -29.8‰

Sample: 38199–1000–1, submitted on 2 March 2004 by D Robertson

Material: wood (waterlogged): *Salix/Populus sp.* (500g) (M Taylor 2004)

Initial comment: the sample comes from a horizontally laid timber (context 1) that formed part of structure Norfolk HER 38199, a possible timber platform or trackway located on Holme Beach. The structure was made from at least 23 pieces of timber laid horizontally in a rectilinear fashion, within an area of about 36m². Located in the south-east of the structure, the timber measured 1.7m long with a diameter of 0.07m. A 0.2m long piece of the timber was collected (sample 1000). This provided a full cross-section of the timber and showed that it had been half split. It had a lot of woody structure and was fibrous; this suggested it maybe of post-prehistoric date.

This timber was firmly embedded in a peat bed. As tidal and wave erosion had destroyed the stratigraphy above the timber it is not possible to say whether it was originally laid on the surface of the peat after the peat had formed or whether it was placed within the peat whilst it was forming. The monument was first observed in March 2003; it is not clear how long its first exposure this was. As the timber was firmly embedded in the peat it is very likely that it was found in its original position and, other than it being eroded and smoothed by the sea, it is unlikely it had been disturbed from it original position.

Objectives: as tidal erosion has removed deposits that once covered the structure (HER 38199), no stratigraphic information is available. As yet no artefactual material has been recovered from the peat in the vicinity of the monument. In the absence of stratigraphic and artefactual information, radiocarbon dating of timbers from the structure will provide an absolute date for the structure.

Dating of the timber, in conjunction with a second sample from another timber (38199–1002–2), will provide the absolute date. This will then provide useful information to enable the fitting of the monument in the known environmental sequence for Holme Beach. It will also allow the structure to be placed more firmly within the Holme Beach archaeological sequence.

Calibrated date: 1σ: 1950–1770 cal BC
                    2σ: 2030–1740 cal BC

Final comment: D Robertson (9 May 2005), as the structure of the timber was more or less intact, it was thought that it would be post-prehistoric in date. However, this was not the case, with the sample dating to the early Bronze Age. This date was statistically consistent with that of the second sample.

The dating of the two samples demonstrated the use of Holme Beach in the few centuries after the construction of the Holme I timber circle (NHER 33771; dated by dendrochronology to 2049 BC). It will allow broader understandings of the local Bronze Age landscape to be developed and will enable comparisons to be made with other known Bronze Age structures.

Laboratory comment: English Heritage (November 2014), the two measurements from this structure (GU-6020 and GU-6021) are statistically consistent (T′=1.0; T′(5%)=3.8; ν=1; Ward and Wilson 1978).

References: Ward and Wilson 1978

GU–6021 3480 ±50 BP

$\delta^{13}C$: -30.3‰

Sample: 38199–1002–2, submitted on 2 March 2004 by D Robertson

Material: wood (waterlogged): *Salix/Populus sp.* (800g) (M Taylor 2004)

Initial comment: as GU-6020, except that the timber was located in the east of the structure, the timber measured 3.8m long. Two notches were cut into it, each of which probably held a crossing timber (although neither cross piece survived). A 0.3m long piece of the timber was collected; this was sub-sampled (sample 1000; a 0.055m long piece). Both the original sample and sub-sample provided a full cross-section of the timber. They showed that it had been made from roundwood and was trapezoidal in cross-section,

measuring 0.14m by 0.12m. It was fibrous; this suggested it may be of post-prehistoric date. During sampling the bark on the lower face became detached from the sapwood (as it was stuck firmly to the peat below).

Objectives: as GU-6020

Calibrated date: 1σ: 1890–1690 cal BC
                    2σ: 1940–1660 cal BC

Final comment: see GU-6020

Laboratory comment: see GU-6020

GU–6022 3520 ±50 BP

$\delta^{13}C$: -29.1‰

Sample: 38200–2002–2, submitted on 2 March 2004 by D Robertson

Material: wood (waterlogged): *Alnus glutinosa* (525g) (M Taylor 2004)

Initial comment: the sample comes from a horizontally laid timber (context 2) that formed part of structure Norfolk HER 38200, a possible timber platform or trackway located on Holme Beach. The structure was made from at least 24 pieces of timber laid horizontally in rectilinear fashion, within an area of about 68m². Located in the centre of the structure, the timber measured 6.9m long with a maximum diameter of 0.15m. A sample was collected from the centre of the timber to give a full cross-section (sample 2002). The timber was originally made from roundwood but had been compressed so it had a roughly trapezoidal cross-section. It was very fibrous (suggesting that it may have been post-prehistoric in date), had a maximum diameter of 0.15m and was firmly embedded in a peat bed. As tidal and wave erosion had destroyed the stratigraphy above the timber it is not possible to say whether it was originally laid on the surface of the peat after the peat had formed or whether it was placed within the peat whilst it was forming.

The monument was first observed in March 2003; it is not clear how long after its first exposure this was. As the timber was firmly embedded in the peat it is very likely that it was found in its original position and, other than it being eroded and smoothed by the sea, it is unlikely it had been disturbed from its original position.

Objectives: as tidal erosion has removed deposits that once covered the structure (HER 38200), no stratigraphic information is available. As yet no artefactual material has been recovered from the peat in the vicinity of the monument. In the absence of stratigraphic and artefactual information, radiocarbon dating of timbers from the structure will provide an absolute date for the structure.

Dating of the timber, in conjunction with a second sample from another timber (38200–2003–3), will provide the absolute date. This will then provide useful information to enable the fitting of the monument in the known environmental sequence for Holme Beach. It will also allow the structure to be placed more firmly within the Holme Beach archaeological sequence.

Calibrated date: 1σ: 1920–1760 cal BC
                    2σ: 2010–1690 cal BC

Final comment: see GU-6020

Laboratory comment: English Heritage (November 2014), the two measurements from this structure (GU–6022–3) are statistically consistent (T'=0.0; T'(5%)=3.8; ν=1; Ward and Wilson 1978).

References: Ward and Wilson 1978

GU–6023 3530 ±50 BP

δ¹³C: -29.3‰

Sample: 38200–2002–3, submitted on 2 March 2004 by D Robertson

Material: wood (waterlogged): *Alnus glutinosa* (320g) (M Taylor 2004)

Initial comment: as GU-6022, except the sample comes from the centre of the timber to give a full cross-section (sample 2003). The timber was originally made from roundwood but had been compressed so it had a roughly semi-circular cross section. It was very fibrous (suggesting a post-prehistoric date) and was trimmed tangentially on two sides.

Objectives: as GU-6020

Calibrated date: 1σ: 1940–1770 cal BC
2σ: 2020–1690 cal BC

Final comment: see GU-6020

Laboratory comment: see GU-6022

GU–6024 3530 ±50 BP

δ¹³C: -30.1‰

Sample: 38205–2401–1, submitted on 25 February 2004 by D Robertson

Material: wood (waterlogged): *Alnus glutinosa* (240g) (M Taylor 2004)

Initial comment: the sample comes from a timber (context 1) that formed part of a collection of seven timbers located on Holme Beach. Located with an area of 0.2m², they ranged in size from 0.05m by 0.09m to 0.13m by 0.1m. All were roughly oval in cross-section, were surrounded by silt, patches of peat and recently deposited sand, and stood proud of the beach surface by up to 0.1m. It was unclear if they formed part of a larger structure. The upper part of an upright quarter split timber was collected from the north-west of the monument (context 1; sample 2301). The timber would originally have had a diameter of 0.12m and was firmly embedded in the surrounding silt deposits. As tidal and wave erosion have destroyed the stratigraphy above the timber it was not possible to firmly establish its stratigraphic relationship with the silt.

The monument was first observed in February 2003; it is not clear how long after its first exposure this was. As the timber was firmly embedded in the peat it is very likely that it was found in its original position and, other than it being eroded and smoothed by the sea, it is unlikely it had been disturbed from its original position.

Objectives: as tidal erosion has removed deposits that once covered the structure (HER 38205), no stratigraphic information is available. As yet no artefactual material has been recovered from the peat in the vicinity of the

monument. In the absence of stratigraphic and artefactual information, radiocarbon dating of timbers from the structure will provide an absolute date for the structure.

Dating of the timber, in conjunction with a second sample from another timber (38205–2402–2), will provide the absolute date. This will then provide useful information to enable the fitting of the monument in the known environmental sequence for Holme Beach. It will also allow the structure to be placed more firmly within the Holme Beach archaeological sequence. At present Bronze Age and Saxon monuments have been identified. As the closeness of the timbers is reminiscent of timbers in Holme I and Holme II, the structure has the potential to be early Bronze Age in date. If it is, the dating of the sample would contribute to the understanding of the landscape around both Holme I and Holme II timber circles. If it is of earlier or later date the work will provide details on how the local settlement and land-use evolved over time. It will also help structural and constructional comparisons to be drawn with other British and north-western European monuments of similar date.

Calibrated date: 1σ: 1940–1770 cal BC
2σ: 2020–1690 cal BC

Final comment: D Robertson (9 May 2005), the sample successfully dated the possible platform to the early Bronze Age. Although the second sample (GU-6025) was also early/middle Bronze Age in date, the two dates are not statistically consistent. As the timbers were definitely part of the same structure, it is possible that the monument was constructed between 2020–1690 cal BC and then repaired or refurbished between 1620–1400 cal BC. On the other hand, it may mean that an old timber was re-used during its construction. The collection and dating of further samples would test these hypotheses.

The dating of the two samples demonstrated the use of Holme Beach in the few centuries after the construction of the Holme I timber circle (NHER 33771; dated by dendrochronology to 2049 BC). It will allow broader understandings of the local Bronze Age landscape to be developed and will enable comparisons to be made with other known Bronze Age structures.

Laboratory comment: English Heritage (November 2014), the two measurements from this structure (GU-6024–5) are not statistically consistent (T'=19.2; T'(5%)=3.8; ν=1; Ward and Wilson 1978).

References: Ward and Wilson 1978

GU–6025 3220 ±50 BP

δ¹³C: -28.6‰

Sample: 38205–2402–2, submitted on 2 March 2004 by D Robertson

Material: wood (waterlogged): *Salix/Populus sp.* (140g) (M Taylor 2004)

Initial comment: as GU-6024, except that the sample comes from the upper part (all that was visible above the surface of the beach) of an upright roundwood post was collected from the centre-west of the monument (context 2; sample 2402). The timber was distorted, measured 0.065m by 0.05m and was firmly embedded in the surrounding silt deposits.

Objectives: as GU-6024

Calibrated date: 1σ: 1530–1430 cal BC
2σ: 1620–1400 cal BC

Final comment: see GU-6024

GU–6026 3230 ±50 BP

δ¹³C: -29.2‰

Sample: 38212–2401–1, submitted on 2 March 2004 by D Robertson

Material: wood: cf *Alnus/Corylus* sp. (150g) (M Taylor 2004)

Initial comment: the sample comes from an upright post (context 1) that was located in the north-west of an arrangement of eight posts located on Holme beach. The largest post in the cluster had a width of 0.17m and the smallest was 0.06m. The upper part of the post was collected (sample 4201). It was made from a piece of alder/hazel roundwood with a diameter of 0.055m and was firmly embedded in a silt exposure. As tidal and wave erosion destroyed the stratigraphy above the timber it was not possible to firmly establish its stratigraphic relationship with the silt. However, it is most likely that it was driven in from above after the silt had finished forming.

The monument was first observed in February 2003; it is not clear how long after its first exposure this was. As the post was firmly embedded in the peat it is very likely that it was found in its original position and, other than it being eroded and smoothed by the sea, it is unlikely it had been disturbed from its original position.

Objectives: as tidal erosion has removed deposits that once covered the structure (HER 38212), no stratigraphic information is available. As yet no artefactual material has been recovered from the peat in the vicinity of the monument. In the absence of stratigraphic and artefactual information, radiocarbon dating of timbers from the structure will provide an absolute date for the structure.

Dating of the timber, in conjunction with a second sample from another timber (38212–4202–2), will provide the absolute date. This will then provide useful information to enable the fitting of the monument in the known environmental sequence for Holme Beach. It will also allow the structure to be placed more firmly within the Holme Beach archaeological sequence. At present Bronze Age and Saxon monuments have been identified. As two post alignments at Holme Beach have been dated to the Saxon period (HER 38402 and HER 38043), this post arrangement has the potential to be Saxon in date. If it is, the dating of the two samples would contribute to the understanding of the landscape around both the known Saxon monuments. If it is of earlier or later date the work will provide details on how the local settlement and land-use evolved over time. It will also help structural and constructional comparisons to be drawn with other British and north-western European monuments of similar date.

Calibrated date: 1σ: 1600–1430 cal BC
2σ: 1630–1410 cal BC

Final comment: D Robertson (9 May 2005), as arrangement of the eight timbers was reminiscent of the Saxon fish-trap recorded in 1999 and 2003 (Holme fish-trap I, NHER

38042), it was thought that it would be early medieval in date. However, this was not the case, with the sample dating to the early to middle Bronze Age. This date was statistically consistent with that of the second sample.

The dating of the two samples demonstrated the use of Holme Beach in the few centuries after the construction of the Holme I timber circle (NHER 33771; dated by dendrochronology to 2049 BC). It will allow broader understandings of the local Bronze Age landscape to be developed and will enable comparisons to be made with other known Bronze Age structures.

Laboratory comment: English Heritage (November 2014), the two measurements from this structure (GU-6026–7) are statistically consistent (T′=0.0; T′(5%)=3.8; Ward and Wilson 1978).

References: Ward and Wilson 1978

GU–6027 3230 ±50 BP

δ¹³C: -29.1‰

Sample: 38212–2402–2, submitted on 2 March 2004 by D Robertson

Material: wood (waterlogged): *Salix/Populus sp.* (125g) (M Taylor 2004)

Initial comment: as GU-6026, except this sample comes from upper part of a willow or poplar with diameter of 0.042m.

Objectives: as GU-6026

Calibrated date: 1σ: 1600–1430 cal BC
2σ: 1630–1410 cal BC

Final comment: see GU-6026

Laboratory comment: see GU-6026

GU–6028 1310 ±50 BP

δ¹³C: -28.0‰

Sample: 39586–1405–5, submitted on 25 March 2004 by D Robertson

Material: wood (waterlogged): *Alnus glutinosa* (420g) (M Taylor 2004)

Initial comment: the sample comes from a timber (context 5) that formed part a fishtrap located on Holme Beach. It was V-shaped structure made up of over 180 timbers. The upper part of a leaning roundwood post was collected from the centre of the western arm of the structure (context 5; sample 1405). The timber had a diameter of 0.09m and was firmly embedded in a peat bed. As tidal and wave erosion had destroyed the stratigraphic relationship with the peat. However, it is most likely that it was probably driven in from above after the peat had finished forming. This is because two fishtraps (HER 38042 and 38043) dated to between the fifth and tenth centuries cal AD were recorded close by in 1999; it was thought these were used in a beach environment rather than an environment in which peat formed.

The monument was first observed in February 2003; it is not clear how long after its first exposure this was. As the timber was firmly embedded in the peat it is very likely that it was found in its original position and, other than it being eroded and smoothed by the sea, it is unlikely it had been disturbed from its original position.

Objectives: as tidal erosion has removed deposits that once covered the structure (HER 39568), no stratigraphic information is available. As yet no artefactual material has been recovered from the peat in the vicinity of the monument. In the absence of stratigraphic and artefactual information, radiocarbon dating of timbers from the structure will provide an absolute date for the structure.

Dating of the timber, in conjunction with a second sample from another timber (39586–1407–7), will provide the absolute date. This will then provide useful information to enable the fitting of the monument in the known environmental sequence for Holme Beach. It will also allow the structure to be placed more firmly within the Holme Beach archaeological sequence. At present Bronze Age and Saxon monuments have been identified. The structure has the potential to be Saxon in date and if it is the dating of the sample would contribute to the understanding of the landscape round the known Saxon fishtraps. If it is of earlier or later date the work will provide details on how the local settlement and land-use evolved over time. It will also help structural and constructional comparisons to be drawn with other British and north-western European monuments of similar date.

Calibrated date: *1σ:* cal AD 650–770
                        2σ: cal AD 640–780

Final comment: D Robertson (9 May 2005), the sample date is statistically consistent with that of the second sample (GU-6029). These date fish-trap III to the middle Saxon period and are broadly comparable with the dates from adjacent fish-trap I (NHER 38042; although there is some uncertainty regarding its dating). As a result, it seems probable that they were built as part of the same complex of fish-traps.

The dating of the sample also confirmed that saltwater resources were exploited at Holme during the middle Saxon period. It will also allow comparisons to be made with other known middle Saxon fish-traps in the future.

Laboratory comment: English Heritage (November 2014), the two measurements on this structure (GU-6028–9) are statistically consistent (T′=0.0; T′(5%)=3.8; *v*=1; Ward and Wilson 1978).

References: Ward and Wilson 1978

GU–6029 1310 ±50 BP

δ¹³C: -28.5‰

Sample: 39586–1407–7, submitted on 25 March 2004 by D Robertson

Material: wood (waterlogged): cf *Alnus/Corylus* sp. (300g) (M Taylor 2004)

Initial comment: as GU-6028, except that the sample comes from the upper part of an upright ?branchwood post was collected from the northern end of the western arm of the structure (context 7; structure 1407). The timber has an eccentric pith, a diameter of 0.075/0.1m and was firmly embedded in a peat bed.

Objectives: as GU-6028

Calibrated date: *1σ:* cal AD 650–770
                        2σ: cal AD 640–780

Final comment: see GU-6028

Laboratory comment: see GU-6028

GU–6030 1480 ±50 BP

δ¹³C: -25.9‰

Sample: 37613–2803–3, submitted on 2 March 2004 by D Robertson

Material: wood (waterlogged): *Quercus* sp., squashed roundwood (900g) (M Taylor 2004)

Initial comment: the sample comes from a timber (context 3) that formed part a fishtrap located on Holme Beach. It was V-shaped structure made up of at least 39 timber posts. The upper part of a leaning oak roundwood post was collected from towards the southern end of the western arm of the structure (context 3; sample 2803). The timber had a diameter of 0.089m. As tidal and wave erosion had destroyed the stratigraphy above the timber it was not possible to firmly establish its stratigraphic relationship with the silt and the peat beds located to the north. However, it is probable that it was driven in from above after the peat had finished forming. This is because two fishtraps (HER 38042 and 38043) dated to between the fifth and tenth centuries cal AD were recorded close by in 1999; it was thought these were used in a beach environment rather than an environment in which peat formed.

The monument was first observed in 2001; it is not clear how long after its first exposure this was. As the timber was firmly embedded in the peat it is very likely that it was found in its original position and, other than it being eroded and smoothed by the sea, it is unlikely it had been disturbed from its original position.

Objectives: as tidal erosion has removed deposits that once covered the structure (HER 37613), no stratigraphic information is available. As yet no artefactual material has been recovered from the peat in the vicinity of the monument. In the absence of stratigraphic and artefactual information, radiocarbon dating of timbers from the structure will provide an absolute date for the structure.

Dating of the timber, in conjunction with a second sample from another timber (37613–2804–47), will provide the absolute date. This will then provide useful information to enable the fitting of the monument in the known environmental sequence for Holme Beach. It will also allow the structure to be placed more firmly within the Holme Beach archaeological sequence. At present Bronze Age and Saxon monuments have been identified. Of the monuments dated to the Saxon period, one was a V-shaped fishtrap (HER 38042). As structure HER 37613 was broadly comparable in plan to this fishtrap, it has the potential to be Saxon in date. If it is, the dating of the sample would contribute to the understanding of the development of the fishtraps on Holme Beach and the landscape around them. If it is of earlier or later date the work will provide details on how the local settlement and land-use evolved over time. It will also help structural and constructional comparisons to be drawn with other British and north-western European monuments of similar date.

Calibrated date: *1σ:* cal AD 540–640
                        2σ: cal AD 420–660

Final comment: see GU-6028

Laboratory comment: English Heritage (November 2014), the two measurements on this structure (GU-6030–1) are statistically consistent (T'=0.2; T'(5%)=3.8; ν=1; Ward and Wilson 1978).

References: Ward and Wilson 1978

GU–6031 1450 ±50 BP

δ¹³C: -24.2‰

Sample: 37613–2804–4, submitted on 2 March 2004 by D Robertson

Material: wood (waterlogged): *Quercus* sp., roundwood including sapwood (675g) (M Taylor 2004)

Initial comment: as GU-6030, except the sample comes from the upper part of a leaning oak roundwood post which was collected from towards the northern end of the western arm of the structure (context 4; sample 2804). The timber has a diameter of 0.096m.

Objectives: as GU-6030

Calibrated date: *1σ:* cal AD 560–650
                        2σ: cal AD 530–670

Final comment: see GU-6028

Laboratory comment: see GU-6030

GU–6032 2860 ±50 BP

δ¹³C: -26.4‰

Sample: 38221–1204–4, submitted on 25 February 2004 by D Robertson

Material: wood (waterlogged): *Quercus* sp., with bark and sapwood (440g) (M Taylor 2004)

Initial comment: the sample comes from a timber (context 4) that formed part of a trackway. The trackway was slightly sinuous and aligned roughly north-west to south-east. 40m long, it comprised 70 timber posts and a series of laid branches and pieces of wood. The sample is from a horizontally laid timber located towards the south-east of the monument. The timber is a tangentially split piece which measures 122mm by 50mm. It was laid with the outside of the tree facing downwards. A section measuring 100mm long was sawn from the south-western end of timber (sample 1204; the sample is a full cross section through the timber). The timber was firmly embedded in the silt surface of the beach; it is more than probable that it was originally laid on the surface of the silt.

The monument was first observed in February 2001; it is not clear how long after its first exposure this was. As the timber was firmly embedded in the silt it is very likely that it was found close its original position and, other than it being eroded and smoothed by the sea, it is unlikely it had been disturbed greatly.

Objectives: as tidal erosion has removed deposits that once covered the structure (HER 38221), no stratigraphic information is available. As yet no artefactual material has been recovered from the peat in the vicinity of the

monument. In the absence of stratigraphic and artefactual information, radiocarbon dating of timbers from the structure will provide an absolute date for the structure.

Dating of the timber, in conjunction with a second sample from another timber (38221–1206–6), will provide the absolute date. This will then provide useful information to enable the fitting of the monument in the known environmental sequence for Holme Beach. It will also allow the structure to be placed more firmly within the Holme Beach archaeological sequence. At present Bronze Age and Saxon monuments have been identified. The structure has the potential to early Bronze Age in date and if it is the dating of the sample would contribute to the understanding of the landscape around both Holme I and Holme II timber circles. If it is of earlier or later date the work will provide details on how the local settlement and land-use evolved over time. It will also help structural and constructional comparisons to be drawn with other British and north-western European monuments of similar date.

Calibrated date: *1σ:* 1120–930 cal BC
                        2σ: 1210–900 cal BC

Final comment: D Robertson (9 May 2005), the sample date is statistically consistent with that of the second sample (GU-6033). This successfully dated trackway II to the middle to late Bronze Age, at least 700 years later than expected and long after the construction of the Holme II timber circle (NHER 38044; dated to 2400–2030 cal BC). The dating potentially places the construction of the monument around the time of the marine transgression at Holme and opens up the possibility that the trackway was built to give access to freshwater marshes during a period of environmental change. The dating also allows broader understandings of the local Bronze Age landscape to be developed and will enable comparisons to be made with other known Bronze Age structures.

Laboratory comment: English Heritage (November 2014), the two measurements from this structure (GU-6032–3) are statistically consistent (T'=1.3; T'(5%)=3.8; ν=1; ward and Wilson 1978).

References: Ward and Wilson 1978

GU–6033 2950 ±50 BP

δ¹³C: -28.6‰

Sample: 38221–1206–6, submitted on 25 February 2004 by D Robertson

Material: wood (waterlogged): *Fraxinus excelsior*, roundwood (190g) (M Taylor 2004)

Initial comment: as GU-6033, except the sample comes from context 6; sample 1206. The timber is a half split piece of ash which would have had an original diameter of *c* 95mm.

Objectives: as GU-6033

Calibrated date: *1σ:* 1230–1050 cal BC
                        2σ: 1380–1000 cal BC

Final comment: see GU-6032

Laboratory comment: see GU-6032

GU–6034 1090 ±50 BP

δ¹³C: -26.6‰

Sample: 38222–2206–6, submitted on 15 March 2004 by D Robertson

Material: wood (waterlogged): cf *Alnus/Corylus* sp. (600g) (M Taylor 2004)

Initial comment: the sample comes from a timber (context 6) that formed part of post alignment - Holme fishtrap V. Aligned north-east to south-west, 39 posts were irregularly spaced across a distance of 28m. All were oval in cross-section and had widths between 0.04m and 0.19m. The most visible post protruded 0.4m above the level of the surrounding sand. The structure appeared to have been a southerly extension of the eastern arm of the Holme fishtrap I which was identified in 1999. The upper part of a leaning post was collected from the central part of the surviving alignment (sample 2206). It was made from alder or hazel roundwood, slightly distorted, had a diameter of 0.084m and was firmly embedded in a silt exposure. As tidal and wave erosion had destroyed the stratigraphy above the time it was not possible to firmly establish its stratigraphic relationship with the peat. However, it is most likely that it was driven in from above the peat after it had finished forming. This is because between 1999 and 2001 the structure was dated to between the fifth and tenth centuries cal AD (GU-5800 and GU-5201; Bayliss *et al* 2016, 145); it was thought then that it was used in a beach environment rather than an environment in which peat formed. The monument was first observed in February 2003; it is not clear how long after its first exposure this was. As the timber was firmly embedded in the silt it is very likely that it was found close its original position and, other than it being eroded and smoothed by the sea, it is unlikely it had been disturbed greatly.

Objectives: as tidal erosion has removed deposits that once covered the structure (HER 38222), no stratigraphic information is available. As yet no artefactual material has been recovered from the peat in the vicinity of the monument. In the absence of stratigraphic and artefactual information, radiocarbon dating of timbers from the structure will provide an absolute date for the structure.

Dating of the timber, in conjunction with a second sample from another timber (38222–2203–2), will provide the absolute date. This will then provide useful information to enable the fitting of the monument in the known environmental sequence for Holme Beach. Dating will also allow the structure to be placed within the Holme Beach archaeological sequence. The structure appeared to be a southern extension of Holme fishtrap (HER 38042), and dating of the sample has the potential to confirm this. If it is Saxon in date, like Holme fishtraps I and II, the work would contribute to understanding the development of the fishtraps and the landscape around them. If it is of earlier or later date the work will provide details on how the local settlement and land-use evolved over time. It will also help structural and constructional comparisons to be drawn with other British and north-western European monuments of similar date.

Calibrated date: *1σ:* cal AD 890–1020
                       2σ: cal AD 770–1030

Final comment: D Robertson (9 May 2005), the sample date is statistically consistent with that of the second sample (GU-6035). These date fish-trap V to the middle/late Saxon period. They could be interpreted as similar or slightly later than those from fish-trap I (NHER 38042), although there is some uncertainty regarding the dating of fish-trap I. As the locations of the two monuments suggest that they were related, it is probable that they were built at about the same time.

The dating of the sample confirmed that saltwater resources were exploited at Holme during the middle/late Saxon period. It will also allow comparisons to be made with other known middle/late Saxon fish-traps in the future.

Laboratory comment: English Heritage (November 2014), the two measurements on this structure (GU-6034–5) are statistically consistent (T′=1.3; T′(5%)=3.8; ν=1; Ward and Wilson 1978).

References: Ward and Wilson 1978

GU–6035 1170 ±50 BP

δ¹³C: -30.9‰

Sample: 38222–2203–2, submitted on 2 March 2004 by D Robertson

Material: wood (waterlogged): *Alnus glutinosa* (300g) (M Taylor 2004)

Initial comment: as GU-6034, except that this sample comes from context 2; sample 2203. The upper part of a leaning post (2) was collected from the southern part of the surviving alignment. It was made from alder roundwood, was slightly distorted, had a diameter of 0.54m/0.088m and was firmly embedded in a silt exposure.

Objectives: as GU-6034

Calibrated date: *1σ:* cal AD 770–950
                       2σ: cal AD 690–990

Final comment: see GU-6034

Laboratory comment: see GU-6034

Howick, Sea Houses Farm: Environmental, Northumberland

Location: NU 25801628
                     Lat. 55.26.34 N; Long. 01.35.34 W

Project manager: I Boomer (University of Newcastle upon Tyne), 2002

Archival body: University of Newcastle upon Tyne

Objectives: to assess whether the lowest units of Holocene age are contemporary with the occupation of the Mesolithic structure. They will also provide a temporal framework for the environmental changes witnessed in sedimentological and microfossil record from the core.

Final comment: I Boomer (4 July 2004), these dates have permitted us to establish a temporal framework for the environmental changes recorded in core HEX11007.

Although the cored sediments began accumulating before the Mesolithic structure was occupied, it would appear that the period represented by Mesolithic occupation is missing from the core due to erosion. Nevertheless, the dates have allowed us to create a detailed history of local environmental conditions at Howick during the last 8000 years.

The evidence indicates a major hiatus between approximately 11000 and 8000 years BP (including the period of Mesolithic occupation), represented by a 30cm layer of course sands and sandstone pebbles, probably as a result of a significant high-energy event dated to about 8300 cal BP. The age and context suggests that this may be associated with the Storegga Slide event, already well-documented along the eastern coast of Scotland.

Laboratory comment: English Heritage (24 June 2014), nine further samples from this series were dated previously (OxA-11833, -11852, 11858–60, 11870, -11936, and 12824–5; Bayliss *et al* 2016, 153-5).

References: Boomer *et al* 2007
                Waddington 2007

OxA–12944 2982 ±34 BP

δ¹³C: -26.3‰

Sample: Core 11007 11, submitted in August 2003 by I Boomer

Material: waterlogged plant macrofossil: *Phragmites* (I Boomer 2003)

Initial comment: this sample comes from a drill depth of 1.57m (decompacted to 1.35m).

Objectives: this sample and sample 12 date the transitional phase between woodland and grassland.

Calibrated date: 1σ: 1270–1120 cal BC
                    2σ: 1380–1110 cal BC

Final comment: I Boomer (4 July 2004), this youngest dated material within the core provides an important upper 'tie-point' for the age-depth model, and indicates a much reduced sedimentation rate in the upper few metres of the sequence.

OxA–12945 3755 ±35 BP

δ¹³C: -27.2‰

Sample: Core 11007 12, submitted in August 2003 by I Boomer

Material: waterlogged plant macrofossil (1.60g) (herbaceous material, ?leaf) (R Gale 2003)

Initial comment: this sample comes from a drill depth of 1.78m (decompacted to 1.67m).

Objectives: as OxA-12944

Calibrated date: 1σ: 2210–2130 cal BC
                    2σ: 2290–2030 cal BC

Final comment: I Boomer (4 July 2004), this sample dated slightly older than would be expected given the overall age-depth relationship within the core. It is ssumed to have been reworked.

OxA–12946 3820 ±35 BP

δ¹³C: -27.7‰

Sample: Core 11007 14, submitted in August 2003 by I Boomer

Material: waterlogged plant macrofossil (1.80g) (alder roundwood, including bark, diameter 7mm) (R Gale 2003)

Initial comment: this sample comes from a drill depth of 2.13m (decompacted to 2.01m).

Objectives: this sample combines with sample 13 to date the initial phase of clearance and the beginning of the cereal peak.

Calibrated date: 1σ: 2340–2200 cal BC
                    2σ: 2460–2140 cal BC

Final comment: I Boomer (4 July 2004), this sample dates the beginning of forest clearance and the beginning of cereal peak for region. By this time the depositional setting is much like we see it today.

OxA–12947 5846 ±38 BP

δ¹³C: -29.1‰

Sample: Core 11007 15, submitted in August 2003 by I Boomer

Material: wood (2.40g) (waterlogged): *Alnus* sp. (R Gale 2003)

Initial comment: this sample comes from a drill depth of 2.97m (decompacted to 2.97m).

Objectives: this sample would date a period of valley floor woodland disturbance, including alder, oak, and hazel. Many ruderals pick up around this point including a large spike in fern spores. This indicates an opening of the forest canopy, possibly due to a flooding event. This sample also dates the final appearance of calcareous microfossils in the core and the middle of the sequence containing a series of sandy laminae. This will anchor the environmental data at the time of cist burials.

Calibrated date: 1σ: 4770–4680 cal BC
                    2σ: 4800–4600 cal BC

Final comment: I Boomer (4 July 2004), dates a possible flood-event due to occurrence in association with small, repeating sandy horizon. Also dates the fern spike (possibly associated with woodland disturbance including opening of the forest canopy). Also marks the last calcareous microfossil assemblage. Between this and next youngest date (OxA-12946) the depositional setting changes from permanent aquatic setting to temporary/seasonal floodplain.

OxA–12948 6473 ±37 BP

δ¹³C: -25.7‰

Sample: Core 11007 16, submitted in August 2003 by I Boomer

Material: wood (0.40g) (waterlogged; oak/ash knotty fragment from very degraded narrow roundwood) (R Gale 2003)

Initial comment: this sample comes from a drill depth of 3.66m (decompacted to 3.51m).

Objectives: this sample would date a period of forest stability prior to the woodland disturbance (dated by sample 15; OxA-12947). The sample dates the top of the sand body, a significant period of local environmental event.

Calibrated date: *1σ:* 5480–5380 cal BC
                       2σ: 5490–5360 cal BC

Final comment: I Boomer (4 July 2004), date appears anomalously old given the overall pattern of age-depth relationships. The sample's stratigraphical occurrence immediately above the coarse sand body would support the hypothesis of possible reworking.

OxA–12949 5985 ±36 BP

δ¹³C: -31.3‰

Sample: Core 11007 17, submitted in August 2003 by I Boomer

Material: waterlogged plant macrofossil (0.70g) (woody stem, diameter 4mm) (I Boomer 2003)

Initial comment: this sample comes from a drill depth of 4.06m (decompacted to 4.06m).

Objectives: this sample would date the end of a period of pulsed woodland disturbances in all the tree taxa, and associated with ruderal presence in the decline stages. This would provide an end date for what may have been episodic farming activity. The sample also comes from immediately below a significant sand deposit, probably initially erosive so that some sediment will have been lost. Dates immediately above and below will reveal the timing and how much sediment was lost.

Calibrated date: *1σ:* 4940–4800 cal BC
                       2σ: 4990–4780 cal BC

Final comment: I Boomer (4 July 2004), dates the end of woodland disturbance and provides limiting date for the overlying sand deposit, which may prove to be a significant erosional feature. This interpretation is supported by anomalously old macrofossils above the sand (OxA-12948 and OxA-13028).

OxA–12950 6290 ±45 BP

δ¹³C: -30.1‰

Sample: Core 11007 18, submitted in August 2003 by I Boomer

Material: waterlogged plant macrofossil (0.50g) (stem diameter 1mm) (I Boomer 2003)

Initial comment: this sample comes from a drill depth of 4.37m (decompacted too 4.37m).

Objectives: this sample dates a period of forest recovery in this phase of pulsed disturbances. This would date a period of potential abandonment of farming during the Neolithic when the woodland recovers. It coincides with the peak abundance of ostracods in the core indicating establishment of a shallow, increasingly freshwater environment.

Calibrated date: *1σ:* 5320–5210 cal BC
                       2σ: 5370–5200 cal BC

Final comment: I Boomer (4 July 2004), together with OxA-12951 confirms continued sedimentation rate and dates both period of forest recovery and peak abundance of ostracods in an increasingly freshwater setting.

OxA–12951 6274 ±38 BP

δ¹³C: -26.6‰

Sample: Core 11007 19, submitted in August 2003 by I Boomer

Material: waterlogged plant macrofossil (1.20g) (hazelnut roundwood, diameter 5mm) (R Gale 2003)

Initial comment: this sample comes from a drill depth of 4.44m (decompacted to 4.44m).

Objectives: as OxA-12950, but probably closer to the penultimate recovery phase.

Calibrated date: *1σ:* 5310–5210 cal BC
                       2σ: 5330–5200 cal BC

Final comment: I Boomer (4 July 2004), together with OxA-12950, these confirm a relatively stable continuation of sediment rate and date the period of forest recovery within local pollen zone HOW 2–4. Also dates the peak abundance of ostracods.

OxA–12952 6988 ±37 BP

δ¹³C: -26.5‰

Sample: Core 11007 21, submitted in August 2003 by I Boomer

Material: waterlogged plant macrofossil (0.60g) (hazelnut shell) (I Boomer 2003)

Initial comment: sample comes from a drill depth of 5.80m (decompacted to 5.73m). This and sample 22 are from an organic rich horizon (plant macrofossils), barren of calcareous microfossils but immediately below the first 'flood event' in the sequence marked by a thin sand horizon at 5.72–5.57cm.

Objectives: this sample would date the first large peak of alder in the valley, which seems to be quite late in comparison to the alder rise elsewhere in the UK. This would date the onset of warmer Atlantic conditions and this is usually related to the beginnings of farming. This sample and the one below would also provide an important age-depth link between the next oldest sample at 6.30m and the next youngest at 5.73m.

Calibrated date: *1σ:* 5980–5830 cal BC
                       2σ: 5990–5750 cal BC

Final comment: I Boomer (4 July 2004), one of two paired samples from the same hazelnut shell (*see* OxA-12953) and the dates agree with this interpretation. Dates the last appearance of foraminifera in the sequence and therefore subsequently, the salinity maximum is about 8‰.

Laboratory comment: English Heritage (November 2014), the measurements on this hazelnut are not statistically consistent (T′=5.8; T′(5%)=3.8; ν=1; Ward and Wilson 1978), although given they are at 99% confidence (T′=5.8; T′(5%)=6.0; ν=1; Ward and Wilson 1978), a weighted mean (7050 ±27 BP) which calibrates to 6010–5880 cal BC at 95% confidence (Reimer *et al* 2009), provides the best estimate for the date of the nut.

References: Reimer *et al* 2009
                      Ward and Wilson 1978

OxA–12953 7117 ±39 BP

$\delta^{13}C$: -26.1‰

Sample: Core 11007 21, submitted in August 2003 by I Boomer

Material: waterlogged plant macrofossil (0.60g) (hazelnut shell) (I Boomer 2003)

Initial comment: as OxA-12952

Objectives: as OxA-12952

Calibrated date: 1σ: 6030–5980 cal BC
                         2σ: 6060–5910 cal BC

Final comment: see OxA-12952

Laboratory comment: see OxA-12952

OxA–12954 7075 ±37 BP

$\delta^{13}C$: -30.7‰

Sample: Core 11007 22, submitted in August 2003 by I Boomer

Material: wood (0.70g) (waterlogged): bark, sliver (R Gale 2003)

Initial comment: this sample comes from a drill depth of 5.83m (decompacted to 5.76m).

Objectives: as OxA-12952, but possibly closer to the penultimate recovery phase.

Calibrated date: 1σ: 6010–5910 cal BC
                         2σ: 6020–5880 cal BC

Final comment: I Boomer (4 July 2004), submitted as one of two samples from approximately same level to provide a chronological framework for the palaeoenvironmental record (along with OxA-12952 and OxA-12953). The result dates the first large peak of alder locally.

OxA–12967 3311 ±35 BP

$\delta^{13}C$: -25.8‰

Sample: Core 11007 13, submitted in August 2003 by I Boomer

Material: waterlogged plant macrofossils (0.50g) *(Phragmites)* (I Boomer 2003)

Initial comment: this sample comes from a drill depth of 2.00m (decompacted to 2.00m).

Objectives: this sample dates the removal of alder and hazel from the valley floor and sides, along with the decline of other woodland taxa like *Hedera*. This is one of the major zone transitions identified. It dates the beginning of woodland clearance and the rise of grass and bracken together with many ruderals. It is also an important sedimentological level marking the end of sandy laminae and probably indicates the onset of stability in the valley floor. Subsequent sedimentation is likely due to occasional overbank flooding. This would date the first large scale clearance, probably associated with the early Bronze Age, and which would go with the cist burials at the same time.

Calibrated date: 1σ: 1630–1520 cal BC
                         2σ: 1690–1500 cal BC

Final comment: I Boomer (4 July 2004), dates the onset of stability in the valley floor (sedimentologically) and the boundary between local pollen zones HOW 2–4 and 2–5.

OxA–13028 6661 ±38 BP

$\delta^{13}C$: -24.3‰

Sample: Core 11007 16, submitted in August 2003 by I Boomer

Material: waterlogged plant macrofossil (0.40g) (oak/ash knotty fragment from very degraded narrow roundwood) (R Gale 2003)

Initial comment: as OxA-12948

Objectives: as OxA-12948

Calibrated date: 1σ: 5630–5550 cal BC
                         2σ: 5650–5510 cal BC

Final comment: I Boomer (4 July 2004), the date of the sample proves anomalously old given overall age-depth model. Therefore, in conjunction with the sedimentological evidence, the sample is believed to have been reworked.

OxA–13029 6586 ±39 BP

$\delta^{13}C$: -27.9‰

Sample: Core 11007 20, submitted in August 2003 by I Boomer

Material: waterlogged plant macrofossil (28g) (alder; wide roundwood) (R Gale 2003)

Initial comment: this sample comes from a drill depth of 4.85m (decompacted to 4.74m).

Objectives: this sample will date the initial phase of the disturbance sequence and also the main phase of alder growth. Furthermore, it will date the beginning of the woodland interference, possibly the onset of agriculture, and the first appearance of freshwater ostracods (ie the point at which the rate of sedimentation and rebound exceeds sea-level rise).

Calibrated date: 1σ: 5610–5480 cal BC
                         2σ: 5620–5470 cal BC

Final comment: I Boomer (4 July 2004), submitted to establish the chronological framework for the palaeoenvironmental record and the sedimentation rate.

Marks the start of local pollen zone HOW 2–4, which records the beginning of woodland disturbance (this sample comes from a large roundwood fragment), and also the beginning of freshwater ostracod assemblages – the first time that the rate of sea-level rise is exceeded by sedimentation.

OxA–13370 10000 ±90 BP

$\delta^{13}C$: -29.1‰

Sample: Core 11007 A/06, submitted on 26 August 2002 by I Boomer

Material: waterlogged plant macrofossils (2g) (herbaceous stem, indet) (R Gale 2002)

Initial comment: sample 6 was taken from near the base of a dark-grey to black, organic-rich, peaty silt from a depth of about 7.70m. It is considered that this material represents a combination of detrital mineral grains and *in situ* organics – probably the result of deposition in a 'floodplain' environment.

Objectives: this sample is taken from the base of the next youngest sedimentary unit and marks the beginning of the deposition of fine-grained sediments containing important microfossil remains.

Calibrated date: 1σ: 9770–9310 cal BC
                  2σ: 10010–9280 cal BC

Final comment: I Boomer (4 July 2004), dates the earliest unequivocal plant macrofossil remains recovered from the core.

Kinsey Cave, North Yorkshire

Location: SD 80376588
                  Lat. 54.05.12 N; Long. 02.18.00 W

Project manager: T Taylor (University of Bradford), 2005

Description: a multi-period cave site. Historic excavations by W Kinsey-Mattinson in 1925–31 are unpublished, but produced a range of material, especially of a late Roman/Romano-British character, but also mammal bones suggestive of earlier dates. Significant archive material is held in the Lord Cave Archive. The site was scheduled in 1949 and revised in 1992, noting potential palaeolithic importance, as perhaps four late-glacial lithics have been potentially associated with the site (no known stratigraphic relationship or clear provenance). Recent AMS dating on archive antler and bone has produced a range of dates from OIS2 (OxA-2456: 11270 ±110 BP, 11380–10960 cal BC at 95% confidence; Reimer *et al* 2013; Hedges *et al* 1992b), through the Mesolithic/Neolithic boundary (human femur and mandible, 2005 dates: OxA-14798, 5111 ±36 BP (3980–3790 cal BC at 95% confidence; Reimer *et al* 2013) and OxA-14799, 5074 ±36 BP (3970–3770 cal BC at 95% confidence; Reimer *et al* 2013), through to a Romano-British ritually deposited horse (OxA-14797: 1807 ±31 BP (cal AD 120–330 at 95% confidence; Reimer *et al* 2013) and an early medieval lynx (OxA-12026, 1550 ±24 BP (cal AD 420–580 at 95% confidence; Reimer *et al* 2013). In March 2005 onwards, visits to the cave identified human bone in recent badger disturbance in the mouth of the cave, as well as

significant vulnerable bone (such as brown bear: sample F101) present on a surface towards the rear of the cave. Subsequent fieldwork to resurvey the cave and conduct limited excavations in the area of immediate badge damage resulted in a three-week fieldwork campaign in September 2005 which, although post-excavation phase funding is pending, produced significant recognisable bone, including that of wild and domestic animals, and human bone with processing traces (consistent with unpublished observations made on now dated archive human bone). This was concentrated in two 'inclusions' in contexts in which some lithics were present - chert but also early Neolithic style flint (Jacobi pers comm.) - but no ceramics, suggestive perhaps of a Mesolithic-Neolithic interface date. Recovery of surface bone, such as brown bear, from the rear of the cave in the area, where it is assumed that reindeer and bison previously originated, suggests potential late-glacial fauna, though following the recent late dating of lynx (the most recent in Britain: Heatherington *et al* 2006), dating of this material remains an open question.

Objectives: to date the faunal remains from the cave.

Final comment: T Taylor (2011), coupled with the results of the fieldwork, we can now call in to question the primary significance of the site as a late-glacial hunting location modified by Roman cultic activity. The site is multiphase, and has preserved evidence for a long sequence of natural and cultural modifications and changes. Archival evidence certainly supports the idea of activity in both the periods mentioned to some degree, but the 2005 intervention recovered a significantly different data set which, with the assistance of radiocarbon dating, was shown to be consistent with a key sub-section of the curated archive. Notably this included evidence for human bone, dating to what is currently presumed to be the very earliest Neolithic occupation of the region (contemporary with long barrow sites in the south of England, and consistent with the earliest dates in Ireland and Scotland). Significant, however, was the form of partially surviving deposit in which these remains were found (inclusion #2), representing a small and apparently deliberate concentration of human bones, including those of children and infants, with those of domesticated and wild animals, spanning a long but discrete period of 2,000 years, and indicating a *terminus ante quem* for the accumulation in the early Bronze Age. The sequence of wild fauna in the cave, as determined from both the EH-funded work and other analyses is equally significant, with implications for understanding the environment in the early Holocene, and in the historical, early medieval period. The cave has produced definitive evidence for the survival of both lynx and brown bear in the wild in northern England in the post-Roman period, with dates of *c* AD 500; these indicate an extensive black hole in our knowledge of wild faunas, and indeed of local environments, over not just preceding centuries, but preceding millennia.

References: Heatherington *et al* 2006
                       Hedges *et al* 1992b, 141–2
                       Reimer *et al* 2013
                       Taylor *et al* 2011

Kinsey Cave: sample series #1, North Yorkshire

Location: SD 80376568
                   Lat. 51.05.12 N; Long. 02.18.00 W

Project manager: T Taylor (University of Bradford), 2005

Archival body: Cravern Museum

Description: material from inclusion 1, either representing a recent badger scatter from inclusion 2 under the cave wall, or the deflated remains of the original context(s) on the cave floor undisturbed during Mattinson's historic trenching operation.

Objectives: to establish the date coherence or discrepancy with inclusion 2 (sample series #2).

Laboratory comment: English Heritage (13 July 2006), this sample from inclusion 1 dates to near the end of this period of accumulation. The human bone in inclusion 2 (OxA-15790) may thus be residual.

References: Taylor *et al* 2011

SUERC–10515 3740 ±40 BP

$\delta^{13}C$: -22.6‰
$\delta^{15}N$ *(diet):* +5.7‰
C/N ratio: 3.3

Sample: F 127, submitted on 28 March 2006 by T Taylor

Material: animal bone: *Sus* sp., ulna (T O'Connor and A Ogden 2006)

Initial comment: material from inclusion 1: either badger scattered or disturbance from Mattinson's excavations.

Objectives: this may be a domestic pig in a Mesolithic or Neolithic context; a date will help establish the cultural context of the deposit.

Calibrated date: 1σ: 2210–2040 cal BC
                    2σ: 2290–2020 cal BC

Final comment: T Taylor (2011), the sample from inclusion 1 dates to near the end of the period of accumulation of inclusion 2 - towards the early second millennium cal BC.

Kinsey Cave: sample series #2, North Yorkshire

Location: SD 80376568
                   Lat. 54.05.12 N; Long. 02.18.00 W

Project manager: T Taylor (University of Bradford), 2005

Archival body: Cravern Museum

Description: human and animal bone from a tightly packed 'ball' of material designated inclusion 2, which may represent an ancient midden deposit placed at the junction of the cave floor and wall and perhaps encised with a deliberate clast placement.

Objectives: to date inclusion 2; including to establish/reject chronological coherence in the core of inclusion 2; to establish/refute chronological coherence of the material at the top and base of inclusion 2; and, to establish congruence or not with inclusion 1.

Final comment: J Meadows (13 July 2006), the animal bones in inclusion 2 accumulated throughout the third and early second millennium cal BC. The sample from inclusion 1 (SUERC-10515) dates to near the end of this period of accumulation. The human bone in inclusion 2 (OxA-15790) may thus be residual. The cattle bone from the top of inclusion 2 (SUERC-10516) is apparently older than one of the bones from the core of the inclusion (OxA-15792), and therefore also seems to be residual. The scatter of results, even excluding the human bone sample, is perhaps greater than expected. This range does provide a reasonable estimate of the date scatter of the bones in this inclusion, but naturally the bone deposit cold have formed much more rapidly than this, if we accept that some of the bones in it are reworked.

References: Taylor *et al* 2011

OxA–15788 3858 ±32 BP

$\delta^{13}C$: -20.4‰

Sample: F 212, submitted on 28 March 2006 by T Taylor

Material: animal bone: *Sus* sp., ulna (T O'Connor and A Ogden 2006)

Initial comment: it is supposed that inclusion 2 is most likely a deliberate primary prehistoric cultural deposit. It is possible, however, that it represents an agglomeration of material that has concentrated following natural transport down the back of the slope into the cave. This possibility is considered les likely due to the good preservation, lack of sorting by size, and the desposition (like 'spillikins' in a tight knot of material) of the bones.

Objectives: to date inclusion 2. The chronological coherence in the core of inclusion 2 will support the interpretation of a primary cultural deposit (midden-like) while analysis of material from the top and bottom will indicate the extent of coherence/lack of such.

Calibrated date: 1σ: 2460–2230 cal BC
                    2σ: 2470–2200 cal BC

Final comment: see series comments

OxA–15789 3982 ±32 BP

$\delta^{13}C$: -22.8‰

Sample: F 222, submitted on 28 March 2006 by T Taylor

Material: animal bone: *Canis* sp., right humerus (T O'Connor and A Ogden 2006)

Initial comment: as OxA-15788

Objectives: as OxA-15788

Calibrated date: 1σ: 2570–2460 cal BC
                    2σ: 2580–2460 cal BC

Final comment: see series comments

OxA–15790 4472 ±33 BP

$\delta^{13}C$: -20.9‰
$\delta^{15}N$ (diet): +11.3‰
C/N ratio: 3.2

Sample: F 227, submitted on 28 March 2006 by T Taylor

Material: human bone (left patella) (T O'Connor and A Ogden 2006)

Initial comment: as OxA-15788

Objectives: as OxA-15788

Calibrated date: 1σ: 3330–3090 cal BC
                  2σ: 3350–3020 cal BC

Final comment: T Taylor (2011), this human bone may be residual given the dates on the animal bones in inclusions 1 and 2.

OxA–15792 3448 ±31 BP

$\delta^{13}C$: -21.9‰

Sample: F 214, submitted on 28 March 2006 by T Taylor

Material: animal bone: *Cervus* sp., left tibia (T O'Connor 2006)

Initial comment: as OxA-15788

Objectives: as OxA-15788

Calibrated date: 1σ: 1870–1690 cal BC
                  2σ: 1890–1680 cal BC

Final comment: see series comments

SUERC–10516 3725 ±35 BP

$\delta^{13}C$: -21.1‰
$\delta^{15}N$ (diet): +3.7‰
C/N ratio: 3.1

Sample: F 199, submitted on 28 March 2006 by T Taylor

Material: animal bone: *Bos* sp., navicular cuboid (T O'Connor 2006)

Initial comment: as OxA-15788

Objectives: as OxA-15788

Calibrated date: 1σ: 2200–2030 cal BC
                  2σ: 2280–2020 cal BC

Final comment: T Taylor (2011), this bone is apparently older than the dated bone from the core of the inclusion (OxA-15792), and is therefore also likely to be residual.

SUERC–10517 4240 ±35 BP

$\delta^{13}C$: -21.9‰
$\delta^{15}N$ (diet): +4.4‰
C/N ratio: 3.3

Sample: F 261, submitted on 28 March 2006 by T Taylor

Material: animal bone: *Bos* sp., phalanx (T O'Connor 2006)

Initial comment: as SUERC-10516

Objectives: as SUERC-10516

Calibrated date: 1σ: 2900–2870 cal BC
                  2σ: 2910–2700 cal BC

Final comment: see series comments

Kinsey Cave: sample series #3, North Yorkshire

Location: SD 80376568
                 Lat. 54.05.12 N; Long. 02.18.00 W

Project manager: T Taylor (University of Bradford), 2005

Archival body: Cravern Museum

Description: large human bone surface finds disturbed by badgers in spring 2005, from the entrance to the sett in the vicinity of inclusions 1 and 2. They are a right and left distal tibia that, on metrical assessment, come from different adult individuals. One has a squatting facet indicative of squatting posture considered more typical of prehistoric than modern or historic populations.

Objectives: to establish the date and potential congruence with archive human bones dated in 2005 and material from inclusions 1 and 2.

Final comment: J Meadows (13 July 2006), all the human bones date to the early Neolithic, although they are not all of the same date, and thus represent different individuals. It is possible that the archival femur and mandible previously dated (OxA-14798 and OxA-14799) and the right tibia dated by OxA-15791 are of the same radiocarbon age (T'=0.6; T'(5%)=6.0; ν=2; Ward and Wilson 1978), although osteological analysis could eliminate the possibility that these bones are from a single person.

References: Taylor *et al* 2011
                 Ward and Wilson 1978

OxA–15791 5086 ±35 BP

$\delta^{13}C$: -20.5‰
$\delta^{15}N$ (diet): +9.9‰
C/N ratio: 3.2

Sample: F 005, submitted on 28 March 2006 by T Taylor

Material: human bone (right distal tibia) T O'Connor and A Ogden 2006)

Initial comment: a human bone disturbed by badgers in 2005.

Objectives: to establish the coherence or lack of such with new radiocarbon dates on archive human bone (*c* 3900 cal BC) and also attempt to relate chronologically to inclusions 1 and 2. As archive human bone (femur and mandible) could come from a single individual, as could the patella in inclusion 2 (F227), there is sense in dating both surface finds as metrically they derive from different adults and therefore will potentially establish date for a population presence rather than a stray individual.

Calibrated date: 1σ: 3960–3800 cal BC
                 2σ: 3970–3780 cal BC

Final comment: see series comments

SUERC–10518 4820 ±40 BP

δ¹³C: -21.4‰
δ¹⁵N (diet): +8.7‰
C/N ratio: 3.4

Sample: F 004, submitted on 28 March 2006 by T Taylor

Material: human bone (left distal tibia) (T O'Connor 2006)

Initial comment: as OxA-15791

Objectives: as OxA-15791

Calibrated date: 1σ: 3650–3530 cal BC
                      2σ: 3660–3520 cal BC

Final comment: see series comments

Kinsey Cave: sample series #4, North Yorkshire

Location: SD 80376568
                  Lat. 54.05.12 N; Long. 02.18.00 W

Project manager: T Taylor (University of Bradford), 2005

Archival body: Cravern Museum

Description: large mammal fauna of unknown date (intrinsic interest). This is a brown bear vertebra which may have lain *in situ* towards the back of Kinsey Cave for some considerable period.

Objectives: given the known material date range in the cave from the late glacial through to early medieval (and later), there is intrinsic interest in establishing whether OIS2 fauna is preserved on or eroding to, the surface, just as there is interest in establishing the probable natural environment of this human-modified site in later periods. Dating of the brown bear vertebra may also inform REE (rare earth provenancing) taphonomy, which may be trialled on the excavated material corpus.

References: Taylor *et al* 2011

SUERC–10519 1540 ±35 BP

δ¹³C: -20.5‰
δ¹⁵N (diet): +11.4‰
C/N ratio: 3.4

Sample: F101, submitted on 28 March 2006 by T Taylor

Material: animal bone (*Ursus arctos*; cervical vertebra) (T O'Connor 2006)

Initial comment: this is a brown bear vertebra which may have lain *in situ* towards the back of Kinsey Cave for some considerable period. The bone was clearly visible but just inaccessible to inadvertent human reach, surrounded by badger faeces and bedding. It was removed by an expert caver following the removal of one or two surface clasts to facilitate access and replaced with a proxy to be used in the monitoring of future badger disturbance.

Objectives: to establish whether OIS2 fauna is preserved on, or eroding to, the surface, just as there is interest in establishing the probable natural environment of this human-modified site in later periods. Dating of the brown bear vertebra may also inform REE taphonomy.

Calibrated date: 1σ: cal AD 430–570
                      2σ: cal AD 420–610

Final comment: T Taylor (2011), the bear bone could be contemporary with the previously dated lynx bone (OxA-12026; 1550 ±24 BP). This suggests Kinsey Cave to have been a significantly 'wild' place in the early medieval period, perhaps culturally so. In addition, it speaks for the integrity of the material in inclusion 2 until its very recent disturbance by badger setting.

London Shellyware: Billingsgate Lorry Park, Greater London

Location: TQ 33608110
                  Lat. 51.30.44 N; Long. 00.04.30 W

Project manager: D Hall (Scottish Urban Archaeological Unit), 1982–3

Archival body: Museum of London

Description: an excavation was carried out on the site of the Billingsgate Fish Market. Many different phases of use of the site were identified, ranging from the Roman period through to the twentieth century. From the eleventh to early thirteenth centuries the waterfront consisted of an artificial bank revetted by timber waterfronts, which were replaced at intervals. Tree-ring dating of the waterfronts was undertaken and provided a felling date of AD 1039/40 for the first waterfront, with subsequent replacements through to the thirteenth century (Hillam 1988). A substantial pottery assemblage was also retrieved which allowed the chronology of the site, and early medieval London, to be refined.

Objectives: to compare and contrast radiocarbon dates from London Shellyware vessels to those excavated in Perth, to contribute towards further refining the chronology for these vessels.

Final comment: D Hall (2010), research carried out on carbonised food residues on London Sandy Shellyware from Perth, Scotland, Bergen, Norway, and London, have all yielded radiocarbon dates between cal AD 930 and 1020. This is some 100 years or so earlier than the accepted London ceramic chronology, created over a number of years, using coin evidence, small finds, and dendrochronology. The previously widely accepted London dating has been used throughout Britain, Ireland and northern Europe to date local pottery sequences that have then provided a basis for dating other finds categories and archeological sites.

Laboratory comment: English Heritage (2010), the dates are in agreement with the model assumption that they belong to a single phase of activity. This activity dates the start of Sandy Shellyware at Billingsgate Lorry Park, London, to *cal AD 820–1020* (95% probability; *start Sandy Shellyware: Billingsgate Lorry Park, London;* Hall *et al* 2010; fig 2) or *cal AD 900–990* (68% probability; Hall *et al* 2010; fig 2). Sandy Shellyware fell out of use in this area in *cal AD 1020–1220* (95% probability; *end Sandy Shellyware: Billingsgate Lorry Park, London)* or *cal AD 1030–1130* (68% probability; Hall *et al* 2010; fig 2).

References: Blackmore and Pearce 2010
                    Hall *et al* 2010

SUERC–8979 1115 ±35 BP

$\delta^{13}C$: -27.2‰

Sample: BIG82 1; context 5868, submitted on 6 February 2006 by D Hall

Material: carbonised residue (cooking pot with thumbed rim)

Initial comment: the sherd came from dumped deposit 5868 associated with the construction of waterfront 6 (dated by dendrochronology to between AD 1080 and AD 1090).

Objectives: to establish the date of London Shelly Sandy Ware vessels which have been dated to the mid to late 12th century by association with dendrochronologically dated waterfront timbers. Similar fabrics and forms have recently been dated to 1020–1030 AD from the excavations at 75 High Street in Perth, Scotland.

Calibrated date: 1σ: cal AD 880–990
                      2σ: cal AD 770–1020

Final comment: D Hall (22 October 2014), this date is earlier than the currently accepted chronology for London Sandy Shellyware but is of a similar bracket to dates on similar fabrics from Perth and Bergen.

SUERC–8980 1105 ±35 BP

$\delta^{13}C$: -24.5‰

Sample: BIG82 4; context 6548, submitted on 6 February 2006 by D Hall

Material: carbonised residue (cooking pot)

Initial comment: from deposit 6548 associated with the construction of a rubble raft forming part of waterfront 11. It has been dated by association with a dendrochronological date of AD 1172–1216.

Objectives: as SUERC-8979

Calibrated date: 1σ: cal AD 890–990
                      2σ: cal AD 870–1020

Final comment: see SUERC-8979

SUERC–8981 1035 ±35 BP

$\delta^{13}C$: -25.0‰

Sample: BIG82 6; context 6542, submitted on 6 February 2006 by D Hall

Material: carbonised residue (cooking pot)

Initial comment: from context 6542, a deposit associated with the infilling of an inlet into the River Thames as part of a consolidation of the area behind waterfronts 8, 10, and 11. It has been dated to AD 1170–1230 based on the presence of London Rouen/North French style pottery.

Objectives: to establish the date of London Sandy Shellware vessels which have been dated to the late-twelfthth/early-thirteenth century by association with London Rouen/North French style pottery. Similar fabrics and forms have recently been dated to AD 1020–1030 from the excavations at 75 High Street in Perth, Scotland.

Calibrated date: 1σ: cal AD 980–1030
                      2σ: cal AD 900–1040

Final comment: see SUERC-8979

SUERC–8982 1000 ±35 BP

$\delta^{13}C$: -24.8‰

Sample: BIG82 7; context 6507, submitted on 6 February 2006 by D Hall

Material: carbonised residue (cooking pot with thumbed rim)

Initial comment: as SUERC-8981

Objectives: as SUERC-8981

Calibrated date: 1σ: cal AD 1010–1040
                      2σ: cal AD 980–1160

Final comment: see SUERC-8979

SUERC–8983 955 ±35 BP

$\delta^{13}C$: -23.2‰

Sample: BIG82 8; context 5072, submitted on 6 February 2006 by D Hall

Material: carbonised residue (lamp)

Initial comment: as SUERC-8981

Objectives: as SUERC-8981

Calibrated date: 1σ: cal AD 1020–1160
                      2σ: cal AD 1010–1170

Final comment: see SUERC-8979

SUERC–8984 1000 ±35 BP

$\delta^{13}C$: -25.0‰

Sample: BIG82 9; context 6395, submitted on 6 February 2006 by D Hall

Material: carbonised residue (cooking pot)

Initial comment: from context 6395, a deposit associated with the consolidation of the foreshore south of the frontage of waterfronts 8 and 10. It has been dated to the late-twelfthth/early-thirteenth century by association with London Rouen/North French style pottery.

Objectives: to establish the date of London Sandy Shellyware vessels which have been dated to between AD 1172 and 1230 by association with London Rouen/North French-style pottery. Similar fabrics and forms have recently been dated to AD 1020–1030 from the excavations at 75 High Street in Perth, Scotland.

Calibrated date: 1σ: cal AD 1010–1040
                      2σ: cal AD 980–1160

Final comment: see SUERC-8979

SUERC–8985 905 ±35 BP

δ¹³C: -28.4‰

Sample: BIG82 10; context 3299, submitted on 6 February 2006 by D Hall

Material: carbonised residue (cooking pot with applied strip)

Initial comment: from context 3299, a deposit associated with the robbing of the phase 8 revetments and construction of the foreshores. It is associated with a dendrochronological date of AD 1172–1205.

Objectives: as SUERC-8979

Calibrated date: 1σ: cal AD 1040–1170
2σ: cal AD 1020–1220

Final comment: see SUERC-8979

London: River Thames between Putney and Syon skulls, Greater London

Location:	between TQ 24200 75800 and TQ 17200 75200 Lat. ; Long. , Lt. 51.28.00.N; Long. 00.12.44.W and Lat. 51.27.46.N; Long. 00.18.47.W
Project manager:	J Sidell (Institute of Archaeology, London), 2005
Archival body:	Museum of London

Description: in 2005, six bone samples from human skulls were submitted to Oxford Radiocarbon Accelerator Unit, Oxford University, for accelerator mass spectrometry (AMS) radiocarbon dating. These skulls were all recovered from the River Thames (either through dredging or casual finds) between 1912 and 1949, with the Putney skull excavated from the foreshore in 2003. They are all held in the Museum of London collections. The skulls were selected on the basis of morphometric assessment, find-location, and relevant archaeological information. Particular weight was given to skull shape with reference to broad population associations determined by Brodie (1994) and others. The find locations along the Thames from Putney, Barn Elms, Mortlake, Kew, and Syon lie within a 7km linear distance and coincide with a stretch of the Thames which has yielded large quantities of prehistoric metalwork.

Objectives: to establish the dates of the skulls to help determine whether their deposition is temporally clustered or diffuse and whether the dates coordinate with deposition of dated artefacts. A comparison of dates with those obtained by Bradley and Gordon (1988; six skulls from various locations in the River Thames), will help identify particular locations where skulls were deposited at particular periods in history, and will allow an estimate of the duration of deposition practices in this region of the River Thames.

Final comment: D Hamilton (2 August 2005), the skulls date from the early Bronze Age through to the medieval periods, with three samples dating to the Bronze Age, one to the Iron Age, one to the Anglo-Saxon, and one to the medieval period. Therefore, these skulls are temporally dispersed.

References:	Brodie 1994
	Edwards *et al* 2009

OxA–14727 768 ±27 BP

δ¹³C: -18.9‰
δ¹⁵N (diet): +12.3‰
C/N ratio: 3.2

Sample: MOL A20001 [GEN01–43], submitted in May 2005 by Y Edwards and A Weisskopf

Material: human bone (skull) (B White)

Initial comment: a skull submerged in the River Thames water at Barn Elms (Museum ID: MOL 0143). It had probably been subject to post-depositional fluvial transport. The skull was not in association with other skeletal remains when recovered. The cranial index is 85.6, a 'round' head that looks almost square. It has large cranial indices, such as are seen in the Bronze Age and medieval periods in Britain, and was noted as 'Bronze Age' on a note accompanying its 1914 purchase.

Objectives: to date the skull and determine if the skull belongs to a group of deliberately deposited artefacts.

Calibrated date: 1σ: cal AD 1250–1280
2σ: cal AD 1210–1290

Final comment:, the result places this skull in the medieval period.

OxA–14728 3819 ±33 BP

δ¹³C: -21.2‰
δ¹⁵N (diet): +10.8‰
C/N ratio: 3.3

Sample: MOL A13603 [GEN01–52], submitted in May 2005 by Y Edwards and A Weisskopf

Material: human bone (skull) (B White)

Initial comment: an adult male skull submerged in the River Thames water at Syon Reach (Museum ID: MOL 0152; MOL A13603). It had probably been subject to post-depositional fluvial transport, and showed signs of trauma caused by a blow from a sharp instrument. The skull was not in association with other skeletal remains when recovered. The cranial index is 87.5, has a characteristic 'round' head, such as are seen in the Bronze Age and medieval periods in Britain, and was noted as 'was found in association with pile dwelling' on a note accompanying it.

Objectives: as OxA-14727

Calibrated date: 1σ: 2300–2200 cal BC
2σ: 2440–2140 cal BC

Final comment:, this result places the skull in the Beaker period.

OxA–14729 1070 ±29 BP

δ¹³C: -19.0‰
δ¹⁵N (diet): +10.9‰
C/N ratio: 3.3

Sample: MOL A13601 [GEN01–31], submitted in May 2005 by Y Edwards and A Weisskopf

Material: human bone (skull) (B White)

Initial comment: an adult male skull recovered from the River Thames at Kew (Museum ID: MOL 0131; MOL A13601). It had probably been subject to post-depositional fluvial transport, and showed signs of trauma (sword cuts to the left parietal and occipital), fracture, and perforation. The skull was not in association with other skeletal remains when recovered. The cranial index is 75.1, intermediate between 'round' and 'long' head. Round heads are seen in the Bronze Age and medieval periods in Britain, and long heads are characteristic of the Neolithic and Romano-British periods.

Objectives: as OxA-14727

Calibrated date: *1σ:* cal AD 960–1020
                      2σ: cal AD 890–1030

Final comment:, the result places the skull in the late Anglo-Saxon period.

OxA–14730 2232 ±29 BP

δ¹³C: -20.5‰
δ¹⁵N (diet): +11.9‰
C/N ratio: 3.3

Sample: MOL 2004.97 [GEN01–51], submitted in May 2005 by Y Edwards and A Weisskopf

Material: human bone (skull) (B White 2003)

Initial comment: an adult male skull found exposed on the River Thames foreshore in 2003 at Putney (Museum ID: MOL 0151; MOL FSP1). It had probably been subject to post-depositional fluvial transport, and showed signs of a healed injury on a frontal bone. The skull was not in association with other skeletal remains when recovered. The cranial index is 74.7, intermediate between 'round' and 'long' head, tending to 'long'. Round heads are seen in the Bronze Age and medieval periods in Britain, and long heads are characteristic of the Neolithic and Romano-British periods.

Objectives: as OxA-14727

Calibrated date: *1σ:* 380–200 cal BC
                      2σ: 390–200 cal BC

Final comment:, the result places the skull in the middle Iron Age.

OxA–14731 3485 ±33 BP

δ¹³C: -21.0‰
δ¹⁵N (diet): +10.8‰
C/N ratio: 3.3

Sample: MOL A13495 [GEN01–27], submitted in May 2005 by Y Edwards and A Weisskopf

Material: human bone (skull) (B White)

Initial comment: an adult male skull recovered from the River Thames at Mortlake prior to 1914 (Museum ID: MOL 0127; MOL A13495). It had probably been subject to post-depositional fluvial transport, and showed no signs of trauma. The skull was not in association with other skeletal remains when recovered. The cranial index is 85.3, characteristic of a 'round' head. Round heads are seen in the Bronze Age and medieval periods in Britain. The skull was identified as 'prehistoric' by a note accompanying the 1914 purchase.

Objectives: as OxA-14727

Calibrated date: *1σ:* 1890–1740 cal BC
                      2σ: 1900–1690 cal BC

Final comment:, the result places the skull in the early Bronze Age.

OxA–14765 2904 ±33 BP

δ¹³C: -20.6‰
δ¹⁵N (diet): +10.5‰
C/N ratio: 3.2

Sample: MOL A13496 [GEN01–29], submitted in May 2005 by Y Edwards and A Weisskopf

Material: human bone (skull) (B White)

Initial comment: a young female skull recovered from the River Thames at Mortlake prior to 1914 (Museum ID: MOL 0129; MOL A13496). It had probably been subject to post-depositional fluvial transport, and showed no signs of trauma. The skull was not in association with other skeletal remains when recovered. The cranial index is 75.4, an intermediate skull shape. The skull was identified as 'prehistoric' by a note accompanying the 1914 purchase.

Objectives: as OxA-14727

Calibrated date: *1σ:* 1190–1010 cal BC
                      2σ: 1220–1000 cal BC

Final comment:, the result places the skull in the middle to late Bronze Age.

Long Barrows Project: Fussell's Lodge, Wiltshire

Location: SU 1932
                    Lat. 51.05.11 N; Long. 01.43.43 W

Project manager: A Whittle (Cardiff University), 2005–7

Archival body: Natural History Museum and Salisbury and South Wiltshire Museum

Description: an earthen long barrow with lengthy trapezoidal mound. A timber mortuary structure held the remains of some 30 individuals. Twenty-four measurements have been obtained from Fussell's Lodge, to join the single determination obtained previous work (BM-134, 5180 ±50 BP, 4350–3690 cal BC at 95% confidence; Reimer *et al* 2004; Ashbee 1966). Samples were submitted from: human bone from the mortuary deposits - these samples represent bones from each of the five bone groups; antler from the bottom of the barrow ditch; an articulated ox vertebral column discovered on top of the primary silts of the ditch; an ox foot place on top of a flint cairn overlying the mortuary deposits; and an ox skull deposit in the mortuary area.

Objectives: to date the construction of the primary structures (timber revetment and mortuary structure) under the long barrow; to determine the dates of the mortuary deposits and their chronological span; to investigate whether there are any differences in date between the separate bone groups of the mortuary deposits; to determine whether the 'green bone' in group B, which exhibits a fracture morphology of fresh breaks, is earlier in date than the overlying 'dry bone' material, which exhibits a fracture morphology consistent with post mortem breakage; to determine if the weathered ox skull incorporated at the proximal end of the mortuary deposits was older than the human remains; and to establish the date of the construction of the long barrow; to establish the relative position of Fussell's Lodge in the typological sequence of long barrows and long cairns.

Final comment: A Bayliss (2007), a total of 27 radiocarbon results are now available from the Fussell's Lodge long barrow, and are presented within an interpretive Bayesian statistical framework in Wysocki *et al* (2007). Three alternative archaeological interpretations of the sequence are given, each with a separate Bayesian model, the third model is preferred. In the first, the construction is a unitary one, and the human remains included are by definition already old. In the second, the primary mortuary structure is seen as having two phases, and is set within a timber enclosure; these are later closed by the construction of a long barrow. In that model of the sequence, deposition began in the thirty-eighth century cal BC and the mortuary structure was extended probably in the 3660s-3650s cal BC; the long barrow was probably built in the 3630s-3620s cal BC; ancestral remains are not in question; and the use of the primary structure may have lasted for a century or so. In the third, preferred model, a variant of the second, we envisage the inclusion of some ancestral remains in the primary mortuary structure alongside fresh remains. This provides different estimates of the date of initial construction (probably in the last quarter of the thirty-eighth century cal BC or the first half of the thirty-seventh century cal BC) and the duration of primary use, but agrees in setting the date of the long barrow probably in the 3630s-3620s cal BC.

Laboratory comment: English Heritage (2007), following the discovery in the Oxford Radiocarbon Laboratory of a contamination problem associated with the gelatinisation protocol (Bronk Ramsey *et al* 2000), five samples were re-processed, graphitized and dated, as described by Bronk Ramsey *et al* (2004a). The original results on these samples (and the other samples unable to be re-dated) were withdrawn.

Laboratory comment: English Heritage (3 July 2014), 16 further samples from this site were dated prior to 2003 (GrA-23183, -23195, OxA-12277-81, -13173-4, -13185-7, -13205-6, 13326, -13329; Bayliss *et al* 2016, 191-5).

References: Ashbee 1966
Bayliss *et al* 2016
Bronk Ramsey *et al* 2000
Bronk Ramsey *et al* 2004a
Reimer *et al* 2004
Wysocki *et al* 2007

GrA–28174 4940 ±45 BP

δ¹³C: -21.9‰

Sample: FL3, submitted on 28 January 2005 by A Whittle

Material: human bone (disarticulated distal third of left femur (older sub-adult)) (M Wysocki 2004)

Initial comment: from bone group B in the primary mortuary structure under the long barrow mound.

Objectives: to establish the date of deaths of the individuals under the barrow, and to contribute to a new understanding of the sequence of construction and the use of the barrow.

Calibrated date: 1σ: 3780–3650 cal BC
2σ: 3900–3640 cal BC

Final comment: A Bayliss (2007), chronological modelling suggests this skeleton dates to *3725–3650 cal BC (95% probability;* table 1; Wysocki *et al* 2007).

GrA–28175 4850 ±45 BP

δ¹³C: -21.1‰

Sample: FL4, submitted on 28 January 2005 by A Whittle

Material: human bone (1.40g) (disarticulated proximal 3/4 of left femur (adult male)) (M Wysocki 2004)

Initial comment: as GrA-28174

Objectives: as GrA-28174

Calibrated date: 1σ: 3660–3630 cal BC
2σ: 3710–3530 cal BC

Final comment: A Bayliss (2007), chronological modelling suggests this skeleton dates to *3705–3645 cal BC (95% probability;* table 1; Wysocki *et al* 2007).

GrA–28199 4880 ±50 BP

δ¹³C: -23.4‰

Sample: FSA1, submitted on 5 March 2001 by A Whittle

Material: antler: *Cervus elaphus*, tip of pick (1–2g) (M Wysocki 2001)

Initial comment: antler from layer 11 of the ditch. This specimen is from the bottom of the flanking quarry ditch of the long barrow. It is almost certainly a broken tip from an antler pick used in the construction of the ditch and barrow long mound.

Objectives: to establish the date of the construction of the ditch and barrow mound of the long barrow. In comparison with FSA3 (an ox metacarpal from the surface of the flint cairn), whose deposition must pre-date the construction of the mound, it will establish and clarify the chronology of the various architectural components of the monument. To also establish the precise date of the ditch construction in comparison to FSA5 (which seals FSA1). It may also clarify whether the human deposits were of an earlier primary mortuary structure, or whether the monument and burials were all part of a single episode.

Calibrated date: 1σ: 3710–3630 cal BC
2σ: 3770–3530 cal BC

Final comment: A Bayliss (2007), chronological modelling suggests this antler dates to *3650–3630 cal BC (95% probability;* table 1; Wysocki *et al* 2007).

Laboratory comment: English Heritage (2007), the three measurements (GrA-28199, GrA-28218, and OxA-13205) on the antler are statistically consistent (T'=0.3; T'(5%)=6.0; ν=2; Ward and Wilson 1978). Their weighted mean (4866 ±26 BP) calibrates to 3780–3540 cal BC (95% confidence; Reimer *et al* 2004).

References: Reimer *et al* 2004
Ward and Wilson 1978

GrA–28207 4760 ±50 BP

δ¹³C: -21.4‰

Sample: FL5, submitted on 28 January 2005 by A Whittle

Material: human bone (disarticulated segment of distal shaft of left femur (adult female)) (M Wysocki 2004)

Initial comment: as GrA-21874

Objectives: as GrA-21874

Calibrated date: 1σ: 3640–3380 cal BC
2σ: 3650–3370 cal BC

Final comment: A Bayliss (2007), this radiocarbon determination is in poor agreement with the chronological model, and was therefore excluded from the analysis. This individual is rather later than the other dated people in bone group B. This suggests that, contrary to the model (fig 7; Wysocki *et al* 2007), access to the distal portion of the mortuary deposits may have been maintained after the construction of the extension which includes bone groups C and D.

GrA–28208 4940 ±50 BP

δ¹³C: -21.5‰

Sample: FL6, submitted on 28 January 2005 by A Whittle

Material: human bone (disarticulated distal third of left femur (adult, not sexed)) (M Wysocki 2004)

Initial comment: as GrA-21874

Objectives: as GrA-21874

Calibrated date: 1σ: 3780–3650 cal BC
2σ: 3910–3640 cal BC

Final comment: A Bayliss (2007), chronological modelling suggests this skeleton dates to *3725–3650 cal BC (95% probability;* table 1; Wysocki *et al* 2007).

GrA–28209 4860 ±50 BP

δ¹³C: -21.2‰

Sample: FL7, submitted on 28 January 2005 by A Whittle

Material: human bone (disarticulated left femur (adult female, individual 1)) (M Wysocki 2004)

Initial comment: from bone group C in the primary mortuary structure under the long barrow mound. The bones were arranged to give the appearance of articulation.

Objectives: as GrA-21874

Calibrated date: 1σ: 3700–3630 cal BC
2σ: 3720–3530 cal BC

Final comment: A Bayliss (2007), chronological modelling suggests this individual dates to *3660–3635 cal BC (95% probability;* table 1; Wysocki *et al* 2007).

Laboratory comment: English Heritage (2007), the two measurements on individual 1 (GrA-28209 and OxA-13186) are statistically consistent (T′=0.3; T′(5%)=3.8; ν=1; Ward and Wilson 1978). Their weighted mean (4838 ±31 BP) calibrates to 3790–3530 cal BC (95% confidence; Reimer *et al* 2004).

References: Reimer *et al* 2004
Ward and Wilson 1978

OxA–14458 4859 ±35 BP

δ¹³C: -20.7‰
δ¹⁵N (diet): +9.4‰
C/N ratio: 3.2

Sample: FL8, submitted on 28 January 2005 by A Whittle

Material: human bone (proximal half of left femur shaft (probable adult female, individual 2)) (M Wysocki 2004)

Initial comment: as GrA-28209

Objectives: as GrA-28174

Calibrated date: 1σ: 3660–3630 cal BC
2σ: 3710–3530 cal BC

Final comment: A Bayliss (2007), chronological modelling suggests this individual dates to *3665–3635 cal BC (95% probability;* table 1; Wysocki *et al* 2007).

OxA–14480 4865 ±39 BP

δ¹³C: -20.9‰
δ¹⁵N (diet): +9.6‰
C/N ratio: 3.5

Sample: FL1, submitted on 28 January 2005 by A Whittle

Material: human bone (left femur shaft (probable adult male)) (M Wysocki 2004)

Initial comment: from bone group A2 in the primary mortuary structure under the long barrow mound.

Objectives: as GrA-28174

Calibrated date: 1σ: 3690–3630 cal BC
2σ: 3710–3530 cal BC

Final comment: A Bayliss (2007), chronological modelling suggests this individual dates to *3705–33645 cal BC (95% probability;* table 1; Wysocki *et al* 2007).

Long Barrows Project: Wayland's Smithy, Oxfordshire

Location: SU 28118536
Lat. 51.33.57 N; Long. 01.35.40 W

Project manager: A Whittle (Cardiff University), 2005–7

Description: a two-phase Neolithic tomb. Wayland's Smithy I is a small oval barrow with a timber and sarsen mortuary structure, containing 14 inhumations. Wayland's Smithy II is a larger transepted megalithic chambered tomb with large trapezoidal barrow, incorporating the earlier Wayland's Smithy monument (Whittle 1991).

Objectives: to establish the date and span of use of the mortuary deposits in Wayland's Smithy I; to establish the date of construction of Wayland's Smithy II; to clarify the chronological interval between the two monuments; to establish the relative position of Wayland's Smithy I and Wayland's Smithy II in the typological sequence of long barrows and long cairns; and to determine the chronological relationship between Wayland's Smithy II and the transepted chambers at West Kennet.

Final comment: A Bayliss (2007), a total of 23 radiocarbon results are now available from the Wayland's Smithy long barrow, and have been presented within an interpretive Bayesian statistical framework (fully described in Whittle *et al* 2007a). Four alternative archaeological interpretations of the sequence were considered, each with a separate Bayesian model, though only two were presented in detail. The differences are based on different readings of the sequence of Wayland's Smithy II. In the preferred interpretation of the sequence, the primary mortuary structure was some kind of lidded wooden box, accessible for deposition over a period of time, and then closed by the mound of Wayland's Smithy I; Wayland's Smithy II was a unitary construction, with transepted chambers, secondary kerb, and secondary ditches all constructed together. In the Bayesian model for this interpretation, deposition began in the earlier thirty-sixth century cal BC, and probably lasted for a generation. A gap of probably *40–100 years* ensued, before the first small mound was constructed in *3520–3470 cal BC*. After another gap, probably of only *1–35 years*, the second phase of the monument was probably constructed in the middle to later part of the thirty-fifth century cal BC (*3460–3400 cal BC*), and its use probably extended to the middle decades of the thirty-fourth century cal BC.

Laboratory comment: English Heritage (2007), following the discovery in the Oxford Radiocarbon Laboratory of a contamination problem associated with the gelatinization protocol (Bronk Ramsey *et al* 2000), 11 samples were re-processed, graphitized and dated, as described by Bronk Ramsey *et al* (2004a). The original results on these samples were withdrawn.

References: Whittle *et al* 2007a
Bronk Ramsey *et al* 2000
Bronk Ramsey *et al* 2004a
Whittle 1991

Long Barrows Project: Wayland's Smithy I, Oxfordshire

Location: SU 28118536
Lat. 51.33.57 N; Long. 01.35.40 W

Project manager: A Whittle (Cardiff University)

Archival body: Reading Museum

Description: six specimens of human bone from discrete individuals from the Wayland's Smithy I mortuary structure. WS16, WS13, and WS8 were stratified, in a sequence, WS16 being the basal deposit. WS10 was overlain by WS7, but the depositional/stratigraphical relationship of these two specimens with the first groups of three is uncertain. WS6 appears to have been the final deposit in the mortuary structure.

Objectives: the series will date the period of depositional use of the tomb, and will give a good indication of the span of use. As the samples are from articulated individuals, the series will furnish a *terminus ante quem* for the construction of the mortuary structure. In comparison, with dates from

Wayland's Smithy II, the series will establish the length of time between the two monuments and help establish the absolute date of Wayland's Smithy II.

Final comment: A Bayliss (2007), the radiocarbon results have indicated that the start of the sequence at Wayland's Smithy is much later than previously thought. The mortuary structure of Wayland's Smithy I was not the first activity to take place on this site, but compared with what had already happened nearby in the thirty-eighth and thirty-seventh centuries cal BC, the situation at Wayland's Smithy from the thirty-sixth into the thirty-fifth or thirty-fourth centuries cal BC may now appear rather unusual. By the time the first people were deposited in Wayland's Smithy I, probably in the earlier thirty-sixth century cal BC, other long cairns and barrows were already old and had indeed been largely finished (*see* Ascott-under-Wychwood, Hazleton, West Kennet, and Fussell's Lodge), although it does have similarities with the wooden box at Haddenham. The form of Wayland's Smithy I is unusual, and its size modest, and this can no longer be attributed to an early date or putative stage in a developmental sequence, since the preferred chronological model suggests a date from the mid thirty-sixth to mid thirty-fifth century cal BC. It is also striking that the dates indicate there was a single generation of use at Wayland's Smithy I, rather than prolonged use over several generations, perhaps in response to a specific event.

Laboratory comment: English Heritage (3 July 2014), five further samples from this site were previously dated (OxA-13170, -13175–6, -13203, and -13330; Bayliss *et al* 2016, 196-7).

References: Bayliss *et al* 2016
Whittle *et al* 2007a
Whittle 1991

KIA–27623 4750 ±32 BP

$\delta^{13}C$: -10.7 ±0.6‰ (AMS)

Sample: WS12, submitted in July 2005 by A Whittle

Material: human bone (right femur, adult male) (M Wysocki 2001)

Initial comment: from the Wayland's Smithy I mortuary deposits, from bone group F. This bone group consists only of this bone, so residuality/ancestorhood cannot be excluded, and for some reason the exact relationship with the other samples/individuals is not certain.

Objectives: to date every individual in the WS1 mortuary assemblage and to refine the dating model for the site.

Calibrated date: 1σ: 3640–3510 cal BC
2σ: 3640–3370 cal BC

Final comment: A Bayliss (2007), chronological modelling suggests this skeleton dates to *3600–3525 (95% probability*; table 1; Whittle *et al* 2007a).

KIA–27624 4779 ±40 BP

$\delta^{13}C$: -25.7 ±0.4‰ (AMS)

Sample: WS14, submitted in July 2005 by A Whittle

Material: human bone (right femur, adult ?female) (M Wysocki 2001)

Initial comment: from the Wayland's Smithy I mortuary deposits, from bone group Q (the third down of four 'layers' within the mortuary deposit). The skeleton is relatively complete, including hand and foot bones, so while residuality cannot be excluded, it is perhaps not likely. The exact relationship with the other samples/individuals is not certain.

Objectives: as KIA-27623

Calibrated date: 1σ: 3640–3520 cal BC
                       2σ: 3650–3380 cal BC

Final comment: A Bayliss (2007), chronological modelling suggests this skeleton dates to *3600–3525 (95% probability;* table 1; Whittle *et al* 2007a).

KIA–27625 4713 ±37 BP

δ¹³C: -22.7 ±0.2‰ (AMS)

Sample: WS15, submitted in July 2005 by A Whittle

Material: human bone (right femur, adult male) (M Wysocki 2001)

Initial comment: from the Wayland's Smithy I mortuary deposits, from bone group V (the lowest of four 'layers' within the mortuary deposit). Residuality cannot be excluded, though it is thought to be part of an articulated leg.

Objectives: as KIA-27623

Calibrated date: 1σ: 3630–3370 cal BC
                       2σ: 3640–3370 cal BC

Final comment: A Bayliss (2007), chronological modelling suggests this skeleton dates to *3600–3550 cal BC (84% probability)* or *3545–3525 cal BC (11% probability;* table 1; Whittle *et al* 2007a).

KIA–27626 4714 ±36 BP

δ¹³C: -18.7 ±0.1‰ (AMS)

Sample: WS18, submitted in July 2005 by A Whittle

Material: human bone (left humerus, adult ?female) (M Wysocki 2001)

Initial comment: from the Wayland's Smithy I mortuary deposits, from bone group PB8 (the topmost of four 'layers' within the mortuary deposit). Many parts of the skeleton are represented so residuality is unlikely. This could well have been one of the latest depositions, though the proximal position in the mortuary deposits makes this a little uncertain.

Objectives: as KIA-27624

Calibrated date: 1σ: 3630–3370 cal BC
                       2σ: 3640–3370 cal BC

Final comment: A Bayliss (2007), chronological modelling suggests this skeleton dates to *3590–3520 cal BC (95% probability;* table 1; Whittle *et al* 2007a).

OxA–14471 4808 ±38 BP

δ¹³C: -20.9‰
δ¹⁵N (diet): +10.1‰
C/N ratio: 3.2

Sample: WS16, submitted on 28 February 2001 by A Whittle

Material: human bone (0.50–1g) (right femur, adult male) (M Wysocki 2001)

Initial comment: a specimen from the 'undisturbed' chamber, human bone from bone group ᵃ, basal layer. This is from a group of bones in contact with the chamber floor and overlain by other stratified human bones, stone, and chalk rubble. It is a basal deposit and must be early in the sequence of deposition. The material was disarticulated, so one cannot exclude excarnation. The specimen was in contact with the sarsen floor.

Objectives: to establish the span of use of the Wayland's Smithy I mortuary structure for deposition of human remains. This was one of the earliest deposits, and was overlain by bones of other individuals (in sequence WS13 and WS8). These three individuals were part of a stratified grouping of bones in the five parts of the chamber. The sample was sealed by contexts containing WS1 to WS5, and will also provide stable isotope/palaeodiet data. A re-dating of the sample WS16 (OxA-10594) which was withdrawn.

Calibrated date: 1σ: 3650–3530 cal BC
                       2σ: 3660–3520 cal BC

Final comment: see OxA-13170

OxA–14769 4812 ±35 BP

δ¹³C: -20.5‰
δ¹⁵N (diet): +10.5‰
C/N ratio: 3.2

Sample: WS7, submitted on 28 February 2001 by A Whittle

Material: human bone (0.50–1g) (right femur, adult male) (M Wysocki 2001)

Initial comment: skeleton PB2, a largely intact, articulated skeleton in a crouched position. It overlay bones of sample WS10. Part of the bone assemblage was lying on and overlain by other bones. It was sealed by deposits of stone and chalk rubble.

Objectives: to establish the depositional span of use of the Wayland's Smithy I mortuary structure. The later of the two deposits, overlying WS10, placed in the northern part of the chamber. It will also provide stable isotope/palaeodiet data. A re-dating of the sample WS7 (OxA-10590) which was withdrawn.

Calibrated date: 1σ: 3650–3530 cal BC
                       2σ: 3660–3520 cal BC

Final comment: see OxA-13176

OxA–14770 4802 ±35 BP

δ¹³C: -20.7‰
δ¹⁵N (diet): +10.1‰
C/N ratio: 3.3

Sample: WS8, submitted on 28 February 2001 by A Whittle

Material: human bone (0.50–1g) (right femur, adult male, *c* 17–25 years old) (M Wysocki 2001)

Initial comment: the specimen is from a partially articulated skeleton from bone group 7. There had been some slippage and movement of bones, but it is almost certain that the individual was articulated at the time of deposition. WS8 overlies both WS16 and WS13. Part of the bone assemblage was lying on and overlain by other bones. It was sealed by deposits of stone, and chalk rubble.

Objectives: to establish the depositional span of use of the Wayland's Smithy I mortuary structure. This is the latest deposit in the sequence WS16, WS13, and WS8, and was part of a concentration in the southern part of the chamber. It was sealed by contexts containing WS1 to WS5 and will also provide stable isotope/palaeodiet data. A re-dating of the sample WS8 (OxA-10591) which was withdrawn.

Calibrated date: 1σ: 3650–3530 cal BC
                     2σ: 3660–3520 cal BC

Final comment: see OxA-14769

OxA–14771 4749 ±34 BP

$\delta^{13}C$: -20.4‰
$\delta^{15}N$ *(diet):* +10.3‰
C/N ratio: 3.2

Sample: WS9, submitted on 14 March 2002 by A Whittle

Material: human bone (right femur, adult male) (M Wysocki 2001)

Initial comment: a partially articulated skeleton (lower limbs and pelvic girdle) overlain by OxA-14770 and overlaying OxA-14771. From the mortuary assemblage sealed by the collapsed cairn (chalk rubble).

Objectives: the specimen shows clear evidence of carnivore scavenging, which strongly suggests exposure elsewhere before deposition of remains. This specimen will help clarify relationship between it and other individuals (OxA-13203 and OxA-14769) deposited in full articulation with no taphonomic evidence of exposure. A re-dating of the sample WS9 (OxA-11378) which was withdrawn.

Calibrated date: 1σ: 3640–3380 cal BC
                     2σ: 3640–3370 cal BC

Final comment: see OxA-13175

OxA–14772 4787 ±34 BP

$\delta^{13}C$: -20.8‰
$\delta^{15}N$ *(diet):* +9.9‰
C/N ratio: 3.3

Sample: WS10, submitted on 28 February 2001 by A Whittle

Material: human bone (1g) (right femur, adult male) (M Wysocki 2001)

Initial comment: most of the skeleton from bone group 11 is represented and the position of the bones strongly suggested this was an originally articulated skeleton subsequently disturbed during later depositional episodes. It overlay some scattered bones on the chamber floor, and was overlain by skeleton PB2 (sample WS7). It is the earlier of the two stratified deposits in the northern half of the chamber. Part of the bone assemblage was lying on and overlain by other bones. It was sealed by deposits of stone and chalk rubble.

Objectives: to establish the span of use of the Wayland's Smithy I mortuary structure. It is the earlier of the two articulated skeletons deposited, one on top of the other, in the northern part of the chamber (ie was overlain by WS7). It was sealed by contexts containing WS1 to WS5 and will also provide stable isotope/palaeodiet data. A re-dating of the sample WS10 (OxA-10592) which was withdrawn.

Calibrated date: 1σ: 3640–3520 cal BC
                     2σ: 3650–3510 cal BC

Final comment: see OxA-13170

Longstone Edge, Derbyshire

Location: SK 20887341
                     Lat. 53.15.26 N; Long. 01.41.18 W

Project manager: J Last (English Heritage)

Description: two adjacent round barrows, threatened by slippage and collapse, were excavated by the Central Archaeological Service of English Heritage in 1996. Both barrows consisted of mixed stone and earthen mounds covering cists or rock-cut graves.

Objectives: to determine the date and duration of mortuary activity prior to the construction of the barrow 1 mound; to provide a date for human remains below barrow 1; to date the Food Vessel cremation burial associated with the construction of the barrow 1 mound; and to date the mortuary activity from barrow 2 as a comparison with that beneath barrow 1.

Final comment: P Marshall and J Last (18 March 2008), the radiocarbon programme has confirmed the existence of Neolithic mortuary activity pre-dating barrow 1, though the formation processes of this assemblage remain unclear. Subsequent mortuary activity prior to the completion of barrow 1 comprised open cist burials just before 2000 cal BC, and a Food Vessel cremation burial just after 2000 cal BC. The radiocarbon results from graves beneath barrow 1 and 2 suggest that the early Bronze Age burial activity at each site may have been contemporary.

References: Last 2014

Longstone Edge: barrow 1, cremations, Derbyshire

Location: SK 20887341
                     Lat. 53.15.26 N; Long. 01.41.18 W

Project manager: J Last (English Heritage), 1996

Archival body: English Heritage

Description: barrow 1 had not previously been excavated. It comprised of a rock-cut grave containing partially articulated remains of two adults. Other bones of these individuals and others (both burnt and unburnt) were spread on the limestone surface beneath the mound and were associated with Beaker pottery, and had also been incorporated into the mound. Large accumulations of small mammal bones suggest the site was open for a period prior to the construction of the mound, which followed the deposition of a Food Vessel cremation which is believed to immediately pre-date the raising of the barrow mound. Finds of Iron Age and Roman pottery attest to later re-use or disturbance of the site.

Objectives: to date this phase of activity and provide a *terminus ante quem* for the earlier mortuary activity and a *terminus post quem* for the mound construction.

Final comment: J Last and P Marshall (18 March 2008), a Food Vessel cremation burial was subsequently placed on the edge of the mound that covered the cist burials, and was in turn sealed by the main barrow mound.

Laboratory comment: English Heritage (18 March 2008), replicate samples of charred hazelnut fragments and cremated human bone were submitted as part of a wider programme to assess the accuracy of the dating of cremated bone (Lanting *et al* 2001). The four measurements are not statistically consistent (T'=11384.2; T'(5%)=7.8; ν=3, Ward and Wilson 1978) and clearly represent more than one phase of activity.

References: Lanting *et al* 2001
                          Ward and Wilson 1978

GrA–26548 3555 ±40 BP

Sample: 3030, submitted on 25 June 2004 by J Last

Material: calcined human bone (S Mays 2004)

Initial comment: 3030 is the fill of a shallow cut, sealed beneath the barrow mound, which contained the Food Vessel and associated cremation (sample 5136, source of the previously dated material was taken from within this context).

Objectives: to provide a *terminus ante quem* for the period of mortuary activity on the site, and a *terminus post quem* for construction of the barrow mound; to date the Food Vessel with which the cremation was associated and compare with the Beaker associated burials in the grave (OxA-13448 and OxA-13449).

Calibrated date: 1σ: 1950–1830 cal BC
                              2σ: 2030–1760 cal BC

Final comment: P Marshall (18 March 2008), the date is in good agreement with the currency for Food Vessels in England.

Laboratory comment: English Heritage (18 March 2008), the two replicate measurements (GrA-26548 and OxA-14087) on calcined bone [3030] are statistically consistent (T'=0.0; T'(5%)=3.8; ν=1, Ward and Wilson 1978), and thus a weighted mean can be taken before calibration (3558 ±29 BP; 1950–1880 cal BC at 68% confidence and 2010–1770 cal BC at 95% confidence; Reimer *et al* 2004).

References: Reimer *et al* 2004
                          Ward and Wilson 1978

OxA–13393 7519 ±40 BP

$\delta^{13}C$: -22.4‰

Sample: 5136A, submitted on 10 July 2003 by J Last

Material: carbonised plant macrofossil (hazelnut shell fragment (*Corylus* sp.)) (W Smith 2002)

Initial comment: from the total sample of Food Vessel pot, contents, and associated objects including cremated human bone. Collected over a 0.5m by 0.5m area in area 12 (the southern part of the site). The context including the vessel was in a shallow cut beneath the mound, and sealed by it.

Objectives: to provide a *terminus ante quem* for the period of mortuary activity on the site, and a *terminus post quem* for construction of the barrow mound; to date the Food Vessel with which the cremation was associated and compare with the Beaker associated burials in the grave.

Calibrated date: 1σ: 6440–6380 cal BC
                              2σ: 6460–6260 cal BC

Final comment: P Marshall and J Last (18 March 2008), the two measurements on charred hazelnut fragments from the Food Vessel cremation burial were excluded from further analysis because they are clearly residual and probably relate to Mesolithic activity in the vicinity of the site.

Laboratory comment: English Heritage (18 March 2008), the two measurements on charred hazelnut fragments [5136A and B) from the cremation deposit (OxA-13393 and OxA-13394) are not statistically consistent (T'=283.9; T'(5%) =3.8; ν=1, Ward and Wilson 1978), and the material thus represents the remains of two distinct periods of activity.

References: Ward and Wilson 1978

OxA–13394 8475 ±40 BP

$\delta^{13}C$: -24.7‰

Sample: 5136B, submitted on 10 July 2003 by J Last

Material: carbonised plant macrofossil (hazelnut shell fragment (*Corylus* sp.)) (W Smith 2002)

Initial comment: as OxA-13393

Objectives: as OxA-13393

Calibrated date: 1σ: 7580–7520 cal BC
                              2σ: 7590–7480 cal BC

Final comment: see OxA-13393

Laboratory comment: see OxA-13393

OxA–14087 3560 ±40 BP

$\delta^{13}C$: -18.5‰

Sample: 3030, submitted on 25 June 2004 by J Last

Material: calcined human bone (S Mays 2004)

Initial comment: a replicate of GrA-26548.

Objectives: as GrA-26548

Calibrated date: 1σ: 1950–1880 cal BC
                              2σ: 2030–1770 cal BC

Final comment: see GrA-26548

Laboratory comment: see GrA-26548

Longstone Edge: barrow 1, inhumations, Derbyshire

Location: SK 20887341
Lat. 53.15.26 N; Long. 01.41.18 W

Project manager: J Last (English Heritage), 1996

Archival body: English Heritage

Description: barrow 1 had not previously been excavated. It comprised of a rock-cut grave containing partially articulated remains of two adults. Other bones of these individuals and others (both burnt and unburnt) were spread on the limestone surface beneath the mound, were associated with Beaker pottery, and had also been incorporated into the mound. Large accumulations of small mammal bones suggest the site was open for a period prior to the construction of the mound, which followed the deposition of a Food Vessel cremation which is believed to immediately pre-date the raising of the barrow mound. Finds of Iron Age and Roman pottery attest to later reuse or disturbance of the site.

Objectives: to determine the date and duration of mortuary activity prior to the construction of the barrow mound.

Final comment: P Marshall and J Last (18 March 2008), the enclosure defined by a circular dry stone wall contained numerous small human bones and crushed or broken fragments of larger bones in association with Neolithic pottery. Originally interpreted as an excarnation platform, analysis of the human remains suggested they did not represent excarnation. An alternative interpretation was that the remains might be contemporary with the disturbed burials in the rock-cut graves. The enclosure was later re-used as a burial site, marked by the presence of three rock-cut graves, all covered by large limestone flags. Apart from a single fragment in one of the others, only the main or 'central' grave contained human remains.

References: Last 2014

OxA–13447 4832 ±31 BP

$\delta^{13}C$: -20.8‰
$\delta^{15}N$ *(diet):* +9.0‰
C/N ratio: 3.2

Sample: 72572, submitted on 10 July 2003 by J Last

Material: human bone (proximal end of right ulna) (S Mays 2003)

Initial comment: from 'subsoil' deposit 1057 underlying barrow make up 1055 in area 12 (the southern part of the site). Material may originally have been deposited on surface of this deposit, but clearly pre-dates barrow mound, which seals it. The human remains in this deposit are fragmented and disarticulated.

Objectives: to help establish the period of mortuary activity on the site. To test the hypothesis of a possible separate Neolithic mortuary phase, which was proposed by the excavator. To compare the dates of the disarticulated remains with the burials in the grave (75501, 75502). In order to achieve a specimen was selected which is duplicated in the cist skeletons and must therefore belong to a third individual.

Calibrated date: *1σ:* 3650–3630 cal BC
2σ: 3660–3530 cal BC

Final comment: P Marshall (18 March 2008), the date indicates a Neolithic mortuary phase that pre-dates the construction of the barrows some considerable time later.

Laboratory comment: English Heritage (18 March 2008), two measurements on human bone from the purported excarnation platform ([72572] OxA-13447 and [72776] OxA-14086) are not statistically consistent (T'=151.2; T'(5%)=3.8; v=1, Ward and Wilson 1978), and clearly represent material of two different ages. However, both are considerably older than the inhumations from the graves.

References: Ward and Wilson 1978

OxA–13448 3691 ±29 BP

$\delta^{13}C$: -20.6‰
$\delta^{15}N$ *(diet):* +10.4‰
C/N ratio: 3.2

Sample: 75501, submitted on 10 July 2003 by J Last

Material: human bone (left femur base) (S Mays 2003)

Initial comment: the co-mingled remains of two individuals (this one 50% complete), partially articulated, in a cist grave cut into the natural limestone. Disturbed in antiquity and by a recent fissure running through the barrow (the reason for the excavation).

Objectives: to help establish the period of mortuary activity on the site (probably Beaker associated); to compare the date of the burials in the grave with the scattered human remains on the adjacent surface; and to test whether the two individuals (OxA-13449) are likely to have been buried around the same time (the cist was probably reopened more than once).

Calibrated date: *1σ:* 2140–2020 cal BC
2σ: 2200–1970 cal BC

Final comment: P Marshall (18 March 2008), the date has helped to refine the main period of mortuary activity.

Laboratory comment: English Heritage (18 March 2008), measurements on the two inhumations from the grave (skeleton 75502: OxA-13449; skeleton 75501: OxA-13448), the latter largely displaced by a recent fissure through the barrow, are statistically consistent (T'=3.8; T'(5%)=3.8; v=1, Ward and Wilson 1978), and could thus be of the same age.

References: Ward and Wilson 1978

OxA–13449 3771 ±29 BP

$\delta^{13}C$: -20.5‰
$\delta^{15}N$ *(diet):* +10.8‰
C/N ratio: 3.3

Sample: 75502 (marked 72711/2), submitted on 10 July 2003 by J Last

Material: human bone (left femur distal shaft) (S Mays 2003)

Initial comment: as OxA-13448

Objectives: as OxA-13448

Calibrated date: 1σ: 2280–2140 cal BC
2σ: 2290–2050 cal BC

Final comment: P Marshall (18 March 2008), *see* OxA-13448

Laboratory comment: see OxA-13448

OxA–14086 4283 ±32 BP

δ¹³C: -22.1‰
δ¹⁵N (diet): +11.7‰
C/N ratio: 3.2

Sample: 72776, submitted on 25 June 2004 by J Last

Material: human bone (perinatal right sphenoid) (S Mays 2003)

Initial comment: from basal mound deposit 1082 underlying mound make-up and above 'subsoil' (ie sealed by barrow) in area 12 (south part of the site). The human remains in this deposit are fragmented and disarticulated.

Objectives: to confirm a Neolithic phase at the site as implied by OxA-13447 on a related sample); or to assess whether some of the scattered human remains in fact derive from the same phase as the cist grave (dated by OxA-13448 and OxA-13449). As previously suggested, because teeth from one of these individuals were found among the disarticulated remains. The objective was therefore to determine which phase an individual not previously dated (this case a child) belongs to.

Calibrated date: 1σ: 2910–2880 cal BC
2σ: 2930–2870 cal BC

Final comment: P Marshall (18 March 2008), the result suggests that Neolithic mortuary activity prior to the construction of the barrows was intermittent with this sample being deposited sometime later than OxA-13447.

Laboratory comment: see OxA-13447

Longstone Edge: barrow 2, inhumations, Derbyshire

Location: SK 20887341
Lat. 53.15.26 N; Long. 01.41.18 W

Project manager: J Last (English Heritage), 1996

Archival body: English Heritage

Description: barrow 2 is the smaller mound of the two and was excavated by Thomas Bateman in 1848.

Objectives: to spot date the mortuary activity from barrow 2 as a comparison with that beneath barrow 1.

Final comment: P Marshall and J Last (18 March 2008), barrow 2 was the smaller of the two barrows and demonstrated a single phase of construction with a rock-cut grave at its centre, previously excavated by Thomas Bateman. The partial remains of a child skeleton found on the ground surface beneath the mound and outside the grave may be the remains of an individual removed from the cist by Bateman. However, other finds from the deposit suggest the bones may be *in situ*, given they are not in Bateman's backfill.

References: Last 2014

OxA–13450 3704 ±29 BP

δ¹³C: -21.2‰
δ¹⁵N (diet): +12.3‰
C/N ratio: 3.3

Sample: 72634, submitted on 10 July 2003 by J Last

Material: human bone (left femur midshaft fragment) (S Mays 2003)

Initial comment: partial remains of child skeleton (<20% complete) found on ground surface (2058) beneath mound and outside the cist grave excavated by Bateman. The excavator suggests this may be the remains of an individual removed from the cist by Bateman (who mentions a child inhumation within the grave) but other finds from the surface suggest these bones may be *in situ* and they are not within Bateman's backfill.

Objectives: to help establish the period of mortuary activity on the site. To compare the date of barrow 2 burials with barrow 1 burials, since there are no *in situ* burials left in the cist.

Calibrated date: 1σ: 2140–2030 cal BC
2σ: 2200–1980 cal BC

Final comment: P Marshall (18 March 2008), the radiocarbon result suggests that the initial interpretation that this individual represents material removed from the cist could be correct.

Laboratory comment: English Heritage (18 March 2008), the child inhumation (72634; OxA-13450) and measurements from the central grave of barrow 1 are statistically consistent (T'=4.4; T'(5%)=6.0; ν=2, Ward and Wilson 1978), and could all be of the same actual age.

References: Ward and Wilson 1978

Lundenwic, City of London

Location: TQ 30358105
Lat. 51.30.48 N; Long. 00.07.22 W

Project manager: R Cowie and G Malcolm (Museum of London Archaeology Service), 1987–2000

Description: in 1984, the site of the middle Saxon trading port of London (Lundenwic) was finally identified as lying on the north bank of the Thames less than a kilometre upstream from the former Roman town of Londinium. Documentary and archaeological evidence suggests that between the late seventh century and the early-to-mid ninth century AD, Lundenwic was part of an international network of trading centres in north-west Europe. Fieldwork was undertaken between 1987 and 2000 at 18 sites within the area.

Objectives: to refine/confirm the date of the introduction of Ipswich Ware into Lundenwic; to refine the dating of certain seventh-century grave types (eg BOB91 buckle set); and to date the start of the settlement at Lundenwic.

Final comment: R Cowie and L Blackmore (2012), the radiocarbon results suggest that Lundenwic developed very rapidly across most of the study area, probably in the last third of the seventh century AD, which would accord with the earliest known reference to the port in AD 672–4.

There appears, therefore, to have been little or no hiatus between the final use of the cemetery and the development of Lundenwic. Furthermore, it would seem that the settlement did not grow gradually in an uncoordinated fashion, but that its development was prompted and controlled by a central authority.

Laboratory comment: English Heritage (2012), a further 16 dates were obtained prior to 2003 (OxA-12348–54, -12450, -12454, 12559, and UB-4879 and -4881; Bayliss *et al* 2016, 210-11).

References: Cowie *et al* 1988
                 Cowie and Blackmore 2012
                 Malcolm *et al* 2003

Lundenwic: post-Ipswich Ware pit, City of London

Location: TQ 30358105
                Lat. 51.30.48 N; Long. 00.07.22 W

Project manager: B Barker, R Cowie, and G Malcolm (Museum of London Archaeology Service), 1989–99

Archival body: Museum of London

Description: a pit at site N which clearly post-dated the introduction of Ipswich Ware.

Objectives: to refine the development and ceramic chronology of Lundenwic.

Final comment: R Cowie and L Blackmore (2012), given that the pit is thought to have contained sufficient water to erode its sides during the accumulation of the primary fill, it seems most likely that older material had been incorporated into it during its use.

Laboratory comment: English Heritage (2012), the samples from the primary fill N[97] of pit N[18] are not statistically consistent (T′=19.0; T′(5%)=3.8; *v*=1; Ward and Wilson 1978).

References: Cowie and Blackmore 2012
                 Ward and Wilson 1978

OxA–13706 1378 ±25 BP

$\delta^{13}C$: -26.1‰

Sample: DRY 90, fill [97], pit [18], submitted in January 2004 by R Cowie

Material: charcoal: Salicaceae, single fragment (R Gale 2003)

Initial comment: context [97], from which the sample was taken, was the primary fill of pit [18], which contained nine successive fills. One of the later fills contained pottery dated to AD 730/750–850 (sherds of Ipswich Ware). Frequent charred plant remains ('charcoal') were noted by the excavator, suggesting that the material was likely to be contemporaneous rubbish disposed in the pit.

Objectives: to establish a more precise date for the use of the rubbish pit, which was dated to AD 730/750–850 on the basis of the presence of Ipswich Ware in one of the pits middle fills.

This will help to confirm or refine the existing 'ceramic chronology' for middle Saxon London.

Calibrated date: *1σ:* cal AD 640–670
                       2σ: cal AD 630–680

Final comment: R Cowie and L Blackmore (2012), OxA-13706 is thought to represent residual material material incorporated into the primary fill of the pit due to erosion of its sides when filled by water.

OxA–14043 1211 ±29 BP

$\delta^{13}C$: -26.9‰

Sample: DRY 90, fill [97], pit [18], submitted in January 2004 by R Cowie

Material: charcoal: *Corylus avellana*, single fragment (R Gale 2003)

Initial comment: as OxA-13706

Objectives: as OxA-13706

Calibrated date: *1σ:* cal AD 730–890
                       2σ: cal AD 690–900

Final comment: R Cowie (14 July 2013), the charcoal used for the sample came from the fill of a middle Saxon pit that containing sherds of Ipswich Ware and a single sherd of shell-tempered ware. The presence of the latter suggests that the pit must have been open after *c* AD 770. The radiocarbon date would be generally consistent with a feature in use after the introduction of Ipswich Ware.

MARISP, Somerset

Location: see individual sites
                Lat. ; Long.

Project manager: R Brunning (Somerset County Council), 1997–2003

Description: the lowland peatlands of Somerset are famous for their waterlogged prehistoric sites. Many of the sites were discovered during, or just before peat extraction, but now another threat, peat wastage, is occurring on a wider scale. The MARISP (Monuments at Risk in Somerset's Peatlands) project was established in response to this threat, with funding from English Heritage, Somerset County Council, and the Environment Agency. There had previously only been three *in situ* assessments of preservation at waterlogged sites in Somerset, and only a few more from the whole of England. The MARISP project was designed to provide an overview of the scale of this loss by studying a representative sample of the known waterlogged sites in the peat area (Jones *et al* 2007, 71–2).

Objectives: the objectives of the MARISP project were to:
1. assess the condition of the monuments examined;
2. obtain baseline condition data by methods that could be replicated; 3. maximise the research opportunities of the site investigations; 4. develop management proposals for each site; and 5. produce recommendations for developing the methodologies for future *in situ* assessment projects (Jones *et al* 2007, 71–2).

Final comment: P Marshall and D Hamilton (2007), thirty samples were submitted for radiocarbon dating from five sites (Dewar's Track, Street Causeway, Harters Hill, Sharpham Park, and Glastonbury Lake Village) investigated as part of the MARISP Project.

References: Brunning 2013
                   Jones *et al* 2007

MARISP: Dewar's track, Somerset

Location: ST 42724017
                   Lat. 51.09.25 N; Long. 02.49.08 W

Project manager: R Brunning (Somerset County Council),
                   October 2003

Archival body: Somerset Records Office, Somerset
                   County Museum

Description: Dewar's/Viper's trackway is a late Bronze Age trackway built across a raised bog at a time when the bog surface was becoming increasingly wet. The trackway consisted of brushwood laid on transverse rods secured by vertical oak piles at the edges of the structure.

Oak timbers were recovered from the ploughsoil, and an excavation in 2003 failed to find convincing evidence of the trackway. This suggests that the structure may have been totally destroyed by ploughing and associated desiccation.

Objectives: to date the peat below the ploughsoil to determine if the prehistoric trackways have been destroyed by ploughing and associated peat wastage. Oak timbers in the ploughsoil may be the remains of the late Bronze Age trackway.

Final comment: R Brunning (31 January 2005), the combined dates have successfully shown that the top of the surviving peat deposit in the field dates at least 1300 years before the late Bronze Ages trackways that once survived there. This demonstrates that the trackways are now destroyed over most of the field and may only survive where the peat is covered by colluvium at one edge of the field.

Laboratory comment: (16 January 2007), the two measurements (GU-6036 and GU-6037) are statistically consistent (T'=0.0; T' (5%)=3.8; v=1; Ward and Wilson 1978) and allow a weighted mean to be calculated (4080 ±35 BP).

References: Brunning 2013
                   Ward and Wilson 1978

GU–6036 4080 ±50 BP

$\delta^{13}C$: -27.9‰

Sample: DV 03 Peat bulk 53–58cm, submitted on 11 March 2004 by R Brunning

Material: peat (humic acid) (1840g)

Initial comment: the peat sample was collected from 53–58cm below the ground surface. This was below the limit of ploughing and below the well humified and desiccated peat soils that existed above it. The sample is from a black, partly humified moss peat.

Objectives: this sample should be able to provide a date for the top of the peat below the ploughsoil destruction zone. This will help to prove or disprove if the peat contemporary with the wooden trackway (*c* 500–900 cal BC) has been destroyed by ploughing/desiccation. No convincing evidence of the trackway was seen in the excavation and oak timbers found in the ploughsoil may be remnant of the structure.

Calibrated date: *1σ:* 2840–2490 cal BC
                     2σ: 2880–2470 cal BC

Final comment: R Brunning (31 January 2005), this provided the proof that the scheduled late Bronze Age trackway in the field has been destroyed by peat wastage and ploughing.

GU–6037 4080 ±50 BP

$\delta^{13}C$: -27.3‰

Sample: DV 03 Peat bulk 53–58cm, submitted on 11 March 2004 by R Brunning

Material: peat (humin) (1840g)

Initial comment: as GU-6036

Objectives: as GU-6036

Calibrated date: *1σ:* 2840–2490 cal BC
                     2σ: 2880–2470 cal BC

Final comment: see GU-6036

MARISP: Glastonbury Lake Village, Somerset

Location: ST 491408
                   Lat. 51.09.48 N; Long. 02.43.40 W

Project manager: R Brunning (Somerset County Council),
                   September 2003

Archival body: Somerset County Museum

Description: the site of the Iron Age lake village. A trench was excavated close to the village pallisade to examine the peat stratigraphy. Wet detritus peat was recorded at the base, with drier silty peat above. The site was covered by orange/grey clay. Some worked wood, clay sling stones, and bone were identified in the peat.

Objectives: to provide a temporal framework for the palaeoenvironmental analysis.

Final comment: P Marshall (16 January 2007), the results suggest that peat dating to the period of occupation of the site do remain intact.

Laboratory comment: English Heritage (17 November 2014), eight further samples were subsequently dated (OxA-16233–8, and OxA-16245–6; Bayliss *et al* forthcoming).

References: Brunning 2013
                   Ward and Wilson 1978

SUERC–9828 2455 ±35 BP

δ¹³C: -27.3‰

Sample: GLV 4.44–4.45m OD, submitted on 23 January 2006 by J Jones

Material: waterlogged plant macrofossils (*Cladium mariscus* (×6) (great fen sedge); *Carex* (×5) (sedge); *Hydrocotyle vulgaris* (×1) (marsh pennywort); *Rumex* (×1) (dock); *Alnus glutinosa* (×2) (alder); *Ranunculus* (×1) (buttercup); *Oenanthe aquatica* (×1) (water dropwort)) (J Jones 2006)

Initial comment: the sample was taken from 4.44–4.45m at the base of the sequence (and 1.14–1.15m from the ground surface), from a poorly humified peat with abundant plant remains.

Objectives: to date the base of the sequence.

Calibrated date: *1σ:* 750–430 cal BC
                     2σ: 770–400 cal BC

Final comment: P Marshall (November 2014), the samples of waterlogged macrofossils (SUERC-9828) and peat (OxA-16233; 2861 ±30 BP, and OxA-16234; 2869 ±30 BP) from the base of the sequence came from a poorly humified peat associated with worked wood, bone and clay sling stones that are presumably from activity in the Lake Village. The three measurements are not statistically consistent (T'=98.0; T' (5%)=6.0; ν=2; Ward and Wilson, 1978), although the two measurements on the humic and humin peat fractions are (T'=0.1; T' (5%)=3.8; ν=1; Ward and Wilson, 1978). This suggests that the sample of plant macrofossils (SUERC-9828) contains some intrusive, younger material and therefore does not provide an accurate date for this level.

References: Ward and Wilson 1978

SUERC–9829 2615 ±40 BP

δ¹³C: -25.5‰

Sample: GLV 4.80–4.81m OD, submitted on 23 January 2006 by J Jones

Material: waterlogged plant macrofossils (*Carex* (×9) (sedge); *Betula* (×1) (birch); *Ranunculus lingua* (×3) (greater spearwort); *Cladium mariscus* (×4) (great fen sedge); *Mentha aquatica* (×6) (water mint); *Urtica dioica* (×1) (nettle); *Salix* bud (×1) (willow)): *Schoenoplectus lacustris,* (×2).

Initial comment: sample from 4.80–4.81m OD from the top of a poorly humified peat (and 0.78–0.79m from the ground surface).

Objectives: to date the top of the poorly humified peat towards the base of the sequence.

Calibrated date: *1σ:* 810–790 cal BC
                     2σ: 840–760 cal BC

Final comment: P Marshall (November 2014), the samples of waterlogged macrofossils (SUERC-9829) and peat (OxA-16235; 2393 ±30 BP, and OxA-16236; 2355 ±33 BP) from 4.80–4.81m come from the top of the poorly humified peat towards the base of the sequence. The three measurements are again not statistically consistent (T'=28.5; T' (5%)=6.0; ν=2; Ward and Wilson, 1978), although the two measurements on the humic and humin peat fractions are

(T'=0.7; T' (5%)=3.8; ν=1; Ward and Wilson, 1978). This suggests that the sample of plant macrofossils (SUERC-9829) contains some ?older material and therefore does not provide an accurate date for this level.

References: Ward and Wilson 1978

MARISP: Harters Hill, Somerset

Location: ST 534428
                     Lat. 51.10.54 N; Long. 02.40.00 W

Project manager: R Brunning (Somerset County Council), October 2003

Archival body: Somerset County Museum

Description: a north/south alignment of large stakes occurs running from High Ground of Harters Hill onto the levels. A small trench was excavated to examine the peat stratigraphy, and revealed organic silty clay at the base, overlain by well-humified peat with abundant wood fragments and small round wood. Some of the wood was charred.

Objectives: to provide a temporal framework for the palaeoenvironmental analysis.

Final comment: P Marshall (16 January 2007), the results suggest that environmental evidence associated with the pile alignment probably does survive.

Laboratory comment: English Heritage (17 November 2014), four further samples were subsequently dated (OxA-15984, -16024, -16179, and -16247; Bayliss *et al* forthcoming).

References: Brunning 2013

SUERC–9825 3150 ±35 BP

δ¹³C: -27.2‰

Sample: HH 4.36–4.37m OD, submitted on 23 January 2006 by J Jones

Material: waterlogged plant macrofossils (*Rumex* sp. (×1) (dock); *Mentha aquatica* (×3) (water mint); *Sparganium erectus* (×3) (branched bur-reed); *Alisma plantago* (×2) (water plantain); *Urtica dioica* (×1) (nettle); *Eleocharis palustris/uniglumis* (×6) (spike-rush); *Hydrocotyle vulgaris* (×1) (marsh pennyworth)): *Chenopodium* cf *album,* (×1) (fat-hen); *Rubus* subg. *Glandolosus,* (×2) (bramble) (J Jones 2006)

Initial comment: sample from 4.36–4.37m OD at the base of the peaty clay (and 0.65–0.66m below the ground surface).

Objectives: to provide a date for the onset of this more peaty clay. The clay fraction in the peat, which does not occur in the peat below, is thought to be a result of slope wash from higher ground. The evidence for this may be forthcoming from pollen analysis associated with peat at this level. This will therefore provide dating for this possible episode of environmental change.

Calibrated date: *1σ:* 1450–1400 cal BC
                     2σ: 1500–1300 cal BC

Final comment: P Marshall (16 January 2007), this dates the period of environmental change.

SUERC–9826 2935 ±35 BP

$\delta^{13}C$: -28.5‰

Sample: HH 4.54–4.55m OD (B), submitted on 23 January 2006 by J Jones

Material: peat (humic acid)

Initial comment: sample from 4.54–4.55m OD from the top of the peaty clay (and 0.47–0.48m below the ground surface).

Objectives: to provide the most recent date for peat deposition at Harters Hill, and provide a final date for the pollen diagram and associated macrosfossil and beetle analysis.

Calibrated date: 1σ: 1220–1050 cal BC
                       2σ: 1260–1010 cal BC

Final comment: P Marshall (16 January 2007), this dates the top of the sequence.

Laboratory comment: English Heritage (16 January 2007), the two measurements on the humic and humin fractions from 5.54–4.55m (SUERC-9826 and SUERC-9836) are statistically consistent (T'=1.7; T'(5%)=3.8; ν=1; Ward and Wilson, 1978) and allow a weighted mean to be calculated (2968 ±25 BP), which calibrates to 1300–1110 cal BC; Reimer *et al* 2004).

References: Reimer *et al* 2004
                    Ward and Wilson 1978

SUERC–9836 3000 ±35 BP

$\delta^{13}C$: -28.7‰

Sample: HH 4.54–4.55m OD (B), submitted on 23 January 2006 by J Jones

Material: peat (humin)

Initial comment: as SUERC-9826

Objectives: as SUERC-9826

Calibrated date: 1σ: 1280–1130 cal BC
                       2σ: 1390–1120 cal BC

Final comment: see SUERC-9826

Laboratory comment: see SUERC-9826

MARISP: Sharpham, Somerset

Location:	ST 467374 Lat. 51.07.57 N; Long. 02.45.42 W
Project manager:	R Brunning (Somerset County Council), September 2003
Archival body:	Somerset County Museum

Description: a small trench was excavated to examine the peat stratigraphy. Lias was exposed at the base of the trench, with overlying lake marl above silty peat with snails. Above this was peat containing well-preserved timbers, covered by a thin layer of blue-grey silty clay, and then further peat with small wood remains (some beaver chewed). The overlying peat was organic silty colluvium to the surface.

Objectives: to provide a temporal framework for the palaeoenvironmental analysis.

Final comment: P Marshall (16 January 2007), the results suggest that peat and environmental remains associated with the wooden platform do survive.

Laboratory comment: English Heritage (17 November 2014), four further samples were subsequently dated (OxA-16248–51; Bayliss *et al* forthcoming).

References: Brunning 2013

SUERC–9830 2870 ±35 BP

$\delta^{13}C$: -27.9‰

Sample: Sharpham 5.03–5.04m OD, submitted on 23 January 2006 by J Jones

Material: waterlogged plant macrofossils (*Alnus glutinosa* (alder) (×11); *Carex* (sedge) (×1); *Cladium mariscus* (great fen sedge) (×1); *Sparganium erectum* (branched bur-reed) (×1): *Ranunculus* sp., (×1) (buttercup); *Oenanthe* sp., *aquatica* (×8) (water dropwolt)) (J Jones 2006)

Initial comment: sample from 5.03–5.04m OD from the base of silty peat (and 0.90–0.91m from the ground surface).

Objectives: to provide dating for the onset of silty peat deposition near the base of the sequence.

Calibrated date: 1σ: 1120–1000 cal BC
                       2σ: 1190–920 cal BC

Final comment: P Marshall (16 January 2007), the sample dates the onset of silty peat deposition.

SUERC–9834 2325 ±35 BP

$\delta^{13}C$: -28.3‰

Sample: Sharpham 5.52–5.53m OD, submitted on 23 January 2006 by J Jones

Material: peat (humic acid)

Initial comment: sample from 5.52–5.53m OD from the top of the upper peat (and 0.41–0.42m from the ground surface).

Objectives: this is the uppermost sample in the sequence at the top of the woody peat before the onset of possible colluvial (clay silt) deposition. It will therefore provide the most recent date for peat deposition at Sharpham.

Calibrated date: 1σ: 410–380 cal BC
                       2σ: 420–360 cal BC

Final comment: P Marshall (16 January 2007), provides a date for the end of peat deposition.

Laboratory comment: English Heritage (16 January 2007), the two measurements (SUERC-9834 and SUERC-9838) are statistically consistent (T'=1.2; T'(5%)=3.8; ν=1; Ward and Wilson, 1978) and thus allow a weighted mean to be calculated (2298 ±25 BP), whch calibrates to 410–260 cal BC (95% confidence; Reimer *et al* 2004).

References: Reimer *et al* 2004
                    Ward and Wilson 1978

SUERC–9835 2510 ±35 BP

δ¹³C: -28.4‰

Sample: Sharpham 5.41–5.42m OD (B), submitted on 23 January 2006 by J Jones

Material: peat (humic acid)

Initial comment: sample from 5.41–5.42m OD from the base of the upper woody peat (and 0.52–0.53m below the ground surface).

Objectives: this sample will provide dating for the onset of the most recent phase of peat deposition at this site to correlate with the pollen, plant macrofossil and beetle analysis being undertaken at this level.

Calibrated date: 1σ: 780–540 cal BC
2σ: 800–510 cal BC

Final comment: P Marshall (16 January 2007), dates the onset of the most recent phase of peat deposition at the site.

Laboratory comment: English Heritage (16 January 2007), the two measurements (SUERC-9835 and SUERC-9839) are statistically consistent (T′=0.5; T′ (5%)=3.8; ν=1; Ward and Wilson, 1978) and once again allow a weighted mean to be calculated (2528 ±25 BP), which calibrates to 800–540 cal BC (95% confidence; Reimer *et al* 2004).

References: Reimer *et al* 2004
Ward and Wilson 1978

SUERC–9838 2270 ±35 BP

δ¹³C: -28.1‰

Sample: Sharpham 5.52–5.53m OD, submitted on 23 January 2006 by J Jones

Material: peat (humin)

Initial comment: as SUERC-9834

Objectives: as SUERC-9834

Calibrated date: 1σ: 400–250 cal BC
2σ: 410–200 cal BC

Final comment: see SUERC-9834

Laboratory comment: see SUERC-9834

SUERC–9839 2545 ±35 BP

δ¹³C: -28.6‰

Sample: Sharpham 5.41–5.42m OD (B), submitted on 23 January 2006 by J Jones

Material: peat (humin)

Initial comment: as SUERC-9835

Objectives: as SUERC-9835

Calibrated date: 1σ: 800–670 cal BC
2σ: 810–540 cal BC

Final comment: see SUERC-9835

Laboratory comment: see SUERC-9835

MARISP: Street Causeway, Somerset

Location: ST 487375
Lat. 51.08.06 N; Long. 02.44.02 W

Project manager: R Brunning (Somerset County Council), September 2003

Archival body: Somerset Records Office, Somerset County Museum

Description: a stone causeway crossing a narrow neck of wetland between the settlements of Street and Glastonbury leading to a bridge or ford across the River Brue. Beside the river two lines of piles were discovered sealed by the stones of the causeway surface that was over 20m wide at this point.

The causeway was built on an organic detrital muddy peat. The peat and the causeway were both covered by a thick alluvial layer suggesting a major change in the local environment at the time the causeway was built, possibly representing early medieval enclosure of the wetland and embanking of the river, the alluvium being deposited by seasonal flooding.

Objectives: to date the construction of the causeway and its active lifespan and to provide a *terminus post quem* for the major environmental change recorded on the site.

Final comment: R Brunning (31 January 2005), the two dates provide an unexpectedly early date in the mid Saxon period. This is very exciting as they may relate to the early activity of Glastonbury Abbey, which acquired land on Polden ridge in the mid-eighth century AD. One possibility is that the causeway was built by the Abbey to connect it to this new land holding. The dates overlap with radiocarbon dates for a nearby canal and suggest major engineering on the floodplain at this time.

Laboratory comment: English Heritage (1 November 2014), two further samples were subsequently dated (OxA-16180–1; Bayliss *et al* forthcoming).

Laboratory comment: (16 January 2007), the results suggest that a considerable accumulation of sediment was removed prior to construction of the stone causeway.

References: Brunning 2013

GU–6040 1250 ±50 BP

δ¹³C: -24.3‰

Sample: Wood number 23, submitted on 11 March 2004 by R Brunning

Material: wood (waterlogged): *Fraxinus excelsior*, *c* 18 rings with sapwood and bark edge (370g) (R Gale 2003)

Initial comment: the sample is from a vertical pile from one of two lines of piles that ran longitudinally under the stone surface of the causeway. They were not inserted through this surface and appear to related to the initial construction of the causeway where it approached the river and a possible bridge site.

Objectives: this sample should be able to provide a date for the initial construction of the causeway. Current dating evidence from the structure is very limited. One sherd of Roman pottery was found under the road and a twelfth or thirteenth century AD spur on top of the road in previous excavations. A fourteenth-century AD bridge was built next to the line of the causeway and probably replaced it. A major environmental change occurs in the stratigraphy at the level of the causeway. This may represent enclosure of the wetlands and embankment of the River Brue. The sample will provide a *terminus post quem* for this change.

Calibrated date: *1σ:* cal AD 680–860
                   2σ: cal AD 660–900

Final comment: R Brunning (31 January 2005), this date has proved to be consistent with GU-6041 linking the detrital content at the edge of the causeway to the piles underneath it. This suggests that the piles were part of the original structure.

GU–6041 1320 ±50 BP

δ¹³C: -28.3‰

Sample: Wood number 200, submitted on 11 March 2004 by R Brunning

Material: wood (waterlogged): *Acer campestre*, roundwood (17 rings) (180g) (R Gale 2004)

Initial comment: the sample comes from a pieces of wood which formed part of an upper strand deposit beside the causeway.

Objectives: as GU-6040

Calibrated date: *1σ:* cal AD 650–770
                   2σ: cal AD 640–780

Final comment: R Brunning (31 January 2005), this date has helped to successfully provide a date for the causeway and the environmental change associated with it.

SUERC–9827 2015 ±35 BP

δ¹³C: -28.4‰

Sample: 4.02–4.03m OD B, submitted on 23 January 2006 by J Jones

Material: peat (humic acid)

Initial comment: this sample was taken from 4.02–4.03m OD from the top of an organic silty clay. It was 1.12–1.13m below the ground surface, and was overlain by the stone causeway and 1m of alluvium.

Objectives: this sample will provide dating for the top of the organic silty clay onto which the stone causeway is directly laid. This date should be contemporary with the two radiocarbon dates already obtained from this site unless peat cutting occurred prior to construction of the causeway.

Calibrated date: *1σ:* 50 cal BC–cal AD 30
                   2σ: 110 cal BC–cal AD 70

Final comment: P Marshall (16 January 2007), the results therefore suggest that a considerable accumulation of sediment was removed prior to construction of the stone causeway.

Laboratory comment: English Heritage (16 January 2007), the two measurements (SUERC-9827 and SUERC-9837) are statistically consistent (T'=0.2; T' (5%)=3.8; ν=1; Ward and Wilson 1978) and allow a weighted mean to be calculated (2025 ±25 BP), which calibrates to 100 cal BC–cal AD 50 (95% confidence; Reimer *et al* 2004).

References: Reimer *et al* 2004
                   Ward and Wilson 1978

SUERC–9837 2035 ±35 BP

δ¹³C: -28.7‰

Sample: 4.02–4.03m OD B, submitted on 23 January 2006 by J Jones

Material: peat (humin)

Initial comment: as SUERC-9827

Objectives: as SUERC-9827

Calibrated date: *1σ:* 90 cal BC–cal AD 10
                   2σ: 170 cal BC–cal AD 60

Final comment: see SUERC-9827

Laboratory comment: see SUERC-9827

Peak District Mines, Derbyshire

Location: *see* individual sites

Project manager: J Barnatt (Peak District National Park), 2003–4

Description: a research project to investigate fire-setting underground in Peak District lead mines. Much of the evidence found to date is as yet undetermined, but is most likely to be medieval or later, the practice probably ceasing by the end of the seventeenth century with the common adoption of gunpowder.

Objectives: to facilitate research and assessment of the importance of these sites If secure dates can be obtained it will establish if the fire-setting is from the medieval, late medieval, or post-medieval periods.

Final comment: J Barnatt (15 August 2013), the suite of radiocarbon dates were designed to test the date of previously unrecognised fire-setting using coal found in a series of historic lead mines in the Peak District ore-field. Before the dating programme was undertaken it was unclear whether the use of this extraction technique in this context, using this fuel, was medieval or post-medieval in date. Two adjacent sites, Old Ash Mine and Lords and Ladies Mine, both in Northern Dale near Snitterton (centred SK 268607), were chosen for initial dating assessment because of exceptional survival in undisturbed context. Eleven radiocarbon dates were obtained, from six discrete underground sample sites where charcoal from kindling used to light the coal. Taken together, these indicated that the fire-setting at Old Ash Mine dated to the late-sixteenth or seventeenth century and that extraction at Lords and Ladies Mine was seventeenth century in date. The results were published in 2006 and interpreted as reflecting a period of mining where most of the easily accessible ore reserves near surface had already been taken and miners either had to invest in costly drainage schemes to follow rich veins deeper, or to use fire-setting to access lesser reserves where hard limestone had to be broken to follow thin deposits. The technique was superseded by the early eighteenth century as gunpowder became widely available.

References: Barnatt and Worthington 2006

Peak District Mines: Lords and Ladies Mine, Derbyshire

Location: SK 270606
                     Lat. 53.08.29 N; Long. 01.35.47 W

Project manager: J Barnatt (Peak District National Park), April 2004

Archival body: Peak District National Park Authority

Description: two sample sites within different 'flat workings' in the mine, which were selected for recovery of charcoal. The fuel was used for fire-setting of limestone beds associated with galena-rich mineralization within bedding planes. These were demonstrably worked prior to gunpowder trials elsewhere in the mine, a technique commonly introduced in the Derbyshire ore-field around the beginning of the eighteenth century. Both samples were recovered from floor deposits in the 'flat workings', comprising waste rock/mineral with burnt coal and small amounts of charcoal.

Objectives: previously dated samples from the adjacent Old Ash Mine indicate fire-setting probably took place in *c* AD 1640–70 (and possibly earlier). The samples from Lords and Ladies Mine will either be similar and reinforce interpretation of ore extraction in a concerted fashion over a short period, or they will show that it took place over a more extended timeframe in the mines as a whole. There is historical documentation indicating ore extraction somewhere in Northern Dale in the sixteenth century, and a sherd of twelfth- or thirteenth-century pottery has been found within the mine entrance.

Final comment: see Project comments

Laboratory comment: English Heritage (25 November 2004), the five measurements (on four samples) are not statistically consistent (T'=10.7; T'(5%)=9.5; v=4, Ward and Wilson 1978) and probably therefore represent the remains of more than one distinct period of fire setting. It is probable that the fire-setting in this mine is later than that at Old Ash Mine.

References: Barnatt and Worthington 2006
                  Ward and Wilson 1978

OxA–13978 201 ±24 BP

$\delta^{13}C$: -24.5‰

Sample: Series B, LAL04–B2(1), submitted on 9 May 2004 by J Barnatt

Material: charcoal: *Corylus* sp., roundwood, single fragment; 4 growth rings (R Gale 2004)

Initial comment: sample from an *in situ* deposit immediately adjacent to the fire-setting, swept to corner of small working. The deposit comprises small pieces of waste stone/mineral containing much burnt coal and a small quantity of charcoal. This sample is charcoal from throughout the deposit (measuring *c* 30cm ∞ 15cm across and up to 10cm thick).

Objectives: to provide a date for the fire-setting.

Calibrated date: 1σ: cal AD 1660–1955★
                       2σ: cal AD 1650–1955★

Final comment: J Barnatt (15 August 2013), analysis of the series of dates as a whole gave confidence that the samples were not contaminated and gave results that were meaningful. Dating the Lords and Ladies Mine to the seventeenth century was later than was initially hoped for, but in retrospect was of particular interest, for rather than demonstrating extensive medieval mining, it forced a more nuanced understanding of the problems faced by miners in the early post-medieval period and the solutions reached by those without access to financial backers for drainage schemes, who instead re-introduced methods of extraction which, while dangerous because of the problems of suffocation and the need for careful ventilation control, allowed reserves to be exploited.

Laboratory comment: see OxA-13981

OxA–13979 196 ±24 BP

$\delta^{13}C$: -24.8‰

Sample: LAL 04 - B2(2), submitted on 9 May 2004 by J Barnatt

Material: charcoal: *Fraxinus* sp., roundwood, single fragment; 6 growth rings (R Gale 2004)

Initial comment: as OxA-13978

Objectives: as OxA-13978

Calibrated date: 1σ: cal AD 1660–1955★
                       2σ: cal AD 1650–1955★

Final comment: see OxA-13978

OxA–13980 115 ±27 BP

$\delta^{13}C$: -23.8‰

Sample: LAL 04 - A1A, submitted on 9 May 2004 by J Barnatt

Material: charcoal: *Corylus avellana* single fragment (R Gale 2004)

Initial comment: as OxA-13978

Objectives: as OxA-13978

Calibrated date: 1σ: cal AD 1680–1930
                       2σ: cal AD 1670–1950

Final comment: see OxA-13978

OxA–13981 161 ±24 BP

$\delta^{13}C$: -25.1‰

Sample: LAL 04 - A1B, submitted on 9 May 2004
by J Barnatt

Material: charcoal: *Fraxinus* sp., roundwood, single
fragment; 3 growth rings (R Gale 2004)

Initial comment: as OxA-13978

Objectives: as OxA-13978

Calibrated date: 1σ: cal AD 1665–1945
                       2σ: cal AD 1665–1955*

Final comment: see OxA-13978

Laboratory comment: English Heritage (25 November 2004),
the two results on this sample are statistically consistent
(T'=1.4; T'(5%)=3.8; ν=1; Ward and Wilson 1978),
and so a weighted mean (141 ±17 BP) can be taken before
calibration (cal AD 1670–1950 at 95% confidence; Reimer
et al 2004).

References: Reimer *et al* 2004
                  Ward and Wilson 1978

OxA–13982 122 ±23 BP

$\delta^{13}C$: -24.7‰

Sample: LAL 04 - A1B, submitted on 9 May 2004 by
30/04/2008

Material: charcoal: *Fraxinus* sp., roundwood single fragment;
c 3 growth rings (R Gale 2004)

Initial comment: a replicate of OxA-13981. As OxA-13978

Objectives: as OxA-13978

Calibrated date: 1σ: cal AD 1680–1930
                       2σ: cal AD 1675–1945

Final comment: see OxA-13978

Laboratory comment: see OxA-13981

Peak District Mines: Old Ash Mine, Derbyshire

Location: SK 270606
                   Lat. 53.08.29 N; Long. 01.35.47 W

Project manager: J Barnatt (Peak District National Park),
                    Spring 2003

Archival body: Peak District National Park

Description: from four sample sites within different 'flat
workings' in the mine. These were selected for recovery of
charcoal, which was used to light coal, the fuel used for fire-
setting of limestone beds associated with galena-rich
mineralisation within bedding planes. These were
demonstrably worked prior to gunpowder extraction
elsewhere in the mines, a technique commonly introduced in
the Derbyshire ore-field around the beginning of the
seventeenth century. Each sample was recovered from floor
deposits in the 'flat workings', each comprising waste
rock/mineral with burnt coal and small amounts of charcoal.

Objectives: the samples will hopefully provide dating for the
fire-setting employed in Old Ash Mine, the most extensive
working of this type in the Derbyshire ore-field. They are
predicted to be either of medieval or early post-medieval
date (probably thirteenth to seventeenth century). The fire-
setting may have been undertaken over many years and
samples were taken from different parts of the workings and
the dates of different parts. If some or all prove to be
medieval, this will have a significant impact on our
understanding of mining in Britain in this period.

Final comment: see Project comments

Laboratory comment: English Heritage (24 November 2004),
the samples from the south-east cave workings (OAM03 B1
& B2) are not statistically consistent (T'=16.3; T'(5%)=3.8;
ν=1, Ward and Wilson 1978) and clearly represent the
remains of two distinct periods of fire setting. However, one
of these (OxA-12940) is clearly medieval in date. Excluding
this result the other four measurements (OxA-12939 and
OxA-12941-3) are statistically consistent (T'=3.0;
T'(5%)=7.8; ν=3, Ward and Wilson 1978) and could be of
the same age.

References: Barnatt and Worthington 2006
                  Ward and Wilson 1978

OxA–12939 221 ±29 BP

$\delta^{13}C$: -25.6‰

Sample: Series A, OAM03 -A1, submitted on 20 July 2003
by J Barnatt

Material: charcoal: *Corylus* sp., roundwood single fragment;
radius 6mm, 3 growth rings (R Gale 2003)

Initial comment: a sample from a thin lower deposit of small
pieces of waste stone/mineral containing burnt coal and a
small quantity of charcoal. This was sealed by a thin
stalactite coating, within a part mined, part natural roof
passage used as a ventilation duct, later backfilled when out
of use. The lower deposits were derived from elsewhere at
the time the passage was mined/used during fire-setting
period of extraction - hence it is not *in situ* with regard to
fire-setting, but dropped from known fire-set workings a
short distance above (now enterable via a different route).

Objectives: this sample is one of a series of four, which (if not
irreversibly contaminated by coal smoke) will date the use of
fire-setting using coal within Old Ash Mine. The use of coal
(lit by wood) in pre-seventeenth-century mine workings has
only recently been identified archaeologically in Peak
District mines and samples have never been previously
dated. While there is post-medieval documentation of its
use, referring to it being an 'ancient' method of extraction, it
is not known when this practise started. Coal from nearby
coal mines is documented from at least the thirteenth
century. The span of dates for the use of coal in lead mines
is currently unknown. Old Ash Mine is an ideal starting
point for investigation, as the early workings are extensive
and relatively undisturbed (in contrast with all other
identified fire-setting sites in the orefield). The scale of the
Old Ash workings, if medieval, will have significant impact
on our understanding of metal mining in Britain at that date.

Calibrated date: 1σ: cal AD 1650–1955*
                       2σ: cal AD 1640–1955*

Final comment: J Barnatt (15 August 2013), analysis of the series of dates as a whole gave confidence that the samples were not contaminated and gave results that were meaningful. Dating Old Ash Mine to the late sixteenth and/or seventeenth century was later than was initially hoped for, but in retrospect was of particular interest, for rather than demonstrating extensive medieval mining, it forced a more nuanced understanding of the problems faced by miners in the early post-medieval period and the solutions reached by those without access to financial backers for drainage schemes, who instead re-introduced methods of extraction which, while dangerous because of the problems of suffocation and the need for careful ventilation control, allowed reserves to be exploited.

OxA–12940 368 ±29 BP

δ¹³C: -24.5‰

Sample: Series A, OAM03 -B1(1), submitted on 20 July 2003 by J Barnatt

Material: charcoal: *Corylus* sp., roundwood, single fragment; 7 growth rings (R Gale 2003)

Initial comment: a sample from an *in situ* deposit immediately adjacent to the fire-setting, swept to a corner of a small working. The deposit comprises small pieces of waste stone/mineral containing much burnt coal and a small quantity of charcoal. This sample is charcoal from throughout the deposit (measuring *c* 30cm ∞ 15cm across and up to 10cm thick).

Objectives: as OxA-12939

Calibrated date: 1σ: cal AD 1450–1620
2σ: cal AD 1440–1640

Final comment: see OxA-12939

OxA–12941 259 ±29 BP

δ¹³C: -25.6‰

Sample: Series A, OAM03–B2, submitted on 20 July 2003 by J Barnatt

Material: charcoal: *Quercus* sp., roundwood, single fragment; 5+ growth rings (R Gale 2003)

Initial comment: this sample is from an *in situ* deposit immediately adjacent to the fire-setting, swept to a corner of a small working. The deposit comprises small pieces of waste stone/mineral containing much burnt coal and a small quantity of charcoal. Like B1 this sample is charcoal from throughout the deposit (measuring *c* 30cm ∞ 15cm across and up to 10cm thick).

Objectives: as OxA-12939

Calibrated date: 1σ: cal AD 1640–1670
2σ: cal AD 1520–1800

Final comment: see OxA-12939

OxA–12942 280 ±29 BP

δ¹³C: -25.7‰

Sample: Series A, OAM03–C3, submitted on 20 July 2003 by J Barnatt

Material: charcoal: Pomoideae, single fragment (R Gale 2003)

Initial comment: samples taken close together from floor deposits in the end of a flatting with no later mining. While from slightly different parts of the flatting (and C2 might be wood rather than charcoal) there is no reason why these cannot be combined (but *see* below).

Objectives: as OxA-12939

Calibrated date: 1σ: cal AD 1520–1660
2σ: cal AD 1510–1670

Final comment: see OxA-12939

OxA–12943 222 ±30 BP

δ¹³C: -23.7‰

Sample: Series A, OAM03–D1, submitted on 20 July 2003 by J Barnatt

Material: charcoal: *Ilex aquifolium*, single fragment (R Gale 2003)

Initial comment: sample taken from floor deposit near the end of a flatting with no later mining. While there has been no later mining, the possibility that the floor deposits have been turned over by later miners, while exploring the passage to see if further mining would be worthwhile, cannot be fully discounted.

Objectives: as OxA-12939

Calibrated date: 1σ: cal AD 1650–1955★
2σ: cal AD 1640–1955★

Final comment: see OxA-12939

OxA–13977 220 ±23 BP

δ¹³C: -24.4‰

Sample: Series A, OAM03–B1(2), submitted on 9 May 2004 by J Barnatt

Material: charcoal: *Corylus* sp., roundwood, single fragment; 11 growth rings (R Gale 2003)

Initial comment: as OxA-12940

Objectives: two samples from sample site B have previously been dated: OAM03–B1; OxA-12940, 368 ±29 BP; and OAM03–B2; OxA-12941, 259 ±29 BP). Statistically these are not consistent and the former appears to represent an earlier period of fire-setting. However, from an archaeological viewpoint the two samples were from a discrete deposit which can be confidently argued to represent one episode of burning - it is very hard to see how it could be interpreted otherwise given the context of the deposit. One explanation could be that significantly earlier charcoal (in B1) was introduced into the fire, but this seems unlikely - a large amount of special pleading would be needed. Thus a further sample from the charcoal retained from B1 (found scattered throughout the discrete deposit) is submitted to seek clarification. This will hopefully demonstrate whether B1 has charcoal consistent in date with B2 or whether there is indeed a significant earlier component.

Calibrated date: 1σ: cal AD 1650–1955★
2σ: cal AD 1645–1955★

Final comment: see OxA-12939

Predictive modelling of archaeological sites in raised mires: Hatfield Moors, Rawcliffe, South Yorkshire

Location:	SE 7060007200 Lat. 53.33.23 N; Long. 00.56.03 W
Project manager:	B Gearey (University of Birmingham), 2005
Archival body:	Doncaster Museum

Description: a former raised mire site on Hatfield Moors. The samples are from various locations on Hatfield Moors. They include both the base and top of peat deposits.

Objectives: the three main objectives were: to date the onset of peat accumulation in different areas of the moor, determined in relation to m OD (basal samples); to assess the chronological depth of sediment surviving in different parts of the moor and to therefore estimate the potential for the preservation of archaeological remains (top samples); to establish the relationship between sediment accumulation and site construction at the Lindholme Neolithic trackway site, excavated under the auspices of this project. The results of the dating are integral to the project as they will be used to create a 4D model of the evolution of the peatland.

Final comment: B Gearey (26 September 2014), for full discussion of all these determinations and their application in establishing a chronology for peat inception at the Lindholme Neolithic trackway site, modeling peat accumulation, and the age of surviving peat deposits at Hatfield and Thorne Moors, see Chapman and Gearey (2013).

Laboratory comment: English Heritage (2006), only three of the eight bulk peat samples for which replicate humic and humin fraction measurements were obtained produced statistically consistent results. Replicate measurements on the humic and humin fractions of bulk peat samples and single macrofossils were also statistically inconsistent in all cases. These results suggest that many of the basal samples incorporated a range of organic fractions of different ages; for example, all four determinations from HAT4 are statistically inconsistent. The lack of consistency between humin and humic fractions suggests that the humic acids may have been mobile down the deposit, since the latter fraction is often younger than the humin or associated macrofossils (eg HAT2; SUERC-9688 and -9699, HAT4; SUERC-9636 and 9637; HAT23; SUERC-8875 and -8876 and -9647 and -9648). The four samples from HAT23 in particular reflect the incorporation of a variety of material of different ages. Such a lack of sample homogeneity is typical of many peat deposits and reflects in part the complexity of formation processes of mire systems (eg Kilian *et al* 1995). This would perhaps be expected to be a particular problem during the earliest stages of peat formation when there is evidence for fluctuations in hydrological conditions. Identifying which determination might give the 'true' estimate of the age of the deposits with such multiple, statistically inconsistent dates are problematic.

References: Chapman and Gearey 2013

GU–6365 5650 ±50 BP

δ¹³C: -27.2‰

Sample: HAT17_NTRACKS7 (c), submitted in July 2005 by B Gearey

Material: peat (humic acid) (B Gearey 2005)

Initial comment: this sample is from the basal sediment unit in the northern part of Hatfield Moors.

Objectives: this sample is intended to provide a date for peat inception in the northern part of Hatfield Moors.

Calibrated date: 1σ: 4540–4440 cal BC
2σ: 4600–4360 cal BC

Final comment: B Gearey (26 September 2014), this result indicates a Mesolithic date for peat inception at this location, but is not consistent with the humin fraction sample (SUERC-9646) also dated.

Laboratory comment: English Heritage (17 August 2012), the replicate samples (GU-6365 and SUERC-9646) are not statistically consistent (T′=11.7; T′(5%)=3.8; v=1; Ward and Wilson 1978).

References: Ward and Wilson 1978

SUERC–8846 5430 ±35 BP

δ¹³C: -26.0‰

Sample: HAT2_LIND501, submitted in July 2005 by B Gearey

Material: waterlogged plant macrofossils (unidentified charred twigs and stems) (D Robinson 2005)

Initial comment: this sample is from trench 5 of the Lindholme Trackway excavations. This sample is from the base of the sediment that the trackway is located in. The sediment is an organic *limus* or pool mud, which immediately overlies the basal windblown sands at the site.

Objectives: this sample is intended to provide a date for peat inception at this location.

Calibrated date: 1σ: 4340–4250 cal BC
2σ: 4350–4230 cal BC

Final comment: B Gearey (26 September 2014), the result indicates a Mesolithic date for peat inception.

Laboratory comment: English Heritage (2006), the three measurements from this depth (SUERC-9688–9 and SUERC-8846) are statistically inconsistent (T′=919.7, T′(5%)=6.0, v=2; Ward and Wilson 1978).

References: Ward and Wilson 1978

SUERC–8847 3305 ±40 BP

δ¹³C: -27.3‰

Sample: HAT3_SE_C3, submitted in July 2005 by B Gearey

Material: waterlogged plant macrofossils (Ericaceae twigs/roots) (D Robinson 2005)

Initial comment: this sample is from the top of the peat at this location and consists of organic raised mire sediment.

Objectives: this sample is intended to assess the remaining chronological depth of peat remaining in this sampling site in the south-eastern part of the moor.

Calibrated date: 1σ: 1630–1520 cal BC
2σ: 1690–1490 cal BC

Final comment: B Gearey (26 September 2014), the result indicates that deposits dating to the Bronze Age survive at this location.

SUERC–8848 4495 ±35 BP

δ¹³C: -28.5‰

Sample: HAT4_SE_C3 (a), submitted in July 2005 by B Gearey

Material: waterlogged plant macrofossils (fragments of charred grass stems) (D Robinson 2005)

Initial comment: this sample is from the base of the peat at this location and consists of highly humified peat.

Objectives: this sample is intended to determine the age of peat inception at this sampling site in the south-eastern part of the moor.

Calibrated date: 1σ: 3350–3090 cal BC
2σ: 3360–3020 cal BC

Final comment: B Gearey (26 September 2014), this result confirms a Neolithic date for peat inception at this location.

Laboratory comment: English Heritage (2006), the four results from this depth (SUERC-9636–7 and SUERC-8848–9) are statistically inconsistent (T′=48.7; T′(5%)=7.8; ν=3; Ward and Wilson 1978).

References: Ward and Wilson 1978

SUERC–8849 4745 ±35 BP

δ¹³C: -27.1‰

Sample: HAT4_SE_C3 (b), submitted in July 2005 by B Gearey

Material: waterlogged plant macrofossils (*Eriphorum* spindle/Ericaceae stems) (D Robinson 2005)

Initial comment: as SUERC-8848

Objectives: as SUERC-8848

Calibrated date: 1σ: 3640–3380 cal BC
2σ: 3640–3370 cal BC

Final comment: see SUERC-8848

Laboratory comment: see SUERC-8848

SUERC–8850 4630 ±40 BP

δ¹³C: -27.9‰

Sample: HAT5_NROADS6, submitted in July 2005 by B Gearey

Material: waterlogged plant macrofossils, twigs (D Robinson 2005)

Initial comment: this sample is from the base of the peat at this location and consists of highly humified peat.

Objectives: this sample is intended to determine the age of peat inception at this sampling site in the western part of the moor.

Calibrated date: 1σ: 3500–3360 cal BC
2σ: 3520–3340 cal BC

Final comment: see SUERC-8848

SUERC–8851 4410 ±35 BP

δ¹³C: -28.1‰

Sample: HAT6_NROADS6, submitted in July 2005 by B Gearey

Material: waterlogged plant macrofossils (Ericaceae stems) (D Robinson 2005)

Initial comment: this sample is from the top of the peat at this location and consists of organic raised mire sediment.

Objectives: this sample is intended to assess the remaining chronological depth of peat in this area, to the west of Lindholme Island.

Calibrated date: 1σ: 3100–2930 cal BC
2σ: 3320–2910 cal BC

Final comment: B Gearey (26 September 2014), the result indicates that deposits dating to the Neolithic survive at this location.

SUERC–8855 2990 ±40 BP

δ¹³C: -28.7‰

Sample: HAT8_PSOUTHC2A, submitted in July 2005 by B Gearey

Material: waterlogged plant macrofossils (Ericaceae stems) (D Robinson 2005)

Initial comment: this sample is from the top of the peat in the south part of the peatland and consists of raised mire sediment.

Objectives: this sample is intended to assess the remaining chronological depth of peat at this sampling site in the southern part of the moor.

Calibrated date: 1σ: 1280–1120 cal BC
2σ: 1390–1110 cal BC

Final comment: see SUERC-8847

SUERC–8856 3470 ±35 BP

δ¹³C: -27.2‰

Sample: HAT9_MIDDLEC4, submitted in July 2005 by B Gearey

Material: waterlogged plant macrofossils (Ericaceae stems/roots) (D Robinson 2005)

Initial comment: this sample is from the base of the peat sequence at this location and consists of highly humified sediment.

Objectives: this sample is intended to provide a date for peat inception at this sampling site in the southern part of the moor.

Calibrated date: *1σ:* 1880–1700 cal BC
                     2σ: 1890–1680 cal BC

Final comment: B Gearey (26 September 2014), this result confirms a Bronze Age date for peat inception at this location.

SUERC–8857 3200 ±40 BP

δ¹³C: -28.0‰

Sample: HAT10_MIDDLEC4, submitted in July 2005 by B Gearey

Material: waterlogged plant macrofossils (Ericaceae stems/roots) (D Robinson 2005)

Initial comment: this sample is from the top of the peat sequence at this sampling site in the southern part of the peatland and consists of raised mire sediment.

Objectives: this sample is intended to assess the remaining chronological depth of peat remaining at this sampling site in the southern part of the moor.

Calibrated date: *1σ:* 1510–1420 cal BC
                     2σ: 1600–1400 cal BC

Final comment: see SUERC-8847

SUERC–8858 3310 ±35 BP

δ¹³C: -27.1‰

Sample: HAT11_PSOUTHC1, submitted in July 2005 by B Gearey

Material: waterlogged plant macrofossils (Ericaceae stems/roots) (D Robinson 2005)

Initial comment: this sample is from the base of the peat sequence at this location and consists of highly humified sediment.

Objectives: this sample is intended to provide a date for peat inception at this sampling site in the southern part of the moor.

Calibrated date: *1σ:* 1630–1520 cal BC
                     2σ: 1690–1500 cal BC

Final comment: see SUERC-8856

SUERC–8859 2960 ±40 BP

δ¹³C: -28.1‰

Sample: HAT12_PSOUTHC1, submitted on 26 September 2014 by B Gearey

Material: waterlogged plant macrofossils (Ericaceae stems/roots and charred leaf) (D Robinson 2005)

Initial comment: this sample is from the top of the peat sequence at this sampling site in the southern part of the peatland and consists of raised mire sediment.

Objectives: this sample is intended to assess the remaining chronological depth of peat remaining at this sampling site in the southern part of the moor.

Calibrated date: *1σ:* 1260–1110 cal BC
                     2σ: 1290–1040 cal BC

Final comment: see SUERC-8847

SUERC–8860 4410 ±35 BP

δ¹³C: -26.9‰

Sample: HAT3_MIDDLEC5 (a), submitted in July 2005 by B Gearey

Material: carbonised plant macrofossil (twigs) (D Robinson 2005)

Initial comment: this sample is from the base of the peat sequence at this location and consists of highly humified sediment.

Objectives: this sample is intended to provide a date for peat inception at this sampling site in the central part of the moor.

Calibrated date: *1σ:* 3100–2930 cal BC
                     2σ: 3320–2910 cal BC

Final comment: see SUERC-8848

SUERC–8861 4410 ±35 BP

δ¹³C: -24.3‰

Sample: HAT13_MIDDLEC5 (b), submitted in July 2005 by B Gearey

Material: waterlogged plant macrofossils: bark, charred (D Robinson 2005)

Initial comment: this sample is from the base of the peat sequence at this location and consists of highly humified sediment.

Objectives: this sample is intended to provide a date for peat inception at this sampling site in the central part of the moor.

Calibrated date: *1σ:* 3100–2930 cal BC
                     2σ: 3320–2910 cal BC

Final comment: see SUERC-8848

SUERC–8875 4895 ±40 BP

δ¹³C: -27.4‰

Sample: HAT23_SWLINDS4 (a), submitted in July 2005 by B Gearey

Material: waterlogged plant macrofossils: twig, and *Carex* sp. nutlets (D Robinson 2005)

Initial comment: this sample is from the base of the peat sequence at this location and consists of highly humified sediment.

Objectives: this sample is intended to provide a date for peat inception at this sampling site in the central part of the moor.

Calibrated date: *1σ:* 3710–3640 cal BC
                         2σ: 3770–3630 cal BC

Final comment: see SUERC-8848

Laboratory comment: English Heritage (2006), the four results from this depth (SUERC-9647–8 and SUERC-8875–6) are not statistically consistent (T′=116.9; T′(5%)=7.8; *v*=3; Ward and Wilson 1978).

References: Ward and Wilson 1978

SUERC–8876 4730 ±35 BP

δ¹³C: -25.4‰

Sample: HAT23_SWLINDS4 (b), submitted in July 2005 by B Gearey

Material: waterlogged plant macrofossils (*Eriphorum* fibres) (D Robinson 2005)

Initial comment: this sample is from the basal sediment unit in the western part of Hatfield Moors and consists of highly humified sediment.

Objectives: this sample is intended to provide a date for peat inception in the western part of Hatfield Moors.

Calibrated date: *1σ:* 3630–3380 cal BC
                         2σ: 3640–3370 cal BC

Final comment: see SUERC-8848

Laboratory comment: see SUERC-8875

SUERC–8877 4250 ±35 BP

δ¹³C: -26.8‰

Sample: HAT24_LIND103 base, submitted in July 2005 by B Gearey

Material: waterlogged plant macrofossils (Ericacaeae stems/roots) (D Robinson 2005)

Initial comment: this sample is from trench 5 of the Lindholme Trackway excavations. This sample is from the top of the sediment that the trackway is located in.

Objectives: this sample is intended to provide a date for peat inception at this location.

Calibrated date: *1σ:* 2910–2870 cal BC
                         2σ: 2920–2760 cal BC

Final comment: B Gearey (26 September 2014), this result confirms a Neolithic date for the deposit.

SUERC–8878 3690 ±35 BP

δ¹³C: -26.7‰

Sample: HAT25_LIND103 top, submitted in July 2005 by B Gearey

Material: waterlogged plant macrofossils (Ericacaeae stems/roots) (D Robinson 2005)

Initial comment: as OxA-8877

Objectives: this sample is intended to provide a date for the top of the remaining peat at this location and to compare the date of the sediment with that of the archaeological feature, to address issues of sediment compaction and structural movement.

Calibrated date: *1σ:* 2140–2020 cal BC
                         2σ: 2200–1960 cal BC

Final comment: B Gearey (26 September 2014), this determination would appear to be too young given the Neolithic date for the trackway, it is probable that the macrofossils were intrusive.

SUERC–8947 5985 ±35 BP

δ¹³C: -27.5‰

Sample: HAT15_PSOUTHC3 base (b), submitted in July 2005 by B Gearey

Material: waterlogged plant macrofossils: bark, unidentified, charred (D Robinson 2005)

Initial comment: this sample is from the base of the peat sequence at this location and consists of raised mire sediment.

Objectives: this sample is intended to assess the remaining chronological depth of peat remaining at this sampling site in the southern part of the moor.

Calibrated date: *1σ:* 4940–4800 cal BC
                         2σ: 4980–4780 cal BC

Final comment: see SUERC-8846

SUERC–8948 3630 ±35 BP

δ¹³C: -27.0‰

Sample: HAT16_PSOUTHC3 top, submitted in 2005 by B Gearey

Material: waterlogged plant macrofossils (Ericaceae stem/root) (D Robinson 2005)

Initial comment: this sample is from the base of the peat sequence at this location and consists of raised mire sediment.

Objectives: this sample is intended to assess the remaining chronological depth of peat remaining at this sampling site in the southern part of the moor.

Calibrated date: *1σ:* 2040–1940 cal BC
                         2σ: 2140–1890 cal BC

Final comment: see SUERC-8847

SUERC–8949 4905 ±35 BP

δ¹³C: -26.9‰

Sample: HAT18_NTRACKS5, submitted in July 2005 by B Gearey

Material: carbonised plant macrofossils (Ericaceae twig and unidentified twig) (D Robinson 2005)

Initial comment: this sample is from the basal sediment unit in the northern part of Hatfield Moors.

Objectives: this sample is intended to provide a date for peat inception in the western part of Hatfield Moors.

Calibrated date: 1σ: 3710–3640 cal BC
2σ: 3770–3630 cal BC

Final comment: see SUERC-8848

SUERC–8950 4655 ±35 BP

δ¹³C: -25.3‰

Sample: HAT19_SWLINDS1, submitted in July 2005 by B Gearey

Material: waterlogged plant macrofossils (Ericaceae stem/root) (D Robinson 2005)

Initial comment: this sample is from the basal sediment unit in the western part of Hatfield Moors and consists of highly humified sediment.

Objectives: as SUERC-8949

Calibrated date: 1σ: 3510–3360 cal BC
2σ: 3630–3350 cal BC

Final comment: see SUERC-8848

SUERC–8951 4905 ±35 BP

δ¹³C: -25.9‰

Sample: HAT20_SWLINDS3, submitted in July 2005 by B Gearey

Material: waterlogged plant macrofossils (herbaceous, small twigs) (D Robinson 2005)

Initial comment: this sample is from the basal sediment unit in the western part of Hatfield Moors and consists of highly humified sediment.

Objectives: as SUERC-8949

Calibrated date: 1σ: 3710–3640 cal BC
2σ: 3770–3630 cal BC

Final comment: see SUERC-8848

SUERC–8952 6355 ±35 BP

δ¹³C: -26.0‰

Sample: HAT21_NTRACKS6, submitted in July 2005 by B Gearey

Material: waterlogged plant macrofossil (herbaceous, small twigs) (D Robinson 2005)

Initial comment: this sample is from the basal sediment unit in the northern part of Hatfield Moors.

Objectives: this sample is intended to provide a date for peat inception in the northern part of Hatfield Moors.

Calibrated date: 1σ: 5370–5300 cal BC
2σ: 5470–5230 cal BC

Final comment: see SUERC-8846

SUERC–8953 4605 ±35 BP

δ¹³C: -24.6‰

Sample: HAT22_SWLINDS2, submitted in July 2005 by B Gearey

Material: waterlogged plant macrofossils (herbaceous roots) (D Robinson 2005)

Initial comment: this sample is from the basal sediment unit in the western part of Hatfield Moors and consists of highly humified sediment.

Objectives: as SUERC-8949

Calibrated date: 1σ: 3490–3350 cal BC
2σ: 3500–3340 cal BC

Final comment: see SUERC-8848

SUERC–9636 4425 ±35 BP

δ¹³C: -28.6‰

Sample: HAT4_SE_C3 (c), submitted in July 2005 by B Gearey

Material: peat (humic acid) (B Gearey 2005)

Initial comment: this sample is from the base of the peat at this location and consists of highly humified peat.

Objectives: this sample is intended to determine the age of peat inception at this sampling site in the south-eastern part of the moor.

Calibrated date: 1σ: 3270–3010 cal BC
2σ: 3330–2920 cal BC

Final comment: B Gearey (26 September 2014), this result indicates a possible Neolithic date for peat inception at this location, but is inconsistent with the other sample (SUERC-9637) at this location.

Laboratory comment: see SUERC-8848

References: Ward and Wilson 1978

SUERC–9637 4500 ±35 BP

δ¹³C: -27.9‰

Sample: HAT4_SE_C3 (c), submitted in July 2005 by B Gearey

Material: peat (humin) (B Gearey 2005)

Initial comment: as SUERC-9636

Objectives: as SUERC-9636

Calibrated date: 1σ: 3350–3090 cal BC
2σ: 3360–3020 cal BC

Final comment: see SUERC-9636

Laboratory comment: see SUERC-8848

SUERC–9638 4225 ±35 BP

$\delta^{13}C$: -27.5‰

Sample: HAT13_MIDDLEC5 (c), submitted in July 2005 by B Gearey

Material: peat (humic acid) (B Gearey 2005)

Initial comment: this sample is from the base of the peat at this location and consists of highly humified peat.

Objectives: this sample is intended to determine the age of peat inception at this sampling site in the south-eastern part of the moor.

Calibrated date: 1σ: 2900–2870 cal BC
                      2σ: 2910–2690 cal BC

Final comment: B Gearey (26 September 2014), this result indicates a Neolithic date for peat inception, but is inconsistent with a replicate measurement (SUERC-9639) on the humin fraction from the same sample.

Laboratory comment: English Heritage (17 August 2012), the replicate samples (SUERC-9638 and SUERC-9639) are not statistically consistent (T'=11.8, T'(5%)=3.8, v=1; Ward and Wilson 1978).

References: Ward and Wilson 1978

SUERC–9639 4395 ±35 BP

$\delta^{13}C$: -28.0‰

Sample: HAT13_MIDDLEC5 (c), submitted in July 2005 by B Gearey

Material: peat (humin) (B Gearey 2005)

Initial comment: as SUERC-9638

Objectives: as SUERC-9638

Calibrated date: 1σ: 3090–2920 cal BC
                      2σ: 3270–2910 cal BC

Final comment: see SUERC-9638

Laboratory comment: see SUERC-9638

SUERC–9640 5350 ±35 BP

$\delta^{13}C$: -28.3‰

Sample: HAT15_PSOUTHC3 base (c), submitted in July 2005 by B Gearey

Material: peat (humic acid) (B Gearey 2005)

Initial comment: this sample is from the base of the peat sequence at this location and consists of raised mire sediment.

Objectives: this sample is intended to assess the remaining chronological depth of peat remaining at this sampling site in the southern part of the moor.

Calibrated date: 1σ: 4260–4070 cal BC
                      2σ: 4330–4040 cal BC

Final comment: B Gearey (26 September 2014), this result indicates a Mesolithic date for peat inception at this

location, but is inconsistent with result on the replicate measurement on the humin fraction (SUERC-9641) from the same sample.

Laboratory comment: English Heritage (17 August 2012), the replicate samples (SUERC-9640 and SUERC-9641) are not statistically consistent (T'=118.8, T'(5%)=3.8, v=1; Ward and Wilson 1978).

References: Ward and Wilson 1978

SUERC–9641 5890 ±35 BP

$\delta^{13}C$: -28.2‰

Sample: HAT15_PSOUTHC3 base (c), submitted in July 2005 by B Gearey

Material: peat (humin) (B Gearey 2005)

Initial comment: as SUERC-9640

Objectives: as SUERC-9640

Calibrated date: 1σ: 4800–4710 cal BC
                      2σ: 4840–4690 cal BC

Final comment: see SUERC-9640

Laboratory comment: see SUERC-9640

SUERC–9646 5860 ±35 BP

$\delta^{13}C$: -26.7‰

Sample: HAT17_NTRACKS7 (c), submitted in July 2005 by B Gearey

Material: peat (humin) (B Gearey 2005)

Initial comment: as GU-6365

Objectives: as GU-6365

Calibrated date: 1σ: 4790–4700 cal BC
                      2σ: 4800–4610 cal BC

Final comment: see GU-6365

Laboratory comment: see GU-6365

SUERC–9647 4390 ±35 BP

$\delta^{13}C$: -26.9‰

Sample: HAT23_SWLINDS4 (c), submitted in July 2005 by B Gearey

Material: peat (humic acid) (B Gearey 2005)

Initial comment: this sample is from the basal sediment unit in the western part of Hatfield Moors and consists of highly humified sediment.

Objectives: this sample is intended to provide a date for peat inception in the western part of Hatfield Moors.

Calibrated date: 1σ: 3090–2920 cal BC
                      2σ: 3270–2900 cal BC

Final comment: see SUERC-8848

Laboratory comment: English Heritage (17 August 2012), the replicate results on the bulk fractions of this peat

(SUERC-9647 and SUERC-9648) are statistically consistent (T'=0.0; T'(5%)=3.8; v=1; Ward and Wilson 1978). *See also* SUERC-8875.

References: Ward and Wilson 1978

SUERC–9648 4400 ±40 BP

$\delta^{13}C$: -26.3‰

Sample: HAT23_SWLINDS4 (c), submitted in July 2005 by B Gearey

Material: peat (humin) (B Gearey 2005)

Initial comment: as SUERC-9647

Objectives: as SUERC-9647

Calibrated date: *1σ:* 3100–2920 cal BC
                          2σ: 3310–2900 cal BC

Final comment: see SUERC-8848

Laboratory comment: see SUERC-9647 and SUERC-8875

SUERC–9688 3880 ±35 BP

$\delta^{13}C$: -27.6‰

Sample: HAT1_LIND501, submitted in July 2005 by B Gearey

Material: peat (humic acid) (B Gearey 2005)

Initial comment: as SUERC-8846

Objectives: this sample is intended to provide both a date for the top of the remaining peat at this location and to compare the date of the sediment with that of the archaeological feature, to address issues of sediment compaction and structural movement.

Calibrated date: *1σ:* 2470–2290 cal BC
                          2σ: 2470–2200 cal BC

Final comment: B Gearey (26 September 2014), this determination is too young for its stratigraphic context, possibly because humic acids may be highly mobile.

Laboratory comment: English Heritage (17 August 2012), the replicate results on the bulk fractions of this peat (SUERC-9688 and SUERC-9689) are not statistically consistent (T'=6.9, T'(5%)=3.8, v=1; Ward and Wilson 1978). *See also* SUERC-8846.

References: Ward and Wilson 1978

SUERC–9689 4010 ±35 BP

$\delta^{13}C$: -27.1‰

Sample: HAT1_LIND501, submitted in July 2005 by B Gearey

Material: peat (humin) (B Gearey 2005)

Initial comment: as SUERC-8846

Objectives: as SUERC-8846

Calibrated date: *1σ:* 2580–2470 cal BC
                          2σ: 2620–2460 cal BC

Final comment: B Gearey (26 September 2014), this determination is too young for its stratigraphic context.

Laboratory comment: see SUERC-9688 and SUERC-8846

SUERC–9690 4470 ±35 BP

$\delta^{13}C$: -27.7‰

Sample: HAT7_PSOUTHC2A, submitted in July 2005 by B Gearey

Material: peat (humic acid) (B Gearey 2005)

Initial comment: this sample is from the base of the peat in the south part of the peatland and consists of raised mire sediment.

Objectives: this sample is intended to assess the date of peat inception at this sampling site in the southern part of the moor.

Calibrated date: *1σ:* 3330–3030 cal BC
                          2σ: 3350–3010 cal BC

Final comment: B Gearey (26 September 2014), this result indicates a Neolithic date for peat inception at this loation, but is inconsistent with the replicate measurement on the humin fraction of this sample (SUERC-9691).

Laboratory comment: English Heritage (17 August 2012), the replicate samples (SUERC-9690 and SUERC-9691) are statistically consistent (T'=2.9; T'(5%)=3.8; v=1; Ward and Wilson 1978).

References: Ward and Wilson 1978

SUERC–9691 4555 ±35 BP

$\delta^{13}C$: -27.8‰

Sample: HAT7_PSOUTHC2A, submitted in July 2005 by B Gearey

Material: peat (humin) (B Gearey 2005)

Initial comment: as SUERC-9690

Objectives: as SUERC-9690

Calibrated date: *1σ:* 3370–3130 cal BC
                          2σ: 3490–3100 cal BC

Final comment: see SUERC-9690

Laboratory comment: see SUERC-9690

SUERC–9692 3205 ±35 BP

$\delta^{13}C$: -26.9‰

Sample: HAT14_MIDDLEC5, submitted in July 2005 by B Gearey

Material: peat (humic acid) (B Gearey 2005)

Initial comment: this sample is from the top of the peat sequence at this location and consists of raised mire sediment.

Objectives: this sample is intended to assess the remaining chronological depth of peat remaining at this sampling site in the central part of the moor.

Calibrated date: *1σ:* 1510–1430 cal BC
                      2σ: 1600–1410 cal BC

Final comment: B Gearey (26 September 2014), this result confirms that deposits dating to the Bronze Age survive at this location.

Laboratory comment: English Heritage (17 August 2012), the replicate samples (SUERC-9692 and SUERC-9696) are statistically consistent (T'=2.0; T'(5%)=3.8; *v*=1; Ward and Wilson 1978).

References: Ward and Wilson 1978

SUERC–9696 3135 ±35 BP

δ¹³C: -26.4‰

Sample: HAT14_MIDDLEC5, submitted in July 2005 by B Gearey

Material: peat (humin) (B Gearey 2005)

Initial comment: as SUERC-9692

Objectives: as SUERC-9692

Calibrated date: *1σ:* 1440–1390 cal BC
                      2σ: 1500–1300 cal BC

Final comment: see SUERC-9692

Laboratory comment: see SUERC-9692

Predictive modelling of archaeological sites in raised mires: Thorne Moors, South Yorkshire

Location: SE 733158
              Lat. 53.38.00 N; Long. 00.53.29 W

Project manager: B Gearey (University of Birmingham), 2005

Archival body: Doncaster Museum

Description: a former raised mire site on Thorne Moors. Rawcliffe Moors is the deepest remaining area of peat on Thorne Moors. A series of samples from a peat sequence were submitted, which has been analysed for testate amoebae. These data will be used to produce a quantitive reconstruction of changes in the mire watertable over time. This will in turn inform on the relative accessibility of the mire surface to human communities and hence the likelihood of finding wetland archaeological remains at specific stratigraphic horizons within the peatland.

Objectives: to provide an absolute chronology for the watertable reconstruction and therefore to identify during which cultural periods the mire surface was dry or wet. This will indicate the likely antiquity of any archaeological remains associated with different stratigraphic horizons.

Final comment: B Gearey (26 September 2014), Bayesian modelling indicates that there are problems with the sequence of determinations for this sequence. The models show good agreement up until 0.78m, above this point there is poor agreement. In order to assess if this was a result of aberrant dates in the form of intrusive material (such as the

Eriophorum or Ericaceous stems), these samples were removed and the model re-run. Removing the humic acid dates makes no difference to the agreement suggesting that there are other issues affecting the robustness of the top segment of the deposits. Certainly the testates and macrofossils indicate that significant variation in peat composition especially between 0.90–0.72m and 0.32–0.04m - it is possible that the samples are affected by lack of homogeneity. This set of determinations was therefore not included in the final publication of this project (Chapman and Gearey 2013).

References: Chapman and Gearey 2013

OxA–15829 2508 ±31 BP

δ¹³C: -24.4‰

Sample: RAW 0.36m, submitted on 28 March 2006 by B Gearey

Material: waterlogged plant macrofossil (*Sphagnum*) (B Gearey 2006)

Initial comment: a peat sample from an ombrotrophic mire.

Objectives: to establish a dating framework for testate amoebae derived watertable curve. The data is to be used for modelling the implications of palaeohydrological changes for anthropogenic access to the mire surface as part of the project.

Calibrated date: *1σ:* 780–540 cal BC
                      2σ: 800–530 cal BC

Final comment: B Gearey (26 September 2014), the result indicates an Iron Age date for the peat at this level.

OxA–15830 2437 ±29 BP

δ¹³C: -22.9‰

Sample: RAW 0.44m, submitted on 28 March 2006 by B Gearey

Material: waterlogged plant macrofossil (*Eriophorum*) (B Gearey 2006)

Initial comment: as OxA-15829

Objectives: as OxA-15829

Calibrated date: *1σ:* 740–410 cal BC
                      2σ: 760–400 cal BC

Final comment: see OxA-125829

OxA–15831 2694 ±31 BP

δ¹³C: -24.7‰

Sample: RAW 0.68m, submitted on 28 March 2006 by B Gearey

Material: waterlogged plant macrofossil (*Sphagnum*) (B Gearey 2006)

Initial comment: as OxA-15829

Objectives: as OxA-15829

Calibrated date: 1σ: 900–800 cal BC
                       2σ: 910–800 cal BC

Final comment: B Gearey (16 September 2014), the resultt indicates a Bronze Age date for the peat at this level.

OxA–15832 3115 ±34 BP

$\delta^{13}C$: -26.0‰

Sample: RAW 0.88m A, submitted on 28 March 2006 by B Gearey

Material: waterlogged plant macrofossil (*Calluna* stem/roots) (B Gearey 2006)

Initial comment: as OxA-15829

Objectives: as OxA-15829

Calibrated date: 1σ: 1430–1310 cal BC
                       2σ: 1450–1280 cal BC

Final comment: see OxA-15831

Laboratory comment: English Heritage (2006), the three results from this depth are statistically inconsistent (T′=12.0; T′(5%)=6.0; ν=2; Ward and Wilson 1978).

References: Ward and Wilson 1978

OxA–15833 3048 ±33 BP

$\delta^{13}C$: -25.7‰

Sample: RAW 0.88m B, submitted on 28 March 2006 by B Gearey

Material: peat (humin) (B Gearey 2006)

Initial comment: as OxA-125829

Objectives: as OxA-125829

Calibrated date: 1σ: 1390–1230 cal BC
                       2σ: 1420–1210 cal BC

Final comment: B Gearey (26 September 2014), the result confirms a Bronze Age date, but the result for OxA-15993 from the same level is significantly different.

Laboratory comment: see OxA-15832

OxA–15834 4374 ±34 BP

$\delta^{13}C$: -25.2‰

Sample: RAW 1.76m, submitted on 28 March 2006 by B Gearey

Material: peat (humin) (B Gearey 2006)

Initial comment: as OxA-15829

Objectives: as OxA-15829

Calibrated date: 1σ: 3080–2910 cal BC
                       2σ: 3100–2900 cal BC

Final comment: B Gearey (26 September 2014), the result confirms a Neolithic date for the peat at this level, however it is significantly different from OxA-15930 at the same level.

Laboratory comment: English Heritage (2006), the two results on this sample are statistically inconsistent (T′=13.5; T′(5%)=3.8; ν=1; Ward and Wilson 1978).

References: Ward and Wilson 1978

OxA–15835 4462 ±34 BP

$\delta^{13}C$: -26.2‰

Sample: RAW 1.96m, submitted on 28 March 2006 by B Gearey

Material: peat (humin) (B Gearey 2006)

Initial comment: as OxA-15829

Objectives: as OxA-15829

Calibrated date: 1σ: 3330–3020 cal BC
                       2σ: 3350–3010 cal BC

Final comment: B Gearey (26 September 2014), this result confirms a Neolithic date for the peat at this level, but is significantly different from OxA-15946 taken from the same level.

Laboratory comment: English Heritage (2006), the two results on this sample are statistically inconsistent (T′=10.7; T′(5%)=3.8; ν=1; Ward and Wilson 1978).

References: Ward and Wilson 1978

OxA–15836 4732 ±33 BP

$\delta^{13}C$: -26.6‰

Sample: RAW 2.45m, submitted on 28 March 2006 by B Gearey

Material: wood (waterlogged): unidentified, roundwood (B Gearey 2006)

Initial comment: as OxA-15829

Objectives: as OxA-15829

Calibrated date: 1σ: 3630–3380 cal BC
                       2σ: 3640–3370 cal BC

Final comment: B Gearey (26 September 2014), this result confirms a Neolithic date for peat inception at this location.

OxA–15885 2845 ±55 BP

$\delta^{13}C$: -27.5‰

Sample: RAW 0.50m, submitted on 28 March 2006 by B Gearey

Material: waterlogged plant macrofossil (*Sphagnum*) (B Gearey 2006)

Initial comment: as OxA-15829

Objectives: as OxA-15829

Calibrated date: 1σ: 1110–920 cal BC
                       2σ: 1210–850 cal BC

Final comment: see OxA-15831

OxA–15928 1964 ±26 BP

$\delta^{13}C$: -23.5‰

Sample: RAW 0.10m, submitted on 28 March 2006 by B Gearey

Material: waterlogged plant macrofossil (*Eriophorum*) (B Gearey 2006)

Initial comment: as OxA-15829

Objectives: as OxA-15829

Calibrated date: *1σ:* cal AD 1–70
                   2σ: 40 cal BC–cal AD 90

Final comment: B Gearey (26 September 2014), the result confirms a late Iron Age/Roman date for the top of the peat.

OxA–15929 2252 ±27 BP

$\delta^{13}C$: -22.6‰

Sample: RAW 0.24m, submitted on 28 March 2006 by B Gearey

Material: waterlogged plant macrofossil (*Sphagnum*) (B Gearey 2006)

Initial comment: as OxA-15829

Objectives: as OxA-15829

Calibrated date: *1σ:* 390–230 cal BC
                   2σ: 400–200 cal BC

Final comment: see OxA-15829

OxA–15930 4205 ±31 BP

$\delta^{13}C$: -26.2‰

Sample: RAW 1.76m, submitted on 28 March 2006 by B Gearey

Material: peat (humic acid) (B Gearey 2006)

Initial comment: as OxA-15829

Objectives: as OxA-15829

Calibrated date: *1σ:* 2890–2760 cal BC
                   2σ: 2900–2680 cal BC

Final comment: see OxA-15836

Laboratory comment: see OxA-15834

OxA–15944 3688 ±32 BP

$\delta^{13}C$: -24.9‰

Sample: RAW 1.20m, submitted on 28 March 2006 by B Gearey

Material: waterlogged plant macrofossil (*Eriophorum*) (B Gearey 2006)

Initial comment: as OxA-15829

Objectives: as OxA-15829

Calibrated date: *1σ:* 2140–2020 cal BC
                   2σ: 2200–1970 cal BC

Final comment: see OxA-15831

OxA–15945 4141 ±32 BP

$\delta^{13}C$: -23.6‰

Sample: RAW 1.52m, submitted on 28 March 2005 by B Gearey

Material: waterlogged plant macrofossil (*Sphagnum*) (B Gearey 2006)

Initial comment: as OxA-15829

Objectives: as OxA-15829

Calibrated date: *1σ:* 2870–2630 cal BC
                   2σ: 2880–2570 cal BC

Final comment: B Gearey (26 September 2014), the result confirms a late Neolithic/Bronze Age date for the peat at this level.

OxA–15946 4305 ±34 BP

$\delta^{13}C$: -27.3‰

Sample: RAW 1.96m, submitted on 28 March 2006 by B Gearey

Material: peat (humic acid) (B Gearey 2006)

Initial comment: as OxA-15829

Objectives: as OxA-15829

Calibrated date: *1σ:* 2920–2890 cal BC
                   2σ: 3010–2880 cal BC

Final comment: see OxA-15835

OxA–15993 2944 ±36 BP

$\delta^{13}C$: -26.0‰

Sample: RAW 0.88m B, submitted on 28 March 2006 by B Gearey

Material: peat (humic acid) (B Gearey 2006)

Initial comment: as OxA-15829

Objectives: as OxA-15829

Calibrated date: *1σ:* 1220–1110 cal BC
                   2σ: 1270–1010 cal BC

Final comment: see OxA-15831

Laboratory comment: see OxA-15832

Riccall: early medieval cemetery, North Yorkshire

Location: SE 60833736
                  Lat. 53.49.44 N; Long. 01.04.38 W

Project manager: R Hall (York Archaeological Trust), 1985

Archival body: Yorkshire Museum

Description: agricultural and building works some 100m from the River Ouse, at a site that could be associated with references to an encampment by King Harald Hardraada hereabouts in September 1066, have revealed human skeletal remains. Some have been archaeologically excavated; others recovered in an archaeological watching brief.

Objectives: to determine whether the cemetery reflects events of AD 1066; to determine the date range of the cemetery; and to provide chronological precision in relation to other scientific studies, notably oxygen isotope analysis.

Final comment: R Hall (23 February 2005), the series of six dates from the Riccall, North Yorkshire, early medieval cemetery site has provided the chronological framework necessary to understand more fully the nature of this burial ground. Four of the date ranges encompass 1066, the time when Harald Hardraada's invasion forces reportedly encamped at Riccall. These ranges are, however, relatively long; typically they extend across almost a century and a half. A fifth date range, AD 995–1030, covers a much briefer (35 years at 2 sigma) span, which pre-dates AD 1066 by 36 years. The final date range, of just over a century, is considerably earlier than the other five, in the late eigth-early ninth century. In all, the dates suggest that this site was used for burial over several centuries, they indicate that it is not a one event/AD 1066/Harald Hardraada cemetery, and they provide a series of reference points against which to assess the oxygen isotope analysis.

References: Hall *et al* 2008

UB–6179 1017 ±15 BP

$\delta^{13}C$: -22.1 ±0.2‰
$\delta^{13}C$ *(diet):* -20.4 ±0.3‰
$\delta^{15}N$ *(diet):* +10.7 ±0.4‰
C/N ratio: 2.6

Sample: skeleton E, submitted on 1 March 2004 by R Hall

Material: human bone (right femur) (J Buckberry 2004)

Initial comment: not located on L P Wenham's plan, but certainly one of his discoveries.

Objectives: to establish a date for the cemetery and to see if/how this correlates with oxygen isotope analysis suggesting a Norwegian origin for population.

Calibrated date: 1σ: cal AD 1015–1025
2σ: cal AD 990–1030

Final comment: R Hall (23 February 2005), this tightly dated sample suggests that this site was in use before the events of AD 1066.

UB–6180 965 ±15 BP

$\delta^{13}C$: -22.4 ±0.2‰
$\delta^{13}C$ *(diet):* -20.4 ±0.3‰
$\delta^{15}N$ *(diet):* +11.0 ±0.4‰
C/N ratio: 2.6

Sample: skeleton 18, submitted on 1 March 2004 by R Hall

Material: human bone (left femur) (J Buckberry 2004)

Initial comment: upper body removed by disturbance; lower limbs *in situ.*

Objectives: as UB-6179

Calibrated date: 1σ: cal AD 1025–1120
2σ: cal AD 1020–1155

Final comment: R Hall (23 February 2005), a relatively wide date range, which extends well beyond AD 1066 to include a large part of the Anglo-Norman period.

UB–6181 971 ±22 BP

$\delta^{13}C$: -20.5 ±0.2‰
$\delta^{13}C$ *(diet):* -20.0 ±0.3‰
$\delta^{15}N$ *(diet):* +11.1 ±0.4‰
C/N ratio: 2.6

Sample: skeleton 1, submitted on 1 March 2004 by R Hall

Material: human bone (right femur) (J Buckberry 2004)

Initial comment: head and long bones present *in situ,* remainder very fragmentary. There is evidence from other skeletons in this area of disturbance by ploughing.

Objectives: to determine the date of the cemetery use. This group lies 10+ metres north east of the 1974.121 group. To see if/how date correlates with oxygen isotope results, which suggest a Norwegian origin for the cemetery population.

Calibrated date: 1σ: cal AD 1020–1120
2σ: cal AD 1015–1155

Final comment: R Hall (23 February 2005), *see* UB-6180, showing that the cemetery's northern eastern area includes burials of similar date to those found in the original investigation.

UB–6182 972 ±15 BP

$\delta^{13}C$: -21.1 ±0.2‰
$\delta^{13}C$ *(diet):* -19.9 ±0.3‰
$\delta^{15}N$ *(diet):* +9.5±0.4‰
C/N ratio: 2.6

Sample: skeleton 17, submitted on 1 March 2004 by R Hall

Material: human bone (left femur) (J Buckberry 2004)

Initial comment: found in the extended watching brief. The skeleton appeared in the machine clearance - only the legs were recovered. A shallow grave (presumably) although no cut was visible.

Objectives: as UB-6181

Calibrated date: 1σ: cal AD 1020–1040
2σ: cal AD 1020–1150

Final comment: see UB-6181

UB–6183 926 ±16 BP

$\delta^{13}C$: -21.5 ±0.2‰
$\delta^{13}C$ *(diet):* -20.4 ±0.3‰
$\delta^{15}N$ *(diet):* +11.5 ±0.4‰
C/N ratio: 2.6

Sample: skeleton A, submitted on 1 March 2004 by R Hall

Material: human bone (right femur) (J Buckberry 2004)

Initial comment: revealed during building works in Yorkshire Water Authority layout, some 50+ metres from site of 1974.121 (1956–7 excavations). Apparently undisturbed. A shallow grave (grave cut unrecognisable), less than 50cm deep.

Objectives: to establish a date of cemetery use; to see if/how this correlates with results of oxygen isotope analysis, which suggests a Norwegian origin for the cemetery population.

Calibrated date: 1σ: cal AD 1040–1160
2σ: cal AD 1030–1165

Final comment: R Hall (23 February 2005), this shows that this part of the cemetery to the south west of the original excavation also contains burials that embrace the late Saxon or Anglo-Norman period.

UB–6184 1256 ±19 BP

δ¹³C: -19.4 ±0.2‰
δ¹³C (diet): -20.0 ±0.3‰
δ¹⁵N (diet): +11.6 ±0.4‰
C/N ratio: 2.7

Sample: skeleton 4, submitted on 1 March 2004 by R Hall

Material: human bone (left femur) (J Buckberry 2004)

Initial comment: although no grave cuts were identified in the burial soil, this very probably came from an individual grave. That part of the skeleton within the excavated area (ie chest down) was intact.

Objectives: to establish the date of use of the cemetery, and to see how/if this correlates with oxygen isotope analysis results, which suggests a Norwegian origin.

Calibrated date: 1σ: cal AD 690–775
2σ: cal AD 680–775

Final comment: R Hall (23 February 2005), the earliest date burial by a margin of at least a century, this indicates that the graveyard may have originated as early as the middle Saxon period.

Ripon Cathedral, North Yorkshire

Location: SE 31447112
Lat. 54.08.05 N; Long. 01.31.08 W

Project manager: A Bayliss (English Heritage), 2005

Description: the first church on the site of Ripon Cathedral was originally part of a Celtic monastery. This was reorganised by St Wilfrid in AD 660. Between then and AD 1050 it was refounded as a College of secular canons under the patronage of the Archbishop of York. It remained as a parish church even after the dissolution of the College in AD 1547. In AD 1604 the college was refounded under James I, dissolved during the Commonwealth, but founded yet again in AD 1660. It was elevated to Cathedral status in AD 1836. The nave roof consists of 15 'truncated' trusses, consisting alternately of single larger principal rafters (trusses 1, 3, 5, etc, numbering from west to east) or of two very slightly smaller principal rafters in close-set pairs (trusses 2, 4, 6, etc). All such principal rafters are of oak. The apex of each truss seems to have been cut off (if indeed the original ever

went to the ridge) and replaced in softwood. Set to the underside of the principal roof timbers are the beams of the ceiling vault. These consist of ridge and vault ribs, from which spring diagonal and intermediate ribs. All these timbers are of oak.

Objectives: the tree-ring analysis of Ripon Cathedral (Arnold *et al* 2005) was unusual in producing two, well-replicated but undated, site chronologies, each containing more than 100 rings. This provided the opportunity to test the accuracy of the radiocarbon dating produced for the ALSF research programme, using samples whose relative age was known by dendrochronology. The submission of related, and replicate, samples to the laboratories collaborating on the ALSF programme tested the comparability of results produced by different laboratories. Accelerator Mass Spectrometry has only recently achieved the precision needed for wiggle-matching, and so a subsidiary aim was to field test the technique to determine whether it can offer, on a routine basis, the accuracy required for applications relating to historic buildings. Further samples from this site were dated as part of the wider English Heritage radiocarbon dating programme, and as part of the internal quality control procedures of the Scottish Universities Environmental Research Centre.

Final comment: A Bayliss (13 November 2007), of the 20 groups of replicate measurements from Ripon Cathedral, 16 are statistically consistent at two standard deviations (Ward and Wilson 1978; *see* below). In two other cases, the results are consistent at three standard deviations, although in the other two cases they are not. Bayesian wiggle-matching suggests that the timbers in site sequence RIPCSQ01 were felled in *cal AD 1855–1870 (95% probability; RIPCSQ01 bark edge;* Bayliss *et al* 2014, fig 10), and the timbers in site sequence RIPCSQ02 were felled in *cal AD 1850–1870 (95% probability; RIPCSQ02 bark edge;* Bayliss *et al* 2014, fig 11). This suggests that the entire roof structure of the nave was reconstructed as part of the works designed by Sir Gilbert Scott and undertaken between AD 1862 and AD 1872. These date estimates are compatible with tentative tree-ring matches, which would date the final ring of both master sequences to AD 1868 (Bayliss *et al* 2014, table 3).

References: Arnold *et al* 2005
Bayliss *et al* 2014
Ward and Wilson 1978

Ripon Cathedral: RIPCSQ01, North Yorkshire

Location: SE 31447112
Lat. 54.08.05 N; Long. 01.31.08 W

Project manager: A Bayliss (English Heritage), 2005

Archival body: Nottingham Tree-Ring Dating Laboratory, English Heritage

Description: the undated 226–ring tree-ring master sequence, RIPCSQ01, is made up of series from six principal rafters from the nave roof (Arnold *et al* 2005). Five of these timbers are complete to bark edge, including both cores from which radiocarbon samples were dated (RIP-C08 and RIP-C11).

Objectives: to date this undated sequence, and so this section of the nave of the Cathedral.

Final comment: A Bayliss (13 November 2007), the wiggle-matching of site sequence RIPCSQ01 suggests that the last ring of the tree-ring chronology was formed in *cal AD 1855–1870 (95% probability; RIPCSQ01 bark edge;* Bayliss *et al* 2014, fig 10), or *cal AD 1860–1865 (68% probability)*. This model has good overall agreement (A$_{overall}$ = 166.3%, A$_n$= 22.4%; Bronk Ramsey 1995). These date estimates are compatible with a tentative tree-ring match, which would date the final ring of this sequence to AD 1868.

Laboratory comment: English Heritage (25 November 2014), ten further samples were also funded by the Aggregates Levy Sustainability Fund (ALSF) and are published in Bayliss *et al* (2008a, 132–5; SUERC-8963–5, -8969–75).

References: Arnold *et al* 2005
Bayliss *et al* 2008a, 130–2
Bayliss *et al* 2014
Bronk Ramsey 1995

GrA–30635 95 ±30 BP

$\delta^{13}C$: -27.2‰

Sample: RIP-C11 <5>, submitted on 13 December 2005 by D Hamilton

Material: wood: *Quercus* sp. (R Howard 2005)

Initial comment: decadal sample from rings 177–186 of floating tree-ring sequence RIPCSQ01. The sample is from core RIP-C11, which was taken from the north principal rafter of truss 11.

Objectives: to demonstrate the accuracy and inter-laboratory comparability of radiocarbon samples dated under the English Heritage research programme, and to provide calendar dating for the undated tree-ring master sequence, RIPCSQ01, from the nave roof of Ripon Cathedral.

Calibrated date: *1σ:* cal AD 1690–1920
2σ: cal AD 1680–1940

Final comment: A Bayliss (13 November 2007), this result is statistically consistent with two other measurements available for this decade of RIPCSQ01, SUERC-11441 (85 ±35 BP) and SUERC-8970 (155 ±35 BP) (T'=2.4;T'(5%)=6.0; v=2; Ward and Wilson 1978). The weighted mean of these three measurements, 110 ±19 BP, calibrates to cal AD 1680–1955★ (Reimer *et al* 2004). The wiggle-matching of RIPCSQ01 suggests that this sample dates to *cal AD 1810–1825 (95% probability; rings 177–186;* Bayliss *et al* 2014, fig 10).

References: Reimer *et al* 2004
Ward and Wilson 1978

GrA–30753 150 ±30 BP

$\delta^{13}C$: -25.7‰

Sample: RIP-C11 <1>, submitted on 13 December 2005 by D Hamilton

Material: wood: *Quercus* sp. (R Howard 2005)

Initial comment: decadal sample from rings 217–226 of floating tree-ring sequence RIPCSQ01. The sample is from core RIP-C11, which was taken from the north principal rafter of truss 11.

Objectives: as GrA-30635

Calibrated date: *1σ:* cal AD 1670–1950
2σ: cal AD 1660–1955★

Final comment: A Bayliss (13 November 2007), this result is statistically consistent with two other measurements available for this decade of RIPCSQ01, SUERC-11434 (140 ±35 BP) and SUERC-8963 (100 ±35 BP) (T'=1.2;T'(5%)=6.0; v=2; Ward and Wilson 1978). The weighted mean of these three measurements, 132 ±19 BP, calibrates to cal AD 1675–1955★ (Reimer *et al* 2004). The wiggle-matching of RIPCSQ01 suggests that this sample dates to *cal AD 1850–1865 (95% probability; rings 217–226;* Bayliss *et al* 2014, fig 10).

References: Reimer *et al* 2004
Ward and Wilson 1978

GrA–30755 115 ±30 BP

$\delta^{13}C$: -25.9‰

Sample: RIP-C11 <1>, submitted on 13 December 2005 by D Hamilton

Material: wood: *Quercus* sp. (R Howard 2005)

Initial comment: decadal sample from rings 207–216 of floating tree-ring sequence RIPCSQ01. The sample is from core RIP-C011, which was taken from the north principal rafter of truss 11.

Objectives: as GrA-30635

Calibrated date: *1σ:* cal AD 1680–1930
2σ: cal AD 1670–1950

Final comment: A Bayliss (13 November 2007), this result is statistically consistent with two other measurements available for this decade of RIPCSQ01, SUERC-11435 (160 ±35 BP) and SUERC-8954 (135 ±35 BP) (T'=1.0;T'(5%)=6.0; v=2; Ward and Wilson 1978). The weighted mean of these three measurements, 135 ±19 BP, calibrates to cal AD 1670–1955★ (Reimer *et al* 2004). The wiggle-matching of RIPCSQ01 suggests that this sample dates to *cal AD 1840–1855 (95% probability; rings 207–216;* Bayliss *et al* 2014, fig 10).

References: Reimer *et al* 2004
Ward and Wilson 1978

GrA–30756 115 ±30 BP

$\delta^{13}C$: -26.1‰

Sample: RIP-C11 <3>, submitted on 13 December 2005 by D Hamilton

Material: wood: *Quercus* sp. (R Howard 2005)

Initial comment: decadal sample from rings 197–206 of floating tree-ring sequence RIPCSQ01. The sample is from core RIP-C11, which was taken from the north principal rafter of truss 11.

Objectives: as GrA-30635

Calibrated date: *1σ:* cal AD 1680–1930
                      2σ: cal AD 1670–1950

Final comment: A Bayliss (13 November 2007), this result is statistically consistent with two other measurements available for this decade of RIPCSQ01, SUERC-11439 (135 ±35 BP) and SUERC-8965 (145 ±35 BP) (T′=0.5;T′(5%)=6.0; *v*=2; Ward and Wilson 1978). The weighted mean of these three measurements, 130 ±19 BP, calibrates to cal AD 1675–1955* (Reimer *et al* 2004).The wiggle-matching of RIPCSQ01 suggests that this sample dates to *cal AD 1830–1845 (95% probability; rings 197–206*; Bayliss *et al* 2014, fig 10).

References: Reimer *et al* 2004
                Ward and Wilson 1978

GrA–30757 65 ±30 BP

δ¹³C: -25.3‰

Sample: RIP-C11 <4>, submitted on 13 December 2005 by D Hamilton

Material: wood: *Quercus* sp. (R Howard 2005)

Initial comment: decadal sample from rings 187–196 of floating tree-ring sequence RIPCSQ01. The sample is from core RIP-C11, which was taken from the north principal rafter of truss 11.

Objectives: as GrA-30635

Calibrated date: *1σ:* cal AD 1710–1910
                      2σ: cal AD 1690–1920

Final comment: A Bayliss (13 November 2007), this result is statistically consistent with two other measurements available for this decade of RIPCSQ01, SUERC-11440 (110 ±35 BP) and SUERC-8969 (150 ±35 BP) (T′=3.4;T′(5%)=6.0; *v*=2; Ward and Wilson 1978). The weighted mean of these three measurements (104 ±19 BP) calibrates to cal AD 1680–1955* (Reimer *et al* 2004). The wiggle-matching of RIPCSQ01 suggests that this sample dates to *cal AD 1820–1835 (95% probability; rings 187–196*; Bayliss *et al* 2014, fig 10).

References: Reimer *et al* 2004
                Ward and Wilson 1978

GrA–30761 75 ±30 BP

δ¹³C: -26.3‰

Sample: RIP-C11 <6>, submitted on 13 December 2005 by D Hamilton

Material: wood: *Quercus* sp. (R Howard 2005)

Initial comment: decadal sample from rings 167–176 of floating tree-ring sequence RIPCSQ01.The sample is from core RIP-C11, which was taken from the north principal rafter of truss 11.

Objectives: as GrA-30635

Calibrated date: *1σ:* cal AD 1690–1920
                      2σ: cal AD 1680–1930

Final comment: A Bayliss (13 November 2007), this result is not statistically consistent with two other measurements available for this decade of RIPCSQ01, SUERC-11442 (140 ±35 BP) and SUERC-8971 (240 ±35 BP) (T′=12.9; T′(5%)=6.0; *v*=2; Ward and Wilson 1978). SUERC-8971 is slightly older than expected. The weighted mean of the three measurements, 144 ±19 BP, calibrates to cal AD 1665–1955* (Reimer *et al* 2004).The wiggle-matching of RIPCSQ01 suggests that this sample dates *to cal AD 1800–1815 (95% probability; rings 167–176*; Bayliss *et al* 2014, fig 10).

References: Reimer *et al* 2004
                Ward and Wilson 1978

GrA–30762 165 ±30 BP

δ¹³C: -25.6‰

Sample: RIP-C11 <7>, submitted on 13 December 2005 by D Hamilton

Material: wood: *Quercus* sp. (R Howard 2005)

Initial comment: decadal sample from rings 157–166 of floating tree-ring sequence RIPCSQ01. The sample is from core RIP-C11, which was taken from the north principal rafter of truss 11.

Objectives: as GrA-30635

Calibrated date: *1σ:* cal AD 1660–1950
                      2σ: cal AD 1660–1955*

Final comment: A Bayliss (13 November 2007), this result is not statistically consistent with two other measurements available for this decade of RIPCSQ01 at 95% confidence (SUERC-11443, 145 ±35 BP and SUERC-8972, 270 ±35 BP; T′=7.6;T′(5%)=6.0; *v*=2;Ward and Wilson 1978), although it is at 99% confidence. The weighted mean of these three measurements, 191 ±19 BP, calibrates to cal AD 1655–1955* (Reimer *et al* 2004).The wiggle-matching of RIPCSQ01 suggests that this sample dates to *cal AD 1790–1805 (95% probability; rings 157–166*; Bayliss *et al* 2014, fig 10).

References: Reimer *et al* 2004
                Ward and Wilson 1978

GrA–30763 210 ±30 BP

δ¹³C: -25.6‰

Sample: RIP-C11 <8>, submitted on 13 December 2005 by D Hamilton

Material: wood: *Quercus* sp. (R Howard 2005)

Initial comment: decadal sample from rings 147–156 of floating tree-ring sequence RIPCSQ01. The sample is from core RIP-C11, which was taken from the north principal rafter of truss 11.

Objectives: as GrA-30635

Calibrated date: *1σ:* cal AD 1650–1955*
                      2σ: cal AD 1640–1955*

Final comment: A Bayliss (13 November 2007), this result is statistically consistent with two other measurements available for this decade of RIPCSQ01, SUERC-11444 (230 ±35 BP)

and SUERC-8973 (275 ±35 BP)(T'=2.0;T'(5%)=6.0; *ν*=2; Ward and Wilson 1978). The weighted mean of these three measurements, 235 ±19 BP, calibrates to cal AD 1640–1950 (Reimer *et al* 2004).The wiggle-matching of RIPCSQ01 suggests that this sample dates to *cal AD 1780–1795 (95% probability; rings 147–156*; Bayliss *et al* 2014, fig 10).

References: Reimer *et al* 2004
           Ward and Wilson 1978

GrA–30765 165 ±30 BP

δ¹³C: -25.1‰

Sample: RIP-C11 <9>, submitted on 13 December 2005 by D Hamilton

Material: wood: *Quercus* sp. (R Howard 2005)

Initial comment: decadal sample from rings 137–146 of floating tree-ring sequence RIPCSQ01.The sample is from core RIP-C11, which was taken from the north principal rafter of truss 11.

Objectives: as GrA-30635

Calibrated date: 1σ: cal AD 1660–1950
              2σ: cal AD 1660–1955*

Final comment: A Bayliss (13 November 2007), this result is statistically consistent with two other measurements available for this decade of RIPCSQ01, SUERC-11445 (140 ±35 BP) and SUERC-8974 (245 ±35 BP) (T'=5.0;T'(5%)=6.0; *ν*=2; Ward and Wilson 1978). The weighted mean of these three measurements, 182 ±19 BP, calibrates to cal AD 1660–1955* (Reimer *et al* 2004). The wiggle-matching of RIPCSQ01 suggests that this sample dates to *cal AD 1770–1785 (95% probability; rings 137–146*; Bayliss *et al* 2014, fig 10).

References: Reimer *et al* 2004
           Ward and Wilson 1978

GrA–30766 120 ±30 BP

δ¹³C: -25.4‰

Sample: RIP-C11 <10>, submitted on 13 December 2005 by D Hamilton

Material: wood: *Quercus* sp. (R Howard 2005)

Initial comment: decadal sample from rings 127–136 of floating tree-ring sequence RIPCSQ01. The sample is from core RIP-C11, which was taken from the north principal rafter of truss 11.

Objectives: as GrA-30635

Calibrated date: 1σ: cal AD 1680–1940
              2σ: cal AD 1670–1950

Final comment: A Bayliss (13 November 2007), this result is not statistically consistent with two other measurements available for this decade of RIPCSQ01, SUERC-11449 (160 ±35 BP) and SUERC-8975 (275 ±35 BP) (T'=11.7; T'(5%)=6.0; *ν*=2;Ward and Wilson 1978). SUERC-8975 is slightly older than expected. The weighted mean of the three measurements, 179 ±19 BP, calibrates to cal AD 1660–1955* (Reimer *et al* 2004).The wiggle-matching of

RIPCSQ01 suggests that this sample dates to *cal AD 1760–1775 (95% probability; rings 127–136*; Bayliss *et al* 2014, fig 10).

References: Reimer *et al* 2004
           Ward and Wilson 1978

SUERC–11434 140 ±35 BP

δ¹³C: -23.9‰

Sample: RIP-C08 <1>, submitted on 13 December 2005 by D Hamilton

Material: wood: *Quercus* sp. (R Howard 2005)

Initial comment: decadal sample from rings 217–226 of floating tree-ring sequence RIPCSQ01.The sample is from core RIP-C08, which was taken from the north principal rafter of truss 3.

Objectives: as GrA-30635

Calibrated date: 1σ: cal AD 1670–1950
              2σ: cal AD 1660–1955*

Final comment: see GrA-30753

Laboratory comment: SUERC Radiocarbon Dating Laboratory (AMS) (26 November 2008), a re-combustion of GU-6305 (SUERC-8963; published in Bayliss *et al* 2008a, 130). *See* GrA-30753.

References: Bayliss *et al* 2008a, 130

SUERC–11435 160 ±35 BP

δ¹³C: -23.5‰

Sample: RIP-C08 <2>, submitted on 13 December 2005 by D Hamilton

Material: wood: *Quercus* sp. (R Howard 2005)

Initial comment: decadal sample from rings 207–216 of floating tree-ring sequence RIPCSQ01.The sample is from core RIP-C08, which was taken from the north principal rafter of truss 3.

Objectives: as GrA-30635

Calibrated date: 1σ: cal AD 1660–1950
              2σ: cal AD 1660–1955*

Final comment: see GrA-30755

Laboratory comment: SUERC Radiocarbon Dating Laboratory (AMS) (26 November 2008), a re-combustion of GU-6306 (SUERC-8964; published in Bayliss *et al* 2008a, 130). *See* GrA-30755.

References: Bayliss *et al* 2008a, 130

SUERC–11439 135 ±35 BP

δ¹³C: -23.7‰

Sample: RIP-C08 <3>, submitted on 13 December 2005 by D Hamilton

Material: wood: *Quercus* sp. (R Howard 2005)

Initial comment: decadal sample from rings 197–206 of floating tree-ring sequence RIPCSQ01.The sample is from core RIP-C08, which was taken from the north principal rafter of truss 3.

Objectives: as GrA-30635

Calibrated date: 1σ: cal AD 1670–1950
2σ: cal AD 1660–1955*

Final comment: see GrA-30756

Laboratory comment: SUERC Radiocarbon Dating Laboratory (AMS) (26 November 2008), a re-combustion of GU-6307 (SUERC-8965; published in Bayliss *et al* 2008a, 131). *See* GrA-30756.

References: Bayliss *et al* 2008a, 131

SUERC–11440 110 ±35 BP

δ¹³C: -23.6‰

Sample: RIP-C08 <4>, submitted on 13 December 2005 by D Hamilton

Material: wood: *Quercus* sp. (R Howard 2005)

Initial comment: decadal sample from rings 187–196 of floating tree-ring sequence RIPCSQ01.The sample is from core RIP-C08, which was taken from the north principal rafter of truss 3.

Objectives: as GrA-30635

Calibrated date: 1σ: cal AD 1680–1930
2σ: cal AD 1670–1950

Final comment: see GrA-30757

Laboratory comment: SUERC Radiocarbon Dating Laboratory (AMS) (26 November 2008), a re-combustion of GU-6308 (SUERC-8969; published in Bayliss *et al* 2008a, 131). *See* GrA-30757.

References: Bayliss *et al* 2008a, 131

SUERC–11441 85 ±35 BP

δ¹³C: -23.3‰

Sample: RIP-C08 <5>, submitted on 13 December 2005 by D Hamilton

Material: wood: *Quercus* sp. (R Howard 2005)

Initial comment: decadal sample from rings 177–186 of floating tree-ring sequence RIPCSQ01.The sample is from core RIP-C08, which was taken from the north principal rafter of truss 3.

Objectives: as GrA-30635

Calibrated date: 1σ: cal AD 1690–1920
2σ: cal AD 1680–1940

Final comment: see GrA-30635

Laboratory comment: SUERC Radiocarbon Dating Laboratory (AMS) (26 November 2008), a re-combustion of GU-6309 (SUERC-8970; published in Bayliss *et al* 2008a, 131). *See* GrA-30635.

References: Bayliss *et al* 2008a, 131

SUERC–11442 140 ±35 BP

δ¹³C: -23.5‰

Sample: RIP-C08 <6>, submitted on 13 December 2005 by D Hamilton

Material: wood: *Quercus* sp. (R Howard 2005)

Initial comment: decadal sample from rings 167–176 of floating tree-ring sequence RIPCSQ01.The sample is from core RIP-C08, which was taken from the north principal rafter of truss 3.

Objectives: as GrA-30635

Calibrated date: 1σ: cal AD 1670–1950
2σ: cal AD 1660–1955*

Final comment: see GrA-30761

Laboratory comment: SUERC Radiocarbon Dating Laboratory (AMS) (26 November 2008), a re-combustion of GU-6310 (SUERC-8971; published in Bayliss *et al* 2008a, 131). *See* GrA-30761.

References: Bayliss *et al* 2008a, 131

SUERC–11443 145 ±35 BP

δ¹³C: -23.3‰

Sample: RIP-C08 <7>, submitted on 13 December 2005 by D Hamilton

Material: wood: *Quercus* sp. (R Howard 2005)

Initial comment: decadal sample from rings 157–166 of floating tree-ring sequence RIPCSQ01.The sample is from core RIP-C08, which was taken from the north principal rafter of truss 3.

Objectives: as GrA-30635

Calibrated date: 1σ: cal AD 1670–1950
2σ: cal AD 1660–1955*

Final comment: see GrA-30762

Laboratory comment: SUERC Radiocarbon Dating Laboratory (AMS) (26 November 2008), a re-combustion of GU-6311 (SUERC-8972; published in Bayliss *et al* 2008a, 131–2). *See* GrA-30762.

References: Bayliss *et al* 2008a, 131–2

SUERC–11444 230 ±35 BP

δ¹³C: -23.2‰

Sample: RIP-C08 <8>, submitted on 13 December 2005 by D Hamilton

Material: wood: *Quercus* sp. (R Howard 2005)

Initial comment: decadal sample from rings 147–156 of floating tree-ring sequence RIPCSQ01.The sample is from core RIP-C08, which was taken from the north principal rafter of truss 3.

Objectives: as GrA-30763

Calibrated date: 1σ: cal AD 1640–1955*
2σ: cal AD 1630–1955*

Final comment: see GrA-30763

Laboratory comment: SUERC Radiocarbon Dating Laboratory (AMS) (26 November 2008), a re-combustion of GU-6312 (SUERC-8973; published in Bayliss *et al* 2008a, 132). *See* GrA-30763.

References: Bayliss *et al* 2008a, 132

SUERC–11445 140 ±35 BP

$\delta^{13}C$: -23.4‰

Sample: RIP-C08 <9>, submitted on 13 December 2005 by D Hamilton

Material: wood: *Quercus* sp. (R Howard 2005)

Initial comment: decadal sample from rings 137–146 of floating tree-ring sequence RIPCSQ01. The sample is from core RIP-C08, which was taken from the north principal rafter of truss 3.

Objectives: as GrA-30635

Calibrated date: 1σ: cal AD 1670–1950
                          2σ: cal AD 1660–1955*

Final comment: see GrA-30665

Laboratory comment: SUERC Radiocarbon Dating Laboratory (AMS) (26 November 2008), a re-combustion of GU-6313 (SUERC-8974; published in Bayliss *et al* 2008a, 132). *See* GrA-30665.

References: Bayliss *et al* 2008a, 132

SUERC–11449 160 ±35 BP

$\delta^{13}C$: -24.1‰

Sample: RIP-C08 <10>, submitted on 13 December 2005 by D Hamilton

Material: wood: *Quercus* sp. (R Howard 2005)

Initial comment: decadal sample from rings 127–136 of floating tree-ring sequence RIPCSQ01. The sample is from core RIP-C08, which was taken from the north principal rafter of truss 3.

Objectives: as GrA-30635

Calibrated date: 1σ: cal AD 1660–1950
                          2σ: cal AD 1660–1955*

Final comment: see GrA-30766

Laboratory comment: SUERC Radiocarbon Dating Laboratory (AMS) (26 November 2008), a re-combustion of GU-6314 (SUERC-8975; published in Bayliss *et al* 2008a, 132). *See* GrA-30766.

References: Bayliss *et al* 2008a, 132

Ripon Cathedral: RIPCSQ02, North Yorkshire

Location:	SE 31447112
	Lat. 54.08.05 N; Long. 01.31.08 W
Project manager:	A Bayliss (English Heritage), 2005
Archival body:	Nottingham Tree-Ring Dating Laboratory, English Heritage

Description: the undated 117–ring tree-ring master sequence, RIPCSQ02, is made up of series from nine timbers, including five 'double' rafters and two ceiling ribs from each end of the nave (Arnold *et al* 2005, fig 5). Six of these timbers are complete to bark edge, including both cores from which radiocarbon samples were dated (RIP-C14 and RIP-C29).

Objectives: to date this undated sequence, and so this section of the nave of the Cathedral.

Final comment: A Bayliss (13 November 2007), the wiggle-matching of site sequence RIPCSQ02 suggests that the last ring of the tree-ring chronology was formed in *cal AD 1850–1870 (95% probability; RIPCSQ02 bark edge;* Bayliss *et al* forthcoming, fig 11), or *cal AD 1855–1865 (68% probability).* This model has good overall agreement ($A_{overall}$ = 69.7%, A_n= 22.4%; Bronk Ramsey 1995). These date estimates are compatible with a tentative tree-ring match, which would date the final ring of this sequence to AD 1868.

Laboratory comment: English Heritage (25 November 2014), ten further samples were dated through the Aggregates Levy Sustainability Fund (ALSF) and are published in Bayliss *et al* (2008a, 132–5; OxA-15406–14 and -15497).

References:	Arnold *et al* 2005
	Bayliss *et al* 2008a, 132–5
	Bayliss *et al* 2014
	Bayliss *et al* 2014
	Bronk Ramsey 1995

GrA–30767 115 ±30 BP

$\delta^{13}C$: -23.9‰

Sample: RIP-C14 <1>, submitted on 13 December 2005 by D Hamilton

Material: wood: *Quercus* sp. (R Howard 2005)

Initial comment: decadal sample from rings 108–117 of floating tree-ring sequence RIPCSQ02. The sample is from core RIP-C14, which was taken from a northern double rafter in truss 6B.

Objectives: to demonstrate the accuracy and inter-laboratory comparability of radiocarbon samples dated under the English Heritage research programme, and to provide calendar dating for the undated tree-ring master sequence, RIPCSQ02, from the nave roof of Ripon Cathedral.

Calibrated date: 1σ: cal AD 1680–1930
                          2σ: cal AD 1670–1950

Final comment: A Bayliss (13 November 2007), this result is statistically consistent with another measurement for this decade of RIPCSQ02, OxA-15406 (132 ±25 BP) (T'=0.2;

T'(5%)=3.8; *v*=1; Ward and Wilson 1978).The weighted mean of these measurements, 126 ±20 BP, calibrates to cal AD 1675–1955* (Reimer *et al* 2004). The wiggle-matching of RIPCSQ02 suggests that this sample dates *to cal AD 1845–1865 (95% probability; rings 108–117*; Bayliss *et al* 2014, fig 11).

References: Reimer *et al* 2004
Ward and Wilson 1978

GrA–30768 100 ±30 BP

$\delta^{13}C$: -24.5‰

Sample: RIP-C14 <2>, submitted on 13 December 2005 by D Hamilton

Material: wood: *Quercus* sp. (R Howard 2005)

Initial comment: decadal sample from rings 98–107 of floating tree-ring sequence RIPCSQ02. The sample is from core RIP-C14, which was taken from a northern double rafter in truss 6B.

Objectives: as GrA-30767

Calibrated date: 1σ: cal AD 1690–1930
2σ: cal AD 1680–1940

Final comment: A Bayliss (13 November 2007), this result is statistically consistent with another measurement for this decade of RIPCSQ02, OxA-15497 (155 ±23 BP) (T'=2.1; T'(5%)=3.8; *v*=1; Ward and Wilson 1978). The weighted mean of these measurements, 135 ±18 BP, calibrates to cal AD 1675–1940 (Reimer *et al* 2004). The wiggle-matching of RIPCSQ02 suggests that this sample dates to *cal AD 1835–1855 (95% probability; rings 98–107*; Bayliss *et al* 2014, fig 11).

References: Reimer *et al* 2004
Ward and Wilson 1978

GrA–30770 95 ±30 BP

$\delta^{13}C$: -24.1‰

Sample: RIP-C14 <3>, submitted on 13 December 2005 by D Hamilton

Material: wood: *Quercus* sp. (R Howard 2005)

Initial comment: decadal sample from rings 88–97 of floating tree-ring sequence RIPCSQ02. The sample is from core RIP-C14, which was taken from a northern double rafter in truss 6B.

Objectives: as GrA-30767

Calibrated date: 1σ: cal AD 1690–1920
2σ: cal AD 1680–1940

Final comment: A Bayliss (13 November 2007), this result is statistically consistent with another measurement for this decade of RIPCSQ02, OxA-15407 (143 ±26 BP) (T'=1.5; T'(5%)=3.8; *v*=1; Ward and Wilson 1978). The weighted mean of these measurements, 123 ±20 BP, calibrates to cal AD 1680–1955* (Reimer *et al* 2004). The wiggle-matching of RIPCSQ02 suggests that this sample dates to *cal AD 1825–1845 (95% probability; rings 88–97*; Bayliss *et al* 2014, fig 11).

References: Reimer *et al* 2004
Ward and Wilson 1978

GrA–30772 60 ±30 BP

$\delta^{13}C$: -24.5‰

Sample: RIP-C14 <4>, submitted on 13 December 2005 by D Hamilton

Material: wood: *Quercus* sp. (R Howard 2005)

Initial comment: decadal sample from rings 78–87 of floating tree-ring sequence RIPCSQ02. The sample is from core RIP-C14, which was taken from a northern double rafter in truss 6B.

Objectives: as GrA-30767

Calibrated date: 1σ: cal AD 1890–1910
2σ: cal AD 1690–1920

Final comment: A Bayliss (13 November 2007), this result is not statistically consistent with another measurement for this decade of RIPCSQ02 at 95% confidence, although it is at 99% confidence (OxA-15408, 141 ±25 BP;T'=4.3;T'(5%)= 3.8; *v*=1; Ward and Wilson 1978). The weighted mean of these measurements, 109 ±20 BP, calibrates to cal AD 1680–1955* (Reimer *et al* 2004). The wiggle-matching of *RIPCSQ02 suggests that this sample dates to cal AD 1815–1835 (95% probability; rings 78–87*; Bayliss *et al* 2014, fig 11).

References: Reimer *et al* 2004
Ward and Wilson 1978

GrA–30773 85 ±30 BP

$\delta^{13}C$: -24.1‰

Sample: RIP-C14 <5>, submitted on 13 December 2005 by D Hamilton

Material: wood: *Quercus* sp. (R Howard 2005)

Initial comment: decadal sample from rings 68–77 of floating tree-ring sequence RIPCSQ02. The sample is from core RIP-C14, which was taken from a northern double rafter in truss 6B.

Objectives: as GrA-30767

Calibrated date: 1σ: cal AD 1690–1920
2σ: cal AD 1680–1930

Final comment: A Bayliss (13 November 2007), this result is statistically consistent with another measurement for this decade of RIPCSQ02, OxA-15409 (147 ±26 BP) (T'=2.4; T'(5%)=3.8; *v*=1; Ward and Wilson 1978).The weighted mean of these measurements, 121 ±20 BP, calibrates to cal AD 1680–1955* (Reimer *et al* 2004). The wiggle-matching of RIPCSQ02 suggests that this sample dates to *cal AD 1805–1825 (95% probability; rings 68–77*; Bayliss *et al* 2014, fig 11).

References: Reimer *et al* 2004
Ward and Wilson 1978

GrA–30775 155 ±35 BP

δ¹³C: -24.4‰

Sample: RIP-C14 <6>, submitted on 13 December 2005 by D Hamilton

Material: wood: *Quercus* sp. (R Howard 2005)

Initial comment: decadal sample from rings 58–67 of floating tree-ring sequence RIPCSQ02. The sample is from core RIP-C14, which was taken from a northern double rafter in truss 6B.

Objectives: as GrA-30767

Calibrated date: 1σ: cal AD 1660–1950
                      2σ: cal AD 1660–1955★

Final comment: A Bayliss (13 November 2007), this result is statistically consistent with another measurement for this decade of RIPCSQ02, OxA-15410 (171 ±25 BP) (T′=0.1;T′(5%)=3.8; ν=1; Ward and Wilson 1978). The weighted mean of these measurements, 166 ±21 BP, calibrates to cal AD 1665–1955★ (Reimer *et al* 2004). The wiggle-matching of RIPCSQ02 suggests that this sample dates to *cal AD 1795–1815 (95% probability; rings 58–67;* Bayliss *et al* 2014, fig 11).

GrA–30776 170 ±30 BP

δ¹³C: -24.1‰

Sample: RIP-C14 <7>, submitted on 13 December 2005 by D Hamilton

Material: wood: *Quercus* sp. (R Howard 2005)

Initial comment: decadal sample from rings 48–57 of floating tree-ring sequence RIPCSQ02. The sample is from core RIP-C14, which was taken from a northern double rafter in truss 6B.

Objectives: as GrA-30767

Calibrated date: 1σ: cal AD 1660–1955★
                      2σ: cal AD 1660–1955★

Final comment: A Bayliss (13 November 2007), this result is statistically consistent with another measurement for this decade of RIPCSQ02, OxA-15411 (208 ±26 BP) (T′=0.9; T′(5%)=3.8; ν=1; Ward and Wilson 1978). The weighted mean of these measurements, 192 ±20 BP, calibrates to cal AD 1655–1955★ (Reimer *et al* 2004). The wiggle-matching of RIPCSQ02 suggests that this sample dates to *cal AD 1785–1805 (95% probability; rings 48–57;* Bayliss *et al* 2014, fig 11).

References: Reimer *et al* 2004
                    Ward and Wilson 1978

GrA–30777 150 ±30 BP

δ¹³C: -25.4‰

Sample: RIP-C14 <8>, submitted on 13 December 2005 by D Hamilton

Material: wood: *Quercus* sp. (R Howard 2005)

Initial comment: decadal sample from rings 38–47 of floating tree-ring sequence RIPCSQ02. The sample is from core RIP-C14, which was taken from a northern double rafter in truss 6B.

Objectives: as GrA-30767

Calibrated date: 1σ: cal AD 1670–1950
                      2σ: cal AD 1660–1955★

Final comment: A Bayliss (13 November 2007), this result is statistically consistent with another measurement for this decade of RIPCSQ02, OxA-15412 (221 ±25 BP) (T′=3.3; T′(5%)=3.8; ν=1; Ward and Wilson 1978). The weighted mean of these measurements, 192 ±20 BP, calibrates to cal AD 1655–1955★ (Reimer *et al* 2004). The wiggle-matching of RIPCSQ02 suggests that this sample dates to *cal AD 1775–1795 (95% probability; rings 38–47;* Bayliss *et al* 2014, fig 11).

References: Reimer *et al* 2004
                    Ward and Wilson 1978

GrA–30779 190 ±30 BP

δ¹³C: -24.4‰

Sample: RIP-C14 <9>, submitted on 13 December 2005 by D Hamilton

Material: wood: *Quercus* sp. (R Howard 2005)

Initial comment: decadal sample from rings 28–37 of floating tree-ring sequence RIPCSQ02. The sample is from core RIP-C14, which was taken from a northern double rafter in truss 6B.

Objectives: as GrA-30767

Calibrated date: 1σ: cal AD 1660–1955★
                      2σ: cal AD 1650–1955★

Final comment: A Bayliss (13 November 2007), this result is statistically consistent with another measurement for this decade of RIPCSQ02, OxA-15413 (188 ±25 BP) (T′=0.0; T′(5%)=3.8; ν=1; Ward and Wilson 1978). The weighted mean of these measurements, 189 ±20 BP, calibrates to cal AD 1660–1955★ (Reimer *et al* 2004). The wiggle-matching of RIPCSQ02 suggests that this sample dates to *cal AD 1765–1785 (95% probability; rings 28–37;* Bayliss *et al* 2014, fig 11).

References: Reimer *et al* 2004
                    Ward and Wilson 1978

GrA–30780 220 ±30 BP

δ¹³C: -23.7‰

Sample: RIP-C14 <10>, submitted on 13 December 2005 by D Hamilton

Material: wood: *Quercus* sp. (R Howard 2005)

Initial comment: decadal sample from rings 18–27 of floating tree-ring sequence RIPCSQ02. The sample is from core RIP-C14, which was taken from a northern double rafter in truss 6B.

Objectives: as GrA-30767

Calibrated date: 1σ: cal AD 1650–1955★
2σ: cal AD 1640–1955★

Final comment: A Bayliss (13 November 2007), this result is statistically consistent with another measurement for this decade of RIPCSQ02, OxA-15414 (211 ±25 BP) (T′=0.1; T′(5%)=3.8; ν=1; Ward and Wilson 1978). The weighted mean of these measurements, 215 ±20 BP, calibrates to cal AD 1645–1955★ (Reimer *et al* 2004). The wiggle-matching of RIPCSQ02 suggests that this sample dates to *cal AD 1755–1775 (95% probability; rings 18–27;* Bayliss *et al* 2014, fig 11).

References: Reimer *et al* 2004
Ward and Wilson 1978

Ryall Quarry: Saxon's Lode Farm, Worcestershire

Location: SO 865391
Lat. 52.03.02 N; Long. 02.11.51 W

Project manager: A Barber (Cotswold Archaeology), 2001 and 2002

Description: the site comprises of approximately 18ha extension to an active quarry on plateau gravels immediately east of the River Severn. Archaeological evaluation of two northern fields of the extension area in January 1998 was followed by evaluation of the two southern fields in September 1998. The second evaluation identified significant Romano-British remains surviving in the eastern half of the south-western field, with the results elsewhere being largely negative.

Subsequent excavation of the western area in September 2001 revealed little archaeology. Topsoil stripping of the eastern area commenced in September 2002 (area 2) and by October 2002 excavations had revealed the near complete plan of a late second-century AD farmstead, defined by large ditched boundaries on its northern, eastern, and southern sides. By November 2002, further features had appeared through weathering and it became apparent that area 2 also contained a number of Anglo-Saxon sunken-featured buildings.

Objectives: to aid interpretation of the site and its placement within the settlement pattern of the region.

Final comment: A Barber (21 May 2005), the individual dating results provide a broadly consistent range within each date series. Taken together the four date series have proved valuable in helping to determine the chronology of occupation across the site. Three of the dating series have provided accurate and consistent dates identifying middle to late Iron Age and AD 560–670 occupation, whilst a wholly unexpected outcome from the remaining dating series was the identification of an early Bronze Age feature within area 2.

References: Barber and Watts 2008
Ward and Wilson 1978

Ryall Quarry, Saxon's Lode Farm, charred cereal remains within pit 1943, Worcestershire

Location: SO 865391
Lat. 52.03.02 N; Long. 02.11.51 W

Project manager: M Watts (Cotswold Archaeology), 2001 and 2002

Archival body: Worcestershire County Museum

Description: a series of two seed samples taken from a primary deposit 1945, containing well-preserved and abundant charred grain, within pit 1943 (also coded with equivalent cut number 1413). Deposit 1945 appears to represent a single event closely associated with the function and use, or closure, of the pit and was sealed by backfill deposit 1944 (= 1414).

The pit was of an atypical form with area 2, being rectangular rather than circular in shape, and was cut shallowly into the natural gravels to a depth of 0.2m.

Pit 1943/1413 lay immediately to the east of a cluster of intercutting circular pits directly alongside, and south of, the northern boundary of the Roman-British farmstead and within the northernmost area of the central, sub-circular, Roman-British enclosure (partly defined by the northern ditched boundary). Pit 1943/1413 also lay approximately 20m to the north of Anglo-Saxon building remains sited within the area of the former Romano-British enclosure.

Objectives: to indicate whether the pit pre-dates the Romano-British sub-circular enclosure/northern boundary; to aid in the establishment of a secure chronological framework for the analysis of the abundant and well-preserved charred grain from this deposit. To aid interpretation of the site and its placement within the settlement pattern of the region.

Final comment: A Barber (21 January 2005), the results from this series provided relatively accurate and consistent date ranges for the final use or disuse of this pit and, despite its atypical shape (rectangular), proved its contemporaneity with circular storage pit 1032. The dates aid determination of the chronology of occupation of the site.

References: Barber and Watts 2008

OxA–13998 2122 ±27 BP

δ¹³C: -22.5‰

Sample: 41 (i), submitted on 28 June 2004 by M Watts

Material: grain: *Triticum* sp., single carbonised grain (E Pearson 2004)

Initial comment: a sample recovered from the southern half of the primary deposit 1945 within pit 1943 (equivalent to 1413), one of a series of pits lying immediately south of the northern boundary ditch of the Romano-British farmstead, within the northernmost part of its sub-circular enclosure. The rectangular shape of pit 1943 was atypical within excavation area 2, contrasting for example with the circular forms of a group of intercutting pits clustered immediately west of pit 1943. Context 1945, which was almost entirely comprised of charred grain, appears to represent a single depositional event closely associated with the use, or closure, of the pit. It was directly sealed by a homogenous deposit 1944 (=1414) of gravelly silty-sand which appears to represent deliberate backfilling of the pit.

Objectives: to establish a reliable date for the use, or closure, of pit 1943 (=1413), as represented by charred grain deposit 1945, and to determine whether the feature predates, is contemporaneous with, or post-dates the sub-circular enclosure and ditched northern boundary of the Romano-British farmstead. To use the results of the analysis, together with that from pit 1032, to aid more detailed phasing of the site, as well as placing the pitting within a chronological framework for the occupation of the remainder of the site and within the wider, regional, settlement pattern.

Calibrated date: 1σ: 200–100 cal BC
2σ: 350–50 cal BC

Final comment: A Barber (21 January 2005), this result correlates well with those from SUERC-4099, a similar sample of charred seed, and identify the basal charred grain deposit within pit 1943 and dating to the middle-late Iron Age (OxA-13998 and SUERC-4099). The dating result successfully establishes a reliable date for the disuse of pit 1943, establishing that it pre-dates the subcircular enclosure and ditched northern boundary of the Romano-British farmstead. The rectangular shape of pit 1943 was atypical within excavation area 2 but the dating result confirms that it belongs to the same phase of pit digging as the circular storage pits.

Laboratory comment: English Heritage (21 May 2005), two samples of well-preserved charred grain were submitted from pit 1413 within pit group 7 (SUERC-4099 and OxA-13998). These measurements are statistically consistent (T'=0.1; v=1; T'(5%)=3.8; Ward and Wilson 1978) and date this feature to the Iron Age (period 3).

References: Ward and Wilson 1978

SUERC–4099 2110 ±35 BP

δ¹³C: -23.0‰

Sample: 41 (ii), submitted on 28 June 2004 by M Watts

Material: grain: *Triticum spelta*, single carbonised grain (E Pearson 2004)

Initial comment: as OxA-13998

Objectives: as OxA-13998

Calibrated date: 1σ: 200–50 cal BC
2σ: 350–40 cal BC

Final comment: see OxA-13998

Laboratory comment: see OxA-13998

Ryall Quarry: Saxon's Lode Farm, charred cereal remains within pit 1032, Worcestershire

Location: SO 865391
Lat. 52.03.02 N; Long. 02.11.51 W

Project manager: M Watts (Cotswold Archaeology), 2001 and 2002

Archival body: Worcestershire County Museum

Description: a series of two seed samples taken from a primary deposit 1915, containing well-preserved and abundant charred grain, within pit 1032. Deposit 1915 appears to represent a single event closely associated with the function and use, or closure, of the pit, and was sealed by a series of gravelly clay-silt backfill deposit 1918, 1919, 1920, and 1033.

The pit, 1.37m in diameter, was of a form typical within area 2, being circular in shape with near vertical sides and a flat base, and was cut deeply into the natural gravels to a depth of approximately 0.5m.

Pit 1032 was one of a series of pit lying close to the western edge of the central sub-circular enclosure of the Romano-British farmstead, but also only approximately 10m south of the Anglo-Saxon building remains sited within the area of the former enclosure.

Objectives: evidence to indicate whether the pit predates the Romano-British sub-circular enclosure. Evidence for association with the Romano-British sub-circular enclosure. Evidence for association with the Anglo-Saxon building sited within the area of the former Romano-British sub-circular enclosure. To aid in the establishment of a secure chronological framework for the analysis of the abundant and well-preserved charred grain from this deposit. To aid interpretation of the site and its placement within the settlement pattern of the region.

Final comment: A Barber (21 January 2005), the results from this series provided accurate and consistent date ranges for the disuse of the storage pit 1032. They also helped determine the chronology of the occupation of the site, helping to place it within a wider regional settlement pattern.

References: Barber and Watts 2008

OxA–14000 2129 ±26 BP

δ¹³C: -22.3‰

Sample: 39 (i), submitted on 28 June 2004 by M Watts

Material: grain: *Triticum dicoccum, single carbonised grain* (E Pearson 2004)

Initial comment: a sample recovered from the eastern half of the primary deposit 1915 within pit 1032, one of a series of pits lying immediately east of the western side of the central sub-circular enclosure of the Romano-British farmstead. Context 1915, which contained abundant charred grain, appears to represent a single depositional event closely associated with the use, or closure, of the pit. It was directly sealed by homogenous deposits 1918, 1919, 1920, and 1033 of gravely clay-silt which appear to represent the deliberate backfilling of the pit.

Objectives: to establish a reliable date for the use, or closure, of pit 1032, and to determine whether the feature predates, is contemporaneous with, or post-dates the sub-circular enclosure and ditched northern boundary of the Romano-British farmstead. To use the results of the analysis, together with that from pit 1943, to aid more detailed phasing of the site, as well as placing the pitting within a chronological framework for the occupation of the remainder of the site and within the wider, regional, settlement pattern.

Calibrated date: 1σ: 200–110 cal BC
                              2σ: 350–50 cal BC

Final comment: A Barber (21 January 2005), this result correlates well with that from SUERC-4098, a similar sample of charred seed, and identified as the basal deposit 1915 within pit 1032 as dating to the middle-late Iron Age (OxA-14000 with SUERC-4098). The dating result successfully establishes a reliable date from the last use or disuse of pit 1032, establishing that it pre-dates the subcircular enclosure and ditched northern boundary of the Romano-British farmstead.

Laboratory comment: English Heritage (21 May 2005), two samples of well-preserved charred grain were submitted from pit 1032 within pit group 12 (SUERC-4098 and OxA-14000). These measurements are statistically consistent (T′=0.0; ν=1; T′(5%)=3.8; Ward and Wilson 1978) and also date this feature to the Iron Age (period 3). The calibrated dates are remarkably consistent in themselves (all ranging from the mid-fourth to the mid-first century BC, and centring on the second century BC) and the later part of this range overlaps with the main ceramic dating for period 3 (undecorated Malvernian-type rock-tempered wares: *c* first century BC to *c* first century AD).

References: Ward and Wilson 1978

SUERC–4098 2135 ±35 BP

$\delta^{13}C$: -21.7‰

Sample: 39 (ii), submitted on 28 June 2004 by M Watts

Material: grain: *Triticum dicoccum, single carbonised grain* (E Pearson 2004)

Initial comment: as OxA-14000

Objectives: as OxA-14000

Calibrated date: 1σ: 210–110 cal BC
                              2σ: 360–50 cal BC

Final comment: see OxA-14000

Laboratory comment: see OxA-14000

Ryall Quarry: Saxon's Lode Farm, sunken-featured building 4, Worcestershire

Location: SO 865391
                      Lat. 52.03.02 N; Long. 02.11.51 W

Project manager: M Watts (Cotswold Archaeology), 2001–2

Archival body: Worcestershire County Museum

Description: a series of three charcoal samples taken from the primary deposit

1013 within Anglo-Saxon sunken-featured building (SFB) 4. Deposit 1013 was a grey-brown gravely silty-sand, averaging only 0.06m in thickness, which appears to represent the gradual accumulation of fine-grained material within the cut, associated with use of the building. It was sealed by a homogenous dark grey-brown sandy-silt deposit 1967, 0.25m in thickness, which formed the main deposit within SFB 4 cut 1012 and appeared to represent a single backfilling episode. SFB 4 was situated at the southern edge of excavation area 2.

Objectives: to determine the date of occupation of the sunken-featured building and thus to refine the dating of the early-middle Saxon occupation within area 2. This is currently based upon organic-tempered pottery only broadly datable to *c* AD 450–850, and annular loomweight fragments. To aid in the establishment of a secure chronological framework for the analysis of the charred plant remains from this deposit.

Final comment: A Barber (21 January 2005), the results from this series provided relatively accurate and consistent date ranges for the use of sunken featured building 4. They also helped determine the chronology of the occupation for the site, helping to place it within a wider, regional settlement pattern.

References: Barber and Watts 2008

OxA–13999 1431 ±25 BP

$\delta^{13}C$: -24.4‰

Sample: 15 (i), submitted on 28 June 2004 by M Watts

Material: charcoal: Pomoideae, single fragment (R Gale 2004)

Initial comment: a sample recovered from the eastern half of the primary deposit 1013 within cut 1012 at the southern edge of excavation area 2. Context 1013, which contained occasional charred plant remains, consisted of a fine-grained silt-sand which may represent the gradual accumulation of material associated with use of the building. It was sealed directly by a homogenous deposit 1967 of more humic gravelly sand-silt. The absence of recognisable structuring/lensing within this deposit suggests that it represents a single episode of deliberate backfilling of the building following its disuse.

Objectives: to establish a secure date for the use of sunken-featured building 4, and consequently to refine the very broad date of *c* AD 450–850, established from recovered organic-tempered pottery and annular loomweight fragments, for the early-middle Saxon occupation within area 2. To aid in the establishment of a secure chronological framework for the analysis of the charred plant remains from this deposit. To aid in the establishment of a secure chronological framework for ceramic types for the region.

Calibrated date: 1σ: cal AD 600–650
                              2σ: cal AD 570–660

Final comment: A Barber (21 January 2005), this result is broadly consistent with the other two samples recovered from the basal deposit 1013 of sunken-featured building 4. The date from the primary (?use-related) deposit 1013 determined the date of the occupation of the building and thus refines the dating of the early-middle Saxon occupation within area 2. The dating result aids establishment of a secure chronological framework for the analysis of the charred plant remains from this deposit.

Laboratory comment: English Heritage (21 May 2005), three charcoal samples were submitted from SFB 4 (SUERC-4101, OxA-13999, and OxA-14001). These measurements are statistically consistent (T′=0.0; ν=1; T′(5%)=3.8; Ward and Wilson 1978) and date this feature to the early to middle Saxon period (period 5).

References: Ward and Wilson 1978

OxA–14001 1445 ±26 BP

δ¹³C: -25.2‰

Sample: 49 (i), submitted on 28 June 2004 by M Watts

Material: charcoal: *Corylus* sp., single fragment (R Gale 2004)

Initial comment: as OxA-13999

Objectives: as OxA-13999

Calibrated date: 1σ: cal AD 590–650
                       2σ: cal AD 560–660

Final comment: see OxA-13999

Laboratory comment: see OxA-13999

SUERC–4101 1435 ±35 BP

δ¹³C: -25.0‰

Sample: 15 (ii), submitted on 28 June 2004 by M Watts

Material: charcoal: Pomoideae, single fragment (R Gale 2004)

Initial comment: as OxA-13999

Objectives: as OxA-13999

Calibrated date: 1σ: cal AD 590–650
                       2σ: cal AD 560–660

Final comment: see OxA-13999

Laboratory comment: see OxA-13999

Ryall Quarry: Saxon's Lode Farm, sunken-featured building 6, Worcestershire

Location: SO 865391
                     Lat. 52.03.02 N; Long. 02.11.51 W

Project manager: M Watts (Cotswold Archaeology)

Archival body: Worcestershire County Museum

Description: a series of three charcoal samples taken from the primary deposit

2269 within sunken-featured building 6. Deposit 2269 was a grey-brown gravelly silty-sand, averaging only 0.16m in thickness, which appears to represent the gradual accumulation of fine-grained material within the cut, associated with use of the building. It was sealed by a homogenous dark grey-brown sandy-silt deposit 2268, 0.15m in thickness, which formed the main deposit within

the cut and appeared to represent a single backfilling episode. The building was situated within the area of the former Romano-British sub-circular enclosure.

Objectives: to determine the date of occupation of the building and thus to refine the dating of the early-middle Saxon occupation within Area 2. This is currently based upon organic-tempered pottery only broadly datable to *c* AD 450–850, and annular loomweight fragments. To aid in the establishment of a secure chronological framework for the analysis of the charred plant remains from this deposit.

Final comment: A Barber (21 January 2005), the consistent date provided by all three samples in this series, identify this feature as being of early Bronze Age, rather than an Anglo-Saxon sunken feature building. The consistency of all three results precludes the possibility of residuality.

OxA–13997 3895 ±29 BP

δ¹³C: -24.2‰

Sample: 57 (ii), submitted on 28 June 2004 by M Watts

Material: charcoal: *Quercus* sp., sapwood (R Gale 2004)

Initial comment: sample recovered from eastern half of primary deposit 2269 within sunken-featured building 6, cut 2267, situated in the area of the former Romano-British sub-circular enclosure. Context 2269, which contained charred plant remains, consisted of a medium-grained gravelly sandy-silt which may represent the gradual accumulation of material associated with use of the building. It was sealed directly by a homogenous deposit 2268 of less gravely sandy-silt. The absence of recognisable structuring/lensing within this deposit suggests that it represents a single episode of deliberate backfilling of the building following its disuse.

Objectives: to establish a secure date for the use of the building and consequently to refine the very broad date of *c* AD 450–850, established from recovered organic-tempered pottery and annular loomweight fragments, for the early-middle Saxon occupation within area 2. To aid in the establishment of a secure chronological framework for the analysis of the charred plant remains from this deposit. To aid in the establishment of a secure chronological framework for ceramic types for the region.

Calibrated date: 1σ: 2470–2300 cal BC
                       2σ: 2480–2280 cal BC

Final comment: A Barber (21 January 2005), this result correlates closely with those of two similar charcoal samples taken from this deposit. The three calibrated dates together identify an early Bronze Age date for this deposit. The dating results, taken together, indicate that his feature is not a sunken featured building but is conceivably a grave cut associated with an early Bronze Age barrow, or some other feature of early Bronze Age date. The charcoal within primary fill 2269, three samples of which all yielded near identical dates, is not considered likely to be residual.

Laboratory comment: English Heritage (21 May 2005), three charcoal samples were submitted from pit 2267 (SUERC-4102–3 and OxA-13997). These are statistically consistent (T′=0.5, ν=2; T′(5%)=6.0; Ward and Wilson 1978) and date this feature to the early Bronze Age (period 1).

References: Ward and Wilson 1978

SUERC–4102 3875 ±35 BP

δ¹³C: -24.3‰

Sample: 57 (i), submitted on 28 June 2004 by M Watts

Material: charcoal: *Quercus* sp., sapwood (R Gale 2004)

Initial comment: as OxA-13997

Objectives: as OxA-13997

Calibrated date: 1σ: 2470–2280 cal BC
                       2σ: 2470–2200 cal BC

Final comment: see OxA-13997

Laboratory comment: see OxA-13997

SUERC–4103 3865 ±35 BP

δ¹³C: -24.1‰

Sample: 57 (iii), submitted on 28 June 2004 by M Watts

Material: charcoal: Pomoideae (R Gale 2004)

Initial comment: as OxA-13997

Objectives: as OxA-13997

Calibrated date: 1σ: 2460–2280 cal BC
                       2σ: 2470–2200 cal BC

Final comment: see OxA-13997

Laboratory comment: see OxA-13997

Sherborne: Tinney's Lane, Dorset

Location: ST 644170
                       Lat. 50.57.01 N; Long. 02.30.24 W

Project manager: J Valentin (Exeter Archaeology), 2002

Archival body: Dorset County Museum

Description: the earliest recorded activity on site was represented by a small number of features in area 3 with sherds of middle Neolithic Peterborough Ware (*c* 3400–2500 cal BC). This pottery is one of very few such assemblages in Dorset outside the monument zone of the South Dorset ridgeway. One of the Neolithic pits also contained an infant (1–3 years old) inhumation. The majority of the archaeological remains in areas 1, 2, and 3 were associated with the on-site production of pottery vessels and fired-clay artefacts, dating to the early part of the late Bronze Age. The focus of this activity was concentrated within a shallow terrace in the central part of area 2, with evidence for clay extraction, spreads of burnt limestone fragments, pits filled with 'industrial' waste, postholes, scorched bases of bonfires used for firing pottery, and dumps of pottery wasters and mis-fired clay objects. Beyond the central part of area 2, the archaeological remains survived only as features cut into the limestone bedrock. More than 13,500 sherds of Post-Deverel-Rimbury pottery (*c* 1100–750 cal BC) were recovered from a variety of features, forming one of the largest assemblages recorded in England. Over 1200 items of fired-clay were found, including a significant number of 'bullet-shaped' cylindrical 'loom-weights'. Other finds included one of the largest assemblages of late Bronze Age

worked bone tools in South West England, the nature of which suggests that these artefacts were also being produced on site for use either in pottery production or for trade. A number of copper alloy artefacts were found, as were several fragments of fired-clay refractory moulds.

Objectives: radiocarbon dates will be required to date the pottery production site and features located outside the main pottery production area to determine whether these represent contemporaneous anthropogenic activity. In particular: to provide a date for the pottery production activity by dating organic fuel residues from the firing areas (bonfire bases); to address the contemporaneity of the putative adjacent settlement site by dating organic residues from potsherds - if present - or of securely provenanced organic material from archaeological features; to date features associated with the Peterborough Wares.

Final comment: P Marshall (2012), the radiocarbon dating programme at Tinney's Lane has demonstrated that activity occurred for a short period some time within the twelfth or eleventh centuries cal BC. One determination (SUERC-9678) was anomalous providing a date in the fifteenth to thirteenth century cal BC. The sample came from a context underneath one of the bonfire bases. It may have derived from activity relating to an earlier phase of occupation, associated with the small number of middle Bronze Age sherds that were also found in the area.

References: Best *et al* 2013
                       Best and Woodward 2011

SUERC–9652 2920 ±35 BP

δ¹³C: -26.2‰

Sample: DSTL02 4470–300 A, submitted on 2 February 2006 by J Valentin

Material: charcoal: *Ulmus* sp., single fragment (R Gale 2002)

Initial comment: from the basal fill of a small pit/posthole 603, part of 'industrial' structure [626]. The context was formed of a mid-brown silty clay with moderate amounts of charcoal, burnt limestone, ash, and burnt-red silty clay. A backfilled deposit, one of twelve sherds of late Bronze Age pottery, part of the same vessel in context 1046 (area 3; rare fabric).

Objectives: to provide dating for on-site pottery production activities. One of a few 'industrial' features: part of possible structure [626]; to date in conjunction with sample 301; to establish whether pottery production activities were contemporary across the development site; and to establish contemporaneity of activity across the site, a conjoining sherd with pottery from feature 1047 in area 3.

Calibrated date: 1σ: 1200–1040 cal BC
                       2σ: 1230–1000 cal BC

Final comment: see project comments

SUERC–9656 2910 ±35 BP

δ¹³C: -25.7‰

Sample: DSTL02 4470–300 B, submitted on 2 February 2006 by J Valentin

Material: charcoal: *Corylus avellana*, single fragment (R Gale 2002)

Initial comment: as SUERC-9652

Objectives: as SUERC-9652

Calibrated date: *1σ:* 1190–1020 cal BC
                         2σ: 1220–1000 cal BC

Final comment: see project comments

SUERC–9657 2950 ±35 BP

δ¹³C: -25.9‰

Sample: DSTL02 4470–301 A, submitted on 2 February 2006 by J Valentin

Material: charcoal: *Alnus glutinosa*, single fragment (R Gale 2002)

Initial comment: from the basal fill of a small pit/posthole 598, part of 'industrial' structure [626]. A dark grey to reddish brown silty clay, with charcoal, burnt limestone, and patches of burnt-red silty clay. A backfilled deposit which contained ten sherds of late Bronze Age pottery.

Objectives: to provide dating for on-site pottery production activities. One of few 'industrial' features, part of possible structure [626]; to establish whether pottery production activities were contemporary across the development site; to date in conjunction with sample 300.

Calibrated date: *1σ:* 1220–1110 cal BC
                         2σ: 1270–1040 cal BC

Final comment: see project comments

SUERC–9658 2960 ±35 BP

δ¹³C: -26.1‰

Sample: DSTL02 4470–301 B, submitted on 2 February 2006 by J Valentin

Material: charcoal: *Corylus avellana*, single fragment (R Gale 2002)

Initial comment: as SUERC-9657

Objectives: as SUERC-9657

Calibrated date: *1σ:* 1230–1110 cal BC
                         2σ: 1280–1040 cal BC

Final comment: see project comments

SUERC–9659 2915 ±35 BP

δ¹³C: -22.9‰

Sample: DSTL02 4470–302 B, submitted on 2 February 2006 by J Valentin

Material: charcoal: *Quercus* sp., sapwood, single fragment (R Gale 2002)

Initial comment: from the middle fill of pit 616. A mid-brown silty clay with moderate amounts of burnt and unburnt limestone. A mixed and ashy backfilled deposit with 23 sherds of late Bronze Age pottery and wasters.

Objectives: to provide dating for on-site pottery production activities. Pit 616 had been backfilled with 'industrial' waste and burnt silty clay. One of several similar pits in this area, close to bonfire bases. It had probably been used to store water (freshwater snails were identified); to establish whether pottery production activities were contemporary across the development site; to provide a date in conjunction with sample 304.

Calibrated date: *1σ:* 1200–1040 cal BC
                         2σ: 1220–1000 cal BC

Final comment: see project comments

SUERC–9660 2880 ±35 BP

δ¹³C: -26.3‰

Sample: DSTL02 4470–304 B, submitted on 2 February 2006 by J Valentin

Material: charcoal: *Corylus avellana*, single fragment (R Gale 2002)

Initial comment: from the lower fill of pit 616. A dark brown silty clay with frequent charcoal and occasional burnt limestone. A backfilled deposit with late Bronze Age pottery.

Objectives: to provide dating for on-site pottery production activities. Pit 616 had been backfilled with 'industrial' waste and burnt silty clay. One of several similar pits in this area, close to bonfire bases. It had probably been used to store water (freshwater snails were identified); to establish whether pottery production activities were contemporary across the development site; to provide a date in conjunction with sample 302.

Calibrated date: *1σ:* 1120–1000 cal BC
                         2σ: 1200–930 cal BC

Final comment: see project comments

SUERC–9661 2855 ±35 BP

δ¹³C: -25.1‰

Sample: DSTL02 4470–304 C, submitted on 2 February 2006 by J Valentin

Material: charcoal: *Acer campestre*, single fragment (R Gale 2002)

Initial comment: as SUERC-9660

Objectives: as SUERC-9660

Calibrated date: *1σ:* 1060–940 cal BC
                         2σ: 1130–910 cal BC

Final comment: see project comments

SUERC–9662 2875 ±35 BP

δ¹³C: -25.3‰

Sample: DSTL02 4470–305, submitted on 2 February 2006 by J Valentin

Material: charcoal: *Sorbus* sp. Pomoideae, single fragment (R Gale 2002)

Initial comment: from the main fill of pit 612, a dark grey to black silty loam with charcoal and moderate amounts of burnt limestone (fairly tightly packed and becoming redder towards the lower part of the fill, but no evidence for burning *in situ*). A backfilled deposit with 23 sherds of late Bronze Age pottery. It was similar to fill 630 in 'industrial' pit 631 to the south. Pit 612 forms the central feature of the 'industrial' structure [626], surrounded by postholes 623, 625, 619, 621, 598 (sample 301), 604, 603 (sample 300), and 608.

Objectives: to provide dating for on-site pottery production activities. Pit 612 had been backfilled with 'industrial' waste and burnt-red limestone and patches of silty clay. One of several similar pits in this area, close to bonfire bases; to establish whether pottery production activities were contemporary across the development site.

Calibrated date: 1σ: 1120–1000 cal BC
 2σ: 1200–920 cal BC

Final comment: see project comments

SUERC–9666 2805 ±35 BP

δ¹³C: -26.5‰

Sample: DSTL02 4470–311 B, submitted on 2 February 2006 by J Valentin

Material: charcoal: *Ulmus* sp., single fragment (R Gale 2002)

Initial comment: from the main fill of large pit 649, a mid-brown silty clay with frequent burnt limestone and moderate amounts of charcoal. It lay beneath burnt limestone spread [2658]. A backfilled deposit with 107 sherds of late Bronze Age pottery, including wasters (and one cup base).

Objectives: to provide dating for on-site pottery production activities. Pit 649 was probably used to store water for use in pottery manufacture. It was fed by gullies to the north-west; to establish whether pottery production activities were contemporary across development site; to provide a date in conjunction with samples 361 and 362.

Calibrated date: 1σ: 1010–900 cal BC
 2σ: 1050–840 cal BC

Final comment: see project comments

SUERC–9667 2875 ±35 BP

δ¹³C: -24.7‰

Sample: DSTL02 4470–311 C, submitted on 2 February 2006 by J Valentin

Material: charcoal: *Corylus avellana*, single fragment (R Gale 2002)

Initial comment: as SUERC-9666

Objectives: as SUERC-9666

Calibrated date: 1σ: 1120–1000 cal BC
 2σ: 1200–920 cal BC

Final comment: see project comments

SUERC–9668 2755 ±35 BP

δ¹³C: -25.6‰

Sample: DSTL02 4470–312 A, submitted on 2 February 2006 by J Valentin

Material: charcoal: *Sorbus* sp. Pomoideae, single fragment (R Gale 2002)

Initial comment: from the basal fill of a large, sub-circular pit 1029 in area 1: a light greyish brown silt with sparse fragments of burnt and unburnt limestone, with a dump of large sherds of late Bronze Age pottery near the base of the fill. Late Bronze Age sherds were found in 1028, a total of 153 including wasters.

Objectives: to provide dating for on-site pottery production activities; to establish whether pottery production activities were contemporary across the development site. This is the only charcoal sample for area 1.

Calibrated date: 1σ: 930–840 cal BC
 2σ: 1000–810 cal BC

Final comment: see project comments

SUERC–9669 2795 ±35 BP

δ¹³C: -26.5‰

Sample: DSTL02 4470–312 B, submitted on 2 February 2006 by J Valentin

Material: charcoal: *Prunus spinosa*, single fragment (R Gale 2002)

Initial comment: as SUERC-9668

Objectives: as SUERC-9668

Calibrated date: 1σ: 1010–900 cal BC
 2σ: 1030–840 cal BC

Final comment: see project comments

SUERC–9670 2850 ±35 BP

δ¹³C: -24.1‰

Sample: DSTL02 4470–314 A, submitted on 2 February 2006 by J Valentin

Material: charcoal: *Quercus* sp., narrow roundwood, single fragment (R Gale 2002)

Initial comment: from the fill of pit 767. A very dark greyish brown silty clay loam with burnt limestone, large fragments of charcoal, burnt-red silty clay and fragments of calcite. Context 766 had 22 sherds of late Bronze Age pottery, other fills also had late Bronze Age pottery (including wasters). Ashy fills with charcoal and burnt stone were present.

Objectives: to provide dating for on-site pottery production activities; to establish whether pottery production activities were contemporary across the development site: this pit was in the north-east corner of area 3; to establish whether it is of similar date to samples 384 and 385.

Calibrated date: 1σ: 1060–940 cal BC
 2σ: 1120–910 cal BC

Final comment: see project comments

SUERC–9671 2835 ±35 BP

δ¹³C: -26.2‰

Sample: DSTL02 4470–314 C, submitted on 2 February 2006 by J Valentin

Material: charcoal: *Prunus* sp., single fragment (R Gale 2002)

Initial comment: as SUERC-9670

Objectives: as SUERC-9670

Calibrated date: *1σ:* 1050–920 cal BC
                               2σ: 1120–900 cal BC

Final comment: see project comments

SUERC–9672 2870 ±35 BP

δ¹³C: -24.0‰

Sample: DSTL02 4470–319, submitted on 2 February 2006 by J Valentin

Material: charcoal: *Fraxinus excelsior*, single fragment (R Gale 2002)

Initial comment: a bonfire base with heat-discoloured area of surface of [2660]. A mid to dark red-purple silty clay. Moderate amounts of burnt limestone and sparse charcoal were present, along with 86 sherds of late Bronze Age pottery with wasters.

Objectives: to provide dating for on-site pottery production activities: 803 was the remains from the bonfire firing of pottery vessels; to establish whether pottery production activities were contemporary across the development site.

Calibrated date: *1σ:* 1120–1000 cal BC
                               2σ: 1190–920 cal BC

Final comment: see project comments

SUERC–9676 2920 ±35 BP

δ¹³C: -23.6‰

Sample: DSTL02 4470–322 A, submitted on 2 February 2006 by J Valentin

Material: charcoal: *Acer campestre*, single fragment (R Gale 2002)

Initial comment: from the middle fill of a large, sub-circular pit, 949 in area 2. A very dark brown-black silty clay with occasional fragments of unburnt limestone, patches of ash and charcoal lenses in burnt-red silty clay.

Objectives: pit 949 was located immediately north of bonfire base 2190, and had been backfilled with waste from pottery production; to provide dating for on-site pottery production activities: 803 was the remains from the bonfire firing of pottery vessels; to establish whether pottery production activities were contemporary across the development site; and to provide a date in conjunction with sample 392.

Calibrated date: *1σ:* 1200–1040 cal BC
                               2σ: 1230–1000 cal BC

Final comment: see project comments

SUERC–9677 2930 ±35 BP

δ¹³C: -26.8‰

Sample: DSTL02 4470–322 B, submitted on 2 February 2006 by J Valentin

Material: charcoal: *Sorbus* sp. Pomoideae, single fragment (R Gale 2002)

Initial comment: as SUERC-9676

Objectives: as SUERC-9676

Calibrated date: *1σ:* 1210–1050 cal BC
                               2σ: 1260–1010 cal BC

Final comment: see project comments

SUERC–9678 3090 ±35 BP

δ¹³C: -26.2‰

Sample: DSTL02 4470–351, submitted on 2 February 2006 by J Valentin

Material: charcoal: *Sorbus* sp. Pomoideae, single fragment (R Gale 2002)

Initial comment: from a bonfire deposit of uneven ashy spread beneath bonfire deposit 2400. A mottled reddish purple to black silty clay. Charcoal flecks in various concentrations were identified throughout 2448. It overlay heat-discoloured silty clay within the coombe, with sherds of late Bronze Age pottery. Context 2448 was part of activity horizon [2569], which includes bonfires and spreads of trample. This part of [2659] was coeval with spreads of burnt limestone [2658].

Objectives: to provide dating for on-site pottery production activities: 2448 was the remains from the bonfire-firing of pottery vessels; also to establish whether pottery production activities were contemporary across the development site.

Calibrated date: *1σ:* 1420–1290 cal BC
                               2σ: 1440–1260 cal BC

Final comment: see project comments

SUERC–9679 2920 ±35 BP

δ¹³C: -24.8‰

Sample: DSTL02 4470–361, submitted on 2 February 2006 by J Valentin

Material: charcoal: *Fraxinus excelsior*, single fragment (R Gale 2002)

Initial comment: from the middle fill of pit 649, a dark greyish-brown to black fine silty loam with scattered burnt limestone forming a well-packed layer. The pottery was mainly sloping into the centre of the pit from the north-eastern edge. Fragments of fired-clay objects were also found in the backfilled deposit, along with 23 sherds of late Bronze Age pottery.

Objectives: to provide dating for on-site pottery production activities: 2461 had been dumped into pit 649 from the north-east as waste material from pottery manufacture on the site; also to establish whether pottery production activities were contemporary across the development site; and to provide a date in conjunction with samples 311 and 362.

Calibrated date: 1σ: 1200–1040 cal BC
2σ: 1230–1000 cal BC

Final comment: see project comments

SUERC–9680 2935 ±35 BP

$\delta^{13}C$: -26.8‰

Sample: DSTL02 4470–362, submitted on 2 February 2006 by J Valentin

Material: charcoal: *Alnus glutinosa*, single fragment (R Gale 2002)

Initial comment: from the basal fill of pit 649, a yellowish brown silty clayey silt with small fragments of burnt and unburnt limestone and occasional charcoal.

Objectives: to provide dating for on-site pottery production activities: pit 649 was probably used to store water for us in pottery manufacture on the site; also to establish whether pottery production activities were contemporary across the development site; and to provide a date in conjunction with samples 311 and 361.

Calibrated date: 1σ: 1220–1050 cal BC
2σ: 1260–1010 cal BC

Final comment: see project comments

SUERC–9681 2845 ±35 BP

$\delta^{13}C$: -26.2‰

Sample: DSTL02 4470–384 B, submitted on 2 February 2006 by J Valentin

Material: charcoal: *Acer campestre*, single fragment (R Gale 2002)

Initial comment: from the fill of large pit 1181. Context 1180 contained 46 fragments of fired-clay 'loom-weights', worked flints, and 144 late Bronze Age pottery sherds (including 28 wasters).

Objectives: to provide dating for on-site pottery production activities: the feature contained deliberately dumped waste products from pottery and fired-clay object manufacture; also to establish whether pottery production activities were contemporary across the development site and area 3; and to provide a date in conjunction with sample 385.

Calibrated date: 1σ: 1050–930 cal BC
2σ: 1120–910 cal BC

Final comment: see project comments

References: Ward and Wilson 1978

SUERC–9682 2795 ±40 BP

$\delta^{13}C$: -25.9‰

Sample: DSTL02 4470–385 A, submitted on 2 February 2006 by J Valentin

Material: charcoal: *Ulmus* sp., single fragment (R Gale 2002)

Initial comment: from the basal fill of large pit 1181, overlain by fill 1180. Fill 1200 contained 11 late Bronze Age pottery sherds (including 5 wasters), and around 12 fired-clay objects (miss- and partially-fired 'bullet-shaped loom weights').

Objectives: to provide dating for on-site pottery production activities: the feature contained deliberately dumped waste products from pottery and fired-clay object manufacture; also to establish whether pottery production activities were contemporary across the development site and area 3; and to provide a date in conjunction with sample 384.

Calibrated date: 1σ: 1010–900 cal BC
2σ: 1050–830 cal BC

Final comment: see project comments

References: Ward and Wilson 1978

SUERC–9686 2885 ±40 BP

$\delta^{13}C$: -23.2‰

Sample: DSTL02 4470–385 B, submitted on 2 February 2006 by J Valentin

Material: charcoal: *Acer campestre*, single fragment (R Gale 2002)

Initial comment: as SUERC-9682

Objectives: as SUERC-9682

Calibrated date: 1σ: 1130–1000 cal BC
2σ: 1210–930 cal BC

Final comment: see project comments

SUERC–9687 2805 ±35 BP

$\delta^{13}C$: -23.6‰

Sample: DSTL02 4470–392, submitted on 2 February 2006 by J Valentin

Material: charcoal: *Acer campestre*, single fragment (R Gale 2002)

Initial comment: from the upper fill of large sub-circular pit 949 in area 2. A mid-reddish brown silty clay with moderate amounts of burnt limestone. It contained late Bronze Age pottery, including wasters and formed part of layer [2659]: an activity horizon coeval with bonfire firings and spreads of burnt limestone [2658].

Objectives: pit 949 was located immediately north of bonfire base 2190 and had been backfilled with waste from pottery production; to provide dating for on-site pottery production activities: the feature contained deliberately dumped waste products from pottery and fired-clay object manufacture; also to establish whether pottery production activities were contemporary across the development site and area 3; and to provide a date in conjunction with sample 322.

Calibrated date: 1σ: 1010–900 cal BC
2σ: 1050–840 cal BC

Final comment: see project comments

Silbury Hill, Wiltshire

Location: SU 09976854
Lat. 51.24.55 N; Long. 01.51.24 W

Project manager: R Atkinson (Cardiff University), 1968–9

Archival body: Alexander Keiller Museum

Description: the monumental mound at Silbury. A central feature of the later Neolithic complex around Avebury. These samples derive from the 1968–9 tunnel dug by R J C Atkinson in the BBC-sponsored excavation.

Objectives: to provide a chronology for the construction of Silbury Hill.

Final comment: A Bayliss (2007), these radiocarbon dates were incorporated into a chronological model along with other dates from Silbury Hill, and indicate that the raising of the primary mound occurred in the twenty-fourth or twenty-third century cal BC (Bayliss *et al* 2007d).

Laboratory comment: English Heritage (24 June 2014), three samples were funded prior to 2003 (OxA-13210–1, and OxA-13333; Bayliss *et al* 2016, 240-1). A further 24 dates from these investigations were subsequently funded (OxA-17470–4, -20805–9, SUERC-24081–2, -24086–91, -24828–9, and -27238–41).

References: Bayliss *et al* 2007d
Bayliss *et al* 2016
Marshall *et al* 2013
Whittle 1997

GrA–27332 4015 ±45 BP

$\delta^{13}C$: -21.4‰

Sample: Sample 5, submitted on 13 October 2004 by A Whittle

Material: animal bone: *Sus* sp. (P Baker 2002)

Initial comment: a replicate of OxA-13333 (3916 ±28 BP). From the old land surface at ring 4 of the western lateral tunnel in the area of the primary mound (Whittle 1997, fig 12).

Objectives: this sample provides a *terminus post quem* for the construction of the primary mound.

Calibrated date: 1σ: 2580–2470 cal BC
2σ: 2840–2460 cal BC

Final comment: P Marshall (2015), the date provides a reliable *terminus post quem* for the construction of the primary mound.

Laboratory comment: English Heritage (2013), the two results on this sample are statistically consistent (T'=3.5; T'(5%)=3.8; ν=1; Ward and Wilson), and the weighted mean (3944 ±24 BP) calibrates to 2550–2345 cal BC (at 95% confidence; Reimer *et al* 2004).

References: Reimer *et al* 2004
Ward and Wilson 1978

GrA–27335 3630 ±45 BP

$\delta^{13}C$: -23.7‰

Sample: Sample 2, submitted on 13 October 2004 by A Whittle

Material: antler: *Cervus elaphus* (P Baker 2002)

Initial comment: a replicate of OxA-11970 (3634 ±30 BP). From clean chalk material above the floor of the tunnel at ring 12 on the west side of the tunnel, in the outer part of the mound (Whittle 1997, figs 10–11).

Objectives: as GrA-27332

Calibrated date: 1σ: 2120–1930 cal BC
2σ: 2140–1880 cal BC

Final comment: P Marshall (2015), the date provides an estimate for the construction of the mound.

Laboratory comment: English Heritage (2013), the two results on this sample are statistically consistent (T'=0.0; T'(5%)=3.8; ν=1; Ward and Wilson), and the weighted mean (3633 ±25 BP) calibrates to 2130–1920 cal BC (at 95% confidence; Reimer *et al* 2004).

References: Reimer *et al* 2004
Ward and Wilson 1978

GrA–27336 3390 ±40 BP

$\delta^{13}C$: -23.7‰

Sample: Sample 1, submitted on 13 October 2004 by A Whittle

Material: antler: *Cervus elaphus* (P Baker 2002)

Initial comment: a replicate of OxA-13210 (3401 ±36 BP). This sample is from the early part of the tunnel excavation of April 1968. It is therefore not far into the tunnel, but had not been given a precise location other than 'e side of chalk block wall'. There are chalk block walls around rings 11–13/14 on both sides of the tunnel (Whittle 1997, figs 10–11) about 14–18m into the mound, in the makeup of the chalk mound, and the sample should belong here.

Objectives: to date the construction of the secondary mound. The sample is a fragment of antler tine probably broken from a pick during the quarrying of the chalk making up the mound. This sample should date the completion of the secondary mound (phase III in Atkinson's terms).

Calibrated date: 1σ: 1750–1620 cal BC
2σ: 1860–1610 cal BC

Final comment: P Marshall (2015), the date provides an estimate for the construction of the mound.

Laboratory comment: English Heritage, the two results on this sample are statistically consistent (T'=0.0; T'(5%)=3.8; ν=1; Ward and Wilson), and the weighted mean (3396 ±27 BP) calibrates to 1750–1620 cal BC (at 95% confidence; Reimer *et al* 2004).

References: Reimer *et al* 2004
Ward and Wilson 1978

OxA–11970 3634 ±30 BP

$\delta^{13}C$: -23.3‰

Sample: Sample 2, submitted on 16 October 2004 by A Whittle

Material: antler: *Cervus elaphus* (P Baker 2002)

Initial comment: a replicate of GrA-27335.

Objectives: as GrA-27332

Calibrated date: 1σ: 2040–1940 cal BC
                     2σ: 2130–1910 cal BC

Final comment: see GrA-27335

Laboratory comment: see GrA-27335

Silbury Hill: 2000–2001 investigations, Wiltshire

Location: SU 09976854
                     Lat. 51.24.55 N; Long. 01.51.24 W

Project manager: F McAvoy (English Heritage), 2000–1

Archival body: Alexander Keiller Museum

Description: the monumental mound at Silbury. A central feature of the later Neolithic complex around Avebury. These samples derive from the 2000–2001 remedial works and excavation that followed the collapse of the top of the mound in 2000.

Objectives: to provide a date for activity associated with construction of the summit of Silbury Hill.

Final comment: A Bayliss (2007), these radiocarbon dates were incorporated into a chronological model along with other dates from Silbury Hill, and indicate that the raising of the primary mound occurred in the 24th or 23rd century cal BC (Bayliss *et al* 2007d).

References: Bayliss *et al* 2007d
                     Marshall *et al* 2013

GrA–27331 3655 ±45 BP

δ¹³C: -23.2‰

Sample: 661–2001 00864, submitted on 25 September 2001 by F McAvoy

Material: antler: *Cervus elaphus* (P Baker 2001)

Initial comment: the antler fragment was recovered from a layer of large chalk blocks with voids (context 30) recorded in the sides of the hole at the top of Silbury Hill. It was probably used in quarrying chalk at the base of the hill and brought up during construction. The chalk layer has the same characteristics as that containing the other antler (661–851) but the exact relationship between the layers could not be established.

Objectives: to establish a date for the construction activity on the summit of Silbury Hill.

Calibrated date: 1σ: 2130–1950 cal BC
                     2σ: 2200–1890 cal BC

Final comment: P Marshall (5 November 2014), the result provides a reliable indication of the date of the chalk mound.

OxA–13228 3856 ±39 BP

δ¹³C: -22.6‰

Sample: 661– 851, submitted on 28 June 2001 by F McAvoy

Material: antler: *Cervus elaphus* (P Baker 2001)

Initial comment: the antler was found lying against the outer face of a very substantial chalk wall (context 7) within/below a layer of loosely packed chalk rubble (context 11). It was probably used in quarrying chalk at the base of the hill and brought up to the top during construction.

Objectives: to establish a date for the construction activity on the summit of Silbury Hill.

Calibrated date: 1σ: 2460–2210 cal BC
                     2σ: 2470–2190 cal BC

Final comment: P Marshall (5 November 2014), the date for antler 661– 851 is poor agreement with its apparent stratigraphic position on top of, ie later than, the antler samples from within the makeup of the chalk mound. The antler is therefore either redeposited from an earlier context or the dated samples from the chalk mound do not relate to its primary construction but a later episode of modification.

Laboratory comment: English Heritage (5 November 2014), the two measurements (OxA-13228 and OxA-14118 are statistically consistent (T′=0.2; T′(5%)= 3.8; ν=1; Ward and Wilson 1978). The weighted mean (3870 ±24 BP) calibrates to 2465–2210 cal BC (Reimer *et al* 2004).

References: Reimer *et al* 2004
                     Ward and Wilson 1978

OxA–14118 3878 ±50 BP

δ¹³C: -22.5‰

Sample: 661– 851, submitted on 28 June 2001 by F McAvoy

Material: antler: *Cervus elaphus* (P Baker 2001)

Initial comment: as OxA-13228

Objectives: as OxA-13228

Calibrated date: 1σ: 2470–2280 cal BC
                     2σ: 2480–2200 cal BC

Final comment: see OxA-13228

Laboratory comment: see OxA-13228

Silbury Hill: primary mound turves, Wiltshire

Location: SU 09976854
                     Lat. 51.24.55 N; Long. 01.51.24 W

Project manager: A Whittle (Cardiff University), 1968–9

Archival body: Cardiff University

Description: these samples derive from three blocks of turf from the primary mound recovered by John Evans during the 1968–9 excavations.

Objectives: to date the construction of the primary Silbury Hill mound.

Final comment: A Bayliss (2007), the new radiocarbon results from samples of mosses retrieved from the surfaces of turves (*see also* OxA-11647 and OxA-11663) incorporated in the primary mound are statistically significantly different, from each other (T'=129.3; T'(5%)=16.9; ν=9; Ward and Wilson 1978), from the series of results from bulk organic soil fractions measured by the Smithsonian Institution in the 1960s (excluding SI-910AH) (T'=365.1; T'(5%)=23.7; ν=14; Ward and Wilson 1978), and from the sample of plant material dated from a similar context in 1968 (T'=143.3; T'(5%)=18.3; ν=10; Ward and Wilson 1978). The bulk soil samples probably contained organic matter dating from the whole period of the soil's formation.

The consistency of the results of the different particle size fractions suggests that these measurements are probably reliable estimates of the radiocarbon content of this material, although the resultant date only provides a *terminus post quem* for the mound. The wide variation in the new measurements appears to relate to contamination by a younger humic component in the samples. The four measurements on the acid/base insoluble fraction of this material are statistically consistent (T'=6.4; T'(5%)=7.8; ν=3; Ward and Wilson 1978) and are considered to be the most reliable for determining the age of the mosses. Since this material was growing on the surface of the turf, all the fragments of moss must have grown in the few years before the mound was raised. The results on the alkali-soluble (humic fraction) are also statistically consistent but are significantly younger (T'=5.8; T'(5%)=9.5; ν=4; Ward and Wilson 1978). Since humics can be mobilised and remobilised within sediments, particularly in an alkaline environment such as the chalk mound of Silbury, it is likely that these results relate to later contaminants. The varying results on OxA-11663 and OxA-11647 also seem to relate to the incomplete removal of a younger humic component, as these samples were fragile and light and so were only treated with acid. For these samples, this appears to have been insufficient to remove all contaminants. The four new results on the acid/alkali insoluble fraction of the moss are, however, significantly later than that of the bulk sample of equivalent material previously dated (I-4136; T'=13.2; T'(5%)=9.5; ν=4; Ward and Wilson 1978). As this sample consisted of unburnt plant material from within the turves of the primary mound, rather than from their surface, the incorporation of some slightly earlier material preserved in the turf is perhaps not unexpected. For these reasons the four measurements on the acid/ alkali insoluble fraction of the moss are believed to provide the best estimate for the construction of the primary mound.

Laboratory comment: Oxford Radiocarbon Accelerator Unit (2007), the humic acid and humin fractions on the Oxford laboratory moss samples were both dated. The results strongly suggest that there was a younger ^{14}C-labelled contaminant derived from the humic component within the moss samples from Silbury Hill. The humin fraction determinations are likely to be the most reliable of these pairs of samples for the age of this material, as they are usually considered to be in radiocarbon dating, since humics can be mobilised and remobilised within sediments, and cam be of younger or older age. The shift in the δ^{13}C values is indicative of a small humic contaminant.

References: Bayliss *et al* 2007d
                   Reimer *et al* 2004
                   Ward and Wilson 1978

GrA–28465 3770 ±40 BP

$\delta^{13}C$: -28.9‰

Sample: Sample 6 (TS1a), submitted on 13 March 2005 by A Whittle

Material: plant macrofossils (mainly *Rhytidiadelphus squarrosus* (Hedw.) Warnst., with some cf. *Eurynchium* sp(p), *Neckara complanata* (Hedw.) H√°b., *Pseudoscleropodium purum* (Hedw.) Fleisch, cf. *Plagiomnium* sp(p), and some moss indet (mainly leafless or otherwise eroded shoots) (A Hall 2005)

Initial comment: a replicate of OxA-14641. The sample, from a block of soil and turf, was collected by John Evans in 1968/9 from spoil removed by the excavation team from the centre of the mound. Its precise location within the primary mound is therefore not known, however, the soil is calcareous and is therefore from the primary mound, and not from the OLS (Whittle 1997). Further proof of the sample being derived from the primary mound consists of visible layering of turves, face to face, within the matrix block.

Objectives: to provide a date for the construction of the primary or secondary mound at Silbury Hill.

Calibrated date: 1σ: 2280–2130 cal BC
                        2σ: 2300–2030 cal BC

Final comment: see series comments

References: Whittle 1997

GrA–28466 3840 ±40 BP

$\delta^{13}C$: -28.9‰

Sample: Sample 7 (TS2) A, submitted on 13 March 2005 by A Whittle

Material: plant macrofossils (<2g) (mainly *Rhytidiadelphus squarrosus* (Hedw.) Warnst., with some *Calliergon cuspidatum* (Hedw.) Kindb., cf. *Plagiomnium* sp(p), and some moss indet (mainly leafless or otherwise eroded shoots)) (A Hall 2005)

Initial comment: the sample material, collected in 1968/9 by John Evans from spoil removed by the excavation team from the centre of the mound is from a block of non-calcareous soil comprising at least five stacked turves. This stacking of the turves demonstrates that this sample came from the make-up of the mound and not the old land surface.

Objectives: to provide a date for the construction of the primary or secondary mound at Silbury Hill.

Calibrated date: 1σ: 2400–2200 cal BC
                        2σ: 2470–2140 cal BC

Final comment: see series comments

GrA–28467 3585 ±40 BP

$\delta^{13}C$: -29.9‰

Sample: Sample 7 (TS2a), submitted on 13 March 2005 by A Whittle

Material: plant macrofossils (mainly *Rhytidiadelphus squarrosus* (Hedw.) Warnst., with some *Calliergon cuspidatum* (Hedw.) Kindb., cf. *Plagiomnium* sp(p), and some moss indet (mainly leafless or otherwise eroded shoots)) (A Hall 2005)

Initial comment: a replicate of OxA-14642.

Objectives: as GrA-28465

Calibrated date: 1σ: 2020–1880 cal BC
2σ: 2040–1780 cal BC

Final comment: see series comments

Laboratory comment: Groningen (28 June 2005), the alkali-soluble fraction was dated.

GrA-28555 3710 ±80 BP

δ¹³C: -29.9‰

Sample: Sample 6 (TS1) A, submitted on 13 March 2005 by A Whittle

Material: plant macrofossils (<2g) (mainly *Rhytidiadelphus squarrosus* (Hedw.) Warnst., with some cf. *Eurynchium* sp(p), *Neckara complanata* (Hedw.) H√°b., *Pseudoscleropodium purum* (Hedw.) Fleisch, cf. *Plagiomnium* sp(p), and some moss indet (mainly leafless or otherwise eroded shoots) (A Hall 2005)

Initial comment: a replicate of OxA-14640. The sample, from a block of soil and turf, was collected by John Evans in 1968/9 from spoil removed by the excavation team from the centre of the mound. Its precise location within the primary mound is therefore not known, however, the soil is calcareous and is therefore from the primary mound, and not from the OLS (Whittle 1997). Further proof of the sample being derived from the primary mound consists of visible layering of turves, face to face, within the matrix block.

Objectives: as GrA-28465

Calibrated date: 1σ: 2210–1970 cal BC
2σ: 2350–1890 cal BC

Final comment: see series comments

Laboratory comment: Groningen (28 June 2005), the alkali-soluble fraction was dated. This was a very small sample, hence the large uncertainty.

OxA-11647 3746 ±40 BP

δ¹³C: -30.4‰

Sample: Sample 6B (SILB5), submitted on 29 May 2002 by A Whittle

Material: plant macrofossils (mainly *Calliergon cuspidatum* (Hedw.) Kindb., with some moss indet (mainly leafless or otherwise eroded shoots)) (A Hall 2002)

Initial comment: from the surface of a turf from the primary mound.

Objectives: to provide a date for the construction of the primary mound at Silbury Hill.

Calibrated date: 1σ: 2200–2135 cal BC
2σ: 2200–2135 cal BC

Final comment: see series comments

Laboratory comment: Oxford Radiocarbon Accelerator Unit (24 August 2004), acid-wash only.

OxA-11663 3295 ±60 BP

δ¹³C: -28.1‰

Sample: Sample 6A (SILB3), submitted on 29 May 2002 by A Whittle

Material: plant macrofossils (mainly *Rhytidiadelphus squarrosus* (Hedw.) Warnst., with some *Calliergon cuspidatum* (Hedw.) Kindb., cf. *Plagiomnium* sp(p), and some moss indet (mainly leafless or otherwise eroded shoots)) (A Hall 2002)

Initial comment: from the surface of a turf from the primary mound.

Objectives: to provide a date for the construction of the primary mound at Silbury Hill.

Calibrated date: 1σ: 1640–1500 cal BC
2σ: 1740–1430 cal BC

Final comment: see OxA-11647

Laboratory comment: Oxford Radiocarbon Accelerator Unit (24 August 2004), acid-wash only.

OxA-14640 3735 ±50 BP

δ¹³C: -28.9‰

Sample: Sample 6 (TS1) B, submitted on 15 March 2005 by A Whittle

Material: plant macrofossils (mainly *Rhytidiadelphus squarrosus* (Hedw.) Warnst., with some cf. *Eurynchium* sp(p), *Neckara complanata* (Hedw.) H√°b., *Pseudoscleropodium purum* (Hedw.) Fleisch, cf. *Plagiomnium* sp(p), and some moss indet (mainly leafless or otherwise eroded shoots) (humic acid)) (A Hall 2005)

Initial comment: a replicate of GrA-28555.

Objectives: as GrA-28555

Calibrated date: 1σ: 2210–2030 cal BC
2σ: 2290–1970 cal BC

Final comment: see series comments

Laboratory comment: Oxford Radiocarbon Accelerator Unit (30 September 2005), the alkali-soluble fraction was dated.

OxA-14641 3898 ±31 BP

δ¹³C: -28.1‰

Sample: Sample 6 TS1b, submitted on 15 March 2005 by A Whittle

Material: plant macrofossils (<2g) (humin; mainly *Rhytidiadelphus squarrosus* (Hedw.) Warnst., with some cf. *Eurynchium* sp(p), *Neckara complanata* (Hedw.) H, *Pseudoscleropodium purum* (Hedw.) Fleisch, cf. *Plagiomnium* sp(p), and some moss indet (mainly leafless or otherwise eroded shoots)) (A Hall 2005)

Initial comment: a replicate of GrA-28465.

Objectives: as GrA-28465

Calibrated date: 1σ: 2470–2300 cal BC
                     2σ: 2480–2280 cal BC

Final comment: see series comments

OxA–14642 3612 ±31 BP

δ¹³C: -28.8‰

Sample: Sample 7 TS2b, submitted on 15 March 2005
by A Whittle

Material: plant macrofossils (<2g) (humic acid; mainly
Rhytidiadelphus squarrosus (Hedw.) Warnst., with some
Calliergon cuspidatum (Hedw.) Kindb., cf. *Plagiomnium* sp(p),
and some moss indet (mainly leafless or otherwise eroded
shoots)) (A Hall 2005)

Initial comment: a replicate of GrA-28467.

Objectives: as GrA-28466

Calibrated date: 1σ: 2030–1920 cal BC
                     2σ: 2120–1880 cal BC

Final comment: see series comments

Laboratory comment: Oxford Radiocarbon Accelerator Unit
(30 September 2005), the alkali-soluble fraction was dated.

OxA–14643 3848 ±31 BP

δ¹³C: -27.8‰

Sample: Sample 7 (TS2) B, submitted on 15 March 2005 by
A Whittle

Material: plant macrofossils (<2g) (humin; mainly
Rhytidiadelphus squarrosus (Hedw.) Warnst., with some
Calliergon cuspidatum (Hedw.) Kindb., cf. *Plagiomnium* sp(p),
and some moss indet (mainly leafless or otherwise eroded
shoots) (A Hall 2005)

Initial comment: a replicate of GrA-28466.

Objectives: as GrA-28466

Calibrated date: 1σ: 2400–2210 cal BC
                     2σ: 2470–2200 cal BC

Final comment: see series comments

Stannon Down, Devon

Location: SX 132803
                   Lat. 50.35.30 N; Long. 04.38.19 W

Project manager: A Jones (Cornwall Archaeological Unit),
                   1998–2000

Description: Stannon Down is situated within a region that
contains many important archaeological remains, including
prehistoric settlements, field systems, and burial cairns.
Roger Mercer excavated part of the middle Bronze Age
settlement with a substantial umber of roundhouses and
field systems in the late 1960s (Mercer 1970). Field
surveyby the Cornwall Archaeological Unit in 1985, revealed
further roundhouses, field walls, and prehistoric burial cairns.
Survey and excavation conducted by the CAU in 1998–2000,
revealed limited Neolithic artefacts, a range of ceremonial
cairns of varying forms, further Bronze Age field boundaries,
and some limited Iron Age activity. The area is on eof
moorland pasture which fell within the pit worklings and
waste disposal area of the Stannon Clay Works.

Objectives: to provide a chreonological framework for the sites
on Stannon Down.

References: Jones 2004
                   Mercer 1970

Stannon Down: site 2, Cornwall

Location: SX 132803
                   Lat. 50.35.30 N; Long. 04.38.19 W

Project manager: A Jones (Cornwall County Council), 1998

Archival body: Royal Cornwall Museum

Description: a small circular cairn, 3.5m in diameter, with a
'tail' extending from its western side. A second phase seems
to have been marked by the digging of a line of postholes
and two pits on its north-eastern side. Five samples from
two major phases of the use of the site were submitted:
layer 27, a placed deposit under the cairn; and pit 30 and
postholes 32, 6, and 11, from the alignment on the north-east
side of the cairn.

Objectives: to confirm the suggested Bronze Age date which
has been inferred from the artefacts; and to confirm the
suggested phasing of the monument as indicated by the
stratigraphy.

Final comment: A Jones (24 October 2013), the radiocarbon
determinations from the cairn confirmed that the site was of
early Bronze Age origin and that it had been the focus for
renewed activity in the form of pits and postholes around the
middle centuries of the second millennium cal BC.

References: Jones 2004

OxA–13385 3385 ±30 BP

δ¹³C: -23.8‰

Sample: Posthole 32 Sample 1035, submitted on
1 December 2003 by A Jones

Material: charcoal: *Quercus* sp., sapwood single fragment
(R Gale 2003)

Initial comment: material was collected from posthole 32.
It was sealed by layer 3 which covered the prehistoric features
on the site.

Objectives: to date the alignment of pits and postholes which
run parallel to site 2. The dates will provide a comparison
with layer 27 which is being obtained from the cairn itself.

Calibrated date: 1σ: 1740–1630 cal BC
                     2σ: 1750–1610 cal BC

Final comment: A Jones (24 October 2013), the resulting
determination was consistent with what was expected and was of
interest as it supported evidence for a subsequent phase of
activity beside the cairn in the later part of the early Bronze Age.

OxA–13386 3254 ±31 BP

δ¹³C: -23.0‰

Sample: Pit 30, Fill 31 Sample 1033, submitted on 24 October 2013 by A Jones

Material: charcoal: *Corylus* sp., single fragment (R Gale 2003)

Initial comment: material was collected from pit 30, which formed part of a structured deposit sealed by a 'capstone'. It contained much charcoal, a snapped flint knife, and an amber bead.

Objectives: to date the alignment of pits and postholes which run parallel to site 2. The dates will provide a comparison with layer 27 which is being obtained from the cairn itself.

Calibrated date: 1σ: 1600–1490 cal BC
  2σ: 1620–1440 cal BC

Final comment: A Jones (24 October 2013), the resulting determination was of interest as it supported evidence for a subsequent phase of activity beside the cairn in the middle centuries of the second millennium cal BC, and provided secure dating evidence for a range of artefacts.

OxA–13387 3919 ±31 BP

δ¹³C: -25.2‰

Sample: Layer 27 Sample 1027, submitted on 1 December 2003 by A Jones

Material: charcoal: *Corylus* sp., single fragment (R Gale 2003)

Initial comment: the sample was sealed beneath a large slab under the body of the cairn. It was a primary deposit and did not appear to have been disturbed.

Objectives: to date the primary use of the monument, which on artefactual grounds is assumed to be early Bronze Age in date. It is also hoped to confirm the phasing of the site by comparing it with dates obtained from pit 30 and postholes 32, 11, and 6, which formed part of an alignment on the north-eastern side of the site.

Calibrated date: 1σ: 2470–2340 cal BC
  2σ: 2490–2290 cal BC

Final comment: A Jones (24 October 2013), the resulting determination was important because it helped establish the date for the construction of the cairn in the early Bronze Age and demonstrated that it was the earliest feature in the cairn complex.

OxA–13388 3223 ±30 BP

δ¹³C: -26.7‰

Sample: Posthole 11 Sample 1014, submitted on 1 December 2003 by A Jones

Material: charcoal: *Corylus* sp., single fragment (R Gale 2003)

Initial comment: material was collected from posthole 11. It was sealed by layer 3 which covered the prehistoric features on the site.

Objectives: to date the alignment of pits and postholes which run parallel to site 2. The dates will provide a comparison with layer 27 which is being obtained from the cairn itself.

Calibrated date: 1σ: 1510–1440 cal BC
  2σ: 1600–1420 cal BC

Final comment: A Jones (24 October 2013), the resulting determination was not consistent with the other dates from the pits and postholes beside site 2. However, it did support the evidence for renewed activity around the cairn in the middle centuries of the second millennium cal BC.

OxA–13389 3274 ±31 BP

δ¹³C: -24.2‰

Sample: Posthole 6 Sample 1011, submitted on 1 December 2003 by A Jones

Material: charcoal: *Corylus* sp., single fragment (R Gale 2003)

Initial comment: material was collected from posthole 6. It was sealed by layer 3 which covered the prehistoric features on the site.

Objectives: as OxA-13388

Calibrated date: 1σ: 1620–1500 cal BC
  2σ: 1630–1450 cal BC

Final comment: A Jones (24 October 2013), the determination was not consistent with the other dates from the pits and postholes beside site 2. However, as with the other dates from the pits and postholes beside the cairn, it did support the evidence for renewed activity around the cairn in the middle centuries of the second millennium cal BC.

Stannon Down: site 6, Cornwall

Location: SX 132802
Lat. 50.35.26 N; Long. 04.38.19 W

Project manager: A Jones (Cornwall County Council), 1998

Archival body: Royal Cornwall Museum

Description: a circular ring-cairn circa 6m in diameter, associated with three phases of use: a primary phase ring-cairn and infilling of pits; infilling of the interior of the site until almost flush with the wall; and erection of a timber ring inside the cairn. The dating of the site is currently based on artefacts. Samples from three postholes from the final phasing of the monument were submitted: posthole 57 was sampled twice.

Objectives: to confirm the suggested Bronze Age date which has been inferred from the artefacts; and to confirm the timber ring was not associated with any later occupation (eg Iron Age) of the site.

Final comment: A Jones (24 October 2013), the four radiocarbon determinations from the ring-cairn were all from the secondary post-ring phase. Three of the dates were not statistically consistent with one another, although two suggested that this phase of the site dated towards the middle centuries of the second millennium cal BC. Taken together the dating suggests that the second phase of the site dates to the end of the early Bronze Age.

References: Jones 2004

OxA–13390 3267 ±31 BP

$\delta^{13}C$: -24.9‰

Sample: Posthole 57, Sample 1010, submitted on 1 December 2003 by A Jones

Material: charcoal: *Corylus* sp., single fragment (R Gale 2003)

Initial comment: material was collected from within the cairn. It was sealed by layer 51 which covered the post-ring phase.

Objectives: to date the timber ring phase of site 6. Three others are being submitted.

Calibrated date: 1σ: 1610–1500 cal BC
                        2σ: 1630–1450 cal BC

Final comment: A Jones (24 October 2013), the determination was broadly what was expected, and was statistically consistent with one of the other dates (OxA-13446), suggesting a phase of activity in the middle centuries in the second millennium cal BC.

OxA–13391 3215 ±30 BP

$\delta^{13}C$: -25.5‰

Sample: Posthole 53 Sample 1007, submitted on 1 December 2003 by A Jones

Material: charcoal: *Corylus* sp., single fragment (R Gale 2003)

Initial comment: material was collected from posthole 53 within the cairn. It was sealed by layer 51 which covered the post-ring phase.

Objectives: as OxA-13390

Calibrated date: 1σ: 1510–1440 cal BC
                        2σ: 1600–1420 cal BC

Final comment: A Jones (24 October 2013), the determination was not statistically consistent with the others in the post-ring but it did confirm a broadly middle second millennium cal BC date for this phase of activity within the ring-cairn.

OxA–13392 3076 ±32 BP

$\delta^{13}C$: -25.2‰

Sample: Posthole 76 Sample 1023, submitted on 1 December 2003 by A Jones

Material: charcoal: *Quercus* sp., sapwood single fragment (R Gale 2003)

Initial comment: material was collected from posthole 76 within the cairn. It was sealed by layer 51 which covered the post-ring phase.

Objectives: to date the timber ring phase of site 6. Three others are being submitted.

Calibrated date: 1σ: 1410–1280 cal BC
                        2σ: 1430–1230 cal BC

Final comment: see OxA-13391

OxA–13446 3200 ±28 BP

$\delta^{13}C$: -25.8‰

Sample: Posthole 57 Sample 1008, submitted on 1 December 2003 by A Jones

Material: charcoal: *Corylus* sp., single fragment (R Gale 2003)

Initial comment: material was collected from posthole 57 within the cairn. It was sealed by layer 51 which covered the post-ring phase.

Objectives: as OxA-13390

Calibrated date: 1σ: 1510–1430 cal BC
                        2σ: 1520–1410 cal BC

Final comment: A Jones (24 October 2013), the determination was broadly what was expected, and was statistically consistent with one of the other dates (OxA-13390) from the post-ring, suggesting a phase of activity in the middle centuries in the second millennium cal BC.

Stannon Down: site 86, Cornwall

Location: SX 143821
                     Lat. 50.36.29 N; Long. 04.37.27 W

Project manager: A Jones (Cornwall County Council), 1998–2000

Archival body: Royal Cornwall Museum

Description: on the north-eastern edge of Stannon Down, a small stream flows through the shallow valley which separates this area from Roughter Moors draining the intervening saddle. At site 86, the stream has cut a peat face some 95cm deep, and at the base of the face a gritty, organic sub-peat soil was exposed with *in situ* tree roots.

Objectives: to provide a chronology for the pollen diagram from site 86, by providing a date for the earliest peat development at the site, and dating three pollen zone boundaries. This pollen diagram provides information about the changing environment of Stannon Down throughout the prehistoric period.

Final comment: A Jones (16 November 2004), the radiocarbon dates from the Stannon pollen column are extremely valuable, as environmental data from the uplands has been limited to a few samples and some site-specific material from other excavations. The length of the sequence (Mesolithic to post-Roman) is also significant because it is the longest to have been obtained from the moor. The inversion in the dates from the early to middle Bronze Age is unfortunate, but this is the period when most activity was taking place - firstly by cairn construction and later by settlement.

References: Jones 2004
                    Mercer 1970

SUERC–3623 6520 ±40 BP

δ¹³C: -29.1‰

Sample: Stan86.1, submitted on 25 May 2004 by H Tinsley

Material: peat (humic acid) (H Tinsley 2004)

Initial comment: the sample is a 10mm slice of peat taken from a horizontal tin extracted from the bottom of a 0.95m stream cut section. The horizontal tin covered the sub-peat soil peat interface at 0.915m. The sample is taken from the base of the peat. 0.895–0.905m below the ground surface.

Objectives: this sample is one of a series for dating critical horizons in the pollen diagram from Stannon 86, which provides a site record for south Bodmin Moor from Mesolithic times to the Dark Ages, and provides environmental context for the recent Bronze Age excavations on Stannon Down. The date for the sub-peat A horizon established a Mesolithic date for the base of this section, and this is an early date for Bodmin. The whole peat sequence at Stannon 86 extends undisturbed up to the Dark Ages. The pollen data demonstrates that, prior to peat formation; woods of hazel and oak were established in this area, these were later replaced by local alder. The objective for dating this specific sample is to establish when peat itself began to form.

Calibrated date: 1σ: 5510–5470 cal BC
2σ: 5550–5380 cal BC

Final comment: A Jones (16 November 2004), evidence dating to the Mesolithic period is rare on Bodmin Moor. There are large flint scatters around Dozmany Pool, but evidence from Stannon is limited to a few unstratified flints. Our understanding of this period is therefore almost entirely dependent on the dating of environmental material. For this reason, the date from the pollen column is significant.

SUERC–3624 3550 ±35 BP

δ¹³C: -29.0‰

Sample: Stan86.2, submitted on 25 May 2004 by H Tinsley

Material: peat (humic acid) (H Tinsley 2004)

Initial comment: this sample is a 10mm slice of herbaceous, topogenous peat taken from a vertical monolith tin which was used to sample a 0.95m stream cut section, with peat overlying a gritty sub-peat soil at 0.915m. This sample is taken at 0.625–0.635m below the ground surface.

Objectives: this sample is one of a series for dating critical horizons in the pollen diagram from Stannon 86, which provides a site record for south Bodmin Moor from Mesolithic times to the Dark Ages, and provides environmental context for the recent Bronze Age excavcations on Stannon Down. This sample is from the horizon of the first sustained expansion of grass pollen, the LPAZ boundary ST86–2b/ST86–2c.

Calibrated date: 1σ: 1940–1830 cal BC
2σ: 2020–1760 cal BC

Final comment: A Jones (16 November 2004), the later Neolithic period/early Bronze Age was a time when monuments appeared across the moor. Site 2 has already been radiocarbon dated to this time, though most of the cairns on the moor are likely to be firmly in the early Bronze

Age (1800–1500 cal BC). The pollen data is therefore significant as it may be linked to the expansion of the ceremonial landscape. The inversion in the later date above is unfortunate as it makes it difficult to separate the later settlement horizon associated with the roundhouses and field systems, and the impact that these events had on woodland and the establishment of grassland. The presence of substantial amounts of mature oak charcoal in the sample from the excavated cairns suggests woodland was a major component of the early Bronze Age landscape.

SUERC–3625 2915 ±35 BP

δ¹³C: -29.0‰

Sample: Stan86.3, submitted on 25 May 2004 by H Tinsley

Material: peat (humic acid) (H Tinsley 2004)

Initial comment: this sample is a 10mm slice of herbaceous, topogenous peat taken from a vertical monolith tin which was used to sample a 0.95m stream cut section, with peat overlying a gritty sub-peat soil at 0.915m. This sample is taken at 0.54–0.55m below the ground surface.

Objectives: this sample is one of a series for dating critical horizons in the pollen diagram from Stannon 86, which provides a site record for south Bodmin Moor from Mesolithic times to the Dark Ages, and provides environmental context for the recent Bronze Age excavcations on Stannon Down. This sample is from the start of maximum values for grass pollen, it is the LPAZ boundary ST86–2c/ST86–3.

Calibrated date: 1σ: 1200–1040 cal BC
2σ: 1220–1000 cal BC

Final comment: A Jones (16 November 2004), the middle Bronze Age (1500–1000 cal BC) is associated with extensive field systems and roundhouse settlements. The recent excavation did not investigate many features dating to this period (half a roundhouse and nine boundaries), but the extent of the field systems means that the dating of the associated environmental horizon is important.

SUERC–3626 2370 ±35 BP

δ¹³C: -28.8‰

Sample: Stan86.4, submitted on 25 May 2004 by H Tinsley

Material: peat (humic acid) (H Tinsley 2004)

Initial comment: this sample is a 10mm slice of herbaceous, topogenous peat taken from a vertical monolith tin which was used to sample a 0.95m stream cut section, with peat overlying a gritty sub-peat soil at 0.915m. This sample is taken at 0.40–0.41m below the ground surface.

Objectives: this sample is one of a series for dating critical horizons in the pollen diagram from Stannon 86, which provides a site record for south Bodmin Moor from Mesolithic times to the Dark Ages, and provides environmental context for the recent Bronze Age excavcations on Stannon Down. This sample is from the horizon where on site alder wood declines, there is increased site wetness and a temporary decline in anthropogenic herbs. This horizon is also the LPAZ boundary ST863/ST86–4.

Calibrated date: 1σ: 420–390 cal BC
2σ: 540–380 cal BC

Final comment: A Jones (16 November 2004), the Iron Age determination is significant as up until the excavation at Stannon there was virtually no evidence for Iron Age occupation on the moor at all. At Stannon site 9 there was evidence for a structure which was constructed inside an earlier ring-cairn. Occupation is likely to have been seasonal and associated with transhumance. The pollen evidence is indicative of a wetter period during the Iron Age. This is perhaps indicated at site 9 by the excavation of a perimeter ditch and a central drain.

Stannon Down: site 9, Cornwall

Location: SX 132802
Lat. 50.35.26 N; Long. 04.38.19 W

Project manager: A Jones (Cornwall County Council), 1999

Archival body: Royal Cornwall Museum

Description: a complex variant ring-cairn which had been reoccupied in the Iron Age. Four phases were identified: a simple ring-cairn possibly associated with two cist graves; covering of the ring-cairn and graves by a double bank and ditch; erection of orthostats on the bank; and Iron Age reuse (post-built structure). Bronze Age and Iron Age finds were recovered. Samples were submitted from two of the site's four phases: 213, the primary deposit of placed charcoal under the kerb; and 238, 258, 226, and 224, Iron Age postholes belonging to a structure.

Objectives: to obtain a date for the primary phase of the monument and to date the Iron Age phase of occupation.

Final comment: A Jones (24 October 2013), the five radiocarbon determinations from the cairn came from two identified phases of activity. The first came from a placed deposit of charcoal under the cairn, and this fell late in the early Bronze Age, towards the middle centuries of the second millennium cal BC from the secondary post-ring phase. The remaining determinations all came from the postholes within the cairn. Two of the determinations fell in the middle centuries of the second millennium cal BC and although not statistically consistent suggest renewed interest in the site in the middle Bronze Age. However, two further determinations fell in the Iron Age, and these are consistent with one another. Taken together, the dating suggests that following the construction of the cairn, there was later interest the site the in the middle Bronze Age, followed by reuse of the site in the Iron Age, when a structure was erected within it. This sequence was also supported by the ceramics from the site.

References: Jones 2004

OxA–13380 2987 ±30 BP

δ¹³C: -25.2‰

Sample: Posthole 224 Sample 1078, submitted on 1 December 2003 by A Jones

Material: charcoal: *Corylus* sp., single fragment (R Gale 2003)

Initial comment: material was collected from posthole 224. It was sealed by layers 201 and 202 which covered the prehistoric phases of the site.

Objectives: to date the timber-ring phase of site 9. It is hoped to confirm the Iron Age date which is indicated by the artefacts.

Calibrated date: 1σ: 1270–1130 cal BC
2σ: 1370–1110 cal BC

Final comment: A Jones (24 October 2013), the determination was broadly what was expected, and was statistically consistent with one of the other dates (OxA-13383) from the post-setting, suggesting a later phase of late Iron Age activity, associated with the construction of a structure within the cairn. This is supported by the ceramic assemblage from the site.

OxA–13381 3127 ±31 BP

δ¹³C: -24.1‰

Sample: Posthole 258 Sample 1085, submitted on 1 December 2003 by A Jones

Material: charcoal: *Quercus* sp., roundwood single fragment (R Gale 2003)

Initial comment: material was collected from posthole 258. It was sealed by layers 201 and 202 which covered the prehistoric phases of the site.

Objectives: as OxA-13380

Calibrated date: 1σ: 1440–1320 cal BC
2σ: 1490–1300 cal BC

Final comment: A Jones (24 October 2013), the determination was later than anticipated and was not statistically consistent with the others in the post-setting. However, it did suggest a broadly middle second millennium cal BC date for some of the postholes within the cairn, which post-dated the primary phase. This is also supported by the ceramic analysis which identified middle Bronze Age ceramics in some of the postholes.

OxA–13382 2183 ±29 BP

δ¹³C: -22.9‰

Sample: Posthole 238 Sample 1084, submitted on 24 October 2013 by A Jones

Material: charcoal: *Quercus* sp., roundwood single fragment (R Gale 2003)

Initial comment: material was collected from posthole 238. It was sealed by layers 201 and 202 which covered the prehistoric phases of the site.

Objectives: as OxA-13380

Calibrated date: 1σ: 360–190 cal BC
2σ: 370–160 cal BC

Final comment: A Jones (24 October 2013), the determination was later than anticipated and was not statistically consistent with the others in the post-setting. However, it did suggest a broadly middle second millennium cal BC date for some of the postholes within the cairn, which post-dated the primary phase. This is also supported by the ceramic analysis which identified middle Bronze Age pottery in some of the postholes.

OxA–13383 2118 ±28 BP

$\delta^{13}C$: -25.0‰

Sample: Posthole 226 SS 1083, submitted on 1 December 2003 by A Jones

Material: charcoal: *?Quercus* sp. (R Gale 2003)

Initial comment: material was collected from posthole 226. It was sealed by layers 201 and 202 which covered the prehistoric phases of the site.

Objectives: as OxA-13380

Calibrated date: 1σ: 200–100 cal BC
2σ: 350–40 cal BC

Final comment: A Jones (24 October 2013), the determination was broadly what was expected, and was statistically consistent with one of the other dates (OxA-13380) from the post-setting, suggesting a later phase of late Iron Age activity, associated with the construction of a structure within the cairn. This is supported by the artefactual patterning.

OxA–13384 3326 ±31 BP

$\delta^{13}C$: -26.6‰

Sample: Layer 213 Sample 1106, submitted on 1 December 2003 by A Jones

Material: charcoal: *Quercus* sp., roundwood single fragment (R Gale 2003)

Initial comment: a deposit of placed charcoal sealed beneath the large kerbstone in the ring-cairn (site 9), in turn sealed by bank 205.

Objectives: to date the primary phase of the monument and this phase of activity.

Calibrated date: 1σ: 1640–1540 cal BC
2σ: 1690–1520 cal BC

Final comment: A Jones (24 October 2013), the resulting determination was important because it helped establish the date for the construction of the site 9 in the latter part of the early Bronze Age, and it demonstrated that the cairn was latter than site 2 which was constructed earlier in the second millennium cal BC.

Suffolk Coastal Survey: Holbrook Bay fishtrap, Suffolk

Location: TM 1715433636
Lat. 51.57.29 N; Long. 01.09.37 E

Project manager: J Plouviez (Suffolk County Council Archaeological Service), 2006

Archival body: Suffolk County Council Archaeological Service

Description: a multi-phased fish weir (STU054) situated in the intertidal zone at Holbrook. The site consists of multiple roundwood stakes, with some evidence of structural wattle. It has been suggested this is one of the largest wooden archaeological features in the UK, certainly in Suffolk.

Objectives: to confirm the date of the fishtrap as Saxon, and to identify phasing within the structure.

Final comment: W Fletcher (20 October 2014), the structure in Holbrook bay is one of the largest known fish weirs in East Anglia and was one considered to be one of the most important structures of its type discovered in recent years. It is very inaccessible, visible only at very low tides and situated some distance from the shore on an extensive shallow mudflat. It was a priority for sampling and further research through dating. Given the size and scale of the structure and the numbers of posts that make up in the structure as a whole, the sampling should be seen as indicative rather than definitive. The results, however, suggest that although the structure may have had been long lived and may have had many phases of use, construction in the seventh century is most likely. The date for this structure is consistent with a number of other sites discovered on the foreshore and dated as part of this programme, in particular two of the site from Awarton, which is a kilometre to the east. The date potentially indicates an increase in the use of fish weirs in the seventh century, and the size of the operation represents an organised approach to its construction.

Laboratory comment: English Heritage (26 September 2007), chronological modelling was undertaken for this fishtrap on the seven samples from STU054. Two samples, UB-5224 and -5225, were excluded from the model as they may represent later re-use of the feature. The model has good overall agreement ($A_{overall}$ =91.8%). The model indicates that construction of this fishtrap occurred in *cal AD 680–850* (*95% probability*) and probably in *cal AD 630–690* (*68% probability*).

References: Everett 2007

UB–5224 1135 ±17 BP

$\delta^{13}C$: -28.2 ±0.2‰

Sample: <1>, submitted in 2006 by J Plouviez

Material: wood (waterlogged): *Alnus glutinosa*, roundwood, 110mm diameter, no bark (175g) (R Gale 2006)

Initial comment: an upright timber from the basket end (eastern) end of the more ephemeral of the two identified features.

Objectives: to establish the date of the structure and to establish phasing within the site as a whole.

Calibrated date: 1σ: cal AD 885–965
2σ: cal AD 880–975

Final comment: W Fletcher (20 October 2014), one of a number of dates taken from the structure with a view to establishing the period of construction and any subsequent re-use. A seventh-century date would be consistent with other dated fish weirs in the area.

UB–5225 1029 ±17 BP

$\delta^{13}C$: -29.1 ±0.2‰

Sample: <3>, submitted in 2006 by J Plouviez

Material: wood (waterlogged): *Alnus glutinosa*, roundwood, 110mm diameter, no bark (191g) (R Gale 2006)

Initial comment: an upright timber from the mid section of the more ephemeral of the two features/phases identified.

Objectives: as UB-5224

Calibrated date: *1σ:* cal AD 990–1025
               2σ: cal AD 985–1030

Final comment: see UB-5224

UB–5227 1312 ±16 BP

δ¹³C: -28.6 ±0.2‰

Sample: <5>, submitted in 2006 by J Plouviez

Material: wood (waterlogged): *Corylus avellana*, roundwood, 45mm diameter, no bark (413g) (R Gale 2006)

Initial comment: the sample was taken from an *in situ* wattle structure associated with the main system of uprights. It was taken roughly from the middle of the southern arm.

Objectives: to establish the date of the structure and to establish phasing within the site as a whole; to establish if the wattle was contemporary with the uprights.

Calibrated date: *1σ:* cal AD 665–690
               2σ: cal AD 660–765

Final comment: see UB-5224

UB–5228 1260 ±20 BP

δ¹³C: -29.5 ±0.2‰

Sample: <6>, submitted in 2006 by J Plouviez

Material: wood (waterlogged): *Corylus avellana*, roundwood, 30mm diameter, no bark (194g) (R Gale 2006)

Initial comment: from the wattle of the main structure at the south-east end.

Objectives: as UB-5224

Calibrated date: *1σ:* cal AD 685–775
               2σ: cal AD 675–775

Final comment: see UB-5224

UB–5229 1269 ±16 BP

δ¹³C: -27.7 ±0.2‰

Sample: <7>, submitted in 2006 by J Plouviez

Material: wood (waterlogged): *Fraxinus excelsior*, sapwood, no bark (247g) (R Gale 2006)

Initial comment: an upright timber from the main structure at the east end.

Objectives: as UB-5224

Calibrated date: *1σ:* cal AD 685–770
               2σ: cal AD 675–775

Final comment: see UB-5224

UB–5230 1287 ±20 BP

δ¹³C: -28.4 ±0.2‰

Sample: <8>, submitted in 2006 by J Plouviez

Material: wood (waterlogged): *Fraxinus excelsior*, sapwood, no bark (146g) (R Gale 2006)

Initial comment: an upright timber from the middle of the main structure.

Objectives: as UB-5224

Calibrated date: *1σ:* cal AD 675–770
               2σ: cal AD 665–775

Final comment: see UB-5224

UB–5231 1323 ±16 BP

δ¹³C: -28.4 ±0.2‰

Sample: <9>, submitted in 2006 by J Plouviez

Material: wood (waterlogged): *Salix/Populus sp.*, roundwood, 90mm diameter, no bark (425g) (R Gale 2006)

Initial comment: an upright timber from the west end of the southern arm of the main structure.

Objectives: to date the main structure; to establish if the main structure is contemporary with the wattle; to establish if the main structure is contemporary with the more ephemeral structure.

Calibrated date: *1σ:* cal AD 660–685
               2σ: cal AD 655–765

Final comment: see UB-5224

Sutton Common: cremations, South Yorkshire

Location: SE 563122
            Lat. 53.36.14 N; Long. 01.08.59 W

Project manager: R Van de Noort (University of Exeter), 1998–9, and 2002–3

Archival body: Doncaster Museum

Description: a 'cemetery' provisionally dated to the late Iron Age or early Roman period, which comprises of *c* 30 geometrically shaped 'mini enclosures'. The shapes were at maximum 3m long and 2m wide, with circular, semi-circular, near square, rectangular, and oblong shapes present. The cemetery was interpreted as a re-use of the marsh fort.

Objectives: to date the cemetery of presumed late Iron Age or early Roman date.

Final comment: R Van de Noort (2007), initially it was thought that the oval-shaped feature containing the cremations was one of the 'mortuary rings' of Iron Age date, despite its large size. The radiocarbon dating has proved beyond doubt that this feature pre-dates the construction of the marsh-fort by 1300 years.

References: Van de Noort *et al* 2007

OxA–14608 3445 ±31 BP

δ¹³C: -19.9‰

Sample: 7074a, submitted on 10 March 2005 by R Van de Noort

Material: calcined human bone (J McKinley 2005)

Initial comment: this sample comes from a closed context thought to represent the secondary deposition of cremated remains.

Objectives: to establish the date of the (secondary) deposition of cremated human and animal bones in what can only be described as a cemetery within the confounds of the earlier marsh fort.

Calibrated date: *1σ:* 1870–1690 cal BC
                       2σ: 1880–1660 cal BC

Final comment: R Van de Noort (2007), the human cremation dates to the early Bronze Age rather than the Iron Age as expected. The radiocarbon dates prove beyond doubt that this feature pre-dates the construction of the marsh-fort by 1300 years, and that it does not belong to the second phase of activity.

Laboratory comment: English Heritage (2007), the two measurements on the cremation (OxA-14608 and SUERC-6147) are statistically consistent (T'=1.1; T'(5%)=3.8; *v*=1; Ward and Wilson 1978) and their weighted mean (3467 ±23 BP) calibrates to 1885–1690 cal BC (95% confidence; Reimer *et al* 2004).

References: Reimer *et al* 2004
                  Ward and Wilson 1978

OxA–14609 2229 ±27 BP

δ¹³C: -17.8‰

Sample: 7246a, submitted on 10 March 2005 by R Van de Noort

Material: animal bone (calcined) (A Outram 2004)

Initial comment: as OxA-14608

Objectives: as OxA-14608

Calibrated date: *1σ:* 370–200 cal BC
                       2σ: 390–200 cal BC

Final comment: P Marshall (November 2014), the date places the animal bone in the Iron Age period and is considerably later than the human cremation.

Laboratory comment: English Heritage (2007), the two measurements (OxA-14609 and SUERC-6147) on the cremated bone are statistically consistent (T'=0.1; T'(5%)=3.8; *v*=1; Ward and Wilson 1978) and their weighted mean (2235 ±21 BP) calibrates to 390–200 cal BC (95% confidence; Reimer *et al* 2004).

References: Reimer *et al* 2004
                  Ward and Wilson 1978

SUERC–6143 3495 ±35 BP

δ¹³C: -19.9‰

Sample: 7074b, submitted on 10 March 2005 by R Van de Noort

Material: calcined human bone (J McKinley 2005)

Initial comment: a replicate of OxA-14608.

Objectives: as OxA-14608

Calibrated date: *1σ:* 1890–1750 cal BC
                       2σ: 1920–1690 cal BC

Final comment: see OxA-14608

Laboratory comment: see OxA-14608

SUERC–6147 2229 ±35 BP

δ¹³C: -22.9‰

Sample: 7246b, submitted on 10 March 2005 by R Van de Noort

Material: calcined animal bone: unidentifiable (A Outram 2004)

Initial comment: a replicate of OxA-14609.

Objectives: as OxA-14608

Calibrated date: *1σ:* 380–200 cal BC
                       2σ: 400–190 cal BC

Final comment: see OxA-14609

Laboratory comment: see OxA-14609

Sutton Common: pollen core, South Yorkshire

Location: SE 564120
                     Lat. 53.36.05 N; Long. 01.08.52 W

Project manager: B Gearey (University of Hull), 2003

Archival body: Doncaster Museum

Description: a series of samples from a sediment core, in the vicinity of an Iron Age marsh fort, which have been analysed palynologically.

Objectives: to determine the ages of the pollen assemblage zones and thereby relate the evidence for environmental change to the archaeological record.

Final comment: B Gearey (25 September 2014), this series of determinations from the pollen core from the palaeochannel was re-analysed using a Bayesian approach and correlated with the archaeological chronology from Sutton Common (Gearey *et al* 2009). This demonstrates that the sequence of radiocarbon determinations is robust but with four determinations removed (SUERC-7612, -7616–7, and -8169).

References: Gearey *et al* 2009
                  Van de Noort *et al* 2007

SUERC–5697 3310 ±40 BP

δ¹³C: -27.6‰

Sample: SC1 0.50m, submitted on 13 December 2004 by B Gearey

Material: wood (waterlogged): *Alnus glutinosa*, narrow roundwood; single fragment (1.49g) (R Gale 2004)

Initial comment: the sample is from a palaeochannel peat deposit (well-humified and woody) adjacent to the Sutton Common enclosure. The context appears undisturbed.

Objectives: to establish the age of the upper sediment of the core, and to determine whether more detailed pollen analysis on these layers is warranted.

Calibrated date: *1σ:* 1640–1520 cal BC
           2σ: 1690–1500 cal BC

Final comment: B Gearey (25 September 2014), this result confirms a Bronze Age date for the deposit.

SUERC–5698 3520 ±35 BP

δ¹³C: -27.8‰

Sample: SC1 0.54m, submitted on 13 December 2004 by B Gearey

Material: wood (waterlogged): *Alnus glutinosa*, narrow roundwood; single fragment (R Gale 2004)

Initial comment: as GU-5697

Objectives: as GU-5697

Calibrated date: *1σ:* 1900–1770 cal BC
           2σ: 1950–1740 cal BC

Final comment: see GU-5697

SUERC–7608 2465 ±35 BP

δ¹³C: -26.3‰

Sample: SC1 0.14m, submitted on 14 July 2005 by B Gearey

Material: waterlogged plant macrofossil: single fragment, bark (R Gale 2004)

Initial comment: as SUERC-8168

Objectives: to establish a chronology for the sequence of environmental changes interpreted from the palynological analysis of the palaeochannel core, and to hence place the archaeological information from the Sutton Common excavations within this picture. This sample will provide a date for the base of the pollen zone SC5, which is thought to reflect the local clearance of woodland during the Iron Age.

Calibrated date: *1σ:* 760–500 cal BC
           2σ: 780–400 cal BC

Final comment: B Gearey (25 September 2014), this result confirms an Iron Age date for the deposit.

SUERC–7612 2495 ±35 BP

δ¹³C: -28.0‰

Sample: SC1 0.58m, submitted on 14 July 2005 by B Gearey

Material: wood (waterlogged): *Alnus* sp., roundwood; single fragment (R Gale 2004)

Initial comment: this sample is from a palaeochannel peat (well-humified and woody) deposit adjacent to the Sutton Common Iron Age enclosure. The context appears undisturbed.

Objectives: to establish a chronology for the sequence of environmental changes interpreted from the palynological analysis of the palaeochannel core, and to hence place the archaeological information from the Sutton Common excavations within this picture. This sample will provide a date for the base of the pollen zone SC4, which is thought to reflect a Bronze Age phase of environmental change.

Calibrated date: *1σ:* 770–540 cal BC
           2σ: 800–480 cal BC

Final comment: B Gearey (25 September 2014), this date is too young for its stratigraphic context and the macofossil is probably intrusive.

SUERC–7613 5140 ±35 BP

δ¹³C: -29.3‰

Sample: SC1 1.30m, submitted on 14 July 2005 by B Gearey

Material: waterlogged plant macrofossil: bark, single fragment (R Gale 2004)

Initial comment: as SUERC-7612

Objectives: to establish a chronology for the sequence of environmental changes interpreted from the palynological analysis of the palaeochannel core, and to hence place the archaeological information from the Sutton Common excavations within this picture. This sample will provide a date for the lower part of pollen zone SC3, which may record the *Ulmus* decline.

Calibrated date: *1σ:* 3980–3950 cal BC
           2σ: 4040–3800 cal BC

Final comment: B Gearey (25 September 2014), this result confirms a Neolithic date for the deposit.

SUERC–7614 4480 ±35 BP

δ¹³C: -28.1‰

Sample: SC1 1.12m, submitted on 14 July 2005 by B Gearey

Material: waterlogged plant macrofossil: *Corylus* sp., single fragment (R Gale 2004)

Initial comment: as SUERC-7612

Objectives: as SUERC-7613

Calibrated date: *1σ:* 3340–3090 cal BC
           2σ: 3360–3020 cal BC

Final comment: see SUERC-7613

SUERC–7615 3880 ±35 BP

δ¹³C: -25.6‰

Sample: SC1 0.70m, submitted on 14 July 2005 by B Gearey

Material: waterlogged plant macrofossil: cf *Alnus* sp., single fragment (R Gale 2004)

Initial comment: as SUERC-7612

Objectives: to establish a chronology for the sequence of environmental changes interpreted from the palynological analysis of the palaeochannel core, and to hence place the archaeological information from the Sutton Common excavations within this picture. This sample will provide a date for the stratigraphic change from organic sediment to woody peat at the top of pollen zone SC3.

Calibrated date: 1σ: 2470–2290 cal BC
                       2σ: 2470–2200 cal BC

Final comment: see SUERC-5697

SUERC–7616 7985 ±35 BP

δ¹³C: -27.6‰

Sample: SC1 1.56m, submitted on 14 July 2005 by B Gearey

Material: waterlogged plant macrofossil: *Betula* sp., single fragment (R Gale 2004)

Initial comment: as SUERC-7612

Objectives: to establish a chronology for the sequence of environmental changes interpreted from the palynological analysis of the palaeochannel core, and to hence place the archaeological information from the Sutton Common excavations within this picture. This sample will provide a date for the base of pollen zone SC6, which may record the alder rise.

Calibrated date: 1σ: 7050–6820 cal BC
                       2σ: 7060–6700 cal BC

Final comment: see SUERC-7613

SUERC–7617 5370 ±35 BP

δ¹³C: -29.2‰

Sample: SC1 1.80m, submitted on 14 July 2005 by B Gearey

Material: wood (waterlogged): *Alnus* sp., roundwood; single fragment (R Gale 2004)

Initial comment: as SUERC-7612

Objectives: to establish a chronology for the sequence of environmental changes interpreted from the palynological analysis of the palaeochannel core, and to hence place the archaeological information from the Sutton Common excavations within this picture. This sample will provide a date for the top of pollen zone SC2, which records the replacement of *Salix* by *Alnus*.

Calibrated date: 1σ: 4320–4170 cal BC
                       2σ: 4340–4050 cal BC

Final comment: B Gearey (25 September 2014), this sample is too young for its stratigraphic context and the macrofossil is probably intrusive.

SUERC–7618 8935 ±35 BP

δ¹³C: -31.0‰

Sample: SC1 3.34m, submitted on 14 July 2005 by B Gearey

Material: waterlogged plant macrofossil: *Corylus* sp., single fragment (R Gale 2004)

Initial comment: from the base of the pollen core.

Objectives: to establish a chronology for the sequence of environmental changes interpreted from the palynological analysis of the palaeochannel core, and to hence place the archaeological information from the Sutton Common excavations within this picture. This sample will provide a date for the base of pollen zone SC1, which defines the base of the pollen diagram.

Calibrated date: 1σ: 8240–8000 cal BC
                       2σ: 8260–7960 cal BC

Final comment: B Gearey (25 September 2014), this result confirms a Mesolithic date for peat inception at this location.

SUERC–7622 8860 ±40 BP

δ¹³C: -30.1‰

Sample: SC1 2.63m, submitted on 14 July 2005 by B Gearey

Material: sediment (11g) (humin) (B Gearey 2004)

Initial comment: this sample is from a water-lain *Limus* or organic mud deposit. The organic matter percentage of this sample has been calculated using LOI as 42%. The deposit appears undisturbed.

Objectives: to establish a chronology for the sequence of environmental changes interpreted from the palynological analysis of the palaeochannel core, and to hence place the archaeological information from the Sutton Common excavations within this picture. This sample will provide a date for the base of pollen zone SC2, which records the *Quercus* rise.

Calibrated date: 1σ: 8210–7940 cal BC
                       2σ: 8230–7790 cal BC

Final comment: B Gearey (25 September 2014), this result confirms a Mesolithic date for this deposit.

Laboratory comment: English Heritage (2006), the two measurements (SUERC-7622 and -8018) on this sediment are statistically consistent (T'=0.7; T'(5%)=3.8; v=1; Ward and Wilson 1978), and so a weighted mean can be taken (8879 ±33 BP).

References: Ward and Wilson 1978

SUERC–8018 8920 ±60 BP

δ¹³C: -29.8‰

Sample: SC1 2.63m, submitted on 14 July 2005 by B Gearey

Material: sediment (11g) (humic acid) (B Gearey 2004)

Initial comment: as SUERC-7622

Objectives: as SUERC-7622

Calibrated date: *1σ:* 8250–7960 cal BC
                      2σ: 8280–7830 cal BC

Final comment: see SUERC-7622

Laboratory comment: see SUERC-7622

SUERC–8168 1730 ±35 BP

δ¹³C: -29.2‰

Sample: SC1 0.0–0.02m, submitted on 14 July 2005 by B Gearey

Material: peat (humic acid) (B Gearey 2004)

Initial comment: this sample is from a palaeochannel deposit adjacent to the Sutton Common Iron Age enclosure.

Objectives: to establish a chronology for the sequence of environmental changes interpreted from the palynological analysis of the palaeochannel core, and to hence place the archaeological information from the Sutton Common excavations within this picture. This sample will provide a date for the top of the pollen zone SC5, which is thought to reflect a later Iron Age period of landscape development.

Calibrated date: *1σ:* cal AD 250–390
                      2σ: cal AD 230–400

Final comment: B Gearey (25 September 2014), this determination was used in the final chronological model (Gearey *et al* 2009).

Laboratory comment: English Heritage (2006), the two measurements (SUERC-8168–9) on this sediment are not statistically consistent (T'=116.6; T'(5%)=3.8; ν=1; Ward and Wilson 1978).

References: Ward and Wilson 1978

SUERC–8169 2265 ±35 BP

δ¹³C: -29.9‰

Sample: SC1 0.0–0.02m, submitted on 14 July 2005 by B Gearey

Material: peat (humin) (B Gearey 2004)

Initial comment: as SUERC-8168

Objectives: as SUERC-8168

Calibrated date: *1σ:* 400–230 cal BC
                      2σ: 400–200 cal BC

Final comment: B Gearey (25 September 2014), this determination is too old for its stratigraphic context and was removed from the final chronological model (Gearey *et al* 2009).

Laboratory comment: see SUERC-8168

Sutton Courtenay, Drayton Road, Oxfordshire

Location: SU 491933
               Lat. 51.38.08 N; Long. 01.17.26 W

Project manager: C Dennis (Oxford Archaeology), 2002

Archival body: Ashmolean Museum

Description: excavations were carried out to investigate a Saxon settlement around Sutton Courtenay on The Thames Valley Second Gravel Terrace. Features examined included pits containing early Saxon pottery, a punishment burial, a timber hall, an early-mid Saxon waterhole, a group of Mesolithic pits, and several Roman ditches.

Objectives: to determine if the burial is contemporary with the early Saxon features in the vicinity.

Final comment: G Hey (2007), the burial dates from the very end of the middle Bronze Age or the beginning of the late Bronze Age, and ore-dates the establishment of the Iron Age settlement by at least 200 years.

References: Hayden and Hey 2008

UB–6031 2890 ±22 BP

δ¹³C: -24.5 ±0.2‰
δ¹³C (diet): -18.8 ±0.3‰
δ¹⁵N (diet): +8.9 ±0.4‰
C/N ratio: 3.4

Sample: DRSCT02 163, submitted in November 2003 by C Dennis

Material: human bone (320g) (right and left humerii, ulnae, and radii) (A Witkin 2003)

Initial comment: the human skeleton 163 was found in an oval cut (162) within a test pit. The skeleton was tightly flexed with the head beneath the thigh bones as though the individual was tightly bound when buried.

Objectives: the very unusual position of the skeleton in this burial indicates that this is a deviant form of interment. All the features in the adjacent trench were of eearly Anglo-Saxon date, but there were no finds in the pit with which to date the burial. It is also possible that it is contemporary with the nearby middle Saxon occupation. Other deviant burials are known from middle Saxon contexts at Yarnton and Higham Ferrers. Dating would help establish the relationship of the burial to nearby features and enable a more useful contribution to be made to an understanding of deviant burials in the early medieval period.

Calibrated date: *1σ:* 1115–1015 cal BC
                      2σ: 1190–1000 cal BC

Final comment: G Hey (2007), the burial is clearly prehistoric rather than Anglo-Saxon in origin as expected. Given that we lack any closely comparable examples, and any very clear picture of the wider context of this burial, it is difficult to arrive at any definite conclusions regarding its significance. It is nonetheless exceptional; firstly because it was articulated, secondly it was tightly bound and flexed, and thirdly it had been placed on its back rather than on its side - another inversion of what may have been more common burial customs in the middle to late Bronze Age.

Taplow Court, Buckinghamshire

Location:	SU 907823
	Lat. 51.31.53 N; Long. 00.41.33 W
Project manager:	T Allen (Oxford Archaeology), 1999–2000
Archival body:	Buckinghamshire County Museums Service

Description: a multi-period hilltop site with Mesolithic and Neolithic struck flint, and several phases of a Bronze Age hilltop enclosure, one ditched. The Bronze Age defences silted up and were overlain by the timber-laced rampart of an Iron Age hillfort, whose ditch lay adjacent. Part of the entrance of the Bronze Age and Iron Age forts lay within the excavations. The Iron Age rampart was fired, and then replaced, the ditch silting up in the Roman and Saxon periods. Saxon domestic material came from the ditch, and a burial and a possible building lay adjacent. This activity is broadly contemporary with the Taplow burial mound close by.

Objectives: to date the first enclosure of the hilltop and clarify the sequence of defensive ditches and palisades; to establish whether internal activity is related to the main sequence; to provide a precise date for the Saxon activity.

Final comment: T Allen (13 October 2014), the Anglo-Saxon dates have demonstrated occupation during the early and middle Saxon periods, significantly augmenting the limited evidence provided by the finds, and most importantly, have confirmed the presence of a double-ditched hillfort surviving into the seventh century AD.

References: Allen *et al* 2009

OxA–14267 2390 ±27 BP

$\delta^{13}C$: -24.0‰

Sample: sample 41, submitted in November 2004 by H Lamdin-Whymark

Material: charcoal: *Quercus* sp., sapwood single fragment (D Challinor and R Gale 2004)

Initial comment: charred timbers 624, part of rampart structure 1113. The timber appears to have been charred *in situ* within the gravels of the rampart core.

Objectives: to establish, in association with samples from contexts 618 and 1000, the date of the rampart structure, thereby dating the construction of the Iron Age hillfort. Furthermore, these dates form part of a run of dates through a well-stratified construction sequence.

Calibrated date: 1σ: 490–400 cal BC
2σ: 540–390 cal BC

Final comment: T Allen (13 October 2014), this sample provides a date for the charred timber.

Laboratory comment: English Heritage (9 January 2007), three samples of oak sapwood were submitted from the charred timbers, two (OxA-14267 and OxA-14295) produced statistically consistent radiocarbon measurements

(T'=1.0; T'(5%)=3.8; v=1, Ward and Wilson 1978) and one (SUERC-4972) produced a modern result (from the 1950 levelling of the site).

References: Ward and Wilson 1978

OxA–14268 3356 ±29 BP

$\delta^{13}C$: -23.4‰

Sample: sample 35, submitted in November 2004 by H Lamdin-Whymark

Material: charcoal (Maloideae), single fragment (D Challinor 2004)

Initial comment: from fill 553 of hollow 642, part of group 1119. The stratigraphically latest pit within a sequence of intercutting pits containing Collared Urn. The pit group is tentatively recorded as cutting post alignment 1107 and is cut by ditch 574.

Objectives: to establish, in association with a date on a charred residue on a Collared Urn sherd, the date of pit 642. The date will also provide a *terminus ante quem* for other pits in the groups and post-alignment 1107. The date will also provide a *terminus post quem* for the cutting of ditch 574.

Calibrated date: 1σ: 1690–1610 cal BC
2σ: 1740–1550 cal BC

Final comment: T Allen (13 October 2014), this sample had no direct functional relationship to hollow 642, stratigraphically the latest in group 1119, and therefore only provides a *terminus post quem* for the context.

OxA–14292 2736 ±26 BP

$\delta^{13}C$: -23.4‰

Sample: sample 4, submitted in November 2004 by H Lamdin-Whymark

Material: grain: *Triticum* sp., single carbonised grain (M Robinson 2004)

Initial comment: from fill 246 of posthole 245, part of posthole alignment 935 in the interior of the hillfort.

Objectives: to establish a date for posthole 245, and in association with a sample from context 244, a date for post alignment 935. This date also forms part of a suite of dates through a well-stratified sequence.

Calibrated date: 1σ: 910–830 cal BC
2σ: 930–810 cal BC

Final comment: T Allen (1 October 2014), this sample helps to date the post-built structures in the interior of the hillfort.

OxA–14293 1224 ±24 BP

$\delta^{13}C$: -25.2‰

Sample: sample 27, submitted in November 2004 by H Lamdin-Whymark

Material: charcoal (Maloideae), single fragment (D Challinor 2004)

Initial comment: from fill 549 of posthole 548. The posthole was part of post alignment 1104 and other postholes in this alignment clearly cut the Bronze Age preserved soil horizon 1105 and were sealed by stabilisation layer 1102 filling the late Bronze Age ditch 574.

Objectives: to establish, in association with a sample from context 535, the date of post alignment 1104 and confirm the stratigraphic relationship with Bronze Age soil 1105 and the stabilisation layer 1102, in ditch 574. Furthermore, to establish a chronological sequence for the construction, modification, and abandonment of palisades, post alignments and ditches.

Calibrated date: 1σ: cal AD 725–860
                     2σ: cal AD 690–890

Final comment: T Allen (13 October 2014), this charcoal is thought to be intrusive in posthole 548.

Laboratory comment: English Heritage (9 January 2007), the two measurements from posthole 548 (SUERC-4969 and OxA-14293) are not statistically consistent (T′=170.5; T′(5%)=3.8; ν=1, Ward and Wilson 1978) and therefore contain material of different ages. The hazelnut shell (SUERC-4969) in posthole 537 is clearly residual, while the fragment of Maloideae (OxA-14293) from posthole 548 is thought to be intrusive.

References: Ward and Wilson 1978

OxA–14294 2851 ±26 BP

δ¹³C: -22.7‰

$\delta^{13}C$: -22.7‰

Sample: sample 46, submitted in November 2004 by H Lamdin-Whymark

Material: charcoal: *Quercus* sp., sapwood single fragment (D Challinor and R Gale 2004)

Initial comment: from postpipe fill 168 of posthole 166, part of post alignment 1108. The postholes in this alignment clearly cut the Bronze Age preserved soil horizon 1105.

Objectives: to establish, in association with samples from contexts 200 and 170, the date of post-alignment 1108. Furthermore, these dates are intended to establish the chronological relationship of post-alignment 1108 to post-alignments 1104 and 1107, palisade 1106, and the late Bronze Age ditch 574.

Calibrated date: 1σ: 1050–970 cal BC
                     2σ: 1120–920 cal BC

Final comment: T Allen (13 October 2014), this material from posthole 166 is thought to be intrusive or residual.

Laboratory comment: English Heritage (9 January 2007), measurements from two of the postholes forming 1108, posthole 166 (OxA-14294) and posthole 198 (SUERC-4967) are not statistically consistent (T′=170.5; T′(5%)=3.8; ν=1, Ward and Wilson 1978) and therefore contain material of different ages.

References: Ward and Wilson 1978

OxA–14295 2428 ±26 BP

δ¹³C: -24.4‰

$\delta^{13}C$: -24.4‰

Sample: sample 40, submitted in November 2004 by H Lamdin-Whymark

Material: charcoal: *Quercus* sp., sapwood single fragment (D Challinor and R Gale 2004)

Initial comment: charred timbers 618, part of rampart structure 1113. The timbers appear to have been charred *in situ* within the gravels of the rampart core.

Objectives: to establish, in association with samples from contexts 624 and 1000, the date of timber rampart structure, thereby dating the construction of the Iron Age hillfort. Furthermore, these dates form part of a run of dates through a well-stratified construction sequence.

Calibrated date: 1σ: 730–410 cal BC
                     2σ: 750–400 cal BC

Final comment: see OxA-14267

Laboratory comment: see OxA-14267

OxA–14296 2508 ±27 BP

δ¹³C: -26.9‰

$\delta^{13}C$: -26.9‰

Sample: sample 67, submitted in November 2004 by H Lamdin-Whymark

Material: charcoal: *Corylus/Alnus* sp., single fragment (D Challinor 2004)

Initial comment: from fill 1018 of posthole 1016. The substantial posthole forms part of the entrance structure of the Iron Age hillfort. The posthole contained two fills. Charcoal was recovered from the lower fill 1018 in sample 67.

Objectives: to establish the date of the posthole 1016 and in relation to samples from the charred timbers from the rampart, date the construction of the second phase hillfort.

Calibrated date: 1σ: 780–540 cal BC
                     2σ: 790–530 cal BC

Final comment: T Allen (13 October 2014), this sample provides a date for the context.

OxA–14297 1258 ±24 BP

δ¹³C: -26.1‰

$\delta^{13}C$: -26.1‰

Sample: sample 21, submitted in November 2004 by H Lamdin-Whymark

Material: charcoal: *Corylus/Alnus* sp., single fragment (D Challinor 2004)

Initial comment: from fill 511 of posthole 510, part of post alignment 1107. The post-alignment also appeared to be sealed by the fills of the early Bronze Age pit group 1119. The two potsherds from the post-alignment, however, included one indeterminate sherd, and one probable Iron Age sherd. The alignment also ran parallel and adjacent to a further post-alignment and palisade of the Bronze Age.

Objectives: to establish, in association with a sample from context 507, the date of post-alignment 1107 and confirm the stratigraphic relationship with early Bronze Age pit group 1119 and overlying Bronze Age soil 1105. Furthermore, to establish a chronological sequence for the construction, modification, and abandonment of the palisades, post-alignments, and ditches.

Calibrated date: 1σ: cal AD 685–775
2σ: cal AD 675–780

Final comment: T Allen (13 October 2014), the measurements on charcoal from the two postholes of the alignment (OxA-14297 and SUERC-5150) are Anglo-Saxon in date rather than Iron Age as thought. It is therefore suggested that the material was intrusive and originated from Saxon activity in the vicinity of the posthole row.

Laboratory comment: English Heritage (9 January 2007), the measurements on charcoal from two different postholes [510 and 506] of this alignment (OxA-14297 and SUERC-5150) are statistically consistent (T'=1.2; T'(5%)=3.8; ν=1, Ward and Wilson 1978), however, they are Anglo-Saxon in date.

References: Ward and Wilson 1978

OxA–14357 2803 ±27 BP

$\delta^{13}C$: -21.7‰

Sample: sample 3, submitted in November 2004 by H Lamdin-Whymark

Material: grain: *Triticum dicoccum/spelta*, single carbonised grain (M Robinson 2004)

Initial comment: from fill 244 of posthole 244, part of posthole row 935 in the interior of the hillfort.

Objectives: to establish a date for posthole 243, and in association with a sample from context 246, a date for post-alignment 935. This date also forms part of a suite of dates through a well-stratified sequence.

Calibrated date: 1σ: 1010–910 cal BC
2σ: 1020–890 cal BC

Final comment: T Allen (13 October 2014), this sample provided a date for the post-built feature in the interior of the hillfort.

OxA–14358 3120 ±30 BP

$\delta^{13}C$: -23.0‰

Sample: sample 10, submitted in November 2004 by H Lamdin-Whymark

Material: grain: *Triticum* sp., single carbonised grain (M Robinson 2004)

Initial comment: from fill 283 of posthole 279, part of pit alignment 1132 in the interior of the hillfort.

Objectives: to establish a date for posthole 279 and indirectly for post-alignment 1132. This date also forms part of a suite of dates through a well-stratified sequence.

Calibrated date: 1σ: 1430–1320 cal BC
2σ: 1450–1290 cal BC

Final comment: T Allen (13 October 2014), this sample provided a date for the post-built feature in the interior of the hillfort. However, the evidence suggests that the posthole row 1132 and palisade trench 1106 date to the late Bronze Age, implying that the material dated here to the middle Bronze Age was residual.

OxA–14359 2687 ±27 BP

$\delta^{13}C$: -23.0‰

Sample: sample 49, submitted in November 2004 by H Lamdin-Whymark

Material: grain: *Triticum spelta*, single carbonised grain (M Robinson 2004)

Initial comment: from the sole fill 207 of posthole 206, part of roundhouse 1117 in the interior of the hillfort.

Objectives: to establish the date of the *Spelta* wheat, a late Bronze Age date would represent an early occurrence. The date will also function to date the posthole and roundhouse.

Calibrated date: 1σ: 840–800 cal BC
2σ: 910–800 cal BC

Final comment: see OxA-14357

OxA–14432 1451 ±30 BP

$\delta^{13}C$: -22.6‰

Sample: context no. 711, submitted in November 2004 by H Lamdin-Whymark

Material: animal bone: *Equus* sp., articulating radius and ulna (E Evans 2004)

Initial comment: layer 711 from upper fill of the hillfort ditch 460. Layer 711 contained Saxon pottery and a lava quern. The silting of the ditch may have incorporated residual finds, but in general the finds assemblage was consistent with a Saxon date. The horse radius and ulna articulate and clearly originated from the same animal, though they were not identified as articulated during excavation.

Objectives: to establish a date for layer 711 in the hillfort ditch and indirectly provide a date for Saxon activity on-site. Furthermore, this layer is dated as part of a dating programme of a stratigraphic sequence of the late Bronze Age and Iron Age defences.

Calibrated date: 1σ: cal AD 580–650
2σ: cal AD 550–660

Final comment: T Allen (13 October 2014), this sample provides a date for the Saxon silting of the ditch.

SUERC–4963 1390 ±40 BP

$\delta^{13}C$: -20.3‰
$\delta^{15}N$ *(diet):* +9.9‰
C/N ratio: 3.4

Sample: skeleton 107, submitted in November 2004 by H Lamdin-Whymark

Material: human bone (left femur) (A Boyle 2004)

Taplow Court, Buckinghamshire

Initial comment: from grave 105 in the hillfort entrance. The skeleton was fully articulated and had not been dispersed, bar the removal of the head by ploughing.

Objectives: to establish the date of the skeleton, particularly given the proximity to the 'princely' Saxon burial mound. Furthermore, in its association with dates on the upper fills of the ditch, this sample will indirectly function to broadly indicate the date of Saxon activity on site.

Calibrated date: 1σ: cal AD 630–670
2σ: cal AD 590–680

Final comment: T Allen (13 October 2014), the result shows the burial is Saxon in date.

SUERC–4967 3415 ±35 BP

δ¹³C: -25.7‰

Sample: sample 45, submitted in November 2004 by H Lamdin-Whymark

Material: charcoal: *Corylus/Alnus* sp., single fragment (D Challinor 2004)

Initial comment: from postpipe fill 200 of posthole 198, part of post alignment 1108. Posthole 198 is, however, part of the alignment and other postholes in this alignment clearly cut the Bronze Age preserved soil horizon.

Objectives: to establish, in association with samples from contexts 168 and 170, the date of post-alignment 1108. Furthermore, these dates are intended to establish the chronological relationship of post-alignment 1108 to post-alignments 1104 and 1107, palisade 1106, and the late Bronze Age ditch 574.

Calibrated date: 1σ: 1750–1660 cal BC
2σ: 1880–1620 cal BC

Final comment: T Allen (13 October 2014), this material from posthole 198 is thought to be residual.

Laboratory comment: see OxA-14294

SUERC–4968 2700 ±40 BP

δ¹³C: -23.8‰

Sample: sample 49, submitted in November 2004 by H Lamdin-Whymark

Material: grain: *Triticum spelta*, single carbonised grain (M Robinson 2004)

Initial comment: from fill 207 of posthole 206, part of roundhouse 1117 in the interior of the hillfort.

Objectives: to establish the date of the *Spelta* wheat, as a late Bronze Age date would represent an early occurrence. The date will also function to date the posthole and roundhouse 1117.

Calibrated date: 1σ: 900–800 cal BC
2σ: 930–790 cal BC

Final comment: see OxA-14357

SUERC–4969 9220 ±40 BP

δ¹³C: -24.9‰

Sample: sample 25, submitted in November 2004 by H Lamdin-Whymark

Material: carbonised plant macrofossil (*Corylus* sp., single hazelnut shell) (D Challinor 2004)

Initial comment: from fill 535 of posthole 537, part of post-alignment 1104. Posthole 537 is part of post-alignment 1104, and other postholes in this alignment clearly cut the Bronze Age preserved soil horizon 1105, and were sealed by stabilisation layer 1102 filling the late Bronze Age ditch 574.

Objectives: to establish, in association with a sample from context 548, the date of post-alignment 1104 and confirm the stratigraphic relationship with Bronze Age soil 1105 and the stabilisation layer 1102, in ditch 574. Furthermore, to establish a chronological sequence for the construction, modification, and abandonment of palisades, post-alignments, and ditches.

Calibrated date: 1σ: 8540–8320 cal BC
2σ: 8570–8290 cal BC

Final comment: T Allen (13 October 2014), this hazelnut shell was clearly residual, attesting to Mesolithic activity in the area.

Laboratory comment: see OxA-14293

SUERC–4970 3020 ±40 BP

δ¹³C: -24.0‰

Sample: sample 30, submitted in November 2004 by H Lamdin-Whymark

Material: charcoal: *Corylus/Alnus* sp., single fragment (D Challinor 2004)

Initial comment: from post-packing 571 within intervention 569, in palisade trench 1106. The palisade trench clearly cuts preserved soil 1105 and post-pipes within the palisade are abutted by the rampart gravels 1120 associated with the cutting of ditch 574.

Objectives: to establish the date of palisade trench 1106 and, in association with the dates from ditch 574 and adjacent stratified palisades, establish a chronology for the construction sequence for the enclosures and activity on the hilltop.

Calibrated date: 1σ: 1380–1210 cal BC
2σ: 1410–1120 cal BC

Final comment: T Allen (13 October 2014), the date is inconsistent with its stratigraphic position, and it seems likely that the charcoal was residual.

SUERC–4971 1165 ±35 BP

δ¹³C: -22.9‰

Sample: sample 51, submitted in November 2004 by H Lamdin-Whymark

Material: grain: *Secale cereale*, single fragment (D Challinor 2004)

Initial comment: from upper fill 579 of hillfort ditch 460. Layer 579 contained Saxon pottery, a lava quern, and was rich in charred remains. The silting of the ditch may have incorporated residual finds, but in general the composition of the charred assemblage was consistent with a Saxon date.

Objectives: to establish the date of the grain assemblage recovered from the ditch and indirectly date the Saxon activity. Furthermore, this date is the latest in the stratigraphic sequence of late Bronze Age and Iron Age defences.

Calibrated date: 1σ: cal AD 770–950
                      2σ: cal AD 770–980

Final comment: see OxA-14432

SUERC–4972 1.2073 ±0.0053 fM

δ¹³C: -25.2‰

Sample: sample 68, submitted in November 2004 by H Lamdin-Whymark

Material: charcoal: *Quercus* sp., sapwood single fragment (D Challinor and R Gale 2004)

Initial comment: charred timber 1000, part of rampart structure 1113. The timber appears to have been charred *in situ* within the gravels of the rampart core. It appeared to represent a halved timber set directly over the silt stabilisation layer 1102.

Objectives: to establish, in association with samples from contexts 618 and 624, the date of the timber rampart structure, thereby dating the construction of the Iron Age hillfort. Furthermore, these dates form part of a run of dates through a well-stratified construction sequence.

Calibrated date: 1σ: cal AD 1960–1986
                      2σ: 1 cal AD 1959–1987

Final comment: T Allen (13 October 2014), this sample obviously derived from the 1950 levelling of the site, and not the Iron age rampart as thought.

Laboratory comment: see OxA-14267

Laboratory comment: English Heritage (4 June 2015), this result has been calibrated using the post-1950 calibration curve for the northern hemisphere (zone 1) compiled by Hua *et al* (2013). The resultant distribution is extremely bimodal, suggesting that the sample dates either to cal AD 1959–1961 (32% probability) or to cal AD 1983–1986 (63% probability) (Stuiver and Reimer 1993).

References: Hua *et al* 2013
                Stuiver and Reimer 1993

SUERC–5150 1305 ±35 BP

δ¹³C: -24.4‰

Sample: sample 23, submitted in November 2004 by H Lamdin-Whymark

Material: charcoal (Maloideae), single fragment (D Challinor 2004)

Initial comment: from fill 507 of posthole 506, part of post-alignment 1107. The posthole cuts natural gravel and was sealed by preserved soil horizon 1105 of Bronze Age date. The post-alignment also appeared to be sealed by the fills of the early Bronze Age pit group 1119. The two pot sherds from the post-alignment, however, included one indeterminate sherd and one probable Iron Age sherd. The alignment also ran parallel and adjacent to a further post-alignment and palisade of the Bronze Age.

Objectives: to establish the date of post-alignment 1107 and confirm the stratigraphic relationship with early Bronze Age pit group 1119 and overlying Bronze Age soil 1105. Furthermore, to establish a chronological sequence for the construction, modification, and abandonment of palisades, post-alignments, and ditches.

Calibrated date: 1σ: cal AD 660–770
                      2σ: cal AD 650–780

Final comment: see OxA-14297

Laboratory comment: see OxA-14297

Tower of London: Royal Menagerie, Greater London

Location: TQ 33438053
               Lat. 51.30.25 N; Long. 00.04.40 W

Project manager: H O'Reegan (University of Nottingham), 1936–7 and 1999

Archival body: Natural History Museum

Description: the Royal Menagerie was one of the earliest, and the longest-running, animal collections in the British Isles. It held lions, bears, and other exotic beasts and was moved from Woodstock, near Oxford, to the Tower of London by King John (AD 1199–1216 (Parnell 1999). The Menagerie was finally closed on the orders of the Duke of Wellington in 1835 after the remaining animals had been moved to the newly created Zoological Gardens in Regents' Park in 1831 (Keeling 2001). Although the Royal Menagerie is known from documentary evidence, few physical remains survive and the whereabouts of the early Menagerie within the Tower are uncertain. The only clues to its fourteenth century location are references to the creation of a lock and key 'towards the lions and leopard' and the 'lion turret' (Parnell 1999). These documents may refer to the semi-circular structure built by Edward I in

AD 1276–7 in the south-western corner of the Tower of London and referred to as the 'Lion Tower' from AD 1531, but this cannot be confirmed. Certainly by the sixteenth century the Menagerie was housed there, but following the closure of the collections the Lion Tower itself was demolished in the 1850s. It was partially uncovered by an excavation conducted by the Ministry of Works in 1936–7 (Harvey 1947) and again in a small excavation in 1999 (Keevill 2004).

Some of the animals are known through illustrations and documentary evidence, but we have no information about their physical condition. However, during the 1937 excavation a quantity of animal skulls were recovered and deposited in the Natural History Museum, London.

Amongst these specimens are three big cats, two lions (*Panthera leo*) and one leopard (*Panthera pardus*). These specimens are extremely important as they are potentially the only remains of big cats to be recovered from the medieval period in the UK. Nineteen dog (*Canis familiaris*) crania were also recovered from the same deposit.

Objectives: the three big cat specimens represent the only potentially medieval big cat remains in the UK, and as such they are very important for our understanding of early menageries and the wild beast trade. However, very few records survive from the 1936–7 excavation, and only a few rudimentary plans exist in the NMR, Swindon. In the absence of a published report or excavation notes, radiocarbon dating the cat skulls is the only way of determining the true age of the remains. Once these dates have been obtained they will be used to guide the documentary researches. This will greatly improve our current scant knowledge of early zoo history in the UK.

Final comment: H O'Regan (20 December 2005), the radiocarbon dating has established that the animal remains range in date from the thirteenth to seventeenth centuries, making them the earliest post-Pleistocene big cat remains in Britain.

We have established through the radiocarbon dates that the two lions are the only medieval big cat remains to have been found in the UK, and all three cats are amongst only a handful of post-Pleistocene big cats to have been uncovered in this country. As such they provide an important insight into how exotic animals were kept, and people's attitudes to their disposal in the medieval period.

There was a partially occluded foramen magnum of one of the lions. This anomaly has also been noted in captive and unprovenanced cat skulls from the early twentieth century, indicating that it is a condition with a long history.

It is obvious that the specimens at the National History Museum are from a number of different time periods and cannot be considered to be associated other than through the mention of 'bones' on an unpublished plan dated June 1936, and that they were found during the same excavation. Based on the stable isotope analyses of nitrogen all animals appear to have had a protein rich (meat-based) diet, including the dog. This suggests that medieval and later menagerie animals were fed a meat rich diet, unlike those omnivorous mammals, particularly bears, which historical records report were fed on bread in the eighteenth and nineteenth centuries (McCann 1994; Keeling 2001).

References: Harvey 1947
Keeling 2001
Keevill 2004
O'Regan *et al* 2006
Parnell 1999

OxA–13495 386 ±25 BP

$\delta^{13}C$: -18.6‰
$\delta^{15}N$ *(diet):* +12.6‰
C/N ratio: 3.2

Sample: leopard, 2001.28, submitted on 15 March 2004 by H O'Regan

Material: animal bone: Panthera pardus, leopard maxilla (H O'Regan 2002)

Initial comment: very few records survive with the exception of a paper by Harvey (1947) which draws on the 1930s excavations to outline the history of the western entrance of the Tower, and shows that the area in which the skulls were found. It was originally part of the Spur Ditch portion of the Tower Moat, which encircled the Barbican and was in-filled *c* AD 1685. The Spur Guard was then built on this area and an extension to it together with the Spur Barracks added in the AD 1780s. A hand-annotated plan of the 1936–7 excavation mentions that bones were found adjacent to the Spur Guard extension, in a layer 1–4 feet 'above general modern ditch level'. Since no other bones are mentioned it is logical to assume that these are the cat and dog skulls in question, but the reference to depth does not make it clear whether they came from the medieval moat sediments or the post-medieval infill, and secure dating may resolve this issue. There is a possibility that the skulls were deposited from elsewhere as part of the infilling material, but they do not exhibit abrasion or rolling and several crania are associated with their mandibles, suggesting that they are from a primary context.

Objectives: this is potentially the only medieval leopard specimen to have been found in archaeological deposits in the UK. This specimen is not as complete as the lion skull (only the forepart of the cranium remains), but it is currently unique in the UK and as such it is important to obtain a date on it. The date will help to direct the documentary studies, as it will allow us to narrow down the dates of interest when examining the historical records. It will also form part of our analysis of the health of animals in early menageries in the UK and provide an important date for at least one inhabitant of England's longest running menagerie. The associated delta carbon and delta nitrogen values may also yield interesting information on the animal's diet, as it appears that carnivores were quite commonly fed on bread in the medieval period. Documentary evidence to this effect exists for bears and dogs but the diet of the cats is not known.

Calibrated date: 1σ: cal AD 1450–1620
2σ: cal AD 1440–1630

Final comment: H O'Regan (20 December 2005), given the level of damage to the skeleton it is possible this specimen was redeposited from elsewhere. The radiocarbon date covers the end of the Plantagenet reign, through the Tudors, to the beginnings of the Stuart dynasty and the post-medieval period. It has a high nitrogen isotope signature indicating it has a protein-rich (meat) diet.

OxA–13496 450 ±24 BP

$\delta^{13}C$: -19.5‰
$\delta^{15}N$ *(diet):* +9.9‰
C/N ratio: 3.3

Sample: Lion, 1952.10.20.16, submitted on 15 March 2004 by H O'Regan

Material: animal bone: *Panthera leo*, buccal surface from left mandible (H O'Regan 2002)

Initial comment: as OxA-13495

Objectives: to obtain a date on what is potentially one of only two lion skulls from the medieval period known from UK deposits. Lion 16 comprises an almost complete cranium and associated mandibles. The radiocarbon date will help to direct the documentary studies, as it will allow us to narrow down the dates of interest when examining the historic records. This specimen exhibits interesting non-metric traits, including extra cusps on the lower M1s and a pathological bone thickening around the foramen magnum. An analysis and description of these features will form an important part of our analysis on the health of animals in early menageries. The associated delta carbon and delta nitrogen values may also yield interesting information on the animal's diet, as it appears that carnivores were quite commonly fed on bread in the medieval period. Documentary evidence to this effect exists for bears and dogs but the diet of the cats is not known.

Calibrated date: 1σ: cal AD 1435–1450
                       2σ: cal AD 1425–1460

Final comment: H O'Regan (20 December 2005), this sample dated to a period from the end of the reign of Henry V to the start of the reign of Richard III, covering much of the War of the Roses. In AD 1496, the Chronicles of London report that all the lions in the Tower died (Thomas 1996), suggesting there was some ill-health in the population, and clearly this specimen may have been among that group. The high nitrogen value suggests that the cat was fed on a protein-rich diet (meat).

OxA–13497 228 ±24 BP

$\delta^{13}C$: -19.9‰
$\delta^{15}N$ *(diet):* +10.6‰
C/N ratio: 3.2

Sample: Domestic dog: 1969.399.3, submitted on 15 March 2004 by H O'Regan

Material: animal bone: *Canis familiaris*, zygomatic arch (H O'Regan 2002)

Initial comment: as OxA-13495

Objectives: to obtain a date on one of 19 dog skulls which were recovered from the same site as the three big cat skulls. Dog specimen 3 is an almost complete cranium with associated mandibles, which suggests it has not been redeposited. The dog specimen will make a useful comparison to the cats, and will also potentially form part of a later study of oxygen isotopes in medieval dog specimens, to see if they can be provenanced, as historical records exist which suggest that they may have been collected from all over England to be taken to London for animal baiting. A date on this specimen will also form part of our analysis of the health of animals in early menageries, this study will include both the dog and big cat remains. At least four of the 19 dog specimens exhibit pathologies. The associated delta carbon and delta nitrogen values may also yield interesting information on the animal's diet, as it appears that carnivores were quite commonly fed on bread (particularly bears and dogs) in the medieval period.

Calibrated date: 1σ: cal AD 1650–1800
                       2σ: cal AD 1640–1955*

Final comment: H O'Regan (20 December 2005), this radiocarbon date on the most complete dog skeleton has a range of 300 years because it hit a plateau on the radiocarbon calibration curve. Such a range offers little insight, and it is not clear from the excavation records whether all these dogs are contemporaneous, or what their temporal relationship was with the big cats. Given the variety of preservation and colouration it is unliley they all date from the same period. This animal has a high nitrogen signature, suggesting that it also had a diet rich in protein.

OxA–13498 686 ±25 BP

$\delta^{13}C$: -19.4‰
$\delta^{15}N$ *(diet):* +11.1‰
C/N ratio: 3.2

Sample: Lion, 1952.10.20.15, submitted on 15 March 2004 by H O'Regan

Material: animal bone: *Panthera leo*, lingual surface of right mandible (H O'Regan 2002)

Initial comment: as OxA-13495

Objectives: lions are often discussed in historical documents, but this sample represents one of only two potentially medieval specimens that have been recovered in the UK. This specimen comprises a completely cranium and associated mandibles and is in excellent condition. The radiocarbon date will help to direct the documentary studies, as it will allow us to narrow down the dates of interest when examining the historical records. Analysis of this specimen will form part of our analysis of the health of animals in early menageries in the UK and provide an important date for at least one inhabitant of the England's longest running menagerie. The associated carbon and nitrogen stable isotope values may also yield interesting information on the animal's diet, as it appears that carnivores were quite commonly fed on bread in the medieval period. Documentary evidence to this effect exists for bears and dogs but the diet of the cats is not known.

Calibrated date: 1σ: cal AD 1280–1300
                       2σ: cal AD 1270–1390

Final comment: H O'Regan (20 December 2005), this result confirms that this specimen is the earliest medieval big cat in the country. This period covers the reigns of Edward I, II, and III. It is a complete skull, retaining even delicate bones such as the ethmoturbinals and the pterygoid processes, suggesting that it was found close to where it was deposited. The animal was a young adult when it died, based on tooth wear comparisons with modern lions, and it had no obvious abnormalities. Records from AD 1314 show that the cats were provided with one quarter of mutton per day (Parnell 1999), and the high delta nitrogen value for this specimen does indicate that it had a protein-rich (meat) diet.

Trevelgue Head, Cornwall

Location: SW 827630
                     Lat. 50.25.33 N; Long. 05.03.35 W

Project manager: J Nowakowski (Cornwall Archaeology Unit), 2003

Archival body: Cornwall Royal Museum

Description: Trevelgue Head is an impressive promontory cliff-castle located on the north Cornish coast on the outskirts of Newquay. The site is defined by a spectacular series of large earth and stone ramparts, which embrace the remains of an Iron Age settlement.

Objectives: to contribute to the overall site chronology; to test the stratigraphic and phasing sequence suggested for the formation of the middens in trench 61; to refine the dating of the sequence within Middle Iron Age South Western Decorated ware for which data from Trevelgue allows for the first time to be proposed; to date a major phase of ironworking activity; to maximise the contribution of the data from Trevelgue to our understanding of the first millennuim cal BC in South West Britain.

Final comment: P Marshall (2011), the radiocarbon dating prgramme has been successful in contributing to the construction of the overall site chronology and in providing estimates for the dates of key contexts and structures. For the first time scientific dating has been used to support the typological development of styles within the South Western Decorated ceramic group. The dating of carbonised residues on sites with good pottery sequences may in the futue therefore provide more precise estimates for the development of pottery styles. The radiocarbon dating programme has also helped further our understanding of the first millennium cal BC in South West Britain.

References: Christie 1985
                  Christie 1988
                  Nowakowski and Quinnell 2011

OxA–13191 2165 ±50 BP

δ¹³C: -25.9‰

Sample: 90257B, submitted on 3 December 2003 by J A Nowakowski

Material: carbonised residue (H Quinnell 2002)

Initial comment: two conjoining sherds with residues recovered from a low layer within midden 3 (251, trench 61. Deeply buried and pottery unabraided therefore highly likely to be *in situ*. The layer was well sealed and deeply buried brown earth midden layer.

Objectives: a significant vessel typologically. A successful date from this sample will enhance current understanding of ceramics for the Iron Age period and contribute new date regionally as well as nationally. To provide spot date to help date the midden formation in trench 61, and to contribute to a series of dates of majority iron activity on the cliff-castle. Part of a paired sample of ceramic residues from midden 3 (251) trench 61.

Calibrated date: 1σ: 360–160 cal BC
                       2σ: 380–50 cal BC

Final comment: P Marshall (2011), one of three samples (OxA-13191, -13192, and -14012) from midden 3 [251] from an assemblage of Outline vessels. The results suggest this sample was not contemporary with midden 3.

Laboratory comment: English Heritage (2011), the three measurements from midden 3 [251] are not statistically consistent (T'=19.6; T'(5%)=6.0; v=2; Ward and Wilson

1978), however, OxA-13191 and OxA-13192 are (T'=1.2; T'(5%)=3.8; v=1; Ward and Wilson 1978), suggesting the residue on sherd 90275B (OxA-14012) is not contemporary with the formation of midden 3.

References: Ward and Wilson 1978

OxA–13192 2225 ±24 BP

δ¹³C: -27.2‰

Sample: 90325, submitted on 3 December 2003 by J A Nowakowski

Material: carbonised residue (H Quinnell 2002)

Initial comment: as OxA-13191

Objectives: as OxA-13191

Calibrated date: 1σ: 365–205 cal BC
                       2σ: 385–200 cal BC

Final comment: see OxA-13191

Laboratory comment: see OxA-13191

OxA–13224 2377 ±31 BP

δ¹³C: -26.4‰

Sample: 90463b, submitted on 3 December 2003 by J A Nowakowski

Material: charcoal: *Corylus* sp., roundwood single fragment (R Gale 2002)

Initial comment: pit (249) was a massive feature cut into the bedrock. It was deep (up to 1m) and well-sealed. It has been interpreted as a furnace used in metalworking processes. This sample was recovered from the base of the feature and one of a pair of samples submitted for dating (the other is 90316).

Objectives: to confirm a major phase of activity on the cliff-castle during the middle Iron Age (sixth to second century BC). This date is one of a pair submitted from this furnace and is likely to be contemporary with other dates submitted as part of this dating programme - especially those from area 62 and 62A.

Calibrated date: 1σ: 420–400 cal BC
                       2σ: 540–390 cal BC

Final comment: P Marshall (2011), the ceramics associated with the industrial features in trench 62 comprise a mixture of all South Western Decorative styles.

Laboratory comment: English Heritage (2011), the two charcoal samples (OxA-13312 and OxA-13224) from furnace [249] are statistically consitent (T'=1.7; T'(5%)=3.8; v=1; Ward and Wilson 1978).

References: Ward and Wilson 1978

OxA–13277 2198 ±27 BP

δ¹³C: -27.6‰

Sample: 90453, submitted on 3 December 2003 by J A Nowakowski

Material: carbonised residue (H Quinnell 2002)

Initial comment: the sample comprises two joining vessel sherds with carbonised residue and came from a small probable domes hearth [293]. This was a sealed deposit found under a wall and midden - the midden contained Iron Age pottery. The hearth lay in a rock-cut hollow.

Objectives: to confirm the major phase of activity on the cliff-castle during the middle Iron Age. A date here is likely to be contemporary with other sample submitted for dating from trenches 61 and 62. Successful dates from key ceramics groups from Trevelgue Head will enhance research of regional and national understanding of ceramic manufacture and technology for this period.

Calibrated date: 1σ: 360–200 cal BC
                     2σ: 370–170 cal BC

Final comment: P Marshall (2011), this context comes from a feature stratified below midden [296] with Outline pottery.

OxA–13278 2144 ±27 BP

δ¹³C: -27.4‰

Sample: 90361, submitted on 3 December 2003 by J A Nowakowski

Material: carbonised residue (H Quinnell 2002)

Initial comment: carbonised residue on single sherd of a significant vessel type (typologically). Found in a sealed posthole cut deep into the house floor. Well-sealed by levelling and abandonment layers during the late Iron Age and Roman period.

Objectives: a significant vessel typologically. A successful date from this sample will enhance current understanding of ceramics for the Iron Age period and contribute new data regionally as well as nationally. To confirm a major phase of activity on the cliff-castle during the middle Iron Age period, as well as give some indication of history of occupation in house 1.

Calibrated date: 1σ: 210–160 cal BC
                     2σ: 360–90 cal BC

Final comment: P Marshall (2011), this result provides a date for a South Western Decorated vessel of Accomplished type.

OxA–13304 2284 ±27 BP

δ¹³C: -24.3‰

Sample: 90054, submitted on 3 December 2003 by J A Nowakowski

Material: charcoal: *Corylus* sp., roundwood single fragment (R Gale 2002)

Initial comment: the sample was recovered from the fill of a well-defined sealed posthole, which had been dug into a former land surface, later sealed by a substantial rampart.

Objectives: to confirm a chronological phase of activity on the cliff-castle, which was identified during the ceramics assessment. This feature appears to be associated with others, which have produced pottery dating to the fourth to first centuries BC. These pre-date the rampart construction in trench 71.

Calibrated date: 1σ: 400–360 cal BC
                     2σ: 400–230 cal BC

Final comment: P Marshall (2011), this result provides a *terminus post quem* for the rampart.

OxA–13305 2467 ±27 BP

δ¹³C: -25.1‰

Sample: 90383, submitted on 3 December 2003 by J A Nowakowski

Material: charcoal: *Corylus* sp., roundwood single fragment (R Gale 2002)

Initial comment: charcoal recovered from a lower layer of midden 2 (253) - trench 61. Well buried and stratified in a layer in the midden which was undisturbed and *in situ.*

Objectives: to provide spot date to help date the midden formation in trench 61.

Calibrated date: 1σ: 760–510 cal BC
                     2σ: 770–410 cal BC

Final comment: P Marshall (2011), this charcoal is thought to be residual within its context. The small group of ceramics in the middens appear to be Outline, the earliest proposed for the middle Iron Age South Western Decorated sequence.

Laboratory comment: English Heritage (2011), the two measurements on samples from midden 2 (OxA-13305 and OxA-13416) are not statistically consistent (T'=36.4; T'(5%)=3.8; *v*=1; Ward and Wilson 1978) suggesting that the charcoal is residual material incorporated into the midden during its formation.

References: Ward and Wilson 1978

OxA–13306 2352 ±29 BP

δ¹³C: -27.8‰

Sample: 90384, submitted on 3 December 2003 by J A Nowakowski

Material: carbonised residue (H Quinnell 2002)

Initial comment: one sherd with carbonised residue excavated from one of the lowest layers, which form part of midden 1 in trench 61. Deeply buried and highly likely to be *in situ.* Recovered from a well-sealed and stratified brown earth midden layer which formed one of the lowest fills of midden 1 in trench 61. Midden 1 was formed against the inner face of rampart (256) and contained by an orthostat set within a rock-cut hollow.

Objectives: a successful date will enhance current understanding of ceramics for the Iron Age period in the south west as well as nationally. To provide a spot date for the formation of midden 1 and the midden sequence in trench 61. To contribute a series of dates for majority Iron Age activity on the cliff-castle. Part of a paired sample of ceramic residues from midden 1 (255) trench 61.

Calibrated date: 1σ: 410–390 cal BC
                     2σ: 430–380 cal BC

Final comment: P Marshall (2011), the difference in ages of the radiocarbon measurements from the individual middens can be expected as they are likely to have formed over a period of time rather than during one discrete event.

Laboratory comment: English Heritage (2011), the two measurements (OxA-13306 and OxA-14032) on sherds from the earliest midden 1, both from a group with broad 'transitional' affinities, considered to be intermediate between Late early Iron Age and middle Iron Age ceramic groups, are not statistically consistent (T'=52.9; T'(5%)=3.8; v=1; ward and Wilson 1978) meaning that the two residues are not of the same actual age.

References: Ward and Wilson 1978

OxA–13307 2163 ±28 BP

$\delta^{13}C$: -24.9‰

Sample: 90445, submitted on 3 December 2003 by J A Nowakowski

Material: carbonised residue (H Quinnell 2002)

Initial comment: carbonised residue attached to plain ceramic sherd found in sealed posthole in house 1. Cut features in the house floor were deep and well-sealed by levelling layers deposited during the late Iron Age/Roman period.

Objectives: to confirm the major phase of activity on the cliff-castle during the middle Iron Age; to give an indication of history of occupation in house 1. A date here is likely to be contemporary with other samples submitted for dating from trenches 61 and 62 and house 1. Successful dates from key ceramic groups from Trevelgue Head will enhance research of regional and national understanding of ceramic manufacture and technology for this period.

Calibrated date: 1σ: 350–170 cal BC
2σ: 360–110 cal BC

Final comment: P Marshall (2011), this result provides an estimate for the date of use of house 1.

OxA–13308 2009 ±28 BP

$\delta^{13}C$: -24.7‰

Sample: 90433A, submitted on 3 December 2003 by J A Nowakowski

Material: carbonised residue (H Quinnell 2002)

Initial comment: carbonised residue on seven joining sherds of a single vessel found in a sealed posthole flanking a doorway in house 1. Cut features in the house floor were deep and well-sealed by levelling layers deposited during the late Iron Age/Roman period.

Objectives: a significant vessel typologically. A successful date from this sample will enhance current understanding of ceramics for the Iron Age period and contribute new data regionally as well as nationally. To confirm a major phase of activity on the cliff-castle during the middle Iron Age period, as well as give some indication of history of occupation in house 1.

Calibrated date: 1σ: 50 cal BC–cal AD 30
2σ: 90 cal BC–cal AD 70

Final comment: P Marshall (2011), this result provides a date for a distinctive South western Decorated vessel of Accomplished type. It was also used to estimate the date of use of house 1.

OxA–13309 2268 ±26 BP

$\delta^{13}C$: -26.9‰

Sample: 90424, submitted on 3 December 2003 by J A Nowakowski

Material: carbonised residue (H Quinnell 2002)

Initial comment: as OxA-13307

Objectives: as OxA-13307

Calibrated date: 1σ: 390–260 cal BC
2σ: 400–210 cal BC

Final comment: see OxA-13307

OxA–13310 2582 ±28 BP

$\delta^{13}C$: -26.9‰

Sample: 90035B, submitted on 3 December 2003 by J A Nowakowski

Material: carbonised residue (H Quinnell 2002)

Initial comment: the vessel was in the fill of a sealed pit ('cooking hole') [257] which had been dug into a former sealed land surface.

Objectives: to confirm a chronological phase of activity o the cliff-castle, which on the basis of the ceramics would appear to be one of the earliest phases in this dating programme.

Calibrated date: 1σ: 800–780 cal BC
2σ: 810–760 cal BC

Final comment: P Marshall (2011), the results from cooking pit [257] suggest there were two distinct episodes of activity.

Laboratory comment: English Heritage (2011), the two samples from cooking pit [257] (OxA-13310 and OxA-13311) are not statistically consistent (T'=13.6; T'(5%)=3.8; v=1; Ward and Wilson 1978).

References: Ward and Wilson 1978

OxA–13311 2728 ±28 BP

$\delta^{13}C$: -24.8‰

Sample: 90035A, submitted on 3 December 2003 by J A Nowakowski

Material: charcoal: *Quercus* sp., sapwood single fragment (R Gale 2002)

Initial comment: as OxA-13310

Objectives: to confirm a chronological phase of activity of the cliff-castle, which was identified during the ceramics assessment. Potentially may produce the earliest scientific date in this series from Trevelgue Head.

Calibrated date: 1σ: 910–830 cal BC
2σ: 930–810 cal BC

Final comment: see OxA-13310

Laboratory comment: see OxA-13310

OxA–13312 2323 ±28 BP

δ¹³C: -25.9‰

Sample: 90463a, submitted on 3 December 2003 by
J A Nowakowski

Material: charcoal: *Corylus* sp., roundwood single fragment
(R Gale 2002)

Initial comment: as OxA-13224

Objectives: as OxA-13224

Calibrated date: *1σ:* 410–380 cal BC
                     2σ: 410–370 cal BC

Final comment: P Marshall (2011), the posterior density
estimate for this sample is *410–350 cal BC (88% probability)*
or *280–250 cal BC (7% probability*; table 6.1; Nowakowski
and Quinnell 2011).

Laboratory comment: see OxA-13224

OxA–13313 2274 ±26 BP

δ¹³C: -23.4‰

Sample: 90272A, submitted on 3 December 2003 by
J A Nowakowski

Material: charcoal: *Quercus* sp., sapwood single fragment
(R Gale 2002)

Initial comment: the sample was recovered from the fill of a
well-defined and well-sealed ovoid cut interpreted as a
furnace (245). This is one of several similar features found
at some depth in the area 62. These have all bee regarded as
metalworking features. This sample is one of a pair
submitted from furnace (245).

Objectives: as OxA-13224

Calibrated date: *1σ:* 400–360 cal BC
                     2σ: 400–230 cal BC

Final comment: P Marshall (2011), the ceramics associated
with the industrial features commmprise a mixture of all South
Western Decorative styles.

Laboratory comment: English Heritage (2011), the
measurements from furnace [245] are statistically consistent
(T'=2.7; T'(5%)=3.8; ν=1; Ward and Wilson 1978).

References: Ward and Wilson 1978

OxA–13314 2212 ±27 BP

δ¹³C: -23.9‰

Sample: 90272B, submitted on 3 December 2003 by
J A Nowakowski

Material: charcoal: *Quercus* sp., sapwood single fragment
 (R Gale 2002)

Initial comment: as OxA-13313

Objectives: as OxA-13224

Calibrated date: *1σ:* 370–200 cal BC
                     2σ: 380–190 cal BC

Final comment: see OxA-13313

Laboratory comment: see OxA-13313

OxA–13416 2120 ±50 BP

δ¹³C: -26.9‰

Sample: 90448, submitted on 3 December 2003 by
J A Nowakowski

Material: carbonised residue (H Quinnell 2002)

Initial comment: one sherd with carbonised residue excavated
from midden 2 (253), trench 61. A layer in the midden
which was undisturbed and *in situ.* Recovered from brown
earth, which formed the lowest deposit of midden 2 in
trench 61.

Objectives: the sherd is diagnostic and a successful date will
enhance our current understanding for ceramics of the Iron
Age period. Whether adequate residue is present for dating
requires assessment. To provide a spot date to help date the
midden formation sequence in trench 61 and to contribute
to a series of paired dates submitted as part of current dating
programme on Trevelgue Head.

Calibrated date: *1σ:* 210–50 cal BC
                     2σ: 360 cal BC–cal AD 1

Final comment: see OxA-13305

Laboratory comment: see OxA-13305

OxA–14012 2021 ±39 BP

δ¹³C: -26.2‰

Sample: 90257B (2), submitted on 29 July 2004 by
J A Nowakowski

Material: carbonised residue (H Quinnell 2002)

Initial comment: six conjoining sherds with carbonised
residues on a typologically vessel recovered from a lower
layer within midden deposit 3 (251), trench 61. Deeply
buried and recovered with diagnostic Iron Age pottery.
Contamination of deposit *in situ* unlikely. Pottery
unabraded therefore highly likely to have been deposited
and recovered *in situ.*

Objectives: a significant vessel typologically. A successful date
from this sample will enhance current understanding of
ceramics for the Iron Age period and contribute new date
regionally as well as nationally. To provide spot date to help
date the midden formation in trench 61, and to contribute
to a series of dates of majority iron activity on the cliff-castle.
Part of a paired sample of ceramic residues from midden 3
(251) trench 61.

Calibrated date: *1σ:* 60 cal BC–cal AD 30
                     2σ: 160 cal BC–cal AD 70

Final comment: see OxA-13191

Laboratory comment: English Heritage (2011), *see* OxA-
13191. This sherd has likely been contaminated with fish
glue resulting in a measurement that is too young, although
probably only by a relatively small amount.

References: Ward and Wilson 1978

OxA–14032 2059 ±28 BP

$\delta^{13}C$: -26.3‰

Sample: 90326, submitted on 29 July 2004 by
J A Nowakowski

Material: carbonised residue (H Quinnell 2002)

Initial comment: one sherd with carbonised residue excavated from one of the lowest layers, which form midden 1 (255) in trench 61. Deeply buried, well-sealed, unabraded and highly likely to be *in situ*. Recovered from one of the lowest fills on midden 1 in trench 61. Midden 1 was formed against the inner face of rampart (256) and contained by an orthostat set within a rock-cut hollow.

Objectives: a successful date will enhance current understanding of ceramics for the Iron Age period in the south west as well as nationally. To provide a spot date for the formation of midden 1 and the midden sequence in trench 61. To contribute a series of dates for majority Iron Age activity on the cliff-castle. Part of a paired sample of ceramic residues from midden 1 (255) trench 61.

Calibrated date: *1σ:* 110–40 cal BC
                 2σ: 170 cal BC–cal AD 10

Final comment: P Marshall (2011), the midden shows no evidence of disturbance or later activity such as pits being cut into it to suggest how an intrusive sherd could have been incorporated. The only plausible explanantion is that the sample has been contaminated at some point following excavation. It was therefore exckuded from the chronological modelling.

Laboratory comment: English Heritage (2011), *see* OxA-13306. It is possible that this sherd was also contaminated by conservation processes although there is no direct evidence of this.

Upland Peats, Cumbria and Lancashire

Location: SD 4062367880centred on
                 Lat. 54.06.12 N; Long. 02.54.29 W

Project manager: J Quartermaine (Oxford Archaeology North), 2004

Description: this study examined four areas of North West England (West Cumbria, Great Langdale, Central Lake District, Forest of Bowland, and the Anglezarke/Rivington Massif, both in Lancashire).

Objectives: the principal aim of this study was to develop a baseline survey and management tool that will be applicable for all upland peat regions in England. The objectives towards this included the implementation of a programme of experimentation in identification and recording techniques of the archaeological and palaeoecological resource, and to assess the nature of the threats to the peatlands.

Final comment: D Druce (20 October 2014), the study has demonstrated that the fate of the underlying archaeology was closely entwined with that of the overlying peats, and therefore the prime recommendation is that a partnership should be established between all environmental and

archaeological agencies so as to afford the protection of the peatlands. Despite this, it was recognised that there was a need to provide preferential management to those areas of greatest archaeological potential, and that there was a need for a scheme comparable to the one successfully tested in Dartmoor (Premier Archaeological Landscapes (PALs). This would entail selecting areas of archaeological importance on the basis of their observed resource and their potential for buried remains, and by working with partners, particularly Natural England, to put forward such areas for Higher Level Stewardship (HLS) schemes. This approach was perceived as a first stage in addressing the long-term issues that have potential to impact on the archaeology preserved within the upland peats.

Upland Peats: Anglezarke, site 1, Lancashire

Location: SD 6465918318
                 Lat. 53.39.37 N; Long. 02.32.11 W

Project manager: J Quartermaine (Oxford Archaeology North), 2004

Archival body: Lancashire County Council

Description: site 1 was situated in the central area of Anglezarke where the blanket peat reaches depths of up to 300cm in places. Site 1 was a 200cm section of peat exposed in a tributary stream of a main channel that runs adjacent to the plateau of Black Hill upper. The site slopes very gently to the south-east.

Objectives: the lowermost sample was to date the inception of the deepest, most central area of blanket peat on Anglezarke Moor. The uppermost sample was to determine when the peat ceased to develop and/or to determine the loss of peat since work that was carried out in the area in the early 1990s (Bain 1991).

References: Bain 1991

SUERC–4511 845 ±35 BP

$\delta^{13}C$: -28.1‰

Sample: 12.5–14.5cm depth, submitted on 9 June 2004 by D Druce

Material: peat (humic acid)

Initial comment: the sample is a 2cm slice of peat taken from near the top of a peat layer *c* 300cm deep. The peat overlies a very organic sandy soil *c* 30cm in depth that overlies bedrock. The top 4cm of the section consisted of charred peat and charred heather. This overlaid *c* 4cm of very loose peat. The sample was taken from immediately below this and consisted of compact humified peat. The site is waterlogged and the sample comes from the exposed section of a peat hagg *c* 300cm deep. The peat hagg has been exposed by a small tributary steam that has cut through the peat down to bedrock. The stream channel was clean and U-shaped and did not show any signs of having meandered in the past.

Objectives: to date the top of the peat in the central area of the Anglezarke area of moorland where the heather is frequently burnt to encourage grouse. The date will inform us of how much peat has been lost and/or the period of time when peat ceased to develop due to the action of frequent burning.

Calibrated date: 1σ: cal AD 1160–1230
                     2σ: cal AD 1050–1270

Final comment: D Druce (20 October 2014), the date suggests that either truncation of the peat has taken place in the central area of Anglezarke, or, indeed, it represents a time when agricultural practices on Anglezarke Moor, such as the burning of heather to provide improved forage for livestock, became more intensive, and restricted the production of surface peat.

SUERC–4512 4945 ±35 BP

$\delta^{13}C$: -28.7‰

Sample: 231–233 cm depth, submitted on 28 August 2004 by D Druce

Material: peat (humic acid)

Initial comment: the sample is a 2cm slice of peat taken from near the base of a peat layer *c* 300cm deep. The peat overlies a very organic sandy soil *c* 30cm in depth that overlies bedrock. The peat hagg has been exposed by a small tributary steam that has cut through the peat down to bedrock. The stream channel was clean and U-shaped and did not show any signs of having meandered in the past.

Objectives: to date peat inception in the central part of the Anglezarke area of moorland.

Calibrated date: 1σ: 3770–3660 cal BC
                     2σ: 3800–3640 cal BC

Final comment: D Druce (20 October 2014), the date is in agreement with others taken from the base of the deep peat on the central plateau of Anglezarke Moor (Bain 1991). The pollen records suggest that peat inception occurred here during a time of increased woodland disturbance and evidence for burning (Bain 1991).

References: Bain 1991

Upland Peats: Anglezarke, site 2, Lancashire

Location: SD 6308417990
                  Lat. 53.39.26 N; Long. 02.33.36 W

Project manager: J Quartermaine (Oxford Archaeology North), 2004

Archival body: Lancashire County Council

Description: site 2 is situated on the edge of the central plateau of Anglezarke near to Hurst Hill. The area is relatively flat but has been subjected to a number of threats including fire. The depth of peat here is considerably lower than in the central area (Anglezarke site 1).

Objectives: the lowermost sample is to date the inception of blanket peat near the periphery of Anglezarke Moor. The uppermost sample is to determine when the peat ceased to develop and/or to determine the loss of peat since work that was carried out in the area in the early 1990s (Bain 1991).

References: Bain 1991

SUERC–4514 210 ±40 BP

$\delta^{13}C$: -29.7‰

Sample: 7–9cm, submitted on 9 June 2004 by D Druce

Material: peat (humic acid)

Initial comment: the sample is a 2cm slice of peat taken from near the top of a peat layer *c* 45cm deep. The peat overlies a very organic sandy soil *c* 2cm in depth that overlies bedrock. The sample taken for dating consisted of compact humified peat, immediately below the charred layer.

Objectives: to date the top of the peat on the edge of the central plateau of the Anglezarke area of moorland where the heather is frequently burnt to encourage grouse. The date will inform us of how much peat has been lost and/or the period of time when peat ceased to develop due to the action of frequent burning.

Calibrated date: 1σ: cal AD 1650–1955★
                     2σ: cal AD 1640–1955★

Final comment: D Druce (20 October 2014), this area of Anglezarke Moor has suffered the most severe erosion where peat has been completely removed in places. Ironically, however, the surface of the peat here is younger that that on the central plateau. In addition, although dates indicate peat inception occurred here during the late Neolithic/early Bronze Age (SUERC-4515 below) a thickness of only *c* 45cm remains. The peat that does survive has been subjected to a number of erosion and regrowth episodes, illustrated by hiatuses in both the stratigraphy and pollen record.

SUERC–4515 3685 ±35 BP

$\delta^{13}C$: -29.0‰

Sample: 41–43cm depth, submitted on 9 June 2004 by D Druce

Material: peat (humic acid)

Initial comment: the sample is a 2cm slice of peat taken from near the base of a peat *c* 45cm deep. The peat overlies an organic sandy soil *c* 2cm in depth that overlies bedrock.

Objectives: to date peat inception on the edge of the central plateau of Anglezarke Moor.

Calibrated date: 1σ: 2140–2020 cal BC
                     2σ: 2200–1950 cal BC

Final comment: D Druce (20 October 2014), the date is consistent with that from another site on the periphery of Anglezarke Moor (Bain 1991), and both suggest that peat inception occurred here later than on the central plateau. This period of peat encroachment was accompanied by a significant rise in plants of disturbed ground and ericaceous pollen, and appears to signify a period of increased human activity in the area (Bain 1991).

References: Bain 1991

Upland Peats: Forest of Bowland, site 1, Lancashire

Location:	SD 5764250853
	Lat. 53.57.07 N; Long. 02.38.49 W
Project manager:	J Quartermaine (Oxford Archaeology North), 2004
Archival body:	Lancashire County Council

Description: site 1 is on the most central areas of the Forest of Bowland. The site, known as White Moss, is situated in a slight basin surrounded by gentle slopes. Due to the nature of the terrain there is a tendency for the site to become waterlogged, hence over 200cm of blanket peat has developed in the area. At present, however, fluvial activity appears to be actively eroding large areas of this peat.

Objectives: the lowermost sample is to date the inception of the deepest, most central, area of peat in the Forest of Bowland transect. The uppermost sample is to date the top of the peat in this location of deepest peat to help to determine current peat growth.

SUERC–4504 2350 ±35 BP

δ¹³C: -28.2‰

Sample: 1–3cm Depth, submitted on 27 August 2004 by D Druce

Material: peat (humic acid)

Initial comment: the 2cm slice of humified peat was taken from near the top of a peat layer *c* 200cm deep.

Objectives: to date the top of the peat in the central area of the Forest of Bowland where the deposits are the deepest and probably most intact compared to those on the periphery of the area.

Calibrated date: 1σ: 410–390 cal BC
2σ: 510–380 cal BC

Final comment: D Druce (20 October 2014), the date suggests that over 2000 years of palaeoecological history has been lost. It is highly unlikely that natural drainage erosion has truncated the peat to this depth, and it is most probably, albeit in part, attributable to peat cutting.

SUERC–4505 6720 ±35 BP

δ¹³C: -27.8‰

Sample: 193.5–195.5cm Depth, submitted on 27 August 2004 by D Druce

Material: peat (humic acid)

Initial comment: the 2cm slice of humified peat was taken from near the base of a peat layer *c* 200cm deep.

Objectives: to date the inception of the peat in the central area of the Forest of Bowland where the deposits are the deepest and probably most intact compared to those on the periphery of the area.

Calibrated date: 1σ: 5660–5620 cal BC
2σ: 5710–5560 cal BC

Final comment: D Druce (20 October 2014), site 1, which forms one of the highest points in the study area, provided the earliest date of peat inception from the whole project. Although the pollen evidence indicates small-scale clearance in the area, there is little to indicate why peat developed so early here when compared to other study areas. One explanation could be that, as topographically this forms a col, the area was one where water collected, although it also formed a watershed. It is possible that this type of topography, coupled with a shift to warmer and wetter conditions in the Atlantic Period (at *c* 5500 cal BC), was instrumental in this very early blanket peat development, particularly when coupled with some loss of tree cover through anthropogenic activity.

Upland Peats: Forest of Bowland, site 2, Lancashire

Location:	SD 5568649901
	Lat. 53.56.36 N; Long. 02.40.36 W
Project manager:	J Quartermaine (Oxford Archaeology North), 2004
Archival body:	Lancashire County Council

Description: site 2 is situated on a spur of gently sloping ground on Stakehouse Fell, Forest of Bowland. Much of the area near to the site has undergone considerable erosion, be it through peat cutting or fire, in the past. It appears that whatever the initial cause, current fluvial activity appears to be removing a bulk of the remaining peat.

Objectives: to date the peat inception at this location in the Forest of Bowland transept in order to compare with the dates from the other two sampling sites.

SUERC–4506 4645 ±35 BP

δ¹³C: -28.8‰

Sample: 79.5–81.5cm Depth, submitted on 27 August 2004 by D Druce

Material: peat (humic acid)

Initial comment: a 2cm slice of humified peat was taken from near the base of a peat layer *c* 80cm deep. The peat overlays a deposit of sandy soil with organic remains. The site is waterlogged and the sample comes from the exposed section of peat hagg. The peat face was almost vertical and clean and did not appear to show any signs of slumping or redeposition of material.

Objectives: to date the inception of the peat in an intermediate area of peat development in the Forest of Bowland. The result will be compared to the dates taken from the other two sampling sites: site 1 where the peat is the deepest and site 3 where the peat is the shallowest.

Calibrated date: 1σ: 3500–3360 cal BC
2σ: 3520–3350 cal BC

Final comment: D Druce (20 October 2014), peat inception occurred at site 2 in a relatively open landscape with less than 50% tree/shrub pollen. It is possible that the spread of peat in the Forest of Bowland was a consequence of the progressive clearance of the area, which, based on the pollen evidence, ultimately developed into a heather-rich landscape at some time during the Neolithic period. It is also possible, however, that the acidification of ground conditions destroyed and naturally inhibited the growth of trees, thus accentuating the spread of peat.

Upland Peats: Forest of Bowland, site 3, Lancashire

Location: SD 5499349415
Lat. 53.56.20 N; Long. 02.41.13 W

Project manager: J Quartermaine (Oxford Archaeology North), 2004

Archival body: Lancashire County Council

Description: site 3 is situated on an area of sloping ground on the periphery of the Forest of Bowland near Stakehouse Farm. The area has undergone considerable erosion in the past and is covered with peat haggs *c* 50cm deep. Bare ground is visible in large areas between the peat haggs.

Objectives: to date the peat inception at this location in the Forest of Bowland transept in order to compare with the dates from the other two sampling sites.

SUERC–4507 2365 ±40 BP

$\delta^{13}C$: -29.1‰

Sample: 49–51cm Depth, submitted on 27 August 2004 by D Druce

Material: peat (humic acid)

Initial comment: a 2cm slice of humified peat was taken from near the base of a peat layer *c* 50cm deep. The peat overlays a deposit of sandy soil with organic remains.

Objectives: to date the inception of the peat on the periphery of the Forest of Bowland. The result will be compared to the dates taken from the other two sampling sites: site 1 where the peat is the deepest and site 2 where the peat is *c* 80cm deep. Site 3 is from an area of immense erosion and it is crucial to understand these erosive processes in relation to the site's geographical location, history, and inception and depth of peat.

Calibrated date: 1σ: 420–390 cal BC
2σ: 540–380 cal BC

Final comment: D Druce (20 October 2014), peat development at site 3 occurred much later than at the other two sites (SUERC–4505 and -4506 above), but earlier than expected. It is possible that an acceleration in peat growth occurred in the Forest of Bowland at around this time following a shift to wetter conditions recorded in the peat at sampling site 1 and at nearby Fenton Cottage (Middleton *et al* 1995).

References: Middleton *et al* 1995

Upland Peats: Langdale, site 1, Cumbria

Location: . NY 2743407469
Lat. 54.27.27 N; Long. 03.07.15 W

Project manager: J Quartermaine (Oxford Archaeology North), 2004

Archival body: Cumbria County Council

Description: site 1 is an area of deep peat situated on a plateau on the slopes below Pike Of Stickle on Langdale Fell. The site itself is gently sloping and the peat has been eroded in places down to bedrock probably by fluvial activity, thus leaving peat haggs up to 200cm deep in places.

Objectives: to date inception of the peat in the southern end of the Langdale transect where the peat is deepest.

SUERC–4516 3620 ±40 BP

$\delta^{13}C$: -27.9‰

Sample: 145–147cm depth, submitted on 16 June 2004 by D Druce

Material: peat (humic acid)

Initial comment: the 1cm sample is from the base of *c* 145cm of peat. The peat overlies *c* 10cm of humic soil that lies directly on the weathered bedrock. The bottom *c* 50cm of peat contains wood fragments. The site is waterlogged and the sample comes from the face of a peat hagg *c* 150cm deep. The peat hagg appears to have been formed through drainage erosion that has left a clean vertical face.

Objectives: to date peat inception at the southern end of Langdale Fell.

Calibrated date: 1σ: 2040–1920 cal BC
2σ: 2140–1880 cal BC

Final comment: D Druce (20 October 2014), the date is very simlar to peat inception at site 4 (SUERC-4521 below), and at Red Tarn Moss, to the south-west of the Langdale Pikes (Pennington 1975). The results reinforce the emerging picture that peat inception on the high fells around Great Langdale took place following a long history of human modification of the landscape. The reason for the date of peat inception at these three sites is not entirely clear, but it is broadly consistent with a number of other dates from this project.

References: Pennington 1975

Upland Peats: Langdale, site 2, Cumbria

Location: NY 2783709020
Lat. 54.28.17 N; Long. 03.06.54 W

Project manager: J Quartermaine (Oxford Archaeology North), 2004

Archival body: Cumbria County Council

Description: site 2 is situated at the northern end of the Langdale transect where the peat is shallower. The site is on a relatively gentle slope that becomes much steeper downwards to the west. Much of the peat appears to have been eroded by fluvial activity and the samples were taken from an eroded edge of peat *c* 110cm deep.

Objectives: to date inception of the peat at the northern end of the Langdale transect where the peat is shallower. The result will be compared to the dates taken from the southern end of the transect.

SUERC–4517 2980 ±35 BP

δ¹³C: -28.2‰

Sample: 100–102 cm depth, submitted on 16 June 2004 by D Druce

Material: peat (humic acid)

Initial comment: the 2cm sample is from the base of a deposit of peat c 110cm deep. The peat overlies *c* 10cm of silty sandy humic soil that lies directly on weathered bedrock. The bottom *c* 25cm of peat contains wood fragments. The site is waterlogged and the sample comes from the face of a peat hagg *c* 110cm deep. The peat hagg appears to have been formed through drainage erosion that has left a clean vertical face.

Objectives: to date peat inception at the northern end of Langdale Fell.

Calibrated date: 1σ: 1270–1120 cal BC
                     2σ: 1380–1110 cal BC

Final comment: D Druce (20 October 2014), peat development in the Langdale Fells encroached northwards as far as High Raise (sampling site 2) at this time. Although the pollen records from many of the tarn deposits surrounding Great Langdale show a recovery of woodland after the Neolithic period (eg at Blea Tarn and Red Tarn, Pennington 1975), the higher fells around Angle Tarn, Red Tarn (Helvellyn) and Great Langdale show an expansion in heathland and the persistence of open conditions following the Neolithic period.

References: Pennington 1975

Upland Peats: Langdale, site 4, Cumbria

Location: NY 2756007370
                   Lat. 54.27.23 N; Long. 03.07.08 W

Project manager: J Quartermaine (Oxford Archaeology North), 2004

Archival body: Cumbria County Council

Description: site 4 is from an area of deep peat situated on a small plateau on the slopes below Pike Of Stickle on Langdale Fell. The site is relatively flat and adjacent to a floor of chippings associated with Langdale axe production.

Objectives: to date inception of the peat at the southern end of the Langdale transect where the peat is deepest and evidence for prehistoric activity the greatest.

SUERC–4521 3865 ±35 BP

δ¹³C: -28.5‰

Sample: 145–147cm depth, submitted on 16 June 2004 by D Druce

Material: peat (humic acid)

Initial comment: the 2cm sample is from the base of a peat layer *c* 150cm deep. The peat overlies *c* 5cm of humic soil, which overlies and is incorporated into an axe-factory chipping floor. The peat immediately overlying the humic soil is very humified. The site is waterlogged and the sample comes from a 100cm ∞ 100cm pit *c* 150cm deep at its northern face.

Objectives: to date peat inception at the southern end of Langdale Fells near to the concentration of axe factory working sites. The date from this site will be compared to the other two peat inception dates put forward from Langdale to determine the chronology of peat development in relation to topography, drainage patterns and archaeological activity.

Calibrated date: 1σ: 2460–2280 cal BC
                     2σ: 2470–2200 cal BC

Final comment: D Druce (20 October 2014), the date is very simlar to peat inception at site 1 (SUERC-4516 above), and at Red Tarn Moss, to the south-west of the Langdale Pikes (Pennington 1975). The results reinforce the emerging picture that peat inception on the high fells around Great Langdale took place following a long history of human modification of the landscape. The reason for the date of peat inception at these three sites is not entirely clear, but it is broadly consistent with a number of other dates from this project.

References: Pennington 1975

Upland Peats: South-west Fells transect, site 1b, Cumbria

Location: SD 1374195509
                   Lat. 54.20.52 N; Long. 03.19.43 W

Project manager: J Quartermaine (Oxford Archaeology North), 2004

Archival body: Cumbria County Council

Description: the site is situated in a small valley bounded to the east by the high ridges of Hesk Fell and to the west by coastal land renowned for its concentration of archaeological sites. The site is waterlogged, being adjacent to a small stream. Roughly 100cm of peat has developed in the bottom of the valley, which is encroaching upslope and, in places, covers some of the archaeological sites.

Objectives: to date the inception of the peat at three altitudinally and archaeologically differing locations in the south-west Fells study area. Also to provide a date for possible early prehistoric activity in the area in relation to vegetation change and peat development.

SUERC–4523 4490 ±40 BP

δ¹³C: -29.4‰

Sample: 78–80 cm depth, submitted on 22 October 2004 by D Druce

Material: peat (humic acid)

Initial comment: the 2cm slice of peat was taken from near the base of a peat layer *c* 100cm deep. The peat overlies a very organic sandy soil *c* 20cm in depth. Woody remains were visible in the lower half of the peat.

Objectives: to date the timing of peat inception on the lowest plateau in the south-west Fells study area. This area is also the richest archaeologically.

Calibrated date: *1σ:* 3350–3090 cal BC
            2σ: 3360–3020 cal BC

Final comment: D Druce (20 October 2014), the dating evidence from this site and from Site 2b (SUERC-4524 above), situated at the lowest and intermediate altitudes of the south-west Fells, suggests that the peat developed here more-or-less simultaneously. It is possible that woodland clearance was instrumental in its development, as the pollen assemblage in the base of the peat at both sites indicates a landscape of relatively open shrub/woodland with disturbance indicators

Upland Peats: South-west Fells transect, site 2b, Cumbria

Location: SD 1465595410
            Lat. 54.20.49 N; Long. 03.18.52 W

Project manager: J Quartermaine (Oxford Archaeology North), 2004

Archival body: Cumbria County Councill, Cumbria County Council

Description: the site is situated on a relatively flat plateau bounded on the north by the steep crags on White Pike, and on the south by the peak of the Knott. The terrain falls away fairly steeply on the west to the lower plateau of Birkby Fell renowned for its wealth of archaeological sites. The sampling site appears to be one of two shallow basins separated by a ridge of slightly higher land where a prehistoric axe was recently discovered through test-pitting. Up to 100cm of peat has developed in places.

Objectives: to date the inception of the peat at three altitudinally and archaeologically differing locations in the south-west Fells study area. Also to provide a date for possible early prehistoric activity in the area in relation to vegetation change and peat development.

SUERC–4524 4530 ±35 BP

δ¹³C: -30.5‰

Sample: 98–100 cm depth, submitted on 22 October 2004 by D Druce

Material: peat (humic acid)

Initial comment: the sample is a 2cm slice of peat taken from near the base of peat 100cm deep. The peat overlies a very organic soil *c* 4cm in depth, which, in turn, overlies *c* 20cm of sandy silt with stones.

Objectives: to date the timing of peat inception on the higher plateau behind 'The Knott' in the south-west Fells study area.

Calibrated date: *1σ:* 3360–3110 cal BC
            2σ: 3370–3090 cal BC

Final comment: D Druce (20 October 2014), the dating evidence from this site and from site 1b (SUERC-4523 below), situated at the lowest and intermediate altitudes of the south-west Fells, suggests that the peat developed here more-or-less simultaneously. It is possible that woodland clearance was instrumental in its development, as the pollen assemblage in the base of the peat at both sites indicates a landscape of relatively open shrub/woodland with disturbance indicators.

Upland Peats: South-west Fells transect, site 3, Cumbria

Location: SD 1632195153
            Lat. 54.20.42 N; Long. 03.17.19 W

Project manager: J Quartermaine (Oxford Archaeology North), 2004

Archival body: Cumbria County Council

Description: the site is situated in a trough of gently sloping land between the two peaks of Woodend Height and Hesk Fell on the south west Cumbrian Fells. The area is covered by blanket peat that reaches up to 200cm in depth in places. Occasional bedrock is exposed due to a combination of drainage erosion and peat cutting. Site 3 consisted of a face of peat exposed through the latter.

Objectives: to date the inception of the peat at three altitudinally and archaeologically differing locations in the SW Fells study area. Also to provide a date for possible early prehistoric activity in the area in relation to vegetation change and peat development.

SUERC–4522 3480 ±40 BP

δ¹³C: -28.0‰

Sample: 85–87 cm depth, submitted on 9 June 2004 by D Druce

Material: peat (humic acid)

Initial comment: the 2cm slice of peat was taken from near the base of a peat layer c 100cm deep. The peat overlies a very organic sandy soil *c* 2cm deep, which overlies *c* 10cm sandy soil with stones. The site is waterlogged and the sample comes from the face of a peat hagg in a large area of peat on Hesk Fell that reaches up to *c* 300cm in depth. The peat hagg appears to be exposed through drainage erosion, although there is clear evidence for small-scale peat cutting in the area. Hesk Fell is one of the altitudinally highest locations in the south-west Fells study area.

Objectives: to date the timing of peat inception on Hesk Fell. This may be compared with the two, altitudinally lower (and archaeologically differing), sampling sites taken from the south-west Fells study area.

Calibrated date: *1σ:* 1890–1700 cal BC
                2σ: 1920–1680 cal BC

Final comment: D Druce (20 October 2014), the development of the blanket peat at sampling site 3, dated to the early Bronze Age, is broadly consistent with other peat inception dates in this study and appears to have occurred at a time of a general acceleration in peat growth in central Cumbria.

Warkworth Castle: Grey Mare's Tail Tower, Northumberland

Location: NU 247057
                    Lat. 55.20.44 N; Long. 01.36.38 W

Project manager: J Goodall (English Heritage), 2004

Archival body: English Heritage

Description: the tower is thought to date from the late-thirteenth or early-fourteenth century on stylistic grounds. Tree-ring dating failed to produce a date for primary lintel timbers in the window openings. Therefore one of these samples was submitted for wiggle-match dating. Six contiguous blocks of ten rings each were sampled from an oak dendrochronology core, a 106–ring sequence ending in the heartwood/sapwood boundary (WKW-A01).

Objectives: to determine whether the surviving timber lintels in the tower relate to the original construction, or to a later phase of repair or modification.

Final comment: A Bayliss (24 August 2014), Bayesian wiggle-matching (Christen and Litton 1995; Bronk Ramsey *et al* 2001) suggests that this timber was felled in the early fourteenth century and is a survival from the primary phase of construction. Grey Mare's Tail tower was probably built between Edward I's building campaign at Berwick-upon-Tweed, which began in AD 1296, and the inheritance of John Fitz Robert to a debt-ridden estate in AD 1310.

Laboratory comment: English Heritage (September 2005), the chronological model for this timber has good overall agreement (Arnold *et al* 2006, fig 8), only when SUERC-6564 is excluded as an early outlier. By offsetting the estimated date for the heartwood/sapwood boundary by the probability distribution of the number of sapwood rings likely to be present on the timber (Bayliss and Tyers 2004), we can suggest that this timber was felled in *cal AD 1295–1340 (95% probability; bark edge*; Arnold *et al* 2006, fig 9).

References: Arnold *et al* 2006
                    Bayliss and Tyers 2004
                    Bronk Ramsey *et al* 2001
                    Christen and Litton 1995
                    Honeyman and Hunter Blair 1990

SUERC–6540 700 ±20 BP

δ¹³C: -25.1‰

Sample: WKW-A01 Block 1, submitted on 8 March 2005 by D Hamilton

Material: wood: *Quercus* sp., 10 rings (R Howard 2005)

Initial comment: rings 93–102 from floating tree-ring sequence WKW-A01. The final ring of this sample is the heartwood/sapwood boundary.

Objectives: to wiggle-match this sample and provide a date for this undated site sequence.

Calibrated date: *1σ:* cal AD 1275–1290
                    2σ: cal AD 1270–1300

Final comment: A Bayliss (24 August 2014), the weighted mean for block 1 has good agreement with its position in the sequence.

Laboratory comment: English Heritage (24 August 2014), the two results from block 1 are statistically consistent (T'=0.0; T'(5%)=3.8; *v*=1; Ward and Wilson 1978), so a weighted mean has been taken before further analysis (698 ±14 BP).

References: Ward and Wilson 1978

SUERC–6541 695 ±20 BP

δ¹³C: -25.3‰

Sample: WKW-A01 Block 1, submitted on 8 March 2005 by D Hamilton

Material: wood: *Quercus* sp., 10 rings (R Howard 2005)

Initial comment: a replicate of SUERC-6540.

Objectives: as SUERC-6540

Calibrated date: *1σ:* cal AD 1275–1295
                    2σ: cal AD 1270–1380

Final comment: see SUERC-6540

Laboratory comment: see SUERC-6540

SUERC–6542 715 ±20 BP

δ¹³C: -24.8‰

Sample: WKW-A01 Block 2, submitted on 8 March 2005 by D Hamilton

Material: wood: *Quercus* sp., 10 rings (R Howard 2005)

Initial comment: rings 83–92 from floating tree-ring sequence WKW-A01.

Objectives: as SUERC-6540

Calibrated date: *1σ:* cal AD 1270–1290
                    2σ: cal AD 1265–1295

Final comment: A Bayliss (24 August 2014), the weighted mean for block 2 has good agreement with its position in the sequence.

Laboratory comment: English Heritage (28 August 2014), the two results from block 2 are statistically consistent (T'=0.0; T'(5%)=3.8; *v*=1; Ward and Wilson 1978), so a weighted mean has been taken before further analysis (713 ±14 BP).

References: Ward and Wilson 1978

SUERC–6543 710 ±20 BP

δ¹³C: -24.4‰

Sample: WKW-A01 Block 2, submitted on 8 March 2005 by D Hamilton

Material: wood: *Quercus* sp., 10 rings (R Howard 2005)

Initial comment: a replicate of SUERC-6542.

Objectives: as SUERC-6540

Calibrated date: *1σ:* cal AD 1275–1290
                     2σ: cal AD 1265–1295

Final comment: see SUERC-6542

Laboratory comment: see SUERC-6542

SUERC–6547 760 ±20 BP

δ¹³C: -24.0‰

Sample: WKW-A01 Block 3, submitted on 8 March 2005 by D Hamilton

Material: wood: *Quercus* sp., 10 rings (R Howard 2005)

Initial comment: rings 73–82 from floating tree-ring sequence WKW-A01.

Objectives: as SUERC-6540

Calibrated date: *1σ:* cal AD 1255–1275
                     2σ: cal AD 1225–1285

Final comment: A Bayliss (24 August 2014), the weighted mean for block 3 has good agreement with its position in the sequence.

Laboratory comment: English Heritage (24 August 2014), the two results from block 3 are statistically consistent (T′=0.0; T′(5%)=3.8; *v*=1; Ward and Wilson 1978), so a weighted mean has been taken before further analysis (763 ±14 BP).

References: Ward and Wilson 1978

SUERC–6548 765 ±20 BP

δ¹³C: -24.7‰

Sample: WKW-A01 Block 3, submitted on 8 March 2005 by D Hamilton

Material: wood: *Quercus* sp., 10 rings (R Howard 2005)

Initial comment: a replicate of SUERC-6547.

Objectives: as SUERC-6540

Calibrated date: *1σ:* cal AD 1255–1275
                     2σ: cal AD 1220–1280

Final comment: see SUERC-6547

Laboratory comment: see SUERC-6547

SUERC–6550 775 ±20 BP

δ¹³C: -26.0‰

Sample: WKW-A01 Block 4, submitted on 8 March 2005 by D Hamilton

Material: wood: *Quercus* sp., 10 rings (R Howard 2005)

Initial comment: rings 63–72 from floating tree-ring sequence WKW-A01.

Objectives: as SUERC-6540

Calibrated date: *1σ:* cal AD 1250–1275
                     2σ: cal AD 1220–1280

Final comment: A Bayliss (24 August 2014), the weighted mean for block 4 has good agreement with its position in the sequence.

Laboratory comment: English Heritage (24 August 2014), the two results from block 4 are statistically consistent (T′=0.3; T′(5%)=3.8; *v*=1; Ward and Wilson 1978), so a weighted mean has been taken before further analysis (783 ±14 BP).

References: Ward and Wilson 1978

SUERC–6551 790 ±20 BP

δ¹³C: -25.4‰

Sample: WKW-A01 Block 4, submitted on 8 March 2005 by D Hamilton

Material: wood: *Quercus* sp., 10 rings (R Howard 2005)

Initial comment: a replicate of SUERC-6550.

Objectives: as SUERC-6540

Calibrated date: *1σ:* cal AD 1220–1265
                     2σ: cal AD 1215–1275

Final comment: see SUERC-6550

Laboratory comment: see SUERC-6550

SUERC–6556 800 ±20 BP

δ¹³C: -25.8‰

Sample: WKW-A01 Block 5, submitted on 8 March 2005 by D Hamilton

Material: wood: *Quercus* sp., 10 rings (R Howard 2005)

Initial comment: rings 53–62 from floating tree-ring sequence WKW-A01.

Objectives: as SUERC-6540

Calibrated date: *1σ:* cal AD 1220–1265
                     2σ: cal AD 1210–1270

Final comment: A Bayliss (24 August 2014), the weighted mean for block 5 has good agreement with its position in the sequence.

Laboratory comment: English Heritage (24 August 2014), the two results from block 5 are statistically consistent (T′=0.1; T′(5%)=3.8; *v*=1; Ward and Wilson 1978), so a weighted mean has been taken before further analysis (805 ±14 BP).

References: Ward and Wilson 1978

SUERC–6557 810 ±20 BP

$\delta^{13}C$: -25.9‰

Sample: WKW-A01 Block 5, submitted on 8 March 2005 by D Hamilton

Material: wood: *Quercus* sp., 10 rings (R Howard 2005)

Initial comment: a replicate of SUERC-6556.

Objectives: as SUERC-6540

Calibrated date: 1σ: cal AD 1215–1260
                       2σ: cal AD 1205–1270

Final comment: see SUERC-6556

Laboratory comment: see SUERC-6556

SUERC–6558 810 ±20 BP

$\delta^{13}C$: -24.2‰

Sample: WKW-A01 Block 6, submitted on 8 March 2005 by D Hamilton

Material: wood: *Quercus* sp., 10 rings (R Howard 2005)

Initial comment: rings 43–52 from floating tree-ring sequence WKW-A01.

Objectives: as SUERC-6540

Calibrated date: 1σ: cal AD 1215–1260
                       2σ: cal AD 1205–1270

Final comment: A Bayliss (24 August 2014), the weighted mean for block 6 has good agreement with its position in the sequence.

Laboratory comment: English Heritage (24 August 2014), the two results from block 6 are statistically consistent (T′=0.0; T′(5%)=3.8; ν=1; Ward and Wilson 1978), so a weighted mean has been taken before further analysis (810 ±14 BP).

References: Ward and Wilson 1978

SUERC–6559 810 ±20 BP

$\delta^{13}C$: -24.4‰

Sample: WKW-A01 Block 6, submitted on 8 March 2005 by D Hamilton

Material: wood: *Quercus* sp., 10 rings (R Howard 2005)

Initial comment: a replicate of SUERC-6558.

Objectives: as SUERC-6540

Calibrated date: 1σ: cal AD 1215–1260
                       2σ: cal AD 1205–1270

Final comment: see SUERC-6558

Laboratory comment: see SUERC-6558

SUERC–6563 850 ±20 BP

$\delta^{13}C$: -24.3‰

Sample: WKW-A01 Block 7, submitted on 8 March 2005 by D Hamilton

Material: wood: *Quercus* sp., 10 rings (R Howard 2005)

Initial comment: rings 33–42 from floating tree-ring sequence WKW-A01.

Objectives: as SUERC-6540

Calibrated date: 1σ: cal AD 1165–1220
                       2σ: cal AD 1155–1245

Final comment: A Bayliss (24 August 2014), both measurements from block 7 have poor agreement in the model, being rather earlier than expected from their position in the sequence.

Laboratory comment: English Heritage (24 August 2014), the two results from block 7 are statistically consistent (T′=1.1; T′(5%)=3.8; ν=1; Ward and Wilson 1978), so a weighted mean has been taken before further analysis (865 ±14 BP). *See* series comments.

References: Ward and Wilson 1978

SUERC–6564 880 ±20 BP

$\delta^{13}C$: -24.4‰

Sample: WKW-A01 Block 7, submitted on 8 March 2005 by D Hamilton

Material: wood: *Quercus* sp., 10 rings (R Howard 2005)

Initial comment: a replicate of SUERC-6563.

Objectives: as SUERC-6540

Calibrated date: 1σ: cal AD 1155–1210
                       2σ: cal AD 1050–1220

Final comment: see SUERC-6563

Laboratory comment: see SUERC-6563

SUERC–6566 850 ±20 BP

$\delta^{13}C$: -24.4‰

Sample: WKW-A01 Block 8, submitted on 8 March 2005 by D Hamilton

Material: wood: *Quercus* sp., 10 rings (R Howard 2005)

Initial comment: rings 23–32 from floating tree-ring sequence WKW-A01.

Objectives: as SUERC-6540

Calibrated date: 1σ: cal AD 1165–1220
                       2σ: cal AD 1155–1245

Final comment: A Bayliss (24 August 2014), the weighted mean for block 8 has good agreement with its position in the sequence.

Laboratory comment: English Heritage (24 August 2014), the two results from block 8 are statistically consistent (T′=0.5; T′(5%)=3.8; ν=1; Ward and Wilson 1978), so a weighted mean has been taken before further analysis (860 ±14 BP).

References: Ward and Wilson 1978

SUERC–6567 870 ±20 BP

$\delta^{13}C$: -24.4‰

Sample: WKW-A01 Block 8, submitted on 8 March 2005 by D Hamilton

Material: wood: *Quercus* sp., 10 rings (R Howard 2005)

Initial comment: a replicate of SUERC-6566.

Objectives: as SUERC-6540

Calibrated date: *1σ:* cal AD 1160–1215
                  2σ: cal AD 1055–1220

Final comment: see SUERC-6566

Laboratory comment: see SUERC-6566

West Heslerton, Yorkshire (East Riding)

Location: SE 917765
                     Lat. 54.10.33 N; Long. 00.35.42 W

Project manager: D Powlesland (Landscape Research Centre), 1978–82, 1985, 1987, 1989–92, 1995

Description: West Heslerton and its hinterland have been the subject of continued archaeological research for more than three decades, including major excavations at Cook's Quarry, West Heslerton (site 1), which started as a rescue excavation between 1978 and 1982 and currently are continuing as the quarry continues to expand. Site 1 proved to be most important on account of the presence of extensive deposits of blown sand sealing evidence dating from the late Mesolithic to early medieval periods, with the key phases including a late Neolithic and early Bronze Age barrow cemetery, parts of a late Bronze Age settlement, and early Anglo-Saxon or Anglian cemetery. Following the publication of the excavations at site 1, excavation continued covering sites 2, 6, and 8, between 1984 and 1986, examining most of the remainder of the Anglian cemetery which had been superimposed on a late Neolithic and early Bronze Age monument complex. Small-scale evaluation excavations at sites 20 and 21 in 1984 were concerned with the sampling within a crop-mark complex, interpreted as a 'ladder settlement' dating from the middle Iron Age to post-Roman period. From 1986 until the end of 1995 work was concentrated on the rescue excavation of an Anglian settlement (sites 2, 11, 12, and 13), associated with the previously excavated cemetery but occupied from the fifth to ninth centuries, a longer duration than the cemetery which ceased to be used by the mid-seventh century. The evidence gathered both from excavation and very intensive aerial and ground-based survey has revealed the most detailed picture of an archaeological landscape for its scale in Britain; providing context for the excavations and an unparalleled insight into the evolution of settlement covering several thousand hectares.

Objectives: the excavations and the associated dating programmes at West Heslerton have covered an important period in the development of radiocarbon dating and its application to excavated datasets. During the early years of the project the objectives were simply to secure dates for

material where we were unsure of the date or to assist in defining the overall chronological sequence. As the precision of the dates returned has increased and the size of the samples required has reduced the dating programmes have been much more precisely targeted. Recently the dating programme has been directed towards two main objectives, the dating of the important prehistoric ceramic assemblage and the dating and sequencing of the vast excavation of the Anglian settlement.

References: Haughton and Powlesland 1999
                  Powlesland *et al* 1986

West Heslerton: environmental, North Yorkshire

Location: SE 9123277879
                     Lat. 54.11.19 N; Long. 00.36.07 W

Project manager: J Rackham (Environmental Archaeology Consultancy) and D Powlesland (The Landscape Research Centre), 2003

Archival body: The Landscape Research Centre

Description: the site constitutes the floor of the valley of West and East Heslerton parishes and Yedingham. It is a landscape covered with numerous ancient field systems revealed in cropmarks and through geophysical survey, as well as old river palaeochannels and landscapes buried by wind blown sands.

The fieldwork constituted a series of trenches laid over the locations of ancient field, trackway, and enclosure ditches revealed by aerial photographs and two trenches specifically targeted on the late-glacial Holocene (?) peat deposits buried beneath wind blown sands. In addition, a series of auger transects were laid out across the courses of palaeochannels revealed in crop marks. 27 trenches were excavated and revealed 29 ditches and 5 locations where the underlying buried peats were exposed. The auger survey comprised three auger transects across the palaeochannels of old courses of the River Derwent.

These investigations were carried out across an area of approximately 15 square kilometres, comprising several farms and different landowners. The valley floor is approximately 26m OD and is marked by a slight ridge along its centre with lower ground to the south before rising onto the Wolds and the present course of the River Derwent to the north before the land rises onto the Yorkshire Moors. The archaeology includes Neolithic, Bronze Age, Iron Age, Roman, Anglo-Saxon, and medieval features including cemeteries, funerary monuments, settlements, shrines, trackways, and field systems.

The present project is primarily targeted at the field systems and natural features of the landscape.

Objectives: the project objective is to establish whether or not deposits suitable for the reconstruction of the palaeoenvironmental history of the valley floor and its archaeological field systems survive in sufficiently good condition to warrant a research project on this rich archaeological landscape. The dating programme is designed to supply a chronology for the deposits and features investigated during this pilot project and being assessed for their palaeoenvironmental potential.

Each date from the field ditches can be attached to the results of the environmental assessment and used to date the fills of that particular ditch and possibly by inference and associated field system. The dating of the peat deposits will establish the age of this extensive deposit and a chronology for the environmental results from the assessment, whether the peats are contemporary across a significant area and possibly whether late Palaeolithic or Mesolithic activity could be expected around its perimeter or both.

The dating of the palaeochannel fills will establish whether the organic deposits are contemporary and directly relatable to any of the phases of archaeology already recorded in the valley as well as affording a chronology for the environmental results assessed within this project from the cores through the channel fills.

Final comment: D Powlesland (27 October 2014), OxA-13118, OxA-13053, GU-5996, and GU-5997 were derived from samples from an extensive peat deposit buried beneath blown and alluvial deposits which have been observed through auger survey and observation of drainage ditches over a large area of the research zone between the River Derwent and the edge of the former wetlands in West Heslerton. This peat deposit is clearly discontinuous and the auger survey and other observations alone cannot be used to infer that the deposit is the same in all locations although there does seem to be a degree of uniformity and there is some indication that the area was subject to a massive alluvial event which buried peat over a wide area. Two of the dates are from the late Palaeolithic and are supported by the pollen analysis also undertaken within the sampling programme.

References: Hunn and Rackham forthcoming
                    Powlesland *et al* 1986
                    Powlesland 2003a
                    Powlesland 2003b
                    Powlesland and Rackham 2007
                    Rackham *et al* in press

GU–5996 10830 ±120 BP

$\delta^{13}C$: -29.0‰

Sample: HPP/DS/132/50–52cm, submitted on 19 November 2003 by J Rackham

Material: peat (humic acid) (J Rackham 2003)

Initial comment: this sample was taken from the upper peat horizon at 50–52cm of a series of peats and organic muds underlying wind blown sands in Trench AA, Field 132. The organic deposits overlay glacial sands and gravels.

Objectives: the sample is designed to give a date for the top of the organic sediments occurring at this location that can be used to give a chronological context to any palaeoenvironmental results, particularly pollen analysis, forthcoming from these deposits.

Calibrated date: *1σ:* 10840–10730 cal BC
                      2σ: 11040–10610 cal BC

Final comment: D Powlesland (27 October 2014), this sample is medieval and derived from towards the base of the peat can only be intrusive, most likely derived from a deep penetrating root fragment.

Laboratory comment: English Heritage (November 2014), the two measurements are statistically consistent (T'=1.2; T'(5%)=3.8; v=1; Ward and Wilson 1978) and the weighted mean (10898 103 BP) calibrates to 11100–10830 cal BC (95% confidence; Reimer *et al* 2004).

References: Ward and Wilson 1978

GU–5997 11080 ±200 BP

$\delta^{13}C$: -30.0‰

Sample: HPP/DS/132/50–52cm, submitted on 19 November 2003 by J Rackham

Material: peat (humin) (J Rackham 2003)

Initial comment: as GU-5996

Objectives: as GU-5996

Calibrated date: *1σ:* 11170–10770 cal BC
                      2σ: 11370–10720 cal BC

Final comment: see GU-5996

OxA–13034 6414 ±38 BP

$\delta^{13}C$: -28.4‰

Sample: HPP/DS/76/AD/35–40cm (a), submitted on 19 November 2003 by J Rackham

Material: wood (waterlogged): *Alnus glutinosa*, roundwood single fragment (R Gale 2003)

Initial comment: this sample was taken from just above (35–40cm) the lowest organic sediment of a 45cm thick peat and organic sediment sequence beneath the archaeological ditches in Trench AD, Field 76. The deposit overlies glacial sands and gravels and appears to be a continuation of the peats found several hundred metres to the west (HPP/DS/132/AA/10–15cm and HPP/DS/132/AA/50–52cm) and in adjacent fields. The deposits were overlain by blown sands until these were dug out for the archaeological ditches at this location.

This portion of the peats lay below the present water table, and underlies the sequence of archaeological ditches above. It represents the first organic horizon in the sequence with sufficient organic content for a radiocarbon date at this location. The peat occurs over quite a wide area in this and adjacent fields. The organic deposits form a 0.45m thick band and this sample comes from the lower 35–40cms in the sequence.

Objectives: the sample is designed to give a date for the base of the organic sediments occurring at this location that can be used to give a chronological context to any palaeoenvironmental results, particularly pollen analysis, forthcoming from these deposits. It is also intended to establish whether the peats which are extensive across several fields in this area of the valley floor are contemporary.

Calibrated date: *1σ:* 5470–5320 cal BC
                      2σ: 5480–5310 cal BC

Final comment: D Powlesland (27 October 2014), despite serious peat loss owing to desiccation resulting from the combined impacts of drainage and irrigation throughout the

Vale of Pickering, early peat deposits still survive in some areas, particularly as here buried beneath blown sand and alluvial deposits. The lower part of the peat profile here, with a later Mesolithic date, and others recovered from ditches sectioned in this trench with early Bronze Age dates (OxA-13049, OxA-13050, OxA-13051, and OxA-13116) provide critical dates for the pollen analysis undertaken as part of this particular sampling project. Both this and the later early Bronze Age dates relate to material for which the pollen analysis which indicates very limited tree cover; this is particularly significant for the late Mesolithic context, but suggests that the preliminary sampling undertaken here should be followed up by a more detailed and targeted programme of sampling and analysis.

OxA–13035 371 ±27 BP

$\delta^{13}C$: -28.1‰

Sample: HPP/DS/76/AD/0–5 cm (a), submitted on 19 November 2003 by J Rackham

Material: wood (waterlogged): Salicaceae, roundwood single fragment (R Gale 2003)

Initial comment: as OxA-13034. This sample was taken from the top 5cm of a 45cm thick peat and organic sediment sequence beneath the archaeological ditches in Trench AD, Field 76.

Objectives: the sample is designed to give a date for the top of the organic sediments occurring at this location that can be used to give a chronological context to any palaeoenvironmental results, particularly pollen analysis, forthcoming from these deposits. It is also intended to establish whether the peats which are extensive across several fields in this area of the valley floor are contemporary.

Calibrated date: 1σ: cal AD 1450–1620
                       2σ: cal AD 1440–1640

Final comment:, the two anomalous dates OxA-13035 and OxA-13036 recovered from the top of the peat deposit here with Medieval/post-medieval dates indicate contamination from later ditch digging and interference at the limit of the water-table at the upper limit of this peat deposit. The samples recovered from site 76, which lies towards the southern edge of the former wetland area in the centre of the Vale, not only indicate that woodland was cleared but also that ditch systems were established as early as the early Bronze Age, perhaps within a larger landscape management regime, and that during the late medieval period new enclosures were being established within what must still have been seasonal wetlands.

OxA–13036 408 ±26 BP

$\delta^{13}C$: -27.8‰

Sample: HPP/DS/76/AD/0–5cm (b), submitted on 19 November 2003 by J Rackham

Material: wood (waterlogged): Salicaceae, roundwood (humic acid) (R Gale 2003)

Initial comment: as OxA-13035.

Objectives: as OxA-13035.

Calibrated date: 1σ: cal AD 1440–1480
                       2σ: cal AD 1430–1620

Final comment: see OxA-13035

OxA–13038 1820 ±31 BP

$\delta^{13}C$: -26.3‰

Sample: HPP/DS/108/AA/NORTH (b), submitted on 19 November 2003 by J Rackham

Material: wood (waterlogged): Salicaceae, roundwood single fragment (R Gale 2003)

Initial comment: the wood was found at the base of the fills of the north ditch of a small rectilinear enclosure. The wood came from the rich peaty fills of the north ditch in a fairly shallow context (approximately 0.5m deep).

Objectives: the sample is designed to give a date to the early fills of this ditch that can be used to give a context to any palaeoenvironmental results, particularly pollen analysis, forthcoming from these deposits.

Calibrated date: 1σ: cal AD 130–250
                       2σ: cal AD 90–320

Final comment: D Powlesland (27 October 2014), this sample and OxA-13223 were derived from the peat filled ditch of a small rectangular enclosure. This was documented from the air and partially through geomagnetic survey which revealed only the southern half of the enclosure (*c* 20m square). The enclosure was located some 210m to the west of a strongly magnetic feature, possibly a kiln site. The importance of the Roman date of this feature is that it indicates use of the land on the very margins of the former wetlands during the Roman period, where it was situated to the north of the linear/ladder settlement following the edge of the wetland and bounded in the late Roman period at least by a large flood-defence ditch. The pollen recovered from this dated sample indicated a locally pastoral, grassland habitat with some evidence of cereal cultivation (or crop processing) and waste ground with bracken nearby. There is also evidence of a background of deciduous trees. It is not possible to state how close because the pollen catchment and influx to such field boundary ditches is likely to be small with pollen assemblages derived from the vegetation on and very near site. *Salix*, however, is a poor pollen producer and was probably growing within the ditch. The presence of *Calluna* (ling) and *Erica* (heather) suggests that some soil deterioration had occurred by this date with areas of heathland growing on acidified soils (R Scaife, pers com).

OxA–13049 3600 ±31 BP

$\delta^{13}C$: -28.7‰

Sample: HPP/DS/76/AD/CENTRE (a), submitted on 19 November 2003 by J Rackham

Material: wood (waterlogged): *Alnus glutinosa*, single fragment (R Gale 2003)

Initial comment: the wood was found in the basal 5cm of the fills of the centre ditch in the east section of Trench AD, Field 76, part of a rectilinear enclosure at a depth of approximately 0.7m.

Objectives: the sample is designed to give a date to the early fills of this ditch that can be used to give a chronological context to any palaeoenvironmental results, particularly pollen analysis, forthcoming from these deposits.

Calibrated date: 1σ: 2020–1910 cal BC
                      2σ: 2040–1880 cal BC

Final comment: see OxA-13034

OxA–13050 3704 ±30 BP

δ¹³C: -29.1‰

Sample: HPP/DS/76/AD/CENTRE (b), submitted on 19 November 2003 by J Rackham

Material: wood (waterlogged): *Alnus glutinosa*, single fragment (R Gale 2003)

Initial comment: as OxA-13049

Objectives: as OxA-13049

Calibrated date: 1σ: 2140–2030 cal BC
                      2σ: 2200–1980 cal BC

Final comment: see OxA-13034

OxA–13051 3353 ±28 BP

δ¹³C: -27.6‰

Sample: HPP/DS/76/AD/SOUTH (a), submitted on 19 November 2003 by J Rackham

Material: wood (waterlogged): *Alnus glutinosa*, roundwood single fragment (R Gale 2003)

Initial comment: the wood was found at the base of the fills of the south ditch of a rectilinear enclosure at a depth of approximately 0.8m.

Objectives: as OxA-13049

Calibrated date: 1σ: 1690–1610 cal BC
                      2σ: 1740–1550 cal BC

Final comment: see OxA-13034

OxA–13052 315 ±24 BP

δ¹³C: -25.9‰

Sample: HPP/DS/76/AB/WEST (b), submitted on 19 November 2003 by J Rackham

Material: wood (waterlogged): Salicaceae, roundwood single fragment (R Gale 2003)

Initial comment: the wood was found at the base of the fills of the west ditch in Trench AB, Field 76, of a rectilinear enclosure at a depth of approximately 0.8m.

Objectives: as OxA-13049

Calibrated date: 1σ: cal AD 1515–1645
                      2σ: cal AD 1485–1650

Final comment: D Powlesland (27 October 2014), this sample and OxA-13117 were recovered to secure a date for a small rectangular enclosure measuring *c* 85m ∞ 55m

situated within the area of the former wetlands. This feature was unusual in that half appeared clearly in crop-mark images whilst the other half was better revealed in the geomagnetic survey results. Although no clearly defined entrance is seen in the images, a gap in the north-east corner is the probable location. There is no information from remote sensing to indicate internal structures, which would probably not show anyway, but the feature does reflect medieval use of the wetland margins, it may simply have been a stock management enclosure.

OxA–13053 6390 ±32 BP

δ¹³C: -27.7‰

Sample: HPP/DS/132/10–15cm (b), submitted on 19 November 2003 by J Rackham

Material: wood: *Alnus glutinosa*, roundwood single fragment (R Gale 2003)

Initial comment: these fragments were extracted from the organic sands at the base (10–15cm) of a sequence of peat and organic mud deposits, lying beneath wind blown sands in Field 132. the plant material probably derives from reed stems or reed roots of vegetation growing on the sands, but might have penetrated several centimetres through the lower organic deposits above. The sands from which this material has been extracted for the lowest organic deposit in a series of peats and organic muds over 0.55m thick.

Objectives: as OxA-13049

Calibrated date: 1σ: 5470–5320 cal BC
                      2σ: 5480–5300 cal BC

Final comment: D Powlesland (27 October 2014), this anomalous date sample from the late Mesolithic can similarly not be relied upon perhaps indicating that this deposit was subject to extensive later disturbance through root action, or influenced by high material mobility when the peat itself was un-desiccated, at which point the now thin deposit would have been over 2m thick.

OxA–13116 3391 ±34 BP

δ¹³C: -27.2‰

Sample: HPP/DS/AD/SOUTH (b), submitted on 19 November 2003 by J Rackham

Material: wood (waterlogged): *Alnus glutinosa*, roundwood single fragment (R Gale 2003)

Initial comment: as OxA-13051

Objectives: as OxA-13051

Calibrated date: 1σ: 1740–1630 cal BC
                      2σ: 1760–1610 cal BC

Final comment: see OxA-13034

OxA–13117 314 ±26 BP

δ¹³C: -26.1‰

Sample: HPP/DS/76/AB/WEST (a), submitted on 19 November 2003 by J Rackham

Material: wood (waterlogged): Salicaceae, roundwood single fragment (R Gale 2003)

Initial comment: as OxA-13052

Objectives: as OxA-13052

Calibrated date: *1σ:* cal AD 1510–1650
                       2σ: cal AD 1480–1650

Final comment: see OxA-13052

OxA–13118 933 ±28 BP

δ¹³C: -26.1‰

Sample: HPP/DS/132/10–15cm (a), submitted on 19 November 2003 by J Rackham

Material: wood (waterlogged; epidermal cylinder from herbaceous stem/root, unidentified monocotyledon), single fragment (R Gale 2003)

Initial comment: these fragments were extracted from the organic sands at the base (10–15cm) of a sequence of peat and organic mud deposits, lying beneath wind blown sands in Field 132.

Objectives: as OxA-13049

Calibrated date: *1σ:* cal AD 1030–1160
                       2σ: cal AD 1020–1170

Final comment: see GU-5996

OxA–13223 1873 ±32 BP

δ¹³C: -26.6‰

Sample: HPP/DS/108/AA/NORTH (a), submitted on 19 November 2003 by J Rackham

Material: wood (waterlogged): Salicaceae, roundwood single fragment (R Gale 2003)

Initial comment: as OxA-13038

Objectives: as OxA-13038

Calibrated date: *1σ:* cal AD 80–210
                       2σ: cal AD 60–240

Final comment: see OxA-13038

OxA–13353 2519 ±27 BP

δ¹³C: -28.3‰

Sample: HPP/A2/BH2A/100–110 A, submitted on 10 December 2003 by J Rackham

Material: wood (waterlogged): *Alnus glutinosa,* single fragment (R Gale 2003)

Initial comment: this wood has been extracted from organic sediments in then upper unhumified part of an organic and peat sequence infilling an old course of the River Derwent at a depth of 1.0m below the present ground surface.

Objectives: the sample is designed to give a date to the later fills of this old river channel that can be used to give a chronological context to any palaeoenvironmental results, particularly pollen analysis, forthcoming from these deposits.

This sample and HPP/A2/BH2A/140–145cm bracket the palaeoenvironmentally important deposits in this channel giving a near start and near end date for the sequence.

Calibrated date: *1σ:* 780–560 cal BC
                       2σ: 800–540 cal BC

Final comment: D Powlesland (27 October 2014), three samples designed to date the upper and lower deposits of a relict channel of the River Derwent indicate that this channel holds an environmental resource that documents the environmental sequence in the centre of the Vale of Pickering from the very end of the Palaeolithic until the middle Iron Age. As such the potential of this resource deserves more extensive investigation, not only because it will allow much of the dry land archaeology investigated at West Heslerton to be viewed within a broader environmental context, but because we are very aware of the rapid loss of this key resource as a consequence of increasing drainage, more intensive and invasive agriculture, and climate change.

OxA–13354 2541 ±27 BP

δ¹³C: -27.5‰

Sample: HPP/A2/BH2A/100–110 B, submitted on 10 December 2003 by J Rackham

Material: wood (waterlogged): *Alnus glutinosa,* single fragment (R Gale 2003)

Initial comment: as OxA-13353

Objectives: as OxA-13353

Calibrated date: *1σ:* 800–760 cal BC
                       2σ: 800–550 cal BC

Final comment: see OxA-13353

OxA–13355 9712 ±40 BP

δ¹³C: -28.4‰

Sample: HPP/A2/BH2A/140–145 A, submitted on 10 December 2003 by J Rackham

Material: wood (waterlogged): *Salix/Populus* sp., single fragment (R Gale 2003)

Initial comment: this wood has been extracted from organic sediments in the lower part of an organic and peat sequence infilling an old course of the River Derwent at a depth of 1.4m below the present ground surface.

Objectives: the sample is designed to give a date to the early fills of this old river channel that can be used to give a chronological context to any palaeoenvironmental results, particularly pollen analysis, forthcoming from these deposits.

This sample and HPP/A2/BH2A/100–110cm bracket the palaeoenvironmentally important deposits in this channel giving a near start and end date for the sequence.

Calibrated date: *1σ:* 9260–9190 cal BC
                       2σ: 9280–8950 cal BC

Final comment: see OxA-13353

OxA–13356 9390 ±40 BP

δ¹³C: -27.9‰

Sample: HPP/A2/BH7A/160–165 A, submitted on 10 December 2003 by J Rackham

Material: wood (waterlogged): *Salix/Populus* sp., single fragment (R Gale 2003)

Initial comment: this wood has been extracted from organic sediments in the upper unhumified part of an organic and peat sequence infilling an old course of the River Derwent at a depth of 1.6m below the present ground surface.

Objectives: the sample is designed to give a date to the later fills of this old river channel that can be used to give a chronological context to any palaeoenvironmental results, particularly pollen analysis, forthcoming from these deposits.

This sample and HPP/A2/BH7A/210–213cm bracket the palaeoenvironmentally important deposits in this channel giving a near start and near end date for the sequence.

Calibrated date: *1σ:* 8740–8620 cal BC
                        2σ: 8780–8560 cal BC

Final comment: see OxA13353

OxA–13357 4723 ±32 BP

δ¹³C: -28.9‰

Sample: HPP/A3/BH8A/100–105 A, submitted on 10 December 2003 by J Rackham

Material: wood (waterlogged): *Alnus glutinosa*, stem, single fragment (R Gale 2003)

Initial comment: the wood was extracted from the upper organic fills of an ancient course of the River Derwent at a depth of 1.0m below the present ground surface.

Objectives: the sample is designed to give a date to the later fills of this ancient river channel that can be used to give a chronological context to any palaeoenvironmental results, particularly pollen analysis, forthcoming from these deposits.

This sample and HPP/A3/BH8A/210–215cm bracket the palaeoenvironmentally important deposits in this channel giving a near start and near end date for the sequence.

Calibrated date: *1σ:* 3630–3380 cal BC
                        2σ: 3640–3370 cal BC

Final comment: D Powlesland (27 October 2014), this date, one of a series returned from auger transects across relict river channels in the centre of the Vale of Pickering collectively, were intended to allow us to assess the potential for the recovery of the environmental narrative from the relict channels and in particular attempt to identify areas for future research which could inform the results of the very extensive dry-land excavations in and around west Heslerton where environmental evidence has tended not to survive. They also served as a point in time sample set which might facilitate future monitoring of the rate of environmental evidence decay as the peat-lands associated with the River Derwent continue to suffer desiccation from aggressive agricultural drainage and climate change. This date confirms the potential for the recovery of late Neolithic evidence. The

date for the lower paired sample suggest that this part of the channel augered holds environmental data which spans the period from the late Mesolithic to mid to late Neolithic.

OxA–13358 4640 ±30 BP

δ¹³C: -29.8‰

Sample: HPP/A3/BH8A/100–105 B, submitted on 10 December 2003 by J Rackham

Material: wood (waterlogged): *Alnus glutinosa*, stem, single fragment (R Gale 2003)

Initial comment: as OxA-13357

Objectives: as OxA-13357

Calibrated date: *1σ:* 3500–3360 cal BC
                        2σ: 3520–3350 cal BC

Final comment: see OxA-13357

OxA–13359 6264 ±33 BP

δ¹³C: -28.9‰

Sample: HPP/A3/BH8A/210–215 A, submitted on 10 December 2003 by J Rackham

Material: wood (waterlogged): *Alnus glutinosa*, single fragment (R Gale 2003)

Initial comment: the wood was found in the basal organic fills of an ancient course of the River Derwent at a depth of 2.1m below the present ground surface.

Objectives: the sample is designed to give a date to the early fills of this ancient river channel that can be used to give a chronological context to any palaeoenvironmental results, particularly pollen analysis, forthcoming from these deposits.

Calibrated date: *1σ:* 5310–5210 cal BC
                        2σ: 5320–5200 cal BC

Final comment: D Powlesland (27 October 2014), *see* OxA-13357 This date and OxA-13360 give the lower date range of a single channel which seems to have infilled completely during the late Mesolithic to late Neolithic period.

OxA–13360 6173 ±37 BP

δ¹³C: -27.0‰

Sample: HPP/A3/BH8A/210–215 B, submitted on 10 December 2003 by J Rackham

Material: wood (waterlogged): *Alnus glutinosa*, single fragment (R Gale 2003)

Initial comment: as OxA-13359

Objectives: as OxA-13359

Calibrated date: *1σ:* 5220–5050 cal BC
                        2σ: 5230–5000 cal BC

Final comment: see OxA-13359

OxA–13361 77 ±25 BP

$\delta^{13}C$: -27.6‰

Sample: HPP/DS/102/AB/0–10 A, submitted on 19 November 2003 by J Rackham

Material: wood (waterlogged): Pomoideae, stem, single fragment (R Gale 2003)

Initial comment: the wood was extracted from a sample taken from the basal 10cm of the organic ditch fill.

Objectives: the sample is designed to give a date to the early fills of this ditch that can be used to give a context to any palaeoenvironmental results, particularly pollen analysis, forthcoming from these deposits.

Calibrated date: 1σ: cal AD 1700–1920
2σ: cal AD 1690–1930

Final comment: D Powlesland (27 October 2014), the post-medieval to modern dates secured from this sample and OxA-13362 confirmed suspicions made in the field that the ditch sectioned here was most likely recent; it was examined because there was some possibility that it may have related to a ditch complex examined in 102AA which was considered likely to be of Roman date. The presence of this modern feature, which remains waterlogged following the alignment of earlier and larger features shows a high degree of landscape continuity on the edge of the wetland even if it did not provide securely datable archaeo-environmental material.

OxA–13362 103 ±24 BP

$\delta^{13}C$: -27.7‰

Sample: HPP/DS/102/AB/0–10 B, submitted on 19 November 2003 by J Rackham

Material: wood (waterlogged): *Ulex/Cytisus* sp., stem, single fragment (R Gale 2003)

Initial comment: as OxA-13361

Objectives: as OxA-13361

Calibrated date: 1σ: cal AD 1690–1925
2σ: cal AD 1680–1940

Final comment: see OxA-13361

OxA–13363 5673 ±31 BP

$\delta^{13}C$: -26.7‰

Sample: HPP/DS/139/AA/0–5 A, submitted on 10 December 2003 by J Rackham

Material: wood (waterlogged): bark, unidentified, single fragment (R Gale 2003)

Initial comment: the wood was found in the basal 5cms of the fills of the sectioned ditch in Trench AA, Field 139, at the base of a humified peat layer, at a depth of approximately 0.6m.

Objectives: as OxA-13361

Calibrated date: 1σ: 4550–4450 cal BC
2σ: 4560–4450 cal BC

Final comment: D Powlesland (27 October 2014), this date is at odds with the analysis of the pollen recovered from the same feature which suggests a much earlier date. It may indicate that the feature (a shallow ditch) is Neolithic and that the pollen recovered is contamination from lower deposits, something which could have occurred when both were waterlogged. The two consistent dates from bark could be used to demonstrate a Neolithic date for the feature; this would be important in terms of understanding activity in the former wetlands during the middle of theNeolithic. The area deserves further examination.

OxA–13364 5189 ±33 BP

$\delta^{13}C$: -29.7‰

Sample: HPP/DS/139/AA/0–5 B, submitted on 10 December 2003 by J Rackham

Material: wood (waterlogged): bark, unidentified, single fragment (R Gale 2003)

Initial comment: as OxA-13361

Objectives: as OxA-13363

Calibrated date: 1σ: 4040–3960 cal BC
2σ: 4050–3950 cal BC

Final comment: see OxA-13363

OxA–13365 347 ±24 BP

$\delta^{13}C$: -26.8‰

Sample: HPP/DS/143/AA/SOUTH/0–10 A, submitted on 10 December 2003 by J Rackham

Material: wood (waterlogged): *Prunus spinosa*, stem, single fragment (R Gale 203)

Initial comment: these wood fragments were extracted from a bulk sample taken from the basal 10cms of the ditch fills of the southern ditch in Trench AA, Field 143. They occurred stratigraphically below a secondary fill containing a flint tool. The ditch is cut into sands and gravels.

Objectives: as OxA-13661

Calibrated date: 1σ: cal AD 1475–1635
2σ: cal AD 1455–1645

Final comment: D Powlesland (27 October 2014), this sample and OxA-13366 were intended to date a potential ditched trackway towards the centre of the Vale of Pickering. The dates recovered show that this was a late medieval feature an interpretation broadly supported by the pollen analysis which indicated a Roman or post-Roman date.

OxA–13366 319 ±23 BP

$\delta^{13}C$: -27.1‰

Sample: HPP/DS/143/AA/SOUTH/0–10 B, submitted on 10 December 2003 by J Rackham

Material: wood (waterlogged): *Ulex/Cytisus* sp., single fragment (R Gale 2003)

Initial comment: as OxA-13361

Objectives: as OxA-13365

Calibrated date: *1σ:* cal AD 1515–1645
                       2σ: cal AD 1480–1650

Final comment: see OxA-13365

OxA–13428 6620 ±45 BP

δ¹³C: -27.1‰

Sample: HPP/A2/BH7A/210–213 A, submitted on 10 December 2003 by J Rackham

Material: wood (waterlogged; 2 fragments): *Alnus glutinosa* (R Gale 2003)

Initial comment: as OxA-13359

Objectives: as OxA-13361

Calibrated date: *1σ:* 5620–5510 cal BC
                       2σ: 5640–5480 cal BC

Final comment: see OxA-13356

Wissett: Church of St Andrews, Suffolk

Location: TM 366792
                       Lat. 52.21.33 N; Long. 01.28.28 W

Project manager: D Hamilton (English Heritage), 2004–5

Archival body: English Heritage

Description: the parish church of Wissett with a round tower at the west end. Plans to replace the existing bellframe have raised the possibility of removing the second floor, which is thought to date to the twelfth century. Dendrochronological analysis has failed to provide a date for the floor, so the wiggle-matching has been requested as an alternative.

Objectives: wiggle-matching of this timber in the bell tower floor to provide a precise date for its construction.

Final comment: P Marshall (November 2014), a series of six radiocarbon measurements was undertaken on samples from the mean chronology once tree-ring analysis had failed to produce absolute dating. Wiggle-matching of these results suggests that these timbers were felled in *cal AD 1095–1135 (13% probablilty)* or *cal AD 1145–1205 (82% probability)*. Consequently, it appears that the second floor of the bell tower is a survival of the original twelfth-centry construction.

References: Bayliss *et al* 2006

OxA–14541 951 ±28 BP

δ¹³C: -24.0‰

Sample: SAW03 (2), submitted on 23 February 2005 by D Hamilton

Material: wood: *Quercus* sp., 10 rings (M Bridge 2005)

Initial comment: rings 11–20 from core SAW03, starting from the interior of the sequence. Core SAW03 comes from an eastern floorbeam. The innermost 3 rings and outermost 4 rings were removed prior to separating into 10–year blocks, due to them being degraded by insect attack.

Objectives: AMS wiggle-matching of this timber in the bell tower floor.

Calibrated date: *1σ:* cal AD 1020–1160
                       2σ: cal AD 1020–1170

Final comment: see series comments

OxA–14542 966 ±28 BP

δ¹³C: -24.0‰

Sample: SAW03 (3), submitted on 23 February 2005 by D Hamilton

Material: wood: *Quercus* sp., 10 rings (M Bridge 2005)

Initial comment: as OxA-14541. Rings 21–30 from core SAW03, starting from the interior of the sequence.

Objectives: as OxA-14541

Calibrated date: *1σ:* cal AD 1020–1150
                       2σ: cal AD 1010–1160

Final comment: see series comments

OxA–14543 918 ±28 BP

δ¹³C: -23.6‰

Sample: SAW03 (4), submitted on 23 February 2005 by D Hamilton

Material: wood: *Quercus* sp., 10 rings (M Bridge 2005)

Initial comment: as OxA-14541. Rings 31–40 from core SAW03, starting from the interior of the sequence.

Objectives: as OxA-14541

Calibrated date: *1σ:* cal AD 1040–1170
                       2σ: cal AD 1020–1210

Final comment: see series comments

OxA–14544 968 ±28 BP

δ¹³C: -24.9‰

Sample: SAW03 (5), submitted on 23 February 2005 by D Hamilton

Material: wood: *Quercus* sp., 10 rings (M Bridge 2005)

Initial comment: as OxA-14541. Rings 41–50 from core SAW03, starting from the interior of the sequence.

Objectives: as OxA-14541

Calibrated date: *1σ:* cal AD 1020–1150
                       2σ: cal AD 1010–1160

Final comment: see series comments

OxA–15035 941 ±24 BP

$\delta^{13}C$: -24.3‰

Sample: SAW01 (1), submitted on 5 August 2005 by D Hamilton

Material: wood: *Quercus* sp., 10 rings (M Bridge 2005)

Initial comment: as OxA-14541. Rings 1–10 from core SAW03, starting from the interior of the sequence.

Objectives: as OxA-14541

Calibrated date: 1σ: cal AD 1030–1155
2σ: cal AD 1020–1165

Final comment: see series comments

OxA–15036 972 ±26 BP

$\delta^{13}C$: -24.5‰

Sample: SAW01 (1), submitted on 5 August 2005 by D Hamilton

Material: wood: *Quercus* sp., 10 rings (M Bridge 2005)

Initial comment: a replicate of OxA-15035.

Objectives: as OxA-14541

Calibrated date: 1σ: cal AD 1020–1120
2σ: cal AD 1010–1160

Final comment: see series comments

OxA–X–2128–16 >58000 BP

$\delta^{13}C$: -22.7‰

Sample: SAW03 (EvoStik), submitted on 23 February 2005 by D Hamilton

Material: organic matter (Evostik wood glue (Batch no. 411340)) (D Hamilton 2005)

Initial comment: EvoStik used to attach a core sample to a lathe, which needs to be removed to avoid contamination.

Objectives: to see if EvoStik is geological or modern in age. This is to aid the interpretation of the AMS wiggle-matching of a core sample from the church that was attached to a lathe with EvoStik.

Final comment: P Marshall (November 2014), the sample of EvoStik was not pre-treated but was graphitised and measured by AMS. The result demonstrates that this product is petroleum-based and cannot have introduced modern contamination into the samples.

Wolverton Turn Enclosure, sunken-featured building, Buckinghamshire

Location: SP 80254066
Lat. 52.03.31 N; Long. 00.49.51 W

Project manager: S Preston (Thames Valley Archaeological Services), 1972, 1991, 1992, and 1994

Archival body: Buckinghamshire Museum

Description: the Wolverton Turn enclosure has been subject to a number of archaeological investigations since the early 1970s. This large sub-rectangular enclosure was identified initially from aerial photographs in 1969, and a part of it was first excavated in advance of construction in 1972, while an adjacent Bronze Age ring-ditch and associated burials were also excavated. Initially it was thought that the enclosure ditches were of Roman date. In 1991, a large area was evaluated in advance of construction, and a number of trenches excavated across the line of the enclosure ditches. This led to an excavation the following year which indicated a middle Saxon date for the main enclosure ditches, which seem to have been backed by a palisade. Except for one sunken-featured building well to the north, however, there was no evidence for structures which might have represented settlement. A geophysical survey of part of the site in 1992 proved inconclusive. Excavation in 1994 extended the 1992 Bronze Age settlement. The main period of use for the enclosure ditches was Saxon but the possibility of a Roman origin cannot be ruled out. The animal bones include a high proportion of horse, perhaps suggesting a specialist breeding or training centre.

Objectives: to establish the correlation with the pottery fabric, which is only putatively early or middle Saxon.

Final comment: S Preston (24 February 2005), a very useful confirmation of an early Saxon date (the ceramics gave no closer than early-to-middle Saxon). It was useful to get two dates so close in age from the same feature. The pottery dating was dependent on the presence/absence of decoration, so this secure date is a great improvement on what was, here, an argument for absence.

Laboratory comment: English Heritage (9 February 2005), the two measurements from this feature are statistically consistent (T'=0.0; T'(5%)=3.8; ν=1; Ward and Wilson 1978), and suggests that no residual material was dated from this context. This feature, therefore dates to cal AD 420–620.

References: Preston 2007
Ward and Wilson 1978

GrA–27202 1540 ±35 BP

$\delta^{13}C$: -22.2‰

Sample: WMC 91 169–1, submitted on 13 October 2004 by S Preston

Material: animal bone: *Bos* sp., carpal (S Anthony 2004)

Initial comment: from the bottom fill of sunken-featured building 168. It was possibly cut by posthole 174, but it is more likely they were contemporary.

Objectives: to establish if the sunken-featured building was contemporary with the ditched enclosure (series WMC92 ditch 1); to prrovide a date for the associated pottery, currently only vaguely 'early to middle Saxon'.

Calibrated date: 1σ: cal AD 430–570
2σ: cal AD 420–610

Final comment: S Preston (24 February 2005), this result confirms an early Saxon date and is clearly separate from the dated material in the ditch.

GrA–27203 1245 ±35 BP

δ¹³C: -23.0‰

Sample: WMC 92 088, submitted on 13 October 2004
by S Preston

Material: animal bone: *Equus* sp., metacarpal
(S Anthony 2004)

Initial comment: the sample came from the only fill of the
ditch segment. No other feature was cutting the ditch or was
or cut by it at this point.

Objectives: to provide a date for the enclosure and its
associated pottery.

Calibrated date: *1σ:* cal AD 690–780
                        2σ: cal AD 670–890

Final comment: S Preston (24 February 2005), this result
confirms the middle Saxon date as suspected, and provides
useful corroboration for the weak ceramic chronology.

OxA–14199 1541 ±28 BP

δ¹³C: -20.8‰

Sample: WMC 91 169–2, submitted on 13 October 2004
by S Preston

Material: animal bone (sheep/goat, metacarpal)
(S Anthony 2004)

Initial comment: as GrA-27202

Objectives: as GrA-27202

Calibrated date: *1σ:* cal AD 430–570
                        2σ: cal AD 420–600

Final comment: see GrA-27202

OxA–14200 1223 ±28 BP

δ¹³C: -19.4‰

Sample: WMC92 037, submitted on 13 October 2004
by S Preston

Material: animal bone: *Sus* sp., skull (petrous)
(S Anthony 2004)

Initial comment: from the primary fill of the ditch segment
where no other feature was identified. From a deposit of a
concentration of pig bones, including several skulls (several
animals) and very little other bone. There was one Roman
sherd amongst a small collection of Saxon pottery, but
residuality is unlikely. The records indicate the bone
clustered towards base of feature (up to 0.9m deep).

Objectives: as GrA-27203

Calibrated date: *1σ:* cal AD 720–870
                        2σ: cal AD 680–890

Final comment: see GrA-27203

Yarnton Neolithic and Bronze Age, Oxfordshire

Location: SP 470110
                     Lat. 51.47.41 N; Long. 01.19.07 W

Project manager: G Hey (Oxford Archaeology), 1989–2006

Description: the Yarnton Project provided a unique
opportunity to trace the development of settlement and
landscape change in a major river valley setting over the
entire span of the Neolithic and Bronze Age (*c* 4000–750 cal
BC). Scientific dating has been integral to the production of
an absolute chronological sequence, providing the
framework for understanding the changes observed over this
period of time.

A total of 127 radiocarbon determinations have been
obtained on samples of Neolithic and Bronze Age date from
Yarnton. Of these, 110 results were produced by the Oxford
Radiocarbon Accelerator Unit between 1994 and 2005; 74
on samples of charred plant remains, 29 on waterlogged
plant macrofossils, five on animal and human bone from the
second gravel terrace, and two on calcined human bone.
Two further samples of calcined human bone and six
samples of charred plant remains were dated using
Accelerator Mass Spectrometry (AMS) by the Scottish
Universities Environmental Research Centre between 2003
and 2005. Six samples of waterlogged wood were dated at
East Kilbride by Liquid Scintillation Spectrometry between
1998 and 2001. Two samples of waterlogged wood were also
dated conventionally by the Queen's University Belfast
Radiocarbon Dating Laboratory in 1996. Finally, a single
sample of carbonised bread was dated using AMS by the
Rafter Radiocarbon Laboratory, New Zealand in 1998. Two
optically-stimulated luminescence measurements of earlier
prehistoric date were also produced by the Oxford
University Research Laboratory in 1992.

Objectives: to establish the earliest date of Neolithic activity
at Yarnton, and assess whether use of the Yarnton landscape
was persistent or intermittent throughout the Neolithic and
Bronze Age periods;

to ascertain the date at which different parts of the Yarnton
floodplain were first cleared and used;

to assist in the identification of contemporary feature groups,
especially in relation to domestic activity;

to date the structures recovered, determine periods when
they are present/absent, and chart changes in form and use;

to date monuments in the landscape and assess the period of
time over which they remained significant;

to date human burials, thus charting changes in burial
practices over time. Particularly, to ascertain the period of
time over which unmarked inhumed and cremated
individuals were buried in the wider landscape, and the dates
at which small quantities of human bone were deposited and
mixed human and animal deposits were made. Also, to
establish the chronological relationship between human
burials and potentially associated monuments and
structures;

to date special or 'votive' activity;

to trace environmental change across the landscape, and chart changing land-use strategies and farming practices through time;

to understand the period at which different crafts and 'industrial' processes were undertaken and date other activity away from settlement;

to establish the chronological relationships of major pottery styles; and

to identify periods at which long-distance exchange networks are in evidence.

Final comment: A Bayliss and G Hey (28 November 2012), the scientific dating programme for Neolithic and Bronze Age remains at Yarnton has made a vital contribution to understanding the development of settlement and landscape change in the study area. The radiocarbon programme was severely restricted by the scarcity of datable material and by the spatially discrete character of many of the archaeological features. Thus, formal Bayesian modelling enabling greater chronological precision could be not be undertaken, except in a small number of cases. Nonetheless, the results acquired through scientific techniques have enabled us to compare developments in different parts of the landscape and construct a meaningful narrative of change through time from the fourth to the mid-first millennia cal BC. Without scientific methods, features would have remained undated or would have been misattributed, for example the small circular early Neolithic roundhouse, 5816, which was initially believed to be late Bronze Age in date.

The results have demonstrated human activity in the Yarnton study area from the early fourth millennium to the mid-first millennium cal BC, and the repeated use of this landscape over this period of time. A particular contribution has been made to the understanding of the date of early Neolithic activity and the diversity of the evidence present for this period, including structures, burials, and cereal production. The hypothesis that there was little evidence for house building in this area over the course of the second half of the fourth millennium and throughout the third millennium cal BC was supported by the results, but deposition within pits over this period of time was clearly demonstrated, with one possible exception.

An explosion of activity in the early-to-mid Bronze Age, principally in the fifteenth and fourteenth centuries cal BC, is suggested by the dating programme, including the start of a long tradition of roundhouse construction and evidence for a range of domestic activities including the digging of waterholes, the creation of burnt stone pits, and evidence for craft activities.

Evidence is present for cremation burial from the earliest Neolithic period through into the later Bronze Age at Yarnton. The scientific dating provided conclusive evidence for the longevity of use of the area around the Neolithic long enclosure for acts of formal deposition, especially those associated with human remains, and the importance of this place from the fourth to the end of the second millennium cal BC. The long-lived significance of a U-shaped enclosure on the second gravel terrace at Cresswell Field may similarly be demonstrated by the presence of three late-second millennium inhumation burials surrounding the monument.

The dating programme has provided a time-frame for landscape change and, in particular, the increasing extent of grazed grassland at the expense of tree cover on many parts of the floodplain in the fifteenth century cal BC, and a sharp rise in the water table at this period of time. The dating of features containing charred food remains has also been of great interest. In general terms, the results confirm the hypothesis that Neolithic assemblages are dominated by gathered plant foods, principally hazelnut shells, with few cereals present (Moffett *et al* 1989), and suggest that this pattern begins to be reversed in the first half of the second millennium cal BC. Cereals become more common in samples dating from the fifteenth century cal BC, and wild foods are very rare by this time. However, radiocarbon dating of features on the floodplain with comparatively large numbers of cereals showed, surprisingly, that a number of these were early Neolithic in date and, overall, it can be suggested that there was a period in the early fourth millennium when gathered foods were in use, but cereals were an important element of food assemblages.

| *References:* | Hey *et al* 2016 |
| | Moffett *et al* 1989 |

Yarnton Neolithic and Bronze Age: activity in early Neolithic rectangular structure 3871, Oxfordshire

Location:	SP 474109
	Lat. 51.47.39 N; Long. 01.18.45 W
Project manager:	G Hey (Oxford Archaeology), 1996
Archival body:	Oxfordshire County Museums Service

Description: an area of approximately 3ha was excavated on the Thames floodplain to the north of the A40 road between Yarnton and Cassington. The site consisted mainly of Neolithic and Bronze Age features cut into the natural gravels. These included a large Neolithic rectangular post-built structure, two waterholes, pit alignment, numerous small circular structures and fencelines, a ring-ditch, as well as a large number of scattered pits and postholes.

Objectives: to date the rectangular structure.

Final comment: G Hey and A Bayliss (28 November 2012), chronological modelling of the results suggests that the building was in use in the later part of the first quarter of the fourth millennium cal BC, probably in the 38th century (*c* 3800 cal BC). It should be noted, however, that material from the postholes of this structure was extremely scarce, and so our estimate for the dating of this building relies on only four measurements. The dates indicate that a cremation burial was placed in the top of a pit to the east of the structure some 100 years after it fell into disuse. Two further main phases of activity were evidenced in this area.

Three radiocarbon results show that one of these later periods of activity took place in the first half of the third millennium cal BC. One sample is from a posthole, 4391, lying within the western side of rectangular structure 3871 (OxA-6773), and two are residual in an area of disturbance (4591) in the south-west of this building (OxA-11881; 4241 ±34 BP, 2910–2710 cal BC at 95% confidence; Reimer *et al* 2004, and OxA-11919; 4193 ±34 BP, 2900–2660 cal BC at

95% confidence; Reimer *et al* 2004). This is believed to be a tree-throw hole, in which a series of burning events took place in the last quarter of the third millennium cal BC (OxA-11877, 3703 ±34 BP, 2210–1970 cal BC; OxA-11880, 3737 ±34 BP, 2280–2030 cal BC; OxA-11934, 3673 ±29 BP, 2140–1950 cal BC; OxA-11935 3732 ±29 BP, 2270–2030 cal BC; and OxA-11920, 3779 ±33 BP, 2300–2050 cal BC; at 95% confidence; Reimer *et al* 2004).

Laboratory comment: English Heritage (2005), four samples from two postholes on the main wall lines of a rectangular structure on site 7 were dated. These produced statistically consistent radiocarbon results (T′=7.7; T′(5%)=7.8; *v*=3; Ward and Wilson 1978), suggesting that this material probably does date to the construction and use of the building.

Laboratory comment: English Heritage (2014), two further radiocarbon measurements were obtained from this structure prior to 1998 and were published in Bayliss *et al* (2015, 243; OxA-6772–3). Ten further samples were dated before 2003 (OxA-11460, -11875–7, -11880–1, -11919–20, and -11934–5; Bayliss *et al* 2016, 313-4 (LHS)).

References: Bayliss *et al* 2015
Hey *et al* 2016
Reimer *et al* 2004
Ward and Wilson 1978

OxA–14479 4867 ±35 BP

Sample: YPFB96 3814 a, submitted in December 2004 by G Hey and C Hayden

Material: human bone (calcined; lower limb) (A Boyle 2004)

Initial comment: from human cremation deposit 3814 in the top of pit 3815. It had been placed in the top of a deep pit with a wooden container to the east of the rectangular structure.

Objectives: to establish to which phase the cremated remains belong. It is possible that they were related to the early Neolithic house nearby. If it can be dated, this deposit would add a significant element to the understanding of the character of activity within the house. It would also provide a *terminus ante quem* for the unusual remains of the wooden container in the base of the pit. In addition, the date will elucidate the period at which the Yarnton floodplain was used for burial, and the extent of such remains at its particular period.

Calibrated date: 1σ: 3660–3630 cal BC
2σ: 3710–3540 cal BC

Final comment: G Hey (28 November 2012), the result showed that the individual, who was cremated and placed in the top of a substantial pit just outside the east wall of the rectangular building, had lived in the early Neolithic but at a time after the house had gone out of use (around 200 years).

Laboratory comment: English Heritage (2005), two statistically-consistent measurements (OxA-14479 and SUERC-5689) were obtained on this individual (T′=3.5; T′(5%)=3.8; *v*=1; Ward and Wilson 1978). The weighted mean (4821 ±25 BP) indicates cremation occurred in 3655–3535 cal BC (95% confidence; Reimer *et al* 2004).

References: Reimer *et al* 2004
Ward and Wilson 1978

SUERC–5689 4775 ±35 BP

Sample: YPFB96 3814 b, submitted in December 2004 by G Hey and C Hayden

Material: human bone (calcined; lower limb) (A Boyle 2004)

Initial comment: as OxA-14479. A replicate measurement.

Objectives: as OxA-14479

Calibrated date: 1σ: 3640–3520 cal BC
2σ: 3650–3380 cal BC

Final comment: see OxA-14479

Laboratory comment: see OxA-14479

Yarnton Neolithic and Bronze Age: Bronze Age activity on sites 4c, 4e, 9, and 10, Oxfordshire

Location: SP 474109
Lat. 51.47.38 N; Long. 01.18.46 W

Project manager: G Hey (Oxford Archaeology), 1998

Archival body: Oxfordshire County Museums Service

Description: an unusual bark vessel was found in clay-lined pit 13058 on site 9. An area of Bronze Age activity (sites 4c and 4e) was located on a small gravel island in the centre of the floodplain. A circular structure (16209) was found in the north of site 4c. Pit 25045, lying 85m to the east of structure 16209 on site 4e, originally contained a sizeable wooden post, part of which remained *in situ*. The pit was recut and a substantial deposit of charred grain, animal bone, worked flint, and Deverel-Rimbury pottery was placed in the top. Waterhole 16010 lay to the south-east of structure 16209. Other features indicated that flax retting was taking place at Yarnton in the Bronze Age. Small pit 14034 containing part of a cremated adult human body was found on the north bank of a palaeochannel, 120m south of structure 16209 and on the same gravel island (site 10).

Objectives: the objectives in dating this sequence of samples were to place the activity within the overall chronology of settlement at Yarnton, to date the first use of the areas on which these sites were set, and to date important changes in the surrounding vegetation and the rising water table that occurred in the middle Bronze Age.

Final comment: G Hey (28 November 2012), dated activity on this central gravel island appears to have occurred from *c* 1750 cal BC to the later Bronze Age, although the majority of dated features fall in the third quarter of the second millennium cal BC.

Laboratory comment: English Heritage (24 June 2014), one further sample was dated prior to 2003 and is published in Bayliss *et al* (2016, 315; OxA–8929).

References: Hey *et al* 2016

OxA–12721 3204 ±30 BP

$\delta^{13}C$: -24.6‰

Sample: YFPB98 25046 <25009> A, submitted in April 2003 by C Chissell

Material: grain: *Hordeum* sp., charred single carbonised grain (M Robinson 2003)

Initial comment: this sample of charred grain came from the upper fill of pit 25045, isolated in the centre of site 4e. A posthole was cut into the base of this pit, and the lower part of the post remained *in situ*. This fill 25046 was clearly deposited as a discrete and intentional backfill to this pit.

Objectives: deposit 25046 contained a total of 153 carbonised cereal grains, the only significant deposit of cereal grain on the Yarnton floodplain, and more than the total found in all the other Neolithic/Bronze Age features. The pit itself contained an interesting series of deposits, including what appeared to be a large wooden post driven into the base of the pit. The deposit containing the grain also contained burnt stone, flint, pottery, and fragments of loom weight. Obtaining an accurate date for this large deposit of cereal grains will help us to date the increased use of cereals on the Yarnton floodplain for which there is little other evidence.

Calibrated date: 1σ: 1510–1430 cal BC
                      2σ: 1530–1410 cal BC

Final comment: G Hey (28 November 2012), this important deposit of charred grain is middle Bronze Age in date and it is likely to be contemporary with deposition structure 16209.

Laboratory comment: English Heritage (2005), the two cereal grains from this pit produced statistically consistent results (T'=0.1; T'(5%)=3.8; ν=1; Ward and Wilson 1978) which date this deposit to 1530–1410 cal BC (95% confidence; OxA-12721–2; Reimer *et al* 2004).

References: Reimer *et al* 2004
                Ward and Wilson 1978

OxA–12722 3220 ±30 BP

$\delta^{13}C$: -21.9‰

Sample: YFPB98 25046 <25009> B, submitted in April 2003 by C Chissell

Material: grain: *Triticum dicoccum*, single charred (M Robinson 2003)

Initial comment: as OxA-12721

Objectives: as OxA-12721

Calibrated date: 1σ: 1510–1440 cal BC
                      2σ: 1600–1420 cal BC

Final comment: see OxA-12721

Laboratory comment: see OxA-12721

OxA–12865 3087 ±29 BP

$\delta^{13}C$: -26.7‰

Sample: YFPB98 16017 <16085> A, submitted in April 2003 by C Chissell

Material: wood (waterlogged): *Quercus* sp., twig (M Robinson 2003)

Initial comment: from one of the primary fills of waterhole 16010.

Objectives: waterhole 16010 may have been a short-lived feature as no evidence of recutting was observed. It therefore offers a good opportunity to date the use of this waterhole, and its association with surrounding features. It will also provide a date for the environmental picture created from the organic deposits. The waterhole will provide further evidence of the overall density of the character of the prehistoric settlement on the floodplain.

Calibrated date: 1σ: 1420–1290 cal BC
                      2σ: 1430–1260 cal BC

Final comment: G Hey (28 November 2012), the use of this waterhole appears to post-date the structure, 16209, which lay nearby and suggests some longevity to the use of this part of the floodplain in the Bronze Age.

Laboratory comment: English Heritage (2005), the basal fill 16017 of waterhole 16010 produced two consistent measurements (T'=0.2; T'(5%)=3.8; ν=1; Ward and Wilson 1978) which show that this waterhole was in use at a slightly later date, probably in the fourteenth century cal BC.

References: Ward and Wilson 1978

OxA–12866 3068 ±28 BP

$\delta^{13}C$: -27.2‰

Sample: YFPB98 16017 <16085> B, submitted in April 2003 by C Chissell

Material: wood (waterlogged): *Quercus* sp., twig (M Robinson 2003)

Initial comment: as OxA-12865

Objectives: as OxA-12865

Calibrated date: 1σ: 1400–1280 cal BC
                      2σ: 1420–1230 cal BC

Final comment: see OxA-12865

Laboratory comment: see OxA-12865

OxA–12880 2820 ±110 BP

$\delta^{13}C$: -29.0‰

Sample: YFPB98 25024 <25004> A, submitted in April 2003 by C Chissell

Material: plant macrofossils (*Linum ustatissimum* seeds and capsules) (M Robinson 2002)

Initial comment: from the primary fill of isolated pit 25014. This deposit was waterlogged, contained organic material, and was sealed beneath four other deposits, the uppermost of which contained large quantities of finds.

Objectives: obtaining a date for the flax seeds and capsule fragments, as well as spelt wheat from the well-preserved organic remains within pit 25014, will contribute to our understanding of the development of crop processing activities across the floodplain.

Calibrated date: *1σ:* 1130–830 cal BC
                      2σ: 1290–790 cal BC

Final comment: G Hey (28 November 2012), the small sample of flax from pit 25014 indicates that flax retting was taking place at Yarnton in the later Bronze Age, and that this part of the gravel island was used from the early to late Bronze Age.

OxA–12938 3158 ±31 BP

δ¹³C: -25.5‰

Sample: YFPB98 16077 <16034> a, submitted in July 2003 by C Chissell

Material: grain: indeterminate single cereal grain, carbonised (M Robinson 2003)

Initial comment: this cereal grain was found within the only fill of posthole 16076 which formed part of circular structure 16209 in the north of site 4c.

Objectives: dating this post-built structure, and by association the features in its immediate vicinity, will allow us to confidently place the activity in this part of the site into the chronological framework for the Yarnton floodplain as a whole.

Calibrated date: *1σ:* 1490–1410 cal BC
                      2σ: 1500–1320 cal BC

Final comment: G Hey (28 November 2012), it is likely that the structure was in use during the fifteenth century cal BC. Structure 16209 is likely to be contemporary with pit 25045 lying 85m to its east on site 4e.

Laboratory comment: English Heritage (2005), the four samples from a circular structure 16209 (OxA-12938, -12964, 12965, and -12966), found in the north of site 4c, produced statistically inconsistent radiocarbon results (T′=17.0; T′(5%)=7.8; ν=3; Ward and Wilson 1978).

References: Ward and Wilson 1978

OxA–12964 3141 ±31 BP

δ¹³C: -25.1‰

Sample: YFPB98 16069 <16030>, submitted in July 2003 by C Chissell

Material: charcoal: *Prunus* sp., single fragment (R Gale 2003)

Initial comment: from within the only fill from posthole 16068 which formed part of circular structure 16209 in the north of site 4c.

Objectives: as OxA-12938

Calibrated date: *1σ:* 1440–1400 cal BC
                      2σ: 1500–1300 cal BC

Final comment: see OxA-12938

Laboratory comment: see OxA-12938

OxA–12965 3236 ±30 BP

δ¹³C: -24.6‰

Sample: YFPB98 16069 <16030> B, submitted in July 2003 by C Chissell

Material: charcoal: *Quercus* sp., sapwood single fragment (R Gale 2003)

Initial comment: as OxA-12938

Objectives: as OxA-12938

Calibrated date: *1σ:* 1530–1450 cal BC
                      2σ: 1610–1430 cal BC

Final comment: see OxA-12938

Laboratory comment: see OxA-12938

OxA–12966 3300 ±31 BP

δ¹³C: -25.1‰

Sample: YFPB98 16077 <16034> b, submitted in July 2003 by C Chissell

Material: charcoal: *Prunus* sp., single fragment (R Gale 2003)

Initial comment: as OxA-12938

Objectives: as OxA-12938

Calibrated date: *1σ:* 1630–1520 cal BC
                      2σ: 1660–1500 cal BC

Final comment: G Hey (28 November 2012), it is likely that this sample was residual.

Laboratory comment: see OxA-12938

OxA–14492 3136 ±29 BP

Sample: YPFB98 14033a = 14007, submitted in December 2004 by G Hey and C Hayden

Material: human bone (calcined) (A Boyle 2004)

Initial comment: this bone is from the single fill 14033 of shallow pit 14034, part of a small deposit (49g). There were no other finds.

Objectives: to establish to which phase these cremated remains belong. The cremation lies away from the main areas of prehistoric activity, and a date will shed light on contemporary activity and help date the period at which isolated cremation burials were made. It will also broaden our understanding of the spatial organisation of activity in that phase, in particular the extent of the burials.

Calibrated date: *1σ:* 1440–1400 cal BC
                      2σ: 1500–1300 cal BC

Final comment: G Hey (28 November 2012), this person could have drawn water from waterhole 16010.

Laboratory comment: English Heritage (2005), pit 14034 containing part of a cremated adult human body produced two statistically consistent radiocarbon determinations (T′=2.1; T′(5%)=3.8; ν=1; Ward and Wilson 1978), suggesting that this person was cremated in 1440–1310 cal BC (95% confidence; Reimer *et al* 2004).

References: Reimer *et al* 2004
                      Ward and Wilson 1978

SUERC–5695 3070 ±35 BP

Sample: YPFB98 14033 b, submitted in December 2004 by 09/10/2008

Material: human bone (calcined) (A Boyle 2004)

Initial comment: as OxA-14492

Objectives: as OxA-14492

Calibrated date: 1σ: 1410–1270 cal BC
2σ: 1430–1220 cal BC

Final comment: see OxA-14492

Laboratory comment: see OxA-14492

Yarnton Neolithic and Bronze Age: early and middle Neolithic pits, Oxfordshire

Location: SP 474109
Lat. 51.47.39 N; Long. 01.18.45 W

Project manager: G Hey (Oxford Archaeology), 1992

Archival body: Oxfordshire County Museums Service

Description: of the widespread scatter of pits across the Yarnton landscape, four have been dated by radiocarbon to the early and middle Neolithic.

Objectives: this series of measurements will hopefully provide dates for the individual features and help to establish whether this group of pits is genuinely contemporary and whether the Neolithic occupation on the site represents a single phase of occupation or several separately defined phases.

Final comment: G Hey (28 November 2012), the dating of pits across the floodplain provided important information on the period over which the Yarnton floodplain was occupied, the dates at which different parts of this area were brought into use, and the persistence of this activity.

Laboratory comment: English Heritage (19 June 2014), six further samples from this series were dated prior to 1998 and were published in Bayliss *et al* (2015, 249; OxA-4661–2, -6412–3, -7716, and NZA-8679). Two further samples were dated prior to 2003 (OxA-11513–4; Bayliss *et al* 2016, 317-8).

References: Bayliss *et al* 2015
Hey *et al* 2016

OxA–14447 4957 ±34 BP

δ¹³C: -23.9‰

Sample: YFP92 2351 a, submitted in December 2004 by G Hey and C Hayden

Material: grain: *Triticum spelta*, or *dioccum* single carbonised grain (M Robinson 2004)

Initial comment: from the primary fill 2351of pit 2349. The pit cut into the top of a tree throw hole adjacent to a midden deposit on site 2.

Objectives: to establish the chronological relationship between the deposits of animal bone, pottery, and cereals in the pit and other activity on the site, especially other middle Neolithic features. The deposit of animal bone, pottery, and grain is substantial, and has significant implications for the character of the activity on the site.

Calibrated date: 1σ: 3780–3690 cal BC
2σ: 3800–3650 cal BC

Final comment: G Hey (28 November 2012), although the Neolithic result for the cereals in this pit was not a surprise, their early date had not been expected. They form part of the earliest activity on this gravel island, along with the early Neolithic midden deposits, and are contemporary with the Neolithic rectangular house to the west.

Laboratory comment: English Heritage (2005), these samples (OxA-14447 and SUERC-5686) produced two statistically consistent radiocarbon measurements (T'=0.0; T'(5%)=3.8; ν=1; Ward and Wilson 1978), which suggest that the cereal was grown in the 38th century cal BC during the use of the rectangular structure 3871.

References: Ward and Wilson 1978

SUERC–5686 4965 ±35 BP

δ¹³C: -23.7‰

Sample: YFP92 2351 b, submitted in December 2004 by G Hey and C Hayden

Material: grain: indeterminate single cereal grain, carbonised (M Robinson 2004)

Initial comment: as OxA-14447

Objectives: as OxA-14447

Calibrated date: 1σ: 3790–3700 cal BC
2σ: 3900–3650 cal BC

Final comment: see OxA-14447

Laboratory comment: see OxA-14447

Yarnton Neolithic and Bronze Age: early Neolithic circular structure 5816, Oxfordshire

Location: SP 473106
Lat. 51.47.28 N; Long. 01.18.51 W

Project manager: G Hey (Oxford Archaeology), 1992

Archival body: Oxfordshire County Museums Service

Description: two samples from the postpipes of two postholes (5060 in posthole 5059, and 5077 in posthole 5076) in small circular structure 5816 excavated on site 3 were dated.

Objectives: to date the use of structure 5816, and thus allow comparison with: a nearby late Bronze Age waterhole (together the features would imply a significant focus of activity in this period); other late Bronze Age activity (eg structures in the palaeochannel); and earlier or later activity (eg building 9568) and thus the continuity or discontinuity of activity across the site.

Final comment: G Hey (28 November 2012), the date of this structure, based on its morphology, was thought to be Bronze Age, and nearby features suggested it may have been late in that period. The early Neolithic results were a complete surprise. This building has no close parallels.

Laboratory comment: English Heritage (2005), the four results form a coherent group and suggest that the structure was in use in the second half of the second quarter of the fourth millennium cal BC, probably in the second half the 37th century and 36th century BC (*c* 3600 cal BC). Thus, the use of this structure post-dates the rectangular building by *c* 200 years. It is likely that the individual (3814) whose cremated remains were placed in the top of pit 3815 next to the rectangular building was cremated during the period that structure 5816 was in use (*75.7% probable*).

References: Hey *et al* 2016

OxA–14444 4700 ±0 BP

δ¹³C: -22.7‰

Sample: YFP92 5077 b, submitted in December 2004 by G Hey and C Hayden

Material: carbonised plant macrofossil (single nutshell, *Corylus avellana*) (M Robinson 2004)

Initial comment: from the postpipe of posthole 4302, associated with post-Deverel-Rimbury pottery. The debris was presumed to have been incorporated into the posthole during the use of the building.

Objectives: to date the use of structure 5816 of which this posthole forms a part.

Calibrated date: 1σ: 3520–3380 cal BC
                          2σ: 3620–3375 cal BC

Final comment: G Hey (28 November 2012), the early Neolithic date of this sample was not anticipated. The pottery has now been reassessed and is now believed to be Neolithic Bowl.

OxA–14446 4850 ±0 BP

δ¹³C: -23.7‰

Sample: YFP92 5060 b, submitted in December 2004 by G Hey and C Hayden

Material: charcoal: Pomoideae, single fragment (M Robinson 2004)

Initial comment: from the postpipe of posthole 5059, associated with post-Deverel-Rimbury pottery. The debris was presumed to have been incorporated into the posthole during the use of the building.

Objectives: as OxA-14444

Calibrated date: 1σ: 3650–3640 cal BC
                          2σ: 3650–3635 cal BC

Final comment: see OxA-14444

SUERC–5687 4795 ±40 BP

δ¹³C: -27.3‰

Sample: YFP92 5060 a, submitted in December 2004 by G Hey and C Hayden

Material: carbonised plant macrofossil (single nutshell, *Corylus avellana*) (M Robinson 2004)

Initial comment: as OxA-14446

Objectives: as OxA-14444

Calibrated date: 1σ: 3640–3520 cal BC
                          2σ: 3660–3380 cal BC

Final comment: see OxA-14444

SUERC–5688 4780 ±35 BP

δ¹³C: -23.5‰

Sample: YFP92 5077 a, submitted in December 2004 by G Hey and C Hayden

Material: grain: *Triticum dicoccum,* or *spelta,* single carbonised grain (M Robinson 2004)

Initial comment: as OxA-14444

Objectives: as OxA-14444

Calibrated date: 1σ: 3640–3520 cal BC
                          2σ: 3650–3380 cal BC

Final comment: see OxA-14444

Yarnton Neolithic and Bronze Age: late Neolithic structure 4291, Oxfordshire

Location: SP 469107
                   Lat. 51.47.32 N; Long. 01.19.12 W

Project manager: G Hey (Oxford Archaeology), 1996

Archival body: Oxfordshire County Museums Service

Description: samples from two postpits (4263 and 4302) of Neolithic structure 4291 were submitted for radiocarbon dating.

Objectives: the recovery of late Neolithic structures not associated with henges in England is highly unusual. Its date will add to our knowledge of the periods at which more substantial buildings were constructed at Yarnton, and will aid our understanding of the character of late Neolithic settlement and its complexity. It will also be possible to compare this occupation area to that of other late Neolithic activity on the floodplain (eg in the area of the early Neolithic house) with the aim of determining the possible duration of later Neolithic activity. Comparisons with earlier and later activity to assess chronological continuity/discontinuity will also be possible.

Final comment: G Hey (28 November 2012), the radiocarbon results support the tentative date provided by nine sherds of Grooved Ware pottery from the features which made up the structure, although they sit on the cusp of the late Neolithic and early Bronze Age. Such structures are rare and this provides a well-dated example. It also provides and good and interestingly late date for Grooved Ware pottery.

Laboratory comment: English Heritage (2005), these samples produced four statistically consistent radiocarbon measurements (T′=3.8; T(5%)=7.8; *v*=3; ward and Wilson 1978), which indicate that the structure was occupied for a relatively short period in the third quarter of the third millennium cal BC (probably in the 25th or the 24th century cal BC).

| *References:* | Hey *et al* 2016 |
| | Ward and Wilson 1978 |

OxA–14445 3869 ±32 BP

δ¹³C: -25.8‰

Sample: YFPB96 4300 b, submitted in December 2004 by G Hey and C Hayden

Material: carbonised plant macrofossil (nutshell, *Corylus avellana*) (M Robinson 2004)

Initial comment: a nut shell from the upper fill of posthole 4302, 0.69m wide and 0.2m deep. The debris was presumed to have been incorporated into the posthole during the use of the building.

Objectives: to establish whether the probable structure is late Neolithic in date.

| *Calibrated date:* | *1σ:* 2460–2280 cal BC |
| | *2σ:* 2470–2200 cal BC |

Final comment: G Hey (28 November 2012), this sample contributes to an overall date for the structure in the third quarter of the second millennium.

OxA–14448 3919 ±33 BP

δ¹³C: -25.5‰

Sample: YFPB96 4264 b, submitted in December 2004 by G Hey and C Hayden

Material: carbonised plant macrofossil (*Malus/Prunus* sp., pip) (M Robinson 2004)

Initial comment: from the single fill of postpit 4263, 0.8m wide and 0.19m deep. The debris was presumed to have been incorporated into the posthole during the use of building 4291.

Objectives: as OxA-14445

| *Calibrated date:* | *1σ:* 2470–2340 cal BC |
| | *2σ:* 2490–2290 cal BC |

Final comment: see OxA-14445

SUERC–5690 3900 ±40 BP

δ¹³C: -24.5‰

Sample: YFPB96 4264 a, submitted in December 2004 by G Hey and C Hayden

Material: carbonised plant macrofossil (nutshell, *Corylus avellana*) (M Robinson 2004)

Initial comment: as OxA-14448

Objectives: as OxA-14445

| *Calibrated date:* | *1σ:* 2470–2290 cal BC |
| | *2σ:* 2480–2210 cal BC |

Final comment: see OxA-14445

SUERC–5694 3960 ±35 BP

δ¹³C: -25.6‰

Sample: YFPB96 4300 a, submitted in December 2004 by G Hey and C Hayden

Material: carbonised plant macrofossil (nutshell, *Corylus avellana*) (M Robinson 2004)

Initial comment: as OxA-14445

Objectives: as OxA-14445

| *Calibrated date:* | *1σ:* 2550–2460 cal BC |
| | *2σ:* 2570–2340 cal BC |

Final comment: see OxA-14445

Yarnton Neolithic and Bronze Age: waterholes and burnt stone features, sites 17 and 21, Oxfordshire

Location:	SP 474109
	Lat. 51.47.39 N; Long. 01.18.45 W
Project manager:	G Hey (Oxford Archaeology), 1998
Archival body:	Oxfordshire County Museums Service

Description: features uncovered on site 21 revealed a variety of activity taking place during the early Bronze Age. A waterhole, 15014, was dug into the base of the channel and must pre-date the reactivation of this channel. An important deposit of worked wood was dumped into the top of the waterhole after it went out of use, which included a log ladder and a wooden bowl, both made of alder. North of the waterhole was a spread of burnt stone and a pit packed with burnt stone and charcoal. Another waterhole, 15072, was dug on the bank of the channel. A waterhole, 10159, a square-sided pit filled with burnt stone and charcoal, 10022, and spreads of burnt stone were found on the north bank of a palaeochannel on site 17.

Objectives: the objectives in dating this sequence of samples were to place the activity within the overall chronology of settlement at Yarnton, to date the first use of the areas on which these sites were set, to date a series of unusual artefacts, to date important changes in the surrounding vegetation and the rising water table that occurred in the middle Bronze Age.

Final comment: G Hey (28 November 2012), activity on site 17 probably began in the fifteenth century cal BC, overlapping in date with the later features on site 21.

Laboratory comment: English Heritage (6 October 2014), two further samples were dated before 2003 (OxA-8673 and OxA-9779; Bayliss *et al* 2016, 321-2).

| *References:* | Hey *et al* 2016 |

OxA–12838 3053 ±31 BP

$\delta^{13}C$: -26.0‰

Sample: YFPB97 10020 <10005> B, submitted in April 2003 by C Chissell

Material: charcoal: Pomoideae, single carbonised grain (M Robinson 2003)

Initial comment: from a single fill in pit 10022. The pit was square-sided and filled with burnt stone and charcoal, and situated to the north of waterhole 10159.

Objectives: the pit was packed with burnt stone and charcoal and would appear to be associated with burnt mound activity. The activity represented by such features is poorly understood. Often thought to be associated with food preparation, the Yarnton burnt stone deposits are not associated with domestic refuse, and as such may offer a different picture. The dating of this burnt mound is important as this form of activity is often Bronze Age in date, however, the arrangement of burnt mounds adjacent to waterholes may be earlier in date. This pit did not contain any datable artefacts, so radiocarbon dating would allow the burnt mound activity to be placed in its immediate context of other securely dated features on the Yarnton site, and regional parallels can also be sought. The quantity and range of burnt mound deposits within the excavation area also provides the opportunity to examine the chronological span of this activity and to explore any changes in the character of the activity over time.

Calibrated date: 1σ: 1390–1260 cal BC
2σ: 1420–1210 cal BC

Final comment: G Hey (28 November 2012), the later date for this pit, in the fourteenth or thirteenth century cal BC, seems most likely, and also its use with the final cut of waterhole 10159. However, there is a strong likelihood that the residual charcoal is associated with long-lived burnt-mound activity in this area, as evidenced by spreads and dumps of this material seen in evaluation along the northern edge of this palaeochannel.

Laboratory comment: English Heritage (2005), a bulk sample of short-lived charcoal from the pit produced a radiocarbon measurement (SUERC-1146) which is significantly earlier than that on the single fragment from the same context (OxA-12838; T'=7.4; T'(5%)=3.8; ν=1; Ward and Wilson 1978). There are two possible explanations for this. Either OxA-12838 is intrusive, in which case SUERC-1146 provides the best estimate of the date of this feature, falling in the fifteenth century cal BC, and so contemporary with waterhole 10159. More probably, sample SUERC-1146 contained reworked material, and OxA-12838 more accurately reflects the date of the pit, in the fourteenth or thirteenth centuries cal BC. In the latter case, the activity might be related to the undated fourth cut of waterhole 10159, which contained the majority of the burnt stone from that feature. The activity which produced burnt stone and charcoal, and which might be associated with the use of the waterhole, may have been long-lived.

References: Ward and Wilson 1978

OxA–12839 3304 ±29 BP

$\delta^{13}C$: -26.1‰

Sample: YFPB98 15034 <21056> A, submitted in April 2003 by C Chissell

Material: charcoal: Pomoideae, single carbonised grain (M Robinson 2003)

Initial comment: from the upper fill of sub-rectangular pit 15010. The fill comprised 50–60% charcoal and 35–40% burnt stone which is likely to represent redeposited material from a burnt mound.

Objectives: a large number of features containing burnt stone were found along the banks of the channels in the western parts of the Yarnton floodplain. It is important to establish a date for these features in order to better interpret the activity represented by these deposits, and to clarify the relationship between them, the channels, causeways, waterholes, and other domestic features in the vicinity.

Calibrated date: 1σ: 1630–1520 cal BC
2σ: 1660–1500 cal BC

Final comment: G Hey (28 November 2012), this charcoal is part of the burnt-mound material found near to the palaeochannel and waterholes on this gravel island, and the date shows that the activity which generated these deposits was taking place here at the early stages of the middle Bronze Age.

Laboratory comment: English Heritage (2005), the short-lived fragments of wood charcoal suggest that the feature was filled around 1600 cal BC, and is broadly contemporary with waterhole 15014.

OxA–12840 3364 ±28 BP

$\delta^{13}C$: -26.3‰

Sample: YFPB98 15034 <21056> B, submitted in April 2003 by C Chissell

Material: charcoal: Pomoideae, single carbonised grain (M Robinson 2003)

Initial comment: as OxA-12839

Objectives: as OxA-12839

Calibrated date: 1σ: 1690–1620 cal BC
2σ: 1740–1610 cal BC

Final comment: see OxA-12839

Laboratory comment: see OxA-12839

OxA–12841 3198 ±28 BP

$\delta^{13}C$: -28.5‰

Sample: YFPB97 10144 <10015> A, submitted in April 2003 by C Chissell

Material: waterlogged plant macrofossil: *Crataegus monogyna* stone (M Robinson 2003)

Initial comment: from the primary fill of the second recut 10165 of waterhole 10159. The waterhole was found on the north bank of a palaeochannel on site 17. The large waterhole had been recut on four occasions.

Objectives: the material within context 10144 provides an opportunity to date the use of waterhole 10159, assess the frequency of the recutting, and also date the character of the environment at the time the waterhole was in use. This waterhole may have been associated with burnt mound activity. It was located immediately south of pit 10020, filled by burnt stone and charcoal, and the fills of the final waterhole recut contained large quantities of burnt stone. The waterhole may therefore have been abandoned at the same time as the burnt mound, and the burnt mound activity could have been contemporary with the use of the waterhole. The lack of occupation features and debris further points to this waterhole representing off-site activity.

Calibrated date: 1σ: 1500–1430 cal BC
                    2σ: 1520–1410 cal BC

Final comment: G Hey (28 November 2012), these results place the filling of the waterhole in the fifteenth century cal BC, and suggest that the fills of the second and third recuts accumulated over a relatively short period of time (probably less than a century).

Laboratory comment: English Heritage (2005), two samples from the primary fill of the second recut, 10165, and two samples from a basal fill of the third recut, 10154, were dated, producing statistically consistent measurements (T′=1.0; T′(5%)=7.8; ν=3; Ward and Wilson 1978). As these samples were waterlogged plant remains they are unlikely to have been reworked, and indeed the measurements show good agreement with their relative stratigraphic positions.

References: Ward and Wilson 1978

OxA–12862 3199 ±29 BP

δ¹³C: -26.2‰

Sample: YFPB97 10144 <10015> B, submitted in April 2003 by C Chissell

Material: waterlogged plant macrofossil: *Prunus spinosa*, stone (M Robinson 2003)

Initial comment: as OxA-12841

Objectives: as OxA-12841

Calibrated date: 1σ: 1510–1430 cal BC
                    2σ: 1530–1410 cal BC

Final comment: see OxA-12841

Laboratory comment: see OxA-12841

OxA–12863 3198 ±30 BP

δ¹³C: -26.0‰

Sample: YFPB97 10173 <10020> A, submitted in April 2003 by C Chissell

Material: waterlogged plant macrofossil: *Crataegus monogyna* stone (M Robinson 2003)

Initial comment: as OxA-12841, from the third fill of the third recut 10154 of waterhole 10159.

Objectives: as OxA-12841. This fill is also of particular importance as it provides the opportunity to date the deposition of the wooden artefact.

Calibrated date: 1σ: 1510–1430 cal BC
                    2σ: 1530–1410 cal BC

Final comment: see OxA-12841

Laboratory comment: see OxA-12841

OxA–12864 3165 ±29 BP

δ¹³C: -27.8‰

Sample: YFPB97 10173 <10020> B, submitted in April 2003 by C Chissell

Material: waterlogged plant macrofossil: *Prunus spinosa*, stone (M Robinson 2003)

Initial comment: as OxA-12863

Objectives: as OxA-12863

Calibrated date: 1σ: 1500–1410 cal BC
                    2σ: 1500–1390 cal BC

Final comment: see OxA-12841

Laboratory comment: see OxA-12841

OxA–12885 3169 ±33 BP

δ¹³C: -26.9‰

Sample: YFPB98 15076, submitted in February 2003 by C Chissell

Material: carbonised residue (interior) (A Barclay 2002)

Initial comment: from deposit 15076 within waterhole 15072 on site 21. This deposit was the fourth fill (of 12) from the base of the waterhole, which had a high organic content, and also contained bone and flint. This fill probably represents a period of natural accumulation just after/during the abandonment of the waterhole.

Objectives: the pottery is from a rare form of vessel which is difficult to date conventionally. The fabric indicates a middle Bronze Age date, but the form suggests a late Bronze Age date, therefore radiocarbon dating will help to identify to which period it belongs, as well as a date for the waterhole falling out of use. It will then be possible to determine whether this waterhole was likely to have been contemporary with the other Bronze Age waterholes and domestic features in the area, or whether it is an indication of some later activity.

Calibrated date: 1σ: 1500–1410 cal BC
                    2σ: 1510–1390 cal BC

Final comment: G Hey (28 November 2012), the probability distribution of this calibrated date suggests that this Globular Urn vessel is most likely to have been used in the fifteenth century cal BC, and belongs to the Deverel-Rimbury tradition. Waterhole 15072 is likely to have been in use after waterhole 15014 had filled, by which time the channel would have become active.

OxA–12886 3437 ±35 BP

δ¹³C: -26.1‰

Sample: YFPB98 15113 <21063> A, submitted in April 2003 by C Chissell

Material: wood (waterlogged): *Quercus* sp., twig (M Robinson 2003)

Initial comment: from the lowest fill of waterhole 15014 on site 21. Context 15113 was a waterlogged deposit derived from natural silting, organic accumulation, and deliberate deposition. Some of the material undoubtedly derived from the partial decomposition of timbers.

Objectives: it is possible that this waterhole pre-dates others found in the vicinity due to its location in the base of the palaeochannel. A date for this feature combined with other dates for other waterholes on the floodplain will enhance our understanding of the environment and landuse of the area and provide information on changing hydrological conditions.

Calibrated date: 1σ: 1860–1690 cal BC
2σ: 1880–1640 cal BC

Final comment: G Hey (28 November 2012), the three samples (OxA-12886–8) dated from the basal silt, 15113, suggest that the feature started to infill in the seventeenth century cal BC.

Laboratory comment: English Heritage (2005), a replicate measurement (OxA-12887) was taken on this sample. These two results are statistically consistent (T′=2.9; T′(5%)=3.8; ν=1; Ward and Wilson 1978), and so a weighted mean can be taken before calibration (3396 ±25 BP; 1750–1620 cal BC at 95% confidence; Reimer *et al* 2004).

References: Reimer *et al* 2004
Ward and Wilson 1978

OxA–12887 3352 ±36 BP

δ¹³C: -26.5‰

Sample: YFPB98 15113 <21063> A, submitted in April 2003 by C Chissell

Material: wood (waterlogged): *Quercus* sp., twig (M Robinson 2003)

Initial comment: as OxA-12886. A replicate measurement.

Objectives: as OxA-12886

Calibrated date: 1σ: 1690–1610 cal BC
2σ: 1750–1530 cal BC

Final comment: see OxA-12886

Laboratory comment: see OxA-12886

OxA–12888 3286 ±35 BP

δ¹³C: -28.7‰

Sample: YFPB98 15113 <21063> B, submitted in April 2003 by C Chissell

Material: wood (waterlogged): Pomoideae, twig (<5g) (M Robinson 2003)

Initial comment: as OxA-12886

Objectives: as OxA-12886

Calibrated date: 1σ: 1620–1500 cal BC
2σ: 1650–1460 cal BC

Final comment: see OxA-12886

Laboratory comment: see OxA-12886

SUERC–1146 3190 ±40 BP

δ¹³C: -25.9‰

Sample: YFPB97 10020 <10005> A, submitted in April 2003 by C Chissell

Material: charcoal: Pomoideae (3.66g); *Prunus* sp. (2.18g) (M Robinson 2003)

Initial comment: as OxA-12838

Objectives: as OxA-12838

Calibrated date: 1σ: 1510–1420 cal BC
2σ: 1530–1400 cal BC

Final comment: G Hey (28 November 2012), the later date for this pit, in the fourteenth or thirteenth century cal BC, seems most likely, and also its use with the final cut of waterhole 10159. However, there is a strong likelihood that the residual charcoal is associated with long-lived burnt-mound activity in this area, as evidenced by spreads and dumps of this material seen in evaluation along the northern edge of this palaeochannel.

Laboratory comment: see OxA-12838

Bibliography

Aerts-Bijma, A T, Meijer, H A J, and van der Plicht, J, 1997 AMS sample handling in Groningen, *Nuclear Instruments and Methods in Physics Research B*, **123**, 221–5

Aerts-Bijma, A T, van der Plicht, J, and Meijer, H A J, 2001 Automatic AMS sample combustion and CO_2 collection, *Radiocarbon*, **43(2A)**, 293–8

Allen, M J, Leivers, M, and Ellis, C, 2008 Neolithic causewayed enclosures and later prehistoric farming: duality, imposition and the role of predecessors at Kingsborough, Isle of Sheppey, Kent, UK, *Proc Prehist Soc*, **74**, 235–322

Allen, T, Hayden, C, and Lamdin-Whymark, H, 2009 *From Bronze Age enclosure to Anglo-Saxon settlement: archaeological excavations at Taplow hillfort, Buckinghamshire, 1999–2005*, Thames Valley Landscapes Monogr, **30**, Oxford: Oxford Archaeol Unit Ltd

Ambers, J, and Bowman, S, 1998 Radiocarbon measurements form the British Museum: datelist XXIV, *Archaeometry*, **40**, 413–35

Ambers, J, Matthews, K, and Bowman, S, 1989 British Museum radiocarbon measurements XXI, *Radiocarbon*, **31**, 15–32

Ambers, J, Matthews, K, and Bowman, S, 1991 British Museum natural radiocarbon measurements XXII, *Radiocarbon*, **33**, 51–68

Anon, 1990 Avebury: Windmill Hill, *Wiltshire Natur Hist Archaeol Mag*, **83**, 218–23

Armour-Chelu, M, 1998 The animal bone, in *Etton: excavations at a Neolithic causewayed enclosure near Maxey, Cambridgeshire* (F Pryor), Engl Heritage Archaeol Rep, **18**, 273–88 London: Engl Heritage

Arneborg, J, Heinemeier, J, Lynnerup, N, Nielsen, H L, Rud, N, and Sveinbjörnsdóttir, Á E, 1999 Change of diet of the Greenland Vikings determined from stable carbon isotope analysis and ^{14}C dating of their bones, *Radiocarbon*, **41**, 157–68

Arnold, A J, Howard, R E , and Litton, C D, 2005 *Tree-ring analysis of timbers from the nave roof and ceiling of the cathedral church of St Peter and Wilfred, Ripon, North Yorkshire*, Engl Heritage Centre for Archaeol Rep, **44/2005**

Arnold, A, Bayliss, A, Cook, G, Goodall, J, Hamilton, W D, Howard, R E, Litton, C D, and van der Plicht, J, 2006 *Scientific dating of timbers from Grey Mare's Tail Tower, Warkworth Castle, Warkworth, near Alnwick, Northumberland*, English Heritage Research Department Research Series, **34/2006**

Ashbee, P, 1966 The Fussell's Lodge Long Barrow excavations 1957, *Archaeologia*, **100**, 1–80

Ashmore, P, 1999 Radiocarbon dating: avoiding errors by avoiding mixed samples, *Antiq*, **73**, 124–30

Bain, M G, 1991 *Palaeoecological studies in the Rivington Anglezarke Uplands Lancashire, Volume I, II, III*, unpubl PhD thesis, Univ Salford

Barber, A, and Watts, M, 2008 Excavations at Saxon's Lode Farm, Ryall Quarry, Ripple, 2001–2: Iron Age, Romano-British and Anglo-Saxon rural settlement in the Severn Valley, *Trans Worcestershire Archaeol Soc*, **21**, 1–90

Barclay, A, and Bayliss, A, 1999 Cursus monuments and the radiocarbon problem, in *Pathways and ceremonies. The cursus monuments of Britain and Ireland* (A Barclay and J Harding), Nelothic Studies Seminar Pap, **4**, 11–29 Oxford: Oxbow Books

Barfoot, J F, and Price-Williams, D, 1976 *The Saxon barrow at Galley Hills, Banstead Down, Surrey*, Res Vols Surrey Archaeol Soc, **3**, Guildford: Surrey Archaeol Soc

Barker, H, and Mackey, J, 1961 British Museum natural radiocarbon measurements III, *Radiocarbon*, **3**, 39–45

Barker, H, and Mackey, J, 1968 British Museum natural radiocarbon measurements V, *Radiocarbon*, **10**, 1–7

Barker, H, Burleigh, R, and Meeks, N, 1971 British Museum Natural Radiocarbon Measurements VII, *Radiocarbon*, **13**, 157–88

Barnatt, J, and Worthington, T, 2006 Using coal to mine lead: firesetting at Peak District mines, *Mining Hist*, **16**, 1–92

Bayliss, A, 2007 Bayesian Buildings: an introduction for the numerically challenged, *Vernacular Architect*, **38**, 75–86

Bayliss, A, 2016 Radiocarbon Dating and Bayesian Chronological Modelling in Early Medieval England, in *The Evidence of Material Culture: Studies in Honour of Professor Vera Evison* (eds I Riddler, J Soulat, and L Keys), Europe Médiévale **10**, 239–56, Autun (Éditions Mergoil)

Bayliss, A, and Bronk Ramsey, C, 2004 Pragmatic Bayesians: a decade integrating radiocarbon dates into chronological models, in *Tools for constructing chronologies: tools for crossing disciplinary boundaries* (eds C E Buck and A R Millard), 25–41, London (Springer)

Bayliss, A, and Tyers, I, 2004 Interpreting radiocarbon dates using evidence from tree rings, *Radiocarbon*, **46**, 957–64

Bayliss, A, and Whittle, A (eds) 2007 Histories of the dead: building chronologies for five southern British long barrows, *Cantab Archaeol J*, **17/1 suppl**

Bayliss, A, Benson, D, Galer, D, Humphrey, L, McFadyen, L, and Whittle, A, 2007e One thing after another: the date of the Ascott-under-Wychwood long barrow, *Cambridge Archaeol J*, **17 (suppl)**, 29–44

Bayliss, A, Beavan, N, Hines, J, Høilund Nielsen, K, and McCormac, G, 2012b Radiocarbon Dating and Chronological Modelling in *Buckland Anglo-Saxon Cemetery, Dover: excavations 1994* (K Parfitt and T Anderson), 360–6, Canterbury (Canterbury Archaeological Trust)

Bayliss, A, Boomer, I, Bronk Ramsey, C, Hamilton, D, and Waddington C, 2007c Chapter 6: Absolute Dating, in *Mesolithic Settlement in the North Sea Basin: a case study from Howick, North-East England* (ed C Waddington), 65–74, Oxford (Oxbow)

Bayliss, A, Bronk Ramsey, C, Cook, G, Marshall, P, and van der Plicht, J, forthcoming *Radiocarbon dates from samples funded by English Heritage 2006–10*, Historic Engl, Swindon

Bayliss, A, Bronk Ramsey, C, Cook, G, McCormac, G, and Marshall, P, 2016 *Radiocarbon dates from samples funded by English Heritage between 1993 and 1998*, Swindon: Hist England

Bayliss, A, Bronk Ramsey, C, Cook, G, McCormac, F G, Otlet, R, and Walker, A J, 2013a *Radiocarbon dates from samples funded by English Heritage between 1988 and 1993*, Swindon: Engl Heritage

Bayliss, A, Bronk Ramsey, C, Cook, G and van der Plicht, J, 2007a *Radiocarbon Dates from samples funded by English Heritage under the Aggregates Levy Sustainability Fund 2002–4*, English Heritage, Swindon

Bayliss, A, Bronk Ramsey, C, van der Plicht, J, and Whittle, A, 2007b Bradshaw and Bayes: towards a timetable for the Neolithic, *Cantab Archaeol J*, **17(1) suppl**, 1–28

Bayliss, A, Cook, G, Bronk Ramsey, C, van der Plicht, J, and McCormac, G, 2008a *Radiocarbon dates from samples funded by English Heritage under the Aggregates Levy Sustainability Fund 2004–7*, Swindon: Engl Heritage

Bayliss, A, Bronk Ramsey, C, Hamilton, D, and van der Plicht, J, 2006 *Radiocarbon wiggle-matching of the second floor of the bell tower, Church of St Andrew, Wissett, Suffolk*, Engl Heritage Res Dept Rep Ser, **32/2006**

Bayliss, A, Bronk Ramsey, C, Hamilton, D, van der Plicht, J, Cook, G T, Tyers, C, and Freeman, S, 2014 *Cathedral church of St Peter and St Wilfrid, Ripon, North Yorkshire: nave roof and ceiling, wiggle-match radiocarbon dating of timbers*, Research Dept Rep Ser, **73/2014**

Bayliss, A, Groves, C, McCormac, F G , Bailie, M G L, Brown, D, and Brennand, M, 1999 Precise dating of the Norfolk timber circle, *Nature*, **402**, 479

Bayliss, A, Healy, F, van der Plicht, J, Bronk Ramsey, C, Reimer P, Shand, G, Weekes, J, and Whittle, A, 2019 Radiocarbon dating, in *Chalk Hill: Neolithic and Bronze Age discoveries at Ramsgate, Kent* (P Clark, G Shand, and Weekes, J), 78-85, Leiden (Sidestone Press)

Bayliss, A, Healy, F, Whittle, A, and Darvill, T, 2011 Radiocarbon dating, in Excavations at a Neolithic enclosure on The Peak, near Birdlip, Gloucestershire, *Proc Prehist Soc*, 77, 187–94

Bayliss, A, Hedges, R, Otlet, R, Switsur, R, and Walker, A J, 2012a *Radiocarbon dates from samples funded by English Heritage between 1981 and 1988*, Swindon: Engl Heritage

Bayliss, A, Hines, J, Hoilund Nielsen, K, McCormac, G, and Scull, C, 2013b *Anglo-Saxon graves and grave goods of the 6th and 7th centuries AD: a chronological framework*, Soc Med Archaeol Monogr, **33**, London: Soc Med Archaeol

Bayliss, A, McAvoy, F, and Whittle, A, 2007d The world recreated: redating Silbury Hill in its monumental landscape, *Antiq*, **81**, 26–53

Bayliss, A, McCormac, F G, van der Plicht, J, 2004a An illustrated guide to measuring radiocarbon from archaeological samples, *Physics Education*, **39**, 137–44

Bayliss, A, Popescu, E, Athfield-Beavan, N, Bronk Ramsey, C, Cook G T, and Locker, A, 2004b The potential significance of dietary offsets for the interpretation of radiocarbon dates: an archaeologically significant example from medieval Norwich, *J Archaeol Sci*, **31**, 563–75

Bayliss, A, Whittle, A, and Healy, F, 2008b Timing, tempo and temporalities in the early Neolithic of southern Britain, in Between foraging and farming. An extended broad spectrum of papers presented to Leendert Louwe Kooijmans, *Analecta Praehist Leidensia*, **40**, 25–42

Beadsmore, E, Garrow, D, and Knight, M, 2010 Refitting Etton: space, time and material culture within a causewayed enclosure in Cambridgeshire, *Proc Prehist Soc*, **76**, 115–34

Beavan, N, Mays, S, Bayliss, A, Hines, J, and McCormac, F G, 2011 *Amino acid and stable isotope analysis of skeletons dated for the Anglo-Saxon chronology project*, Research Dept Rep Ser, **88/2011**

Bedwin, O, 1981a Excavations at the Neolithic enclosure on Bury Hill, Houghton, West Sussex 1979, *Proc Prehist Soc*, **47**, 69–86

Bedwin, O R, 1981b Excavations at The Trundle, *Sussex Archaeol Coll*, **119**, 208–14

Bedwin, O, 1984 The excavation of a small hilltop enclosure on Court Hill, Singleton, West Sussex, *Sussex Archaeol Coll*, **122**, 13–22

Benson, D, and Clegg, I N I, 1978 Cotswold Burial Rites, *Man*, **13**, 134–137

Benson, D, and Whittle, A (eds) 2007 *Building memories: the Neolithic Cotswold long barrow at Ascott-under-Wychwood, Oxfordshire*, Oxford (Oxbow)

Berridge, P J, 1986 Mesolithic evidence from Hembury, *Proc Devon Archaeol Soc*, **44**, 163–6

Best, J, and Woodward, A, 2011 Late Bronze Age pottery production: evidence from a 12th–11th century cal BC settlement at Tinney's Lane, Sherborne, Dorset, *Proc Prehist Soc*, **78**, 207–61

Best, J, Woodward, A, and Tyler, K, 2013 *Late Bronze Age pottery production: evidence from a 12th to 11th century BC settlement at tinney's Lane, Sherborne, Dorset*, Dorset Natur Hist Archaeol Soc Monogr Ser, **21**, Dorset: Dorset Natur Hist Archaeol Soc

Blackmore, L, and Pearce, J, 2010 *Shelly-Sandy Ware and the Greyware Industries A Dated Type Series of London Medieval Pottery: Part 5, Shelly-Sandy Ware and the Greyware Industries*, MOLAS Monogr, **49**, London: Museum of London

Boomer, I, Waddington, C, Stevenson, T, and Hamilton, D, 2007 Holocene coastal change and geoarchaeology at Howick, Northumberland, UK, *The Holocene*, **17**, 89–104

Bowman, S, 1990 *Radiocarbon dating*, London (British Museum)

Boyle, A, Jennings, D, Miles, D, and Palmer, S, 1998 *The Anglo-Saxon cemetery at Butler's Field, Lechlade, Gloucestershire. Vol 1: prehistoric and Roman activity and Anglo-Saxon grave catalogue*, Thames Valley Landscapes Monogr, **10**, Oxford: Oxford Archaeol Unit

Boyle, A, Jennings, D, Miles, D, and Palmer, S, 2011 *The Anglo-Saxon cemetery at Butler's Field, Lechlade, Gloucestershire. Vol 2: the Anglo-Saxon grave goods specialist reports, phasing, and discussion*, Thames Valley Landscapes Monogr, **33**, Oxford: Oxford Archaeol Unit

Bradley, P, 2004 Causewayed enclosures: monumentality, architecture, and spatial distribution of artefacts - the evidence from Staines, Surrey, in *Towards a New Stone Age: aspects of the Neolithic in South-east England* (J Cotton and D Field), CBA Res Rep, **137**, 115–23 York: CBA

Bradley, R, 1969 The Trundle revisited, *Sussex Notes Quer*, **17**, 133–4

Bradley, R, 1983 The bank barrows and related monuments of Dorset in the light of recent fieldwork, *Proc Dorset Natur Hist Archaeol Soc*, **105**, 15–20

Branch, N P, Batchelor, C R, Cameron, N G, Coope, G R, Densem, R, Gale, R, Green, C P, and Williams, A N, 2012 Holocene environmental changes in the Lower Thames Valley, London, UK: implications for understanding the history of Taxus woodland, *The Holocene*, **22**, 1143–58

Brennand, M, and Taylor, M, 2003 The survey and excavation of a Bronze Age timber circle at Holme-next-the-Sea, Norfolk, 1998–9, *Proc Prehist Soc*, **69**, 1–84

Brock, F, Bronk Ramsey, C, and Higham, T, 2007 *Radiocarbon dating bone samples recovered from gravel sites*, Engl Heritage Res Dept Rep Ser, **30/2007**

Brodie, N, 1994 *The Neolithic-Bronze transition in Britain. A critical review of some archaeological and craniological concepts*, BAR Brit Ser, **238**

Brock, F, Bronk Ramsey, C, and Higham, T, 2007 Quality assurance of ultrafiltered bone dating, *Radiocarbon*, **49**, 187–92

Brock, F, Higham, T, Ditchfield, P, and Bronk Ramsey, C, 2010 Current pretreatment methods for AMS radiocarbon dating at the Oxford Radiocarbon Accelerator Unit (ORAU), *Radiocarbon*, **52**, 103–12

Bronk Ramsey, C, 1995 Radiocarbon calibration and analysis of stratigraphy: the OxCal program, *Radiocarbon*, **36**, 425–30

Bronk Ramsey, C, 1998 Probability and dating, *Radiocarbon*, **40**, 461–74

Bronk Ramsey, C, 2000 Comment on 'The Use of Bayesian Statistics for ^{14}C dates of chronologically ordered samples: a critical analysis', *Radiocarbon*, **42**, 199–202.

Bronk Ramsey, C, 2001 Development of the radiocarbon calibration program OxCal, *Radiocarbon*, **43**, 355–63

Bronk Ramsey, C, 2008 Deposition models for chronological records, *Quaternary Science Review*, **27**, 42–60

Bronk Ramsey, C, 2009a Bayesian analysis of radiocarbon dates, *Radiocarbon*, **51**, 37–60

Bronk Ramsey, C, 2009b Dealing with outliers and offsets in radiocarbon dating, *Radiocarbon*, **51**, 1023–45

Bronk Ramsey, C, and Hedges, R E M, 1997 Hybrid ion sources: radiocarbon measurements from microgram to milligram, Nuclear *Instruments and Methods in Physics Research B*, **123**, 539–45

Bronk Ramsey, C, Higham, T, Bowles, A, and Hedges, R E M, 2004a Improvements to the pre-treatment of bone at Oxford, *Radiocarbon*, **46**, 155–63

Bronk Ramsey, C, Higham, T, and Leach, P, 2004b Towards high precision AMS: progress and limitations, *Radiocarbon*, **46**, 17–24

Bronk Ramsey, C, and Lee, S, 2013 Recent and planned developments of the program OxCal, *Radiocarbon*, **55**, 720–30

Bronk Ramsey, C, Pettitt, P B, Hedges, R E M, Hodgins, G W L, and Owen, D C, 2000 Radiocarbon dates from the Oxford AMS system: Archaeometry datelist 30, *Archaeometry*, **42**, 459–79

Bronk Ramsey, C, van der Plicht, J, and Weninger, B, 2001 'Wiggle-matching' radiocarbon dates, *Radiocarbon*, **43**, 381–90

Brothwell, D R, 1965 Description of the bones, in *Windmill Hill and Avebury: Excavations by Alexander Keiller, 1925–1939* (I F Smith), 138–40, Oxford: Clarendon

Brothwell, D R, 1971 Forensic aspects of the so-called Neolithic skeleton Q1 from Maiden Castle, Dorset, *World Archaeol*, **3**, 233–41

Brown, A, 1989 The social life of flint at Neolithic Hembury, *Lithics*, **10**, 46–9

Bruhn, F, Durr, A, Grootes, P M, Mintrop, A, and Nadeau, M, 2001 Chemical removal of conservation substances by "Soxhlet"-type extraction, *Radiocarbon*, **43**, 229–37

Brunning, R, 2013 *Somerset's peatland archaeology: managing and investigating a fragile resource*, Oxford: Oxbow Books

Buck, C E, Cavanagh, W G, and Litton, C D, 1996 *Bayesian Approach to Interpreting Archaeological Data*, Chichester

Burstow, G P, 1942 The stone age camp at Whitehawk, *Sussex County Mag*, **16**, 314–9

Campbell, G, 2008 Charred plant remains, in *The Danebury environs programme. A Wessex landscape in the Roman era, volume 2, part 2: Grateley South, Grateley, Hants, 1998 and 1999* (B Cunliffe and C Poole), Engl Heritage and Oxford Univ School of Archaeol Monogr, **71**, 166–74 Oxford: School of Archaeol

Carlton, R, 2011 Archaeological excavations at Harehaugh hillfort in 2002, *Archaeol Aeliana*, **60**, 5 ser, 85–115

Carr, R, 1992 The Middle-Saxon settlement at Staunch Meadow, Brandon, Suffolk - a final update, *The Quarterly: J Norfolk Archaeol Res Group*, **5**, 16–22

Carr, R D, Tester, A, and Murphy, P, 1988 The middle Saxon settlement at Staunch Meadow, Brandon, *Antiq*, **62**, 371–7

Carver, M, 2005 *A seventh-century princely burial ground and its context*, Rep Res Comm Soc Antiq London, **69**, London: Brit Museum Press

Chapman, H P, and Gearey, B R, 2013 *Modelling archaeology and palaeoenvironments: the hidden landscape archaeology of Hatfield and Thorne Moors, eastern England*, Oxford: Oxbow Books

Christen, J A, and Litton, C D, 1995 A Bayesian approach to wiggle-matching, *J Archaeol Sci*, **22**, 719–25

Christie, P M L, 1985 Barrows on the North Cornish Coast: wartime excavations by C K Croft Andrew 1939–44, *Cornish Archaeol*, **24**, 23–121

Christie, P M L, 1988 A barrow cemetery on Davidstow Moor, Cornwall: wartime excavations by C K Croft Andrew, *Cornish Archaeol*, **27**, 27–169

Clark, A J, Tarling, D H, and Noël, M, 1988 Developments in archaeomagnetic dating in Britain, *J Archaeol Sci*, **15**, 645–67

Clark, P, Shand, G, and Weekes, J, 2019 *Chalk Hill: Neolithic and Bronze Age discoveries at Ramsgate, Kent*, Leiden: Sidestone Press

Cleal, R, 1991 Earlier prehistoric pottery, in *Maiden Castle: excavations and field survey 1985–6* (ed N Sharples), 171–85, microfiche M9: A4–E4, London: Engl Heritage

Cleal, R M J, 2004 The dating and diversity of the earliest ceramics of Wessex and south-west England, in *Monuments and material culture. Papers in honour of an Avebury archaeologist: Isobel Smith* (R M J Cleal and J Pollard), 166–92, East Knoyle: Hobnob Press

Connah, G, 1965 Excavations at Knap Hill, Alton Priors, *Wiltshire Archaeol Natur Hist Mag*, **60**, 1–23

Connah, G, 1969 Radiocarbon dating for Knap Hill, *Antiq*, **43**, 304–5

Cook, G T, Bonsall, C, Hedges, R E M, McSweeney, K, Boroneant, V, and Pettitt, P B, 2001 A freshwater diet-derived ^{14}C reservoir effect at the stone age sites in the Iron Gates gorge, *Radiocarbon*, **43**, 453–60

Cowie, R, and Blackmore, L, 2012 *Lundenwic: excavations in Middle Saxon London, 1987–2000*, MOLA Monogr, **63**, London: MOLA

Cowie, R, Layard Whytehead, R, and Blackmore, L, 1988 Two Middle Saxon occupation sites: excavations at Jubilee Hall and 21–22 Maiden Lane, *Trans London Middlesex Archaeol Soc*, **39**, 47–163

Cunliffe, B, and Poole, C, 2008 *The Danebury environs programme. A Wessex landscape in the Roman era, volume 2, part 2: Grateley South, Grateley, Hants, 1998 and 1999*, Engl Heritage and Oxford Univ School of Archaeol Monogr, **71**, Oxford: School of Archaeol

Cunnington, M E, 1909 On a remarkable feature in the entrenchments of Knap Hill Camp, Wiltshire, *Man*, **9**, 49–52

Cunnington, M E, 1912 Knap Hill Camp, Wilts Archaeol Mag, **37**, 42–65

Curwen, E C, 1929 Excavations in The Trundle, Goodwood, 1928, *Sussex Archaeol Coll*, **70**, 33–85

Curwen, E C, 1930 Neolithic camps, *Antiq*, **4**, 22–54

Curwen, E C, 1931 Excavations in the Trundle, Goodwood (second season) 1930, *Sussex Archaeol Coll*, **72**, 33–85

Curwen, E C, 1934 Excavations at Whitehawk Camp, Brighton 1932–3, *Antiquity J*, **14**, 99–113

Curwen, E C, 1935 Whitehawk 1935 Field-Book. Manuscript, unpubl rep, Dept Local Hist Archaeol, Brighton Museum Art Gallery

Curwen, E C, 1936 Excavations in Whitehawk Camp, Brighton, third season, 1935, *Sussex Archaeol Coll*, **77**, 60–92

Curwen, E C, 1954 *The archaeology of Sussex*, London: Methuen and Co. Ltd

Darvill, T, 1981 Excavations at the Peak Camp, Cowley: an interim report, *Glevensis*, **15**, 52–6

Darvill, T, 1982a Excavations at the Peak Camp, Cowley, Gloucestershire, *Glevensis*, **16**, 20–5

Darvill, T, 1986 Prospects for dating Neolithic sites and monuments in the Cotswolds and adjacent areas, in *Archaeological Results from Accelerator Dating* (J A J Gowlett and R E M Hedges), OUCA Monogr, **11**, 119–24 Oxford: Oxford Univ

Darvill, T, 2011 Excavations at a Neolithic enclosure on The Peak, near Birdlip, Gloucestershire, *Proc Prehist Soc*, **77**, 139–204

Darvill, T, 1982b The megalithic chambered tombs of the Cotswold-Severn region: an assessment of certain architectural elements and their relation to ritual practice and Neolithic Society, *VORDA*, **5**, 142

Darvill, T, and Fulton, A K, 1998 *MARS: the Monuments at Risk Survey of England, 1995: main report*, Bournemouth and London: Bournemouth Univ and Engl Heritage

Davies, G H, 1956 Maiden Bower near Dunstable, *Bedfordshire Archaeol*, **12**, 98–101

Dee, M, and Bronk Ramsey, C, 2000 Refinement of graphite target production at ORAU, *Nuclear Instruments and Methods in Physics Research B*, **172**, 449–53

DeNiro, M J, 1985 Post-mortem preservation and alteration of *in vivo* bone collagen isotope ratios in relation to palaeodietary reconstruction, *Nature*, **317**, 806–9

Dixon, P, 1971 *Crickley Hill. Third report 1971*, Cheltenham: Gloucestershire College of Art and Design

Dixon, P, 1972a Excavations at Crickley Hill, *Antiq*, **46**, 49–52

Dixon, P, 1972b *Crickley Hill. Fourth Report 1972*, Cheltenham: Gloucestershire College of Art and Design

Dixon, P, 1979 A Neolithic and Iron Age site on a hilltop in southern England, *Sci American*, **241**, 142–50

Dixon, P, 1981 Crickley Hill, *Curr Archaeol*, **76**, 145–7

Dixon, P, 1988a The Neolithic settlements on Crickley Hill, in *Enclosures and defences in the Neolithic of Western Europe* (C Burgess, P Topping, C Mordant, and M Maddison), BAR Int Ser, **403**, 75–88 Oxford: BAR

Dixon, P, 1988b Crickley Hill 1969–1987, *Curr Archaeol*, **110**, 73–8

Dixon, P, 1994 *Crickley Hill. Volume 1. The hillfort defences*, Nottingham: Crickley Hill Trust and Univ Nottingham

Dixon, P, 2005 Thirty-five years at Crickley, *Curr Archaeol*, **200**, 390–5

Dixon, P, and Borne, P, 1977 *Crickley Hill and Gloucestershire Prehistory*, Gloucester: Crickley Hill Archaeol Trust

Down, A, and Welch, M, 1990 *Chichester excavations 7 [Appledown and The Mardens]*, Chichester Excavations Ser, 7, Chichester: Chichester District Council

Down, C, 1997 *The Archaeological Monitoring of Groundworks at The Trundle, Singleton, West Sussex*, Chichester: Southern Archaeol

Drewett, P, 1977 The excavation of a Neolithic causewayed enclosure at Offham Hill, East Sussex, 1976, *Proc Prehist Soc*, **43**, 201–42

Drewett, P, 2003 Taming the wild: the first farming communities in Sussex, in *The Archaeology of Sussex to AD 2000* (D Rudling), 39–46, King's Lynn: Heritage Marketing and Publ for Univ Sussex

Drewett, P, Rudling, D, and Gardiner, M, 1988 *The South East to AD 1000*, London: Longman

Drinkall, G, and Foreman, M, 1998 *The Anglo-Saxon cemetery at Castledyke South, Barton-on-Humber*, Sheffield Excavation Rep, **6**, Sheffield: Sheffield Academic Press

Duncan, H, Duhig, C, and Phillips, M, 2003 A late migration/final phase cemetery at Water Lane, Melbourn, *Proc Cambridge Antiq Soc*, **92**, 57–134

Dyer, J, 1955 Maiden Bower near Dunstable, *Bedfordshire Archaeol*, **1**, 47–52

Dyer, J, 1961 Maiden Bower, *Bedfordshire Mag*, **7**, 320

Dyson, L, Shand, G, and Stevens, S, 2000 Causewayed enclosures, *Curr Archaeol*, **168**, 470–2

Edwards, Y H, Weisskopf, W, and Hamilton, D, 2009 Age, taphonomic history and mode of deposition of human skulls in the River Thames, *Trans London Middlesex Archaeol Soc*, **60**, 35–51

Edwards, Y, Paton, A, Wells, M, Gover, J, and Birbeck, V, 2010 Chesham Bois Manor, home to the Cheyne family for 350 Years. Historical and archaeological investigation, *Rec Buckinghamshire*, **50**, 53–82

Evans, C, and Hodder, I, 2006 *A woodland archaeology: the Haddenham Project*, MacDonald Inst Archaeol Res, **1**, Cambridge: MacDonald Inst Archaeol Res

Evans, J G, 1966 Land mollusca from the Neolithic enclosure on Windmill Hill, *Wiltshire Archaeol Natur Hist Mag*, **61**, 91–2

Evans, J G, 1971 Notes on the environment of early farming communities in Britain, in *Economy and settlement in Neolithic and Early Bronze Age Britain and Europe* (ed D D A Simpson), 23–73, Leicester University Press

Everett, L, 2007 *Targeted inter-tidal survey, SCCAS Rep No 2007/192*, http://content.historicengland.org.uk/images-books/publications/suffolk-rczas-targeted-inter-tidal-survey-report/2007192targetedinter-tidalsurveyreport.pdf/,

Evison, V, 1987 *Dover: Buckland Anglo-Saxon cemetery*, Hist Bldg Comm Engl Archaeol Rep, **3**

Fernandes, R, Millard, A R, Brabeck M, Nadeau, M-J, and Grootes, P, 2014 Food Reconstruction Using Isotopic Transferred Signals (FRUITS): a Bayesian model for diet reconstruction, *PLoS ONE*, **9**, e87436

Ferris, I, 2010 *The beautiful rooms are empty: excavations at Binchester Roman fort, County Durham, 1976–1981 and 1986–1991*, Durham: Durham County Council

Finlayson, R, 2006 *Ainsbrook, North Yorkshire: a report on an archaeological assessment*, unpubl typescript rep, York Archaeol Tr, Rep, 2006/51

Finn, N, 2011 *Bronze Age ceremonial enclosures and cremation cemetery at Eye Kettleby, Leicestershire: the development of a prehistoric landscape*, Leicester Archaeol Monogr, **20**, Leicester: Univ Leicester Archaeol Services

Fox, A, 1963 Neolithic charcoal from Hembury, *Antiq*, **37**, 228–9

Freeman, S, Bishop, P, Bryant, C, Cook, G, Fallick, A, Harkness, D, Metcalfe, S, Scott, M, Scott, R, and Summerfield, M, 2004 A new environmental sciences AMS laboratory in Scotland, *Nuclear Instruments and Methods in Physics Research B*, **223-4**, 31–4

French, C , and Pryor, F, 2005 *Archaeology and Environment of the Etton Landscape*, E Anglian Archaeol Rep, **109**, Peterborough: Fenland Archaeol Trust

Funnell, B M, and Pearson, I, 1984 A guide to the Holocene geology of the north Norfolk, *Bull Geolog Soc Norfolk*, **34**, 123–40

Fyfe, R, 2009 *Roman Lode, Burcombe, Exmoor, North Devon: pollen analysis of blanket peat deposits*, Engl Heritage Res Rep Ser, **24–2009**

Gale, J D, 1986 *Leaf arrowheads: the evidence from Crickley Hill and its wider implications*, unpubl BA thesis, Univ Nottingham

Gearey, B G, Marshall, P, and Hamilton, D, 2009 Correlating archaeological and palaeoenvironmental records using a Bayesian approach: a case study from Sutton Common, South Yorkshire, England, *J Archaeol Sci*, **36**, 1477–87

Gelfand, A E, and Smith, A F M, 1990 Sampling approaches to calculating marginal densities, *J Amer Stat Assoc*, **85**, 398–409

Geophysical Surveys of Bradford, 1989 Report No 89/38, unpubl rep, Geophysical Surv Bradford

Geophysical Surveys of Bradford, 1993 Whitehawk Camp, Brighton. Report number 93/118, unpubl typescript rep, Geophysical Surv of Bradford

Geophysical Surveys of Bradford, 1995 Whitesheet Hill, Wiltshire. Survey no. 92/95, unpubl rep, Geophysical Surv Bradford

Geophysical Surveys of Bradford Prospection, 2004 Northborough Causewayed Enclosure, Peterborough, unpubl geophysics rep, GSB Prospection

Gilks, W R, Richardson, S, and Spiegelhalther, D J, 1996 *Markov Chain Monte Carlo in practice*, London: Chapman and Hall

Gillespie, R, Gowlett, J A J, Hall, E T, Hedges, R E M, and Perry, C, 1985 Radiocarbon dates from the Oxford AMS system: Archaeometry datelist 2, *Archaeometry*, **27**, 237–46

Glamorgan-Gwent Archaeol Trust, 2000 Beech Court Farm Enclosure, Ewenny Quarry, Vale of Glamorgan: Archaeological Field evaluation stage 1b Internal trenches. GGAT Report 2000/038

Glamorgan-Gwent Archaeol Trust, 2001 Beech Court Farm Enclosure, Ewenny Quarry, Vale of Glamorgan: palaeoenvironmental assessment. GGAT Report

Gossip, J, 2004 *Boden Vean, St Anthony-in-Meneage, Cornwall: Archaeological evaluation, archive summary*, unpubl rep, Truro Hist Environ Serv

Gossip, J, 2013 The evaluation of a multi-period prehistoric site and fogou at Boden Vean, St Anthony-in-Menuage, Cornwall, 2003, *Cornish Archaeol*, **52**, 1–98

Gossip, J, and Johns, C, undated *Boden Vean, St Anthony-in-Meneage, Cornwall: Assessment and updated project design*, unpubl rep, Truro Hist Environ Serv

Graves-Brown, P, 1998 Ewenny, Beech Court Farm prehistoric enclosure (SS 9040 7660), *Archaeol in Wales*, **38**, 111–2

Griffith, F M, 2001 Recent work on Neolithic enclosures in Devon, in *Neolithic enclosures in Atlantic Northwest Europe* (T Darvill and J Thomas), Neolithic Studies Grp Seminar Pap, **6**, 66–77 Oxford: Oxbow Books

Grigson, C, 1982 Sexing Neolithic cattle skulls and horncores, in *Ageing and sexing animal bones from archaeological sites* (B Wilson, C Grigson, and S Payne), BAR Brit Ser, **109**, 24–35 Oxford: BAR

Grigson, C, 1984 The domestic animals of the earlier Neolithic in Britain, in *Die Anfänge des Neolithikums vom Orient bis Nordeuropa. IX. Der Beginn der Haustierhaltung in der 'Alten Welt'* (G Nobis), 205–20, Köln: Böhlau

Grigson, C, 1999 The mammalian remains, in *The harmony of symbols: the Windmill Hill causewayed enclosure, Wiltshire* (A Whittle, J Pollard, and C Grigson), 164–252, Oxford: Oxbow Books

Grootes, P M, Nadeau, M-J, and Rieck, A, 2004 ^{14}C-AMS at the Leibniz-Labor: radiometric dating and isotope research, *Nuclear Instruments and Methods in Physics Research B*, **223–224**, 55–61

Groves, C, Locatelli, C, and Nayling, N, 2004 *Tree-ring analysis of oak samples from Stert Flats fish weirs, Bridgwater Bay, Somerset*, CfA Rep, **43/2004**

Guthrie, A, 1969 Excavation of a settlement at Goldherring, Sancreed, 1958–61, *Cornish Archaeol*, **8**, 5–39

Hall, D W, Cook, G T, Hall, M A, Muir, G K P, Hamilton, W D, and Scott, E M, 2007 The early Medieval origin of Perth, Scotland, *Radiocarbon*, **49**, 639–44

Hall, D W, Cook, G T, and Hamilton, W D, 2010 New dating evidence for North Sea trade between England, Scotland and Norway in the 11th century AD, *Radiocarbon*, **52**, 331–6

Hall, R A, 2005 *The Ainsbrook Site, Yorkshire*, unpubl typescript rep, York Archaeol Tr, Rep 2005/15

Hall, R A, Buckberry, J, Storm, R, Budd, P, Hamilton, W D, and McCormac, G, 2008 The medieval cemetery at Riccall Landing: a reappraisal, *Yorkshire Archaeol J*, **80**, 55–92

Hamilton, W D, 2006 Appendix 1. Radiocarbon dates from a possible buried land surface (context 1318) in trench 3, in *Groundwell Ridge Roman villa, Swindon: excavations 2003–2005* (R Brickstock, K Brown, G Campbell, D Dungworth, A Hammon, D Hamilton, P Harding, N Hembrey, B Hill, S Jennings, N Linford, P Linford, L Martin, G Morley, J Timby, and P Wilson), Engl Heritage Res Dept Rep Ser, **69/2010**, Swindon: Engl Heritage

Hamilton, W D, Bayliss, A, Menuge, A, Bronk Ramsey, C, and Cook, G, 2007 'Rev Thomas Bayes: get ready to wiggle' - Bayesian modelling, radiocarbon wigglematching, and the north wing of Baguley Hall, *Vernacular Architecture*, **38**, 87–97

Hammerow, H, Holleveot, Y, and Vince, A, 1994 Migration period settlement and Anglo-Saxon pottery from Flanders, *Medieval Archaeol*, **38**, 1–18

Hanson-James, N, 1993 *An Investigation of the Neolithic central area on Crickley Hill*, unpubl BA thesis, Univ Nottingham

Hardiman, M A, Fairchild, J E, and Longworth, G, 1992 Harwell radiocarbon measurements XI, *Radiocarbon*, **34**, 47–70

Harkness, D D, 1983 The extent of natural ^{14}C deficiency in the coastal environment of the United Kingdom, *PACT*, **8**, 351–64

Harris, O, 2006 Agents of identity: performative practice at the Etton causewayed enclosure, in *Elements of being: identities, mentalities and movements* (A Cochrane, D Hofmann, and J Mills), BAR Int Ser, **1437**, 40–9 Oxford: BAR

Harvey, J H, 1947 The western entrance of the Tower, *Trans London Middlesex Archaeol Soc*, **9**, 20–35

Haughton, C, and Powlesland, D, 1999 *West Heslerton: the Anglian cemetery*, Landscape Res Centre Monogr, **1**, Yedingham: Landscape Res Centre

Hayden, C, and Hey, G, 2008 Anglo-Saxon and earlier settlement near Drayton Road, Sutton Courtenay, Berkshire, *Archaeol J*, **164**, 109–96

Head, K, Turney, C S M, Pilcher, J R, Palmer, J G, and Bailie, M G L, 2007 Problems with identifying the '8200-year cold event' in terrestrial records of the Atlantic seaboard: a case study from Dooagh, Achill Island, Ireland, *J Quaternary Sci*, **22**, 65–75

Heatherington, D A, Lord, T C, and Jacobi, R M, 2006 New evidence for the occurrence of Eurasian lynx (*Lynx lynx*) in medieval Britain, *J Quat Sci*, **21**, 3–8

Hedges, R E M, Bronk, C R and Housley, R A 1989a The Oxford Accelerator Mass Spectrometry facility: technical developments in routine dating, *Archaeometry*, **31**, 99–113

Hedges, R E M, Housley, R A, Law, I A, and Bronk Ramsey, C, 1989b Radiocarbon dates from the Oxford AMS system: Archaeometry datelist 9, *Archaeometry*, **31**, 207–34

Hedges, R E M, Tiemei, C, and Housley, R A, 1992a
Results and methods in the radiocarbon dating of pottery,
Radiocarbon, **34**, 906–15

Hedges, R E M, Housley, R A, Bronk, C R, and van
Klinken, G J, 1992b Radiocarbon dates from the Oxford
AMS system: Archaeometry datelist 14, *Archaeometry*, **34**,
141–59

Hedges, R E M, Humm, M J, Foreman, J, Klinken, G J van,
and Bronk, C R, 1992c Developments in sample
combustion to carbon dioxide, and in the Oxford AMS
carbon dioxide ion source system, *Radiocarbon*, **34**, 306–11

Hedges, R E M, Housley, R, Petitt, P B, Bronk Ramsey, C,
and van Klinken, G J, 1996 Radiocarbon dates from the
Oxford AMS system: Archaeometry datelist 21,
Archaeometry, **38**, 181–207

Hedges, R E M, and Reynard, L M, 2007a Nitrogen
isotopes and the trophic level of humans in archaeology,
Journal of Archaeological Science, **34**, 1240–51

Hedges, R E M, Clement, J G, Thomas, C D L, and
O'Connell, T C, 2007b Collagen turnover in the adult
femoral mid-shaft: modelled from anthropogenic
radiocarbon tracer measurements, *American Journal of
Physical Anthropology*, **133**, 808–16

Hey, G, 2004 *Yarnton Saxon and Medieval Settlement and
Landscape*, Thames Valley Landscape Monogr, **20**, Oxford
(Oxford Archaeology)

Hey, G, Booth, P, and Timby, J, 2011 *Yarnton: Iron Age and
Romano-British settlement and landscape*, Thames Valley
Landscape Monogr, **35**, Oxford (Oxford Archaeology)

Hey, G, Bell, C, Dennis, C, and Robinson, M, (eds) 2016
Yarnton: Neolithic and Bronze Age settlement and landscape,
Thames Valley Landscapes Monogr, **39**, Oxford: Oxford
Univ School Archaeol

Hills, C, and O'Connell, T, 2009 New light on the Anglo-
Saxon succession: two cemeteries and their dates, *Antiq*, **83**,
1096–1108

Holden, E W, 1951 Earthworks on Court Hill, *Sussex Notes
Queries*, **13**, 183–5

Hollos, D B, 1999 *The Long Mound Sequence at Crickley
Hill*, unpubl draft PhD thesis, Univ Nottingham

Honeyman, H L, and Hunter Blair, H, 1990 *Warkworth
Castle and Hermitage, Northumberland*, English Heritage
Guidebooks

Hoper, S T, McCormac, F G, Hogg, A G, Higham, T F G,
and Head, M J, 1998 Evaluation of wood pretreatments on
oak and cedar, *Radiocarbon*, **40**, 45–50

Hua, Q, Barbetti, M, and Rakowski, A Z, 2013
Atmospheric radiocarbon for the period 1950–2010,
Radiocarbon, **55**, 2059–72

Hunn, J R, and Rackham, D J, forthcoming *Rectory Farm,
West Deeping, Lincolnshire, a reading of lost landscapes*, BAR
Brit Ser

Institute of Archaeology, 1987 The Trundle, West Sussex: a
Geophysical Survey of the Proposed British Telecom
Goodwood Radio Station, unpubl rep, Instit Archaeol
Sussex Archaeol Fld Unit

International Study Group, 1982 An inter-laboratory
comparison of radiocarbon measurements in tree-rings,
Nature, **298**, 619–23

Jacobi, R M, Higham, T F G, and Bronk Ramsey, C, 2006
AMS radiocarbon dating of Middle and Upper Palaeolithic
bone in the British Isles: improved reliability using
ultrafiltration, *J Quaternary Sci*, **21**, 557–73

Jones, A M, 2004 Settlement and ceremony; archaeological
investigations at Stannon Down, St Breward, Cornwall,
Cornish Archaeol, **43/44**, 1–141

Jones, D, 2003 *Fiskerton, Lincolnshire*, Engl Heritage Aerial
Survey Rep, **2/2003**

Jones, J, Tinsley, H, and Brunning, R, 2007 Methodologies
for assessment of the state of preservation of pollen and
plant macrofossil remains in waterlogged deposits,
Environmental Archaeol, **12**, 71–86

Jordan, D, Haddon-Reece, D, and Bayliss, A, 1994
*Radiocarbon dates from samples funded by English Heritage and
dated before 1981*, London: Engl Heritage

Juleff, G, and Bray, L, 2007 Minerals, metal, colour and
landscape: Exmoor's Roman Lode in the early Bronze Age,
Cambridge Archaeol J, **17**, 285–96

Keeling, C H, 2001 Zoological gardens of Great Britain, in
*Zoo and aquarium history - ancient animal collection to
zoological gardens* (Jr V N Kisling), 49–74, Boca Raton:
CRC Press

Keevill, G, 2004 *The Tower of London Moat, archaeological
excavations 1995–9*, Historic Royal Palaces Monogr, **1**,
Oxford: Oxford Archaeol

Keiller, A, 1934 Excavation at Windmill Hill, in *Proceedings
of the First International Congress of Prehistoric and Protohistoric
Sciences. London, August 1–6 1932* (H Milford), 135–8,
Oxford: Oxford Univ Press

Kendall, H G O, 1923 Excavations conducted on the NE
side of Windmill Hill, in *Report of the Earthworks Committee:
Accounts. Reports of the Council and of the Congress for the Year
1922*, 25–6, London: Congress Archaeol Soc/Soc Antiq
London

Kenny, J, 1994 *Archaeological Investigation: Trundle Triangle
Car Par*, Chichester: Chichester District Archaeol Unit

Kinnes, I, 1998 The pottery, in *Etton: Excavations at a
Neolithic causewayed enclosure near Maxey, Cambridgeshire,
1982–87* (F Pryor), Engl Heritage Archaeol Rep, **18**, 161–
214 London: Engl Heritage

Lanting, J N, and Brindley, A L, 1998 Dating cremated
bone: the dawn of a new era, *J Irish Archaeol*, **9**, 151–65

Lanting, J N, and van der Plicht, J, 1998 Reservoir effects
and apparent ages. The *Journal of Irish Archaeology*, **9**,
151–65

Lanting, J N, Aerts-Bijma, A T, and van der Plicht, J, 2001
Dating of cremated bones, *Radiocarbon*, **43**, 249–54

Larsen, G, and Eiriksson, J, 2008 Late Quaternary terrestrial tephrochronology of Iceland - frequency of explosive eruptions, type and volume of tephra, *J Quaternary Sci*, **23**, 109–20

Last, J, 2014 The excavation of two round barrows on Longstone Edge, Derbyshire, *Derbyshire Archaeol J*, **134**, 81–174

Lawrence, S, and Smith, A, 2009 *Between villa and town: excavations of a Roman roadside settlement and shrine at Higham Ferrers, Northamptonshire*, Oxford Archaeol Monogr, 7, Oxford: Oxford Archaeol Unit

Lewis, C, 2005 My time, *Curr Archaeol*, **196**, 198–9

Liddell, D M, 1930 Report on the excavations at Hembury Fort, Devon, 1930, *Proc Devon Archaeol Soc*, **1.2**, 39–63

Liddell, D M, 1931 Report on the excavations at Hembury Fort, Devon: second season 1931, *Proc Devon Archaeol Soc*, **1.3**, 90–120

Liddell, D M, 1932 Report on the excavations at Hembury Fort, Devon, third season 1932, *Proc Devon Archaeol Soc*, **1.4**, 162–90

Liddell, D M, 1935 Report on the excavations at Hembury Fort, Devon, 4th and 5th seasons 1934 and 1935, *Proc Devon Archaeol Soc*, **2.3**, 135–75

Lindley, D V, 1985 *Making decisions*, 2nd edn, London (Wiley)

Linford, N T, 1998 Geophysical survey at Boden Vean, Cronwall, including an assessment of the microgravity technique for the location of suspected archaeological features, *Archaeometry*, **40**, 187–216

Linford, N T, 2004 *Boden Vean, St Anthony-in-Meneage, Cornwall: Report on geophysical survey, October 2003*, CfA Rep, **11/2004**

Linford, P K, 2010 *Groundwell Ridge, Swindon: archaeomagnetic dating report, July 2005*, Engl Heritage Res Dept Rep Ser, **69/2010**

Longin, R, 1971 New method of collagen extraction for radiocarbon dating, *Nature*, **230**, 241–2

Loveday, R, Gibson, A, Marshall, P D, Bayliss, A, Bronk Ramsey, C, and van der Plicht, H, 2007 The antler maceheads dating project, *Proc Prehist Soc*, **73**, 381–92

Lucy, S, Tipper, J, and Dickens, A, 2009 *The anglo-Saxon settlement and cemetery at Bloodmoor Hill, Carlton Colville, Suffolk*, E Anglian Archaeol Rep, **131**, Cambridge: Cambridge Archaeol Unit

Lyons, A, 2011 *Life and afterlife at Duxford, Cambridgeshire: archaeology and history in a chalkland community*, E Anglian Archaeol Rep, **141**, Bar Hill: Oxford Archaeol E

Malcolm, G, Bowsher, D, and Cowie, R, 2003 *Middle Saxon London: excavations at the Royal Opera House 1989–1999*, MoLAS Monogr, **15**, London: Museum London Archaeol Service

Malim, T, and Hines, J, 1998 *The Anglo-Saxon cemetery at Edix Hill (Barrington A)*, Cambridgeshire, CBA Res Rep, **112**

Malone, C, 1989 *The English Heritage Book of Avebury*, London: Batsford and Engl Heritage

Maltby, J M, 2004 Animal bones, in Investigation of the Whitesheet Down environs 1989–90: Neolithic causewayed enclosure and Iron Age settlement, *Wiltshire Studies*, **97**, 167–71

Marshall, P, 2013 Appendix 1. Radiocarbon dating protocol and Bayesian chronological modelling methodology, in *Modelling archaeology and environments in a wetland landscape: the hidden landscape archaeology of Hatfield and Thorne Moors* (eds H Chapman and B Gearey), Oxbow Books, Oxford, 167–184

Marshall, P, and Bayliss, A, 2014 Radiocarbon dating and Bayesian modelling, in *Staunch Meadow, Brandon, A high status Middle Saxon settlement* (A Tester, S Anderson, I Riddler and B Carr), East Anglian Archaeol, **151**, 16–19

Marshall, P, Bayliss, A, Leary, J, Campbell, G, Worley, F, Bronk Ramsey, C, and Cook, G, 2013 The Silbury Chronology, in *Silbury Hill: the largest prehistoric mound in Europe* (J Leary, D Field, and G Campbell), 97–116, Swindon: Engl Heritage

Marshall, P, Bayliss, A, McCormac F G, and Bronk Ramsey, C, 2012 Radiocarbon dating, in *Lundenwic: excavations in Middle Saxon London, 1987-2000* (R Cowie, L Blackmore, A Davis, J Keily, and K Rielly), MoLAS Monograph Ser, **63**, 307–12

Marshall, P, Tipper, J, Bayliss, A, McCormac, F G, van der Plicht, J, Bronk Ramsey, C, and Beavan-Athfield, N, 2009 Absolute Dating, in *The Anglo-Saxon Settlement and Cemetery at Bloodmoor Hill, Carlton Colville, Suffolk* (S Lucy, J Tipper, and A Dickens), East Anglian Archaeol, **131**, 322–9

Marshall, P D, Meadows, J, Bayliss, A, Sparks, R, Bronk Ramsey, C, and Beavan-Athfield, N, 2010 Scientific Dating, in *The Beautiful Rooms are Empty: excavations at Binchester Roman Fort, County Durham, 1976-1981 and 1986-1991* (I Ferris), 527–38, Durham (Durham County Council)

Martin, L, 2002 *Fiskerton, Witham Valley, Lincolnshire. Report on geophysical surveys, August 2002*, Centre for Archaeol Rep, **100/2002**

Matthews, C L, 1976 *Occupation sites on the Chiltern Ridge: excavation at Puddlehill and sites near Dunstable, Bedfordshire. Part 1: Neolithic, Bronze Age and early Iron Age*, BAR Brit Ser, **29**, Oxford: BAR

Matthews, I, 2008 *Roman Lode, Exmoor, Dartmoor: Tephrochronology*, Engl Heritage Res Rep Ser, **26–2008**

Mays, S, 1998 *The archaeology of human bones*, London (Routledge)

McAvoy, F, Morris, E L, and Smith, G H, 1980 The excavation of a multi-period site at Cargoon Bank, Lizard, Cornwall, 1979, *Cornish Archaeol*, **19**, 31–62

McCormac, F G, 1992 Liquid scintillation counter characterisation, optimisation, and benzene purity correction, *Radiocarbon*, **34**, 37–45

McCormac, F G, Bayliss, A, Baillie, M G L, and Brown, D M, 2004 Radiocarbon calibration in the Anglo-Saxon period: AD 495–725, *Radiocarbon*, **46**, 1123–5

McCormac, F G, Bayliss, A, Brown, D M, Reimer, P J, and Thompson, M M, 2008 Extended radiocarbon calibration in the Anglo-Saxon period, AD 395–485 and AD 735–805, *Radiocarbon*, **50**, 11–7

McCormac, F G, Kalin, R M, and Long, A, 1993 Radiocarbon Dating beyond 50,000 years by liquid scintillation counting, in *Liquid Scintillation Spectrometry 1992* (eds J E Noakes, F Schönhofer, and H Polach), 125–33, Tucson: Radiocarbon

McCormac, F G, Thompson, M, and Brown, D, 2001 Characterisation, optimisation and standard measurements for two small-sample high-precision radiocarbon counters, *Centre for Archaeol Rep*, **8/2001**

McCormac, F G, Thompson, M, Brown, D, Bayliss, A, Beavan, N, Reimer, P J, and Hoper, S T, 2011 *Laboratory and quality assurance procedures and the Queen's University, Belfast Radiocarbon Dating Laboratory for samples dated for the Anglo-Saxon Chronology Project*, Engl Heritage Res Dept Rep Ser, **89/2011**

McOmish, D, Field, D, and Brown, G, 2002 *The field archaeology of the Salisbury Plain Training Area*, Swindon: Engl Heritage

Meadows, J, Bayliss, A, Bronk Ramsey, C, Cook, G, Hamilton, D, Marshall, P, Morley, G, and Wilson, P, 2012 *Groundwell Ridge, Swindon, Wiltshire: radiocarbon dating and chronological modeling*, Engl Heritage Res Dept Rep Ser, **24/2012**

Meddens, F M, 1996 Sites from the Thames Estuary Wetlands, England and their Bronze Age use, *Antiq*, **70**, 325–34

Mercer, R J, 1970 The excavation of a Bronze Age hut-circle settlement, Stannon Down, St Breward, Cornwall, 1968, *Cornish Archaeol*, **9**, 17–46

Mercer, R J, 1999 The origins of warfare in the British Isles, in *Ancient Warfare* (J Carman and A Harding), 143–56, Stroud: Sutton Publishing Ltd

Mercer, R J, 2003 The early farming settlement of south western England in the Neolithic, in *Neolithic Settlement in Ireland and Western Britain* (I Armit, E Murphy, E Nelis, and D Simpson), 56–70, Oxford: Oxbow Books

Mercer, R J, and Healy, F, 2008 *Hambledon Hill, Dorset, England: excavation and survey of a Neolithic monument complex and its surrounding landscape*, Swindon (English Heritage)

Middleton, R, Wells, C E, and Huckerby, E, 1995 *The wetlands of North Lancashire*, North West Wetlands Survey, **3**, Lancaster: Lancaster Univ Archaeol Unit

Miles, D, and Palmer, D, 1983 Claydon Pike, *Curr Archaeol*, **8**, 88–92

Miles, D, and Palmer, S, 1982 *Figures in a landscape: archaeological investigations at Claydon Pike, Fairford/Lechlade: an interim report 1979–82*, Oxford: Oxford Archaeol Unit

Miles, D, and Palmer, S, 1990 Thornhill Farm and Claydon Pike, *Curr Archaeol*, **121**, 19–23

Miles, D, Palmer, S, and Perpetua Jones, G, 2007 *Iron Age and Roman settlement in the upper Thames valley: excavations at Claydon Pike and other sites within the Cotswold Water Park*, Thames Valley Landscapes Monogr, **26**, Oxford: Oxford Univ School Archaeol

Miles, H, 1975 Barrows on the St Austell Granite, Cornwall, *Cornish Archaeol*, **14**, 5–82

Moffett, L, Robinson, M A, and Straker, V, 1989 Cereals, fruit and nuts: charred plant remains from Neolithic sites in England and Wales and the Neolithic economy, in *The beginnings of agriculture* (A Miles, D Williams, and N Gardener), BAR Internat Ser, **496**, 243–61 Oxford: BAR

Momber, G, 2000 Drowned and deserted: a submerged prehistoric landscape in the Solent, England, *Nautical Archaeol*, **29**, 86–99

Momber, G, Tomalin, D, Scaife, R, Satchell, J, and Gillespie, J, 2011 *Bouldnor Cliff and the submerged Mesolithic landscape of the Solent*, CBA Monogr Ser, **164**, York: CBA

Mook, W G, 1986 Business meeting: Recommendations/Resolutions adopted by the Twelfth International Radiocarbon Conference, *Radiocarbon*, **28**, 799

Mook, W G, and van der Plicht, J, 1999 Reporting ^{14}C activities and concentrations, *Radiocarbon*, **41**, 227–40

Mook, W G, and Streurman, H J, 1983 Physical and chemical aspects of radiocarbon dating, *PACT*, **8**, 31–55

Mook, W G, and Waterbolk, H T, 1985 *Radiocarbon Dating: European Science Foundation Handbook for Archaeologists 3*, Strasbourg

Musty, J, 1969 The excavation of two barrows, one of Saxon date at Ford, Laverstock, near Salisbury, Wiltshire, *Antiquity J*, **49**, 98–117

Nadeau, M-J, Grootes, P M, Schleicher, M, Hasselberg, P, Rieck, A, and Bitterling, M, 1998 Sample throughput and data quality at the Leibniz-Labor AMS facility, *Radiocarbon*, **40**, 239–45

Nadeau, M-J, Schleicher, M, Grootes, P M, Erlenkeuser, H, Gottdang, A, Mous, D J W, Sarnthein, J M, and Willkomm, H, 1997 The Leibniz-Labor AMS facility at the Christian-Albrechts University, Kiel, Germany, *Nuclear Instruments and Physics Research B*, **123**, 22–30.

Nakamura, T, Taniguchi, Y, Tsuji, S, and Oda, H, 2001 Radiocarbon dating of charred residues on the earliest pottery in Japan, *Radiocarbon*, **43**, 1129–38

Nayling, N, 2003 *Tree-ring analysis of timbers from Baguley Hall, Greater Manchester, CfA report, 101/2003*, Centre for Archaeol Rep, **101/2003**, Portsmouth: English Heritage

Nayling, N, 2005 *Tree-ring analysis of timbers from Baguley Hall, Greater Manchester*, Centre for Archaeol Rep, **10/2005**, Portsmouth: English Heritage

Naysmith, P, Cook, G T, Freeman, S P H T, Scott, E M, Anderson, R, Xu, S, Dunbar, E, Muir, G K P, Dougans, A, Wilcken, K, Schnabel, C, Russell, N, Ascough, P L, and Maden, C, 2010 ^{14}C AMS at SUERC: improving QA data with the 5MV Tandem and 250 kV SSAMS, *Radiocarbon*, **52**, 263–71

Noakes, J E, Kim, S M, and Stipp, J J, 1965 Chemical and counting advances in Liquid Scintillation Age dating, in *Proceedings of the Sixth International Conference on Radiocarbon and Tritium Dating* (eds E A Olsson and R M Chatters), 68–92, Washington DC

Nowakowski, J A, and Quinnell, H, 2011 *Trevelgue Head, Cornwall: the importance of CK Ctroft Andrew's 1939 excavations for prehistoric and Roman Cornwall*, Cornwall: Cornwall Council

Olsen J, Heinemeier J, Hornstrup K M, Bennike P, Thrane, H, 2012 "Old wood" effect in radiocarbon dating of prehistoric cremated bones? *Journal of Archaeological Science*, 40, 30–4

Olsson, I U, 1979 The importance of the pretreatment of wood and charcoal samples, in *Radiocarbon Dating: Proceedings of the 9th International Radiocarbon Conference*, 135–46, Los Angeles and San Diego (Univ California Press)

O'Regan, H J, Turner, A, and Sabin, R, 2006 Medieval big cat remains from the Royal Menagerie at the Tower of London, *Internat J Osteoarchaeol*, 16, 385–394

Oswald, A, Dyer, C, and Barber, M, 2001 *The creation of monuments: Neolithic causewayed enclosures in the British Isles*, Swindon: Engl Heritage

Otlet, R L, Walker, A J, Hewson, A D, and Burleigh, R, 1980 ^{14}C interlaboratory comparison in the UK: experiment design, preparation, and preliminary results, *Radiocarbon*, 22, 936–46

Palmer, R, 1976 Interrupted ditch enclosures in Britain: the use of aerial photography for comparative studies, *Proc Prehist Soc*, 42, 161–86

Parfitt, K, and Anderson, T, 2012 *Buckland Anglo-Saxon cemetery, Dover: excavations 1994*, Archaeol Canterbury New Ser, 6, Canterbury: Canterbury Archaeol Trust

Parfitt, K, and Brugmann, B, 1997 *The Anglo-Saxon cemetery on Mill Hill, Deal, Kent*, Soc Med Archaeol Monogr, 14, London: Soc Med Archaeol

Parnell, A C, Inger, R, Bearhop S, and Jackson, A L, 2010 Source partitioning using stable isotopes: coping with too much variation, *PLoS ONE*, 5, e9672

Parnell, G, 1999 *The Royal Menagerie at the Tower of London*, Leeds: Royal Armouries Museum

Pearsall, D M, 1989 *Palaeoethnobotany: a handbook of procedures*, San Diego: Academic Press

Pearson, A, 2003 Beech Court Farm Enclosure, Ewenny, Vale of Glamorgan: post-excavation summary. GGAT Project A835, unpubl rep, Glamorgan-Gwent Archaeol Trust

Pearson, G W, 1984 *The development of high-precision ^{14}C measurement and its application to archaeological time-scale problems*, unpubl PhD thesis, Queens Univ Belfast

Pearson, G W, 1987 How to cope with calibration, *Antiq*, 61, 98–103

Pearson, G W, and Stuiver, M, 1986 High-precision calibration of the radiocarbon timescale, 500–2500 BC, *Radiocarbon*, 28, 839–62

Pennington, W, 1975 The effect of man on the environment in North-West England: the use of absolute pollen diagrams, in *The effect of man on the landscape: the Highland Zone* (J G Evans, S Limbrey, and H Cleere), CBA Res Rep, 11, 74–86 CBA

Phillips, D, and Gregg, J W, 2003 Source partitioning using stable isotopes: coping with too many sources, *Oecologia*, 136, 261–9

Piggott, S, 1931 The Neolithic pottery of the British Isles, *Archaeol J*, 88, 67–158

Piggott, S, 1952 The Neolithic camp on Whitesheet Hill, Kilmington parish, *Wiltshire Archaeol Natur Hist Mag*, 54, 404–10

Piggott, S, 1954 *Neolithic cultures of the British Isles*, Cambridge: Cambridge Univ Press

van der Plicht, J, Wijma, S, Aerts, A T, Pertuisot, M H, and Meijer, H A J, 2000 Status report: the Groningen AMS facility, *Nuclear Instruments and Methods in Physics Research B*, 172, 58–65

Plunkett, G, Pilcher, J R, McCormac, F G, and Hall, V A, 2004 New dates for first millennium BC tephra isochrones in Ireland, *The Holocene*, 14, 780–6

Pollard, J, 1999 The Keiller excavations, in *The harmony of symbols: the Windmill Hill causewayed enclosure, Wiltshire* (A Whittle, J Pollard, and C Grigson), 24–72, Oxford: Oxbow Books

Pollard, J, and Hamilton, M, 1994 Recent fieldwork at Maiden Bower, *Archaeol J*, 21, 10–18

Powlesland, D J, Haughton, C A, and Hanson, J H, 1986 Excavations at Heslerton, North Yorkshire 1978–82, *Archaeol J*, 143, 53–173

Powlesland, D J, 2003a The Heslerton Parish Project: 20 years of archaeological research in the Vale of Pickering, in *The Archaeology of Yorkshire: an assessment at the beginning of the 21st century* (T G Manby, S Moorhouse, and P Ottaway), York Archaeol Soc Occas Pap, 3, 275–291 Leeds: York Archaeol Soc

Powlesland, D J, 2003b *25 years of archaeological research on the sands and gravels of Heslerton*, Yedingham: Landscape Res Centre

Powlesland, D J, and Rackham, J, 2007 *LRC Archive Pilot Project: Environmental Assessment Project for the central Vale of Pickering (English Heritage Project 3038)*, unpubl typescript rep, Engl Heritage

PPG16 1990 *Planning Policy Guidance: Archaeology and Planning*, Department of the Environment

Preston, S, 2007 Bronze Age occupation and Saxon features at the Wolverton Turn enclosure, near Stony Stratford, Milton Keynes: investigations by Tim Schadla-Hall, Philip Carstairs, Jo Lawson, Hugh Beamish, Andrew Hunn, Ben Ford and Tess Durden, 1972 to 1994, *Rec Buckinghamshire*, 47, 81–117

Pryor, F M M, 1987 Etton 1986: Neolithic metamorphoses, *Antiq*, 61, 78–81

Pryor, F M M, 1988 Etton, near Maxey, Cambridgeshire; a causewayed enclosure on the fen edge, in *Enclosures and defences in the Neolithic of Western Europe* (C Burgess, P Topping, C Mordant, and M Maddison), BAR Int Ser, **403(i)**, 107–26 Oxford: BAR

Pryor, F M M, 1998 *Excavations at a Neolithic Causewayed Enclosure near Maxey, Cambridgeshire, 1982–87*, Engl Heritage Rep, **18**

Pryor, F M M, and Kinnes, I A, 1982 A waterlogged causewayed enclosure in the Cambridgeshire Fens, *Antiq*, **56**, 124–6

Pryor, F M M, French, C A I, and Taylor, M, 1985 An interim report on excavations at Etton, Maxey, Cambridgeshire, *Antiquity J*, **65**, 275–311

Quinnell, H, 2004 *Trethurgy: excavations at Trethurgy Round, St Austell: community and status in Roman and post-Roman Cornwall*, Truro: Cornwall County Council

Rackham, D J, French, C A I, Fryer, V, Murphy, P, Scaife, R G, Smith, D, and Taylor, M, in press *A multiperiod landscape at Rectory Farm, West Deeping*

Rackham, J, 2004 *Fiskerton Auger Survey, Fiskerton, report EAC 27/04*, Unpubl rep, Environmental Archaeol Consultancy

Rawlings, M, Allen, M J, and Healy, F, 2004 Investigation of the Whitesheet Down environs 1989–90: Neolithic causewayed enclosure and Iron Age settlement, *Wiltshire Studies*, **97**, 144–96

RCHME, 1995a A causewayed enclosure and The Trundle hillfort on St Roche's Hill, Singleton, West Sussex. An earthwork survey, unpubl rep, RCHME

RCHME, 1995b *A survey of earthworks at Whitehawk Camp, Brighton, East Sussex*, Cambridge: RCHME

Reimer, P J, Baillie, M G L, Bard, E, Bayliss, A, Beck, J W, Bertrand, C J H, Blackwell, P G, Buck, C E, Burr, G S, Cutler, K B, Damon, P E, Edwards, R L, Fairbanks, R G, Friedrich, M, Guilderson, T P, Hogg, A G, Hughen, K A, Kromer, B, McCormac, F G, Manning, S, Bronk Ramsey, C, Reimer, R W, Remmele, S, Southon, J R, Stuiver, M, Talamo, S, Taylor, F W, van der Plicht, J, and Weyhenmeyer, C E, 2004 IntCal04 Terrestrial Radiocarbon Age Calibration, 0–26 cal kyr BP, *Radiocarbon*, **46**, 1029–58

Reimer, P J, Baillie, M G L, Bard, E, Bayliss, A, Beck, J W, Blackwell, P G, Bronk Ramsey, C, Buck, C E, Burr, G S, Edwards, R L, Friedrich, M, Grootes, P M, Guilderson, T P, Hajdas, I, Heaton, T J, Hogg, A G, Hughen, K A, Kaiser, K F, Kromer, B, McCormac, F G, Manning, S W, Reimer, R W, Richards, D A, Southon, J R, Talamo, S, Turney, C S M, van der Plicht, J, and Weyhenmeyer, C E, 2009 IntCal09 and Marine09 radiocarbon age calibration curves, 0–50,0000 Years cal BP, *Radiocarbon*, **51**, 1111–50

Reimer, P J, Bard, E, Bayliss, A, Beck, W, Blackwell, P G, Bronk Ramsey, C, Buck, C E, Cheng, H, Edwards, R L, Friedrich, M, Grootes, P M, Guilderson, T P, Haflidason, H, Hajdas, I, Hatte, C, Heaton, T J, Hoffman, D L, Hogg, A G, Hughen, K A, Kaiser, K F, Kromer, B, Manning, S W,

Niu, M, Reimer, R W, Richards, D A, Scott, E M, Southon, J R, Staff, R A, Turney, C S M, and van der Plicht, J, 2013 INTCAL13 and MARINE13 radioacrbon age calibration curves 0–50,000 years cal BP, *Radiocarbon*, **55**, 1869–87

Richards, J C, 1990 *The Stonehenge Environs Project*, Engl Heritage Archaeol Rep, **16**

Robertson, D, and Ames, J, 2015 Timber Monuments and Coastal Processes: Recording and Monitoring of Archaeological Remains at Holme Beach, Norfolk, UK 2003–2008, *Journal Wetland Archaeol*, **15**, 34–56

Robertson-Mackay, R, 1962 The excavation of the causewayed camp at Staines, Middlesex, *Middlesex Archaeol News*, **7**, 131–4

Robertson-Mackay, R, 1965 The primary Neolithic settlement in southern England; some new aspects, in *Atti del VI Congresso Internazionale delle Scienze Preistoriche e Protoistoriche, Roma 1962: II Communicazioni, Sezione I-IV*, 319–23, Rome: UISPP

Robertson-Mackay, R, 1987 The Neolithic causewayed enclosure at Staines, Surrey: excavations 1961–63, *Proc Prehist Soc*, **53**, 23–128

Robertson-Mackay, R, Blackmore, L, Hurst, J G, Jones, P, Moorhouse, S, and Webster, L, 1981 A group of Saxon and medieval finds from the site of the Neolithic causewayed enclosure at Staines, Surrey, with a note on the topography of the area, *Trans London Middlesex Archaeol Soc*, **32**, 107–31

Rose, P, and Preston-Jones, A, 1991 *Boden St Anthony (SW 76842404): Report of site visit and survey, September 1991*, unpubl rep, Cornwall Archaeol Unit

Ross Williamson, R P, 1930 Excavations in Whitehawk Neolithic camp, near Brighton, *Sussex Archaeol Coll*, **71**, 56–96

Rouse, A, and Rowland, S, 1999 Small vertebrates, in *The harmony of symbols: the Windmill Hill causewayed enclosure* (A Whittle, J Pollard, and C Grigson), 253–6, Oxford: Oxbow Books

Rozanski, K, 1991 *Report on the International Atomic Energy Agency consultants' group meeting on C-14 reference materials for radiocarbon laboratories, February 18-20, 1991, Vienna, Austria*, unpubl report, IAEA (Vienna)

Rozanski, K, Stichler, W, Gonfiantini, R, Scott, E M, Beukens, R P, Kromer, B, and van der Plicht, J, 1992 The IAEA ^{14}C intercomparison exercise 1990, *Radiocarbon*, **34**, 506–19

Russell, M, 1996 Discussion, in Excavations at Whitehawk Neolithic enclosure, Brighton, East Sussex, 1991–1993, *Sussex Archaeol Coll*, **134**, 56–60

Russell, M, and Rudling, D, 1996 Excavations at Whitehawk Neolithic enclosure, Brighton, East Sussex, 1991–1993, *Sussex Archaeol Coll*, **134**, 39–61

Rylatt, J, Carruthers, W, Darling, M, Fell, V, Field, N, Kenward, H, Meadows, J, Panter, I, Rackham, J, Richardson, J, Scaife, R, Steane, K, Taylor, M, Tyers, I, Vince, A, and Williams, J, 2011 The Iron Age causeway, Fiskerton, Lincolnshire: investigation of preservation, Unpubl rep, The Environ Archaeol Consultancy

Salin, B, 1904 *Die altgermanische Thierornamentik*, Stockholm: Wahlstrom and Widstrand

Savage, R, 1988 *Village, fortress, shrine. Crickley Hill Gloucesteshire 3500 BC-AD 500*, Cheltenham: Crickley Hill Archaeol Trust

Scott, E M, 2003 The Third International Radiocarbon Intercomparison (TIRI) and the Fourth International Radiocarbon Intercomparision (FIRI) 1990 – 2002: results, analyses, and conclusions, *Radiocarbon*, **45**, 135–408

Scott, E M, Aitchison, T C, Harkness, D D, Cook, G T, and Baxter, M S, 1990 An overview of all three stages of the international radiocarbon intercomparison, *Radiocarbon*, **32**, 309–19

Scott, E M, Cook, G T, Naysmith, P, Bryant, C, O'Donnell, D, 2007 A report on Phase 1 of the fifth international radiocarbon inter-comparison (VIRI), *Radiocarbon*, **49**, 409–26

Scott, E M, Cook, G T, and Naysmith, P, 2010a A report on phase 2 of the fifth international radiocarbon intercomparison (VIRI), *Radiocarbon*, **52**, 846–58

Scott, E M, Cook, G T, and Naysmith, P, 2010b The fifth international radiocarbon intercomparison (VIRI): an assessment of laboratory performance in stage 3, *Radiocarbon*, **52**, 859–65

Scull, C, 2009 *Early medieval (late 5th-early 8th centuries AD) cemeteries at Boss Hall and Buttermarket, Ipswich, Suffolk*, Soc Med Archaeol Monogr, **27**, Leeds: Soc Med Archaeol

Selkirk, A, 1971 Ascott-under-Wychwood Long Barrow, *Current Archaeology*, **3**, 7–10

Shand, G, 1998 A Neolithic causewayed enclosure in Kent, *PAST*, **29**, 1

Shand, G, 2001 *Archaeological excavations at Chalk Hill, Ramsgate Harbour Approach Road 1997/8*, unpubl rep, Canterbury Archaeol Trust

Sharples, N M, 1986 Maiden Castle project 1985: an interim report, *Proc Dorset Natur Hist Archaeol Soc*, **107**, 111–9

Sharples, N M, 1987 Maiden Castle project 1986: an interim report, *Proc Dorset Natur Hist Archaeol Soc*, **108**, 53–61

Sharples, N M, 1991a *Maiden Castle. Excavations and field survey 1985–6*, Engl Heritage Archaeol Rep, **19**

Sharples, N M, 1991b *English Heritage Book of Maiden Castle*, London: Batsford and Engl Heritage

Sharples, N M, n d Maiden Castle. Excavations 1985, 1986. Archive report, unpubl doc, Dorset County Museum 1992.91.26.1.645/1

Sidell, J, Wilkinson, K, Scaife, R, and Cameron, N, 2000 *The Holocene evolution of the London Thames*, Museum London Archaeol Service Monogr, **5**, London: Museum of London Archaeol Service

Siegmund, F, 1998 *Merowingerzeit am Niederrhein: Die fruhmittelalterlichen Funde aus dem Regierungsbezirk Dusseldorf und dem Kreis Heinsburg*, Rheinland-Verlag: Cologne

Simpson, D D A, 1996 Crown antler maceheads and the later Neolithic of Britain, *Proc Prehist Soc*, **62**, 293–310

Slota Jr, P J, Jull, A J T, Linick, T W and Toolin, L J, 1987 Preparation of small samples for ^{14}C accelerator targets by catalytic reduction of CO, *Radiocarbon*, **29**, 303–6

Smith, I F, 1959 Excavations at Windmill Hill, Avebury, Wilts, 1957–8, *Wiltshire Archaeol Natur Hist Mag*, **57**, 149–62

Smith, I F, 1965 *Windmill Hill and Avebury, excavations by Alexander Keiller 1925–1939*, Oxford: Clarendon Press

Smith, I F, 1966 Windmill Hill and its implications, *Palaeohistoria*, **12**, 469–81

Smith, I F, 1971 Causewayed enclosures, in *Economy and settlement in Neolithic and early Bronze Age Britain and Europe* (D D A Simpson), 89–112, Leicester: Leicester Univ Press

Smith, W G, 1894 *Man the primaeval savage: his haunts and relics from the hill-tops of Bedfordshire to Blackwall*, London: Edward Stanford

Smith, W G, 1904a Early man, in *The Victoria History of the Counties of England. A History of Bedfordshire. Volume I*, 145–74, Westminster: Archibald Constable and Company Ltd

Smith, W G, 1904b *Dunstable: its history and surroundings. the homeland library III*, London: Homeland Assoc for the Encouragement of Touring in Great Britain

Smith, W G, 1915 Maiden Bower, Bedfordshire, *Proc Prehist Soc Antiq London Ser 2*, **27**, 143–61

Snashall, N, 1997 *The Neolithic Shrine at Crickley Hill*, unpubl BA dissert, Univ Nottingham

Snashall, N, 1998 The Interior of the Neolithic enclosures at Crickley Hill, unpubl MA dissert, Univ Nottingham

Snoeck, C, Brock, F, and Schulting, R J, 2014 Carbon exchanges between bone apatite and fuels during cremation: impact on radiocarbon dates, *Radiocarbon*, **56**, 591–602

Staff, R A, Reynard, L, Brock, F, and Bronk Ramsey, C, 2014 Wood pretreatment protocols and measurement of tree-ring standards at the Oxford Radiocarbon Accelerator Unit (ORAU), *Radiocarbon*, **56**, 709–15

Stafford, T W, Brendal, K, and Duhamel, R C, 1988 Radiocarbon, ^{13}C and ^{15}N analysis of fossil bone: removal of humates with SAD-2 resin, *Geochimica et Cosmochimica Acta*, **52**, 2257–67

Stein, F, 1967 Adelsgraber des achten Jahrhunderts in Deutschland, in *Germanische Denkmaler der Volkerwanderungszeit A*, 9, Berlin: De Gruyter

Stenhouse, M J, and Baxter, M S, 1983 ^{14}C dating reproducibility: evidence from routine dating of archaeological samples, *PACT*, **8**, 147–61

Stuiver, M, and Kra, R S, 1986 Editorial comment, *Radiocarbon*, **28(2B)**, ii

Stuiver, M, and Polach, H A, 1977 Reporting of ^{14}C data, *Radiocarbon*, **19**, 355–63

Stuiver, M, and Pearson, G W, 1986 High-precision calibration of the radiocarbon timescale, AD 1950–2500 BC, *Radiocarbon*, **28**, 805–38

Stuiver, M, and Reimer, P J, 1986 A computer program for radiocarbon age calculation, *Radiocarbon*, **28**, 1022–30

Stuiver, M, and Reimer, P J, 1993 Extended ^{14}C data base and revised CALIB 3.0 ^{14}C age calibration program, *Radiocarbon*, **35**, 215–30

Stuiver, M, Reimer, P J, Bard, E, Beck, J W, Burr, G S, Hughen, K A, Kromer, B, McCormac, F G, van der Plicht, J, and Spurk, M, 1998 INTCAL98 radiocarbon age calibration, 24,000–0 cal BP, *Radiocarbon*, **40**, 1041–84

Stukely, W, 1743 *Abury, a Temple of the British druids*, London: printed and sold by W Innys, R Manby, B Dod and J Brindley

Tallis, J, 1991 Forest and moorland in the South Pennine Uplands in the mid-Flandrian period III: the spread of moorland, local, regional and national, *J Ecology*, **79**, 749–56

Taylor, M, 1998 Wood and bark from the enclosure ditches, in *Excavations at a Neolithic causewayed enclosure near Maxey, Cambridgeshire, 1982–87* (F Pryor), Engl Heritage Archaeol Rep, **18**, 115–59 London: Engl Heritage

Taylor, T, Lord, T C, and O'Connor, T P, 2011 *Recent work at Kinsey Cave*, unpubl rep, Univ of Bradford for Engl Heritage and Yorkshire Dales National

Tester, A, Anderson, S, Riddler, I, and Carr, R, 2014 *Staunch Meadow, Brandon, Suffolk: a high status middle Saxon Settlement*, E Anglian Archaeol, **151**, Suffolk: Suffolk County Council Archaeol Service

Thomas, C, and Rackham, D J, 1996 Bramcote Green, Bermondsey: a Bronze Age trackway and palaeoenvironmental assessment, *Proc Prehist Soc*, **61**, 221–53

Thomas, C, Thorpe, C, and Quinnell, H, 2004 *Post-Roman pottery - initial appraisal* in Archaeology beneath the Towans. Excavations at Gwithian, Cornwall 1949–1969, unpubl rep, Truro: Historic Environ Serv

Thomas, K, 1996 A contribution to the environmental history of Whitehawk Neolithic enclosure, in Excavations at Whitehawk Neolithic enclosure, Brighton, East Sussex, 1991–1993, *Sussex Archaeol Coll*, **134**, 51–6

Thomas, K D, 1982 Neolithic enclosures and woodland habitats on the south downs in Sussex, in *Archaeological Aspects of Woodland Ecology* (M Bell and S Limbrey), BAR Int Ser, **146**, 147–70 Oxford: BAR

Thomas, N, 1964 The Neolithic causewayed camp at Robin Hood's Ball, Shrewton. Wiltshire, *Wiltshire Archaeol Natur Hist Mag*, **59**, 1–27

Time Team, 2005 http://www.channel4.com/history/timeteam/2005_north_found.html, unpubl rep, Time Team

Todd, M, 1984a Excavations at Hembury, Devon, 1980–3: a summary report, *Antiquity J*, **64**, 251–68

Todd, M, 1984b Hembury (Devon): Roman troops in a hillfort, *Antiq*, **58**, 171–4

Underwood, D, 1996 The worked flint, in Excavations at Whitehawk Neolithic enclosure, Brighton, East Sussex, 1991–1993, *Sussex Archaeol Coll*, **134**, 49–50

Vandeputte, K, Moens, L, and Dams, R, 1996 Improved sealed-tube combustion of organic samples to CO_2 for stable isotopic analysis, radiocarbon dating and percent carbon determinations, *Analytical Letters*, **29**, 2761–73

Van de Noort, R, Chapman, H, and Collis, J, 2007 *Sutton Common: the excavation of an Iron Age 'marsh-fort'*, CBA Res Rep, **154**

van den Bogaard, C, and Schmincke, H U, 2002 Linking the North Atlantic to central Europe: a high-resolution Holocene tephrochronological record from northern Germany, *The Holocene*, **2002**, 3–20

Vyner, B, and Wall, I, 2011 A Neolithic cairn at Whitwell, Derbyshire, *Derbyshire Archaeol J*, **131**, 1–132

Waddington, C, 2007 *Mesolithic settlement in the North Sea basin: a case study from Howick, north-east England*, Oxford: Oxbow Books

Wainwright, G J, and Cunliffe, B W, 1985 Maiden Castle: excavation, education and entertainment?, *Antiq*, **59**, 97–100

Ward, G K, and Wilson, S R, 1978 Procedures for comparing and combining radiocarbon age determinations: a critique, *Archaeometry*, **20**, 19–31

Waterbolk, H T, 1971 Working with radiocarbon dates, *Proc Prehist Soc*, **37**, 15–33

West, S E, 1988 *Westgarth Gardens Anglo-Saxon cemetery, Suffolk*, E Anglian Archaeol Monogr, **38**, Ipswich: Suffolk County Council

Wheeler, R E M, 1943 *Maiden Castle, Dorset.*, Rep Res Comm Soc Antiq London, **12**

Whittle, A, 1991 Wayland's Smithy, Oxfordshire: excavations at the Neolithic Tomb in 1962–63 by R. J. C. Atkinson and S. Piggott, *Proc Prehist Soc*, **57**, 61–101

Whittle, A, 1997 *Sacred mound. Holy rings*, Oxbow Monogr, **74**, Oxford: Oxbow

Whittle, A, Bayliss, A, and Wysocki, M, 2007a Once in a lifetime: the date of the Wayland's Smithy long barrow, *Cantab Archaeol J*, **17(1) suppl**, 103–21

Whittle, A, Barclay, A, Bayliss, A, McFadgen, L, Schulting, R, and Wysocki, M, 2007b Building for the dead: events, processes and changing worldviews from the 38th to the 34th centuries cal BC in southern Britain, *Cantab Archaeol J*, **17**, 123–47

Whittle, A, and Pollard, J, 1998 Windmill Hill causewayed enclosure: the harmony of symbols, in *Understanding the Neolithic of North-Western Europe* (M Edmonds and C Richards), 231–47, Glasgow: Cruithne Press

Whittle, A, Healy, F, and Bayliss, A, 2011 *Gathering time: dating the early Neolithic enclosures of southern Britain and Ireland*, Oxford: Oxbow Books

Whittle, A, Pollard, J, and Grigson, C, 1999 *The harmony of symbols: the Windmill Hill causewayed enclosure, Wiltshire*, Oxford: Oxbow Books

Wilkinson, K N, Scaife, R J, and Sidell, E J, 2000 Environmental and sea-level changes in London from 10500 BP to the present: a case study from Silvertown, *Proc Geolog Assoc*, **111**, 41–54

Williams, C, and Switsur, V R, 1985 *Mesolithic exploitation patterns in the central Pennines: a palynological study of Soyland Moor*, BAR Brit Ser, **139**, Oxford: BAR

Wysocki, M, Bayliss, A, and Whittle, A, 2007 Serious mortality: the date of the Fussell's Lodge Long Barrow, *Cambridge Archaeol J*, **17 (suppl)**, 65–84

Xu, S, Anderson, R, Bryant, C, Cook, G T, Dougans, A, Freeman, S, Naysmith, P, Schnabel, C, and Scott, E M, 2004 Capabilities of the new SUERC 5MV AMS facility for ^{14}C dating, *Radiocarbon*, **46**, 59–64

Yates, A, 2000a Ewenny, Beech Court Farm prehistoric enclosure (SS 904 766), *Archaeol Wales*, **40**, 89

Yates, A, 2002 A prehistoric enclosure at Beech Court Farm, Ewenny Quarry, Ewenny, Vale of Glamorgan, June 1998–present, unpubl rep, Glamorgan-Gwent Archaeol Trust

Zhao, Z, and Pearsall, D M, 1998 Experiments in improving phytolith extraction from soils, *J Archaeol Sci*, **25**, 587–98

Zienkiewicz, L, and Hamilton, M, 1999 Pottery, in *The harmony of symbols: the Windmill Hill causewayed enclosure, Wiltshire* (A Whittle, J Pollard, and C Grigson), 257–317, Oxford: Oxbow Books

Index of laboratory codes

GrA–23828 23

GrA–23829 23

GrA–23831 23

GrA–23927 23–4

GrA–23933 24

GrA–25292 24

GrA–25294 24

GrA–25295 24

GrA–25296 24–5

GrA–25304 25

GrA–25305 25

GrA–25306 25

GrA–25367 146

GrA–25368 140

GrA–25379 137

GrA–25389 146

GrA–25391 137–8

GrA–25545 146–7

GrA–25546 147

GrA–25549 147

GrA–25550 147

GrA–25553 147–8

GrA–25554 141

GrA–25555 141

GrA–25556 141

GrA–25558 138

GrA–25559 141

GrA–25560 138

GrA–25563 59

GrA–25589 59

GrA–25590 59–60

GrA–25592 60

GrA–25706 142

GrA–25707 142

GrA–25821 148

GrA–25923 60

GrA–25925 60–1

GrA–25926 61

GrA–25927 61

GrA–25929 61–2

GrA–25931 62

GrA–25935 62

GrA–25936 62

GrA–25937 62–3

GrA–25949 63

GrA–25950 63

GrA–26355 63–4

GrA–26357 64

GrA–26548 201

GrA–26817 122–3

GrA–26819 123–4

GrA–26962 125

GrA–26963 125

GrA–26965 125

GrA–26966 127

GrA–26967 128

GrA–26969 128

GrA–26971 128

GrA–26972 132

GrA–26973 132

GrA–26975 128–9

GrA–26976 129

GrA–26977 129

GrA–27093 25–6

GrA–27094 26

GrA–27096 26

GrA–27098 26

GrA–27099 26

GrA–27100 26–7

GrA–27102 27

GrA–27202 284

GrA–27203 285

GrA–27318 72

GrA–27320 72

GrA–27321 76

GrA–27322 116

GrA–27325 129

GrA–27326 129

GrA–27327 129–30

GrA–27328 130

GrA–27330 130

GrA–27331 242

GrA–27332 241

GrA–27335 241

GrA–27336 241

GrA–27417 22

GrA–27806 84

GrA–27808 85

GrA–27809 85

GrA–27810 85

GrA–27813 87

GrA–27814 80

GrA–27815 80–1

GrA–27816 81

GrA–27818 81

GrA–27820 81

GrA–27821 81–2

GrA–27828 82

GrA–27911 78

GrA–27914 78

GrA–28174 196

GrA–28175 196

GrA–28199 196

GrA–28207 197

GrA–28208 197

GrA–28209 197

GrA–28465 243

GrA–28466 243

GrA–28467 243–4

GrA–28555 244

GrA–29107 106

GrA–29108 106

GrA–29109 107

GrA–29111 107

GrA–29112 107

GrA–29113 112

GrA–29120 112

GrA–29141 114

GrA–29142 114–5

GrA–29143 107

GrA–29145 112–3

GrA–29146 111

GrA–29147 111

GrA–29207 107

GrA–29209 107–8

GrA–29210 108

GrA–29211 108

GrA–29213 113

GrA–29336 111

GrA–29353 90

GrA–29354 90

GrA–29355 90

GrA–29357 90

GrA–29358 90

GrA–29359 93

GrA–29362 88–9

GrA–29363 130

GrA–29364 132–3

GrA–29367 91

GrA–29368 91

GrA–29369 91

GrA–29372 91

GrA–29551 99

GrA–29553 99

GrA–29554 99–100

GrA–29555 100–101

GrA–29557 101

GrA–29706 142

GrA–29707 138

GrA–29708 138–9

GrA–29711 148

GrA–29712 148

GrA–29713 148–9

GrA–29714 149

GrA–29743 108

GrA–29744 108

GrA–29746 139

GrA–29808 102

GrA–29809 102

GrA–29810 102

GrA–29891 104

GrA–29892 104

GrA–30026 105

GrA–30028 117

GrA–30029 117

GrA–30030 117

GrA–30031 117

GrA–30033 120

GrA–30035 120

GrA–30036 120

GrA–30038 118–9

GrA–30066 120

GrA–30067 133–4

GrA–30068 134

GrA–30072 135

GrA–30073 135

GrA–30074 135

GrA–30076 115

GrA–30176 121

GrA–30197 121–2

GrA–30368 87

GrA–30635 225

GrA–30753 225

GrA–30755 225

GrA–30756 225–6

GrA–30757 226

GrA–30761 226

GrA–30762 226

GrA–30763 226–7

GrA–30765 227

GrA–30766 227

GrA–30767 229–30

GrA–30768 230

GrA–30770 230

GrA–30772 230

GrA–30773 230

GrA–30775 231

GrA–30776 231

GrA–30777 231

GrA–30779 231

GrA–30780 231–2

GrA–30880 73

GrA–30882 74

GrA–30884 74

GrA–30885 74

GrA–30886 74

GrA–30888 74

GrA–31094 95

GrA–31100 82

GrA–31101 82

GrA–31103 87–8

GrA–31105 78

GrA–31106 78

GrA–31110 85

GrA–31111 85–6

GrA–31113 86

GrA–31114 86

GrA–31184 94

GrA–31185 94

GrA–31201 95

GrA–31204 95

GrA–31205 96

GrA–31206 96

GrA–31207 96

GrA–31209 96

GrA–31210 96–7

GrA–31211 97

GrA–31213 97

GrA–31463 97

GrA–31466 97

GrA–31467 97

GrA–31544 98

GrA–31545 98

GrA–31546 98

GrA–31548 98

GrA–31550 98–9

GrA–31559 99

GrA–32367 125–6

GU–5817 157

GU–5818 157

GU–5919 158

GU–5920 158

GU–5996 277

GU–5997 277

GU–5999 155

GU–6000 155

GU–6001 155

GU–6002 53–4

GU–6003 54

GU–6004 54

GU–6005 54

GU–6006 54

GU–6007 54–5

GU–6008 55

GU–6009 55

GU–6010 55

GU–6011 55

GU–6012 176–7

GU–6013 177

GU–6014 177

GU–6015 177–8

GU–6016 178

GU–6017 178

GU–6018 178–9

GU–6019 179

GU–6020 179–80

GU–6021 180

GU–6022 180–1

GU–6023 181

GU–6024 181

GU–6025 181–2

GU–6026 182

GU–6027 182

GU–6028 182–3

GU–6029 183

GU–6030 183–4

GU–6031 184

GU–6032 184

GU–6033 184

GU–6034 185

GU–6035 185

GU–6036 205

GU–6037 205

GU–6038 55

GU–6039 56

GU–6040 208–9

GU–6041 209

GU–6050 51

GU–6365 213

KIA–27623 198

KIA–27624 198–9

KIA–27625 199

KIA–27626 199

OxA–11647 244

OxA–11663 244

OxA–11970 241–2

OxA–12370 37

OxA–12371 37

OxA–12372 37

OxA–12675 27

OxA–12676 27

OxA–12677 27

OxA–12678 27

OxA–12679 27–8

OxA–12680 28

OxA–12721 288

OxA–12722 288

OxA–12838 293

OxA–12839 293

OxA–12840 293

OxA–12841 293–4

OxA–12862 294

OxA–12863 294

OxA–12864 294

OxA–12865 288

OxA–12866 288

OxA–12868 163

OxA–12869 163

OxA–12870 163

OxA–12880 288–9

OxA–12885 294

OxA–12886 295

OxA–12887 295

OxA–12888 295

OxA–12892 164

OxA–12914 164

OxA–12915 164

OxA–12916 164

OxA–12917 164–5

OxA–12918 165

OxA–12919 165

OxA–12920 165–6

OxA–12938 289

OxA–12939 211–2

OxA–12940 212

OxA–12941 212

OxA–12942 212	OxA–13307 265	OxA–13394 201	OxA–13715 139
OxA–12943 212	OxA–13308 265	OxA–13400 29	OxA–13726 65–6
OxA–12944 186	OxA–13309 265	OxA–13401 29	OxA–13727 66
OxA–12945 186	OxA–13310 265	OxA–13402 30	OxA–13728 66
OxA–12946 186	OxA–13311 265	OxA–13403 30	OxA–13730 143–4
OxA–12947 186	OxA–13312 266	OxA–13404 30	OxA–13732 139
OxA–12948 186–7	OxA–13313 266	OxA–13416 266	OxA–13752 66
OxA–12949 187	OxA–13314 266	OxA–13428 283	OxA–13753 66
OxA–12950 187	OxA–13315 28	OxA–13446 247	OxA–13754 67
OxA–12951 187	OxA–13316 28	OxA–13447 202	OxA–13755 67
OxA–12952 187–8	OxA–13317 28	OxA–13448 202	OxA–13756 67
OxA–12953 188	OxA–13318 28–9	OxA–13449 202–3	OxA–13757 67
OxA–12954 188	OxA–13319 29	OxA–13450 203	OxA–13759 150
OxA–12964 289	OxA–13320 29	OxA–13461 170	OxA–13760 139
OxA–12965 289	OxA–13353 280	OxA–13462 170	OxA–13773 160–1
OxA–12966 289	OxA–13354 280	OxA–13463 170	OxA–13774 161
OxA–12967 188	OxA–13355 280	OxA–13464 170	OxA–13775 161
OxA–13028 188	OxA–13356 281	OxA–13465 170	OxA–13776 161
OxA–13029 188–9	OxA–13357 281	OxA–13466 170–1	OxA–13777 161
OxA–13034 277–8	OxA–13358 281	OxA–13476 169	OxA–13778 161
OxA–13035 278	OxA–13359 281	OxA–13495 261	OxA–13779 161–2
OxA–13036 278	OxA–13360 281	OxA–13496 261–2	OxA–13780 162
OxA–13038 278	OxA–13361 282	OxA–13497 262	OxA–13781 162
OxA–13049 278–9	OxA–13362 282	OxA–13498 262	OxA–13812 144
OxA–13050 279	OxA–13363 282	OxA–13499 149	OxA–13813 144
OxA–13051 279	OxA–13364 282	OxA–13500 149	OxA–13814 144
OxA–13052 279	OxA–13365 282	OxA–13501 149	OxA–13815 139
OxA–13053 279	OxA–13366 282–3	OxA–13502 150	OxA–13871 158
OxA–13080 166	OxA–13370 189	OxA–13503 150	OxA–13882 67–8
OxA–13116 279	OxA–13380 249	OxA–13504 150	OxA–13883 68
OxA–13117 279–80	OxA–13381 249	OxA–13505 142	OxA–13884 162
OxA–13118 280	OxA–13382 249	OxA–13561 150	OxA–13885 162
OxA–13135 28	OxA–13383 250	OxA–13609 169	OxA–13890 158
OxA–13191 263	OxA–13384 250	OxA–13679 142–3	OxA–13891 162
OxA–13192 263	OxA–13385 245–6	OxA–13680 143	OxA–13892 68
OxA–13223 280	OxA–13386 246	OxA–13706 204	OxA–13935 124
OxA–13224 263	OxA–13387 246	OxA–13707 64	OxA–13966 68
OxA–13228 242	OxA–13388 246	OxA–13708 64–5	OxA–13967 68
OxA–13277 263–4	OxA–13389 246	OxA–13709 65	OxA–13977 212
OxA–13278 264	OxA–13390 247	OxA–13710 65	OxA–13978 210
OxA–13304 264	OxA–13391 247	OxA–13711 65	OxA–13979 210
OxA–13305 264	OxA–13392 247	OxA–13713 143	OxA–13980 210
OxA–13306 264–5	OxA–13393 201	OxA–13714 143	OxA–13981 211

OxA–13982 211
OxA–13997 235
OxA–13998 232–3
OxA–13999 234–5
OxA–14000 233–4
OxA–14001 235
OxA–14004 69
OxA–14005 69
OxA–14006 69
OxA–14007 69
OxA–14008 69–70
OxA–14009 123
OxA–14012 266
OxA–14016 70
OxA–14017 70
OxA–14018 70
OxA–14019 70
OxA–14024 123
OxA–14030 126
OxA–14031 127
OxA–14032 267
OxA–14039 126
OxA–14040 126
OxA–14041 130–1
OxA–14043 204
OxA–14044 70
OxA–14061 128
OxA–14062 131
OxA–14063 131
OxA–14064 133
OxA–14065 133
OxA–14086 203
OxA–14087 201
OxA–14118 242
OxA–14126 126
OxA–14142 72
OxA–14143 131
OxA–14144 131
OxA–14145 131
OxA–14157 126–7
OxA–14175 73
OxA–14176 76–7
OxA–14177 116

OxA–14178 131–2
OxA–14199 285
OxA–14200 285
OxA–14204 132
OxA–14232 71
OxA–14244 71
OxA–14267 256
OxA–14268 256
OxA–14292 256
OxA–14293 256–7
OxA–14294 257
OxA–14295 257
OxA–14296 257
OxA–14297 257–8
OxA–14311 86
OxA–14312 86
OxA–14313 86
OxA–14314 82
OxA–14315 82
OxA–14321 78
OxA–14322 78–9
OxA–14354 83
OxA–14357 258
OxA–14358 258
OxA–14359 258
OxA–14386 88
OxA–14413 83
OxA–14414 83
OxA–14415 83
OxA–14416 79
OxA–14417 79
OxA–14418 86–7
OxA–14428 79
OxA–14432 258
OxA–14444 291
OxA–14445 292
OxA–14446 291
OxA–14447 290
OxA–14448 292
OxA–14458 197
OxA–14469 115
OxA–14470 115
OxA–14471 199

OxA–14479 287
OxA–14480 197
OxA–14486 41
OxA–14487 41
OxA–14492 289
OxA–14497 87
OxA–14507 156
OxA–14508 156–7
OxA–14509 157
OxA–14510 157
OxA–14511 157
OxA–14512 157
OxA–14514 41
OxA–14515 41
OxA–14516 41
OxA–14517 39
OxA–14518 41
OxA–14519 44
OxA–14520 42
OxA–14521 42
OxA–14522 42–3
OxA–14523 43
OxA–14541 283
OxA–14542 283
OxA–14543 283
OxA–14544 283
OxA–14560 44–5
OxA–14567 39
OxA–14569 52
OxA–14584 56
OxA–14586 33
OxA–14587 33
OxA–14588 33
OxA–14593 52
OxA–14594 33–4
OxA–14595 34
OxA–14596 34
OxA–14597 34
OxA–14598 34
OxA–14599 34
OxA–14600 34
OxA–14601 34–5
OxA–14602 35

OxA–14603 52
OxA–14604 52–3
OxA–14605 53
OxA–14606 53
OxA–14607 53
OxA–14608 252
OxA–14609 252
OxA–14640 244
OxA–14641 244–5
OxA–14642 245
OxA–14643 245
OxA–14727 194
OxA–14728 194
OxA–14729 194–5
OxA–14730 195
OxA–14731 195
OxA–14732 101
OxA–14733 108–9
OxA–14734 109
OxA–14765 195
OxA–14766 100
OxA–14767 100
OxA–14768 100
OxA–14769 199
OxA–14770 199–200
OxA–14771 200
OxA–14772 200
OxA–14790 101
OxA–14791 101
OxA–14792 109
OxA–14793 113
OxA–14794 113
OxA–14808 56
OxA–14831 111
OxA–14832 109
OxA–14833 109
OxA–14834 109
OxA–14835 109–10
OxA–14836 113
OxA–14837 114
OxA–14838 111
OxA–14881 112
OxA–14883 91

OxA–14965 150–1

OxA–14966 151

OxA–14967 144

OxA–14968 140

OxA–14969 91–2

OxA–14970 92

OxA–14971 92

OxA–14972 92

OxA–14973 92

OxA–14974 93

OxA–14975 140

OxA–14995 92–3

OxA–14996 89

OxA–15033 93

OxA–15034 89

OxA–15035 284

OxA–15036 284

OxA–15039 93

OxA–15075 144–5

OxA–15076 145

OxA–15079 104–5

OxA–15088 145

OxA–15096 110

OxA–15097 110

OxA–15098 105

OxA–15177 145

OxA–15199 102–3

OxA–15200 103

OxA–15249 118

OxA–15250 118

OxA–15251 118

OxA–15252 121

OxA–15253 121

OxA–15254 119

OxA–15284 118

OxA–15290 134

OxA–15291 134

OxA–15292 135

OxA–15293 136

OxA–15305 103

OxA–15319 122

OxA–15320 119

OxA–15322 136

OxA–15323 136

OxA–15324 136

OxA–15325 115–6

OxA–15390 74

OxA–15447 75

OxA–15448 75

OxA–15449 75

OxA–15509 75

OxA–15543 75

OxA–15544 75–6

OxA–15574 83

OxA–15575 83–4

OxA–15695 47

OxA–15696 47

OxA–15697 47

OxA–15698 47–8

OxA–15699 48

OxA–15704 88

OxA–15714 167

OxA–15715 167

OxA–15716 48

OxA–15717 48

OxA–15718 48

OxA–15719 49

OxA–15720 49

OxA–15721 49

OxA–15722 49

OxA–15723 49

OxA–15750 159

OxA–15788 190

OxA–15789 190

OxA–15790 191

OxA–15791 191

OxA–15792 191

OxA–15825 159

OxA–15826 159

OxA–15827 160

OxA–15828 167

OxA–15829 220

OxA–15830 220

OxA–15831 220–1

OxA–15832 221

OxA–15833 221

OxA–15834 221

OxA–15835 221

OxA–15836 221

OxA–15865 160

OxA–15885 221

OxA–15923 167

OxA–15924 167

OxA–15925 168

OxA–15926 168

OxA–15927 168

OxA–15928 222

OxA–15929 222

OxA–15930 222

OxA–15944 222

OxA–15945 222

OxA–15946 222

OxA–15993 222

OxA–17122 76

OxA–30071 134

OxA–X–2128–16 . 284

OxA–X–2135–46 . 110

SUERC–10161. . . 1

SUERC–10162. . . 1

SUERC–10163. . . 1–2

SUERC–10164. . . 2

SUERC–10168. . . 2

SUERC–10169. . . 2

SUERC–10170. . . 2

SUERC–10171. . . 3

SUERC–10172. . . 3

SUERC–10173. . . 3

SUERC–10182. . . 151–2

SUERC–10183. . . 152

SUERC–10184. . . 152

SUERC–10188. . . 152

SUERC–10511. . . 3–4

SUERC–10515. . . 190

SUERC–10516. . . 191

SUERC–10517. . . 191

SUERC–10518. . . 192

SUERC–10519. . . 192

SUERC–10642. . . 4

SUERC–11434. . . 227

SUERC–11435. . . 227

SUERC–11439. . . 227–8

SUERC–11440. . . 228

SUERC–11441. . . 228

SUERC–11442. . . 228

SUERC–11443. . . 228

SUERC–11444. . . 228–9

SUERC–11445. . . 229

SUERC–11449. . . 229

SUERC–1146. . . . 295

SUERC–3623. . . . 248

SUERC–3624. . . . 248

SUERC–3625. . . . 248

SUERC–3626. . . . 248–9

SUERC–4098. . . . 234

SUERC–4099. . . . 233

SUERC–4101. . . . 235

SUERC–4102. . . . 236

SUERC–4103. . . . 236

SUERC–4504. . . . 269

SUERC–4505. . . . 269

SUERC–4506. . . . 269–70

SUERC–4507. . . . 270

SUERC–4511. . . . 267–8

SUERC–4512. . . . 268

SUERC–4514. . . . 268

SUERC–4515. . . . 268

SUERC–4516. . . . 270

SUERC–4517. . . . 271

SUERC–4521. . . . 271

SUERC–4522. . . . 272–3

SUERC–4523. . . . 272

SUERC–4524. . . . 272

SUERC–4963. . . . 258–9

SUERC–4967. . . . 259

SUERC–4968. . . . 259

SUERC–4969. . . . 259

SUERC–4970. . . . 259

SUERC–4971. . . . 259–60

SUERC–4972. . . . 260

SUERC–5150. . . . 260

SUERC–5686. . . . 290

SUERC–5687. . . . 291

SUERC–5688. . . . 291	SUERC–6572. . . . 35	SUERC–8855. . . . 214	SUERC–9661. . . . 237
SUERC–5689. . . . 287	SUERC–6573. . . . 35	SUERC–8856. . . . 214–5	SUERC–9662. . . . 237–8
SUERC–5690. . . . 292	SUERC–6574. . . . 35	SUERC–8857. . . . 215	SUERC–9666. . . . 238
SUERC–5694. . . . 292	SUERC–6575. . . . 36	SUERC–8858. . . . 215	SUERC–9667. . . . 238
SUERC–5695. . . . 290	SUERC–6579. . . . 36	SUERC–8859. . . . 215	SUERC–9668. . . . 238
SUERC–5697. . . . 253	SUERC–6580. . . . 36	SUERC–8860. . . . 215	SUERC–9669. . . . 238
SUERC–5698. . . . 253	SUERC–6581. . . . 36	SUERC–8861. . . . 215	SUERC–9670. . . . 238
SUERC–5817. . . . 157	SUERC–6582. . . . 36	SUERC–8875. . . . 215–6	SUERC–9671. . . . 239
SUERC–5818. . . . 157	SUERC–6583. . . . 36	SUERC–8876. . . . 216	SUERC–9672. . . . 239
SUERC–5819. . . . 158	SUERC–6865. . . . 57	SUERC–8877. . . . 216	SUERC–9676. . . . 239
SUERC–5820. . . . 158	SUERC–6932. . . . 57	SUERC–8878. . . . 216	SUERC–9677. . . . 239
SUERC–5821. . . . 158	SUERC–7354. . . . 31	SUERC–8947. . . . 216	SUERC–9678. . . . 239
SUERC–6137. . . . 56–7	SUERC–7355. . . . 31	SUERC–8948. . . . 216	SUERC–9679. . . . 239–40
SUERC–6138. . . . 57	SUERC–7356. . . . 31	SUERC–8949. . . . 216–7	SUERC–9680. . . . 240
SUERC–6139. . . . 57	SUERC–7360. . . . 31	SUERC–8950. . . . 217	SUERC–9681. . . . 240
SUERC–6143. . . . 252	SUERC–7361. . . . 32	SUERC–8951. . . . 217	SUERC–9682. . . . 240
SUERC–6147. . . . 252	SUERC–7362. . . . 32	SUERC–8952. . . . 217	SUERC–9686. . . . 240
SUERC–6168. . . . 43	SUERC–7363. . . . 32	SUERC–8953. . . . 217	SUERC–9687. . . . 240
SUERC–6169. . . . 40	SUERC–7364. . . . 32	SUERC–8979. . . . 193	SUERC–9688. . . . 219
SUERC–6170. . . . 40	SUERC–7560. . . . 45	SUERC–8980. . . . 193	SUERC–9689. . . . 219
SUERC–6171. . . . 43	SUERC–7561. . . . 45	SUERC–8981. . . . 193	SUERC–9690. . . . 219
SUERC–6172. . . . 43	SUERC–7562. . . . 45	SUERC–8982. . . . 193	SUERC–9691. . . . 219
SUERC–6173. . . . 43	SUERC–7579. . . . 46	SUERC–8983. . . . 193	SUERC–9692. . . . 219–20
SUERC–6177. . . . 43	SUERC–7580. . . . 46	SUERC–8984. . . . 193	SUERC–9696. . . . 220
SUERC–6178. . . . 43–4	SUERC–7608. . . . 253	SUERC–8985. . . . 194	SUERC–9825. . . . 206
SUERC–6540. . . . 273	SUERC–7612. . . . 253	SUERC–9636. . . . 217	SUERC–9826. . . . 207
SUERC–6541. . . . 273	SUERC–7613. . . . 253	SUERC–9637. . . . 217	SUERC–9827. . . . 209
SUERC–6542. . . . 273	SUERC–7614. . . . 253	SUERC–9638. . . . 218	SUERC–9828. . . . 206
SUERC–6543. . . . 274	SUERC–7615. . . . 254	SUERC–9639. . . . 218	SUERC–9829. . . . 206
SUERC–6547. . . . 274	SUERC–7616. . . . 254	SUERC–9640. . . . 218	SUERC–9830. . . . 207
SUERC–6548. . . . 274	SUERC–7617. . . . 254	SUERC–9641. . . . 218	SUERC–9834. . . . 207
SUERC–6550. . . . 274	SUERC–7618. . . . 254	SUERC–9646. . . . 218	SUERC–9835. . . . 208
SUERC–6551. . . . 274	SUERC–7622. . . . 254	SUERC–9647. . . . 218–9	SUERC–9836. . . . 207
SUERC–6552. . . . 35	SUERC–8018. . . . 254–5	SUERC–9648. . . . 219	SUERC–9837. . . . 209
SUERC–6556. . . . 274	SUERC–8157. . . . 46	SUERC–9649. . . . 174	SUERC–9838. . . . 208
SUERC–6557. . . . 275	SUERC–8168. . . . 255	SUERC–9650. . . . 174–5	SUERC–9839. . . . 208
SUERC–6558. . . . 275	SUERC–8169. . . . 255	SUERC–9651. . . . 175	UB–4896 153
SUERC–6559. . . . 275	SUERC–8846. . . . 213	SUERC–9652. . . . 236	UB–4897 153
SUERC–6563. . . . 275	SUERC–8847. . . . 213–4	SUERC–9656. . . . 236–7	UB–4898 153
SUERC–6564. . . . 275	SUERC–8848. . . . 214	SUERC–9657. . . . 237	UB–4920 15–20
SUERC–6566. . . . 275	SUERC–8849. . . . 214	SUERC–9658. . . . 237	UB–4921 17
SUERC–6567. . . . 276	SUERC–8850. . . . 214	SUERC–9659. . . . 237	UB–4922 14
SUERC–6568. . . . 35	SUERC–8851. . . . 214	SUERC–9660. . . . 237	UB–4923 14

UB–4924 17–18	UB–4985 12	UB–6034 7	UB–6348 15
UB–4925 18	UB–5208 5	UB–6035 7	UB–6394 79
UB–4926 18	UB–5215 171	UB–6036 7	UB–6395 79
UB–4928 18–19	UB–5216 171–2	UB–6037 7	UB–6396 80
UB–4929 19	UB–5217 172	UB–6038 7–8	UB–6397 80
UB–4930 19	UB–5218 172	UB–6039 13	UB–6472 6
UB–4931 19–20	UB–5219 172–3	UB–6040 8	UB–6473 6
UB–4958 154	UB–5220 173	UB–6041 8	UB–6474 11
UB–4959 154	UB–5221 173	UB–6042 8	UB–6475 11
UB–4960 154	UB–5222 173	UB–6179 223	UB–6476 6
UB–4961 20	UB–5223 173–4	UB–6180 223	UB–6477 11
UB–4962 20	UB–5224 250	UB–6181 223	UB–6478 20–21
UB–4963 10	UB–5225 250–1	UB–6182 223	UB–6479 17
UB–4964 8–9	UB–5227 251	UB–6183 223–4	UB–6534 21
UB–4965 4–5	UB–5228 251	UB–6184 224	UB–6858 50
UB–4975 5	UB–5229 251	UB–6185 71	UB–6859 50
UB–4976 15	UB–5230 251	UB–6186 145	UB–6860 50
UB–4981 12	UB–5231 251	UB–6345 16	UB–6861 50
UB–4982 13	UB–6031 255	UB–6346 20	UB–6862 50
UB–4984 9	UB–6032 10	UB–6347 14	UB–6863 50–1

General Index

Page numbers in **bold** refer to figures, page numbers in *italic* refer to tables

accelerator mass spectrometry viii, x, **x**, 194, 224, 285
Acer campestre, Sherborne: Tinney's Lane 237, 239, 240
Acer sp., Street Causeway 209
aerial photography 175
Aggregates Levy Sustainability Fund ix, 229
agriculture, onset of 188
Ainsbrook, North Yorkshire xxx, 1
  Bronze Age 2
  charcoal 1–3
  field systems 1
  geophysical survey 1
  human skeletal remains *xxviii*, 3–4
  pottery 2
Albion Archaeology 16
Aldsworth, F 122
Alexander Keiller Museum 137, 140, 145, 241
Allen, T 256–60
Alnus glutinosa
  Aveley Marshes 31–2
  Bouldnor Cliff: BCII and BCIV 45, 46, 47, 48, 49
  Bridgwater Bay 54, 55, 56
  Etton Woodgate 89
  Fiskerton: auger survey 163–4, 165–6
  Harehaugh Hillfort 170
  Holbrook Bay fishtrap 250–1
  Holme-next-the-Sea 177–9, 180–1, 182–3, 185
  Northborough 114–5
  Sherborne: Tinney's Lane 237, 240
  Sutton Common 253
  West Heslerton 277–8, 278–9, 279, 280, 281, 283
Alnus sp.
  Boden Vean 42
  Howick, Sea Houses Farm 186
  Sherracombe Ford 161, 162
  Sutton Common 253, 254
Alnus/Corylus sp., Holme-next-the-Sea 182, 183, 185
Anderson, Robert **xv**
Anderson, S 8
Anderson, T 11
Anglezarke, Lancashire
  site 1 267–8
  site 2 268
Anglo-Saxon chronology project xviii, xxx
Anglo-Saxon graves and grave goods xviii, xxx, *xviii*
Anglo-Saxon graves and grave goods (female graves) 4
  Apple Down, Compton 4–5
  Aston Clinton, Tring Hill 5
  Binchester 38
  Buckland Dover, Kent 5–6
  Castledyke South, Barton-on-Humber 6–8
  Coddenham, 8–9
  Dover Buckland 154
  Lechlade, Butler's Field 9
  St Peter's Tip, Broadstairs *xviii*, 9–10
Anglo-Saxon graves and grave goods (male graves) 8, 10
  Buckland Dover 11
  Bury St Edmunds, Westgarth Gardens 12
  Butler's Field, Lechlade 12–3
  Castledyke South, Barton-on-Humber 13

Dover Buckland 153–4
Edix Hill (Barrington A) xviii, 13–4
Eriswell, Lakenheath 14–5
Ford, Laverstock 15
Gally Hills, Banstead xxiv, *xviii*, 15–6
Melbourn, Water Lane *xviii*, *xxiv*, 16
Mill Hill, Deal 17
St Peter's Tip *xxiv*, *xxviii*, 17 –21
Anglo-Saxon period
  Ainsbrook 1–4
  barrows 15, 15–6
  Brandon: Staunch Meadow 51–3
  Bridgwater Bay 53–6
  Carlton Colville: Bloodmoor Hill 58–71
  coins 52, 53
  diet 14
  Dover Buckland 153–4
  fish weirs 54, 55
  fishing structures 53–6
  human skeletal remains 153–4, 194, 195, *xxviii*
  Lundenwic 203–4
  pottery 52, 58, 59–63, 64–7, 67–8, 68–71, 203–4,
    234, 235
  Ryall Quarry: Saxon's Lode Farm 232, 233, 234–6
  textile working industry 52–3
animal baiting 262
antler xi, **xi**, xv, xvi
  Ascott under Wychwood 23, 26–7
  Binchester xiv, xxiv
  Crickley Hill 83, 86–7
  Fussell's Lodge 196
  Haddenham 94
  Knap Hill *xix*, 102
  Maiden Bower 104, 104–5, 105
  Maiden Castle 110, 111
  radiocarbon offsets xii
  Silbury Hill 241, 242
  Thames Valley 22
  Whitehawk Hill *xx*, 125, 126, 133
  Whitesheet Hill 133–4
  Windmill Hill 138–9, 139, 141, 142, 143, 144–5, 145,
    xiii
Antler Maceheads Project 21
  Burwell Fen, Cambridgeshire 22
  Thames Valley, Greater London 22
Apple Down, Compton, West Sussex, Anglo-Saxon graves
    and grave goods (female graves) 4–5
Archaeology South-East 99
arrowheads
  Crickley Hill 77
  Hembury 98
  Peak Camp 116
Arts and Humanities Research Board viii
Arundel Castle 72
Ascott under Wychwood: long barrow, Oxfordshire
    vii, **ix**, *xxiv*, 22–30
  animal bone *xviii*, 23–4, 28–9
  antler 23
  carbonised residue 28
  charcoal 27–8
  human skeletal remains 24, 25, 29–30
  postholes 27

Ashmolean Museum 255
Ashmore, P xi
Aston Clinton, Tring Hill, Buckinghamshire, Anglo-Saxon
    graves and grave goods (female graves) 5
Atkinson, R 241
Avebury henge 136
Aveley Marshes *xviii*
Aveley Marshes, Essex 30–2
axes 26, 90, 92

Baguley Hall: north wing, Greater Manchester
    xviii, xxx, 32–6
Barber, A 232, 234
Barbican House Museum 116, 122, 123, 125, 127, 128,
    132
Barfoot, J 15
Barker, B 204
Barn Elms, Greater London 194
Barnard Castle 37, 38
Barnatt, J 209–12
barrows **ix**
    Ascott under Wychwood vii, *xviii*, *xxiv*, 22–30
    Beckhampton Road Long Mound 23
    Crickley Hill 77, 80–4
    dating viii, xxix
    endings **ix**
    Ford, Laverstock 15
    Fussell's Lodge 195–7, *xxi*, *xxvi*
    Gally Hills, Banstead 15–6
    Longstone Edge, Derbyshire 200–3
    Wayland's Smithy xvi, 197–200
Batchelor, R 30–2
Bateman, T 203
Baxter, M S xv
Bayes' theorem xxviii, **xxix**
Bayesian chronological modelling viii
Bayesian chronological models viii, x, xxviii–xxix, **xi**, **xxix**
Bayliss, A xxix, 4, 5, 6, 7–8, 9, 10, 11, 12, 13, 14, 15, 16,
    17, 22, 22–3, 23–4, 25, 26, 27, 28, 29, 153–4, 196–7,
    198–200, 224–32, 241, 242, 243, 273–6, 286–95
beads 6, 7, 9, 10, 40, 154
Beaker Period 88, 146, 202
Beckhampton Road Long Mound, Wiltshire 23
Beckton 31
Bedfordshire, Maiden Bower 103–5
Bedwin, O 72, 76, 122
Beech Court Farm, Vale of Glamorgan, causewayed
    enclosure 71–2
Belfast Radiocarbon Dating Laboratory xv, 16
Benson, D 22
Bergen 194
Betula sp.
    Blacklake Wood 157, 158
    Bouldnor Cliff: BCII and BCIV 47–8, 49
    Bridgwater Bay 54
    Harehaugh Hillfort 170
    Hembury 98–9
    Holme-next-the-Sea 177
    Sherracombe Ford 160–1, 161
    Sutton Common 254
Billingsgate Lorry Park, Greater London xxix
Binchester, Durham xiv, *xviii–xix*, xxiv, xxx, 36–7, 38
    Saxon burial *xix*, 38
Blacklake Wood, Somerset *xxi*, *xxvi*, xxx, 156–8,
Blackmore, L 203–4

Boden Vean, Cornwall xxx, 38
    Bronze Age structure *xix*, *xxv*, 38, 39–40
    curvilinear ditch/creep 40, 44
    enclosure ditch 38
    Iron Age activity *xix*, *xxv*, 40–4
    post-Roman activity 44
    pottery 38, 39, 40, 41, 44
    stone-box sequence 40, 43
Bodmin Moor 248
Boismier, W 175
bone xi, **xi**
    radiocarbon offsets xii
bone, animal xv, xvi
    Ascott under Wychwood *xviii*, 23–4, 25–7, 28–9
    Binchester *xviii–xix*, 37
    Bos sp. 23, 24–5, 37, 73, 74, 75, 79–80, 81, 82, 83, 83–4,
        84, 86, 87, 91, 102–3, 104, 105, 106, 107, 112–3,
        113, 117, 118, 121–2, 125–6, 130, 131, 132, 134,
        136, 138, 141, 142, 143, 144, 145, 146, 147–8,
        148–9, 149, 150, 191, 284
    Bury Hill 72–3
    Canis familiaris 261, 262
    Canis sp. 28–9, 138, 142, 149, 190
    Capra sp. 92, 130, 131–2
    Capreolus capreolus 26
    Carlton Colville: Bloodmoor Hill 63–4, 67, 68, 70,
        71, *xix*
    Cervus elaphus 23–4, 26, 28, 150
    Cervus sp. 106, 191
    Chalk Hill 73–4, 75, 75–6
    Court Hill, enclosure 76–7
    Crickley Hill *xix*, 79, 79–80, 80–1, 81–2, 82, 83, 83–4,
        84, 84–5, 85, 86, 87
    Equus sp. 14, 15, 258, 285
    Eriswell, Lakenheath 14, 15
    Etton *xix*, 91, 91–2, 93
    Fussell's Lodge 195
    Haddenham 94
    Kingsborough 1 99
    Kinsey Cave 189, 190, 191, 192
    Knap Hill *xx*, 102–3
    leopard 261
    lion 261, 261–2, 262
    Longstone Edge 202
    Maiden Bower 104, 105
    Maiden Castle 106–7, 109–10, 112, 112–3, 113, **xiii**, *xx*
    Ovis sp. 63–4, 67, 68, 71, 73, 75, 112–3, 134
    Panthera leo 261, 261–2, 262
    Panthera pardus 261
    Peak Camp 117–8, 118, *xx*
    Roman period *xviii–xix*
    sheep/goat 25–6, 91, 92, 109–10, 118, 139, 285
    Silbury Hill 241
    Staines 121–2
    Sus sp. xviii, 24, 67, 73, 74, 79, 81–2, 83, 85, 86, 87,
        91–2, 93, 109, 117, 135, 140, 147, 190, 241, 285
    Sutton Common 252
    Taplow Court 258
    toad 140, 144
    Tower of London: Royal Menagerie 260–2
    Ursus arctos 192
    Whitehawk Hill *xx*, 125–6, 130, 131, 131–2, 132, 132–3
    Whitesheet Hill 134, 135–6, *xx*
    Windmill Hill 138, 139, 140, 141–2, 142–4, 145, 146–7,
        147–8, 148–9, 149–50

Wolverton Turn Enclosure 284–5
Yarnton 290
Boomer, I 185–9
Bouldnor Cliff, Isle of Wight *xix*
　BCII and BCIV *xxv,* 44–6
　BCII, BCIV and BCV 46–9
　wiggle-matching 50–1
Bourne Hall Museum, Ewell 15
Bowes Museum 37, 38
Bradley, R 105
Brandon: Staunch Meadow, Suffolk *xxv,* xxx, 51
　cemetery 2 51
　mortuary structure 51
　waterfront activity 52–3
　wood, waterlogged *xix*
Bridgwater Bay, Somerset 53–6
Brighton and Hove Museum 125, 127, 128, 132
British Museum 9, 17, 88–9, 89, 93, 120, 121, 153
Bronk Ramsey, C xvi
Bronze Age
　Boden Vean xxx, 38, 39–40, *xix, xxv*
　burials 146, 255
　cairn 245–6
　Fiskerton: auger survey 165–6
　Hatfield Moors, Rawcliffe 214, 215, 220
　Holme-next-the-Sea 175–85
　Howick, Sea Houses Farm 188
　human skeletal remains 194, 195, 252, 289–90, **xii**
　Kinsey Cave 189
　pottery vii, 2, 38, 39, 40, 138, 236–40, 249, 294
　ring-cairn 246–7
　Ryall Quarry: Saxon's Lode Farm 232, 235–6
　Sherborne: Tinney's Lane 236–40
　Stannon Down 245–50
　Sutton Common 252, 253
　Sutton Courtenay, Drayton Road 255
　Taplow Court 257, 257–8, 259, 260
　Thorne Moors 221, 222
　trackways 205
　waterholes 288, 292–5
　West Heslerton 276, 278
　Yarnton *xxii,* 285–6, 287–90, 292–5
brooches 6, 7, 8, 9, 40
Brue, River 208–9
Brunning, R 53–6, 204, 205, 206, 207, 208–9
Bryggen, Bergen, Norway xxix
Buck, C E xxviii
Buckinghamshire
　Aston Clinton, Tring Hill 5
　Chesham Bois House 151–2
　Taplow Court xxx, 256–60
　Wolverton Turn Enclosure, sunken-featured building
　　284–5
Buckinghamshire County Council 5
Buckinghamshire County Council Museum 151
Buckinghamshire County Museums Service 256
Buckinghamshire Museum 284–5
Buckland Dover, Kent
　Anglo-Saxon graves and grave goods (female graves) 5–6
　Anglo-Saxon graves and grave goods (male graves) 11
buckles 6, 8
　buckle-type BU9–a 154
　English Heritage 16
　type BU2–d 17, 154
　type BU2–h 17, 154

type BU3–a 18
type BU3–g 13
type BU3–h 19
type BU6 13
type BU7 15, 18, 19, 20, 21
bulk materials xi, **xi**
burials xiii, xiv–xv, 13
　Ainsbrook 3–4
　Anglo-Saxon viii, xxiv, 51
　Anglo-Saxon (female) 4–10, 38
　Anglo-Saxon (male) 10–21
　animal 63–4
　Apple Down, Compton 4–5
　Aston Clinton, Tring Hill 5
　Binchester 38, *xix*
　Brandon: Staunch Meadow 51
　Bronze Age 146, 255
　Buckland Dover 5–6
　Callington: St Sampson's Church 56–7
　Carlton Colville: Bloodmoor Hill 58
　Castledyke South, Barton-on-Humber 6–8
　Christian 56–7
　dating viii
　Higham Ferrers: Kings Meadow Lane 171–5
　Iron Age 124
　Lechlade, Butler's Field 9
　Medieval period 56–7
　Offham Hill 116
　punishment 255
　Roman period 171–5
　Saxon period *xix*
　Sutton Courtenay, Drayton Road 255
　The Trundle 124
　weapon 14
　Whitehawk 129
　Windmill Hill 146
　Yarnton 285
burnt stone deposits, Yarnton 293, 294, 295
Burwell Fen, Cambridgeshire, Antler Maceheads Project 22
Bury Hill, West Sussex, causewayed enclosure 72–3
Bury St Edmunds, Westgarth Gardens, Suffolk, Anglo-
　Saxon graves and grave goods (male graves) 12
Butler's Field, Lechlade, Gloucestershire, Anglo-Saxon
　graves and grave goods (male graves) 12–3

C4 plants xii
cairn, Stannon Down 245–6
Calgon, chemical contamination 167
calibration xvii, **xvii**
　maximum intercept method xvii
calibration curves x, xvii
Callington: St Sampson's Church, Cornwall *xix, xxv,*
　xxviii, 56–7
Cambridge Archaeological Unit 94
Cambridge University 22
Cambridgeshire
　Burwell Fen 22
　Duxford: Hinxton Road 154–5
　Edix Hill (Barrington A) 13–4
　Etton *xxv,* 89–93, 114
　Etton Woodgate, ditch 88–9, 114
　Haddenham 94
　Melbourn, Water Lane *xxiv, xxviii,* 16
　Northborough 114–6
Cambridgeshire County Council 13, 154–5

Campbell, G 166–8
Canterbury Archaeological Trust 73
carbonised residue xxiv, 59
  Ascott under Wychwood 28
  Billingsgate Lorry Park 193–4
  Boden Vean xix, 39, 40, 41, 44
  Carlton Colville: Bloodmoor Hill xix, 59–63, 64–7,
    67–8, 68–71
  Chalk Hill xix, 74, 75, 76
  Crickley Hill xix, 81, 82, 83, 87–8
  Etton 90, 91, 92, 92–3
  Etton Woodgate 89
  Hembury 95, 98
  Knap Hill 103
  Maiden Castle xx, 107–8, 108–9, 110, 112, 113
  Robin Hood's Ball 118–9
  Staines xx, 120–1, 122
  Trevelgue Head 263, 263–4, 264–5, 266–7
  The Trundle 122–3, 123
  Whitehawk Hill xx, 125, 126, 126–7, 127, 128, 129,
    130–1
  Whitesheet Hill xx, xxi
  Windmill Hill 137–8, 139, 146, 147, 148, 149, 150
  Yarnton 294
Cardiff University 242
Carlton, R 169–71
Carlton Colville: Bloodmoor Hill, Suffolk xiii, xxx,
  58–71, xxv
  animal bone 63–4, 67, 68, 70, 71, xix
  carbonised residue 59–63, 64–7, 67–8, 68–71, xix
  Grubenhäuser 58, 59, 59–63, 64–7, 67–8, 68–71
  middens 60, 62–3, 65, 67
  pits 59–60, 61–2, 63, 64, 64–6, 67, 68, 69, 70, 71
  postholes 60, 62, 63–4, 68, 69, 70, 71
  pottery 59–63, 64–7, 67–8
  Roman period 63, 68, 69
Carr, R 52
Caruth, J 14
Castledyke South, Barton-on-Humber, Humberside 6–8, 13
causeway, timber, Fiskerton: auger survey 162–6
causewayed enclosures
  Beech Court Farm, Vale of Glamorgan 71–2
  Bury Hill 72–3
  Chalk Hill xix, xxv, 73–6, 99, 100
  Court Hill 76–7
  Crickley Hill xix, xxv, 77–88, 117
  dating viii, xxix, 72
  Etton xix, xxv, 89–93, 114
  Etton Woodgate, ditch, Cambridgeshire 88–9, 114
  Haddenham 93
  Hembury 94–9
  Iron Age 72
  Kingsborough 1 99–100
  Kingsborough 2 99, 100–1
  Knap Hill xix–xx, xxv, 101–3
  Maiden Bower 103–5
  Maiden Castle xx, xxiv, xxv, 105–14
  Neolithic 72–3, 76–7, 94, 94–9, 99–101, 101–3, 103–5,
    105–14, 114–6, 116, 116–8, 118–9, 122–4, 124–33,
    133–6, 136–51
  Northborough 114–6
  Offham Hill xx, xxv, 116
  Peak Camp xx, xxvi, 116–8
  Robin Hood's Ball 118–9
  Staines xx, xxvi, 119–22

  The Trundle xx, xxvi, 122–4
  Whitehawk Hill xx, xxvi, xxviii, 124–33
  Whitesheet Hill xx–xxi, xx, xxvi, 133–6
  Windmill Hill ix, xx–xxi, xxvi, xxviii, 136–51
cemeteries
  Anglo-Saxon 51, 203–4
  Anglo-Saxon (female) 4–10
  Anglo-Saxon (male) 10–21, 153–4
  Apple Down, Compton 4–5
  Binchester 37
  Brandon: Staunch Meadow 51
  Buckland Dover 5, 11
  Bury St Edmunds, Westgarth Gardens 12
  Butler's Field, Lechlade 12–3
  Carlton Colville: Bloodmoor Hill 58
  Castledyke South, Barton-on-Humber 6–8, 13
  Claydon Pike, Lechlade 152–3
  dating xxx
  Edix Hill (Barrington A) 13–4
  Eriswell, Lakenheath 14–5
  Iron Age xxx, 155
  Lechlade, Butler's Field 9
  Lundenwic 203–4
  Medieval period 222–4
  Melbourn, Water Lane 16
  Mill Hill, Deal 17
  post-Roman 152–3
  Riccall xxx, 222–4
  Roman period xxx, 251–2
  St Peter's Tip, Broadstairs 9–10, 17–21
  Sutton Common 251–2
Centre for Archaeology vii
cereals, Ryall Quarry: Saxon's Lode Farm 232–4
Chalk Hill, Kent 99, 100
  animal bone 73–4, 75, 75–6
  carbonised residue xix, 74, 75, 76
  causewayed enclosure xxv, 73–6
  pottery 74, 75, 76
charcoal xi, xi
  Ainsbrook 1–3
  Ascott under Wychwood 27–8
  Beech Court Farm, Vale of Glamorgan 72
  Blacklake Wood xxi, 156–8
  Boden Vean xix, 39, 40, 41–4, 44
  Chesham Bois House 151–2
  Crickley Hill xix, 78–9, 79, 84
  Groundwell Ridge 169
  Haddenham 94
  Harehaugh Hillfort 170–1
  Hembury 97, 98–9
  Kingsborough 2 100–1
  Lords and Ladies Mine xxi, 210–1
  Lundenwic 204
  Maiden Castle 108, 110
  Northborough 114–5
  Old Ash Mine 211–2
  Peak Camp 117, 118
  radiocarbon offsets xii
  Roman Lode 158
  Ryall Quarry: Saxon's Lode Farm 234–6
  Sherborne: Tinney's Lane 236–40
  Sherracombe Ford 160–2
  Stannon Down 245–6, 247, 249–50
  Taplow Court 256, 256–8, 259, 259–60
  Trevelgue Head 263, 264, 265–6

ultrafiltration protocol xvi
  Whitehawk 129–30, 131, 132
  Windmill Hill 137, 139, 140, 149, 151
  Yarnton 289, 291, 293
chemical contamination, grain 166–8
Chesham Bois House, Buckinghamshire 151–2
Chess Valley Archaeological and Historical Society 151
Chichester District Museum 4, 76
Chissell, C 288–9, 293–5
Christian-Albrechts-Universität zu Kiel, Germany viii, **ix**
Chronicles of London 262
chronology building x, **xi**
cist graves 56–7, 186, 188
City of London, Lundenwic 204
Clark, P 73
Claydon Pike, Lechlade, Gloucestershire 152–3
cobbles 3
Coddenham, Suffolk, Anglo-Saxon graves and grave goods
  (female graves) 8–9
coins
  Brandon: Staunch Meadow 52, 53
  Chesham Bois House 151, 152
combs 8
cooking pits, Hembury 95–7
Corinium Museum 9, 12, 77–80, 80, 84, 87, 116, 152
Cornwall
  Boden Vean · xxv, xxx, 38–44
  Callington: St Sampson's Church xxv, xxviii, 56–7
  Iron Age promontory fort, Trevelgue Head
  Trevelgue Head vii, **vii**, xxx, 262–7
Cornwall Archaeological Society 39
Cornwall Archaeological Unit 245
Cornwall Royal Museum 263
Corylus avellana
  Ascott under Wychwood 27–8
  Boden Vean 43, 43–4
  Bouldnor Cliff: BCII and BCIV 48
  Brandon: Staunch Meadow 52, 53
  Bridgwater Bay 53–4, 55
  Crickley Hill 78, 79
  Groundwell Ridge 169
  Harehaugh Hillfort 170–1
  Hembury 97, 98
  Holbrook Bay fishtrap 251
  Lords and Ladies Mine 210–1
  Lundenwic 204
  Peak Camp 117
  Sherborne: Tinney's Lane 237, 238
  Whitehawk 131, 132
  Windmill Hill 137, 139, 140
Corylus sp. 149, 235
  Ainsbrook 2, 3
  Ascott under Wychwood 27
  Boden Vean 40, 41, 44
  Crickley Hill 78, 85–6
  Hembury 99
  Maiden Castle 110
  Old Ash Mine 211–2, 212
  Sherracombe Ford 161, 162
  Stannon Down 246, 247, 249
  Sutton Common 253, 254
  Trevelgue Head 263, 264, 266
  Windmill Hill 151
Corylus Viburnum sp., Whitehawk 130

Corylus/Alnus sp. 259
  Brandon: Staunch Meadow 53
  Fiskerton: auger survey 163–6
  Haddenham 94
  Taplow Court 257–8, 259
Court Hill, enclosure, West Sussex 76–7
Cowie, R 203
Cravern Museum 190, 191, 192
cremations and cremation deposits xiv–xv
  animal 58, 63–4
  Carlton Colville: Bloodmoor Hill 58, 63–4
  Higham Ferrers: Kings Meadow Lane 174–5
  Longstone Edge *xxi*, 200–1
  Sutton Common 251–2
  Yarnton 286–7, 289–90
Crickley, battle of 77
Crickley Hill, Gloucestershire 86–7, 117
  animal bone *xix*, 79, 79–80, 80–1, 81–2, 82, 83, 83–4,
    84, 84–5, 85, 86, 87
  antler 83, 86–7
  arrowheads 77
  banana barrow 77, 80–4
  burnt stone 78–9
  carbonised residue *xix*, 81, 82, 83, 87–8
  causewayed enclosure *xxv*, 77–88
  charcoal *xix*, 78–9, 79, 84, 85–6
  continuous ditch 77–80
  entrance 80, 84, 87
  hillfort 77
  inner causewayed ditch 80–4, *xxv*
  long mound valley sequence 84–7
  Mesolithic activity 84
  outer causewayed ditch 87–8, *xxv*
  pits 80
  plant macrofossil 85, 86
  postholes 84, 85, 86
  pottery 81, 82, 83, 84, 85, 86, 87–8
  stone circle 84, 85–6
  timber structure 84
Cumbria
  Antler Maceheads Project 21
  Langdale, site 1 270, 271
  Langdale, site 2 270–1
  Langdale, site 4 270, 271
  South-west Fells transect, site 1b 271–2
  South-west Fells transect, site 2b 272
  South-west Fells transect, site 3 272–3
  Upland Peats 267, 270–3
Cumbria County Council 270, 271, 272
cups 8, 9
Curwen, E 122, 123, 124, 125, 127, 128, 132

Dagobert I (King of the Merovingian Franks) 9
Darvill, T 116
date ranges, calibration xvii, **xvii**
dendrochronology viii, 33, 177, 273–6, 283–4
Dennis, C 255
Derbyshire
  Antler Maceheads Project 22
  Longstone Edge xxx, 200–3
  Lords and Ladies Mine 210–1, *xxvi*
  Old Ash Mine 210, 211–2
  Peak District Mines 209–12
Derwent, River 280–1

Devon
  Hembury 94–9
  Roman Lode 158–60
  Sherracombe Ford xxx, 160–2
  Stannon Down 245–50
Dewar's track, Somerset xvii, **xvii,** xxi, xxx, 205
Dickens, A 58
dietary sources xii, 14
ditched enclosures, Taplow Court xxx
Dixon, P 77, 80, 84, 87
Doncaster Museum 213, 220, 251–2, 252–4
Dorset
  Hambledon Hill viii
  Maiden Castle xiii, xx, xxiv, xxv,105–14
  Sherborne: Tinney's Lane vii, xxx, 236–40
Dorset County Museum 106, 110, 112, 236
Dover Buckland, Kent 153–4
Dover Museum 5, 11, 17
Down, A 4
Drewett, P 116
Druce, D 267, 267–70
Duncan H 16
Durham, Binchester xiv, xxiv, xxx, 36–7
Duxford: Hinxton Road, Cambridgeshire 154–5

earthworm activity 29
East Sussex
  Offham Hill xiii, xxv, 116
  Whitehawk xxvi, xxviii, 124–33
  Whitehawk Hill
Edix Hill (Barrington A), Cambridgeshire, Anglo-Saxon
      graves and grave goods (male graves) xviii, 13–4
Edward I, King 273
Edwards, Y 151–2, 194–5
English Heritage vii, viii, ix, 1, 2, 4, 5, 10, 15, 16, 18, 19,
      21–33, 37–9, 41–4, 48, 51–6, 58–64, 69, 70, 82, 89,
      120, 121, 140, 153, 155, 157–85, 188, 190, 192,
      196–202, 204, 206, 207, 208–11, 213, 214, 216,
      218–20, 221, 224, 225, 229, 233, 234, 235, 241,
      242, 250–1, 252, 254–5, 257, 258, 260, 263, 264–7,
      273–7, 283–4, 287–95
Environment Agency 204
environmental changes, Howick, Sea Houses Farm 185–9
Eriswell, Lakenheath, Suffolk, Anglo-Saxon graves and
      grave goods (male graves) 14–5
erosion damage vii
Essex, Aveley Marshes 30–2
Etton, Cambridgeshire xxv, 114
  animal bone xix, 91, 91–2, 93,
  axe handle 90, 92
  carbonised residue 90, 91, 92, 92–3
  ditch 89–93
  internal features 89
  pottery 90, 91, 92, 92–3
  wood, waterlogged xix
Etton Woodgate, ditch, Cambridgeshire 88–9, 114
Evans, C 94
Evans, J 243
Evostik wood glue 284
excarnation platform, Longstone Edge 202
Exmoor, Devon, Sherracombe Ford xxx
Exmoor Iron Project 155–6
  Blacklake Wood xxi, xxvi, 156–8
  Roman Lode xxi, 158–60
  Sherracombe Ford 160–2

Fagus sp.
  Ascott under Wychwood 27
  Chesham Bois House 152
Ferris, I 37–8
Finlayson, R 1–4
fish weirs 54, 55, 250–1
fishing structures, Bridgwater Bay 53–6
fishtraps xxx
  Holbrook Bay fishtrap 250–1
  Holme-next-the-Sea 175, 181–4, 185
Fiskerton: auger survey, Lincolnshire 162–6
Fitz Robert, J 273
Fletcher, W 250–1
flint, knapped, Bouldnor Cliff: BCII and BCIV
      45, 46, 47, 49
flint assemblages, Ascott under Wychwood 22, 26
flint scatters, Maiden Bower 103
fogou, Boden Vean 38, 40–4
Food Vessel cremation burial, Longstone Edge 200–1
Ford, Laverstock, Wiltshire, Anglo-Saxon graves and grave
      goods (male graves) 15
Foreman, M 6, 13
Forest of Bowland, Lancashire 269
  site 1 296
  site 2 269–70
  site 3 270
Fourier Transform Infra Red Spectroscopy 16
fractionation xvi–xvii
Fraxinus excelsior
  Bridgwater Bay 54–5
  Crickley Hill 78, 78–9
  Groundwell Ridge 169
  Holbrook Bay fishtrap 251
  Holme-next-the-Sea 184
  Sherborne: Tinney's Lane 239, 239–40
  Street Causeway 208–9
Fraxinus sp.
  Ainsbrook 2
  Lords and Ladies Mine 210, 211
funding vii, ix
furnaces 156–8, 161, 169, 266
Fussell's Lodge, Wiltshire **ix** , xxi, xxvi, 195–7
Fyfe, R 158

Gally Hills, Banstead, Surrey, Anglo-Saxon graves and
      grave goods (male graves) xviii, xxiv, 15–6
gate structures, Whitehawk 124
Gearey, R 213–22, 252–4
Ghebru, Fsaha **xvi**
Gibson, A 22
Gillespie, J 45–6, 47–9
Glastonbury Abbey 208
Glastonbury Lake Village, Somerset xxi, xxx, 205–6
Gloucestershire
  Butler's Field, Lechlade 9, 12–3
  Claydon Pike, Lechlade 152–3
  Crickley Hill 77–88, 117, xxv
  Peak Camp 116–8, xxvi
Goodall, J 273
Gossip, J 38, 39, 41, 42, 43, 44, 56–7
GR99 dating experiment, Hampshire 166–8
grain
  chemical contamination 166–8
  GR99 dating experiment 166–8
  Hordeum sp. 288

Ryall Quarry: Saxon's Lode Farm 232–4
Secale cereale 259–60
Taplow Court 256, 258, 259, 259–60
Triticum dicoccum 233–4, 288, 291
Triticum dicoccum/Spelta 258
Triticum sp. 232–3, 256, 258
Triticum spelta 233, 258, 259
Yarnton 288, 289, 290, 291
grave-assemblages *see* Anglo-Saxon graves and grave goods
Greater London
    Barn Elms 194
    Billingsgate Lorry Park xxix, 192–4
    Kew 195
    Mortlake 195
    Putney 195
    Syon Reach 194
    Tower of London: Royal Menagerie 260–2
Greater Manchester, Baguley Hall xxx, 32–6
Greatorex, R 100
Groningen 244
Groundwell Ridge, Wiltshire 168–9
Grubenhäuser, Carlton Colville: Bloodmoor Hill 58, 59, 59–63, 64–7, 67–8, 68–71

Haddenham, Cambridgeshire, causewayed enclosures 93
Hall, D 192–4
Hall, R 1, 222–4
Hambledon Hill, Dorset viii
Hamilton, D 32–6, 194, 205, 224–9, 229–32, 283–4
Hamilton, M 103
Hampshire
    GR99 dating experiment 166–8
    Grateley South 166, 167
Hampshire and Wight Trust for Maritime Archaeology 44, 46–9, 50
Hardraada, King Harald 222, 223
Harehaugh Hillfort, Northumberland vii, 169–71
Harters Hill, Somerset *xxi*, xxx, 206
Hatfield Moors *xxi*
    plant macrosfossils *xxi*
Hatfield Moors, Rawcliffe, South Yorkshire xxx
    Bronze Age 214, 215, 220
    carbonised plant macrofossil 215, 216–7
    Mesolithic 213, 218
    Neolithic 214, 216, 217–8, 219
    peat 213, 217–20
    raised mires 213–20
    waterlogged plant macrofossils 213–5, 215–6, 217
hawthorn, Brandon: Staunch Meadow 52
Hayden, C 289, 292
hazelnut shells 84, 85, 86, 95–7, 99–100, 115–6, 135, 136, 187–8, 201
Healy, F 72, 73–6, 76–7, 77–88, 88–9, 89–93, 94, 94–9, 99–100, 100–1, 101–3, 103–5, 106–14, 114–6, 116, 117–8, 118–9, 119–22, 122–4, 124–33, 133–6, 137–51
hearths xxx, 23–4
    Bouldnor Cliff: BCII and BCIV 48, 49
    Chesham Bois House 151–2
    Harehaugh Hillfort 170
    Hembury 95, 98–9
    Trevelgue Head 264
Hedera decline 188
Hedges, R E M xvi
Hekla eruption AD 1947 159

Hembury, Devon
    arrowheads 98
    carbonised plant macrofossil 95–7, 99
    carbonised residue 95, 98
    causewayed enclosures 94–9
    charcoal 98–9
    cooking pits 95–7
    hearths 95, 98–9
    Iron Age rampart 94, 95, 96–7, 97, 98
    pits 95, 95–7, 97, 98–9
    pottery 98
Henry V, King 262
Hey, G 255, 285–95
Higham Ferrers: Kings Meadow Lane, Northamptonshire xii, xxx
    burial group 1 171–4
    burial group 2 171, 174–5
Higher Level Stewardship (HLS) schemes 267
high-precision liquid scintillation spectrometry **x**
hillforts xxx
    Crickley Hill 77
    Harehaugh 169–71
    Iron Age 105, 122, 123, 124, 169–71, 256, 257
    Maiden Bower 103, 105
    Maiden Castle 105
    Taplow Court 256–60
Hines, J 4, 7–8, 10, 13
Historic Environment Enabling Programme vii
Historic Scotland xxix
Hodder, I 94
Hogarth, A 9, 17
Holbrook Bay fishtrap, Suffolk xxx, 250–1
Holme I timber circle 175, 176, 177, 178, 179, 180, 181, 182, 184
Holme II timber circle 175, 176, 177, 178, 179, 181, 184
Holme-next-the-Sea, Norfolk 175–85, 176
    fishtraps 175, 181–4, 185
    Holme I timber circle 175, 176, 177, 178, 179, 180, 181, 182, 184
    Holme II timber circle 175, 176, 177, 178, 179, 181, 184
    posts 176
    Saxon period 176, 177, 181–4, 185
    tidal erosion 176, 177, 178, 179, 179–80, 180, 182, 183, 184, 185
    trackway II 184
Hoper, S T xv
Hornchurch Marshes 31
Howick, Sea Houses Farm *xxi*, xxx, *xxvi*, 185–9
human skeletal remains 17
    Ainsbrook 3–4
    Anglo-Saxon *xvii*, *xxviii i*, 51, 194, 195
    Anglo-Saxon (female) 4–10
    Anglo-Saxon (male) 10–21
    Apple Down, Compton 4–5
    Ascott under Wychwood 24, 25, 29–30
    Aston Clinton, Tring Hill 5
    Binchester 38
    Brandon: Staunch Meadow 51
    Bronze Age 194, 195, 289–90, **xii**
    Buckland Dover 5–6, 11
    Bury St Edmunds, Westgarth Gardens 12
    Butler's Field, Lechlade 12–3
    Callington: St Sampson's Church 56–7
    Castledyke South, Barton-on-Humber 6–8, 13

child 148, 151, 203
Claydon Pike, Lechlade 152–3
Coddenham 8–9
contexts xiv–xv
dating viii
Dover Buckland 153–4
Duxford: Hinxton Road 155
Edix Hill (Barrington A) xviii, 13–4
Eriswell, Lakenheath 14–5
female 4–10, 38, 123–4, 128, 129, 131, 153, 154, 195,
    197, 198–9, 199
Ford, Laverstock 15
Fussell's Lodge xxi 195–7
Gally Hills, Banstead xviii, 15–6
Higham Ferrers: Kings Meadow Lane xii, 171–5
infant 146, 150
Iron Age 155, 194, 195, **xii**
juvenile 29
Kinsey Cave 189, 190, 191, 191–2
Lechlade, Butler's Field 9
Longstone Edge xxi, 201, 202–3
Maiden Castle 109, 112, 113, 114, xx
male 8, 127, 128, 153, 153–4, 195, 196, 197, 198,
    199, 199–200
Medieval period 56, 194, 222–4, **xii**
Melbourn, Water Lane 16, xviii
Neolithic 109, 127, 128, 129, 131, 148, 149, 150, 151,
    191–2, 195–7, 197–200, 201, 202–3, 286–7, **xii**
neonate 174
nitrogen values xii, **xii**
Norwegian origin 222–4
Offham Hill xiii, xx, 116
post-Roman 152–3
Riccall: early medieval cemetery 222–4
River Thames vii, 194–5
Roman period xii, 171–5, 252, **xii**
Romano-British period 155
St Peter's Tip, Broadstairs xviii, 9–10, 17–21
St Sampson's Church, Callington xix
Saxon period **xii**
stable isotope values xii, **xii**
Staines 122
Sutton Common 252
Sutton Courtenay, Drayton Road 255
Taplow Court 258–9
teenage 7, 10
The Trundle xx, 123–4
Wayland's Smithy 197–200
Whitehawk Hill xx, 127, 128, 129, 131
Whitesheet Hill xx
Windmill Hill 146, 148, 150, 151
Yarnton 285, 286–7, 289–90
Humberside, Castledyke South, Barton-on-Humber 6–8, 13
hydrogen peroxide, chemical contamination 168

Ilex aquifolium, Old Ash Mine 212
International Radiocarbon Inter-comparison study (VIRI)
    xxviii
Iron Age
    Beech Court Farm, Vale of Glamorgan 72
    Boden Vean xix, xxv, 40–4
    burials 124
    causewayed enclosures 72
    cemeteries xxx, 155
    Duxford: Hinxton Road 154–5

Fiskerton: auger survey 162–6
hillforts xxx, 103, 105, 122, 123, 124, 169–71, 256, 257
human skeletal remains **xii,** 155, 194, 195
lake village 205–6
pottery 40, 41, 112, 122, 249, 260, 263, 264–5, 266,
    266–7
Stannon Down 245, 249, 249–50
Taplow Court 256, 257
Thorne Moors 220, 222
Trevelgue Head 262–7
West Heslerton 276, 280
iron extraction activities, Roman Lode 158–60
iron working *see* Exmoor Iron Project
Isle of Wight
    Bouldnor Cliff: BCII and BCIV 44–6, 46–9, xxv
    Bouldnor Cliff: wiggle-matching 50–1
    Mesolithic period 46–9
Isotope Ratio Mass Spectrometry xxiv–xxviii, xvii
isotopic composition xvii

John, King 260
Johns, C 38–44
Jones, J 206, 206–7, 207–8
Jones A 245–50
Juleff, G 156–8, 158, 160–2

Keiller, A 136, 137, 140, 145
Kendall, H 136, 145
Kent
    Buckland Dover 5–6, 11
    Chalk Hill 73–6, 99, 100, xxv
    Dover Buckland 153–4
    Kingsborough 1 99–100
    Kingsborough 2 99, 100–1
    Mill Hill, Deal 17
    St Peter's Tip, Broadstairs 9–10, 17–21, xxiv, xxviii
Kew, Greater London 195
Kingsborough 1, Kent, causewayed enclosure 99–100
Kingsborough 2, Kent, causewayed enclosure 99, 100–1
Kinsey Cave, North Yorkshire 189
    animal bone 189, 190, 191, 192
    Bronze Age 189
    human skeletal remains 189, 190, 191, 191–2
    Medieval period 192
    Neolithic 189, 191–2
    Roman cultic activity 189
    sample series #1 190
    sample series #2 190–1
    sample series #3 191–2
    sample series #4 192
Kinsey-Mattinson, W 189, 190
Knap Hill, Wiltshire, causewayed enclosure xix, xx,
    xxv, 101–3

laboratory codes xxxi
laboratory methods xv–xvi, **xv**, **xvi**
    benzene synthesis xv, **xv**
    international inter-comparison exercises xxviii
    quality assurance xviii, *xviii–xxii*, xxiii–xxiv, **xxiii,**
        xxiv–xxviii, xxviii
    variations in the pretreatment protocols xxviii
lake village, Iron Age 205–6
Lamdin-Whymark, H 256–60
Lancashire
    Anglezarke, site 1 267–8

Anglezarke, site 2 268
  Forest of Bowland, site 1 269
  Forest of Bowland, site 2 269–70
  Forest of Bowland, site 3 270
  Upland Peats 267, 267–70
Lancashire County Council 267–8, 268, 270
landscape, West Heslerton 276–83
Landscape Research Centre, The 276–83
Langdale, Cumbria
  site 1 271
  site 2 270–1
  site 3 270, 271
Last, J 200–3
Lawrence, S 171–5
Lechlade, Butler's Field, Gloucestershire, Anglo-Saxon
    graves and grave goods (female graves) 9
Leibniz Labor für Altersbestimmung und Isotopenforschung,
    Christian-Albrechts Universität, Kiel xvi, xvii
Lewis, R 71
Liddell, D 94, 95
lime kiln, Duxford: Hinxton Road 155
Lincolnshire, Fiskerton: auger survey 162–6
Lincolnshire County Museum 162–6
Lindholme Island 214
Lindholme Neolithic trackway 213, 216
Lindley, D V xxix
liquid scintillation spectrometry viii, x, xv, 285
London Shellyware: Billingsgate Lorry Park, Greater
    London, carbonised residue xxix, 192–4
Long, N 5
Long Barrows Project
  Fussell's Lodge
  Fussell's Lodge, Wiltshire xxi, xxvi, 195–7
  Wayland's Smithy, Oxfordshire 197–200
Longin xv
Longstone Edge, Derbyshire xxx
  barrow 1 xxi, 200, 200–3
  barrow 2 200, 203
  cremations 200, 200–1
loomweights 235, 236, 240, 288
Lords and Ladies Mine, Derbyshire xxi, xxvi, 210–1
Loveday, R 21, 22
Lundenwic, City of London 203–4
  post-Ipswich Ware pit 204
Lyons, A 155

McAvoy, F 242
McCormac, F G 4, 10
maceheads, antler 21–2
McFadyen, L 23–4, 24–5, 27–8
Maiden Bower, Bedfordshire
  causewayed enclosure 103–5
  ditch 103, 104–5
  Iron Age hillfort 103, 105
  pit 103, 104–5
Maiden Castle, Dorset
  animal bone xiii, xx, 106–7, 109–10, 112, 112–3, 113
  antler 110
  carbonised residue xx, 107–8, 108–9, 110, 112, 113
  causewayed enclosure xx, xxiv, xxv, 105–14
  charcoal 108, 110
  human skeletal remains 109, 112, 113, 114, xx
  inner ditch xx, xxv 106, 106–10
  Iron Age hillfort 105
  Long Mound 105, 106, 110–2

Neolithic 105–14
  outer ditch xx, xxv, 106, 112–4
  pits 105
  pottery xiii, **xiv**, 105, 107–8, 108–9, 110, 112, 113
Malcolm, G 203
Malim, T 13
MARISP, Somerset 204–5
  Dewar's track xvii, **xvii**, xxi, 205
  Glastonbury Lake Village xxi, 205–6
  Harters Hill xxi, 206–7
  Sharpham 207–8
  Street Causeway xxi, 208–9
Markov Chain Monte Carlo sampling xxix
Marshall, P 1, 2, 3, 22, 37, 51, 52, 53, 58–71, 156, 158,
    159–62, 170, 200–3, 205–9, 236, 241, 263, 264–5,
    266, 283–4
Mary Rose xviii, xxviii
Masefield, R 5
Matthews, C L 103, 104, 105
maximum intercept method xvii
Meadows, J 191
Medieval period
  burials 56–7
  Callington: St Sampson's Church 56
  cemeteries 222–4
  human skeletal remains 56, 194, 222–4, **xii**
  Kinsey Cave 192
  mines and mining 210, 211
  Riccall 222–4
  Tower of London: Royal Menagerie 262
  West Heslerton 278
Melbourn, Water Lane, Cambridgeshire, Anglo-Saxon
    graves and grave goods (male graves) xxiv, xxviii, 16,
Mercer, R 245
Mesolithic period
  Bouldnor Cliff: BCII and BCIV 46–9
  Crickley Hill 84
  Fiskerton: auger survey 163–6
  Hatfield Moors, Rawcliffe 213, 218
  Hembury 94
  Howick, Sea Houses Farm 185–9
  Stannon Down 248
  Sutton Common 254
  Taplow Court 259
  West Heslerton 276, 277, 278
metal detectorists 1
middens
  Ascott under Wychwood 23, 26
  Carlton Colville: Bloodmoor Hill 60, 62–3, 65, 67
  Etton Woodgate 88
  Trevelgue Head 263, 264–5, 266–7
Miles, D 9, 12, 152
Mill Hill, Deal, Kent, Anglo-Saxon graves and grave goods
    (male graves) 17
mines and mining
  Derbyshire 209–12, xxvi
  Medieval period 210, 211
molluscs 116, 127
Momber, G 44, 45–6, 46–9, 50–1
Mook, W G xvii
Mortlake, Greater London 195
mortuary rings 251
mortuary structure, Brandon: Staunch Meadow 51
Moyses Hall Museum, Bury St Edmunds 12
Museum of London 22, 192–4, 194, 204
Musty, J 15

National Museums and Galleries of Wales 71
Natural History Museum, The 22, 88–9, 89, 93, 120, 121,
    125, 127, 128, 132, 195, 260, 261
Nayling, N 50–1
Neolithic
  causewayed enclosures 72–7, 94–9, 99–119, 122–51
  Crickley Hill 77–88
  Fiskerton: auger survey 163–6
  Hatfield Moors, Rawcliffe 214, 216, 217–8, 219
  Howick, Sea Houses Farm 187
  human skeletal remains **xii,** 109, 127, 128, 129, 131,
      148, 149, 150, 151, 191–2, 195–7, 197–200, 201,
      202–3, 286–7
  Kinsey Cave 189, 191–2
  Maiden Castle 105–14
  pottery xiii, **xiv,** 74, 75, 76, 81, 82, 83, 87–8, 89, 90,
      91, 92, 92–3, 98, 101, 103, 105, 107–8, 108–9, 110,
      113, 116, 118–9, 120–1, 122, 122–3, 123, 125, 126,
      126–7, 127, 128, 129, 130–1, 133, 136, 139, 140,
      146, 147, 148, 149, 150, 236, 290, 291, 291–2
  Stannon Down 245
  Thorne Moors 221, 222
  trackways 213, 216
  West Heslerton 276, 281, 282
  Yarnton 286–7, 290–5, *xxii*
  see also barrows, Neolithic
Noakes, J E xv
Norfolk, Holme-next-the-Sea 175–85
Norfolk Museums and Archaeology Service 175
North Yorkshire
  Ainsbrook xxx, 1–4
  Kinsey Cave 189–92
  Riccall: early medieval cemetery xxx, 222–4
  Ripon Cathedral *xxvi–xxvii,* 224–32
Northamptonshire, Higham Ferrers: Kings Meadow Lane
    xxx, 171–5
Northborough, Cambridgeshire
  causewayed enclosures 114–6
  inner ditch 114–5
  outer ditch 115–6
  pottery 114
Northumberland
  Harehaugh Hillfort vii, 169–71
  Howick, Sea Houses Farm *xxvi,* xxx,185–9,
  Warkworth Castle viii, **viii,** *xxvii–xxviii,* xxx, 273–6
Norway xxix
Nottingham Tree-Ring Dating Laboratory 224, 229
Nottinghamshire, Antler Maceheads Project 22
Nowakowski, J 262–7

occupation deposits xiii
Offham Hill, East Sussex xiii, *xx, xxv,* 116
Old Ash Mine, Derbyshire 210, 211–2
O'Reegan, H 260–2
ostracods 188
Ouse, River 51
OxCal software xxix
Oxford Archaeology 152, 171
Oxford City and County Museum 22
Oxford Radiocarbon Accelerator Unit viii, xvi, xviii, xxiv,
    xxviii, 21, 88, 119, 145, 159–60, 194, 196, 243,
    243–5, 285, **ix**
Oxford Radiocarbon Laboratory 198
Oxfordshire
  Ascott under Wychwood: long barrow vii, **ix,** *xxiv,* 22–30

Sutton Courtenay, Drayton Road 255
  Wayland's Smithy 197–200
  Yarnton Neolithic and Bronze Age vii, xxviii, xxx,
      285–95
Oxfordshire County Museums Service 286, 287, 290,
    291–2, 292

paleoenvironmental proxies xxx
Palmer, S 9, 12
Parfitt, K 5, 11, 17, 153
Parrett, River 53
Peak Camp, Gloucestershire *xx, xxvi,* 77, 116–8,
Peak District Mines, Derbyshire 209–10
  Lords and Ladies Mine *xxi, xxvi,* 210–1
  Old Ash Mine 211–2
Peak District National Park Authority 210
peat xi, xiii, xiii–xiv, **xi**
  accumulation rates 31–2
  Anglezarke, site 1 267–8
  Anglezarke, site 2 268
  Aveley Marshes 30–1, 31
  Bouldnor Cliff: BCII and BCIV 44–5
  Dewar's track xvii, **xvii,** 205
  Fiskerton: auger survey 163, 166
  Forest of Bowland, site 1 269
  Forest of Bowland, site 2 269–70
  Forest of Bowland, site 3 270
  Harters Hill 207
  Hatfield Moors, Rawcliffe 213, 217–20
  Holme-next-the-Sea 175, 180
  Langdale, site 1 270, 271
  Langdale, site 2 270–1
  Langdale, site 4 270, 271
  MARISP, Somerset xvii, **xvii,** 204–9
  Roman Lode 158–60
  Sharpham 207–8
  South-west Fells transect, site 1b 271–2
  South-west Fells transect, site 2b 272
  South-west Fells transect, site 3 272–3
  Stannon Down 247–9
  Street Causeway 209
  Sutton Common 255
  Thorne Moors 221, 222
  upland 267–73
  wastage 204–9
  West Heslerton 276, 277
pendants 6, 9, 10
Perth
Perth, Scotland xxix, 192, 194
Phragmites, Howick, Sea Houses Farm 186, 188
Piggott, S 133
pits xiii
  Ascott under Wychwood 22, 24
  Bouldnor Cliff: BCII and BCIV 48, 49
  Carlton Colville: Bloodmoor Hill 59–60, 61–2, 63, 64,
      64–6, 67, 68, 69, 70, 71
  Chesham Bois House 151–2
  cooking 95–7, 265
  Crickley Hill 80
  Etton 92, 93
  Etton Woodgate 88
  Haddenham 94
  Hembury 95, 95–7, 97, 98–9
  Lundenwic 204
  Maiden Bower 103, 104–5

Maiden Castle 105
Ryall Quarry: Saxon's Lode Farm 232–4
Sherborne: Tinney's Lane 236–40
Trevelgue Head 265
Whitesheet Hill 135–6
Yarnton 288, 290, 293
plant macrofossil, carbonised
Crickley Hill 85, 86
Hatfield Moors, Rawcliffe 215, 216–7
Hembury 95–7, 99
Kingsborough 1 99–100
Kingsborough 2 101
Longstone Edge 201
Northborough 115–6
Taplow Court 259
Whitesheet Hill *xx*, 135, 136
Yarnton 291, 292
plant macrofossil, waterlogged
Aveley Marshes 31–2
Bouldnor Cliff: BCII and BCIV 46
Fiskerton: auger survey 163
Glastonbury Lake Village *xxi*, 206
Harters Hill 206
Hatfield Moors, Rawcliffe 213–5, 215–6, 217
Howick, Sea Houses Farm 186, 187–8, 188–9
Sharpham 207
Sutton Common 253, 253–4
Thorne Moors 220–1, 221–2, 222
Yarnton 293–4
plant macrofossils xi, **xi**, xvi
Silbury Hill 243–5
Yarnton 288–9
plant remains, charred xi, xiii, xv, **xi**
plant remains, waterlogged xi, xiv, xv, **xi**
Plouviez, J 250–1
Pollard, J 103
pollen core, Sutton Common *xxii*
pollen diagram, Stannon Down 247–9
Pomoideae 162
Ainsbrook 2
Ascott under Wychwood 27, 28
Boden Vean 42, 43
Brandon: Staunch Meadow 52
Kingsborough 2 100–1, 101
Maiden Castle 108
Northborough 115
Old Ash Mine 212
Ryall Quarry: Saxon's Lode Farm 234–5, 235, 236
West Heslerton 282
Whitehawk 129
Yarnton 291, 293, 295
Portable Antiquities Scheme 1
posterior density estimates xxviii, xxix
post-excavation analysis vii
post-glacial deposits, Fiskerton: auger survey 162–6
postholes 86, 94
Ascott under Wychwood 27
Carlton Colville: Bloodmoor Hill 60, 62, 63–4, 68, 69,
70, 71
Crickley Hill 84, 85
Etton 93
Taplow Court 256–8, 259, 260
Trevelgue Head 264, 265
Yarnton 289, 290
post-nuclear bomb atmospheric calibration curve 159

post-Roman period
Boden Vean 44
cemeteries 152–3
Claydon Pike, Lechlade 152–3
Fiskerton: auger survey 166
pottery xi, **xi,** xiii
Abingdon Ware 116
Ainsbrook 2
Anglo-Saxon period 52, 58, 59–63, 64–7, 67–8, 68–71,
203–4, 234, 235
Billingsgate Lorry Park 192–4
biotite and quartz tempered 61, 62, 62–3, 65, 67,
67–8, 70
biotite-tempered 58
Boden Vean 38, 39, 40, 41, 44
Brandon: Staunch Meadow 52
Bronze Age vii, 2, 38, 39, 40, 138, 236–40, 249, 294
Carlton Colville: Bloodmoor Hill 58, 59–63, 64–7,
67–8, 68–71
chaff-tempered 59, 60, 62, 65, 65–6, 66, 67, 68–9, 70
Chalk Hill, Kent 74, 75, 76
Collared Urn 256
cooking pots 193–4
Cord Impressed Ware 2
Crickley Hill 81, 82, 83, 84, 85, 86, 87–8
dating 59, 63
decoration 58, 59, 60, 62, 64, 66, 68, 69, 71, 114
Duxford: Hinxton Road 154–5
Ebbsfleet Ware 90
Etton 90, 91, 92, 92–3
Etton Woodgate 89
flint-tempered 120, 137–8, 139
Food Vessel 201
gabbroic 110
Globular Urn 294
grog-tempered 58, 64, 70, 84
Grooved Ware 291–2
Gwithian 38, 44
Harehaugh Hillfort 169, 170
Hembury 98
Ipswich Ware xxx, 52, 203–4
Iron Age 40, 41, 112, 122, 249, 260, 263, 264–5, 266,
266–7
Kingsborough 1 99
Kingsborough 2 101
Knap Hill 103
London Rouen/North French 194
London Shellyware xxix, 192–4
Longstone Edge 201, 202
lugged 98
Lundenwic 203–4
Maiden Bower 105
Maiden Castle xiii, 105, 107–8, 108–9, 110, 112,
113, **xiv**
Neolithic xiii, 74, 75, 76, 81, 82, 83, 87–8, 89, 90, 91,
92, 92–3, 98, 101, 103, 105, 107–8, 108–9, 110,
113, 116, 118–9, 120–1, 122, 122–3, 123, 125, 126,
126–7, 127, 128, 129, 130–1, 133, 136, 137–8, 139,
140, 146, 147, 148, 149, 150, 236–40, 290, 291,
291–2, **xiv**
Northborough 114
organic-tempered 234
Outline 264
Peak Camp 116
Peterborough Ware 92, 105, 138, 140, 236–40

Post-Deverel-Rimbury xxx, 236, 291
post-Roman period 44
ripple-burnished 120, 121
Robin Hood's Ball 118–9
Roman period 209
Ryall Quarry: Saxon's Lode Farm 234, 235
Saxon period 260, 284–5
shell-tempered 90, 91, 92, 147, 149, 204
Sherborne: Tinney's Lane vii, 236–40
South Western Decorative 263, 264, 265, 266
Staines 120–1, 122
Stannon Down 249
Street Causeway 209
Taplow Court 256, 257–8, 258, 260
Trevelgue Head 263, 264, 266, 266–7
Trevisker Ware 38, 39
The Trundle 122–3, 123
Western style Bowl 133
Whitehawk 125, 126, 126–7, 127, 128, 129, 130–1
Whitehawk Style 99
Whitesheet Hill 133, 136
Windmill Hill 137–8, 138, 139, 140, 146, 147, 148, 149, 150
Wolverton Turn Enclosure 284–5
Yarnton 290, 291, 291–2, 294
Powlesland, D 276–83
Preston, S 284–5
Price-Williams, D 15
Prunus sp.
Beech Court Farm, Vale of Glamorgan 72
Sherborne: Tinney's Lane 239
Yarnton 289
Prunus spinosa
Kingsborough 2 101
Sherborne: Tinney's Lane 238
West Heslerton 282
Pryor, F 88–9, 89, 93
publication programme x
Putney, Greater London 195

Quartermaine, J 267, 268, 270, 271, 272
Queen's University, Belfast viii, **ix**, x, 285
Quercus sp. 256
Ainsbrook 1, 2
Baguley Hall: north wing 33–6
Blacklake Wood 156–7, 157–8, 158
Boden Vean 41–2, 42–3, 43
Bouldnor Cliff 46, 49, 50–1
Bridgwater Bay 54, 55
Chesham Bois House 151–2, 152
Dewar's track 205
Etton 93
Haddenham 94
Holme-next-the-Sea 183–4
Maiden Castle 108, 110
Old Ash Mine 212
Peak Camp 117, 118
Ripon Cathedral 224–32
Roman Lode 158
Ryall Quarry: Saxon's Lode Farm 235–6
Sherborne: Tinney's Lane 237, 238
Sherracombe Ford 161–2, 162
Stannon Down 245, 247, 249–50
Taplow Court 257, 260
Trevelgue Head 265, 266

Warkworth Castle: Grey Mare's Tail Tower 273–6
Whitehawk 129, 130, 131
Wissett: Church of St Andrews 283–4
Yarnton 288, 289, 295

Rackham, J 162–6, 276–83
radiocarbon ages
accuracy xxiv
calibration xvii, **xvii**
divergent groups xxiii–xxiv
international inter-comparison exercises xxviii
offsets xxiii, **xxiii**
quoting conventions xvi–xvii
radiocarbon dating facilities viii–ix, **ix**
radiocarbon dating techniques ix, x, **x**
accuracy xxiv
radiocarbon offsets xii
Rafter Radiocarbon Laboratory, New Zealand
xv, xxiv, 5, 6, 8, 9, 10, 11, 12, 13, 14, 15, 16, 17, 18, 19, 20, 21, 22, 154
raised mires
predictive modelling 213–22
Hatfield Moors 213–20
Thorne Moors 213, 220–2
Rawlings, M 133, 135
Reading Museum 198
Reimer, P J xvii
replicate measurements x, *xviii–xxii*, xxiii–xxiv, **xxiii**, xxviii, *xxiv–xxviii*
weighted means xxviii
Riccall: early medieval cemetery, North Yorkshire
xxx, 222–4
Richard III, King 262
Richards, J 118, 133, 135
Rijksuniversitat Groningen 21
Rijkuniversiteit Groningen, the Netherlands
viii, **ix**, xvi, **xvi**, 21, 24, 127, 146, 148
ring-cairn, Stannon Down 246–7, 249–50
rings 8, 9, 10, 154
Ripon Cathedral, North Yorkshire xxiv
RIPCSQ01 224–9, *xxvi–xxvii*
RIPCSQ02 229–32, *xxvii*
structural timbers *xxii*, 224–32
Roberts, J 154–5
Robertson, D 175–85
Robertson-Mackay, R 119, 120, 121
Robin Hood's Ball, Wiltshire, causewayed enclosure 118–9
Robinson, D E 159–60
Roman cultic activity, Kinsey Cave 189
Roman Lode, Devon 158
peat sequence 158–60
Roman period
animal bone *xviii–xix*
bath-house xxx
burials 171–5
Carlton Colville: Bloodmoor Hill 63–4, 68, 69
cemeteries xxx, 251–2
Groundwell Ridge 168–9
Higham Ferrers: Kings Meadow Lane 171–5
human skeletal remains xii, **xii**, 171–5, 252
iron working xxx
pottery 209
villa buildings 168–9
West Heslerton 278, 282

Romano-British period
  Carlton Colville: Bloodmoor Hill 58
  Duxford: Hinxton Road 155
  human skeletal remains 155
  iron working 160–2
  Kinsey Cave 189
  Ryall Quarry: Saxon's Lode Farm 232
Rosaceae, Kingsborough 2 101
Ross Williamson, R 124, 125, 127, 128
Royal Albert Memorial Museum and Art Gallery, Exeter 95
Royal Cornwall Museum 39, 40, 44, 56–7, 245, 246,
        247, 249
Royal Holloway College 30
Rudling, D 124
Russell, M 124
Ryall Quarry: Saxon's Lode Farm, Worcestershire 232
  Anglo-Saxon period 232, 233, 234–6
  Bronze Age 232, 235–6
  charred cereal remains within pit 1032 233–4
  charred cereal remains within pit 1943 232–3
  pottery 234, 235
  Romano-British period 232
  sunken-featured building 4 234–5
  sunken-featured building 6 235–6

St Peter's Tip, Broadstairs, Kent *xxiv*
  Anglo-Saxon graves and grave goods (female graves)
        xviii, 9–10
  Anglo-Saxon graves and grave goods (male graves)
        *xxviii*17–21
St Sampson's Church, Callington, human skeletal remains
        xix
Salicaceae
  Ainsbrook 1, 3
  Lundenwic 204
  West Heslerton 278, 279, 280
Salisbury and South Wiltshire Museum 118, 133, 135, 195
Salix/Populus sp.
  Brandon: Staunch Meadow 52–3, 53
  Holbrook Bay fishtrap 251
  Holme-next-the-Sea 179–80, 181–2
  West Heslerton 280–1
sample characterisation xi–xv, **xi**
sample selection xi–xv, **xiii, xiv**
samples
  bulk materials xi, **xi**
  contexts xiii–xv, **xiii**
  eligibility vii
  identification xi–xii
  quantity x, **x**, xi
sampling strategies x, **x, xi**
Saxon period
  burials 255, *xix*
  fishtraps 175, 176, 181–4, 185
  Holme-next-the-Sea 175, 176, 177, 181–4, 185
  human skeletal remains **xii**
  pottery 260, 284–5
  Sutton Courtenay, Drayton Road 255
  Taplow Court 258–9
  Wolverton Turn Enclosure, sunken-featured building
        284–5
Scientific Dating Team of the Centre for Archaeology vii
Scottish Universities Environmental Research Centre
        viii, **ix**, xv, **xv**, xviii, xxiv, xxviii, 224, 227–9, 285
Scull, C 4–5, 9, 10, 12, 12–3, 14, 17, 17–21

Scunthorpe Museum 6, 13
Seahenge, Holme-next-the-Sea 175–85
sea-level change 31–2, 45–6, 46–9, 166
seax
  type SX1–a 13
  type SX1–b 12, 18, 19
  type SX1–c 15
  type SX2–a 13
  type SX3–a 18, 21
  type SX4–d 13
sediments xi, xiii, **xi,** xxx
  Aveley Marshes 30–2
  Sutton Common 254–5
sequential sampling strategies x
Severn, River 232
Severn Estuary 53
Shand, G 73
Sharpham, Somerset Levels xxx
Sharples, N 105, 106, 110, 112
Sherborne: Tinney's Lane, Dorset vii, xxx, 236–40
Sherracombe Ford: Exmoor, Devon xxx
shield bosses
  type SB1–b 14
  type SB3–b2 12
  type SB3–b3+4 11, 14
  type SB3–c 17
  type SB4– b1+2 20
  type SB4–b1+2 18
  type SB4–b2 154
  type SB5–b+c 15
Sidell, J 194
Silbury Hill, Wiltshire xi, xxx, 136, 241–2
  2000–2001 investigations 242
  animal bone 241
  antler 241, 242
  bulk soil samples 243
  contamination 243
  plant macrofossils 243–5
  primary mound turves 241, 242–5
  remedial works 242
  western lateral tunnel 241
simulation models x
sites, threats to vii–viii
slagheaps, Sherracombe Ford 160–2
smelting slag 156–8
Smith, A 152–3, 171–5
Smith, I 136, 137, 140, 145, 146
Smith, W G 103, 104
smithing slag 151, 152
sodium bicarbonate, chemical contamination 167
sodium hexametaphosphate, chemical contamination 168
soil xiii
Somerset xxiii–xxiv
  Blacklake Wood 156–8, *xxvi*
  Bridgwater Bay 53–6
  Dewar's track xvii, **xvii**, xxx, 205
  de-watering viii
  Glastonbury Lake Village xxx, 205–6
  Harters Hill xxx, 206–7
  MARISP project 204–9
  Sharpham xxx, 207–8
  Street Causeway xxx, 208–9
Somerset County Council 204
Somerset County Museum 205, 206, 207, 208
Somerset Heritage Centre 53–6

Somerset Records Office 205, 208
Sorbus sp., Sherborne: Tinney's Lane 237–8, 238, 239
South Yorkshire
    Hatfield Moors, Rawcliffe xxx, 213–20
    Sutton Common 251–4
    Thorne Moors 213, 220–2
spearheads
    type SP1–a2 13, 19
    type SP1–a3 17, 20
    type SP1–a4 11, 20
    type SP2–a1a2 14
    type SP2–a1b1 15
    type SP2–a2c 12, 18
    type SP2–a3 19
    type SP2–B1a2 14
    type SP2–b1a3 12
    type SP3–a 154
    type SP3–b 11
    type SP4 16, 17, 19
Staines, Surrey xxvi
    causewayed enclosure xx, 119–22
    inner ditch 120–1
    outer ditch 121–2
Stannon Clay Works 245
Stannon Down, Devon 245
    Bronze Age 245–50
    cairn 245–6
    charcoal 245–6, 247, 249–50
    Iron Age 245, 249, 249–50
    Mesolithic period 248
    Neolithic 245
    pollen diagram 247–9
    pottery 249
    ring-cairn 246–7, 249–50
    site 2 245–6
    site 6 246–7
    site 9 249–50
    site 86 247–9
    timber ring 246–7
statistical modelling xxviii–xxxi, **xxix**
statistical simulation x
Stenhouse, M J xv
Stevens, S 99
Stone, J 133
stone circles, Crickley Hill 84, 85–6
Stonehenge 118
Storegga Slide event 186
Street Causeway, Somerset xxx, 208–9
structural timbers xi, xiv, **xi**
    Baguley Hall: north wing 32–6
    Ripon Cathedral 224–32
    Warkworth Castle: Grey Mare's Tail Tower 273–6
    Wissett: Church of St Andrews 283–4
Suffolk
    Brandon: Staunch Meadow xxv, xxx, 51–3
    Bury St Edmunds, Westgarth Gardens 12
    Carlton Colville: Bloodmoor Hill xiii, xxv, xxx, 58–71
    Coddenham 8–9
    Eriswell, Lakenheath 14–5
    Holbrook Bay xxx, 250–1
    Wissett: Church of St Andrews 283–4
Suffolk Coastal Survey: Holbrook Bay fishtrap, Suffolk xxx, 250–1
Suffolk County Council 8, 14
Suffolk County Council Archaeological Service 51, 52, 250–1

Suffolk Museums 58
Surrey
    Gally Hills, Banstead xxiv, 15–6
    Staines xxvi, 120–1
Sutton Common: South Yorkshire xxx
    cremations 251–2
    pollen core xxii, 252–4
Sutton Courtenay, Drayton Road, Oxfordshire 255
Swindon Museum and Art Gallery 168
swords
    type SW1–a 14
    type SW3–a 11
    type SW4 11, 17, 154
Syon Reach, Greater London 194

Taplow Court, Buckinghamshire xxx, 256–60
Taylor, T 189–92
Tester, A 51, 52–3
tetrasodium pyrophosphate decahydrate, chemical contamination 168
textile working industry, Anglo-Saxon period 52–3
Thames, River, 30
    human skeletal remains vii, 194–5
Thames Valley, Greater London, Antler Maceheads Project 21–2
Thomas, N 118
Thorne Moors, South Yorkshire, raised mires xxi, xxii, xxx, 213, 220–2
tidal erosion 176, 177, 178, 179, 179–80, 180, 182, 183, 184, 185
timber circle, Holme-next-the-Sea 175–85
timber ring, Stannon Down 246–7
Time Team 114, 115
Tinsley, H 248–9
Tipper, J 59–71
Todd, M 94, 95
Tower of London: Royal Menagerie, Greater London 260–2
trackways
    Bronze Age 205
    Neolithic 213, 216
tree-ring analysis, Ripon Cathedral 224–32
tree-ring sequences xii, xviii, xxx
tremissis accessory-type 9
Trevelgue Head, Cornwall xxx, 262–7
    carbonised residue 263, 263–4, 264–5
    charcoal 263, 265–6
    hearth 264
    Iron Age promontory fort vii, **vii**
    middens 264–5, 266–7
    pits 265
    pottery 263, 264, 265, 266, 266–7
Trondheim Convention, the xvi–xvii
Trundle, The, West Sussex
    causewayed enclosures xx, xxvi, 122–4
    ditch 2 122–3
    hillfort 122, 123, 124
    inner ditch 123
    outer ditch 123–4
Tyne and Wear Archives and Museums 169

Ulex sp., Boden Vean 43
Ulex/Cytisus sp.
    Ainsbrook 1–2
    Boden Vean 39

West Heslerton 282, 282–3
Ulmus sp.
  Sherborne: Tinney's Lane 236, 238, 240
  Sutton Common 253
University of Exeter 156, 158, 160
University of Newcastle upon Tyne 185–9
Upland Peats
  Anglezarke, site 1 267–8
  Anglezarke, site 2 268
  Cumbria 267, 270–3
  Forest of Bowland, site 1 269
  Forest of Bowland, site 2 269–70
  Forest of Bowland, site 3 270
  Lancashire 267, 267–70
  Langdale, site 1 270, 271
  Langdale, site 2 270–1
  Langdale, site 4 270, 271
  South-west Fells transect, site 1b 271–2
  South-west Fells transect, site 2b 272
  South-west Fells transect, site 3 272–3

Vale of Glamorgan, Beech Court Farm 71–2
Valentin, J 236–40
Van de Noort, R 251–2
Viking period, Ainsbrook 1–4

walk-over survey, Holme-next-the-Sea 175–85
Ward, G K xvii
Wardown Park Museum, Luton 104, 105
Warkworth Castle: Grey Mare's Tail Tower,
      Northumberland viii, **viii**, *xxii, xxvii–xxviii*, xxx,
      273–6
waterholes, Yarnton *xxviii*, 288, 292–5
Watts, M 232–3, 233–4, 234–6
Wayland's Smithy, Oxfordshire xvi, 197–200
weapon burials 14
Weekes, J 73
weighted means, replicate measurements xxviii
Weisskopf, A 194–5
Wellington, Duke of 260
Wessex Archaeology 99, 100, 114, 115
West, C 12
West Heslerton, Yorkshire (East Riding) *xxii*, 276–83
  Bronze Age 276, 278
  Iron Age 276, 280
  Medieval period 278
  Mesolithic period 277, 278
  Neolithic 276, 281, 282
  peat 276, 277
  Roman period 278, 282
  wood, waterlogged 277–83
West Kennet Farm 136
West Kennet long barrow **ix**
West Sussex
  Apple Down, Compton 4–5
  Bury Hill 72–3
  Court Hill, enclosure 76–7
  The Trundle *xxvi*, 122–4
Wheeler R E M 105, 106, 110, 112
Whitehawk Hill, East Sussex
  animal bone 125–6, 130, 131, 131–2, 132, *xx*
  antler 125, 126, 132–3, 133, *xx*
  carbonised residue 125, 126, 126–7, 127, 128, 129,
      130–1, *xx*
  causewayed enclosures *xx*, 124–33

charcoal 129–30, 131, 132
ditch I *xx, xxvi*, 125–7
ditch II *xxvi* , 127–8
ditch III *xxvi, xxviii*, 128–32
ditch IV *xxvi*, 132–3
gate structures 124
human skeletal remains *xx*, 127, 128, 129, 131
pottery 125, 126, 126–7, 127, 128, 129, 130–1
Whitesheet Hill, Wiltshire
  animal bone *xx*, 134, 135–6
  antler 133–4
  carbonised plant macrofossil *xx, xxi*, 135, 136
  causewayed enclosures *xxvi*, 133–6
  ditch 133–4
  human skeletal remains *xx*
  internal features 135–6
  pits 135–6
  plant macrofossil *xx*
  pottery 133, 136
Whittle, A 24, 25–7, 29–30, 136, 137, 140, 145, 195–7,
      197–200, 241, 242–5
wiggle-matching viii, xxx, 283–4
  Baguley Hall: north wing 32, 33, 35, 36
  Bouldnor Cliff: wiggle-matching 50–1
  Ripon Cathedral 224–32
  Warkworth Castle: Grey Mare's Tail Tower 273–6
Wilfrid, St 224
Wilson, P 168–9
Wiltshire
  Beckhampton Road Long Mound 23
  Ford, Laverstock 15
  Fussell's Lodge *xxvi*, 195–7
  Groundwell Ridge 168–9
  Knap Hill *xxv* , 101–3
  Robin Hood's Ball 118–9
  Silbury Hill xxx, 241–5
  Whitesheet Hill *xxvi*, 133–6
  Windmill Hill **ix**, xiii, *xxvi, xxviii*, 136–51
Wiltshire Heritage Museum 133, 135, 137, 145–6
Windmill Hill Culture 101
Windmill Hill, Wiltshire
  animal bone 138, 139, 140, 141–2, 142–4, 145, 146–7,
      147–8, 148–9, 149–50
  antler xiii, 138–9, 139, 141, 142, 143, 144–5, 145
  carbonised residue 137–8, 139, 146, 147, 148, 149, 150
  causewayed enclosures **ix**, *xx–xxi, xxvi, xxviii*, 136–51
  charcoal 137, 139, 140, 149, 151
  human skeletal remains 146, 148, 150, 151
  inner ditch *xxvi* , 137, 140–5
  middle ditch *xxvi, xxviii*, 137, 140–5
  outer ditch *xxvi, xxviii*, 137, 145–51
  pottery 137–8, 139, 140, 146, 147, 148, 149, 150
Wissett: Church of St Andrews, Suffolk 283–4
Witham, River 163, 165
Wolverton Turn Enclosure, sunken-featured building,
      Buckinghamshire 284–5
wood
  Baguley Hall: north wing 32–6
  radiocarbon offsets xii
  Ripon Cathedral *xxii*, 224–32
  Warkworth Castle: Grey Mare's Tail Tower *xxii*, 273–6
  Wissett: Church of St Andrews 283–4
wood, waterlogged xv, xvi
  Bouldnor Cliff: BCII and BCIV 45, 46, 47–9
  Bouldnor Cliff: wiggle-matching 50–1

Brandon: Staunch Meadow *xix*, 52–3
Bridgwater Bay 53–6
Etton *xix*, 90, 93
Etton Woodgate 88–9, 89
Fiskerton: auger survey 163–6
Holbrook Bay fishtrap 250–1
Holme-next-the-Sea 175–85
Howick, Sea Houses Farm 186–7, 188
radiocarbon offsets xii
Street Causeway 208–9
Sutton Common 253
Thorne Moors 221
West Heslerton 277–83
Yarnton Neolithic and Bronze Age *xxii*, 288, 295
Worcestershire, Ryall Quarry: Saxon's Lode Farm 232–6
Worcestershire County Museum 232–3, 233, 234–6, 235

Yarnton Neolithic and Bronze Age, Oxfordshire
    vii, xxviii, xxx, 285–6
  animal bone 290
  Bronze Age 287–90
  burials 285
  burnt stone deposits 293, 294, 295
  carbonised plant macrofossil 291, 292

carbonised residue 294
charcoal 289, 291, 293
cremations 286–7, 289–90
grain 288, 289, 290, 291
human skeletal remains 285, 286–7, 289–90
Neolithic 286–7
Neolithic circular structure 5816 290–1
Neolithic pits 290
Neolithic structure 4291 291–2
pits 288, 293
plant macrofossil, waterlogged 293
plant macrofossils 288–9
postholes 289, 290
pottery 290, 291, 291–2, 294
rectangular structure 3871 286–7
sites 4c, 4e, 9, and 10 287–90
waterholes *xxviii*, 288, 292–5
wood, waterlogged *xxii*, 288, 295
York Archaeological Trust 1
Yorkshire (East Riding)
  Antler Maceheads Project 22
  West Heslerton 276–83
Yorkshire Museum, The 1, 222